J. Frank Adams was one of the world's leading topologists. He solved a number of celebrated problems in algebraic topology, a subject in which he initiated many of the most active areas of research. He wrote a large number of papers during the period 1955–1988, and they are characterised by elegant writing and depth of thought. Few of them have been superseded by later work.

This selection brings together all his major research contributions. They are organised by subject matter rather than in strict chronological order. The first volume contains papers on: the cobar construction, the Adams spectral sequence, higher order cohomology operations, and the Hopf invariant one problem; applications of *K*-theory; generalised homology and cohomology theories. This second volume is mainly concerned with Adams' contributions to: characteristic classes and calculations in *K*-theory; modules over the Steenrod algebra and their Ext groups; finite *H*-spaces and compact Lie groups; maps between classifying spaces of compact groups.

THE SELECTED WORKS OF J. FRANK ADAMS

THE SELECTED WORKS OF J. FRANK ADAMS
VOLUME II

Edited by

J. P. MAY
Department of Mathematics, University of Chicago

and

C. B. THOMAS
Department of Pure Mathematics and Mathematical Statistics, University of Cambridge

CAMBRIDGE
UNIVERSITY PRESS

CAMBRIDGE UNIVERSITY PRESS
Cambridge, New York, Melbourne, Madrid, Cape Town, Singapore, São Paulo, Delhi

Cambridge University Press
The Edinburgh Building, Cambridge CB2 8RU, UK

Published in the United States of America by Cambridge University Press, New York

www.cambridge.org
Information on this title: www.cambridge.org/9780521110686

First published 1992
This digitally printed version 2009

A catalogue record for this publication is available from the British Library

ISBN 978-0-521-41065-6 hardback
ISBN 978-0-521-11068-6 paperback

Contents

Contents of Volume I

Introduction

Frank Adams was a great mathematician with a fine expository style. These two volumes contain the bulk of his numerous papers, grouped according to subject, and roughly chronologically within groups. We chose this organisation because of Adams' practice of returning periodically to certain subjects dear to his heart, re-examining them in the light of intervening research. We have added no editorial material since we prefer to let Adams' own introductions to his papers speak for themselves. We have also made no attempt to note errors since Adams was a scrupulously careful author. Several of his papers have published corrections; these we have included.

Adams' final bibliography contains 82 published items. These volumes contain 52 of them. The omitted items fall into five categories. There are five early papers (to 1957), an appendix to a paper of another author, and a published letter which we feel are probably not of lasting mathematical interest. There are four announcements of results which were published elsewhere. There are three biographical items. There are fifteen primarily expository articles, most of which were published in conference proceedings; we have chosen to include four of these, which we feel are of more lasting interest than the others.

The five remaining omitted items are books. The first of these, from 1964, is *Stable Homotopy Theory, Volume 3* in the Springer Lecture Notes in Mathematics. It contains material that has been superseded mathematically by results published elsewhere. Still, it has many delightful passages, and comparison of it with later work shows how rapidly algebraic topology in general, and Adams' work in particular, was progressing in those days. A second, *Algebraic Topology: a student's guide* consists primarily of selected reprints. The remaining three are highly recommended reading. Two of them, *Lectures on Lie Groups* and *Stable Homotopy and Generalized Homology* are being kept in print by the University of Chicago Press and remain among the best references on their subjects. The third, *Infinite Loop Spaces* is Study 90 of the Annals of Mathematics, published by Princeton University Press. It is vintage Adams, beautifully and humorously written, and it contains capsule summaries of various topics in algebraic topology other than the one described by the title.

The second volume ends with two expository articles which perhaps were not intended for publication. They well illustrate Adams' interest in the work of young algebraic topologists and his success in illuminating the most recent developments in his subject.

Biographical data

J. Frank Adams was born on 5 November 1930, in Woolwich, London, and died on 7 January 1989. He was married in 1953 and had one son and three daughters.

The main dates in his education were Bedford School until 1948, followed by military service in the Royal Engineers (1948–9) and attendance at Trinity College, Cambridge (1949–55), obtaining first class honours in Part II of the mathematical tripos (1951) and a distinction in Part III a year later. He was awarded his B.A. in 1952, his Ph.D. in 1955, his M.A. in 1956 and the higher degree of Sc.D. in 1982.

Following the completion of his Ph.D., he was appointed a Junior Lecturer at Oxford (1955–6). and he held a Research Fellowship at Trinity College, Cambridge, until 1958. During this time he also made his first visit to the University of Chicago (Summer 1957) and held a Commonwealth Fund Fellowship at the Institute of Advanced Study in Princeton (1957–8).

On returning to England he became an Assistant Lecturer at Cambridge, combining this with the Directorship of Studies in Mathematics at Trinity Hall. He was again a visiting member of the Institute at Princeton in the fall of 1961.

In 1962 Adams moved to the University of Manchester first as a Reader and then as Fielden Professor of Pure Mathematics (1964–71). In 1970 he returned to Cambridge as Lowndean Professor of Astronomy and Geometry, and he was also active as a Fellow of Trinity College.

His professional affiliations included the Association of University Teachers, the American Mathematical Society (since 1957) and the London Mathematical Society (since 1958). He was also elected a Fellow of the Royal Society in 1964 and a Foreign Associate of the National Academy of Sciences of the United States in 1985.

His frequent trips to the United States included ten lengthy visits to the University of Chicago, the last in 1985. Adams was in great demand as a speaker – his more important lectures included addresses to the International Congress of Mathematicians in Stockholm (1962) and in Moscow (1966), the Herman Weyl Lectures at the Institute in Princeton (1975) and an address to the American Mathematical Society as Bicentennial Exchange Lecturer (1976).

Over many years he contributed to the *Mathematical Reviews*, and in addition acted as referee for numerous other journals. He served as Editor, Member of the Editorial Board, or Editorial Advisor for three journals of the London Mathematical Society, the *Annals of Mathematics*, *Inventiones Mathematicae*, the *Journal of Pure and Applied Algebra*, and the *Mathematical Proceedings of the Cambridge Philosophical Society*. He also served on subcommittees to choose speakers in algebraic topology for the International Congress (once as Chairman), on the committee to choose Fields Medallists, and on the Consultative (programme) committee of the ICM. He gave similar organisational help to other conferences.

As a Fellow of the Royal Society he served on the mathematical sectional committee (two terms, one as chairman) and on the Council. He was also a member of the Council of the London Mathematical Society, and of the mathematics committee of the Science and Engineering Research Council (two terms).

Adams' honours include the Junior Berwick Prize (1963), the Senior Whitehead Prize (1974) and the Sylvester Medal (1982). Besides the Sc.D. from Cambridge he was also made a Doctor (h.c.) by the Universität Heidelberg in 1986.

Acknowledgements

The following papers appear in these volumes with grateful thanks to the original publishers:

On the chain algebra of a loop space, *Comment. Math. Helv.* **30** (1956) 305–330
On the structure and applications of the Steenrod algebra, *Comment. Math. Helv.* **32** (1958) 180–214
Reproduced here by kind permission of Birkäuser-Verlag.

Lectures on generalised cohomology, Lecture Notes in Mathematics 99 (1969) 1–138
Maps between classifying spaces, *Invent. Math.* **35** (1976) 1–41
Maps between classifying spaces II, *Invent. Math.* **49** (1978) 1–65
2-Tori in E_8, *Math. Ann.* **278** (1978) 29–39
Reproduced here by kind permission of Springer-Verlag.

The sphere, considered as an H-space mod p, *Quart. J. Math.* **12** (1961) 52–60
K-theory and the Hopf invariant, *Quart. J. Math.* **17** (1966) 31–38
Primitive elements in the K-theory of BSU, *Quart. J. Math.* **27** (1976) 253–262
Reproduced here by kind permission of Oxford University Press.

Atomic spaces and spectra, *J. Edinburgh. Math. Soc.* **32** (1989) 473–481
Reproduced here by kind permission of the Edinburgh Mathematical Society.

Finite H-spaces and Lie groups, *J. Pure Appl. Algebra* **19** (1980) 1–8
Reproduced here by kind permission of Elsevier Science Publishers.

Operations of the nth kind in K-theory and what we don't know about RP^∞, in *New developments in topology*, London Math. Soc. Lecture Notes Series. 11 (1974) 1–9
Maps between classifying spaces, III, London Math. Soc. Lecture Note Series. 86 (1983) 136–153
On formulae of Thom and Wu, *Proc. London Math. Soc.* **11** (1961) 741–752
Hopf algebras of cooperations for real and complex K-theory, *Proc. London Math. Soc.* **23** (1971) 385–408
The Hurewicz homomorphism for MU and BP, *J. London Math. Soc.* **5** (1972) 539–545
Reproduced here by kind permission of the London Mathematical Society.

Maps between p-completed classifying spaces
Reproduced by kind permission of the Royal Society of Edinburgh from *Proc. Roy. Soc. Edinburgh*, **112A**, 231–235.

On matrices whose real linear combinations are nonsingular, *Proc. Amer. Math. Soc.* **16** (1965) 318–322 & Correction, *ibid.* **17** (1966) 193–222

Algebraic topology in the last decade, *Proc. Sympos. Pure Math.* **22** (1971) 1–22
Graeme Segal's Burnside ring conjecture, *Bull. Amer. Math. Soc.* **6** (1982) 201–210
The fundamental representations of E_8, *Contemp. Math.* **3** (1985) 1–10
Reproduced here by kind permission of the American Mathematical Society.

H-spaces with few cells, *Topology* **1** (1962) 67–72
On the groups $J(X)$—I, *Topology* **2** (1963) 181–195
On the groups $J(X)$—II, *Topology* **3** (1965) 137–171
On the groups $J(X)$—III, *Topology* **3** (1965) 193–222
On the groups $J(X)$—IV, *Topology* **5** (1966) 21–71 & Correction, *ibid.* **7** (1968) 331
Modules over the Steenrod algebra, *Topology* **10** (1971) 271–282
A variant of E. H. Brown's representability theorem, *Topology* **10** (1971) 185–198
The Segal conjecture for elementary abelian p-groups, *Topology* **24** (1985) 435–460
A generalization of the Atiyah–Segal completion theorem, *Topology* **27** (1988) 1–6
A generalization of the Segal conjecture, *Topology* **27** (1988) 7–21
Reproduced here by kind permission of Pergamon Press.

On the non-existence of elements of Hopf invariant one, *Ann. of Math.* **72** (1960) 20–104
Vector fields on spheres, *Ann. of Math.* **75** (1962) 603–632
Finite H-spaces and algebras over the Steenrod algebra, *Ann. of Math.* **111** (1980) 95–143 & Correction, *ibid.* **113** (1981) 621–622
Reproduced here by kind permission of The Annals of Mathematics.

Chern characters revisited, *Illinois J. Math.* **17** (1973) 333–336 & Addendum, *ibid.* **20** (1976) 372
Stable operations on complex K-theory, *Illinois J. Math.* **21** (1977) 826–829
Reproduced here by kind permission of the Illinois Journal of Mathematics.

An example in homotopy theory, *Proc. Cambridge Philos. Soc.* **53** (1957) 922–923
A finiteness theorem in homological algebra, *Proc. Cambridge Philos. Soc.* **57** (1961) 52–60
On Chern characters and the structure of the unitary group, *Proc. Cambridge Philos. Soc.* **57** (1961) 189–199
An example in homotopy theory, *Proc. Cambridge Philos. Soc.* **60** (1964) 699–700
On complex Stiefel manifolds, *Proc. Cambridge Philos. Soc.* **61** (1965) 81–103
A periodicity theorem, *Proc. Cambridge Philos. Soc.* **62** (1966) 365–377
The Kahn–Priddy theorem, *Proc. Cambridge Philos. Soc.* **73** (1973) 43–55
Sub-Hopf-algebras of the Steenrod algebra, *Proc. Cambridge Philos. Soc.* **76** (1974) 45–52
Operations on K-theory of torsion-free spaces, *Math. Proc. Cambridge Philos. Soc.* **79** (1976) 483–491
Uniqueness of BSO, *Math. Proc. Cambridge Philos. Soc.* **80** (1976) 475–509
Calculation of Lin's Ext groups, *Math. Proc. Cambridge Philos. Soc.* **87** (1980) 459–469
Reproduced here by kind permission of the Cambridge Philosophical Society.

ON FORMULAE OF THOM AND WU

By J. F. ADAMS

[Received 5 January 1961]

1. RECENTLY, in preparing a set of lectures on characteristic classes, I had occasion to consider the formulae of Thom and Wu (**10**) which relate Stiefel–Whitney classes to Steenrod squares. Briefly, they are as follows. Let M be a compact differentiable n-manifold, not necsssarily orientable, with fundamental class $\mu \in H_n(M; Z_2)$. Then there is a unique class $v_i \in H^i(M; Z_2)$ such that

$$\langle Sq^i x, \mu \rangle = \langle v_i x, \mu \rangle$$

for each $x \in H^{n-i}(M; Z_2)$; and the Stiefel–Whitney classes $w_k \in H^k(M; Z_2)$ satisfy

$$w_k = \sum_{i+j=k} Sq^j v_i.$$

Although these formulae are simple and attractive, I did not feel that they gave me that complete understanding which I sought. For example, they raise a problem first recorded by Thom (**9**); briefly, it is as follows. One may use these formulae to define Stiefel–Whitney classes w_k in the cohomology of a manifold which is not necessarily differentiable, or indeed, to define Stiefel–Whitney classes in any algebra over Z_2 which admits operations Sq^i and satisfies suitable axioms. Do these generalized Stiefel–Whitney classes satisfy every formula which holds in the differentiable case? In particular, in the differentiable case we have

$$Sq^i w_k = Q(w_1, w_2, \ldots, w_{i+k})$$

for a certain polynomial $Q = Q(i, k)$; does this formula hold for the generalized Stiefel–Whitney classes?

We shall prove that the answer to this question of Thom is in the affirmative (see Corollary 3 below). The paper is arranged as follows. In §§ 2, 3 we consider abstract algebras of the sort indicated above, and attempt to obtain a better understanding of their theory. § 2 contains sufficient to answer Thom's question, and § 3 contains the remainder of the work. In § 4 we make some supplementary remarks.

It is a pleasure to express my gratitude to R. Thom for a helpful letter. A similar acknowledgement to F. Hirzebruch appears in context in § 4.

2. We have said that generalized Stiefel–Whitney classes are defined in any algebra over Z_2 which admits operations Sq^i and satisfies suitable axioms. In this section, therefore, the initial object of study will be a

Proc. London Math. Soc. (3) 11 (1961) 741–52

graded, anticommutative algebra $H = \sum_{0 \leqslant i \leqslant n} H^i$ over the field Z_p which admits operations Sq^i (if $p = 2$) or P^k (if $p > 2$). More precisely, if $p = 2$ we define A to be the mod 2 Steenrod algebra (**3, 6**); if $p > 2$ we define A to be that subalgebra of the mod p Steenrod algebra which is generated by the P^k; we now assume that H is a graded left module over the graded algebra A. We impose the following axioms.

(*a*) (Cartan formula.) Let $\Delta \colon A \to A \otimes A$ be the diagonal map (**6**); as a standard convention, we will write

$$\Delta a = \sum_r a'_r \otimes a''_r.$$

Then we have $$a(hk) = \sum_r (a'_r h)(a''_r k) \quad (h, k \in H).$$

(We need no signs in this formula, because either $p = 2$ or the elements of A are of even degree.)

(*b*) (Dimension axiom.) If $p = 2$, $h \in H^i$ and $i < j$, then $Sq^j h = 0$. If $p > 2$, $h \in H^i$, and $i < 2k$, then $P^k h = 0$.

(*c*) (Poincaré duality.) There is given an element μ in the vector-space dual of H^n. We write $\langle h, \mu \rangle$ for $\mu(h)$, to preserve the analogy with the topological case. The bilinear function $\langle hk, \mu \rangle$ of the variables $h \in H^i$, $k \in H^{n-i}$ gives a dual pairing from the finite-dimensional vector spaces H^i and H^{n-i} to Z_p.

For example, let M be a compact topological n-manifold, without boundary; and if $p > 2$, let M be oriented. Then the cohomology ring $H = H^*(M; Z_p)$ satisfies all these axioms, provided that we take μ to be the fundamental class.

If H is an algebra satisfying the axioms we have given, then we can make H into a graded right module over A; in fact, if $h \in H^i$, $a \in A^j$ we define ha by the equation

$$\langle ha\,.\,k, \mu \rangle = \langle h\,.\,ak, \mu \rangle \quad (k \in H^{n-i-j}).$$

However, we cannot assert that these operations of A on the right commute with those on the left; nor can we assert that they satisfy the Cartan formula or the dimension axiom.

In particular, we shall have much to do with the classes $\mathscr{E}_H a$, where $\mathscr{E}_H$ is the unit in H. The characteristic property of $\mathscr{E}_H a$ is

$$\langle \mathscr{E}_H a\,.\,k, \mu \rangle = \langle ak, \mu \rangle.$$

If we take $p = 2$, then $\mathscr{E}_H Sq^i$ is the class v_i which appears in the formulae of Thom and Wu.

In any algebra H, we can define various classes by starting from the unit $\mathscr{E}_H$ and iterating the operations we have mentioned above. (For example,

$$w_k = \sum_{i+j=k} Sq^i(\mathscr{E}_H\, Sq^j)$$

is a class of this sort, and so is $Sq^l w_k$.) We wish to study how many classes we can obtain in this way, and what universal formulae they satisfy; that is, what formulae hold in every H. We therefore proceed as follows.

We first define a class of 'words' W, by laying down the following four inductive rules:

(i) The letter $\mathscr{E}$ is a word.

(We emphasize that here $\mathscr{E}$ is merely a formal symbol; in particular, it should not be confused with the unit of any particular algebra H.)

(ii) If W is a word and $a \in A$, then aW and Wa are words.
(iii) If W and W' are words, then the 'cup-product' WW' is a word.
(iv) If W and W' are words and λ, $\mu \in Z_p$, then $\lambda W + \mu W'$ is a word.

For example, if $p = 2$, the following formula is a word:

$$Sq^2\{[(\mathscr{E} Sq^1)(\mathscr{E} Sq^2)] + (\mathscr{E} Sq^3)\}.$$

And, in general, a formula W is a word if and only if it is shown to be such by a finite number of applications of the four given rules.

If H is an algebra, satisfying the axioms we have given above, then we can regard each word W as a formula defining a specific element of H. More formally, we can define a function θ_H which assigns to each word W an element of H, by giving the following four inductive rules.

(i) $\theta_H(\mathscr{E}) = \mathscr{E}_H$.
(ii) $\theta_H(aW) = a(\theta_H(W))$, $\quad \theta_H(Wa) = (\theta_H(W))a$.
(iii) $\theta_H(WW') = (\theta_H(W))(\theta_H(W'))$.
(iv) $\theta_H(\lambda W + \mu W') = \lambda(\theta_H(W)) + \mu(\theta_H(W'))$.

We may refer to $\theta_H(W)$ as 'the value of W in H'.

We now divide the words W into equivalence classes, putting W and W' into the same class if we have $\theta_H(W) = \theta_H(W')$ for every H. We take these equivalence classes as the elements of a 'universal domain' U. The problem mentioned above is therefore equivalent to determining the structure of U.

It is trivial to check that U admits well-defined cup-products, linear combinations, and operations from A (both on the left and right). The operations from A are linear; the operations on the left satisfy the Cartan formula and the dimension axiom.

The ring-structure of U is given by Theorem 1 below; the remaining structure of U will be given in section 3.

THEOREM 1. *U is a polynomial algebra on the generators $u_1, u_2, \ldots, u_i, \ldots$ defined below.*

In order to define u_i, we write χ for the canonical anti-automorphism of A (**6**); this is defined, inductively, by the equations

$$\chi(1) = 1, \qquad \sum_r \chi(a'_r)a''_r = 0 \quad (\dim a > 0)$$

(where $\Delta a = \sum_r a'_r \otimes a''_r$, as always). We now define $u_i = \mathscr{E}(\chi Sq^i)$ if $p = 2$, $u_k = \mathscr{E}(\chi P^k)$ if $p > 2$.

The degree of u_i is thus i if $p = 2$, and $2i(p-1)$ if $p > 2$.

Our next theorem will show that U is faithfully represented in the cohomology of differentiable manifolds.

THEOREM 2. *Suppose given an integer N; let* M *run over those monomials in the u_i $(i > 0)$ whose degree is N or less. Then there is a differentiable manifold D (orientable if $p > 2$) such that the values in $H^*(D; Z_p)$ of the monomials* M *are linearly independent.*

Theorems 1 and 2 lead immediately to the following corollary.

COROLLARY 3. *Let W be a word of the sort considered above. Suppose that the value of W in $H^*(D; Z_p)$ is zero for every differentiable manifold D; then the value of W in H is zero for any H.*

It is clear that this corollary answers Thom's question in the affirmative. In fact, let us take $p = 2$; then

$$w_k = \sum_{i+j=k} Sq^i(\mathscr{E} Sq^j)$$

is a word of the sort considered; hence so is

$$W = Sq^i w_k - Q(w_1, w_2, \ldots, w_{i+k})$$

for any polynomial Q. If we choose Q so that $W = 0$ in every differentiable manifold, then the corollary shows that $W = 0$ in every H.

The remainder of this section will be devoted to proving Theorems 1 and 2. The manifolds D which we exhibit to prove Theorem 2 are cartesian products of projective spaces. We begin work as follows.

LEMMA 4. *Take $p = 2$; let x be the cohomology generator in $H^1(RP^\infty; Z_2)$. If $a \in A^i$ and $i+j+k = 2^s - 1$, then*

$$x^j . ax^k = x^k . (\chi a)x^j.$$

Take $p > 2$; let y be the cohomology generator in $H^2(CP^\infty; Z_p)$. If $a \in A^{2i}$ and $i+j+k = p^s - 1$, then

$$y^j . ay^k = y^k . (\chi a)y^j.$$

In proving this lemma, and later, it will be convenient to make a convention concerning the expansion

$$\Delta a = \sum_r a'_r \otimes a''_r.$$

If $\dim a > 0$, we may assume that this expansion contains the term $a \otimes 1$ (for $r = \alpha$, say) and the term $1 \otimes a$ (for $r = \omega$, say), while the remaining terms have $\dim a'_r > 0$, $\dim a''_r > 0$.

We give the proof of Lemma 4 for the case $p > 2$; the case $p = 2$ is closely similar. We proceed by induction over $\dim a$. The result is trivial if $\dim a = 0$; as an inductive hypothesis, we suppose it true if $\dim a < l$ ($l > 0$); we must deduce it when $\dim a = l$.

It is easy to see that if $N = p^s - 1$ and $m > 0$, then the operation

$$P^m\colon H^{2N-2m(p-1)}(CP^\infty; Z_p) \to H^{2N}(CP^\infty; Z_p)$$

is zero. Hence any operation a taking values in $H^{2N}(CP^\infty; Z_p)$ is zero, at least if $\dim a > 0$. In particular, if $a \in A^{2i}$ and $i+j+k = N$, we have $a(y^k . y^j) = 0$. That is,

$$ay^k . y^j + \sum_{r \neq \alpha} a'_r y^k . a''_r y^j = 0.$$

Using the inductive hypothesis, we have

$$ay^k . y^j + \sum_{r \neq \alpha} y^k . \chi(a'_r) a''_r y^j = 0.$$

Using the characteristic property of χ, we have

$$ay^k . y^j = y^k . \chi(a) y^j.$$

This completes the induction.

From this point onwards, we shall permit ourselves to write $\mathscr{E}$ instead of $\mathscr{E}_H$ for the unit in any algebra H under consideration.

Lemma 5. *Take $p = 2$, $n = 2^s - 2$ ($s \geqslant 2$), $M = RP^n$, and let x be the cohomology generator in $H^1(M; Z_2)$. Then in $H^*(M; Z_2)$ we have $\mathscr{E}(\chi Sq^1) = x$ and $\mathscr{E}(\chi Sq^i) = 0$ for $i > 1$.*

Take $p > 2$, $m = p^s - 2$ ($s \geqslant 2$), $M = CP^m$, and let y be the cohomology generator in $H^2(M; Z_p)$. Then in $H^*(M; Z_p)$ we have $\mathscr{E}(\chi P^1) = y^{p-1}$ and $\mathscr{E}(\chi P^k) = 0$ for $k > 1$.

Proof. According to Lemma 4, the calculation of an operation χa which maps into the top dimension of M reduces to the calculation of ax or ay, as the case may be. The values of $Sq^i x$ and $P^k y$ are well known, and lead to the result stated.

Lemma 6. *In a product manifold $M = M' \times M''$, the values of the classes $\mathscr{E}a$ are given by*

$$\mathscr{E}a = \sum_r \mathscr{E}'a'_r \otimes \mathscr{E}''a''_r,$$

where $\mathscr{E}'b$ and $\mathscr{E}''c$ denote the corresponding classes in M' and M''.

The verification is obvious.

We will now prove Theorem 2. We take the manifold D to be a cartesian product of N factors, where each factor is a projective space RP^n (if $p = 2$) or CP^m (if $p > 2$). We suppose, of course, that $n = 2^s-2 \geqslant N$, or that $m = p^s-2 \geqslant \frac{1}{2}N$, according to the case considered. If $p = 2$, we write $x_1, x_2, \ldots, x_N$ for the cohomology generators in the separate factors; if $p > 2$, we call them $y_1, y_2, \ldots, y_N$.

LEMMA 7. *If $p = 2$, $\mathscr{E}(\chi Sq^i)$ is the i-th elementary symmetric function in $x_1, x_2, \ldots, x_N$.*

If $p > 2$, $\mathscr{E}(\chi p^k)$ is the k-th elementary symmetric function in $(y_1)^{p-1}$, $(y_2)^{p-1}, \ldots, (y_N)^{p-1}$.

This lemma follows immediately from Lemmas 5 and 6; and it completes the proof of Theorem 2.

In order to prove Theorem 1, it is now sufficient to show that U is multiplicatively generated by the elements u_i, $i > 0$. We begin work as follows:

LEMMA 8 (i). *If* $\dim a > 0$ *and* $h \in H$, *then*

$$(\mathscr{E}a).h = ah + \sum_{r \neq \alpha,\omega} (a'_r h)a''_r + ha.$$

(ii) *If* $\dim a > 0$ *and* $u \in U$, *then*

$$(\mathscr{E}a).u = au + \sum_{r \neq \alpha,\omega} (a'_r u)a''_r + ua.$$

(iii) *If* $\dim a > 0$, *then*

$$(\mathscr{E}a).(\mathscr{E}b) = a(\mathscr{E}b) + \sum_{r \neq \alpha,\omega} (a'_r(\mathscr{E}b))a''_r + \mathscr{E}ba.$$

Proof. We begin with (i). If k is also in H and of the appropriate dimension, we have

$$\begin{aligned}\langle(\mathscr{E}a).h.k, \mu\rangle &= \langle a(hk), \mu\rangle \\ &= \langle ah.k, \mu\rangle + \sum_{r \neq \alpha,\omega} \langle a'_r h.a''_r k, \mu\rangle + \langle h.ak, \mu\rangle \\ &= \langle ah.k, \mu\rangle + \sum_{r \neq \alpha,\omega} \langle (a'_r h)a''_r.k, \mu\rangle + \langle ha.k, \mu\rangle.\end{aligned}$$

Since this holds for each k, it establishes (i).

Now take an element $u \in U$. The equation

$$(\mathscr{E}a).u = au + \sum_{r \neq \alpha,\omega} (a'_r u)a''_r + ua$$

holds in every H; therefore it holds in U. This proves part (ii). Part (iii) follows by substituting $u = \mathscr{E}b$.

LEMMA 9. *U is multiplicatively generated by the elements $\mathscr{E}a$, $a \in A$.*

Proof. It is sufficient to prove that if $a \in A$, and W is a polynomial in the elements $\mathscr{E}b$, then aW, Wa may also be written as polynomials in the $\mathscr{E}b$. We will prove this proposition by induction over $\dim a$. The proposition is trivial if $\dim a = 0$; as an inductive hypothesis, we suppose it true when $\dim a < k$ $(k > 0)$; we must deduce it when $\dim a = k$.

We begin with the expression $a(\mathscr{E}b)$. Consider the equation of Lemma 8 (iii). The terms $(\mathscr{E}a).(\mathscr{E}b)$, and $\mathscr{E}ba$ are already polynomials in the $\mathscr{E}c$; and each term $(a'_r(\mathscr{E}b))a''_r$ can be written in that form, by the inductive hypothesis. Hence $a(\mathscr{E}b)$ can be written as a polynomial in the $\mathscr{E}c$.

If W is a polynomial in the $\mathscr{E}b$, then aW can be expanded (by linearity and the Cartan formula) in terms of expressions $c(\mathscr{E}b)$ with $\dim c \leqslant k$. Each of the expressions $c(\mathscr{E}b)$ can be written as a polynomial in the $\mathscr{E}d$, as we have just shown; hence aW can be written as a polynomial in the $\mathscr{E}d$.

This completes the inductive step, so far as aW is concerned; we turn to Wa. By Lemma 8 (ii) we have

$$Wa = (\mathscr{E}a).W - aW - \sum_{r \neq \alpha,\omega} (a'_r W)a''_r.$$

The term $(\mathscr{E}a)W$ is already a polynomial in the $\mathscr{E}b$; aW can be written in that form, as we have just shown; and so can each term $(a'_r W)a''_r$, by the inductive hypothesis. Hence Wa can be written as a polynomial in the $\mathscr{E}b$. This completes the induction, and the proof of Lemma 9.

We have yet to show that U is multiplicatively generated by the u_i $(i > 0)$. Let us write $I(U) = \sum_{j>0} U^j$, and let $D(U)$ be the set of decomposable elements in U, that is, those which can be written in the form $u = \sum_r u'_r u''_r$ with $u'_r \in I(U)$, $u''_r \in I(U)$. Our task amounts to calculating $I(U)/D(U)$.

LEMMA 10. *If* $u \in D(U)$, $a \in A$, *then* $ua \in D(U)$.

The proof is by induction over $\dim a$. The result is trivial when $\dim a = 0$; as an inductive hypothesis, we suppose it true when $\dim a < k$ $(k > 0)$; we must deduce it when $\dim a = k$.

Consider the equation of Lemma 8 (ii). The term $(\mathscr{E}a).u$ is certainly decomposable. Since u is decomposable, au is decomposable, by the Cartan formula; and similarly, each term $a'_r u$ is decomposable. By the inductive hypothesis, each term $(a'_r u)a''_r$ is decomposable. Hence ua is decomposable. This completes the induction.

LEMMA 11. *If* $\dim b > 0$, *then*

$$a(\mathscr{E}b) = \mathscr{E}b\chi(a) \quad \bmod D(U).$$

The proof is again by induction over $\dim a$. The result is trivial when $\dim a = 0$; as an inductive hypothesis, we suppose it true when $\dim a < k$ $(k > 0)$; we must deduce the result when $\dim a = k$.

By Lemma 8 (iii) we have

$$a(\mathscr{E}b) = (\mathscr{E}a).(\mathscr{E}b) - \mathscr{E}ba - \sum_{r \neq \alpha,\omega} (a'_r(\mathscr{E}b))a''_r.$$

Since $\dim b > 0$, the term $(\mathscr{E}a).(\mathscr{E}b)$ is decomposable. By the inductive hypothesis, the term $a'_r(\mathscr{E}b)$ yields $\mathscr{E}b\chi(a'_r)$, modulo $D(U)$. Using Lemma 10, the term $(a'_r(\mathscr{E}b))a''_r$ yields $\mathscr{E}b\chi(a'_r)a''_r$, modulo $D(U)$. Hence

$$a(\mathscr{E}b) = -\mathscr{E}ba - \sum_{r \neq \alpha,\omega} \mathscr{E}b\chi(a'_r)a''_r$$

modulo $D(U)$. Using the characteristic property of χ, we find

$$a(\mathscr{E}b) = \mathscr{E}b\chi(a)$$

modulo $D(U)$. This completes the induction.

LEMMA 12. *$I(U)/D(U)$ is spanned by the elements u_i, $i > 0$.*

We will give the proof for the case $p > 2$. We easily see (using Lemma 9) that $I(U)/D(U)$ is spanned by the elements $\mathscr{E}a$, where $\dim a > 0$. In fact, it is spanned by the elements $\mathscr{E}a$ as a runs over a set of elements which span $I(A) = \sum_{i>0} A^i$.

Consider the set S of elements

$$P^{k_1}P^{k_2}\dots P^{k_l}$$

such that $k_1 \geqslant pk_2$, $k_2 \geqslant pk_3, \dots, k_{l-1} \geqslant pk_l > 0$. It is easy to show from the Adem relations that these elements do span $I(A)$. It is well known (**3**) that they form a base for it; but we do not need this fact here.

The set S contains the elements P^k $(k > 0)$. Every other element in S can be written in the form $P^l c$ with $0 < \dim c < 2l$. As a runs over S, χa runs over a set χS which also spans $I(A)$. The set χS contains the elements χP^k $(k > 0)$; every other element in χS can be written in the form $d\chi(P^l)$ with $0 < \dim d < 2l$. By Lemma 11, we have

$$\mathscr{E}d\chi(P^l) = P^l(\mathscr{E}d) \quad \operatorname{mod} D(U).$$

But $P^l(\mathscr{E}d)$ is zero, because $\dim(\mathscr{E}d) < 2l$. We have shown that $\mathscr{E}d\chi(P^l)$ is decomposable. Hence $I(U)/D(U)$ is spanned by the elements $\mathscr{E}\chi(P^k)$ $(k > 0)$. This completes the proof in case $p > 2$.

The proof in case $p = 2$ is closely similar. We begin with the set S of elements

$$Sq^{i_1}Sq^{i_2}\dots Sq^{i_l}$$

such that $i_1 \geqslant 2i_2$, $i_2 \geqslant 2i_3, \dots, i_{l-1} \geqslant 2i_l > 0$. The set χS contains the elements χSq^i $(i > 0)$; every other element in χS can be written in the form $d\chi(Sq^j)$ with $0 < \dim d < j$; $Sq^j(\mathscr{E}d)$ is zero, and therefore $\mathscr{E}d\chi(Sq^j)$ is decomposable. This completes the proof of Lemma 12.

Lemma 12 shows that U is multiplicatively generated by the elements u_i $(i > 0)$. This completes the proof of Theorem 1.

3. In this section we shall complete the description of U, by describing its operations from A (on the left and on the right). This description is given in Theorem 13 below. After proving this theorem, the section concludes by remarking that U can be given the structure of a Hopf algebra.

THEOREM 13. *Let $P(u_1, u_2,...)$ be a polynomial in the u_i. Then in U we have*

$$a(P(u_1, u_2,...)) = Q(u_1, u_2,...),$$
$$(P(u_1, u_2,...))a = R(u_1, u_2,...),$$

where the polynomials Q and R are constructed according to the method given below.

We now give the method for constructing Q and R. If $p = 2$, take $N \geqslant \dim a+\dim P$, and take a cartesian product of N copies of RP^∞, with fundamental classes $x_1, x_2,..., x_N$. Let σ_i be the ith elementary symmetric function in $x_1, x_2,..., x_N$; set $X = x_1 x_2 ... x_N$. Solve the equations

$$a(P(\sigma_1, \sigma_2,...)) = Q(\sigma_1, \sigma_2,...),$$
$$(\chi a)(XP(\sigma_1, \sigma_2,...)) = XR(\sigma_1, \sigma_2,...)$$

for Q and R.

If $p > 2$, take $2N \geqslant \dim a+\dim P$, and take a cartesian product of N copies of CP^∞, with fundamental classes $y_1, y_2,..., y_N$. Let σ_i be the ith elementary symmetric function in $(y_1)^{p-1}, (y_2)^{p-1},..., (y_N)^{p-1}$; set $Y = y_1 y_2 ... y_N$. Solve the equations

$$a(P(\sigma_1, \sigma_2,...)) = Q(\sigma_1, \sigma_2,...),$$
$$(\chi a)(YP(\sigma_1, \sigma_2,...)) = YR(\sigma_1, \sigma_2,...)$$

for Q and R.

Example. Take $p = 2$, $P = \mathscr{E}$, $a = Sq^3$; we will calculate R. We have

$$\begin{aligned}(\chi a)X &= (Sq^2Sq^1)(x_1 x_2 ... x_N)\\ &= Sq^2(\textstyle\sum x_1^2 x_2 ... x_N)\\ &= \textstyle\sum x_1^4 x_2 ... x_N + \sum x_1^2 x_2^2 x_3^2 x_4 ... x_N\\ &= X(\textstyle\sum x_1^3 + \sum x_1 x_2 x_3).\end{aligned}$$

But $\sum x_1^3 + \sum x_1 x_2 x_3 = \sigma_1^3 + \sigma_1 \sigma_2$; therefore $\mathscr{E} Sq^3 = u_1^3 + u_1 u_2$.

Caution. This representation of U (for $p = 2$) does not throw w_k onto the elementary symmetric function σ_k.

We give the proof of Theorem 13 for the case $p > 2$; the case $p = 2$ is closely similar. Lemma 7 shows that U is faithfully represented (up to any given dimension) in $H^*(D; Z_p)$, where D is a certain cartesian product of projective spaces CP^m with $m = p^s-2$. Moreover, in this representation, u_i becomes the ith elementary symmetric function σ_i in $(y_1)^{p-1}$, $(y_2)^{p-1},..., (y_N)^{p-1}$. If we find formulae $aP = Q$, $Pa = R$ which hold in

$H^*(D; Z_p)$, then these formulae must hold in U. This establishes the construction for Q; it remains to establish the construction for Pa in $H^*(D; Z_p)$.

Let us write the iterated diagonal in A in the form

$$\Delta^{N-1}a = \sum a^{(1)} \otimes a^{(2)} \otimes \ldots \otimes a^{(N)},$$

without indicating the parameter of summation. Then we have

$$\begin{aligned}\langle (y_1^{j_1}y_2^{j_2}\ldots y_N^{j_N})a \,.\, Y \,.\, y_1^{k_1}y_2^{k_2}\ldots y_N^{k_N}, \mu\rangle &\\ &= \langle y_1^{j_1}\ldots y_N^{j_N} \,.\, a(y_1^{k_1+1}\ldots y_N^{k_N+1}), \mu\rangle \\ &= \sum \langle y_1^{j_1}\ldots y_N^{j_N} \,.\, a^{(1)}y_1^{k_1+1}\ldots a^{(N)}y_N^{k_N+1}, \mu\rangle \\ &= \sum \langle y_1^{j_1} \,.\, a^{(1)}y_1^{k_1+1}, \mu_1\rangle \ldots \langle y_N^{j_N} \,.\, a^{(N)}y_N^{k_N+1}, \mu_N\rangle.\end{aligned}$$

Using Lemma 4 and the fact that $m = p^s - 2$, this yields

$$\begin{aligned}\sum \langle (\chi a^{(1)})y_1^{j_1+1} \,.\, y_1^{k_1}, \mu_1\rangle \ldots \langle (\chi a^{(N)})y_N^{j_N+1} \,.\, y_N^{k_N}, \mu_N\rangle &\\ &= \sum \langle (\chi a^{(1)})y_1^{j_1+1}\ldots (\chi a^{(N)})y_N^{j_N+1} \,.\, y_1^{k_1}\ldots y_N^{k_N}, \mu\rangle \\ &= \langle (\chi a)(y_1^{j_1+1}\ldots y_N^{j_N+1}) \,.\, y_1^{k_1}\ldots y_N^{k_N}, \mu\rangle \\ &= \langle (\chi a)(Y y_1^{j_1}y_2^{j_2}\ldots y_N^{j_N}) \,.\, y_1^{k_1}y_2^{k_2}\ldots y_N^{k_N}, \mu\rangle.\end{aligned}$$

Therefore $(Pa).Y = (\chi a)(YP)$. This establishes the construction of R.

Instead of using this representation of U, it would be possible (and almost equivalent) to use the theory of Hopf algebras (**8**). One would first have to make U into a Hopf algebra; this is done below. In order to give the structure maps

$$A \otimes U \to U, \qquad U \otimes A \to U,$$

one would then consider the dual maps

$$A_* \otimes U_* \leftarrow U_*, \qquad U_* \otimes A_* \leftarrow U_*$$

defined on the dual U_* of U. One would remark that these maps are multiplicative, and one would finish by giving their values on the generators of U_*. We will now provide the foundations for this alternative approach.

We first note that if H' and H'' are algebras satisfying the axioms given above, then $H' \otimes H''$ is another such. This leads to the following lemma.

LEMMA 14. *For each u in U there is one and only one element*

$$\Delta u = \sum_s u'_s \otimes u''_s$$

in $U \otimes U$ such that

$$u(H' \otimes H'') = \sum_s u'_s(H') \otimes u''_s(H'')$$

for all H' and H''. The map $\Delta: U \to U \otimes U$ makes U into a Hopf algebra. We have

$$\Delta(au) = \sum_{r,s} a'_r u'_s \otimes a''_r u''_s,$$

$$\Delta(ua) = \sum_{r,s} u'_s a'_r \otimes u''_s a''_r.$$

The proof is obvious.

It follows, in particular, that Δ is given on the generators of U by

$$\Delta u_k = \sum_{i+j=k} u_i \otimes u_j,$$

where $u_0 = 1$. From this it follows that the dual U_* of U is again a polynomial algebra. We may describe generators g_j in U_* as follows: we set $\langle (u_1)^j, g_j \rangle = 1$, $\langle \mathrm{M}, g_j \rangle = 0$ for any other monomial M in the u_i.

4. In the preceding sections, we have been primarily concerned with 'characteristic classes' in abstract cohomology rings. One may ask how this theory applies to the cohomology of differentiable manifolds; the answer is that in this case those classes, which were defined in the abstract case in § 2, can be calculated in terms of the classical characteristic classes.

Let us write O for the 'infinite' orthogonal group, and BO for its classifying space. If M is a differentiable manifold, then its tangent bundle τ induces a map

$$\tau^*: H^*(BO; Z_p) \to H^*(M; Z_p).$$

Let the algebra U be as in §§ 2, 3. Then one can define a (unique) monomorphism

$$\nu: U \to H^*(BO; Z_p)$$

of Hopf algebras, with the following property: if $u \in U$, and M is any differentiable manifold, then the value of u in $H^*(M; Z_p)$ is given by the characteristic class $\tau^*\nu u$. In other words, the (universal) characteristic class νu gives a universal formula for calculating the value of u in differentiable manifolds.

By using classical representations of $H^*(BO; Z_p)$, it is possible to present ν in a form convenient for calculation. In this direction we present the following formulae as a sample; the u_i which occur are the generators of U considered in §§ 2, 3.

If $p = 2$, $\sum_0^\infty \nu(u_i)$ is given by the multiplicative sequence of polynomials (**4, 5**) in the Stiefel–Whitney classes corresponding to the power-series $1/(1+x)$.

If $p > 2$, $\sum_0^\infty \nu(u_i)$ is given by the multiplicative sequence of polynomials in the mod p Pontryagin classes corresponding to the power-series $1/(1+y^{p-1})$.

In order to state our next remark, we recall that in § 2 we made

$H^*(M; Z_p)$ into a right A-module (for each M). There is a unique way of making $H^*(BO; Z_p)$ into a right A-module so that

$$\tau^*: H^*(BO; Z_p) \to H^*(M; Z_p)$$

is always a map of right A-modules. In elementary terms, this means the following. Let $c \in H^*(M; Z_p)$ be a (classical) characteristic class of M, and take $a \in A$. Then there is a universal formula for finding a second characteristic class d in M such that

$$\langle c.ah, \mu\rangle = \langle d.h, \mu\rangle$$

for all h.

It is possible to present the right A-module structure of $H^*(BO; Z_p)$ in a form convenient for calculation.

In the first draft of this paper I developed these remarks at some length, with proofs; I now forbear to do so, for three reasons. The first is space; and the second is that they belong to a cadre of ideas which is much more generally known and understood than that studied in §§ 2, 3 of this paper. The third is that they overlap to some extent with work of F. Hirzebruch (**4**); a paper on the subject is being prepared by Atiyah and Hirzebruch. I am very grateful to Hirzebruch for sending me a copy of an unpublished manuscript of great elegance.

REFERENCES

1. A. Borel, 'Sur la cohomologie des espaces fibrés principaux et des espaces homogènes de groupes de Lie compacts', *Annals of Math.* 57 (1953) 115–207.

2. —— and F. Hirzebruch, 'Characteristic classes and homogeneous spaces I, II', *American J. Math.* 80 (1958) 458–538, 81 (1959) 315–82.

3. H. Cartan, 'Sur l'itération des opérations de Steenrod', *Comment. Math. Helvetici* 29 (1955) 40–58.

4. F. Hirzebruch, 'On Steenrod's reduced powers, the index of inertia, and the Todd genus', *Proc. Nat. Acad. Sci. U.S.A.* 39 (1953) 951–6.

5. —— *Neue topologische Methoden in der algebraischen Geometrie* (Springer-Verlag: Berlin, 1956).

6. J. W. Milnor, 'The Steenrod algebra and its dual', *Annals of Math.* 67 (1958) 150–71.

7. —— *Lectures on characteristic classes* (Princeton; mimeographed notes).

8. —— and J. C. Moore, *On the structure of Hopf algebras* (Princeton; mimeographed notes).

9. R. Thom, 'Espaces fibrés en sphères et carrés de Steenrod', *Ann. Sci. École Norm. Sup.* 69 (1952) 109–82.

10. W.-T. Wu, 'Classes caractéristiques et i-carrés d'une variété', *C.R. Acad. Sci Paris* 230 (1950) 508–11.

Trinity Hall
Cambridge

Reprinted from the *Proceedings of the Cambridge Philosophical Society*,
Volume 57, Part 2, pp. 189–199, 1961

ON CHERN CHARACTERS AND THE STRUCTURE OF THE UNITARY GROUP

BY J. F. ADAMS

Received 28 May 1960

1. *Introduction*. The purpose of this paper is twofold. In order to state our first aim, let U denote the 'infinite' unitary group, and let BU be a classifying space for U. Then Bott (2), (3) has shown that

$$\pi_{2t}(BU) = Z, \quad \pi_{2t+1}(BU) = 0;$$

we propose to investigate the 'Postnikov system' of BU.

Our second aim is to prove certain results about characteristic classes. As an example of these results, we refer the reader to Theorem 1 below; this is an integrality theorem for Chern characters.

It will appear, and will formally be shown in § 5, that the two aims mentioned above are closely related to one another.

In order to state Theorem 1, we need certain standard conventions. Let $k: Z \to Q$ denote the embedding of the integers in the rationals, and let X be a space; then a cohomology class $h^n \in H^n(X; Q)$ is said to be *integral* if it lies in the image of the induced homomorphism

$$k_*: \quad H^n(X; Z) \to H^n(X; Q).$$

By way of data, we assume that Y is a CW-complex, that η is a complex vector bundle over Y, and that the restriction of η to the $(2q-1)$-skeleton Y^{2q-1} of Y is trivial. We write $\mathrm{ch}_n(\eta)$ for the nth component of the Chern character of η, so that

$$\mathrm{ch}_n(\eta) \in H^{2n}(Y; Q).$$

THEOREM 1. (Integrality Theorem for Chern Characters.) *If $m(r)$ denotes the numerical function defined below, then the class $m(r)\,\mathrm{ch}_{q+r}(\eta)$ is integral.*

The numerical function $m(r)$ is defined by

$$m(r) = \prod_p p^{[r/(p-1)]},$$

where p runs over the prime numbers and $[x]$ denotes the integer part of x. The values of $m(r)$ for small values of r are as follows:

$r =$	0	1	2	3	4
$m(r) =$	1	2	12	24	720.

We emphasize that the integer $m(r)$ does not depend on the initial dimension $2q$. In this sense, Theorem 1 is a 'stable' result. Among such results it is 'best possible'; that is, it would not remain true if $m(r)$ were replaced by any smaller number. This will appear during the course of the proof. Alternatively, it is possible (and more elementary) to argue by displaying particular bundles with suitable Chern characters. In this context we remark that the function $m(r)$ is already well-known in the theory of characteristic classes; see (4), p. 16, Lemma 1.7.3.

Theorem 1 will follow as a trivial corollary from Theorem 2, which is more technical but contains more information. In order to state Theorem 2, we need further notation. Following (1), we introduce the group $K(X)$. It is sufficient for our purposes to know that an element of $K(X)$ has two independent components; the first is an integer n; the second is a homotopy class of maps $f\colon X \to BU$. This account may be taken as the definition of $K(X)$ whenever X is a connected CW-complex. If $\xi = (n, f)$ is an element of $K(X)$, then the Chern character of ξ is given by

$$\mathrm{ch}_0(\xi) = n,$$
$$\mathrm{ch}_q(\xi) = f^* \mathrm{ch}_q \quad (q > 0),$$

where ch_q is the 'universal' Chern character in $H^{2q}(BU; Q)$.

Let p be a prime; then $\rho\colon Z \to Z_p$ shall denote the quotient map from the integers to the integers mod p, and

$$\rho_*\colon \quad H^*(X; Z) \to H^*(X; Z_p)$$

shall denote the induced homomorphism of cohomology groups. Let P^k denote the Steenrod cyclic reduced power; if $p = 2$, we interpret P^k as Sq^{2k}. Let χ denote the canonical anti-automorphism of the mod p Steenrod algebra (5), so that we have an operation

$$\chi(P^k)\colon \quad H^n(X; Z_p) \to H^{n+2k(p-1)}(X; Z_p).$$

THEOREM 2. *It is possible to define characteristic classes* $\mathrm{ch}_{q,r}(\xi)$ *which satisfy the following conditions.*

(1) $\mathrm{ch}_{q,r}(\xi)$ *is defined if and only if* $\xi \in K(X)$ *and* X *is a CW-complex which is* $(2q-1)$*-connected.*

(2) $\mathrm{ch}_{q,r}(\xi)$ *lies in* $H^{2q+2r}(X; Z)$.

(3) $k_* \mathrm{ch}_{q,r}(\xi) = m(r)\, \mathrm{ch}_{q+r}(\xi)$.

(4) *Suppose that* $f\colon X \to X'$ *is a map in which* X *and* X' *are* $(2q-1)$*-connected and* $(2q'-1)$*-connected respectively. Assume that* $q \geq q'$ *and* $q+r = q'+r'$; *let* $\xi' \in K(X')$. *Then*

$$f^* \mathrm{ch}_{q',r'}(\xi') = \frac{m(r')}{m(r)} \mathrm{ch}_{q,r}(f^*\xi').$$

(5) *Suppose that* $r = s(p-1)+t$ *with* s, t *integers and* $0 \leq t < p-1$. *Then*

$$\rho_* \mathrm{ch}_{q,r}(\xi) = \frac{m(r)}{p^s m(t)} \chi(P^s) \rho_* \mathrm{ch}_{q,t}(\xi).$$

These conditions call for certain comments. In condition (1), the requirement that X should be $(2q-1)$-connected is vital. Condition (4) has two aspects; on the one hand, if we set $q = q'$, it asserts that $\mathrm{ch}_{q,r}(\xi)$ is natural; on the other hand, if we set $f = 1$, we obtain a statement about the dependence of $\mathrm{ch}_{q,r}(\xi)$ on r. The coefficient $m(r')/m(r)$ which occurs here is an integer. In condition (5), the coefficient $m(r)/p^s m(t)$ is an integer prime to p.

We will now deduce Theorem 1 from Theorem 2. If $q+r = 0$, the result is trivial. Otherwise, let η be a bundle of the sort considered in Theorem 1; then η can be induced by a suitable map

$$f\colon\ Y/Y^{2q-1} \to BU(n).$$

We may compose f with the inclusion $BU(n) \to BU$; since Y/Y^{2q-1} is $(2q-1)$-connected, Theorem 1 follows immediately from condition (3) of Theorem 2.

The author's main interest in condition (5) of Theorem 2 can be explained as follows. By studying the Postnikov system of BU, it is possible to define certain cohomology operations, of kinds higher than the first, which can be computed in suitable spaces X by calculations involving Chern characters. Condition (5) then appears as a formula relating such operations to primary operations. I hope to return to this subject in a further paper.

Theorem 2 is also relevant to recent work of Atiyah; an account will appear in a paper by Atiyah and Hirzebruch, now in preparation. It is a pleasure to acknowledge the stimulus of conversations with Atiyah.

2. *Killing homotopy groups.* We will briefly recall some ideas from homotopy theory, in order to fix the notation needed below.

Let X, Y be connected CW-complexes. We call Y a *space of type* $X(1, \ldots, n)$ if:

(1) There is a map $f\colon X \to Y$ such that $f_*\colon \pi_r(X) \to \pi_r(Y)$ is an isomorphism for $1 \leqslant r \leqslant n$, and

(2) $\pi_r(Y) = 0$ for $r > n$.

For each X and n, there is at least one such space Y. Moreover, any such Y has the universal property indicated by the following diagram.

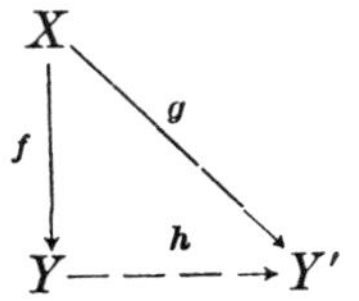

In full, let Y' be another space such that $\pi_r(Y') = 0$ for $r > n$; then any map $g\colon X \to Y'$ can be factored in the form $g \sim hf$, and this equation determines $h\colon Y \to Y'$ up to homotopy. It follows that the homotopy type of Y is determined by that of X.

Let W, X be connected CW-complexes. We call W a *space of type* $X(m, \ldots, \infty)$ if:

(1) There is a map $f\colon W \to X$ such that $f_*\colon \pi_r(W) \to \pi_r(X)$ is an isomorphism for $r \geqslant m$, and

(2) $\pi_r(W) = 0$ for $1 \leqslant r < m$.

This notion is justified in a manner similar to the preceding one.

Now suppose that W, X, Y, Z are connected CW-complexes, and that Y is of type $X(1, \ldots, n)$, W is of type $X(m, \ldots, \infty)$, where $m \leqslant n$. We call Z a space of type $X(m, \ldots, n)$ if it is

(1) a space of type $W(1, \ldots, n)$, or

(2) a space of type $Y(m, \ldots, \infty)$.

These two conditions are equivalent, since each leads to a diagram of the following form.

$$\begin{array}{ccc} X & \leftarrow & W \\ \downarrow & & \downarrow \\ Y & \leftarrow & Z \end{array}$$

For clarity, we display the types of the four spaces involved in the diagram above:

$$\begin{array}{ccc} X(1, \ldots, \infty) & \leftarrow & X(m, \ldots, \infty) \\ \downarrow & & \downarrow \\ X(1, \ldots, n) & \leftarrow & X(m, \ldots, n) \end{array}$$

These notions have been presented for CW-complexes, but by applying them to singular complexes we obtain corresponding singular notions for arbitrary connected spaces. In particular, given X and given $r \leqslant s < t$, we can construct a Serre fibring $F \to E \to B$ so that F, E and B are spaces of singular type $X(s+1, \ldots, t)$, $X(r, \ldots, t)$ and $X(r, \ldots, s)$.

We will allow ourselves to use the symbol $X(m, \ldots, n)$ to denote some space, not specified, of type $X(m, \ldots, n)$.

3. *Reduction of the problem; calculations of cohomology.* If $q+r = 0$, all the conditions of Theorem 2 can be satisfied in a trivial way; we may therefore suppose $q+r > 0$. Since we deal only with spaces X which are $(2q-1)$-connected, we may lift each map $f\colon X \to BU$ and obtain a map $g\colon X \to BU(2q, \ldots, \infty)$, whose homotopy class is determined by that of f. In order to define a natural characteristic class $\mathrm{ch}_{q,r}(\xi)$, then, it is necessary and sufficient to define a universal class

$$\mathrm{ch}_{q,r} \in H^{2q+2r}(BU(2q, \ldots, \infty); Z).$$

These universal classes have, of course, to satisfy certain conditions. In stating these conditions, it is convenient to write

$$n_{q,q'} : BU(2q, \ldots, \infty) \to BU(2q', \ldots, \infty)$$

for the canonical map, which exists if $q \geqslant q'$. Otherwise we use the notation and assumptions of Theorem 2. The conditions we must satisfy are as follows.

$$(3') \qquad k_* \, \mathrm{ch}_{q,r} = m(r) \, n^*_{q,1} \, \mathrm{ch}_{q+r}.$$

$$(4') \qquad n^*_{q,q'} \, \mathrm{ch}_{q',r'} = \frac{m(r')}{m(r)} \, \mathrm{ch}_{q,r}.$$

$$(5') \qquad \rho_* \, \mathrm{ch}_{q,r} = \frac{m(r)}{p^s m(t)} \chi(P^s) \, \rho_* \, \mathrm{ch}_{q,t},$$

at least if $q+t > 0$; if $q+t = 0$, we require

$$\rho_* \, \mathrm{ch}_{q,r} = 0.$$

We now invoke two results of Bott. First, BU is equivalent to the identity-component $\Omega^2 BU$ (2), (3). Secondly, we may take an equivalence between BU and $\Omega^2 BU$ in which the universal Chern character $\mathrm{ch}_n \in H^{2n}(BU; Q)$ corresponds to the double suspension $\sigma^2 \, \mathrm{ch}_{n+1} \in H^{2n}(\Omega^2 BU; Q)$. (This result may be recovered from Proposition 1 of (1).) The first result shows that $BU(2q-2, \ldots, \infty)$ is equivalent to $\Omega^2(BU(2q, \ldots, \infty))$.

We propose to choose the classes $\mathrm{ch}_{q,r}$ so that $\mathrm{ch}_{q-1,r}$ corresponds to $\sigma^2\,\mathrm{ch}_{q,r}$ under this equivalence, at least if $q-1+r > 0$. This means that we need only show how to choose $\mathrm{ch}_{q,r}$ when q is large compared with r. If we choose our classes $\mathrm{ch}_{q,r}$ so that $\mathrm{ch}_{q-1,r}$ corresponds to $\sigma^2\,\mathrm{ch}_{q,r}$, we need only verify properties (3′), (4′) and (5′) when q is large; this is easily shown, granted the second result mentioned above. During the rest of our work, therefore, we may assume that q is as large as we need.

We will now begin calculating $H^*(BU(2q, \ldots, \infty); G)$ for various coefficient groups G. As remarked above, whenever $a \leqslant b < c$, we can construct a fibring $F \to E \to B$ in which F, E and B are spaces of type $BU(2b+2, \ldots, 2c)$, $BU(2a, \ldots, 2c)$ and $BU(2a, \ldots, 2b)$. Each such fibring gives us an exact cohomology sequence, valid in a certain range of dimensions. These exact sequences evidently yield a spectral sequence, which one can use in computing $H^*(BU(2q, \ldots, \infty); G)$ in a certain range of dimensions, this range depending on q. The details of this spectral sequence are analogous to those given in (8). In particular, the differentials in the spectral sequence are obtained from the transgressions in the exact sequences, by passing to subgroups and quotient groups. These differentials therefore commute (up to sign) with stable cohomology operations. The E_1 term of the spectral sequence is

$$\sum_q H^*(BU(2q); G) = \sum_q H^*(Z, 2q; G).$$

(Of course, all the results that follow may be obtained by reasoning directly with the exact cohomology sequences; the use of spectral theory is not essential, although it does give a conceptual grasp of the situation.)

The simplest case is of course $G = Q$; in this case we know ((9), p. 501) that

$$H^{2a+k}(Z, 2a; Q) = \begin{cases} Q & \text{for } k = 0, \\ 0 & \text{for } 0 < k < 2a. \end{cases}$$

We find

LEMMA 3. *If q is large compared with k, we have*

$$H^{2q+k}(BU(2q, \ldots, \infty); Q) = \begin{cases} Q & \text{for } k \text{ even}, \\ 0 & \text{for } k \text{ odd}. \end{cases}$$

Moreover, $\quad n^*_{q,q'}\colon\ H^{2q}(BU(2q', \ldots, \infty); Q) \to H^{2q}(BU(2q, \ldots, \infty); Q)$

is an isomorphism if q and q' are large compared with $q-q'$.

It follows from this that instead of (3′), (4′) it is sufficient to satisfy the following conditions.

$$(3'') \qquad k^*\,\mathrm{ch}_{q,0} = n^*_{q,1}\,\mathrm{ch}_q,$$

$$(4'') \qquad n^*_{q+1,q}\,\mathrm{ch}_{q,r} = \frac{m(r)}{m(r-1)}\,\mathrm{ch}_{q+1,r-1}.$$

We now invoke a third result of Bott; there is a class

$$\mathrm{ch}_{q,0} \in H^{2q}(BU(2q, \ldots, \infty); Z)$$

such that

$$k_*\,\mathrm{ch}_{q,0} = n^*_{q,1}\,\mathrm{ch}_q,$$

and this class (which is clearly unique) is a generator of $H^{2q}(BU(2q, \ldots, \infty); Z)$. (See the end of § 8 of (3).) It is clear that the classes $\mathrm{ch}_{q,0}$ correspond to one another under σ^2.

We now return to our calculations of cohomology, taking $G = Z_p$. In this case we know that $H^*(Z, 2a; Z_p)$ is a module over the (mod p) Steenrod algebra A, and that in a suitable range of dimensions, it is isomorphic to the quotient module $A/A\beta_p$ (7), (10). The first term of our spectral sequence can therefore be described as follows: it is a direct sum of modules $A/A\beta_p$, on generators $g^{2q} \in H^{2q}(BU(2q); Z_p)$. We can take these generators g^{2q} so that they correspond to $\rho_* \operatorname{ch}_{q,0}$ under the map

$$BU(2q, \ldots, \infty) \to BU(2q).$$

We now inquire after the differentials in this spectral sequence. They are zero, for dimensional reasons, until we get to that differential δ which is given by

$$\delta g^{2q} = \lambda \beta_p P^1 g^{2q-2(p-1)}$$

(for p odd) or

$$\delta g^{2q} = \lambda Sq^3 g^{2q-2}$$

(for $p = 2$). Here $\lambda \in Z_p$ is a coefficient which remains to be determined; however, it is independent of q, as we see by applying σ^2; moreover, it is not zero; this is seen as follows. Let g^3 be a generator for $H^3(U; Z_p)$; then $\beta_p P^1 g^3$ is zero, because the homology of U is torsion-free; that is, $\beta_p P^1$ is zero on $H^3(U(3, \ldots, \infty); Z_p)$, although it is non-zero on $H^3(Z, 3; Z_p)$. This shows that the coefficient λ is non-zero. The argument applies equally well to the case $p = 2$, provided we interpret P^1 as Sq^2 in this case.

We now appeal to Milnor's work on the Steenrod algebra (5). Following Milnor, we write

$$\tau_1' = P^1\beta_p - \beta_p P^1$$

if p is odd; if $p = 2$, P^1 is to be interpreted as Sq^2, so that τ_1' is to be interpreted as Milnor's $Sq^{0,1}$. The first non-zero differential mentioned above is therefore given by

$$\delta(ag^{2q}) = \mu(-1)^{|a|} a\tau_1' g^{2q-2(p-1)},$$

where $a \in A$, $|a|$ is the dimension of a and $\mu = -\lambda \neq 0$. Let us define a map

$$d\colon\ A/A\beta_p \to A/A\beta_p$$

by $d(a) = (-1)^{|a|} a\tau_1'$. The elements β_p and τ_1' generate an exterior subalgebra E of A, and A is free *quâ* right module over E. Therefore the sequence

$$A/A\beta_p \xrightarrow{d} A/A\beta_p \xrightarrow{d} A/A\beta_p$$

is exact. Therefore the spectral sequence becomes trivial at this point. We reach the following conclusion.

LEMMA 4. *In a range of dimensions which becomes large with q, $H^*(BU(2q, \ldots, \infty); Z_p)$ is a direct sum of modules $A/(A\beta_p + A\tau_1')$ on generators $g^{2q}, g^{2q+2}, \ldots, g^{2q+2(p-2)}$ which are uniquely determined by the equation*

$$n^*_{q+t,q} g^{2(q+t)} = \rho_* \operatorname{ch}_{q+t,0} \quad (0 \leqslant t < p-1).$$

The definition of these classes $g^{2(q+t)}$ is evidently compatible with σ^2.

We next remark that the Bockstein coboundary β_p may be interpreted as a boundary map on $H^*(BU(2q, \ldots, \infty); Z_p)$, so that this graded group becomes a chain complex.

LEMMA 5. *In a range of dimensions which becomes large with q, the homology of $H^*(BU(2q, \ldots, \infty); Z_p)$ for the boundary β_p has a base consisting of the cosets*

$$\{\chi(P^s)\, g^{2(q+t)}\} \quad (0 \leqslant s,\ 0 \leqslant t < p-1).$$

Proof. This evidently reduces to calculating the homology of $A/(A\beta_p + A\tau_1')$ for the boundary defined by $d(a) = \beta_p a$. Let us call this chain complex C. We have an exact sequence of chain complexes

$$0 \to C \xrightarrow{i} A/A\beta_p \xrightarrow{j} C \to 0$$

defined by $i(a) = a\tau_1'$, $j(a) = a$. The middle term $A/A\beta_p$ is acyclic. Passing to the exact homology sequence, we have

$$\partial\colon \quad H_n(C) \cong H_{n-2(p-1)}(C) \quad (n > 0);$$

also $H_0(C) = Z_p$, generated by the coset containing the identity element $1 \in A_0$. Since $\chi P^0 = 1$, it remains only to show that

$$\partial\{\chi P^{s+1}\} = \{\chi P^s\}.$$

This follows immediately from the following lemma.

LEMMA 6. $\beta_p \chi(P^{s+1}) = \chi(P^{s+1})\,\beta_p + \chi(P^s)\,\tau_1'$.

Proof. If p is odd, Milnor's methods allow one to verify that

$$P^{s+1}\beta_p = \beta_p P^{s+1} + \tau_1' P^s.$$

The required result follows immediately by applying χ.

Similarly, if $p = 2$, one can verify the formula

$$Sq^{2(s+1)} Sq^1 = Sq^1 Sq^{2(s+1)} + Sq^{0,1} Sq^{2s}$$

and apply χ to it.

Lemma 5 will evidently help us to discuss the homomorphism

$$\rho_*\colon \quad H^{2q+n}(BU(2q, \ldots, \infty); Z) \to H^{2q+n}(BU(2q, \ldots, \infty); Z_p)$$

(for a suitable range of n). In fact, we already know from our calculation of $H^{2q+n}(BU(2q, \ldots, \infty); Q)$ that $H^{2q+n}(BU(2q, \ldots, \infty); Z)$ contains a cyclic infinite direct summand when n is even; and when n is even, $H^{2q+n}(BU(2q, \ldots, \infty); Z_p)$ contains just one element of the form $\chi(P^s)\, g^{2(q+t)}$. Therefore the image of ρ_* is exactly the subgroup generated by the elements $\chi(P^s)\, g^{2(q+t)}$ and $\beta_p a g^{2(q+u)}$. Moreover,

$$H^{2q+n}(BU(2q, \ldots, \infty); Z)$$

contains no elements of order p^2. In discussing the situation more precisely later, we shall need the following lemma.

LEMMA 7. *Suppose given a pair E, F, an integer m and classes $e \in H^n(E; Z)$, $f \in H^n(F; Z)$ such that $i^*e = mf$. Then there exists $b \in H^n(E, F; Z_m)$ such that*

$$\rho_* e = j^* b \quad \text{and} \quad \delta_m b = -\delta f,$$

where ρ_* is induced by $\rho\colon Z \to Z_m$ and δ_m is the Bockstein coboundary associated with

$$0 \to Z \xrightarrow{m} Z \xrightarrow{\rho} Z_m \to 0.$$

The proof is immediate, by taking representative cochains.

In the applications, F and E will be the fibre and total space of a fibring; we shall have an isomorphism

$$\pi^*: \quad H^n(B; Z_m) \to H^n(E, F; Z_m)$$

valid in a suitable range of dimensions; we shall therefore obtain a class b in $H^n(B; Z_m)$ such that

$$\rho_* e = \pi^* b \quad \text{and} \quad \delta_m b = -\tau f,$$

where τ denotes the transgression.

This completes the preliminaries to the proof of Theorem 2.

4. *Proof of Theorem* 2. We will now prove the following lemma.

LEMMA 8. *For $r > 0$ and q sufficiently large compared with r, it is possible to choose classes $\mathrm{ch}_{q,r} \in H^{2q+2r}(BU(2q, \ldots, \infty); Z)$ which satisfy the following conditions.*

$$(4'') \qquad n^*_{q+1,q}\,\mathrm{ch}_{q,r} = \frac{m(r)}{m(r-1)}\,\mathrm{ch}_{q+1,r-1} \quad (r > 0).$$

(5″) *If $r = s(p-1)+t$ with $0 \leqslant t < p-1$, then*

$$\rho_*\,\mathrm{ch}_{q,r} = \frac{m(r)}{p^s}(-\mu)^s\,\chi(P^s)\,g^{2(q+t)},$$

where μ is as in § 3. *Moreover, the classes $\mathrm{ch}_{q,r}$ are uniquely determined by these properties, and correspond to one another under σ^2.*

Proof. The proof is by induction over r, beginning from the classes $\mathrm{ch}_{q,0}$ which we have already chosen; these certainly satisfy (5″). Let us suppose that the lemma is true for $r-1$; we thus have classes

$$\mathrm{ch}_{q+1,r-1} \in H^{2q+2r}(BU(2q+2, \ldots, \infty); Z).$$

We have a fibring $F \to E \to B$ in which F, E and B are spaces of type

$$BU(2q+2, \ldots, \infty), \quad BU(2q, \ldots, \infty) \quad \text{and} \quad BU(2q).$$

Since $H^{2q+2r+1}(Z, 2q; Z)$ is finite if q is large compared with r, the element $\tau\,\mathrm{ch}_{q+1,r-1}$ has a finite order; we wish to find this order. Consider the set of homomorphisms

$$\rho_*: \quad H^{2q+2r+1}(Z, 2q; Z) \to H^{2q+2r+1}(Z, 2q; Z_p)$$

as p runs over all primes; it is known that the intersection of their kernels is zero. We have

$$\begin{aligned}\rho_*\tau\,\mathrm{ch}_{q+1,r-1} &= \tau\rho_*\,\mathrm{ch}_{q+1,r-1} \\ &= \tau\frac{m(r-1)}{p^s}(-\mu)^s\,\chi(P^s)\,g^{2(q+1+t)},\end{aligned}$$

where $r-1 = s(p-1)+t$ with $0 \leqslant t < p-1$. Two cases now arise.

Case (i). $t+1 < p-1$ and $r = s(p-1)+(t+1)$. Then $\tau g^{2(q+1+t)} = 0$ and

$$\rho_*\tau\,\mathrm{ch}_{q+1,r-1} = 0.$$

Case (ii). $t+1 = p-1$ and $r = (s+1)(p-1)$. Then $\tau g^{2(q+1+t)} = \mu\tau_1' g^{2q}$ and

$$\rho_*\tau\,\mathrm{ch}_{q+1,r-1} = \frac{m(r-1)}{p^s}\mu(-\mu)^s\,\chi(P^s)\,\tau_1' g^{2q}.$$

Now, $\chi(P^s)\tau_1'$ is not zero in $A/A\beta_p$; for if it were, the method of Lemma 5 would yield $\partial\{\chi P^{s+1}\} = 0$, a contradiction. Thus $\rho_* \tau \operatorname{ch}_{q+1,r-1}$ is not zero in this case.

Therefore the order of $\tau \operatorname{ch}_{q+1,r-1}$ is exactly the product of those primes p such that r is divisible by $(p-1)$; that is, the order of $\tau \operatorname{ch}_{q+1,r-1}$ is $m(r)/m(r-1)$. Hence we can choose a class e in $H^{2q+2r}(BU(2q, \ldots, \infty); Z)$ so that

$$n^*_{q+1,q} e = \frac{m(r)}{m(r-1)} \operatorname{ch}_{q+1,r-1}.$$

However, it may be possible to choose e in more than one way; to make the choice, we study the projections $\rho_* e$ corresponding to the different primes p. As above, we have two cases.

Case (i). $t+1 < p-1$; p does not divide $m(r)/m(r-1)$. In this case we have

$$\rho_* e = \frac{m(r)}{p^s}(-\mu)^s \chi(P^s)\, g^{2(q+1+t)} + ag^{2q}$$

for some $a \in A$, as may be seen by applying $n^*_{q+1,q}$. If $s = 0$, then $a = 0$, for dimensional reasons; we may therefore suppose $s > 0$. We know that

$$\beta_p \rho_* e = 0;$$

we have

$$\beta_p \chi(P^s)\, g^{2(q+1+t)} = 0$$

by Lemma 6; therefore

$$\beta_p a g^{2q} = 0.$$

By Lemma 5, ag^{2q} can be written in the form $\beta_p b g^{2q}$ for some $b \in A$. We can therefore replace e by $e - \delta_p b g^{2q}$, and so ensure that

$$\rho_* e = \frac{m(r)}{p^s}(-\mu)^s \chi(P^s)\, g^{2(q+1+t)}.$$

One should note three points about this replacement process. (i) It is required for only a finite number of primes p, namely, those for which $s > 0$. (ii) The replacement does not alter the value of $n^*_{q+1,q} e$. (iii) The replacement does not alter the projections $\rho_* e$ corresponding to primes other than the one under consideration.

Case (ii). $t+1 = p-1$; p divides $m(r)/m(r-1)$. In this case we apply Lemma 7, with

$$f = \frac{m(r)}{pm(r-1)} \operatorname{ch}_{q+1,r-1}$$

and $m = p$. We find that

$$\rho_* e = \pi^* b,$$

where $b \in H^{2q+2r}(BU(2q); Z_p)$ is a class such that

$$\delta_p b = -\tau f.$$

The last equation implies

$$\begin{aligned}\beta_p b &= -\rho_* \tau f \\ &= \frac{m(r)}{p^{s+1}}(-\mu)^{s+1} \chi(P^s)\, \tau_1' g^{2q}.\end{aligned}$$

Using Lemma 6, we find

$$\beta_p b = \frac{m(r)}{p^{s+1}}(-\mu)^{s+1} \beta_p \chi(P^{s+1})\, g^{2q}.$$

The solution of this equation is

$$b = \frac{m(r)}{p^{s+1}}(-\mu)^{s+1}\chi(P^{s+1})g^{2q} + \beta_p a g^{2q}$$

(where $a \in A$). This shows that

$$\rho_* e = \frac{m(r)}{p^{s+1}}(-\mu)^{s+1}\chi(P^{s+1})g^{2q} + \beta_p a g^{2q}.$$

Arguing as in case (i), we can replace e by $e - \delta_p a g^{2q}$, and so ensure that

$$\rho_* e = \frac{m(r)}{p^{s+1}}(-\mu)^{s+1}\chi(P^{s+1})g^{2q}.$$

We have now shown that it is possible to choose $\mathrm{ch}_{q,r}$ so as to satisfy (4″), (5″); and in this respect our induction is complete. It remains to show that the choice of $\mathrm{ch}_{q,r}$ is unique. As an inductive hypothesis, we may assume that the choice of $\mathrm{ch}_{q+1,r-1}$ is unique. Let e_1, e_2 be two choices of $\mathrm{ch}_{q,r}$; thus we have

(i) $n^*_{q+1,q} e_1 = n^*_{q+1,q} e_2$,

while

(ii) $\rho_* e_1 = \rho_* e_2$ for each prime p.

By (i), we have
$$e_1 - e_2 = \pi^* b$$
for some $b \in H^{2q+2r}(BU(2q); Z)$; hence

$$e_1 - e_2 = \sum_p \delta_p a_p g^{2q}$$

for suitable a_p; here a_p lies in the mod p Steenrod algebra and has dimension $2r-1$. By (ii), we now have
$$\beta_p a_p g^{2q} = 0;$$
hence, by Lemma 5,
$$a_p g^{2q} = \beta_p \alpha_p g^{2q}$$
for some α_p; thus $\delta_p a_p g^{2q} = 0$ and $e_1 = e_2$. The choice of $\mathrm{ch}_{q,r}$ is thus unique. It follows immediately that it is compatible with σ^2. The proof of Lemma 8 is thus complete.

Condition (5″) of Lemma 8 still refers to the class $g^{2(q+t)}$ $(0 \leqslant t < p-1)$. However, we have now set up the class $\mathrm{ch}_{q,t}$, and we easily verify that

$$\rho_* \mathrm{ch}_{q,t} = m(t)\, g^{2(q+t)}.$$

Therefore condition (5″) becomes

$$\rho_* \mathrm{ch}_{q,r} = \frac{m(r)}{p^s m(t)}(-\mu)^s \chi(P^s)\rho_* \mathrm{ch}_{q,t},$$

which implies the formula

$$\rho_* \mathrm{ch}_{q,r}(\xi) = \frac{m(r)}{p^s m(t)}(-\mu)^s \chi(P^s)\rho_* \mathrm{ch}_{q,t}(\xi).$$

It remains only to determine the coefficient μ, which can easily be done by applying the formula to a particular ξ. Let us take ξ to be a line bundle over CP^∞ with Chern character e^x, where x is the generator of $H^2(CP^\infty; Q)$. If we take $q = 1$, $r = p-1$ we easily find $(-\mu) = 1$. This completes the proof of (5′), and thus completes the proof of Theorem 2.

5. *The Postnikov system of BU.* Peterson (6) has already investigated the Postnikov invariants

$$k^{2r+1}(BU) \in H^{2r+1}(BU(2, \ldots, 2r-2); Z).$$

However, since the identity-component of $\Omega^2 BU$ is equivalent to BU, the spaces

$$\Omega^{2q}BU(2q+2, \ldots, \infty), \quad BU(2, \ldots, \infty) = BU$$

are equivalent. The k-invariant $k^{2q+2r+1}$ of $BU(2q+2, \ldots, \infty)$ determines the k-invariant $\sigma^{2q}k^{2q+2r+1}$ of $\Omega^{2q}BU(2q+2, \ldots, \infty)$, and therefore determines the k-invariant $k^{2r+1}(BU)$. It is thus sufficient, in principle, to study the Postnikov system of $BU(2q, \ldots, \infty)$ when q is large; this is a 'stable' problem.

Consider the fibring $F \to E \to B$ in which F, E and B are spaces of type

$$BU(2q+2r, \ldots, \infty), \quad BU(2q, \ldots, \infty) \quad \text{and} \quad BU(2q, \ldots, 2q+2r-2);$$

and suppose q large. The Postnikov invariant $k^{2q+2r+1}$ of E is (up to a sign) the transgression $\tau\, \mathrm{ch}_{q+r,0}$ of the fundamental class in the fibre. According to Lemma 7, we have

$$-\tau\, \mathrm{ch}_{q+r,0} = \delta_m b,$$

where $m = m(r)$ and b is some class in $H^{2q+2r}(B; Z_m)$ such that

$$\pi^* b = \rho_* \mathrm{ch}_{q,r}.$$

Moreover, it is easy to see that the last equation determines b uniquely. The Postnikov invariant $k^{2q+2r+1}$ of E is therefore determined in terms of $\mathrm{ch}_{q,r}$ by the equation

$$k^{2q+2r+1} = \delta_m (\pi^*)^{-1} \rho_* \mathrm{ch}_{q,r}.$$

This gives a formal statement of the relation between the characteristic classes we have investigated and the Postnikov invariants of BU.

REFERENCES

(1) Atiyah, M. F. and Hirzebruch, F. Riemann–Roch theorems for differentiable manifolds. *Bull. Amer. Math. Soc.* 65 (1959), 276–81.

(2) Bott, R. The stable homotopy of the classical groups. *Proc. Nat. Acad. Sci., Wash.*, 43 (1957), 933–5.

(3) Bott, R. The space of loops on a Lie group. *Mich. Math. J.* 5 (1958), 35–61.

(4) Hirzebruch, F. *Neue Topologische Methode in der Algebraischen Geometrie* (Berlin, 1956).

(5) Milnor, J. The Steenrod algebra and its dual. *Ann. Math.* 67 (1958), 150–71.

(6) Peterson, F. P. Some remarks on Chern classes. *Ann. Math.* 69 (1959), 414–20.

(7) Cartan, H. Sur les groupes d'Eilenberg–MacLane $H(\pi, n)$: I, II. *Proc. Nat. Acad. Sci., Wash.*, 40 (1954), 467–71, 704–7.

(8) Eilenberg, S. *Exposé* VIII, *Séminaire des Topologie Algebrique˘E.N.S.* (Cartan seminar notes), III (1950–51).

(9) Serre, J.-P. Homologie singulière des espaces fibrés. *Ann. Math.* 54 (1951), 425–505.

(10) Serre, J.-P. Cohomologie modulo 2 des complexes d'Eilenberg–MacLane. *Comment. Math. Helv.* 27 (1953), 198–232.

Trinity Hall
Cambridge

Reprinted from ILLINOIS JOURNAL OF MATHEMATICS
Vol. 17, No. 2, June 1973

CHERN CHARACTERS REVISITED

BY
J. F. ADAMS

1. Introduction

The title of this paper refers to an earlier one [1]. Although I still feel that the questions studied in the earlier paper were indeed worth study, I have long felt that when I wrote the earlier paper I did not have a satisfactory way of stating the results. Recently I had occasion to reformulate the results of [1], for some lectures I gave at the University of Chicago. This reformulation is given as Theorem 1 below. I then found that results of a more general nature were contained in some work by Larry Smith [3]. I am grateful to Larry Smith and A. Liulevicius for letting me read a copy of [3] before publication. The object of this note, then, is to answer the question raised in the last sentence of [3], by recording a proof of Larry Smith's theorem which seems more elementary and direct than the one in [3].

2. Statement of results

Let **bu** be the connective BU-spectrum. Then $\pi_2(\mathbf{bu})$ is isomorphic to Z; let $u \in \pi_2(\mathbf{bu})$ be a generator. The homotopy ring $\pi_*(\mathbf{bu})$ is the polynomial ring $Z[u]$. We may identify $u \in \pi_2(\mathbf{bu})$ with its image in $H_2(\mathbf{bu}; Z)$ or $H_2(\mathbf{bu}; Q)$. The homology ring $H_*(\mathbf{bu}; Q)$ is the polynomial ring $Q[u]$. As in [1], let $m(r)$ be the numerical function given by

$$m(r) = \prod_p p^{[r/(p-1)]}.$$

THEOREM 1. *The image of $H_*(\mathbf{bu}; Z)$ in $H_*(\mathbf{bu}; Q)$ is the Z-submodule generated by the elements*

$$u^r/m(r), \quad r = 0, 1, 2, 3, \cdots .$$

Let $H(Q, n)$ be the Eilenberg-MacLane spectrum for the group Q of rational numbers in dimension n. The rth component of the Chern character defines an element

$$ch_r \in H^{2r}(\mathbf{bu}; Q)$$

or a map of spectra

$$\mathbf{bu} \longrightarrow H(Q, 2r).$$

This map of spectra induces a homomorphism of homology theories, say

$$ch_r : \mathbf{bu}_n(X) \longrightarrow H_{n-2r}(X; Q).$$

This homomorphism is defined whether X is a space or a spectrum.

Received May 22, 1971.

THEOREM 2 (L. Smith). *The image of*

$$m(r)ch_r : \mathbf{bu}_n(X) \to H_{n-2r}(X; Q)$$

is integral, that is, it is contained in the image of

$$H_{n-2r}(X; Z) \to H_{n-2r}(X; Q).$$

This theorem differs only in minor details from Theorem 3.1 of [3]. That is, I have written $m(r)$ where Smith writes μ_r; the dimensional indexing is slightly different; n may be odd as well as even; and X may be a spectrum as well as a space. Theorem 2.1 of [3] follows, as is remarked at the end of [3].

3. Proof of Theorem 1

The proof proceeds by separating the primes p. Let Q_p be the localisation of Z at p, that is, the subring of fractions a/b with b prime to p. We wish to prove that the image of $H_*(\mathbf{bu}; Q_p)$ in $H_*(\mathbf{bu}; Q)$ is the Q_p-subalgebra generated by u and u^{p-1}/p. We give the proof for the case $p = 2$; the case of an odd prime is similar.

The spectrum **bu** has a (stable) cell decomposition of the form

$$\mathbf{bu} = S^0 \cup_\eta e^2 \cup \cdots$$

where η is the generator for the stable 1-stem, and the cells omitted have (stable) dimension ≥ 4. It follows that the Hurewicz homomorphism

$$Z \cong \pi_2(bu) \to H_2(\mathbf{bu}) \cong Z$$

is multiplication by 2; that is, $H_2(\mathbf{bu})$ is generated by $u/2$. It follows immediately that the image of

$$H_*(\mathbf{bu}) \to H_*(\mathbf{bu}; Q)$$

contains $(u/2)^r$. We wish to prove a result in the opposite direction.

Recall from [1] that we have

$$H^*(\mathbf{bu}; Z_2) = A/(ASq^1 + ASq^{01}),$$

where A is the mod 2 Steenrod algebra. Equivalently, let $\mathbf{HZ}$, $\mathbf{HZ}_2$ be the Eilenberg-MacLane spectra for the groups Z, Z_2 in dimension 0; then the generator in $H^0(\mathbf{bu}; Z_2)$ gives a map of spectra $\mathbf{bu} \to \mathbf{HZ}_2$, which induces a monomorphism

$$H_*(\mathbf{bu}; Z_2) \to H_*(\mathbf{HZ}_2; Z_2).$$

Here $H_*(\mathbf{HZ}_2; Z_2)$ is A_*, the dual of the mod 2 Steenrod algebra [2]. We use this monomorphism to identify $H_*(\mathbf{bu}; Z_2)$ with a subalgebra of A_*; we write ξ_r for the Milnor generators in A_* [2]. Then the image of $u/2 \in H_2(\mathbf{bu}; Z)$ in $H_2(\mathbf{bu}; Z_2)$ is ξ_1^2. The E_2-term of the Bockstein spectral sequence, namely

$$\operatorname{Ker} Sq^1/\operatorname{Im} Sq^1 = \operatorname{Ker} \beta_2/\operatorname{Im} \beta_2,$$

is the polynomial algebra $Z_2[\xi_1^2]$; this fact is essentially in [1], and is easily proved using A_*. The remainder of the argument is obvious from the Bockstein spectral sequence, but I give it in full.

The image of

$$H_{2r}(\mathbf{bu}) \to H_{2r}(\mathbf{bu}; Q)$$

is a finitely-generated abelian group, and since it is non-zero, it is isomorphic to Z. Let $h \in H_{2r}(\mathbf{bu})$ map to a generator. Let $v = u/2$, and let us write $\bar{h}$, $\bar{v}$ for the images of these elements in $H_*(\mathbf{bu}; Z_2)$. Then we have

$$\beta_2 \bar{h} = 0;$$

therefore

$$\bar{h} = \lambda \xi_1^{2r} + \beta_2 k$$

where $\lambda \in Z$ and $k \in H_{2r-1}(\mathbf{bu}; Z_2)$. That is,

$$\bar{h} = \lambda \bar{v}^r + (\delta_2 k)^-,$$

where $\delta_2 : H_{2r-1}(\mathbf{bu}; Z_2) \to H_{2r}(\mathbf{bu}; Z)$ is the integral Bockstein. This gives

$$h = \lambda v^r + \delta_2 k + 2l,$$

where $l \in H_{2r}(\mathbf{bu})$. For the images in $H_{2r}(\mathbf{bu}; Q)$ we have

$$h = \lambda (u/2)^r + 2\mu h$$

where $\mu \in Z$, that is,

$$h = \frac{\lambda}{1 - 2\mu} (u/2)^r$$

where $\lambda/(1 - 2\mu) \in Q_2$. This proves Theorem 1.

4. Proof of Theorem 2

By definition, we have

$$\mathbf{bu}_n(X) = \pi_n(\mathbf{bu} \wedge X).$$

We have therefore to consider the map of homotopy induced by

$$\mathbf{bu} \wedge X \xrightarrow{ch_r \wedge 1} H(Q, 2r) \wedge X.$$

This map evidently factors through

$$\mathbf{HZ} \wedge \mathbf{bu} \wedge X \xrightarrow{1 \wedge ch_r \wedge 1} \mathbf{HZ} \wedge H(Q, 2r) \wedge X \xrightarrow{\mu \wedge 1} H(Q, 2r) \wedge X,$$

where μ is the obvious pairing of Eilenberg-MacLane spectra. Now

$$\pi_n(\mathbf{HZ} \wedge \mathbf{bu} \wedge X)$$

may be interpreted as $H_n(\mathbf{bu} \wedge X)$ and calculated by the ordinary Künneth formula. The terms $\mathrm{Tor}_1^Z\ (H_i(\mathbf{bu}),\ H_j(X))$ evidently map to zero in

$$\pi_n(H(Q,\ 2r) \wedge X) = H_{n-2r}(X;\ Q),$$

since $H_{n-2r}(X;Q)$ is torsion-free. If we consider the term $H_i(\mathbf{bu}) \otimes H_j(X)$, we see that $H_i(\mathbf{bu})$ maps into $\pi_i(H(Q,\ 2r))$, which is zero unless $i = 2r$. There remains the term $H_{2r}(\mathbf{bu}) \otimes H_{n-2r}(X)$. Here $H_{n-2r}(X)$ maps to $H_{n-2r}(X;Q)$ by the canonical map, and $u^r \epsilon H_{2r}(\mathbf{bu})$ maps to $1 \epsilon H_{2r}(Q;2r) = Q$ under ch_r. Using Theorem 1, we see that the image of $H_{2r}(\mathbf{bu}) \otimes H_{n-2r}(X)$ in $H_{n-2r}(X;\ Q)$ is $1/m(r)$ times the image of $H_{n-2r}(X;\ Z)$. This proves Theorem 2.

References

1. J. F. Adams, *On Chern characters and the structure of the unitary group*, Proc. Camb. Phil. Soc., vol. 57 (1961), pp. 189–199.
2. J. W. Milnor, *The Steenrod algebra and its dual*, Ann. of Math., vol. 67 (1958), pp. 150–171.
3. L. Smith, *The Todd character and the integrality theorem for the Chern character*, Illinois J. Math., vol. 17 (1973), pp. 301–310.

Cambridge University
Cambridge, England
The University of Chicago
Chicago, Illinois
The University of Manchester
Manchester, England

ADDENDUM TO "CHERN CHARACTERS REVISITED"

BY

J. F. ADAMS

In the Mathematical Reviews, vol. 48 (1974), no. 7274, the reviewer comments as follows on my paper in the Illinois Journal of Mathematics, vol. 17 (1973), pp. 333–336. "The reviewer's one objection is with the phrase 'the case of an odd prime is similar'; he does not see how to do it by any analogous technique."

The reader may be reassured; it is now agreed by those concerned that the case $p > 2$ is as "similar" to the case $p = 2$ as could be expected. The few hints which follow give all that is needed.

(1) One has to use the fact that at an odd prime p, the spectrum bu splits as the sum of $(p - 1)$ similar summands; this step passes unnoticed for $p = 2$.

(2) The cohomology of one summand is $A/(AQ_0 + AQ_1)$; one needs to describe its homology as a subobject of A_*, the dual of the Steenrod algebra; the canonical anti-automorphism of A_* throws this subobject onto

$$E[\tau_2, \tau_3, \ldots] \otimes Z_p[\xi_1, \xi_2, \ldots]$$

(where the notation for the Milnor generators is as usual.)

(3) The E_2 term for the Bockstein spectral sequence is then $Z_p[\xi_1]$.

(4) Apart from (1) above, one makes only the obvious changes to my argument on pp. 334–335; it is not necessary to change reference [2] into reference [p].

CAMBRIDGE UNIVERSITY
CAMBRIDGE, ENGLAND

Received November 7, 1975.

THE HUREWICZ HOMOMORPHISM FOR MU AND BP

J. FRANK ADAMS AND ARUNAS LIULEVICIUS†

§1. *Statement of results*

For background on the stable category we refer the reader to **[5]**, **[6]** and **[16]**. We shall use the notation of **[4]**: S denotes the sphere spectrum, MU the Thom spectrum for the unitary group, H the Eilenberg–MacLane spectrum for the additive group of integers Z, k the spectrum for connective K-theory, BP the Brown–Peterson spectrum [7] for a given prime p as presented by Quillen **[14]** (see also **[4]**). Each of the spectra E which we will consider comes furnished with a homotopy associative and commutative multiplication

$$m^E : E \wedge E \to E$$

together with a unit

$$i^E : S \to E.$$

We wish to study the Hurewicz homomorphism

$$(i^H \wedge 1)_* : \pi_*(S \wedge E) \to \pi_*(H \wedge E)$$

and the analogous Hattori homomorphism

$$(i^k \wedge 1)_* : \pi_*(S \wedge E) \to \pi_*(k \wedge E).$$

The following is an interpretation of the theorem of Stong **[15]** and Hattori **[9]**:

THEOREM 1. *If $E = MU$ or BP, then $(i^H \wedge 1)_*$ is a monomorphism and $(i^k \wedge 1)_*$ is the injection of a direct summand.*

Our proof is in the spirit of Milnor's original paper on $\pi_*(MU)$ **[12]** in the use of the Adams spectral sequence **[1]**. It differs from **[15]** in that we do not have to construct example manifolds and from **[9]** in that we use no limit arguments or K-cohomology.

Milnor in his original paper **[12]** proved that $(i^H \wedge 1)_*$ is a monomorphism. This implies that $(i^k \wedge 1)_*$ is a monomorphism. The groups $\pi_*(MU)$ and $\pi_*(k \wedge MU)$ are finitely generated free Abelian in each dimension, so to prove that $(i^k \wedge 1)_*$ is an injection onto a direct summand it is sufficient to show that for each prime p the map $(i^k \wedge 1)_* \otimes 1_p$ is a monomorphism, where 1_p is the identity map of Z_p. Let us fix the prime p and denote by Q_p the subring of the rationals of the form a/b, $a, b \in Z$, $(a, b) = 1$, $(p, b) = 1$. We replace MU by MUQ_p and k by kQ_p. Recall that MUQ_p splits as a direct sum of suspensions of BP for the prime p (**[4, 14]**) and

$$HZ_p{}^*(BP) = \frac{A}{A(Q_0, Q_1, \ldots, Q_r, \ldots)} = M, \quad \text{say,}$$

where A is the Steenrod algebra over Z_p and $Q_r \in A^{2p^r-1}$ are the Milnor elements **[11]**.

Received 8 December, 1970.

†The second author was partially supported by NSF grant GP-13168.

[J. LONDON MATH. SOC. (2), 5 (1972), 539–545]

The spectrum kQ_p splits as a direct sum of suspensions of a spectrum G (see [3], Chapter 4, pp 84–92) with

$$HZ_p^*(G) = \frac{A}{A(Q_0, Q_1)} = N, \quad \text{say.}$$

The map $i^G \wedge 1$ on the E_2-terms of the Adams spectral sequences for $\pi_*(BP)$ and $\pi_*(G \wedge BP)$ is the map

$$j^* : \mathrm{Ext}_A^{s,t}(M, Z_p) \to \mathrm{Ext}_A^{s,t}(N \otimes M, Z_p)$$

induced by the obvious A-map

$$j : N \otimes M \to M = Z_p \otimes M$$

defined by $j(a \otimes b) = \varepsilon(a)\, b$, where $\varepsilon : A \to Z_p$ is the augmentation.

We now determine the map in Ext defined by j. Denote by M_* the graded Z_p-dual of M. Since M is a quotient of the Steenrod algebra A over Z_p, M_* is a subcomodule of A_*. Recall [11] that additively

$$A_* = E_p[\tau_0, \ldots, \tau_r, \ldots] \otimes Z_p[\xi_1, \ldots, \xi_r, \ldots]$$

where $\xi_r \in A_{2p^r-2}$, $\tau_t \in A_{2p^r-1}$, and if $p > 2$ the above gives the algebra structure for A_*, whereas if $p = 2$ we have $\tau_r^2 = \xi_{r+1}$. Let $c : A_* \to A_*$ be the conjugation anti-homomorphism and let $\alpha_r = c\tau_r$, $\beta_r = c\xi_r$. Then $M_* = Z_p[\beta_1, \ldots, \beta_r, \ldots]$ as algebra and left comodule over A_*; similarly $N_* = E_p[\alpha_2, \ldots, \alpha_r, \ldots] \otimes Z_p[\beta_1, \ldots, \beta_r, \ldots]$ as comodule over A_*.

The information required to complete the proof of Theorem 1 is given by the following Theorem 2 which will be proved in §2.

THEOREM 2. *For M, N, j^* above*

$$\mathrm{Ext}_A^{*,*}(M, Z_p) = Z_p[q_0, \ldots, q_r, \ldots],$$

$$\mathrm{Ext}_A^{*,*}(M \otimes N, Z_p) = Z_p[q_0, q_1] \otimes M_*$$

as algebra over Z_p; moreover j^* is an algebra homomorphism defined by

$$j^*(q_r) = q_0 \otimes \beta_r + q_1 \otimes \beta_{r-1}^p.$$

The proof of Theorem 1 is now easily completed. Notice that $q_r \in \mathrm{Ext}^{1,2p^r-1}$, $\beta_r \in \mathrm{Ext}^{0,2p^r-2}$; hence in the Adams spectral sequences for $\pi_*(BP)$ and $\pi_*(G \wedge BP)$ we have $E_2 = E_\infty$ since $E_2^{s,t} = 0$ unless $t-s$ is divisible by $2p-2$. Let us write π_m for $\pi_m(BP)$ or $\pi_m(G \wedge BP)$ and $F^s \pi_m$ for the s-th Adams filtration of π_m. Suppose $py \in F^s \pi_m$ for $s > 0$; we claim that $y \in F^{s-1} \pi_m$ (that is, in our situation there are no exotic p-extensions). If $y \in F^t \pi_m$ for all t, the proof is complete [1]; say $[y] \neq 0$ in $E_2^{t,m+t}$ then $[py] = q_0[y] \neq 0$ in $E_2^{t+1,m+t+1}$, since E_2 is free over $Z_p[q_0]$ according to Theorem 2, hence $t \geqslant s-1$, and $y \in F^{s-1} \pi_m$.

Recall that we want to prove that

$$(i^G \wedge 1) \otimes 1_p : \pi_*(BP) \otimes Z_p \to \pi_*(G \wedge BP) \otimes Z_p$$

is a monomorphism.

Suppose $x_s \in F^s \pi_m(BP)$ and $(i^G \wedge 1)_* x_s = py_{s-1}$ (it follows that

$$y_{s-1} \in F^{s-1} \pi_m(G \wedge BP)$$

from our discussion above). We claim that there exists a $w_{s-1} \in F^{s-1}\,\pi_m(BP)$ such that $x_{s+1} = x_s - pw_{s-1} \in F^{s+1}\,\pi_m(BP)$ and $(i^G \wedge 1)_* x_{s+1} = py_s$. Now

$$[py_{s-1}] = q_0[y_{s-1}],$$

so $[x_s]$ lies in the kernel of

$$j^* : \frac{E_2(BP)}{q_0\, E_2(BP)} \to \frac{E_2(G \wedge BP)}{q_0\, E_2(G \wedge BP)},$$

which is a monomorphism according to Theorem 2; so $[x_s] = q_0[w_{s-1}]$ for some $w_{s-1} \in F^{s-1}\,\pi_m(BP)$, that is, $x_{s+1} = x_s - pw_{s-1} \in F^{s+1}\,\pi_m(BP)$, and if we let

$$y_s = y_{s-1} - (i^G \wedge 1)_*\, w_{s-1}$$

the proof of our claim is complete. By an easy induction we obtain the following result: suppose $x_0 \in F^0\,\pi_m(BP)$ and $(i^G \wedge 1)_* x_0 = py$, then for each integer $k \geqslant 1$ there exists $x_k \in F^k\,\pi_m(BP)$ such that $x_0 - x_k$ is divisible by p. By Lemma 8.7 of **[4]** there exists an s such that $F^s\,\pi_m(BP) \subset p\pi_m(BP)$, so by choosing $k \geqslant s$ we have $x_0 = pz$ for some z, and Theorem 1 follows.

Remark. Theorem 1 is equivalent to Hattori's theorem since $k_*(MU)$ is a direct summand of $K_*(MU)$.

§2. *Proof of Theorem* 2

Let p be a fixed prime, A the Steenrod algebra over Z_p, A_* its graded dual. The structure of A_* is exhibited by Milnor **[11]**. We shall henceforth use the notation of $p > 2$ and will let the reader carry out the notational changes for $p = 2$. Recall that

$$A_* = E_p[\tau_0, \ldots, \tau_r, \ldots] \otimes Z_p[\xi_1, \ldots, \xi_r, \ldots]$$

as an algebra, $\tau_r \in A_{2p^r-1}$, $\xi_r \in A_{2p^r-2}$, $r = 0, 1, \ldots$, where $\xi_0 = 1$. Let $c : A_* \to A_*$ be the conjugation anti-automorphism and let $\alpha_r = c\tau_r$, $\beta_r = c\xi_r$, then the coproduct $\phi_* : A_* \to A_* \otimes A_*$ is defined by

$$\phi_*(\alpha_r) = 1 \otimes \alpha_r + \sum_{s=0}^{r} \alpha_s \otimes \beta_{r-s}^{p^s},$$

$$\phi_*(\beta_r) = \sum_{s=0}^{r} \beta_s \otimes \beta_{r-s}^{p^s}.$$

Consider the following subalgebras and subcomodules $M_* \subset N_*$ of A_*:

$$M_* = Z_p[\beta_1, \ldots, \beta_r, \ldots],$$

$$N_* = E_p[\alpha_2, \ldots, \alpha_r, \ldots] \otimes Z_p[\beta_1, \ldots, \beta_r, \ldots].$$

We wish to find resolutions by extended A_*-comodules for the comodules M_* and $N_* \otimes M_*$ and the map in Ext induced by the inclusion $M_* \xrightarrow{\eta \otimes 1} N_* \otimes M_*$, where where $\eta : Z_p \to N_*$ is the unit of N_*.

First, it is easy to write down a resolution for M_* over A_*: let

$$V_{*,*} = Z_p[q_0, q_1, \ldots, q_r, \ldots]$$

where

$$q_r \in V_{1,\,2p^r-1}$$

and

$$d^V : E_p[\alpha_0, \ldots, \alpha_r, \ldots] \otimes M_* \otimes V_{*,*} \circlearrowleft$$

is the A_*-map determined by the Z_p-linear map

$$\bar{d}^V : E_p[\alpha_0, \ldots, \alpha_r, \ldots] \otimes M_* \otimes V_{*,*} \to V_{*,*}$$

defined by

$$\bar{d}^V(\alpha^E \otimes m \otimes v) = \begin{cases} \varepsilon(m) q_r v & \text{if } E = \Delta_r \\ 0 & \text{if } E \neq \Delta_r, \end{cases}$$

where $\varepsilon : M_* \to Z_p$ is the augmentation, $E = (e_0, \ldots, e_s, \ldots)$ is an exponent sequence of finitely non-zero natural numbers and Δ_r is the sequence consisting of $e_r = 1$, $e_s = 0$, $s \neq r$.

Similarly it is trivial to write down a minimal resolution for N_* over A_*. Let $W_{*,*} = Z_p[q_0, q_1]$, and define d^W by the Z_p-linear map

$$\bar{d}^W(\alpha^E \otimes m \otimes w) = \begin{cases} \varepsilon(m) q_r w & \text{if } E = \Delta_r,\ r = 0, 1, \\ 0 & \text{otherwise.} \end{cases}$$

Of course, we have used for the second time that $A_* = E_p[\alpha_0, \ldots, \alpha_r, \ldots] \otimes M_*$ as a left $E_p[\alpha_0, \ldots, \alpha_r, \ldots]$ comodule. The inclusion of N_* is the obvious one.

We easily obtain a resolution of $N_* \otimes M_*$ from this resolution $N_* \xrightarrow{\varepsilon^*} A_* \otimes W_{*,*}$, namely we take

$$N_* \otimes M_* \xrightarrow{\varepsilon_* \otimes 1} A_* \otimes W_{*,*} \otimes M_*$$

for the inclusion and $d^W \otimes 1$ for the differential. It follows that the resulting gadget is a resolution by extended A_*-comodules because of the standard isomorphism of left A_*-comodules

$$_D(A_* \otimes W) \otimes M_* \cong {_L A_*} \otimes (W \otimes M_*)$$

([10], Proposition 1.1). Here the symbol D on the left means that the coaction on the left hand side is the diagonal (tensor product) coaction and the symbol L on the right means that the coaction there is the left coaction, so that the right hand side is a left extended A_*-comodule on $W \otimes M_*$. To describe the isomorphism, let

$$\phi_* : A \to A_* \otimes A_*$$

be the coproduct map,

$$\varepsilon_* : A_* \to Z_p$$

the augmentation,

$$\psi_* : A_* \otimes A_* \to A_*$$

the product,

$$c : A_* \to A_*$$

the conjugation, and

$$\mu : M_* \to A_* \otimes M_*$$

the coaction on M_*. The map

$$h : {}_D(A_* \otimes W) \otimes M_* \to {}_L A_* \otimes (W \otimes M_*)$$

is defined by the Z_p-linear map

$$h = \varepsilon \otimes 1 \otimes 1,$$

that is, h is the composition (tensor signs omitted; T the appropriate twist map):

$$\begin{array}{ccc}
A_* W_* M_* & \xrightarrow{\phi_* 1 \mu_*} & A_*(A_* W_*) A_* M_* \\
 & \searrow & \downarrow 1T1 \\
 & h & A_* A_*(A_* W_*) M_* \\
 & \searrow & \downarrow \psi_* \varepsilon 1 1 \\
 & & A_* W_* M_*
\end{array}$$

In terms of elements: if $\alpha \in A_*$, $w \in W_*$, $m \in M_*$ with $\mu_* m = \sum \gamma_i \otimes m_i$, then

$$h(\alpha \otimes w \otimes m) = \sum_i (-1)^{|w|\,|\gamma_i|} \alpha\gamma_i \otimes w \otimes m_i.$$

The inverse isomorphism of A_*-comodules

$$k : {}_L(A_* \otimes W_* \otimes M_*) \to {}_D(A_* \otimes W_* \otimes M_*)$$

is given by the composition (see **[10]**, Proposition 1.1)

$$\begin{array}{ccc}
A_* W_* M_* & \xrightarrow{1 1 \mu_*} & A_* W_* A_* M_* \\
 & \searrow & \downarrow 11c1 \\
 & & A_* W_* A_* M_* \\
 & k & \downarrow 1T1 \\
 & & A_* A_* W_* M_* \\
 & \searrow & \downarrow \psi_* 1 1 \\
 & & A_* W_* M_*
\end{array}$$

or in terms of elements: if α, w, m are as above

$$k(\alpha \otimes w \otimes m) = \sum (-1)^{|w|\,|\gamma_i|} \alpha c(\gamma_i) \otimes w \otimes m_i.$$

We now wish to compute a covering A_*-comodule map

$$\begin{array}{ccc}
A_* \otimes V_{*,*} & \xrightarrow{I} & {}_L[A_* \otimes (W_{*,*} \otimes M_*)] \\
\uparrow & & \uparrow \\
M_* & \xrightarrow{i} & N_* \otimes M_*
\end{array}$$

where $i(m) = 1 \otimes m$. Now the inclusion of $N_* \otimes M_*$ into ${}_L[A_* \otimes (W_{*,*} \otimes M_*)]$ is given by

$$(n \otimes m) \to \sum_i n\gamma_i \otimes 1 \otimes m_i,$$

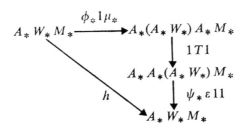

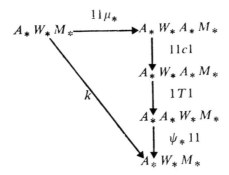

so we can take

$$I_0 : A_* \otimes V_{0,*} \to A_* \otimes (W_{0,*} \otimes M_*)$$

to be the A_*-map defined by the Z_p-linear map

$$\bar{I}_0 : A_* \otimes V_{0,*} \to W_{0,*} \otimes M_*$$

$$\bar{I}_0(\alpha^E \beta^F \otimes 1) = \begin{cases} 1 \otimes \beta^F & \text{if } E = 0, \\ 0 & \text{if } E \neq 0, \end{cases}$$

where E and F are exponent sequences, and $\{\alpha^E \beta^E\}_{E,F}$ forms a basis for A_*. We are interested in a part of the map

$$\bar{I}_1 : A_* \otimes V_{1,*} \to (W_{1,*} \otimes M_*),$$

namely in $\bar{I}_1(1 \otimes q_r)$, $r = 0, 1, 2, \ldots$. Now $1 \otimes q_r$ is the image under $d_0{}^V$ of only the basis elements $\alpha_r \otimes 1$, so we have to compute the composition

$$\bar{d}_0{}^{tW} I_0(\alpha_r \otimes 1),$$

where $\bar{d}_0{}^{tW} = (\bar{d}_0{}^W \otimes 1)k : A_* \otimes W_{0,*} \otimes M_* \to W_{1,*} \otimes M_*$ is the Z_p-linear map defining the differential $d_0{}^{tW}$ on ${}_L[A_* \otimes (W_{0,*} \otimes M_*)]$ which corresponds to $d_0{}^W \otimes 1$ under the isomorphisms h, k. We have

$$I_0(\alpha_r \otimes 1) = \sum_s \alpha_s \otimes 1 \otimes \beta_{r-s}^{p^s},$$

$$kI_0(\alpha_r \otimes 1) = \sum_{s,t} \alpha_s \xi_t^{p^s} \otimes 1 \otimes \beta_{r-s-t}^{p^{s+t}}$$

$$(\bar{d}_0{}^W \otimes 1)kI_0(\alpha_r \otimes 1) = q_0 \otimes \beta_r + q_1 \otimes \beta_{r-1}^p.$$

Let us also notice that

$$d^{tW}(1 \otimes w \otimes m) = 0$$

for all $w \in W_{*,*}$ and $m \in M_*$, since $\mu_* m = \sum_i \gamma_i \otimes m_i$, where $\gamma_i \in M_* \subset A_*$, and we have proved Theorem 2.

References

1. J. F. Adams, "On the structure and applications of the Steenrod algebra", *Comment. Math. Helv.*, 32 (1958), 180–214.
2. ———, *S.P. Novikov's work on operations on complex cobordism* (University of Chicago, 1967.)
3. ———, Lectures on generalised cohomology, in "Category Theory, Homology Theory and their Applications" III, *Lecture Notes in Mathematics*, 99 (1968), 1–138. (Springer–Verlag.)
4. ———, *Quillen's work on formal groups and complex cobordism.* (University of Chicago, 1970.)
5. J. M. Boardman, *Thesis.* (Cambridge, 1964.)
6. ———, *Stable homotopy theory.* (University of Warwick, 1965–1966.)
7. E. H. Brown and F. P. Peterson, "A spectrum whose Z_p-cohomology is the algebra of reduced p-th powers", *Topology* 5 (1966), 149–154.
8. J. P. de Carvalho, F. Clausen, A. Liulevicius, P. Norlamo, "Generators for $\pi_*(MU)$, $\pi_*(BP)$" (in preparation).
9. A Hattori, "Integral characteristic numbers for weakly almost complex manifolds", *Topology* 5 (1966), 259–280.
10. A. Liulevicius, "The cohomology of Massey–Peterson algebras", *Math. Z.*, 105 (1968), 226–256.
11. J. W. Milnor, "The Steenrod algebra and its dual", *Ann. of Math.*, 67 (1958), 150–171.

12. J. W. Milnor, " On the cobordism ring Ω^* and a complex analogue ", *Amer. J. of Math.*, 82 (1960), 505–521.

13. P. Norlamo, *Lecture notes on generators for* $\Omega^* U$. (Nordic Summer School in Mathematics, Aarhus 1968.)

14. D. Quillen, " On the formal group laws of unoriented and complex cobordism theory ", *Bulletin of the A.M.S.*, 75 (1969), 1293–1298.

15. R. E. Stong, " Relations among characteristic numbers, I ", *Topology*, 4 (1965), 267–281.

16. R. Vogt, *Boardman's stable homotopy category.* (Aarhus University lecture notes series No. 21, Aarhus, 1970.)

The Universities of Manchester and Cambridge,
The University of Chicago.

HOPF ALGEBRAS OF COOPERATIONS FOR REAL AND COMPLEX K-THEORY

By J. F. ADAMS, A. S. HARRIS, *and* R. M. SWITZER

[Received 23 July 1970]

1. Introduction

We recall from [**10**] that with each spectrum E there is associated a generalized homology theory and a generalized cohomology theory. We shall write these theories E_*, E^*. Then we may form $E^*(E)$ and interpret it as the algebra of stable cohomology operations on E^*. If we specialize to ordinary cohomology with $\mathbf{Z}_p$ coefficients, we obtain the mod p Steenrod algebra. Unfortunately, if we specialize to real or complex K-theory, the algebra $E^*(E)$ behaves very badly.

We may also form $E_*(E)$. We recall from [**3**] that if the spectrum E satisfies a certain condition, then we may interpret $E_*(E)$ as the algebra of cooperations for E_*. More precisely, $E_*(E)$ is then a Hopf algebra, and for any X, $E_*(X)$ is a comodule with respect to the coalgebra $E_*(E)$. The study of $E_*(X)$ as a comodule over $E_*(E)$ allows one to carry over many proofs and calculations which would ordinarily require stable cohomology operations. If we specialize to ordinary cohomology with $\mathbf{Z}_p$ coefficients, then $E_*(E)$ becomes the dual of the mod p Steenrod algebra; this has been studied by Milnor [**8**]. For the applications, it is desirable to know the structure of $E_*(E)$ in other cases.

The requisite condition on the spectrum E is satisfied in the case of real or complex K-theory. In each case the algebra $E_*(E)$ behaves very well. It is the object of this paper to record the structure of these two algebras. The results are given in §2; see especially Theorems 2.3, 2.5, and 2.8.

By way of comment, and for further motivation, we recall that in [**2**] one of us gave an *ad hoc* construction for certain Ext groups, with applications to real and complex K-theory. We would now suggest that in the main applications, these Ext groups are better replaced by standard Ext groups of comodules over the coalgebras $E_*(E)$ computed in this paper.

Similarly, we recall that in [**9**] § 8 S. P. Novikov performed some rather tricky constructions, by which he sought to cirvumvent the deplorable lack of stable cohomology operations on complex K-theory, so long as the spaces to be considered are finite complexes which are torsion-free. We observe that for such complexes X the duality between $K_*(X)$ and $K^*(X)$

Proc. London Math. Soc. (3) 23 (1971) 385–408
5388.3.23 N

behaves in the best possible way; and we suggest that it would be better to consider $K_*(X)$ as a comodule with respect to one of the coalgebras computed in this paper.

Portions of the material in this paper have appeared in [**4**] and [**7**].

2. Notation and statement of results

In this section we introduce notation, state our main results, and summarize the rest of the paper.

Ordinary homology and cohomology (with integer coefficients) are universally written H_* and H^*. We therefore write H for the corresponding spectrum, which is the Eilenberg–MacLane spectrum for the group $\mathbf{Z}$. This frees the letter K for other uses. We require notation to distinguish between the space BU and the BU-spectrum. Since the homology and cohomology theories corresponding to the BU-spectrum are usually written K_* and K^*, we write K for the BU-spectrum. This allows us to write BU for the space BU. Similarly, we write KO for the BO-spectrum and BO for the space BO.

To avoid holding up the exposition in this section, we defer the proofs of even the easy results.

PROPOSITION 2.1. *The groups $K_*(K)$, $KO_{4n}(KO)$ are torsion-free.*

It follows that the maps

$$K_*(K) \to K_*(K) \otimes \mathbf{Q},$$

$$KO_{4n}(KO) \to KO_{4n}(KO) \otimes \mathbf{Q}$$

are monomorphisms. We propose to describe the structure of $K_*(K)$ and $KO_{4n}(KO)$ by giving their images in $K_*(K) \otimes \mathbf{Q}$ and $KO_{4n}(KO) \otimes \mathbf{Q}$. For this purpose we must first recall the structure of $K_*(K) \otimes \mathbf{Q}$ and $KO_*(KO) \otimes \mathbf{Q}$.

We recall from the Bott periodicity theorem [**5**] that $\pi_*(K)$ is the ring of finite Laurent series $\mathbf{Z}[t, t^{-1}]$ in one generator $t \in \pi_2(K)$. We also recall certain notation from [**3**]. Let S^0 be the sphere-spectrum, and let $i: S^0 \to E$ be the unit map for the ring-spectrum E; we can use i to induce two homomorphisms. First, for any spectrum F we have

$$i_*: F_*(S^0) \to F_*(E);$$

specializing to the case $F = E$, we obtain

$$\eta_L: \pi_*(E) \to E_*(E).$$

Secondly, for any spectrum X we have

$$i_*: (S^0)_*(X) \to E_*(X);$$

specializing to the case $X = E$, we obtain

$$\eta_R\colon \pi_*(E) \to E_*(E).$$

The difference is that the E in $\pi_*(E)$ goes in one case into the left-hand E of $E_*(E)$, and in the other case into the right-hand E. Let us then define

$$u = \eta_L(t), \quad v = \eta_R(t) \in K_2(K).$$

Finally, let $c\colon KO \to K$ be the 'complexification' map; this induces a homomorphism

$$KO_*(KO) \to K_*(K).$$

PROPOSITION 2.2. *$K_*(K)\otimes \mathbf{Q}$ is the ring of finite Laurent series $\mathbf{Q}[u, v, u^{-1}, v^{-1}]$. The map $KO_*(KO) \to K(_*K)$ induces an isomorphism from $KO_*(KO)\otimes \mathbf{Q}$ to the ring of finite Laurent series $\mathbf{Q}[u^2, v^2, u^{-2}, v^{-2}]$.*

We propose to characterize those finite Laurent series which lie in the image of $K_*(K)$ or $KO_*(KO)$ by integrality conditions. The integrality conditions used arise from the existence of the operations Ψ^k, as will appear in § 4.

THEOREM 2.3. *The map*

$$K_*(K) \to K_*(K)\otimes \mathbf{Q}$$

gives an isomorphism between $K_*(K)$ and the set of finite Luarent series $f(u, v)$ which satisfy the following condition.

(2.4) *For any pair of non-zero integers h, k we have*

$$f(ht, kt) \in \mathbf{Z}\left[t, t^{-1}, \frac{1}{hk}\right].$$

THEOREM 2.5. *The map*

$$\sum_n KO_{4n}(KO) \to K_*(K)\otimes \mathbf{Q}$$

gives an isomorphism between $\sum_n KO_{4n}(KO)$ and the set of finite Laurent series $f(u, v)$ which satisfy the following conditions.

(2.6) $f(-u, v) = f(u, v)$; $f(u, -v) = f(u, v)$.

(2.7) *For any pair of non-zero integers h, k we have*

$$f(ht, kt) \in \mathbf{Z}\left[t^4, 2t^2, t^{-4}, \frac{1}{hk}\right].$$

In these theorems, conditions (2.4) and (2.7) have been written in terms of two integers h, k for reasons of symmetry. That is: for any two spectra E, F we have an isomorphism

$$E_*(F) \cong F_*(E),$$

as in [**10**]. If we specialize to the case $E = F$, we obtain a non-trivial anti-automorphism of $E_*(E)$; see [**3**]. In our case, $E = K$, the anti-automorphism sends $f(u, v)$ into $f(v, u)$. Conditions (2.4) and (2.7) have been written in the form given so that they are obviously invariant under this anti-automorphism. If we take condition (2.4) and set $h = 1$, the resulting condition, while apparently weaker, still defines the same set of functions f; this will appear from the proof. Similarly for conditions (2.6) and (2.7). Applying the anti-automorphism, we see that the theorems would also remain true if we set $k = 1$ (leaving h general).

For the rest of this section, we will use Theorem 2.3 or Theorem 2.5 to identify $K_*(K)$ or $\sum_n KO_{4n}(KO)$ with the set of finite Laurent series described in Theorem 2.3 or Theorem 2.5, as the case may be.

Our next task is to describe $KO_m(KO)$ for all values of m. We recall that $KO_*(X)$ is a left module over $\pi_*(KO)$, so that in particular we have a product map

$$\pi_m(KO) \otimes_{\mathbf{Z}} KO_0(KO) \to KO_m(KO).$$

THEOREM 2.8. *This map*

$$\pi_m(KO) \otimes_{\mathbf{Z}} KO_0(KO) \to KO_m(KO)$$

is an isomorphism.

Thus we have

$$KO_m(KO) \cong \begin{cases} \mathbf{Z}_2 \otimes_{\mathbf{Z}} KO_0(KO) & (m = 1, 2 \bmod 8) \\ 0 & (m \equiv 3, 5, 6, 7 \bmod 8). \end{cases}$$

At the risk of labouring the obvious, we make the following result explicit.

PROPOSITION 2.9. *A finite Laurent series $f(u, v) \in KO_0(KO)$ gives the zero element of $\mathbf{Z}_2 \otimes_{\mathbf{Z}} KO_0(KO)$ if and only if it satisfies the following condition.*

(2.10) *For any pair of odd integers h, k we have*

$$f(h, k) \in 2\mathbf{Z}\left[\frac{1}{hk}\right].$$

It is perhaps worth asking whether some analogue of Theorem 2.8 holds before we pass to the limit along the BO-spectrum KO. Is it perhaps true that $KO_*(BO(4n))$ or $KO_*(B\,\mathrm{Spin}(8n))$ is a free module over $\pi_*(KO)$ on generators of dimension 0? We have heard privately from D. W. Anderson that he can answer this sort of question in the affirmative.

As Hopf algebras, $K_*(K)$ and $KO_*(KO)$ have structure maps which we have not yet considered; but with the representation of Theorems 2.3 and 2.5, the behaviour of these structure maps is easily determined.

PROPOSITION 2.11. (i) *In the real case, the generator* $g \in \pi_1(KO)$ *satisfies*

$$\eta_L(g) = \eta_{\mathbf{R}}(g).$$

(ii) *The counit map is given by*

$$\varepsilon(u) = t, \quad \varepsilon(v) = t.$$

(iii) *The canonical anti-automorphism is given by*

$$c(u) = v, \quad c(v) = u.$$

(iv) *The coproduct map is given by*

$$\psi(u) = u \otimes 1, \quad \psi(v) = 1 \otimes v.$$

The rest of this paper is arranged as follows. In § 3 we give some of the easy proofs which were deferred in § 2; we also give the structure of $K_*(BU)$ and $KO_*(B\,\mathrm{Sp})$. In § 4 we use the operations Ψ^k; these yield an easy proof that every finite Laurent series $f(u, v)$ in the image of $K_*(K)$ or $KO_*(KO)$ satisfies conditions (2.4) or (2.7), as the case may be. We also use the operations Ψ^k to deduce Theorem 2.8 from Theorem 2.5. Section 5 is devoted to pure algebra. Observe that the set of finite Laurent series $f(u, v)$ which satisfies condition (2.4) is a subring of $\mathbf{Q}[u, v, u^{-1}, v^{-1}]$, and similarly for those which satisfy conditions (2.6) and (2.7). In § 5 we obtain generators for these subrings. In § 6 we complete the proof of the main theorems. We take the generators for $K_*(BU)$ and $KO_*(B\,\mathrm{Sp})$ obtained in § 3, and locate their images in $K_*(K) \otimes \mathbf{Q}$; it turns out that these images coincide with the generators obtained in § 5.

3. Easy proofs

In this section we give some of the proofs which were deferred in § 2; we also give the structure of $K_*(BU)$ and $KO_*(B\,\mathrm{Sp})$. For this purpose we need some generators.

We identify $BU(1)$ with CP^∞ and $B\,\mathrm{Sp}(1)$ with HP^∞; let ξ be the universal complex line bundle over $BU(1)$, and let η be the universal symplectic line bundle over $B\,\mathrm{Sp}(1)$. Let

$$x = \xi - 1 \in \tilde{K}^0(BU(1)),$$

and similarly let

$$y = \eta - 1 \in \widetilde{K\mathrm{Sp}}^0(B\,\mathrm{Sp}(1)) \cong \widetilde{KO}^4(B\,\mathrm{Sp}(1)).$$

(In the second case, 1 means the trivial symplectic line bundle.)

LEMMA 3.1. $K_0(CP^n)$ *contains one and only one element* β_n *such that*

$$\langle x^i, \beta_n \rangle = \delta_{in} = \begin{cases} 1 & \text{if } i = n \\ 0 & \text{if } i \neq n. \end{cases}$$

Similarly, $KO_{4n}(HP^n)$ *contains one and one only element* γ_n *such that*

$$\langle y^i, \gamma_n \rangle = \delta_{in}.$$

The image of the element β_n in $K_0(CP^\infty) = K_0(BU(1))$ or in $K_0(BU)$ will also be written β_n. Similarly, the image of the element γ_n in $KO_{4n}(HP^\infty) = KO_{4n}(B\,\mathrm{Sp}(1))$ or in $KO_{4n}(B\,\mathrm{Sp})$ will also be written γ_n.

We recall that we have product maps

$$BU \times BU \to BU, \quad B\,\mathrm{Sp} \times B\,\mathrm{Sp} \to B\,\mathrm{Sp}$$

(corresponding to the Whitney sum of virtual bundles); so we have products in $K_*(BU)$ and $KO_*(B\,\mathrm{Sp})$. We pause to emphasize that these products must be distinguished from the products in $K_*(K)$ and $KO_*(KO)$ (see § 2 or [**3**]). The products in $K_*(K)$ and $KO_*(KO)$ arise from the fact that K and KO are ring-spectra; ultimately they derive from the tensor product of virtual bundles. There is a homomorphism $\tilde{K}_q(BU) \to K_q(K)$, but it does not send the one product into the other. We can easily keep the products separate; in $K_*(BU)$ and $KO_*(B\,\mathrm{Sp})$ we always use the 'Whitney sum' product; in $K_*(K)$ and $KO_*(KO)$ we always use the 'tensor' product.

LEMMA 3.2. (i) $K_*(BU(1))$ *is a free module over* $\pi_*(KO)$ *with a base consisting of the elements* $\beta_0, \beta_1, \beta_2, \ldots, \beta_n, \ldots$.

$KO_*(B\,\mathrm{Sp}(1))$ *is a free module over* $\pi_*(KO)$ *with a base consisting of the elements* $\gamma_0, \gamma_1, \gamma_2, \ldots, \gamma_n, \ldots$.

(ii) $K^*(BU(1))$ *is the ring of formal power-series* $\pi_*(K)[[x]]$. $KO^*(B\,\mathrm{Sp}(1))$ *is the ring of formal power-series* $\pi_*(KO)[[y]]$.

(iii) $K_*(BU)$ *is the ring of polynomials*

$$\pi_*(K)[\beta_1, \beta_2, \ldots, \beta_n, \ldots].$$

$KO_*(B\,\mathrm{Sp})$ *is the ring of polynomials*

$$\pi_*(KO)[\gamma_1, \gamma_2, \ldots, \gamma_n, \ldots].$$

(*We have* $\beta_0 = 1$, $\gamma_0 = 1$.)

(iv) *We have*

$$K^*(BU) = \mathrm{Hom}_{\pi_*(K)}(K_*(BU), \pi_*(K)),$$

$$KO^*(B\,\mathrm{Sp}) = \mathrm{Hom}_{\pi_*(KO)}(KO_*(B\,\mathrm{Sp}), \pi_*(KO)).$$

Lemmas 3.1 and 3.2 are easy to prove, by considering the spectral sequences

$$H_*(X; \pi_*(E)) \Rightarrow E_*(X),$$

$$H^*(X; \pi_*(E)) \Rightarrow E^*(X)$$

(and the pairing between these spectral sequences) for the appropriate values of E and X. Compare [**4**] (2.5) (2.14) (4.11). These results are probably known to others.

Proof of Proposition 2.1. We assume known the structure of $\pi_*(K)$ and $\pi_*(KO)$ ([**5**] [**6**]). Then Lemma 3.2(iii) shows that $K_*(BU)$ and $KO_{4n}(B\,\mathrm{Sp})$ are torsion-free. In the spectrum $E = K$, each even term E_{2m} is the space BU; so

$$K_q(K) = \lim_{m\to\infty} \tilde{K}_{q+2m}(BU),$$

and it is torsion-free. Similarly, in the spectrum $E = KO$, each term E_{8m+4} is the space $B\,\mathrm{Sp}$; so

$$KO_{4n}(KO) = \lim_{m\to\infty} \widetilde{KO}_{4n+8m+4}(B\,\mathrm{Sp}),$$

and it is torsion-free. This proves Proposition 2.1.

COROLLARY 3.3. *Every element in $KO_*(KO)$ can be written in the form $ax' + by'$, with $a, b \in \pi_*(KO)$, $x' \in KO_0(KO)$, $y' \in KO_4(KO)$.*

Proof. By Lemma 3.2(iii) the corresponding result is true for $\widetilde{KO}_*(B\,\mathrm{Sp})$. Now Corollary 3.3 follows by passing to the limit, as in the proof of Proposition 2.1.

Proof of Proposition 2.2. It is easy to show by standard methods that for any two spectra E, F, the map

$$\pi_*(E) \otimes \pi_*(F) \otimes Q \to E_*(F) \otimes Q$$

is an isomorphism. (Use the fact that $\otimes Q$ and $\otimes \pi_*(E) \otimes Q$ preserve exactness.)

Proof of Proposition 2.9 *from Theorem* 2.5. A finite Laurent series $f(u, v)$ in $KO_0(KO)$ gives the zero element of $\mathbf{Z}_2 \otimes_{\mathbf{Z}} KO_0(KO)$ if and only if $\frac{1}{2}f(u, v)$ lies in $KO_0(KO)$. By Theorem 2.5 this happens if and only if $\frac{1}{2}f$ satisfies conditions (2.6) and (2.7). Now $\frac{1}{2}f$ clearly satisfies condition (2.6), and it satisfies condition (2.7) if either h or k is even. If we recall that f is homogeneous of degree 0, the remainder of condition (2.7) reduces to condition (2.10).

Proof of Proposition 2.11. Let S^0 be the sphere spectrum, and let $i: S^0 \to KO$ be the unit map for the ring-spectrum KO. Then the generator $g \in \pi_1(KO)$ lies in the image of

$$i_* : \pi_1(S^0) \to \pi_1(KO).$$

Part (i) follows immediately. Parts (ii), (iii), and (iv) follow trivially from the definitions and formal properties given in [**3**].

4. Use of the operations Ψ^k

In this section we give the arguments which depend on the operations Ψ^k. We show that the conditions (2.4) and (2.7) are necessary. We also deduce Theorem 2.8 from Theorem 2.5 and Corollary 3.3.

We recall that the operations Ψ^k were originally constructed as unstable operations. In order to construct Ψ^k as a stable operation, we need to introduce coefficients $\mathbf{Z}\left[\frac{1}{k}\right]$. Crudely speaking, we cannot define a map of spectra from the spectrum $E = K$ to itself by taking each component map $E_{2n} \to E_{2n}$ to be $\Psi^k \colon BU \to BU$, because the following diagram is not commutative.

$$\begin{array}{ccc} S^2 \wedge BU & \xrightarrow{B} & BU \\ \big\downarrow{\scriptstyle 1\wedge\Psi^k} & & \big\downarrow{\scriptstyle \Psi^k} \\ S^2 \wedge BU & \xrightarrow{B} & BU \end{array}$$

Here, of course, the map B is the Bott map, which enters as the map $S^2 \wedge E_{2n} \to E_{2n+2}$ occurring in the spectrum $E = K$; see § 6.

We therefore give some details on homology and cohomology with coefficients. Let G be an abelian group, and let n be a fixed integer with $n \geqslant 1$. Then we can construct a Moore space $M = M(G, n)$ so that

$$\pi_r(M) = 0 \quad \text{for } r < n,$$
$$\pi_n(M) \cong G,$$
$$H_r(M) = 0 \quad \text{for } r > n.$$

Given a spectrum E, we define a 'spectrum with coefficients' EG by

$$(EG)_m = E_{m-n} \wedge M.$$

(The result is essentially independent of n.) Taking $E = K$, $G = \mathbf{Z}\left[\frac{1}{k}\right]$ we construct the spectrum $K\mathbf{Z}\left[\frac{1}{k}\right]$.

The product

$$M\left(\mathbf{Z}\left[\frac{1}{h}\right], n\right) \wedge M\left(\mathbf{Z}\left[\frac{1}{k}\right], n'\right)$$

is a Moore space for the group $\mathbf{Z}\left[\frac{1}{hk}\right]$ in dimension $n+n'$; so we have a pairing of spectra from $K\mathbf{Z}\left[\frac{1}{h}\right]$ and $K\mathbf{Z}\left[\frac{1}{k}\right]$ to $K\mathbf{Z}\left[\frac{1}{hk}\right]$. This pairing induces a homomorphism

$$\mu\colon \left(K\mathbf{Z}\left[\frac{1}{h}\right]\right)_*\left(K\mathbf{Z}\left[\frac{1}{k}\right]\right) \to \pi_*\left(K\mathbf{Z}\left[\frac{1}{hk}\right]\right).$$

After we tensor with $\mathbf{Q}$, μ carries the finite Laurent series $f(u,v)$ into $f(t,t)$.

We can replace the spectrum $K\mathbf{Z}\left[\frac{1}{k}\right]$ constructed above by an Ω-spectrum without changing the corresponding homology and cohomology theories. We take $BU\mathbf{Z}\left[\frac{1}{k}\right]$ to be the $(2n)$th term of this Ω-spectrum. Since $BU\mathbf{Z}\left[\frac{1}{k}\right]$ is a loop-space, it is an H-space. (See § 6 for a discussion showing that the two possible H-structures on BU agree.) Let

$$k\colon BU\mathbf{Z}\left[\frac{1}{k}\right] \to BU\mathbf{Z}\left[\frac{1}{k}\right]$$

be $1+1+\ldots+1$ (k summands). Then the map k induces an isomorphism of homotopy groups, so it has a homotopy inverse

$$\frac{1}{k}\colon BU\mathbf{Z}\left[\frac{1}{k}\right] \to BU\mathbf{Z}\left[\frac{1}{k}\right].$$

We can now construct a map of spectra

$$\Psi^k\colon K \to K\mathbf{Z}\left[\frac{1}{k}\right]$$

by taking its $(2n)$th component map to be

$$\frac{1}{k^n}\Psi^k\colon BU \to BU\mathbf{Z}\left[\frac{1}{k}\right].$$

The induced map

$$(\Psi^k)_*\colon \pi_*(K) \to \pi_*\left(K\mathbf{Z}\left[\frac{1}{k}\right]\right)$$

carries the finite Laurent series $f(t)$ into $f(kt)$.

LEMMA 4.1. *In Theorem* 2.3, *if a finite Laurent series lies in the image of* $K_*(K)$, *then it satisfies condition* (2.4).

LEMMA 4.2. *In Theorem* 2.5, *if a finite Laurent series lies in the image of* $KO_*(KO)$, *then it satisfies conditions* (2.6) *and* (2.7).

Proof of Lemma 4.1. Suppose given an element $e \in K_*(K)$, whose image in $K_*(K)\otimes Q$ is the finite Laurent series $f(u,v)$. The maps Ψ^h and Ψ^k of spectra induce a homomorphism

$$K_*(K) \to \left(KZ\left[\frac{1}{h}\right]\right)_*\left(KZ\left[\frac{1}{k}\right]\right);$$

we then apply

$$\mu\colon \left(KZ\left[\frac{1}{h}\right]\right)_*\left(KZ\left[\frac{1}{k}\right]\right) \to \pi_*\left(KZ\left[\frac{1}{hk}\right]\right).$$

This carries e to an element of

$$\pi_*\left(KZ\left[\frac{1}{hk}\right]\right) = Z\left[t, t^{-1}, \frac{1}{hk}\right].$$

When we tensor with Q, the same homomorphisms carry $f(u,v)$ via $f(hu,kv)$ into $f(ht,kt)$. So we see that

$$f(ht,kt) \in Z\left[t, t^{-1}, \frac{1}{hk}\right];$$

that is, f satisfies (2.4). This proves (4.1).

Proof of Lemma 4.2. The necessity of condition (2.6) follows immediately from Proposition 2.2. The operation Ψ^k exists in the real case [**1**] and gives a map of spectra

$$\Psi^k\colon KO \to KOZ\left[\frac{1}{k}\right].$$

The necessity of condition (2.7) now follows exactly as in the proof of Lemma 4.1.

Proof of Theorem 2.8 *from Theorem* 2.5. We first observe that if $m \equiv 0 \bmod 4$, then the assertion of Theorem 2.8 follows immediately from Theorem 2.5. We may therefore suppose $m \not\equiv 0 \bmod 4$.

We begin by showing that the map

$$\pi_m(KO) \otimes_Z KO_0(KO) \to KO_m(KO)$$

is epimorphic. By Corollary 3.3, every element in $KO_m(KO)$ can be written in the form $ax'+by'$, with $a, b \in \pi_*(KO)$, $x' \in KO_0(KO)$, $y' \in KO_4(KO)$. Using Theorem 2.5 as in the last paragraph, y' can be written in the form cz', where c is the generator of $\pi_4(KO)$ and $z' \in KO_0(KO)$. This shows that the map

$$\pi_m(KO) \otimes_Z KO_0(KO) \to KO_m(KO)$$

is epimorphic.

If $\pi_m(KO) = 0$, this completes the proof. We may therefore suppose $m \equiv 1$ or $2 \bmod 8$, $\pi_m(KO) = Z_2$. We will prove that the map

$$\pi_m(KO) \otimes_Z KO_0(KO) \to KO_m(KO)$$

is monomorphic. Take any element in its kernel; this element may be written $a \otimes x$, where a is the non-zero element of $\pi_m(KO)$ and x is an element in $KO_0(KO)$, corresponding to a finite Laurent series $f(u, v)$. For any pair of odd integers h, k we can use the operations Ψ^h, Ψ^k (as in the proof of Lemmas 4.1 and 4.2) to construct a homomorphism

$$KO_m(KO) \to \pi_m(KO) \otimes \mathbf{Z}\left[\frac{1}{hk}\right]$$

mapping ax into $a \otimes f(h, k)$. Here $\pi_m(KO) \otimes \mathbf{Z}\left[\frac{1}{hk}\right] = \mathbf{Z}_2$, generated by $a \otimes 1$; since we are assuming $ax = 0$, we must have

$$f(h, k) \in 2\mathbf{Z}\left[\frac{1}{hk}\right]$$

for each pair of odd integers h, k. By Proposition 2.9 we have $a \otimes x = 0$ in $\pi_m(KO) \otimes_{\mathbf{Z}} KO_0(KO)$. This proves that the map

$$\pi_m(KO) \otimes_{\mathbf{Z}} KO_0(KO) \to KO_m(KO)$$

is monomorphic, and proves Theorem 2.8.

5. Algebra

In $\mathbf{Q}[u, v, u^{-1}, v^{-1}]$ the finite Laurent series $f(u, v)$ which satisfy condition (2.4) form a subring; similarly, those which satisfy conditions (2.6) and (2.7) form another subring. In this section we will use purely algebraic methods to obtain generators for these two subrings. The main results are Lemmas 5.3 and 5.4. Their topoolgical application will appear at the end of § 6, when we use them to prove Theorems 2.3 and 2.5.

We begin with two results in which the data comes over Z rather than over $\mathbf{Z}\left[\frac{1}{hk}\right]$.

LEMMA 5.1. *Let $f(t)$ be a polynomial in $\mathbf{Q}[t]$ such that*

$$f(k) \in \mathbf{Z}$$

for each integer k. Then $f(t)$ can be written as a $\mathbf{Z}$-linear combination of the polynomials which arise in the binomial theorem

$$p_n(t) = \frac{1}{n!} t(t-1)(t-2)\ldots(t-(n-1)) \quad (n = 0, 1, 2, \ldots).$$

Here $p_0(t)$ is interpreted as 1.

LEMMA 5.2. *Let $f(t)$ be a polynomial in $\mathbf{Q}[t]$ such that*

$$f(-t) = f(t)$$

and

$$f(k) \in \mathbf{Z}$$

for each integer k. Then f(t) can be written as a **Z**-*linear combination of the polynomial* 1 *and the polynomials*

$$q_n(t) = \frac{2}{(2n)!}\,t^2(t^2-1^2)(t^2-2^2)\dots(t^2-(n-1)^2) \quad (n = 1, 2, 3 \dots).$$

We note that $q_0(t)$ would have to be interpreted as 2, but it is excluded from the list.

We also note that the 'binomial' polynomials $p_n(t)$ do satisfy the condition that $p_n(k) \in \mathbf{Z}$ for $k \in \mathbf{Z}$. Since

$$q_n(t) = 2p_{2n}(t+n) - p_{2n-1}(t+n-1),$$

the polynomials $q_n(t)$ satisfy the same condition. It is also clear that they satisfy

$$q_n(-t) = q_n(t).$$

We prove Lemma 5.2; the proof of Lemma 5.1 is similar but simpler.

Proof of Lemma 5.2. We proceed by induction. The result is clearly true for polynomials $f(t)$ of degree 0; suppose it true for polymomials of degree $\leqslant 2(d-1)$. Let $f(t)$ be a polynomial of degree $\leqslant 2d$ which satisfies the conditions of Lemma 5.2. Define a polynomial $\delta^2 f$ by

$$(\delta^2 f)(t) = f(t+1) - 2f(t) + f(t-1).$$

Then $\delta^2 f$ also satisfies the condition of (5.2) and has degree at most $2(d-1)$, so by the inductive hypothesis we can write

$$(\delta^2 f)(t) = n_0 + \sum_{r \geqslant 1} n_r q_r(t),$$

where $n_r \in \mathbf{Z}$. Here n_0 is even, for we have

$$n_0 = (\delta^2 f)(0) = 2(f(1) - f(0)).$$

Now we have

$$\delta^2 q_n = q_{n-1}.$$

So take

$$g(t) = (\tfrac{1}{2}n_0)q_1 + \sum_{r \geqslant 1} n_r q_{r+1}(t);$$

then $g(t)$ is even and $\delta^2 g = \delta^2 f$. So

$$f(t) = a.1 + bt + g(t)$$

with $a, b \in \mathbf{Q}$. Since $f - g$ is even, we have $b = 0$. Since $a = f(0)$, we have $a \in \mathbf{Z}$. So $f(t)$ can be written in the required form; this completes the induction and proves Lemma 5.2.

The proof of Lemma 5.1 is similar, using the first difference δf.

We continue with two results in which the data come over $\mathbf{Z}\left[\frac{1}{hk}\right]$.

LEMMA 5.3. *Let $f(u,v)$ be a finite Laurent series in $\mathbf{Q}[u,v,u^{-1},v^{-1}]$ which satisfies condition (2.4); that is, for any pair of non-zero integers h,k we have*

$$f(ht,kt) \in \mathbf{Z}\left[t,t^{-1},\frac{1}{hk}\right].$$

Then $f(u,v)$ can be written as a $\mathbf{Z}[u,u^{-1},v^{-1}]$-linear combination of the polynomials

$$p'_n(u,v) = \frac{1}{n!}v(v-u)(v-2u)\ldots(v-(n-1)u) \quad (n = 1,2,3,\ldots).$$

LEMMA 5.4. *Let $f(u,v)$ be a finite Laurent series in $\mathbf{Q}[u,v,u^{-1},v^{-1}]$ which satisfies conditions (2.6) and (2.7); that is,*

$$f(-u,v) = f(u,v), \quad f(u,-v) = f(u,v),$$

and for any pair of non-zero integers h,k we have

$$f(ht,kt) \in \mathbf{Z}\left[t^4,t^{-4},2t^2,\frac{1}{hk}\right].$$

Then $f(u,v)$ can be written as a $\mathbf{Z}[u^4,2u^2,u^{-4},v^{-4}]$-linear combination of the polynomials

$$q'_n(u,v) = \frac{2}{(2n+2)!}(v^2-u^2)(v^2-2^2u^2)\ldots(v^2-n^2u^2) \quad (n = 0,1,2,3\ldots).$$

We prove Lemma 5.4; the proof of Lemma 5.3 is similar but simpler.

Proof of Lemma 5.4. Let $f(u,v)$ be a finite Laurent series in $\mathbf{Q}[u,v,u^{-1},v^{-1}]$ which satisfies conditions (2.6) and (2.7). Then $f(u,v)$ is a Laurent series in u^2 and v^2. Separating $f(u,v)$ into homogeneous components, we see that it is sufficient to consider the case when f is homogeneous. Multiplying by a suitable power of u^2, we see that it is sufficient to consider the case in which f has degree zero. Set $w = u^{-1}v$; then

$$f(u,v) = g(w) \in \mathbf{Q}[w^2,w^{-2}].$$

The finite Laurent series g has only a finite number of coefficients; let ν be the highest power to which any prime occurs in the denominator of any such coefficient. Multiplying g by a suitable positive power of $w^4 = u^{-4}v^4$, we see that it is sufficient to consider the case in which g is a polynomial divisible by w and w^ν. Then $g(0) = 0$. If $k \neq 0$, we have

$$g(k) = f(t,kt).$$

Let p be a prime dividing k; then p does not occur in the denominator of $g(k)$, by construction. Let p be a prime not dividing k; then $f(t,kt) \in \mathbf{Z}\left[\frac{1}{k}\right]$ by condition (2.7), and again p does not occur in the denominator of $g(k)$.

So $g(k) \in \mathbf{Z}$ and g satisfies the assumptions of Lemma 5.2. So $f(u,v) = g(u^{-1}v)$ can be written as a $\mathbf{Z}$-linear combination of the polynomial $1 = q'_0$ and the polynomials

$$q_n(u^{-1}v) = \frac{2}{(2n)!}u^{-2n}v^2(v^2-u^2)(v^2-2^2u^2)\dots(v^2-(n-1)^2u^2).$$

But now we have

$$q_{2n}(u^{-1}v) = u^{-4n}[(4n+2)(4n+1)q'_{2n}(u,v)+4n^2u^2q'_{2n-1}(u,v)],$$

$$q_{2n-1}(u^{-1}v) = u^{-4n}[4n(4n-1)u^2q'_{2n-1}(u,v)+(2n-1)^2u^4q'_{2n-2}(u,v)],$$

where $q'_m(u,v)$ is as in the statement of Lemma 5.4. This proves the lemma.

6. Proof of the main theorems

In this section we will prove Theorems 2.3 and 2.5. These theorems will be deduced at the end of the section from Propositions 6.13 and 6.14, whose nature we now explain. We recall that in § 3 we introduced generators

$$\beta_n \in K_0(BU), \quad \gamma_n \in KO_{4n}(B\,\mathrm{Sp});$$

we consider BU as term number 0 in the spectrum K, and $B\,\mathrm{Sp}$ as term number 4 in the spectrum KO, so that we have the following homomorphisms.

$$\tilde{K}_0(BU) \to K_0(K)\otimes\mathbf{Q},$$

$$\widetilde{KO}_{4n}(B\,\mathrm{Sp}) \to KO_{4n-4}(KO)\otimes\mathbf{Q}.$$

Propositions 6.13 and 6.14 give the images of β_n and γ_n under these homomorphisms. For this purpose, of course, we use the description of $K_*(K)\otimes\mathbf{Q}$ and $KO_*(KO)\otimes\mathbf{Q}$ given in Proposition 2.2.

To prove Propositions 6.13 and 6.14 we study the effect on K_* of the Bott map $S^2\wedge BU \to BU$; see Lemmas 6.1, 6.3, and 6.11. To handle the calculations it is convenient to work in the dual, and consider the group of primitive elements $PK^*(BU)$, which we identify as a group of cohomology operations; see Lemmas 6.4 and 6.8 to 6.10.

We continue with a piece of exposition which is not necessary to the proof, in the hope that it will help to explain the line taken in §§ 5 and 6 of this paper. We recall that in the spectrum K, every even term is the space BU, and the maps $S^2\wedge BU \to BU$ come from the Bott periodicity theorem. We therefore have

$$K_n(K) = \lim_{m\to\infty}\tilde{K}_{n+2m}(BU).$$

In order to compute $K_*(K)$, it would be reasonable first to compute

$K_*(BU)$ (which has been done in § 3) and then to compute the homomorphisms

$$\tilde{K}_q(BU) \to \tilde{K}_{q+2}(BU)$$

of the direct system. Lemma 6.3 will show that these homomorphisms annihilate decomposable elements. We may therefore pass to the indecomposable quotient $QK_*(BU)$, and write

$$K_n(K) = \lim_{m\to\infty} Q_{n+2m}(K_*(BU)).$$

As we have seen in § 3, the indecomposable quotient $QK_*(BU)$ is a free module over $\pi_*(K)$ on generators $\beta_1, \beta_2, \ldots, \beta_n, \ldots$. Let us for the moment write $\beta_{n,m}$ for the image in $K_*(K)$ of the class β_n in the $2m$th term of the spectrum K; then it is clear that $K_*(K)$ is generated over $\mathbf{Z}[u, u^{-1}]$ by the classes $\beta_{n,m}$. One can show using Lemma 6.1 that

$$\beta_{n,m} = v^{-m}\beta_{n,0}.$$

Therefore $K_*(K)$ is generated over $\mathbf{Z}[u, u^{-1}, v^{-1}]$ by the classes $\beta_{n,0}$. Compare this with Lemma 5.3 and Proposition 6.13.

Similar arguments apply to the real case, with β_n replaced by γ_n.

From the formal point of view, our proof will not refer much to the direct system mentioned above; but we still need the homomorphism

$$\tilde{K}_q(BU) \to \tilde{K}_{q+2}(BU)$$

to prove Propositions 6.13 and 6.14. We therefore continue by studying this homomorphism.

We first recall that we have a map

$$\mu\colon BU \wedge BU \to BU$$

corresponding to the tensor product of virtual bundles whose virtual dimension is zero. We choose a map

$$f\colon S^2 \to BU$$

representing the generator of $\pi_2(BU)$; we define the Bott map

$$B\colon S^2 \wedge BU \to BU$$

to be the following composite:

$$S^2 \wedge BU \xrightarrow{f\wedge 1} BU \wedge BU \xrightarrow{\mu} BU.$$

We write B_* for the composite

$$\tilde{K}_q(BU) \xrightarrow{\cong} \tilde{K}_{q+2}(S^2 \wedge BU) \longrightarrow \tilde{K}_{q+2}(BU)$$

in which the first factor is the double-suspension isomorphism and the second is induced by B.

Let us regard BU as the $2m$th term of the spectrum K, so that we obtain a homomorphism

$$\tilde{K}_q(BU) \to K_{q-2m}(K).$$

We use this homomorphism for the two horizontal maps in the following diagram; the right-hand vertical map is multiplication by v.

$$\begin{array}{ccc} \tilde{K}_q(BU) & \longrightarrow & K_{q-2m}(K) \\ {\scriptstyle B_*}\downarrow & & \downarrow{\scriptstyle v} \\ \tilde{K}_{q+2}(BU) & \longrightarrow & K_{q-2m+2}(K) \end{array}$$

LEMMA 6.1. *This diagram is commutative.*

The proof is easy.

The adjoint of the map

$$B\colon S^2 \wedge BU \to BU$$

is a map

$$B'\colon BU \to \Omega_0^2 BU,$$

where Ω_0 means the component of the constant map in the loop-space Ω.

LEMMA 6.2. *Under the map $B'\colon BU \to \Omega_0^2 BU$, the product in BU (which represents the Whitney sum of virtual bundles) corresponds up to homotopy with the loop-space product in $\Omega_0^2 BU$.*

LEMMA 6.3. *The homomorphism*

$$B_*\colon \tilde{K}_q(BU) \to \tilde{K}_{q+2}(BU)$$

annihilates decomposable elements.

Proof of (6.2). BU is an H-space; so the loop-space product μ_Ω on $\Omega_0^2(BU)$ is homotopic to the product μ_H induced from the H-space product in BU. The periodicity isomorphism

$$\tilde{K}^0(X) \cong \tilde{K}^0(S^2 \wedge X)$$

is an isomorphism of additive groups; this says that under $B'\colon BU \to \Omega_0^2 BU$, the H-space product in BU corresponds to the product μ_H in $\Omega_0^2 BU$. This proves (6.2).

Proof of Lemma 6.3. We have the following commutative diagram.

$$\begin{array}{ccc} \tilde{K}_q(BU) & & \\ {\scriptstyle B'_*}\downarrow & \searrow^{B_*} & \\ \tilde{K}_q(\Omega_0^2 BU) & \longrightarrow & \tilde{K}_{q+2}(BU) \end{array}$$

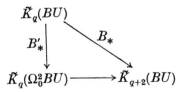

Here the horizontal map is the appropriate double-suspension homomorphism; and it is well known that it annihilates products, providing that the products in $K_*(\Omega_0^2 BU)$ are those induced by the loop-space product; the proof for ordinary homology goes over. Now use Lemma 6.2. This proves Lemma 6.3.

It follows from Lemma 6.3 that B_* induces a homomorphism

$$B_*: Q_q(K_*(BU)) \to Q_{q+2}(K_*(BU)).$$

Here Q occurring at the beginning of a formula means the indecomposable quotient; when Q means $\mathbf{Q}$, the rationals, we will put it at the end. We will compute the effect of B_* on $QK_*(BU)$ by making use of its dual, the group $PK^*(BU)$ of primitive elements.

LEMMA 6.4. *We have*

$$PK^*(BU) = \operatorname{Hom}_{\pi_*(K)}(QK_*(BU), \pi_*(K)),$$
$$PKO^*(B\,\mathrm{Sp}) = \operatorname{Hom}_{\pi_*(KO)}(QKO_*(B\,\mathrm{Sp}), \pi_*(KO)).$$

More precisely, an element in $PK^*(BU)$ is determined by its restriction to $BU(1)$, and this may be any element of $\tilde{K}^*(BU(1))$; an element in $PKO^*(B\,\mathrm{Sp})$ is determined by its restriction to $B\,\mathrm{Sp}(1)$, and this may be any element of $\tilde{K}O^*(B\,\mathrm{Sp}(1))$.

Proof. By Lemma 3.2 (iv) we have

$$\tilde{K}^*(BU) = \operatorname{Hom}_{\pi_*(K)}(\tilde{K}_*(BU), \pi_*(K)).$$

The condition for an element $h \in \tilde{K}^*(BU)$ to be primitive is that its values on all monomials $\beta_{i_1}\beta_{i_2}\ldots\beta_{i_r}$ with $r \geqslant 2$ should be zero. This still allows us to choose its values on the monomials β_i; these values may be arbitrary. Similarly for the real case, replacing β_i by γ_i. This proves (6.4).

We have an alternative description of $PK^*(BU)$. Let A^* be the set of unstable cohomology operations

$$\varphi: K^0(X) \to K^*(X)$$

which are additive:

$$\varphi(y+z) = \varphi(y) + \varphi(z).$$

We may decompose A^* in the form

$$A^* = \pi_*(K) + \tilde{A}^*;$$

here the first summand is generated by the operation Ψ^0 which carries every bundle into a trivial bundle of the same dimensions; the second summand consists of operations which act as zero when X is a point. We may also identify A^* with the set of additive operations

$$\tilde{\varphi}: \tilde{K}^0(X) \to \tilde{K}^*(X)$$

(where X runs over connected spaces), or with $PK^*(BU)$.

COROLLARY 6.5. *An operation $\varphi \in A^*$ is determined by giving its value on the universal line bundle ξ over $BU(1)$; this value may be any element of $K^*(BU(1)) = \pi_*(K)[[x]]$.*

This follows immediately from (6.4). A more direct proof is given in [**3**] pp. 86, 87.

We now introduce the operations we need. In the complex case, we define basic operations φ^n so that

$$\varphi^n(\xi) = x^n$$

for $n \geqslant 1$; compare [**3**] p. 86. In the real case, we have to exploit the fact that real and symplectic objects are self-conjugate.

The operation Ψ^{-1} defined by using complex conjugation is an operation of the sort considered; we have

$$\Psi^{-1}\xi = \bar{\xi} = \xi^{-1} = (1+x)^{-1}.$$

This operation commutes with all other operations in A^0; for suppose that $\varphi(\xi) = f(x)$ for some formal power-series f with integral coefficients; then

$$\begin{aligned}\varphi(\Psi^{-1}\xi) &= f(\bar{\xi}-1) \quad \text{(since } \bar{\xi} \text{ is a line-bundle)}\\ &= f(\Psi^{-1}x);\end{aligned}$$

but also

$$\begin{aligned}\Psi^{-1}(\varphi\xi) &= \Psi^{-1}(fx)\\ &= f(\Psi^{-1}x)\end{aligned}$$

(since Ψ^{-1} commutes with sums, products and limits). So $\Psi^{-1}\varphi = \varphi\Psi^{-1}$.

The operation Ψ^{-1} satisfies $(\Psi^{-1})^2 = 1$, so over $\mathbf{Z}[\frac{1}{2}]$, A^0 splits as the direct sum of the $+1$ and -1 eigenspaces of Ψ^{-1}. In the $+1$ eigenspace we take basic operations S^r such that

$$S^r(\xi) = (x+\bar{x})^r,$$

where we write $\bar{x}$ for $\Psi^{-1}x = (1+x)^{-1}-1$. In the -1 eigenspace we take basic operations A^r such that

$$A^r(\xi) = (x-\bar{x})(x+\bar{x})^{r-1}.$$

LEMMA 6.6. *If $h \in QK_*(BU)$ and either*

(i) $\langle \varphi^r, h\rangle = 0$ *for* $r \geqslant 1$, *or*

(ii) $\langle S^r, h\rangle = 0$ *for* $r \geqslant 1$,

$\langle A^r, h\rangle = 0$ *for* $r \geqslant 1$,

then $h = 0$.

Proof. Part (i) is trivial. For part (ii), we note that

$$S^r(\xi) = x^{2r} + \text{higher powers of } x,$$

$$A^r(\xi) = 2x^{2r-1} + \text{higher powers of } x.$$

This proves Lemma 6.6.

Next we wish to relate the operations φ^r, S^r, and A^r to our homology generators. By means of the complexification maps

$$KO \to K, \quad B\,\mathrm{Sp} \to BU$$

the homology class

$$\gamma_n \in KO_{4n}(B\,\mathrm{Sp})$$

gives an element in $Q_{4n}(K_*(BU))$, provided that $n \geqslant 1$.

LEMMA 6.7.

(i) $\langle \varphi^r, \beta_n \rangle = \delta_{rn} = \begin{cases} 1 & \text{if } r = n, \\ 0 & \text{if } r \neq n. \end{cases}$

(ii) $\langle S^r, \gamma_n \rangle = 2t^{2n}\delta_{rn}$,

$\langle A^r, \gamma_n \rangle = 0.$

Here t stands for the generator of $\pi_2(K)$, as in § 2.

Proof. Part (i) is trivial, and so is the assertion about A^r; for since γ_n is real, we have

$$\langle A^r, \gamma_n \rangle = \langle \Psi^{-1} A^r, \gamma_n \rangle = -\langle A^r, \gamma_n \rangle.$$

For S^r, we consider the following diagram.

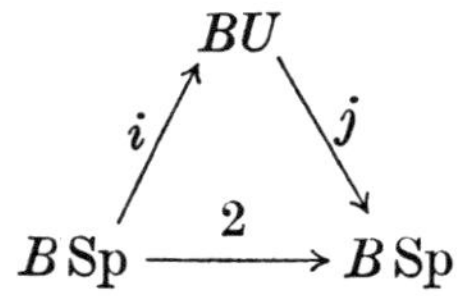

Here the map $2\colon B\,\mathrm{Sp} \to B\,\mathrm{Sp}$ means $1+1$ in the sense of the Whitney sum. By Lemma 6.4 there is for any $r \geqslant 1$ a unique primitive element

$$P^r \in P^{4r}(KO^*(B\,\mathrm{Sp}))$$

whose restriction to $B\,\mathrm{Sp}(1)$ is y^r. The image of y in $\tilde{K}^4(BU(1))$ is

$$t^{-2}(\xi + \bar{\xi} - 2) = t^{-2}(x + \bar{x}),$$

and so the image of y^r in $\tilde{K}^{4r}(BU(1))$ is

$$t^{-2r}(x + \bar{x})^r.$$

The map $j\colon BU \to B\,\mathrm{Sp}$ is an H-map, so it carries primitives to primitives. Therefore the image of P^r in $P^{4r}(K^*(BU))$ is $t^{-2r}S^r$.

Let us write c for the complexification map induced by $KO \to K$. Then we have

$$\begin{aligned} \langle S^r, ci_*\gamma_n\rangle &= \langle t^{2r}cj^*P^r, ci_*\gamma_n\rangle \\ &= t^{2r}c\langle 2^*P^r, \gamma_n\rangle \\ &= 2t^{2r}c\langle y^r, \gamma_n\rangle \\ &= 2t^{2r}\delta_{rn}. \end{aligned}$$

This proves Lemma 6.7.

Let $\varphi \in \tilde{A}^*$ be an additive operation of the sort considered above. Then we can define another additive operation $B^*\varphi \in \tilde{A}^*$ so that the following diagram is commutative.

$$\begin{array}{ccc} \tilde{K}^0(X) & \xrightarrow{\cong} & \tilde{K}^0(S^2 \wedge X) \\ {\scriptstyle B^*\varphi}\downarrow & & \downarrow{\scriptstyle \varphi} \\ \tilde{K}^*(X) & \xrightarrow{\cong} & \tilde{K}^*(S^2 \wedge X) \end{array}$$

Here each horizontal arrow is the periodicity isomorphism.

LEMMA 6.8. *The homomorphisms*

$$\begin{gathered} B^*\colon P^n(K^*(BU)) \to P^n(K^*(BU)), \\ B_*\colon Q_m(K_*(BU)) \to Q_{m+2}(K_*(BU)) \end{gathered}$$

are related by the following formula.

$$\langle \varphi, B_*h\rangle = t\langle B^*\varphi, h\rangle.$$

Here $\varphi \in P^n(K^*(BU))$, $h \in Q_m(K_*(BU))$, and t is the generator in $\pi_2(K)$. The proof is easy.

Given a specific operation φ, we can easily work out $B^*\varphi$ by computing its effect on a line bundle.

LEMMA 6.9. *If $\varphi(\xi) = f(x)$ for some formal power-series f, then*

$$(B^*\varphi)(\xi) = (1+x)f'(x) - f'(0).$$

Proof. Consider the following commutative diagram.

$$\begin{array}{ccccc} \tilde{K}^0(X) & \xrightarrow{\cong} & \tilde{K}^0(S^2 \wedge X) & \longrightarrow & \tilde{K}^0(S^2 \times X) \\ {\scriptstyle B^*\varphi}\downarrow & & \downarrow{\scriptstyle \varphi} & & \downarrow{\scriptstyle \varphi} \\ \tilde{K}^*(X) & \xrightarrow{\cong} & \tilde{K}^*(S^2 \wedge X) & \longrightarrow & \tilde{K}^*(S^2 \times X) \end{array}$$

Let ω be the Hopf bundle over S^2, and set $w = \omega - 1$; then each horizontal map is multiplication by w. Take $X = BU(1)$ and apply the diagram to $\xi - 1$. We find

$$\begin{aligned} w((B^*\varphi)(\xi-1)) &= \varphi((\omega-1)(\xi-1)) \\ &= \varphi(\omega\xi - \xi - \omega + 1) \\ &= \varphi(\omega\xi) - \varphi(\xi) - \varphi(\omega) + \varphi(1) \end{aligned}$$

(since φ is additive). That is,

$$\begin{aligned} w((B^*\varphi)(\xi-1)) &= f(wx+w+x) - f(x) - f(w) + f(0) \\ &= w(1+x)f'(x) - wf'(0) \end{aligned}$$

(since $w^2 = 0$). So

$$(B^*\varphi)(\xi-1) = (1+x)f'(x) - f'(0).$$

This proves Lemma 6.9.

We next observe that $B^*\Psi^{-1} = -\Psi^{-1}$; so B^* maps the $+1$ eigenspace of Ψ^{-1} into the -1 eigenspace, and vice versa.

LEMMA 6.10. *We have*

$$B^*\varphi^r = r\varphi^r + r\varphi^{r-1},$$

$$B^*S^r = rA^r,$$

and

$$B^*A^r = rS^r + (4r-2)S^{r-1},$$

where φ^0 and S^0 are interpreted as 0.

Proof. We apply Lemma 6.9. If we regard $\tilde{A}^*$ as a quotient of A^*, so that the operation $\Psi^0 = \varphi^0 = S^0$ becomes zero in $\tilde{A}^*$, then we can neglect the term $f'(0)$ in Lemma 6.9. For $\varphi = \varphi^r$ we have $f(x) = x^r$ and

$$\begin{aligned} (1+x)f'(x) &= (1+x)rx^{-1} \\ &= rx^r + rx^{r-1}. \end{aligned}$$

This gives $B^*\varphi^r = r\varphi^r + r\varphi^{r-1}$.

Next we observe that

$$(1+x)(1+\bar{x}) = 1,$$

so, differentiating,

$$(1+\bar{x}) + (1+x)\frac{d}{dx}\bar{x} = 0.$$

Thus

$$(1+x)\frac{d}{dx}(x+\bar{x}) = x - \bar{x},$$

$$(1+x)\frac{d}{dx}(x-\bar{x}) = x + \bar{x} + 2.$$

Also

$$(x-\bar{x})^2 = (x+\bar{x})^2 - 4x\bar{x}$$
$$= (x+\bar{x})^2 + 4(x+\bar{x}).$$

Now for S^r we have $f(x) = (x+\bar{x})^r$ and

$$(1+x)f'(x) = r(x+\bar{x})^{r-1}(x-\bar{x}).$$

This gives $B^*S^r = rA^r$.

For A^r we have

$$f(x) = (x-\bar{x})(x+\bar{x})^{r-1}$$

and

$$(1+x)f'(x) = (x+\bar{x}+2)(x+\bar{x})^{r-1} + (r-1)(x-\bar{x})^2(x+\bar{x})^{r-2}$$
$$= r(x+\bar{x})^r + (4r-2)(x+\bar{x})^{r-1}.$$

This gives

$$B^*A^r = rS^r + (4r-2)S^{r-1},$$

and completes the proof of Lemma 6.10.

We now calculate the effect of the homomorphism

$$B_*: Q_q(K_*(BU)) \to Q_{q+2}(K_*(BU))$$

on our homology generators. For this purpose we identify γ_n with its image in $K_{4n}(BU)$, as in Lemma 6.7.

LEMMA 6.11. *We have*

$$B_*\beta_n = t(n\beta_n + (n+1)\beta_{n+1}),$$
$$(B_*)^2\gamma_n = n^2t^2\gamma_n + (2n+1)(2n+2)\gamma_{n+1}.$$

Proof. The first formula comes immediately from the formula for $B^*\varphi_r$ in Lemma 6.10, by dualizing using Lemmas 6.6, 6.7, and 6.8. A slightly different proof is given in [**4**] (13.5). For the second formula we deduce from Lemma 6.10 that

$$(B^*)^2S^r = r^2S^r + 2r(2r-1)S^{r-1}$$

and $(B^*)^2A^r$ is a linear combination of A^r and A^{r-1}. Then we dualize using Lemmas 6.6, 6.7, and 6.8.

COROLLARY 6.12. *In* $QK_*(BU) \otimes \mathbf{Q}$ *we have*

$$t^n\beta_n = \frac{1}{n!}(B_*-t)(B_*-2t)\dots(B_*-(n-1)t)t\beta_1,$$

$$\gamma_n = \frac{2}{(2n)!}(B_*^2-t^2)(B_*^2-2^2t^2)\dots(B_*^2-(n-1)^2t^2)\gamma_1.$$

Proof. Immediate from (6.11), by induction over n. Note that B_* is linear over $\mathbf{Z}[t, t^{-1}]$.

Now we reach our main results. We recall that our generator β_n originally comes from $K_0(BU)$, and our generator γ_n originally comes from $KO_{4n}(B\,\mathrm{Sp})$. We consider BU as term number 0 in the spectrum K, and $B\,\mathrm{Sp}$ as term number 4 in the spectrum KO, so that we have the following homomorphisms.

$$\tilde{K}_{2n}(BU) \to K_{2n}(K),$$

$$\tilde{K}O_{4n+4}(B\,\mathrm{Sp}) \to KO_{4n}(KO).$$

PROPOSITION 6.13. *For $n \geqslant 1$, the image of $t^n\beta_n$ in $K_{2n}(K)\otimes\mathbf{Q}$ is*

$$p'_n(u,v) = \frac{1}{n!}v(v-u)(v-2u)\ldots(v-(n-1)u).$$

PROPOSITION 6.14. *For $n \geqslant 0$, the image of γ_{n+1} in $K_{4n}(K)\otimes\mathbf{Q}$ is*

$$q'_n(u,v) = \frac{2}{(2n+2)!}(v^2-u^2)(v^2-2^2u^2)\ldots(v^2-n^2u^2).$$

In these propositions, the polynomials $p'_n(u,v)$ and $q'_n(u,v)$ are as in Lemmas 5.3 and 5.4.

Proof of Proposition 6.13. Observe that the image of $t^n\beta_n$ in $K_{2n}(K)\otimes\mathbf{Q}$ depends only on its class in $QK_*(BU)\otimes\mathbf{Q}$, by (6.3). For $n = 1$, $t\beta_1$ is the generator of $\tilde{K}_2(CP^1)$, and its image in $K_*(K)$ is certainly v. Now Proposition 6.13 follows from Corollary 6.12 by using Lemma 6.1; the map $QK_*(BU) \to K_*(K)$ sends the action of B_* into multiplication by v, and of course sends the action of t into multiplication by u.

Proof of Proposition 6.14. This is closely similar to the proof of Proposition 6.13, except for one minor point. For $n = 1$, γ_1 is the generator of $\tilde{K}O_4(HP^1)$, and its image in $KO_*(KO)$ or $K_*(K)$ is certainly 1. Now Proposition 6.14 follows from Corollary 6.12 by using Lemma 6.1.

Proof of Theorem 2.3. For ease of writing, let us identify $K_*(K)$ with its image in $K_*(K)\otimes\mathbf{Q}$. We have shown in § 4 that if $f(u,v)$ lies in $K_*(K)$, then f satisfies condition (2.4). To prove the converse, suppose that f satisfies condition (2.4). Then by Lemma 5.3 f can be written as a $\mathbf{Z}[u,u^{-1},v^{-1}]$-linear combination of the polynomials p'_n. But by Proposition 6.13 the polynomials p'_n lie in $K_*(K)$, and clearly $K_*(K)$ is a module over $\mathbf{Z}[u,v,u^{-1},v^{-1}]$. Therefore f lies in $K_*(K)$.

Proof of Theorem 2.5. This is precisely similar to the proof of Theorem 2.3, using Lemma 5.4 instead of Lemma 5.3 and Proposition 6.14 instead of Proposition 6.13.

REFERENCES

1. J. F. ADAMS, 'Vector fields on spheres', *Ann. of Math.* (2) 75 (1962) 603–32.
2. —— 'On the groups $J(X)$, IV', *Topology* 5 (1966) 21–71.
3. —— 'Lectures on generalised cohomology' (Springer Lecture Notes in Mathematics 99, 1969).
4. —— Mimeographed lecture notes (Chicago, 1970).
5. M. F. ATIYAH and R. BOTT, 'On the periodicity theorem for complex vector bundles', *Acta Math.* 112 (1964) 229–47.
6. R. BOTT, 'The stable homotopy of the classical groups', *Ann. of Math.* (2) 70 (1959) 313–37.
7. A. S. HARRIS, M.Sc. thesis (University of Manchester, 1970).
8. J. MILNOR, 'The Steenrod algebra and its dual', *Ann. of Math.* (2) 67 (1958) 150–71.
9. S. P. NOVIKOV, 'The methods of algebraic topology from the viewpoint of cobordism theories', *Izv. Akad. Nauk SSSR* 31 (1967) 855–951.
10. G. W. WHITEHEAD, 'Generalized homology theories', *Trans. Amer. Math. Soc.* 102 (1962) 227–83.

The University
Manchester 13

OPERATIONS OF THE N^{TH} KIND IN K-THEORY, AND WHAT WE DON'T KNOW ABOUT RP^{∞}

J. F. ADAMS

I. Operations of the nth kind in K-theory

In the old days, if you wanted to solve some concrete problem in homotopy theory, you began by calculating the ordinaty cohomology groups of all the spaces involved. Then you used primary cohomology operations, such as cup-products and the Steenrod operations. If, or when, those didn't yield enough information you tried secondary ones, and then tertiary and higher ones. Of course, if the problem needed tertiary operations you didn't publish the argument in that form because it was too nasty. However, it was sometimes possible to avoid some of the nastiness by using suitable formal machinery like the Adams spectral sequence.

A little later we realised, with great pleasure, that sometimes by using a generalised cohomology theory - perhaps with primary operations - you could successfully tackle a geometrical problem which if done by ordinary cohomology would have needed operations of arbitrarily high kind. It was always conceded that the choice of the cohomology theory most useful for a particular problem might take hard work, or luck, or both. But there was a sort of democratic movement, which proclaimed that every generalised cohomology theory deserved equal rights. For example, Atiyah showed that it is technically possible to teach K-theory before ordinary cohomology. Of course all normal people still did their calculations in ordinary cohomology first, but they were made to feel that they were mere slaves of habit. The philosophy prevailed that all the apparatus of calculation, with which we are so familiar in the ordinary case, should be set up for generalised cohomology. It was conceded that some results like the Kunneth theorem might require restrictive hypotheses. On the other hand, basic things like cohomology operations seem to arise from mere category-theory, and one imagined that they would work in fair generality.

It is now time to explain that the work to be presented is joint work of David Baird and myself. David Baird had been studying classical complex K-theory localised at an odd prime p; and by any standards this is a very good cohomology theory. David arrived at a belief which I will state in the the following simplified form: in this theory you may be able to set up the apparatus of stable tertiary and higher operations, but the information yielded will be exactly zero. It is only fair to say that at first I found this suggestion both implausible and unwelcome. However, I have convinced myself that it is well founded.

To make the suggestion precise, we need machinery to estimate the scope and field of action for stable tertiary and higher operations in a given theory. In the classical case one uses homological algebra, introducing Tor and Ext over the Steenrod algebra. It is natural to try and carry this approach over to the generalised case. Given a spectrum E, you can form $E^*(E)$, the E-cohomology of the spectrum E - this is the graded algebra of stable primary operations on E-cohomology. It has to be considered as a topological algebra. Novikov took this line with success in the case of complex cobordism $E = MU$.

In the case of K-theory I think that Graeme Segal made some tentative calculations with rings of operations generated by operations Ψ^k with k prime to p. But here one suffers from a certain lack of confidence that the algebra is well related to any geometry one can do, so that one is not certain one is doing the right calculations. I believe that when the calculations seemed difficult to interpret, they were abandoned.

Later I suggested that for suitable spectra E one should consider $E_*(E)$, the E-homology of X. Under reasonable assumptions on E this behaves like the dual of the Steenrod algebra, and for any X one can make the E-homology groups $E_*(X)$ into a comodule over the coalgebra $E_*(E)$. One can form Ext of comodules over this coalgebra, and one can be sure that this homological algebra is well related to suitable geometry. (See my lectures in Springer Lecture Notes No. 99).

We are now close to a theorem. Let K be classical complex K-theory localised at an odd prime p. Let X and Y be (say) finite CW-complexes. Then we can form the reduced groups $\tilde{K}_*(X)$, $\tilde{K}_*(Y)$

and regard them as comodules over the coalgebra $K_*(K)$. Sufficient information on $K_*(K)$ has been published by Adams, Harris and Switzer. We can form $\mathrm{Ext}^{s,t}_{K_*(K)}\left(\tilde{K}_*(X), \tilde{K}_*(Y)\right)$.

Theorem. $\mathrm{Ext}^{s,t}_{K_*(K)}\left(\tilde{K}_*(X), \tilde{K}_*(Y)\right) = 0$ <u>for</u> $s \geq 3$.

This theorem has something to displease everybody. On the one hand, some of us are to some extent true English problem-solvers, and so we like tertiary operations, and when we are told that in this context they are useless, that is a source of pain and grief. On the other hand, some of us are to some extent true French Bourbakistes, and so we hate these inelegant higher operations and would love a theorem which dispenses us from ever considering the nasty things again. But this theorem won't do that, because it says you may still need secondary operations, and we can give examples where you do need them.

Perhaps it will be best to suggest how we should go on from here. The groups $\mathrm{Ext}^{s,t}_{K_*(K)}\left(\tilde{K}_*(X), \tilde{K}_*(Y)\right)$ should be the E_2 term of an Adams spectral sequence, but it should not converge to ordinary stable homotopy theory. However, there is a plausible candidate for the groups to which it should converge. To construct them, we start from some category C in which we can do stable homotopy theory. We then construct a category of fractions F. More precisely, F comes equipped with a functor $T : C \to F$, which has the following two properties.

(i) If $f : X \to Y$ is a morphism in C such that $K_*(f) : K_*(X) \to K_*(Y)$ is iso, then $Tf : TX \to TY$ is invertible in F.

(ii) $T : C \to F$ is universal with respect to property (i).

Heuristically, the category F would give an account of stable homotopy-theory, so far as it can be seen through the spectacles of K-theory. For example, it is plausible that an Adams spectral sequence starting from $\mathrm{Ext}^{s,t}_{K_*(K)}\left(K_*(X), K_*(Y)\right)$ should converge to the set of morphisms in F from X to Y.

By constructing F we should get a theory with all the same formal properties as stable homotopy theory, but with different coefficient groups. For example, in ordinary stable homotopy theory the group $C[S^{n+r}, S^n]$ contains a summand Z_p if $r = 2(p-1)q - 1$ with

q prime to p, $q > 0$. I presume (but it is not proved) that in the category of fractions the group $F[S^{n+r}, S^n]$ would actually be Z_p if $r = 2(p-1)q - 1$ with q prime to p, whether q is positive or negative. The price for gaining a certain amount of periodicity is that you lose the Hurewicz theorem.

I would hope that to calculate the set of morphisms $F[X, Y]$ in this category of fractions would be quite reasonable, although we do not yet have the theorems which would enable us to do it. If so, then the category of fractions F might become useful as a computational tool. It is possible that F would retain just enough of the phenomena in C to have some interest.

One final question: should there exist also a version of unstable homotopy-theory as seen through the eyes of K-theory?

Math. Proc. Camb. Phil. Soc. (1976), **79**, 483
MPCPS 79–42

Operations on K-theory of torsion-free spaces

BY J. F. ADAMS AND P. HOFFMAN
University of Cambridge, and University of Waterloo, Ontario

(*Received* 5 *June* 1975)

1. *Introduction.* The object of this paper is to study stable operations on the K-cohomology of torsion-free spaces and spectra. Such a study was begun by Novikov ((4), §8). We hope the present work may among other things serve to clarify some of the obscure points in this part of his work.

To proceed to the details, let K be the BU-spectrum, and let K_n, K^n be the corresponding homology and cohomology functors; we need them only for $n = 0$. Let n, m be integers, and let $C(n, m)$ be the full subcategory of CW-spectra X which satisfy the following conditions:

(i) $\pi_r(X) = 0$ for $r < 2n$.

(ii) $H_r(X)$ is free for all r.

(iii) $H_r(X) = 0$ for $r > 2m$.

Let $A(n, m)$ be the algebra of operations

$$a: K^0(X) \to K^0(X)$$

which are defined and natural for X in $C(n, m)$. The object of this paper is to calculate $A(n, m)$.

2. *Results.* In this section we will explain the theory and its results; the proofs will mostly be postponed to the next section.

We propose to describe $A(n, m)$ by duality. The first step is to introduce an object $A_0(n, m)$ such that $A(n, m)$ will turn out to be the dual of $A_0(n, m)$. The subscript zero may be interpreted as 'degree zero'; it would be inappropriate to write $A_*(n, m)$ because no other degree is to be considered.

Let X be a spectrum in $C(n, m)$, let x be an element of $K_0(X)$, and let $f: X \to K$ be a map. Then we can form the element $f_* x$ in $K_0(K)$; we define $A_0(n, m)$ to be the subset of $K_0(K)$ consisting of all such elements $f_* x$, as X, x and f run over all possibilities.

PROPOSITION 2·1. (*a*) *$A_0(n, m)$ is a finitely-generated free Z-module.* (*b*) *The product $K_0(K) \otimes K_0(K) \to K_0(K)$ maps $A_0(n, m) \otimes A_0(n', m')$ into $A_0(n+n', m+m')$.*

Here $\otimes$ means $\otimes_Z$, as usual.

We pause to prove part (*b*), because the ideas will be useful in this section. Take objects $X \in C(n, m)$, $Y \in C(n', m')$, elements $x \in K_0(X)$, $y \in K_0(Y)$, and maps $f: X \to K$, $g: Y \to K$, so that $f_* x$, $g_* y$ are typical elements of $A_0(n, m)$, $A_0(n', m')$. Then we may form the smash product $X \wedge Y$ and it lies in $C(n+n', m+m')$. We may also form the external product $xy \in K_0(X \wedge Y)$, and the map

$$X \wedge Y \xrightarrow{f \wedge g} K \wedge K \xrightarrow{\mu} K,$$

where μ is the product map for the ring-spectrum K. If we recall that f is an element of $K^0(X)$, and g is an element of $K^0(Y)$, then the map just constructed is their external product $fg \in K^0(X \wedge Y)$. The product in $K_0(K)$ of $f_* x$ and $g_* y$ is $(fg)_*(xy)$, and therefore it lies in $A_0(n+n', m+m')$.

Of course, a similar proof with the wedge-sum $X \vee Y$ shows that $A_0(n, m)$ is a Z-module.

Proposition 2·1 allows one to continue with the theory; examples and a precise calculation of $A_0(n, m)$ will be given below – see (2·5) to (2·9).

PROPOSITION 2·2. (*a*) *If X lies in $C(n, m)$, then the coaction map*

$$\nu: K_0(X) \to K_0(K) \otimes K_0(X)$$

maps $K_0(X)$ into $A_0(n, m) \otimes K_0(X)$.

(*b*) *The coproduct map $\psi: K_0(K) \to K_0(K) \otimes K_0(K)$ maps $A_0(n, m)$ into*

$$A_0(n, m) \otimes A_0(n, m).$$

Some comments on the statement are in order. According to the details given in (1), the values of the coaction map ν lie in $K_*(K) \otimes_{\pi_*(K)} K_*(X)$, and the values of the coproduct map ψ lie in $K_*(K) \otimes_{\pi_*(K)} K_*(K)$. However, the 0-dimensional components of these graded groups are $K_0(K) \otimes_Z K_0(X)$ and $K_0(K) \otimes_Z K_0(K)$ (up to obvious isomorphisms). Again, one should check that the maps

$$A_0(n, m) \otimes K_0(X) \to K_0(K) \otimes K_0(X)$$
$$A_0(n, m) \otimes A_0(n, m) \to K_0(K) \otimes K_0(K)$$

are monomorphic; this is immediate, because $K_0(X)$ is torsion-free for X in $C(n, m)$ and $K_0(K)$ is torsion-free by (3).

If X lies in $C(n, m)$ then $K_0(X)$ is a free Z-module and we have

$$K^0(X) \cong \operatorname{Hom}_Z(K_0(X), Z),$$

the isomorphism being defined by the Kronecker product. Let $A^0(n, m)$ be

$$\operatorname{Hom}_Z(A_0(n, m), Z),$$

the dual of $A_0(n, m)$. By duality from Proposition 2·2 we obtain an action map

$$A^0(n, m) \otimes K^0(X) \to K^0(X)$$

and a product map

$$A^0(n, m) \otimes A^0(n, m) \to A^0(n, m).$$

Thus $K^0(X)$ becomes a module over the ring $A^0(n, m)$. We thus obtain a homomorphism of rings

$$\theta: A^0(n, m) \to A(n, m).$$

THEOREM 2·3. *This homomorphism $\theta: A^0(n, m) \to A(n, m)$ is an isomorphism.*

COROLLARY 2·4. *The ring $A(n, m)$ is commutative.*

This follows immediately, because the coproduct ψ in $K_0(K)$ is commutative (3).

In order to proceed further, we need notation for elements of $K_0(K)$. According to (3), $K_0(K)$ is torsion-free, so that we have an embedding

$$K_0(K) \to K_0(K) \otimes Q.$$

Here $K_*(K) \otimes Q$ is the ring of finite Laurent series $Q[u, v, u^{-1}, v^{-1}]$ on two generators u and v, of degree 2, which are described in (3). In particular, $K_0(K) \otimes Q$ is the ring of finite Laurent series $Q[w, w^{-1}]$, where $w = u^{-1}v$. Alternatively, we may introduce w by the method of the following example.

EXAMPLE 2·5. *Let r be an integer (positive, negative or zero) and let X be the sphere-spectrum S^{2r}; it lies in $C(r, r)$. Let x be the generator for $K_0(S^{2r}) \cong \pi_{-2r}(K) \cong Z$, and let $f: S^{2r} \to K$ be the generator for $K^0(S^{2r}) \cong \pi_{2r}(K) \cong Z$. Then $f_* x = w^r$; thus w^r lies in $A_0(r, r)$. Since the objects in $C(r, r)$ are (up to equivalence) wedge-sums of S^{2r}, $A_0(r, r)$ is the Z-module generated by w^r.*

The proof is trivial, and will be omitted.

COROLLARY 2·6. *Multiplication by w^r defines an isomorphism*

$$A_0(n, m) \xrightarrow{\cong} A_0(n + r, m + r)$$

whose inverse is multiplication by w^{-r}.

This follows immediately from (2·1*b*) and (2·5). Of course the result merely reflects the fact that the categories $C(n, m)$ and $C(n + r, m + r)$ are equivalent; the equivalences are given by iterated suspension, that is, multiplication by S^{2r}, S^{-2r}.

EXAMPLE 2·7. *Let ξ be the canonical line bundle over CP^m, so that $K^0(CP^m)$ has a base consisting of the powers $(\xi - 1)^i$ for $0 \leqslant i \leqslant m$; let $\{x_i\}$ be the dual base for $K_0(CP^m)$. Let X be the suspension-spectrum of CP^m, so that X lies in $C(1, m)$; we recall that $K^0(X) = \tilde{K}^0(CP^m)$, and similarly for K_0. Let $f: X \to K$ represent $\xi - 1$. Then*

$$f_* x_m = \frac{w(w-1)(w-2)\ldots(w-m+1)}{1.2.3\ldots m}.$$

Thus

$$\frac{w(w-1)(w-2)\ldots(w-m+1)}{1.2.3\ldots m} = \binom{w}{m}$$

lies in $A_0(1, m)$.

This formula is due to Adams, Harris & Switzer ((3), p. 407, Proposition 6·13); but the methods of the present paper permit a shorter and easier proof.

COROLLARY 2·8. *Let*

$$p_r(w) = \frac{(w-1)(w-2)\ldots(w-r)}{2.3\ldots r+1} = w^{-1}\binom{w}{r+1};$$

then $p_r(w)$ lies in $A_0(0, r)$.

This follows immediately from (2·6) and (2·7).

THEOREM 2·9. *$A_0(n,m)$ is the Z-submodule of $Q[w,w^{-1}]$ generated by the products*

$$w^n p_{r_1}(w)\, p_{r_2}(w) \ldots p_{r_\nu}(w)$$

with $r_1 \geqslant 1, r_2 \geqslant 1, \ldots, r_\nu \geqslant 1, r_1+r_2+\ldots+r_\nu \leqslant m-n$ (and therefore with $\nu \leqslant m-n$).

Of course the empty product is interpreted as 1. The fact that these products do lie in $A_0(n,m)$ follows immediately from (2·1*b*) and (2·8). This completes the statement of results.

3. *Proofs.* In this section we will prove the results stated in section 2.

First we note that while there appears to be no 'universal example' for the problem of computing $A(n,m)$, there is a splendid one for computing $A_0(0,m)$, namely the Thom spectrum MU. More precisely, we filter $K_0(\mathrm{MU})$ in the usual way, taking $K_0(\mathrm{MU})_{2m}$ to be the image of

$$K_0(\mathrm{MU}^{2m}) \to K_0(\mathrm{MU}),$$

where MU^{2m} is the $2m$-skeleton of MU.

LEMMA 3·1. *$A_0(0,m)$ is the image of $K_0(\mathrm{MU})_{2m}$ under the homomorphism*

$$t_*: K_0(\mathrm{MU}) \to K_0(K)$$

induced by the Todd map $t: \mathrm{MU} \to K$.

Proof. The skeleton MU^{2m} qualifies as an object of $C(0,m)$ (and can be chosen finite if required). Therefore the image of

$$K_0(\mathrm{MU}^{2m}) \to K_0(K)$$

is contained in $A_0(0,m)$, by definition.

Conversely, let X be an object of $C(0,m)$, and let $f: X \to K$ be a map. Since X is (-1)-connected, the map f lifts to the connective BU-spectrum k. The Atiyah–Hirzebruch spectral sequences for $[X, \mathrm{MU}]$ and $[X, k]$ are both trivial since $H_*(X)$ is free. The induced homomorphism

$$\pi_*(\mathrm{MU}) \to \pi_*(k)$$

is split epi, so that

$$H^*(X; \pi_*(\mathrm{MU})) \to H^*(X; \pi_*(k))$$

is split epi. Now an easy spectral sequence argument shows that the induced homomorphism

$$[X, \mathrm{MU}] \to [X, k]$$

is epi; so the map f lifts to MU. Since X has homological dimension $\leqslant 2m$, the map f lifts to MU^{2m}. It is now clear that any class $f_* x$ with $x \in K_0(X)$ is contained in the image of $K_0(\mathrm{MU}^{2m}) \to K_0(K)$. This proves Lemma 3·1.

Proof of Proposition 2·1(*a*). By Corollary 2·6 it is sufficient to consider the case $n = 0$. In Lemma 3·1 it is clear that $K_0(\mathrm{MU})_{2m}$ is a finitely-generated Z-module; therefore its image $A_0(0,m)$ is a finitely-generated Z-module. It is torsion-free, since $K_0(K)$ is torsion-free by (3); therefore $A_0(0,m)$ is free. This proves Proposition 2·1(*a*).

To prove Proposition 2·2 we need another lemma.

LEMMA 3·2. *Let $\nu: K_0(X) \to K_0(K) \otimes K_0(X)$ be the coaction map; suppose $x \in K_0(X)$ and $\nu(x) = \sum_i k_i \otimes x_i$, where $k_i \in K_0(K)$ and $x_i \in K_0(X)$; let $f: X \to K$ be a map. Then*

$$f_* x = \sum_i k_i \langle f, x_i \rangle \in K_0(K).$$

On the right-hand side, f is interpreted as an element of $K^0(X)$; since $x_i \in K_0(X)$, we can form the Kronecker product $\langle f, x_i \rangle$, and it lies in Z.

Proof. This is one of the basic properties of ν – see (1), p. 74. For our present purposes we may deduce it from the following commutative diagram, where ϵ is as in (1).

$$\begin{array}{ccc} K_0(X) & \xrightarrow{\nu} & K_0(K) \otimes K_0(X) \\ {\scriptstyle f_*}\downarrow & & \downarrow {\scriptstyle 1 \otimes f_*} \\ K_0(K) & \xrightarrow{\psi} & K_0(K) \otimes K_0(K) \\ & {\scriptstyle 1}\searrow & \downarrow {\scriptstyle 1 \otimes \epsilon} \\ & & K_0(K) \otimes_Z Z \end{array}$$

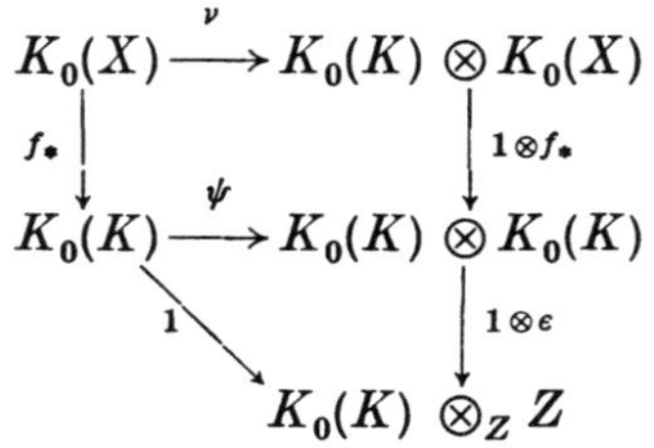

The composite

$$K_0(X) \xrightarrow{f_*} K_0(K) \xrightarrow{\epsilon} Z$$

carries x_i to $\langle f, x_i \rangle$. This proves Lemma 3·2.

Proof of Proposition 2·2. We start with part (*a*). Let X lie in $C(n, m)$. Then $K_0(X)$ is a free Z-module and $K^0(X) \cong \mathrm{Hom}_Z(K_0(X), Z)$; let us take dual bases $\{x_i\}$ in $K_0(X)$ and $\{f_i\}$ in $K^0(X)$. If $x \in K_0(X)$, we may write

$$\nu x = \sum_i k_i \otimes x_i$$

as in Lemma 3·2; and using that lemma,

$$k_j = \sum_i k_i \langle f_j, x_i \rangle = (f_j)_* x \in A_0(n, m).$$

Thus νx lies in $A_0(n, m) \otimes K_0(X)$. This proves part (*a*).

Part (*b*) follows from part (*a*) by naturality, since every element of $A_0(n, m)$ comes via a map $f: X \to K$. This proves Proposition 2·2.

In order to prove Theorem 2·3, we need to make the homomorphism

$$\theta: A^0(n, m) \to A(n, m)$$

more explicit.

LEMMA 3·3. *Let $\alpha \in A^0(n, m)$; then the corresponding operation $a = \theta\alpha \in A(n, m)$ is determined by the equation*

$$\langle af, x \rangle = \alpha(f_* x)$$

valid for each $X \in C(n, m)$, each $f: X \to K$ and each $x \in K_0(X)$.

Proof. Let $\nu x = \sum_i k_i \otimes x_i$ with $k_i \in A_0(n, m)$, $x_i \in K_0(X)$. Then the definition of $a = \theta\alpha$ by duality is

$$\langle af, x \rangle = \sum_i \alpha(k_i) \langle f, x_i \rangle.$$

Since α is linear we have

$$\sum_i \alpha(k_i)\langle f, x_i\rangle = \alpha(\sum_i k_i\langle f, x_i\rangle),$$

and by Lemma 3·2 we have

$$\sum_i k_i\langle f, x_i\rangle = f_* x.$$

This proves Lemma 3·3.

COROLLARY 3·4. The map θ is monomorphic.

Proof. Consider the equation

$$\langle af, x\rangle = \alpha(f_* x).$$

If $a = 0$, then α vanishes on every element $f_* x$ of $A_0(n, m)$.

To prove that θ is epimorphic requires a few preliminaries. We write KQ for the BU-spectrum with rational coefficients.

LEMMA 3·5. *Let $b, c: K^0(X) \to KQ^0(X)$ be operations, defined and natural for X in $C(n, m)$, which agree on S^{2r} for $n \leqslant r \leqslant m$. Then $b = c$.*

Proof. Let $a = b - c$; then $a: K^0(X) \to KQ^0(X)$ is an operation defined and natural for X in $C(n, m)$, and zero on S^{2r} for $n \leqslant r \leqslant m$; it will be sufficient to prove that $a = 0$. For any X in $C(n, m)$, any maps $f: X \to K$ and $g: S^{2r} \to X$, and any $s \in K_0(S^{2r})$ we have

$$\langle af, g_* s\rangle = \langle g^*(af), s\rangle = \langle a(g^* f), s\rangle = 0$$

(since a is zero on $K^0(S^{2r})$). Now

$$K_*(K) \otimes Q \cong \pi_*(K) \otimes \pi_*(X) \otimes Q;$$

therefore such classes $g_* s$ generate $K_0(X) \otimes Q$ as a Q-module. But

$$\begin{aligned} KQ^0(X) &\cong \operatorname{Hom}_Z(K_0(X), Q) \\ &\cong \operatorname{Hom}_Q(K_0(X) \otimes Q; Q); \end{aligned}$$

so we may conclude that $af = 0$. Since this holds for all X and f, we have $a = 0$. This proves Lemma 3·5.

Next we need to describe $KQ^0(KQ)$, the algebra of operations

$$a: KQ^0(X) \to KQ^0(X)$$

defined and natural for all spectra X. For any X there is an isomorphism

$$ch: KQ^0(X) \xrightarrow{\cong} \prod_{r \in Z} H^{2r}(X; Q).$$

Since Q acts on $H^{2r}(X; Q)$, the algebra $\prod_{r \in Z} Q$ acts on $\prod_{r \in Z} H^{2r}(X; Q)$. More precisely, an element of $\prod_{r \in Z} Q$ is a vector $\{\lambda_r\}$ with $\lambda_r \in Q$ for each r; an element of $\prod_{r \in Z} H^{2r}(X; Q)$ is a vector $\{h_r\}$ with $h_r \in H^{2r}(X; Q)$; and the action is given by

$$\{\lambda_r\}\{h_r\} = \{\lambda_r h_r\}.$$

This defines a map of algebras

$$\prod_{r \in Z} Q \to KQ^0(KQ);$$

it is easy to show that this map is an isomorphism, but we will not need this fact. We may write $\sum_r \lambda_r \pi_r$ for the operation corresponding to $\{\lambda_r\}$; here π_r is the idempotent operation corresponding to the projection of $\prod_{s \in Z} H^{2s}(X; Q)$ on its rth factor.

Let $j: K \to KQ$ be the obvious map.

LEMMA 3·6. *Let $a \in A(n, m)$ be an operation. Then there is a map $b: K \to KQ$ such that the composite operation*

$$K^0(X) \xrightarrow{a} K^0(X) \xrightarrow{j} KQ^0(X)$$

is induced by b for all X in $C(n, m)$. That is, for any $f: X \to K$ we have

$$j(af) = bf.$$

Proof. Let $a \in A(n, m)$ be an operation. Then for r in the range $n \leqslant r \leqslant m$, a must act on $K^0(S^{2r})$ by multiplication with a scalar $\lambda_r \in Z$. Choose scalars λ_r arbitrarily for $r < n$ and for $r > m$, and let b be the composite

$$K \xrightarrow{j} KQ \xrightarrow{\sum_r \lambda_r \pi_r} KQ.$$

Then by construction, the composite operation

$$K^0(X) \xrightarrow{a} K^0(X) \xrightarrow{j} KQ^0(X)$$

agrees with that induced by b when $X = S^{2r}, n \leqslant r \leqslant m$; so the two operations are equal by Lemma 3·5. This proves Lemma 3·6.

Proof of Theorem 2·3. After (3·4), it remains to prove that θ is epimorphic. Let $a \in A(n, m)$ be an operation, and let $b: K \to KQ$ be as in Lemma 3·6. We can define a linear function

$$\alpha: A_0(n, m) \to Q$$

by

$$\alpha(f_* x) = \langle b, f_* x \rangle;$$

then α maps $A_0(n, m)$ into Z, and $\theta\alpha = a$. In fact, for any $X \in C(n, m)$, any $f: X \to K$ and any $x \in K_0(X)$ we have

$$\langle b, f_* x \rangle = \langle f^* b, x \rangle = \langle bf, x \rangle = \langle j(af), x \rangle;$$

here $\langle j(af), x \rangle$ is the image in Q of the integer $\langle af, x \rangle$. So $\alpha \in A^0(n, m)$ and

$$\alpha(f_* x) = \langle af, x \rangle;$$

by Lemma 3·3, we have $a = \theta\alpha$. This completes the proof of Theorem 2·3.

Proof of Example 2·7. Let X, f, x_m be as in Example 2·7. For $k \neq 0$ the operation ψ^k may be interpreted as a map $\psi^k: K \to KQ$. By the proof of (2·3) we have

$$\langle \psi^k, f_* x_m \rangle = \langle \psi^k f, x_m \rangle.$$

By the ordinary calculation of ψ^k in CP^m we have

$$\begin{aligned}\psi^k(\xi-1) &= \xi^k - 1\\ &= ((\xi-1)+1)^k - 1\\ &= \sum_{1\leqslant j\leqslant m} \frac{k(k-1)\ldots(k-j+1)}{1.2\ldots j}(\xi-1)^j.\end{aligned}$$

So in this case we have

$$\langle \psi^k f, x_m\rangle = \frac{k(k-1)\ldots(k-m+1)}{1.2\ldots m}.$$

But by a similar calculation in S^{2r}, or directly, we have

$$\langle \psi^k, w^r\rangle = k^r.$$

So for a finite Laurent series

$$q(w) = \sum_r q_r w^r$$

with coefficients $q_r \in Q$, we have

$$\langle \psi^k, q(w)\rangle = \sum_r q_r k^r = q(k).$$

If we set $q(w) = f_* x_m$, we have

$$q(k) = \frac{k(k-1)\ldots(k-m+1)}{1.2\ldots m}.$$

But such a finite Laurent series is determined by its values for non-zero integral k. Therefore

$$q(w) = \frac{w(w-1)\ldots(w-m+1)}{1.2\ldots m}.$$

This proves Example 2·7.

Proof of Theorem 2·9. We rely on Lemma 3·1. It is well known – see for example (2) p. 24, Lemma 4·5 – that $K_0(\mathrm{MU})$ is generated as a ring by elements coming from $\mathrm{MU}(1)$. In order to have a map of spectra $\mathrm{MU}(1)\to\mathrm{MU}$, we must interpret $\mathrm{MU}(1)$ as the suspension-spectrum in which the space $\mathrm{MU}(1)$ occurs as the second term of the spectrum; that is, it is the spectrum $S^{-2}\wedge CP^\infty$. Then the generator x_{r+1} of filtration $r+1$ in $K_0(CP^\infty)$ gives a generator of filtration r in $K_0(\mathrm{MU}(1))$; in (2) this generator is called b_r^K. We have $b_0^K = 1$. By (2·7) the image of b_r^K in $K_0(K)$ is

$$p_r(w) = \frac{(w-1)(w-2)\ldots(w-r)}{2.3\ldots(r+1)} = w^{-1}\binom{w}{r+1}.$$

(The factor w^{-1} comes from the factor S^{-2}.) The proof in (2), by the Atiyah–Hirzebruch spectral sequence, actually shows that $K_0(\mathrm{MU})_{2m}$ is the Z-submodule generated by the products

$$b_{r_1}^K b_{r_2}^K \ldots b_{r_\nu}^K$$

with $\qquad r_1 \geqslant 1,\quad r_2\geqslant 1,\quad \ldots,\quad r_\nu \geqslant 1,\quad r_1+r_2+\ldots+r_\nu \leqslant m.$

The map $t_*: K_0(\mathrm{MU})\to K_0(K)$ preserves products. By Lemma 3·1, $A_0(0,m)$ is the Z-submodule generated by the products

$$p_{r_1}(w)\,p_{r_2}(w)\ldots p_{r_\nu}(w)$$

with $r_1 \geqslant 1, \quad r_2 \geqslant 1, \quad \ldots, \quad r_\nu \geqslant 1, \quad r_1+r_2+\ldots+r_\nu \leqslant m.$

This proves the result for $A_0(0, m)$; the general case follows by (2·6). This completes the proof of Theorem 2·9.

REFERENCES

(1) Adams, J. F. Lectures on generalised cohomology. In *Category Theory, Homology Theory and Their Applications*, III (Lecture Notes in Mathematics no. 99, Springer, 1969).
(2) Adams, J. F. *Stable Homotopy and Generalised Homology*, part II (Chicago University Press, 1974).
(3) Adams, J. F., Harris, A. S. and Switzer, R. M. Hopf algebras of cooperations for real and complex K-theory. *Proc. London Math. Soc.* **23** (1971), 385–408.
(4) Novikov, S. P. The methods of algebraic topology from the view-point of cobordism theories. *Izv. Akad. Nauk SSSR*, Serija Matematiceskaja, **31** (1967), 885–951.

ILLINOIS JOURNAL OF MATHEMATICS
Volume 21, Number 4, December 1977

STABLE OPERATIONS ON COMPLEX K-THEORY

BY

J. F. ADAMS AND F. W. CLARKE

1. Introduction

Let $\mathbf{K}$ be the spectrum representing classical (periodic) complex K-theory. A stable operation (of degree zero) on complex K-theory should then correspond to an element of the K-cohomology group $\mathbf{K}^0(\mathbf{K})$; equivalently, it should correspond to a map of spectra $\mathbf{f}: \mathbf{K} \to \mathbf{K}$. (It will be convenient if the word "map" means a homotopy class, and is restricted to maps of degree zero.) Two maps from $\mathbf{K}$ to $\mathbf{K}$ are well known: Ψ^1, the identity map, and Ψ^{-1}, the map induced by complex conjugation. One may then form integral linear combinations $\lambda\Psi^1 + \mu\Psi^{-1}$, where $\lambda, \mu \in \mathbf{Z}$. It has been conjectured, and some have tried to prove, that in this way one obtains all the maps from $\mathbf{K}$ to $\mathbf{K}$. Although some of our colleagues have found it hard to believe, we will show that this conjecture is false; there are uncountably many maps from $\mathbf{K}$ to $\mathbf{K}$. We deduce this from a result which has other applications in K-theory.

2. Study of K-homology

Let $\mathbf{K}_*(\mathbf{K})$ be the K-homology of the spectrum $\mathbf{K}$. It has been sufficiently described by Adams, Harris, and Switzer [3]; but these authors omit the following fundamental result.

THEOREM 2.1. *$\mathbf{K}_*(\mathbf{K})$, considered as a left module over $\pi_*(\mathbf{K})$, is free on a countably infinite set of generators (of degree zero).*

Because of the structure of $\pi_*(\mathbf{K})$, any graded module M_* over $\pi_*(\mathbf{K})$ which is zero in odd degrees satisfies $M_* \cong \pi_*(\mathbf{K}) \otimes_{\mathbf{Z}} M_0$. So Theorem 2.1 will follow from the following result.

THEOREM 2.2. *$\mathbf{K}_0(\mathbf{K})$ is a free abelian group on a countably infinite set of generators.*

In order to prove this, recall that according to [3] we have an embedding $\mathbf{K}_0(\mathbf{K}) \subset \mathbf{K}_0(\mathbf{K}) \otimes \mathbf{Q} = \mathbf{Q}[w, w^{-1}]$ where $w = u^{-1}v$ (u and v being as in [3]). Let $F(n, m)$ be the intersection of $\mathbf{K}_0(\mathbf{K})$ with the $\mathbf{Q}$-module generated by w^n, $w^{n+1}, \ldots, w^m$.

LEMMA 2.3. *$F(n, m)/F(n, m-1)$ and $F(n, m)/F(n+1, m)$ are free abelian groups of rank 1.*

Received April 20, 1976.

Proof. We give the proof for $F(n, m)/F(n, m-1)$; the proof for $F(n, m)/F(n+1, m)$ is parallel.

An element of $F(n, m)$ may be written in the form $\sum_{n \le r \le m} c_r w^r$, where the coefficients c_r lie in $\mathbf{Q}$. We can define an embedding

$$F(n, m)/F(n, m-1) \to \mathbf{Q}$$

by sending $\sum_{n \le r \le m} c_r w^r$ to the coefficient c_m of w^m. We wish to determine the image I of this embedding. It is a subgroup of $\mathbf{Q}$, and clearly contains $\mathbf{Z}$, since w^m belongs to $F(n, m)$. The result will follow if we show that there is an integer M such that the image I is contained in $(1/M)\mathbf{Z}$. We prove this by localization; it will be sufficient to prove the following.

(i) For each prime p there is a power p^e such that

$$I \subset (1/p^e)\mathbf{Z}_{(p)}$$

(where $\mathbf{Z}_{(p)}$ means the localization of $\mathbf{Z}$ at p, as usual.)

(ii) For all but a finite number of primes p we can take $p^e = 1$.

So let p be a prime. Then in $\mathbf{K}^0(\mathbf{K}; \mathbf{Z}_{(p)})$ we have an element Ψ^k for each integer k prime to p; and we have $\langle \Psi^k, w^r \rangle = k^r$. Let r run over the range $n \le r \le m$, and let k run over an equal number of distinct integers k_n, $k_{n+1}, \ldots, k_m$ prime to p; then the matrix with entries k^n is nonsingular, for we will show that its determinant Δ is nonzero. In fact, by removing from Δ a factor $(k_n k_{n+1} \cdots k_m)^n$, we obtain a Vandermonde determinant, which is nonzero because $k_n, k_{n+1}, \ldots, k_m$ are distinct. We can therefore choose coefficients λ_k in $\mathbf{Z}_{(p)}$ such that

$$\left\langle \sum_k \lambda_k \Psi^k, w^r \right\rangle = \begin{cases} 0 & \text{if } n \le r < m \\ \Delta & \text{if } r = m. \end{cases}$$

In particular, for any element $x = \sum_{n \le r \le m} c_r w^r$ in $F(n, m)$ we have

$$\left\langle \sum_k \lambda_k \Psi^k, x \right\rangle = \Delta c_m.$$

But certainly we have $\langle \sum_k \lambda_k \Psi^k, x \rangle \in \mathbf{Z}_{(p)}$; therefore $c_m \in (1/\Delta)\mathbf{Z}_{(p)}$. Moreover, for $p - 1 \ge m - n + 1$ we can arrange for Δ to be nonzero mod p, for we can arrange for $k_n, k_{n+1}, \ldots, k_m$ to be distinct mod p. This completes the proof.

Proof of Theorem 2.2. This follows immediately from Lemma 2.3. Suppose, as an inductive hypothesis, that we have found a base for $F(n, m)$; we may also suppose that the base contains $m - n + 1$ elements. Then Lemma 2.3 allows one to extend the base to a base for $F(n, m+1)$ or $F(n-1, m)$; we may also assert that this base contains $m - n + 2$ elements. The induction does start, because the case $n = m$ of Lemma 2.3 is to be interpreted as saying that $F(n, n)$ is a free abelian group of rank 1. (The proof even shows that $F(n, n)$ has a base consisting of the element w^n.) It is natural to arrange the induction so that alternate steps increase m and decrease n, but the induction may be conducted in any way provided that $m \to +\infty$ and $n \to -\infty$. The induction constructs a base for $\bigcup F(n, m) = \mathbf{K}_0(\mathbf{K})$. This proves Theorem 2.2, and Theorem 2.1 follows.

3. Maps from K to K

These are described by the following result.

THEOREM 3.1. *The Kronecker product gives an isomorphism*

$$\mathbf{K}^*(\mathbf{K}) \to \mathrm{Hom}_{\pi_*(\mathbf{K})}(\mathbf{K}_*(\mathbf{K}), \pi_*(\mathbf{K})).$$

Proof. This follows immediately from Theorem 2.1, by using the universal coefficient theorem in K-theory. The basic ideas for the proof of such a theorem were given by Atiyah [4], but in the context of the Künneth theorem for spaces. A discussion in the context of the universal coefficient theorem for spectra is given in [1]; it lacks a treatment of the convergence of the spectral sequence, but this may be supplied from the indications given in [2].

COROLLARY 3.2. $\mathbf{K}^1(\mathbf{K}) = 0$; $\mathbf{K}^0(\mathbf{K})$ *is uncountable.*

This follows immediately from Theorems 2.1 and 3.1.

COROLLARY 3.3. $\mathbf{K}^0(\mathbf{K})$ *contains maps not of the form* $\lambda\Psi^1 + \mu\Psi^{-1}$, *where* λ, $\mu \in \mathbf{Z}$.

This follows immediately from Corollary 3.2.

We will now show how to construct a map which is not of the form $\lambda\Psi^1 + \mu\Psi^{-1}$. For a map of the form $\phi = \lambda\Psi^1 + \mu\Psi^{-1}$ we have

$$\langle\phi, 1\rangle = \lambda + \mu, \qquad \langle\phi, w\rangle = \lambda - \mu;$$

so $\langle\phi, 1\rangle = 0$ and $\langle\phi, w\rangle = 0$ imply $\phi = 0$, and in particular $\langle\phi, w^2\rangle = 0$. Let h be the composite

$$F(0, 2) \longrightarrow F(0, 2)/F(0, 1) \xrightarrow{\cong} \mathbf{Z},$$

where the isomorphism comes from Lemma 2.3; then we have $h(1) = 0$, $h(w) = 0$ but $h(w^2) \neq 0$. (In fact calculation shows that $h(w^2) = \pm 24$, but this is irrelevant.) We will now extend h to an element of

$$\mathrm{Hom}_{\mathbf{Z}}(\mathbf{K}_0(\mathbf{K}), \mathbf{Z}) = \mathrm{Hom}^0_{\pi_*(\mathbf{K})}(\mathbf{K}_*(\mathbf{K}), \pi_*(\mathbf{K})).$$

In fact, according to the proof of Theorem 2.2, a base of $F(0, 2)$ may be extended to a base of $\mathbf{K}_0(\mathbf{K})$, and so h may be extended over $\mathbf{K}_0(\mathbf{K})$ by giving it arbitrary values on the remaining basis elements. Applying Theorem 3.1, we obtain a map $\phi \in \mathbf{K}^0(\mathbf{K})$ such that $\langle\phi, 1\rangle = 0$, $\langle\phi, w\rangle = 0$ but $\langle\phi, w^2\rangle \neq 0$; this map ϕ is not of the form $\lambda\Psi^1 + \mu\Psi^{-1}$.

REFERENCES

1. J. F. ADAMS, *Lectures on generalized cohomology*, Lecture Notes in Mathematics, no. 99, Springer 1969, especially pp. 1–45.
2. ———, *Algebraic topology in the last decade*, Proceedings of Symposia in Pure Mathematics, vol. 22, Amer. Math. Soc., 1971, especially p. 11.
3. J. F. ADAMS, A. S. HARRIS, AND R. M. SWITZER, *Hopf algebras of cooperations for real and complex K-theory*, Proc. London Math. Soc., vol. 23 (1971), pp. 385–408.
4. M. F. ATIYAH, *Vector bundles and the Künneth formula*, Topology, vol. 1 (1962), pp. 245–248.

CAMBRIDGE UNIVERSITY
CAMBRIDGE, ENGLAND

UNIVERSITY COLLEGE OF SWANSEA
SWANSEA, WALES

PRIMITIVE ELEMENTS IN THE K-THEORY OF BSU

By J. F. ADAMS

[Received 18 February 1975]

1. Introduction and statement of results

THE question considered in this note arises from the work of Dr. V. P. Snaith. In order to fill in a step in his interesting joint work with Madsen and Tornehave on maps of infinite loop-spaces [**3**], he requires information on the primitive elements in the K-cohomology of BSU. The object of this note is to record such information. A result sufficient for Dr. Snaith's work is given as Corollary 1.4.

I proceed to explain the main results of this note. Let

$$i: BSU \to BU$$

be the usual map. Since CP^∞ is an Eilenberg-MacLane space of type $(\mathbf{Z}, 2)$, maps $f: X \to CP^\infty$ correspond to elements in $H^2(X)$; let

$$\pi: BU \to CP^\infty$$

be the map corresponding to the first Chern class $c_1 \in H^2(BU)$. This map π may also be described by saying that its restriction to $BU(n)$ is the map

$$B\det: BU(n) \to BU(1)$$

induced by the homomorphism

$$\det: U(n) \to U(1).$$

Then the composite

$$BSU \xrightarrow{i} BU \xrightarrow{\pi} CP^\infty$$

is nullhomotopic; indeed, if we wish we may assume that π is a fibering and i is the inclusion of its fibre.

We give BSU the H-space structure corresponding to the Whitney sum of bundles; similarly for BU; the space CP^∞ has a unique H-space structure. The maps i and π are then maps of H-spaces.

If X is an H-space, then the K-homology group $K_0(X)$ becomes a ring, and we write $QK_0(X)$ for its indecomposable quotient (that is, the augmentation ideal I modulo the ideal I^2 of decomposable elements.)

Quart. J. Math. Oxford (2), 27 (1976), 253–262

THEOREM 1.1. *The sequence*

$$0 \longrightarrow QK_0(BSU) \xrightarrow{i_*} QK_0(BU) \xrightarrow{\pi_*} QK_0(CP^\infty) \longrightarrow 0$$

is exact.

(There is a corresponding result in ordinary homology, and the reader may like to consider it, turning to § 2 to check his conclusions; but I feel that in 1975 one cannot justify writing more than is necessary about $H_*(BSU)$.)

I will show that in Theorem 1.1 we have a complete grasp of the map

$$\pi_*: QK_0(BU) \to QK_0(CP^\infty);$$

so the exact sequence determines $QK_0(BSU)$.

The use of this is that the indecomposable quotient group $QK_0(BSU)$ determines the required subgroup of primitive elements. More precisely, let X be any one of the H-spaces BSU, BU, CP^∞; then the Atiyah-Hirzebruch spectral sequence shows that $K_0(X)$ is a free $\mathbf{Z}$-module, $K_1(X)$ is zero, and for any abelian group A of coefficients we have

$$K^0(X; A) \cong \operatorname{Hom}(K_0(X), A)$$

(where the isomorphism is given by the Kronecker product.) We write $PK^0(X; A)$ for the subgroup of primitive elements in $\tilde{K}^0(X: A)$; this subgroup can be identified with the subgroup of those homomorphisms $\tilde{K}_0(X) \to A$ which annihilate decomposable elements, that is, with $\operatorname{Hom}(QK_0(X), A)$.

In particular, the K-theory of CP^∞ is usually described by saying that $K^0(CP^\infty; A)$ is the additive group of formal power-series $A[[\lambda - 1]]$, where λ is the canonical line bundle over CP^∞. One can equally well say that $K_0(X)$ is a free $\mathbf{Z}$-module on generators β_k for $k = 0, 1, 2, \ldots$ [2]. The connection between the two accounts is the Kronecker product formula

$$\left\langle \sum_{j=0}^{\infty} a_j(\lambda - 1)^j, \beta_k \right\rangle = a_k.$$

We also write β_k for the image of the element above under the usual map

$$i': CP^\infty \simeq BU(1) \to BU.$$

Then $K_0(BU)$ is a polynomial ring

$$\mathbf{Z}[\beta_1, \beta_2, \ldots, \beta_k, \ldots]$$

with β_0 as its unit. Thus $QK_0(BU)$ is a free $\mathbf{Z}$-module on generators β_k for $k = 1, 2, 3, \ldots$.

The map

$$\pi_*: K_0(BU) \to K_0(CP^\infty)$$

clearly carries β_k to β_k.

By substituting $x = \lambda - 1$ in the series

$$\log(1+x) = x - \frac{x^2}{2} + \frac{x^3}{3} - \ldots + (-1)^{k-1}\frac{x^k}{k} \ldots,$$

we obtain an element

$$\log \lambda = \sum_{k=1}^{\infty} (-1)^{k-1} \frac{(\lambda-1)^k}{k}$$

in $\tilde{K}^0(CP^\infty; \mathbf{Q})$, and this element is primitive.

THEOREM 1.2. *The homomorphism*

$$j: QK_0(CP^\infty) \to \mathbf{Q}$$

defined by

$$j\kappa = \langle \log \lambda, \kappa \rangle$$

is an isomorphism.

According to the details given above, an explicit formula for j is

$$j(\beta_k) = (-1)^{k-1}\frac{1}{k}.$$

For example, the formula

$$\beta_1^2 = 2\beta_2 + \beta_1$$

shows that if β_1 is "worth 1" in $QK_0(CP^\infty)$, then β_2 is "worth $-\frac{1}{2}$" in $QK_0(CP^\infty)$. This completes the description of the map π_* in Theorem 1.1.

COROLLARY 1.3. *There is an exact sequence*

$$0 \longrightarrow \text{Hom}(\mathbf{Q}, A) \longrightarrow PK^0(BU; A) \xrightarrow{i^*} PK^0(BSU; A) \longrightarrow \text{Ext}(\mathbf{Q}, A) \longrightarrow 0.$$

According to the discussion above, this follows immediately from Theorem 1.1, by applying Hom(, A).

COROLLARY 1.4. *Let A be the ring of p-adic integers; then the induced map of primitive elements*

$$i^*: PK^0(BU; A) \to PK^0(BSU; A)$$

is an isomorphism.

This follows from Corollary 1.3; for if A is the ring of p-adic integers, then Hom $(\mathbf{Q}, A) = 0$ and Ext $(\mathbf{Q}, A) = 0$.

In order to state a final result, let the operations Ψ^k be as in [**1**]. Let $a_0, a_1, a_2, \ldots, a_m$ be any finite set of coefficients in A such that $\sum_{k=0}^{m} a_k = 0$; then the finite linear combination

$$\sum_{k=0}^{m} a_k \Psi^k$$

yields an element in $\tilde{K}^0(BU; A)$, and this element is primitive. Let $FL\Psi$ be the set of such finite linear combinations of the Ψ^k.

PROPOSITION 1.5. *For any A, the image of $FL\Psi$ under*

$$i^*\colon PK^0(BU; A) \to PK^0(BSU; A)$$

is dense in $PK^0(BSU; A)$ (with respect to the filtration topology.)

The remainder of this note is arranged as follows: §§ 2, 3 and 4 contain respectively the proofs of Theorem 1.1, Theorem 1.2 and Proposition 1.5.

2. Proof of Theorem 1.1

First we prove exactness at $QK_0(CP^\infty)$. The usual map

$$i'\colon CP^\infty \simeq BU(1) \to BU$$

satisfies $\pi i' = 1$. It follows from this—and it is already implicit in the account given in § 1—that the map

$$\pi_*\colon K_0(BU) \to K_0(CP^\infty)$$

is epi. Therefore the map

$$\pi_*\colon QK_0(BU) \to QK_0(CP^\infty)$$

is epi.

Next we prove exactness at $QK_0(BU)$. It is clear that the composite $\pi_* i_*$ is zero. For the converse, we form the following homotopy-commutative diagram.

$$\begin{array}{ccccc}
BSU \times CP^\infty & \xrightarrow{i \times i'} & BU \times BU & \xrightarrow{\mu} & BU \\
\Big\downarrow{\scriptstyle c \times 1} & & \Big\downarrow{\scriptstyle \pi \times \pi} & & \Big\downarrow{\scriptstyle \pi} \\
pt \times CP^\infty & \longrightarrow & CP^\infty \times CP^\infty & \xrightarrow{\mu} & CP^\infty
\end{array}$$

Here i, i' and π are as above, and the maps μ are the H-space structure maps; the composite

$$BSU \times CP^\infty \xrightarrow{\mu(i\times i')} BU$$

is a weak equivalence. We obtain the following commutative diagram.

$$\begin{array}{ccc} K_0(BSU)\otimes_{\mathbf{Z}} K_0(CP^\infty) & \xrightarrow{\cong} & K_0(BU) \\ \downarrow{\scriptstyle \epsilon\otimes 1} & & \downarrow{\scriptstyle \pi_*} \\ \mathbf{Z}\otimes_{\mathbf{Z}} K_0(CP^\infty) & \xrightarrow{\cong} & K_0(CP^\infty) \end{array}$$

Take now an element $x \in \tilde{K}_0(BU)$ whose image in $QK_0(CP^\infty)$ is zero; say

$$\pi_* x = \sum_r y_r z_r$$

where $y_r, z_r \in \tilde{K}_0(CP^\infty)$. Since

$$\pi_*: \tilde{K}_0(BU) \to \tilde{K}_0(CP^\infty)$$

is epi, we may lift y_r, z_r to elements y'_r, z'_r in $\tilde{K}_0(BU)$; then

$$x' = x - \sum_r y'_r z'_r$$

represents the same element of $QK_0(BU)$ as x, and satisfies $\pi_* x' = 0$. According to the commutative diagram above, x' corresponds to an element in

$$\tilde{K}_0(BSU)\otimes_{\mathbf{Z}} K_o(CP^\infty).$$

The component of this element in $\tilde{K}_0(BSU)\otimes_{\mathbf{Z}}\tilde{K}_0(CP^\infty)$ is decomposable, while the component in $\tilde{K}_0(BSU)\otimes_{\mathbf{Z}}\mathbf{Z}$ yields an element $x'' \in \tilde{K}_0(BSU)$ whose image in $QK_0(BU)$ is the class of x. This proves exactness at $QK_0(BU)$.

So much will not surprise the reader familiar with [**4**]; such a reader will also expect that to prove exactness at $QK_0(BSU)$ requires more detailed information. In fact it is sufficient to have a suitable result in ordinary homology, such as Lemma 2.4 below. It is also very tempting to declare that at this date all such results on ordinary homology may be assumed known; and if they are not on record, why, that is a defect in the papers written ten or twenty years ago, and not in the present one. Unfortunateiy, a sense of duty impels me to sketch a proof, and I build it up in steps.

For the moment, fix a prime p, and let $\mathbf{k}$ be the field $\mathbf{Z}/p\mathbf{Z}$.

LEMMA 2.1. *The kernel of*

$$\pi_*\colon H_*(BU;\mathbf{k}) \to H_*(CP^\infty;\mathbf{k})$$

may be generated as an ideal by elements of degree 4, 6, 8, 10, ...

Proof. Let $b_n \in H_{2n}(CP^\infty)$ be the generator, with the sign chosen so that

$$\langle (c_1)^n, b_n \rangle = 1.$$

We also write b_n for the image of this element under the usual map

$$i'\colon CP^\infty \simeq BU(1) \to BU.$$

Then $H_*(BU)$ is a polynomial algebra $\mathbf{Z}[b_1, b_2, \ldots, b_n, \ldots]$. It is well known that $H_*(CP^\infty)$ is a divided polynomial algebra, with product given by

$$b_n b_m = \frac{(n+m)!}{n!m!} b_{n+m}.$$

Over $\mathbf{k} = \mathbf{Z}/p\mathbf{Z}$, $H_*(CP^\infty;\mathbf{k})$ is a tensor product of truncated polynomial algebras; as a $\mathbf{k}$-module it has a base consisting of the monomials

$$b_1^{e_0} b_p^{e_1} b_{p^2}^{e_2} \ldots b_{p^\nu}^{e_\nu}$$

with $0 \leqslant e_r \leqslant p-1$ for each r; and the product is given by the relations

$$(b_{p^\nu})^p = 0.$$

So the ideal Ker π_* can be generated by the following elements.

(i) A suitable element

$$b_n - \lambda_n b_1^{e_0} b_p^{e_1} b_{p^2}^{e_2} \ldots b_{p^\nu}^{e_\nu} \qquad (\lambda_n \in k)$$

of degree $2n$ for each n which is not a power of p.

(ii) An element

$$(b_{p^\nu})^p$$

of degree $2p^{\nu+1}$ for $\nu = 0, 1, 2, \ldots$.

This proves Lemma 2.1.

LEMMA 2.2. $QH_*(BSU;\mathbf{k})$ *can be generated as a* $\mathbf{k}$*-module by generators of degree* 4, 6, 8, 10, ...

Proof. Arguing as in the second paragraph of this section, we obtain the following commutative diagram.

$$\begin{array}{ccc}
H_*(BSU;\mathbf{k}) \oplus_{\mathbf{k}} H_*(CP^\infty;\mathbf{k}) & \xrightarrow{\cong} & H_*(BU;\mathbf{k}) \\
\downarrow{\scriptstyle \epsilon\otimes 1} & & \downarrow{\scriptstyle \pi_*} \\
\mathbf{k} \otimes_{\mathbf{k}} H_*(CP^\infty;\mathbf{k}) & \xrightarrow{\cong} & H_*(CP^\infty;\mathbf{k})
\end{array}$$

Write

$$A = H_*(BSU; \mathbf{k})$$
$$B = H_*(BU; \mathbf{k})$$
$$C = H_*(CP^\infty; \mathbf{k}).$$

Then the diagram shows that

$$\tilde{A} \otimes_{\mathbf{k}} C \cong \operatorname{Ker} \pi_*.$$

From this we see that the obvious map

$$\tilde{A} \otimes_A B \to \operatorname{Ker} \pi_*$$

is iso, for it can be factored as

$$\begin{aligned} \tilde{A} \otimes_A B &\cong \tilde{A} \otimes_A (A \otimes_{\mathbf{k}} C) \\ &\cong \tilde{A} \otimes_{\mathbf{k}} C \\ &\cong \operatorname{Ker} \pi_*. \end{aligned}$$

Hence

$$\begin{aligned} (\operatorname{Ker} \pi_*) \otimes_B \mathbf{k} &\cong (\tilde{A} \otimes_A B) \otimes_B \mathbf{k} \\ &\cong \tilde{A} \otimes_A \mathbf{k} \\ &\cong Q(A). \end{aligned}$$

According to Lemma 2.1 $(\operatorname{Ker} \pi_*) \otimes_B \mathbf{k}$ can be generated as a $\mathbf{k}$-module by generators of degree 4, 6, 8, 10, . . . , so the same is true for $Q(A) = QH_*(BSU; \mathbf{k})$. This proves Lemma 2.2.

LEMMA 2.3. *$Q_nH_*(BSU)$ is $\mathbf{Z}$ for $n = 4, 6, 8, 10, \ldots$ and 0 for other values of n.*

Proof. $Q_nH_*(BSU)$ is a finitely-generated abelian group. It is of rank 1 for $n = 4, 6, 8, 10, \ldots$ and of rank 0 for other values of n, because $P^nH^*(BSU; \mathbf{Q})$ is $\mathbf{Q}$ for $n = 4, 6, 8, 10, \ldots$ and 0 for other values of n. Now Lemma 2.3 follows from the structure theorem for finitely-generated abelian groups, plus the information in Lemma 2.2 about

$$Q_nH_*(BSU) \otimes (\mathbf{Z}/p\mathbf{Z}) \cong Q_nH_*(BSU; \mathbf{Z}/p\mathbf{Z}).$$

LEMMA 2.4 *$H_*(BSU)$ is a polynomial algebra over $\mathbf{Z}$ on generators of degree 4, 6, 8, 10, ...*

Proof. Using Lemma 2.3, we can take a polynomial algebra $\mathbf{Z}[y_2, y_3, y_4, y_5, \ldots]$ with generators of degree 4, 6, 8, 10, . . . and construct an epimorphism

$$\theta: \mathbf{Z}[y_2, y_3, y_4, y_5, \ldots] \to H_*(BSU).$$

This epimorphism must be a monomorphism by counting ranks, for in each degree $H_*(BSU)$ has the same rank as

$$H^*(BSU) = \mathbf{Z}[c_2, c_3, c_4, c_5, \ldots].$$

This proves Lemma 2.4.

We now return to the proof of Theorem 1.1. Using Lemma 2.4 and the Atiyah-Hirzebruch spectral sequence, we see that $K_0(BSU)$ is a polynomial algebra over $\mathbf{Z}$, so that $QK_0(BSU)$ is a free $\mathbf{Z}$-module. So the map

$$QK_0(BSU) \to QK_0(BSU) \otimes \mathbf{Q}$$

is mono. On the other hand, in ordinary cohomology

$$PH^*(BU; \mathbf{Q}) \xrightarrow{i^*} PH^*(BSU; \mathbf{Q})$$

is epi, as is well known. Using the Chern character ch, this shows that

$$PK^0(BU; \mathbf{Q}) \xrightarrow{i^*} PK^0(BSU; \mathbf{Q})$$

is epi. By duality, this shows that

$$QK_0(BSU) \otimes \mathbf{Q} \xrightarrow{i_*} QK_0(BU) \otimes \mathbf{Q}$$

is mono. Hence

$$QK_0(BSU) \xrightarrow{i_*} QK_0(BU)$$

is mono. This completes the proof of Theorem 1.1.

3. Proof of Theorem 1.2

The formula

$$j(\beta_k) = (-1)^{k-1} \frac{1}{k}$$

proves that the map

$$j\colon\ QK_0(CP^\infty) \to \mathbf{Q}$$

is epi. To prove that j is mono, we proceed by induction over the filtration. We recall that the subgroup of elements of filtration $\leqslant 2n$ in $\tilde{K}_0(CP^\infty)$ is generated by $\beta_1, \beta_2, \ldots, \beta_n$. We therefore assume, as an inductive hypothesis, that if $\kappa \in \tilde{K}_0(CP^\infty)$ is a linear combination of $\beta_1, \beta_2, \ldots, \beta_{n-1}$ and $j\kappa = 0$, then κ is decomposable. (This is clearly true for $n = 2$, since $j\beta_1 = 1$.) According to the Atiyah-Hirzebruch spectral sequence, the graded ring associated to the filtered ring $K_0(CP^\infty)$ is $H_*(CP^\infty)$. This is a divided polynomial algebra, as stated in § 2. Its indecomposable quotient $Q_{2n}H_*(CP^\infty)$ is $\mathbf{Z}$ for $n=1$; for $n>1$ it follows from standard properties of binomial coefficients that $Q_{2n}H_*(CP^\infty)$ is 0 if n is not a prime power, $\mathbf{Z}/p\mathbf{Z}$ if $n=p^f$.

If n is not a prime power, then this remark allows us to find a decomposable element $\kappa' \in \tilde{K}_0(CP^\infty)$ of filtration $\leq 2n$ such that $\kappa - \kappa'$ has filtration $\leq 2(n-1)$; then the inductive hypothesis applies to $\kappa-\kappa'$, and we see that κ is decomposable.

Suppose then that $n=p^f$, and write

$$\kappa = \sum_{k=1}^{n} a_k \beta_n$$

with $a_k \in \mathbf{Z}$. The equation $j\kappa = 0$ gives

$$\sum_{k=1}^{n} (-1)^{k-1} \frac{a_k}{k} = 0 \quad \text{in } \mathbf{Q}.$$

But this equation shows that

$$a_n \equiv 0 \bmod p,$$

for a_n occurs with denominator p^f and the p-primary factor of every other denominator k is at most p^{f-1}. We can now proceed as before to find a decomposable element $\kappa' \in \tilde{K}_0(CP^\infty)$ of filtration $\leq 2n$ such that $\kappa-\kappa'$ has filtration $\leq 2(n-1)$; the inductive hypothesis applies to $\kappa-\kappa'$, and we see that κ is decomposable.

This completes the induction, and proves Theorem 1.2.

4. Proof of Proposition 1.5

First we need an alternative description of $FL\Psi$.

LEMMA 4.1. *$FL\Psi$ consists of those homomorphisms $PK_0(BU) \to A$ which annihilate all but a finite number of $\beta_1, \beta_2, \beta_3, \ldots$.*

Proof. The restriction of Ψ^k to $BU(1)$ is

$$((\lambda-1)+1)^k = \sum_{i+j=k} \frac{k!}{i!j!}(\lambda-1)^j;$$

so

$$\langle \Psi^k, \beta_k \rangle = 1$$

and

$$\langle \Psi^k, \beta_j \rangle = 0 \qquad \text{if} \qquad j > k.$$

The lemma follows at once.

We also need an alternative description of the filtration topology on $PK^0(BSU; A)$; a typical neighbourhood of zero in $PK^0(BSU; A)$ is the annihilator of a finitely-generated subgroup in $QK_0(BSU)$.

Suppose then that we are given an element of $PK^0(BSU; A)$, or equivalently, a homomorphism $h: QK_0(BSU) \to A$. We wish to show that h can be approximated by elements coming from $FL\Psi$, and so we suppose given a typical neighbourhood U of zero in $PK^0(BSU; A)$, determined by a finitely-generated subgroup $F \subset QK_0(BSU)$.

By choosing n sufficiently large, we can arrange that F is contained in the subgroup of $QK_0(BU)$ generated by $\beta_1, \beta_2, \ldots, \beta_n$; call this subgroup G_n. The image of G_n under

$$j\pi_*: QK_0(BU) \to \mathbf{Q}$$

is clearly isomorphic to $\mathbf{Z}$ (being in fact $(1/m)\mathbf{Z}$ where m is the least common multiple of $1, 2, 3, \ldots, n$). So $G_n \cap QK_0(BSU)$ is a direct summand in G_n, and the restriction of h to $G_n \cap QK_0(BSU)$ admits an extension h' over G_n. Also h' admits an extension h'' over $QK_0(BU)$ such that $h''(\beta_k) = 0$ for $k > n$. Then h'' lies in $FL\Psi$; also $h - h''$ annihilates F, and so lies in the given neighbourhood U of zero. This proves Proposition 1.5.

REFERENCES

1. J. F. Adams, "Vector Fields on Spheres", *Annals of Mathematics* 75 (1962) 603–632.
2. J. F. Adams, "*Stable Homotopy and Generalised Homology*", University of Chicago Press 1974; see especially p. 42.
3. I. Madsen, V. Snaith and J. Tornehave, "*H^∞ endomorphisms of K-theory are infinite loop maps*", manuscript.
4. J. W. Milnor and J. C. Moore, "On the structure of Hopf Algebras", *Annals of Math.* 81 (1965) 211–264; see especially Proposition 3.11, p. 226.

D.P.M.M.S.
16 Mill Lane
Cambridge

Reprinted from the *Proceedings of the Cambridge Philosophical Society*, Volume 57, Part 1, pp. 31–36, 1961

A FINITENESS THEOREM IN HOMOLOGICAL ALGEBRA

By J. F. ADAMS

Received 28 May 1960

In (1), (2), (3) and (4) it is shown that homological algebra (5) can be applied to stable homotopy-theory. In this application, we deal with A-modules, where A is the mod p Steenrod algebra. In the present paper, we shall prove a finiteness theorem for the cohomology of the Steenrod algebra. This theorem is stated as Corollary 2 below. It is purely algebraic, but it is not claimed that it has any algebraic interest; it is inspired solely by the application mentioned above. Here it has the following uses.

(1) When applied in conjunction with the spectral sequence of (1), it allows us to put an explicit upper bound on the order of elements in the stable homotopy group $\pi_{n+r}(S^n)$, $n > r+1$.

(In the first instance, this bound is no better than that previously known; however, it can be improved a little by further calculation.)

(2) It allows one to give a simple discussion of the spectral sequence of (1), including the convergence of this spectral sequence, in the particular case when the space X considered is a sphere.

It is therefore possible that the present finiteness theorem might be useful in simplifying the proof of the convergence of this spectral sequence, even in the general case. Such a simplification would be welcome.

We remark that the methods of the present paper, which are wholly elementary, are quite different from those used in (1), (3), (4) to make a partial computation of the cohomology of the Steenrod algebra. Further progress in the study of this cohomology might result if one were able to relate the two methods in a satisfactory way.

In order to state our first theorem, we need certain notation. The letter K will denote a fixed ground field. The letter A will denote a graded algebra $A = \sum_{t \geqslant 0} A_t$ over K. In the applications mentioned above, we take K to be Z_p, and A to be the mod p Steenrod algebra; however, for the theorems to be proved below, we need only insist that A has the following properties.

(1) *A contains a subalgebra E which is an exterior algebra over K on one generator e of dimension ϵ.*

(2) *There is a set $\{a_i\}$ of homogeneous elements a_i in A such that the elements*

$$1, \quad e, \quad a_i, \quad a_i e, \quad e a_i, \quad e a_i e$$

form a K-base for A.

(3) *If the dimension of a_i is q_i and $q = \mathrm{Min}\,(q_i)$, then $0 < \epsilon < q$.*

In the case of the mod p Steenrod algebra, we take e to be β_p, so that these properties are satisfied, with $\epsilon = 1$ and $q = 2(p-1)$.

The letters L, M, N will denote graded left modules over A; such modules are also left modules over E. We give K the structure of a graded A-module in the trivial way, so that the elements of K have dimension zero. The injection $\iota\colon E \to A$ induces a homomorphism

$$\iota^*\colon \mathrm{Ext}_A^{s,t}(M,K) \to \mathrm{Ext}_E^{s,t}(M,K).$$

(In elementary terms, any resolution C of M by modules C_s which are free over A is also a resolution by modules free over E; and any graded A-map $f\colon C_s \to K$ of degree $-t$ is also an E-map.)

THEOREM 1. *If $M_t = 0$ for $t < m$, then*

$$\iota^*\colon \mathrm{Ext}_A^{s,t}(M,K) \to \mathrm{Ext}_E^{s,t}(M,K)$$

is an isomorphism for $t < m+sq$.

This is our main theorem. Theorem 5 will give means by which Theorem 1 can be strengthened a little in favourable cases.

COROLLARY 2. $\mathrm{Ext}_A^{s,t}(K,K) = 0$ *for $s\epsilon < t < sq$.*

This follows immediately from Theorem 1, since $\mathrm{Ext}_E^{*,*}(K,K)$ is a polynomial algebra on one generator of bidegree $(1,\epsilon)$. It is this corollary which is used in the applications mentioned above.

Since the proof of Theorem 1 is by induction, we introduce an abbreviated notation for our inductive hypotheses. We write $P(m,\sigma,\tau)$ for the following proposition:

'*If $M_t = 0$ for $t < m$, then*

$$\iota^*\colon \mathrm{Ext}_A^{s,t}(M,K) \to \mathrm{Ext}_E^{s,t}(M,K)$$

is an isomorphism provided $s \geqslant \sigma$ and $t < m+\tau$.'

It is easy to see that the truth or falsity of this proposition is independent of m; one has only to regrade the module M. We shall therefore write $P(\sigma,\tau)$ for this proposition in future.

The proposition $P(0,0)$ is true in a trivial fashion.

LEMMA 3. *If $P(\sigma,\tau)$ is true for the special case $M = K$, then it is always true.*

Proof. Assume that $P(\sigma,\tau)$ is true for the special case $M = K$. Consider the case in which $M_t = 0$ for $t < m$ and for $t > m$, so that M is isomorphic to a direct sum of copies of K; we see that $P(\sigma,\tau)$ is true in this case also. As an inductive hypothesis, suppose that $P(\sigma,\tau)$ is true whenever $M_t = 0$ for $t < m$ and $t > m+r-1$. Let M be a new module, such that $M_t = 0$ for $t < m$ and $t > m+r$. Let us write $L = M_{m+r}$, $N = M/M_{m+r}$; then

$$0 \to L \to M \to N \to 0$$

is an exact sequence of A-modules. It yields the following diagram of exact cohomology sequences.

$$\begin{array}{ccccccccc}
\mathrm{Ext}_A^{s+1,t}(N,K) & \leftarrow & \mathrm{Ext}_A^{s,t}(L,K) & \leftarrow & \mathrm{Ext}_A^{s,t}(M,K) & \leftarrow & \mathrm{Ext}_A^{s,t}(N,K) & \leftarrow & \mathrm{Ext}_A^{s-1,t}(L,K) \\
\downarrow{\scriptstyle \iota_1} & & \downarrow{\scriptstyle \iota_2} & & \downarrow{\scriptstyle \iota_3} & & \downarrow{\scriptstyle \iota_4} & & \downarrow{\scriptstyle \iota_5} \\
\mathrm{Ext}_E^{s+1,t}(N,K) & \leftarrow & \mathrm{Ext}_E^{s,t}(L,K) & \leftarrow & \mathrm{Ext}_E^{s,t}(M,K) & \leftarrow & \mathrm{Ext}_E^{s,t}(N,K) & \leftarrow & \mathrm{Ext}_E^{s-1,t}(L,K)
\end{array}$$

Let us suppose that $s \geqslant \sigma$, $t < m+\tau$. Then the maps ι_1 and ι_4 are isomorphisms, by our inductive hypothesis; and ι_2 is an isomorphism, as remarked above. In order to apply the Five Lemma, it is sufficient to show that ι_5 is an epimorphism. But the operations of A and E on L and K are trivial, and therefore compatible with the following diagram of ring-homomorphisms.

$$\begin{array}{ccc} E & \xrightarrow{1} & E \\ & {}_{\iota}\searrow \quad \nearrow_{\lambda} & \\ & A & \end{array}$$

(Here, of course, $\lambda(A_t) = 0$ for $t > \epsilon$.) We thus obtain the following diagram.

$$\begin{array}{ccc} \mathrm{Ext}_E^{s,t}(L,K) & \xleftarrow{1} & \mathrm{Ext}_E^{s,t}(L,K) \\ & {}^{\iota^*}\nwarrow \quad \swarrow^{\lambda^*} & \\ & \mathrm{Ext}_A^{s,t}(L,K) & \end{array}$$

This shows that ι_5 is the projection onto a direct summand.

This completes the induction, and shows that $P(\sigma,\tau)$ is true for modules M which have only a finite number of non-zero components. This proves $P(\sigma,\tau)$ for all modules M, because each homomorphism

$$\iota^*\colon \mathrm{Ext}_A^{s,t}(M,K) \to \mathrm{Ext}_E^{s,t}(M,K)$$

depends on only a finite number of components $M_m, M_{m+1}, \ldots, M_t$. (If the reader wishes, he may give a more formal proof by considering the exact cohomology sequence corresponding to

$$0 \to L = \sum_{u>t} M_u \to M \to M/L \to 0$$

and applying $P(0,0)$ to the module L.)

LEMMA 4. *The proposition* $P(1,q)$ *is true. If* $\sigma > 0$, $P(\sigma,\tau)$ *implies* $P(\sigma+1,\tau+q)$.

Proof. Let M be the submodule of A whose K-base consists of the elements e, a_i, $a_i e$, ea_i, $ea_i e$. We will show that one can replace the map

$$\iota^*\colon \mathrm{Ext}_A^{s+1,t}(K,K) \to \mathrm{Ext}_E^{s+1,t}(K,K)$$

by the map

$$i^*\colon \mathrm{Ext}_A^{s,t}(M,K) \to \mathrm{Ext}_A^{s,t}(Ae,K).$$

This is evident from the following diagram.

$$\begin{array}{ccccc} \mathrm{Ext}_A^{s+1,t}(K,K) & \underset{\cong}{\xleftarrow{\delta}} & \mathrm{Ext}_A^{s,t}(M,K) & \xrightarrow{i^*} & \mathrm{Ext}_A^{s,t}(Ae,K) \\ \downarrow{\scriptstyle \iota^*} & & \downarrow & \swarrow{\scriptstyle f^*,\ \cong} & \\ \mathrm{Ext}_E^{s+1,t}(K,K) & \underset{\cong}{\longleftarrow} & \mathrm{Ext}_E^{s,t}(Ee,K) & & \end{array}$$

In this diagram, the left-hand square derives from the following diagram of exact sequences of modules.

$$\begin{array}{ccccccccc} 0 & \longrightarrow & M & \longrightarrow & A & \longrightarrow & K & \longrightarrow & 0 \\ & & \uparrow & & \uparrow & & \uparrow & & \\ 0 & \longrightarrow & Ee & \longrightarrow & E & \longrightarrow & K & \longrightarrow & 0 \end{array}$$

The vertical arrows are, of course, compatible with the injection

$$\begin{array}{c} A \\ {\scriptstyle\iota}\big\uparrow \\ E \end{array}$$

of rings. The right-hand triangle of the diagram derives from the following triangle.

$$\begin{array}{ccc} M & \overset{i}{\longleftarrow} & Ae \\ \big\uparrow & \nearrow_{f} & \\ Ee & & \end{array}$$

The fact that f^* is an isomorphism may be treated at various levels of sophistication; we give only two. The most elementary treatment is by direct calculation. We can construct a resolution C of Ee by taking C_s to be E-free on one generator c_s of degree $(s+1)\epsilon$ and setting $dc_s = ec_{s-1}$. Similarly, we can construct a resolution C' of Ae by taking C'_s to be A-free on one generator c'_s of degree $(s+1)\epsilon$ and setting $dc'_s = ec'_{s-1}$. We can construct an E-linear chain map $g: C \to C'$ inducing the map $f: Ee \to Ae$ of homology, by setting $g(c_s) = c'_s$. The induced map

$$g^*: \operatorname{Hom}^t_A(C'_s, K) \to \operatorname{Hom}^t_E(C_s, K)$$

is an isomorphism, and the result follows.

The most sophisticated method is to treat the matter as a case of the graded analogue of Prop. 4.1.3, p. 118 of (5). (One substitutes $\iota: E \to A$ for the map $\phi: \Lambda \to \Gamma$ considered there.)

We have now justified the replacement of ι^* by i^*. Let us define $N = M/Ae$; then we have the following exact sequence.

$$\operatorname{Ext}^{s+1,t}_A(N,K) \longleftarrow \operatorname{Ext}^{s,t}_A(Ae,K) \overset{i^*}{\longleftarrow} \operatorname{Ext}^{s,t}_A(M,K) \longleftarrow \operatorname{Ext}^{s,t}_A(N,K).$$

According to our original assumptions about A, N is a free left E-module and $N_t = 0$ for $t < q$. Thus $\operatorname{Ext}^{s,t}_E(N,K) = 0$ for $s > 0$; if we assume $P(\sigma,\tau)$ for some $\sigma > 0$, we find that

$$\operatorname{Ext}^{s,t}_A(N,K) = 0 \quad \text{for} \quad s \geqslant \sigma, \quad t < q+\tau.$$

This shows that i^* is an isomorphism in this range; hence

$$\iota^*: \operatorname{Ext}^{s+1,t}_A(K,K) \to \operatorname{Ext}^{s+1,t}_E(K,K)$$

is an isomorphism for $s+1 \leqslant \sigma+1$, $t < \tau+q$. Invoking Lemma 3, we conclude that the proposition $P(\sigma+1, \tau+q)$ is true.

Similarly, we have

$$\operatorname{Ext}^{s,t}_A(N,K) = 0 \quad \text{for} \quad t < q,$$

which proves the proposition $P(1,q)$.

This completes the proof of Lemma 4. From this lemma, we obtain the proposition $P(\sigma, \sigma q)$ by induction over σ. This proves Theorem 1.

THEOREM 5. *When $S > 0$ and $\sigma > 0$, the propositions $P(S,T)$ and $P(\sigma,\tau)$ imply $P(S+\sigma, T+\tau)$.*

This theorem shows that if we can improve the estimate given by Theorem 1 in one dimension s, then we can extrapolate this improvement to all higher dimensions.

For example, let A be the mod 2 Steenrod algebra, so that $q = 2$. Then the propositions $P(1,2)$, $P(2,4)$ and $P(3,6)$ are best possible results; but for $\sigma = 4$ the best possible result is $P(4,11)$. By Theorem 5, this immediately implies

$$P(4a+b, 11a+2b).$$

There is some reason to conjecture that for $\sigma = 2^n$, $n \geqslant 2$ the best possible result is $P(\sigma, 3\sigma-1)$.

Proof of Theorem 5. We will generalize the proof of Lemma 4. Let C be a minimal resolution of K by free A-modules C_s. Then each C_s will contain a basis element c_s of bidegree $(s, s\epsilon)$ such that $dc_s = ec_{s-1}$. Thus C contains a minimal resolution C' of K by free E-modules; we take C'_s to be E-free on the single generator c_s. Let Z_{S-1}, Z'_{S-1} be the submodules of cycles in C_{S-1}, C'_{S-1}. If we assume the proposition $P(S,T)$, then C_S contains no basis element other than c_S in any dimension less than T. Therefore the quotient

$$N = Z_{S-1}/Aec_{S-1}$$

is such that $N_t = 0$ for $t < T$. Arguing as before, but with iterated connecting homomorphisms, we examine the following diagram.

$$\begin{array}{ccccc}
\mathrm{Ext}_A^{S+s,t}(K,K) & \xleftarrow[\cong]{\delta^S} & \mathrm{Ext}_A^{s,t}(Z_{S-1},K) & \xrightarrow{i^*} & \mathrm{Ext}_A^{s,t}(Aec_{S-1},K) \\
\downarrow{\scriptstyle \iota^*} & & \downarrow & \swarrow{\scriptstyle f^*}\,\cong & \\
\mathrm{Ext}_E^{S+s,t}(K,K) & \xleftarrow[\cong]{\delta^S} & \mathrm{Ext}_E^{s,t}(Z'_{S-1},K) & &
\end{array}$$

By Theorem 1, ι^* is an isomorphism if $t < q(S+s)$. Therefore i^* is an isomorphism in the same range; it follows that

$$\mathrm{Ext}_A^{s+1,t}(N,K) = 0$$

if $t < q(S+s)$. Using Theorem 1 for N, we deduce that

$$\mathrm{Ext}_E^{s+1,t}(N,K) = 0$$

if $t < q(S+s)$ and $t < T+q(s+1)$. But we have

$$\mathrm{Ext}_E^{1,t}(N,K) \cong \mathrm{Ext}_E^{1+s,t+\epsilon s}(N,K).$$

In order to show that $\mathrm{Ext}_E^{1,t}(N,K) = 0$, it is now sufficient to show that we can choose s so large that

$$t+\epsilon s < q(S+s), \quad t+\epsilon s < T+q(s+1).$$

Since $q-\epsilon > 0$, this is always possible. We conclude that

$$\mathrm{Ext}_E^{1,t}(N,K) = 0,$$

whence

$$\mathrm{Ext}_E^{s,t}(N,K) = 0 \quad \text{for} \quad s \geqslant 1.$$

If we now assume the proposition $P(\sigma,\tau)$, we find that

$$\mathrm{Ext}_A^{s,t}(N,K) = 0$$

for $s \geqslant \sigma$, $t < T+\tau$. It follows that i^* is an isomorphism in the same range; therefore

$$\iota^*: \operatorname{Ext}_A^{S+s,t}(K,K) \to \operatorname{Ext}_E^{S+s,t}(K,K)$$

is an isomorphism for $S+s \geqslant S+\sigma$, $t < T+\tau$. Invoking Lemma 3, we conclude that our data imply the proposition $P(S+\sigma, T+\tau)$. This completes the proof of Theorem 5.

REFERENCES

(1) ADAMS, J. F. On the structure and applications of the Steenrod algebra. *Comment. math. helvet.* 32 (1958), 180–214.

(2) ADAMS, J. F. On the non-existence of elements of Hopf invariant one. *Bull. Amer. Math. Soc.* 64 (1958), 279–82.

(3) ADAMS, J. F. On the non-existence of elements of Hopf invariant one. *Ann. Math.* 72 (1960), 20–104.

(4) CARTAN SEMINAR NOTES, 1958/59.

(5) CARTAN, H. and EILENBERG, S. *Homological algebra* (Princeton, 1956).

TRINITY HALL
CAMBRIDGE

Proc. Camb. Phil. Soc. (1966), **62**, 365
PCPS 62–50

A periodicity theorem in homological algebra

By J. F. ADAMS
Department of Mathematics, University of Manchester

(*Received* 29 *September* 1965)

1. *Introduction.* In (**1–3, 6**) it is shown that homological algebra can be applied to stable homotopy-theory. In this application, we deal with A-modules, where A is the mod p Steenrod algebra. To obtain a concrete geometrical result by this method usually involves work of two distinct sorts. To illustrate this, we consider the spectral sequence of (**1, 2**):

$$\mathrm{Ext}_A^{s,t}(H^*(Y; Z_p), H^*(X; Z_p)) \underset{s}{\Rightarrow} {}_p\pi_*^S(X, Y).$$

Here each group $\mathrm{Ext}^{s,t}$ which occurs in the E_2 term can be effectively computed; the process is purely algebraic. However, no such effective method is given for computing the differentials d_r in the spectral sequence, or for determining the group extensions by which ${}_p\pi_*^S(X, Y)$ is built up from the E_∞ term; these are topological problems.

A mathematical logician might be satisfied with this account: an algorithm is given for computing E_2; to find the maps d_r still requires intelligence. The practical mathematician, however, is forced to admit that the intelligence of mathematicians is an asset at least as reliable as their willingness to do large amounts of tedious mechanical work. In fact, when a chance has arisen to show that such a differential d_r is non-zero, it has been regarded as an interesting problem, and duly solved; see (**3, 8, 12**). However, the difficulty of actually computing groups $\mathrm{Ext}_A^{s,t}(L, M)$ has remained the greatest obstacle to the method.

In the circumstances, what we need are theorems to tell us the value of certain groups $\mathrm{Ext}_A^{s,t}$. I have given some results in this direction in lectures delivered at the University of California, Berkeley, in July 1961 ((**5**)). Unfortunately, those lectures contained only the barest hint of proof. It is the object of the present paper to give a proper treatment of these results; I must apologize to my readers for this long delay.

This paper (like the lectures mentioned) deals only with the case $p = 2$. When p is an odd prime, the analogous questions have been investigated by Liulevicius (see (**9**), especially the foot of p. 975).

To indicate the nature of the results, I will show how they apply to the special case

$$H^{s,t}(A) = \mathrm{Ext}_A^{s,t}(Z_2, Z_2),$$

which is relevant in computing the stable homotopy groups of spheres. The groups $H^{s,t}(A)$ are zero for $t < s$ and known for $t = s$.

Theorem 1·1. *We have* $H^{s,t}(A) = 0$ *provided* $0 < s < t < U(s)$, *where* $U(s)$ *is the following numerical function:*

$$U(4s) = 12s - 1, \quad U(4s+1) = 12s+2, \quad U(4s+2) = 12s+4, \quad U(4s+3) = 12s+6.$$

This result is best possible, in the sense that the function $U(s)$ cannot be increased. It supersedes the corresponding result in my earlier note ((4)).

THEOREM 1·2. *For each $r \geq 2$ there is a suitable neighbourhood of the line $t = 3s$ in which we have a 'periodicity' isomorphism*

$$\pi_r: H^{s,t}(A) \xrightarrow{\cong} H^{s+2^r, t+3.2^r}(A)$$

defined by

$$\pi_r(x) = \langle h_{r+1}, h_0^{2^r}, x \rangle.$$

The precise inequalities on s and t for which the isomorphism is proved will be given in section 5; see Corollaries 5·5, 5·8. The symbol $\langle z, y, x \rangle$ means the Massey product, and the element $h_i \in H^{1, 2^i}(A)$ is as in (3).

The 'periodicity' isomorphism π_r increases the total degree $t-s$ by 2^{r+1}. So this result seems to hint that there may be phenomena in the stable homotopy groups of spheres which recur with periods 8, 16, 32, etc. It would be most interesting to have geometric information on this point.

The theorems stated above will be proved by considering $\text{Ext}_{A'}^{s,t}(L; Z_2)$, where A' runs over suitable subalgebras of the Steenrod algebra, and L is a module more general than Z_2. In what follows, all our algebras and modules will be graded, and all their components will be finitely generated over Z_2; their components in sufficiently large negative dimensions will be zero.

2. *The Vanishing Theorem.* With the ordering adopted in (5), the first sort of theorem to be discussed is the Vanishing Theorem ((5), p. 62, Theorem 3).

We shall need some notation. Let A be the mod 2 Steenrod algebra; if r is finite, let A_r be the subalgebra of A generated by $Sq^1, Sq^2, \ldots, Sq^{2^r}$; A_∞ will mean A. We assume that L is a left module over A_r, that L is free *qua* left module over A_0, and that $L_t = 0$ for $t < l$. We define a numerical function by

$$T(4k) = 12k, \quad T(4k+1) = 12k+2, \quad T(4k+2) = 12k+4, \quad T(4k+3) = 12k+7,$$

where k runs over the integers.

THEOREM 2·1. (*Vanishing.*) *$\text{Ext}_{A_r}^{s,t}(L, Z_2)$ is zero if $t < l + T(s)$.*

In (5) the proof of this theorem proceeds hand-in-hand with the proof of the Approximation Theorem ((5), p. 63, Theorem 4). In the present paper, however, the proofs will be separated; I hope this will be found simpler. The proof proceeds in stages. First we remark that A_0 can be regarded as an A_r-module in a unique way.

LEMMA 2·2. *The Vanishing Theorem is true in the special case $r = \infty$, $L = A_0$, $s \leq 4$.*

Proof. This lemma is essentially computational. At least two proofs may be given, depending on how much one is willing to assume known. As a matter of fact, a good deal is known about the groups $\text{Ext}_A^{s,t}(Z_2, Z_2)$ (3, 10); suppose that one is willing to assume as much of this as is needed. Then one simply expresses the module A_0 as an extension with submodule and quotient module both isomorphic to Z_2 (but differently graded); this extension gives rise to an exact sequence of Ext groups, from which one easily computes $\text{Ext}_A^{s,t}(A_0, Z_2)$ in low dimensions, say for $t-s \leq 8$.

On the other hand, some of the methods which have been used in computing $\mathrm{Ext}_A^{**}(Z_2, Z_2)$ are less elementary than others. Since the present lemma can be proved by an elementary and explicit calculation, it might be held that this is the proper way to do it. To do things this way one has to give an explicit A-free resolution of A_0 (preferably a minimal one); at least one must do this in low dimensions. I have done this; it is not prohibitively laborious; and the reader may duplicate the calculation if he wishes. However, it is hardly worth publishing.

LEMMA 2·3. *The Vanishing Theorem is true in the special case $r = \infty$, $s \leqslant 4$.*

Proof. Suppose that L is an A-module which is free over A_0 and such that $L_t = 0$ for $t < l$. Pick an A_0-basis of L, and let $L(\nu)$ be the sub-A_0-module generated by the basis elements of grading $t \geqslant \nu$. Then $L(\nu)$ is actually a sub-A-module of L. The module $L(\nu)/L(\nu+1)$ is an A_0-free module on basis elements all of the same grading, and the present lemma holds for it, by Lemma 2·2 and addition. This allows us to perform an induction. Suppose as an inductive hypothesis that the present lemma is true for the module $L/L(\nu)$. (Since $L = L(l)$, the induction starts with $\nu = l$). Form the exact sequence

$$L(\nu)/L(\nu+1) \to L/L(\nu+1) \to L/L(\nu).$$

This yields an exact sequence

$$\mathrm{Ext}_A^{s,t}(L(\nu)/(L(\nu+1), Z_2) \leftarrow \mathrm{Ext}_A^{s,t}(L/L(\nu+1), Z_2) \leftarrow \mathrm{Ext}_A^{s,t}(L/L(\nu), Z_2).$$

Hence the middle groups are zero for $t < l + T(s)$ $(s \leqslant 4)$ and the present lemma holds for $L/L(\nu+1)$. This completes the induction.

On the other hand, we have

$$\mathrm{Ext}_A^{s,t}(L, Z_2) \cong \mathrm{Ext}_A^{s,t}(L/L(\nu), Z_2)$$

by taking ν sufficiently large compared with t; so what we have proved above is sufficient to prove the lemma.

LEMMA 2·4. *Let*

$$0 \to L' \to L \to L'' \to 0$$

be an exact sequence of A_0-modules. If two of them are A_0-free, then so is the third.

Proof. A module over A_0 is the same thing as a chain complex. It is free over A_0 if and only if its homology is zero. Now the result follows from the exact homology sequence.

PROPOSITION 2·5. *The Vanishing Theorem is true in the special case $r = \infty$.*

Proof. Let L be as in the data, so that L is an A-module, L is free over A_0 and $L_t = 0$ for $t < l$. Form the first four terms of a minimal A-free resolution of L, say

$$L \xleftarrow{\epsilon} C_0 \xleftarrow{d_1} C_1 \xleftarrow{d_2} C_2 \xleftarrow{d_3} C_3 \xleftarrow{d_4} C_4.$$

Since A is A_0-free, and hence each C_i is A_0-free, Lemma 4 shows successively that $\mathrm{Im}\, d_1$, $\mathrm{Im}\, d_2$, $\mathrm{Im}\, d_3$ and $\mathrm{Im}\, d_4$ are A_0-free. Let us write M for $\mathrm{Im}\, d_4$; then Lemma 2·3 (the special case $r = \infty$, $s = 4$) shows that $M_t = 0$ for $t < l + 12$.

The result is true for $k = 0$, by Lemma 2·3; let us suppose, as an inductive hypothesis, that the result is true for some value of k (where k is the integer used in defining T). Then we may apply the inductive hypothesis to M, and we find that the group

$$\mathrm{Ext}_A^{s+4,t}(L, Z_2) \cong \mathrm{Ext}_A^{s,t}(M, Z_2)$$

is zero if

$$s = 4k, \quad t > l+12+12k,$$

$$s = 4k+1, \quad t > l+12+12k+2,$$

$$s = 4k+2, \quad t > l+12+12k+4,$$

or

$$s = 4k+3, \quad t > l+12+12k+7.$$

That is, the result is true for L with k replaced by $k+1$. This completes the induction, and proves the result.

Proof of Theorem 2·1, *the Vanishing Theorem.* For $r = 0$ there is nothing to be proved; and we have already proved the special case $r = \infty$ (Proposition 2·5). So let us assume that $0 < r < \infty$, that L is a (graded) left module over A_r, that L is free over A_0, and that $L_t = 0$ for $t < l$.

By a standard result on Hopf algebras and subalgebras, A is free *qua* right module over A_r. By a standard result on change-of-rings, which is in Cartan–Eilenberg ((7), p. 118) for the ungraded case, we have the following isomorphism:

$$\mathrm{Ext}_{A_r}^{s,t}(L, Z_2) \cong \mathrm{Ext}_A^{s,t}(A \otimes_{A_r} L, Z_2).$$

This allows us to prove that $\mathrm{Ext}_{A_r}^{s,t}(L, Z_2)$ is zero by applying Proposition 2·5 (the case $r = \infty$ of the Vanishing Theorem) to the module $A \otimes_{A_r} L$. Of course we have to verify the assumptions of the Vanishing Theorem for this module. It is easy to see that $(A \otimes_{A_r} L)_t = 0$ for $t < l$. It remains only to check that $A \otimes_{A_r} L$ is free *qua* left module over A_0. Actually we will prove something slightly more general, for use in sections 3, 5.

We assume $0 < r < \rho$; then A_r is a subalgebra of A_ρ. We assume L is a left module over A_r.

Proposition 2·6. *If L is free qua left module over A_0, then so is $A_\rho \otimes_{A_r} L$.*

The special case which we need to prove Theorem 2·1 is the case $\rho = \infty$.

Proof of Proposition 2·6. As in the proof of Lemma 2·3, we can choose an A_0-base of L and filter it by dimensions, thus filtering L by A_r-submodules. Since $A_\rho \otimes_{A_r}$ is an exact functor, this filters $A_\rho \otimes_{A_r} L$. It is now sufficient to prove the result for the special case $L = A_0$.

This requires us to consider the module $A_\rho \otimes_{A_r} A_0$. By using the canonical anti-automorphism of A (which preserves A_ρ, A_r), we may change the question and consider instead the module $A_0 \otimes_{A_r} A_\rho$, considering it as a right module over A_0. The point is that it is easier to write down the Z_2-dual of this module ((11)). We may identify A_ρ^* with the quotient of A^* which has as a base the monomials $\xi_1^{i_1}\xi_2^{i_2}\ldots\xi_q^{i_q}$ such that $i_p < 2^{\rho+2-p}$ for each p, the remaining monomials being zero. Similarly for A_r^*.

The module $A_0 \otimes_{A_r} A_\rho$ is defined so that the following sequence is exact:

$$A_0 \otimes A_r \otimes A_\rho \xrightarrow{\mu\otimes 1 - 1\otimes\mu} A_0 \otimes A_\rho \to A_0 \otimes_{A_r} A_\rho \to 0.$$

Here the map $\mu\colon A_r \otimes A_\rho \to A_\rho$ is the usual product map, but the map $\mu\colon A_0 \otimes A_r \to A_0$ is the map which makes A_0 an A_r-module; that is, the t-dimensional part of A_r annihilates A_0 for $t \geqslant 2$.

The Z_2-dual of the exact sequence displayed above is the following exact sequence.

$$A_0^* \otimes A_r^* \otimes A_\rho^* \xleftarrow{\mu^*\otimes 1 - 1\otimes \mu^*} A_0^* \otimes A_\rho^* \leftarrow (A_0 \otimes_{A_r} A_\rho)^* \leftarrow 0.$$

Here the map $\mu^*\colon A_\rho^* \to A_r^* \otimes A_\rho^*$ is the usual coproduct map, while the map $\mu^*\colon A_0^* \to A_0^* \otimes A_r^*$ is given by

$$\mu^*(1) = 1 \otimes 1, \quad \mu^*(\xi_1) = \xi_1 \otimes 1 + 1 \otimes \xi_1.$$

Next we have to describe the kernel of $\mu^* \otimes 1 - 1 \otimes \mu^*$. First we have the elements

$$1 \otimes \xi_1^{i_1} \xi_2^{i_2} \ldots \xi_q^{i_q}$$

such that $i_p < 2^{\rho+2-p}$ (for each p) and $i_p \equiv 0 \bmod 2^{r+2-p}$ (for each p such that $p \leqslant r+2$). Next we have the elements

$$\xi_1 \otimes \xi_1^{i_1} \xi_2^{i_2} \ldots \xi_q^{i_q} + 1 \otimes \xi_1^{i_1+1} \xi_2^{i_2} \ldots \xi_q^{i_q},$$

where $(i_1, i_2, \ldots, i_q)$ is as before. It is not too hard to verify (using the explicit form of the coproduct in A) that these elements are indeed annihilated by $\mu^* \otimes 1 - 1 \otimes \mu^*$; one can also check (using the fact that A_ρ is a free left module over A_r) that there are the correct number of elements in each dimension. We conclude that the elements given constitute a base of $(A_0 \otimes_{A_r} A_\rho)^*$.

The base elements we have given are of two kinds, and the subset of elements

$$1 \otimes \xi_1^{i_1} \xi_2^{i_2} \ldots \xi_q^{i_q}$$

has an obvious interpretation. We have an epimorphism of A_r-modules

$$A_0 \to Z_2,$$

whence an epimorphism $\quad A_0 \otimes_{A_r} A_\rho \to Z_2 \otimes_{A_r} A_\rho$

and a monomorphism $\quad (A_0 \otimes_{A_r} A_\rho)^* \leftarrow (Z_2 \otimes_{A_r} A_\rho)^*.$

The elements $\quad 1 \otimes \xi_1^{i_1} \xi_2^{i_2} \ldots \xi_q^{i_q}$

represent a base of the submodule $(Z_2 \otimes_{A_r} A_\rho)^*$.

The fact that our basis elements occur in pairs

$$1 \otimes \xi_1^{i_1} \xi_2^{i_2} \ldots \xi_q^{i_q}$$

and $\quad \xi_1 \otimes \xi_1^{i_1} \xi_2^{i_2} \ldots \xi_q^{i_q} + 1 \otimes \xi_1^{i_1+1} \xi_2^{i_2} \ldots \xi_q^{i_q}$

now means that we have found an isomorphism

$$(A_0 \otimes_{A_r} A_\rho)^* \cong A_0^* \otimes (Z_2 \otimes_{A_r} A_\rho)^*.$$

It might be interesting to know if some reason for this isomorphism could be found in the theory of Hopf algebras; but for our purposes this is not essential.

We must now consider the map

$$\partial\colon A_0 \otimes_{A_r} A_\rho \to A_0 \otimes_{A_r} A_\rho$$

defined by operating with S_q^1 on the right; we must show that $A_0 \otimes_{A_r} A_\rho$ is acyclic under ∂. We have the following commutative diagram (in which $dx = xS_q^1$).

$$\begin{array}{ccc} A_0 \otimes A_\rho & \longrightarrow & A_0 \otimes_{A_r} A_\rho \\ {\scriptstyle 1 \otimes d}\downarrow & & \downarrow{\scriptstyle \partial} \\ A_0 \otimes A_\rho & \longrightarrow & A_0 \otimes_{A_r} A_\rho \end{array}$$

On dualizing, therefore, we get the following commutative diagram.

$$\begin{array}{ccc} A_0^* \otimes A_\rho^* & \longleftarrow & (A_0 \otimes_{A_r} A_\rho)^* \\ {\scriptstyle 1 \otimes d^*}\uparrow & & \uparrow{\scriptstyle \partial^*} \\ A_0^* \otimes A_\rho^* & \longleftarrow & (A_0 \otimes_{A_r} A_\rho)^* \end{array}$$

If we examine the effect of $1 \otimes d^*$ on our elements

$$1 \otimes \xi_1^{i_1} \xi_2^{i_2} \ldots \xi_q^{i_q}$$

and

$$\xi_1 \otimes \xi_1^{i_1} \xi_2^{i_2} \ldots \xi_q^{i_q} + 1 \otimes \xi_1^{i_1+1} \xi_2^{i_2} \ldots \xi_q^{i_q}$$

we see that our isomorphism

$$(A_0 \otimes_{A_r} A_\rho)^* \cong A_0^* \otimes (Z_2 \otimes_{A_r} A_\rho)^*$$

expresses $(A_0 \otimes_{A_r} A_\rho)^*$ as the tensor product (in the usual sense) of two chain complexes, of which one (viz. A_0^*) is acyclic. By the Künneth theorem, $(A_0 \otimes_{A_r} A_\rho)^*$ is acyclic. This proves Proposition 2·6, which completes the proof of Theorem 2·1.

3. *The Approximation Theorem.* The second sort of theorem to be discussed is the Approximation Theorem ((5), p. 63, Theorem 4).

We use the same notation as before. We assume $0 < r < \rho$; then we have an injection $i: A_r \to A_\rho$. As before, we assume that L is a (graded) left module over A_ρ, that L is free *qua* left module over A_0, and that $L_t = 0$ for $t < l$. We also assume $s > 0$.

THEOREM 3·1. (*Approximation.*) *The map*

$$i^*: \mathrm{Ext}_{A_r}^{s,t}(L, Z_2) \leftarrow \mathrm{Ext}_{A_\rho}^{s,t}(L, Z_2)$$

is an isomorphism if

$$t < l + 2^{r+1} + T(s-1).$$

The theorem remains true for $s = 0$, provided we interpret $T(-1)$ as 0.

Proof. Let K be the kernel of the obvious map $A_\rho \otimes_{A_r} L \to L$, so that we have the following exact sequence:

$$0 \to K \to A_\rho \otimes_{A_r} L \to L \to 0.$$

Since every element of dimension less than 2^{r+1} in A_ρ is in the subalgebra A_r, it is easy to see that $K_t = 0$ for $t < l + 2^{r+1}$. By Proposition 2·6, $A_\rho \otimes_{A_r} L$ is free over A_0; also L is free over A_0; therefore K is free over A_0, by Lemma 2·4. This will allow us to apply the Vanishing Theorem to K.

We now argue as in the proof of Theorem 2·1. By a standard result on Hopf algebras and subalgebras, A_ρ is free *qua* right module over A_r; by a standard result on change-of-rings, we have the following isomorphism:

$$\mathrm{Ext}_{A_r}^{s,t}(L, Z_2) \cong \mathrm{Ext}_{A_\rho}^{s,t}(A_\rho \otimes_{A_r} L, Z_2).$$

Moreover, we have the following commutative diagram:

$$\begin{array}{ccccccc}
 & & \mathrm{Ext}^{s,t}_{A_r}(L, Z_2) & & & & \\
 & & \cong \uparrow & \overset{i^*}{\nwarrow} & & & \\
\mathrm{Ext}^{s,t}_{A_\rho}(K, Z_2) & \leftarrow & \mathrm{Ext}^{s,t}_{A_\rho}(A_\rho \otimes_{A_r} L, Z_2) & \leftarrow & \mathrm{Ext}^{s,t}_{A_\rho}(L, Z_2) & \leftarrow & \mathrm{Ext}^{s-1,t}_{A_\rho}(K, Z_2).
\end{array}$$

By the Vanishing Theorem, the groups

$$\mathrm{Ext}^{s,t}_{A_\rho}(K, Z_2) \quad \text{and} \quad \mathrm{Ext}^{s-1,t}_{A_\rho}(K, Z_2)$$

are zero for

$$t < l + 2^{r+1} + T(s-1).$$

Therefore the map i^* is an isomorphism for these values of t. This completes the proof.

4. *Construction of periodicity elements.* In this section we shall construct certain elements

$$\varpi_r \in H^{2^r, 3.2^r}(A_r) \quad (r \geqslant 2),$$

which are needed for the statement and proof of our periodicity theorems. We shall also prove some of their properties; see Lemmas 4·3, 4·4 and 4·5. The motivation for this work is to be found in section 5.

We first recall from, for example, (3) that $H^{**}(A)$ can be defined as the cohomology of a ring of cochains, by using the cobar construction $F(A^*)$. For example, $h_{r+1} h_0^{2^r}$ is the cohomology class of the cocycle

$$z_r = [\xi_1^{2^{r+1}} | \xi_1 | \ldots | \xi_1],$$

where the symbol ξ_1 appears 2^r times. We propose to construct cochains c_r in $F(A^*)$ for $r \geqslant 2$ such that

$$\delta c_r = z_r. \tag{4·1}$$

For $r = 2$ we may assume it known by direct calculation that

$$H^{5,12}(A) = 0, \quad H^{4,12}(A) = 0.$$

Therefore it is possible to choose c_2, and c_2 is unique up to a coboundary.

Let us now suppose, as an inductive hypothesis, that we have chosen c_r in such a way that it is defined up to a coboundary. Let the $\cup_1$ product in $F(A^*)$ be as in (3), p. 36. Then we have

$$\delta(c_r \cup c_r + z_r \cup_1 c_r) = z_r \cup_1 z_r.$$

Moreover, the cochain $c_r \cup c_r + z_r \cup_1 c_r$ is defined up to a coboundary. If we evaluate $z_r \cup_1 z_r$ by the explicit formula given in (3), p. 36, we find terms of three sorts.

(i) $[\xi_1^{2^{r+2}} | \xi_1 | \ldots | \xi_1]$ with 2^{r+1} entries ξ_1. This is the cocycle z_{r+1}.

(ii) $[\xi_1^{2^{r+1}} | \xi_1 | \ldots | \xi_1 | \xi_1^{2^{r+1}+1} | \xi_1 | \ldots | \xi_1]$ with a entries ξ_1 in the first batch and b entries ξ_1 in the second batch, where $0 \leqslant a \leqslant 2^r - 1$ and $a + b = 2^{r+1} - 1$. Each such term occurs twice, and these terms cancel.

(iii) $[\xi_1^{2^{r+1}} | \xi_1 | \ldots | \xi_1 | \xi_1^{2^{r+1}} | \xi_1 | \ldots | \xi_1 | \xi_1^2 | \xi_1 | \ldots | \xi_1]$ with a entries ξ_1 in the first batch, b in the second batch and c in the third batch. Here $0 \leqslant a \leqslant 2^r - 1$ and $a + b + c = 2^{r+1} - 2$, so $b + c \geqslant 3$. Since

$$[\xi_1^2 | \xi_1] = \delta[\xi_2], \quad [\xi_1 | \xi_1^2] = \delta[\xi_2 + \xi_1^3],$$

each such term is the boundary of a cochain γ, where γ is either

$$[\xi_1^{2r+1}|\xi_1|\ldots|\xi_1|\,\xi_1^{2r+1}|\xi_1|\ldots|\xi_1|\,\xi_2|\xi_1|\ldots|\xi_1]$$

or

$$[\xi_1^{2r+1}|\xi_1|\ldots|\xi_1|\,\xi_1^{2r+1}|\xi_1|\ldots|\xi_1|\,\xi_2+\xi_1^3|\xi_1|\ldots|\xi_1].$$

If both choices for γ are possible, then they differ by a coboundary.

We now set

$$c_{r+1} = c_r \cup c_r + z_r \cup_1 c_r + \Sigma\gamma. \tag{4·2}$$

Then c_{r+1} is defined up to a coboundary, and $\delta c_{r+1} = z_{r+1}$. This completes the induction.

Remark. What we propose to prove includes the following two facts.

(*a*) $H^{s,t}(A) = 0$ for $s = 2^r+1$, $t = 3.2^r$ (Theorem 1·1). This shows that it is possible to choose c_r satisfying (4·1).

(*b*) $H^{s,t}(A) = 0$ for $s = 2^r$, $t = 3.2^r$ (Corollary 5·6). This shows that (4·1) defines c_r up to a coboundary.

However, one of the properties of c_r (viz. 4·4) seems easiest to prove from a semi-explicit construction.

We will now define ϖ_r. Let $i\colon A_r \to A$ be the injection map. Since the cobar construction is functorial, this induces $i^*\colon F(A^*) \to F(A_r^*)$. Since the dual of i annihilates ξ_1^{2r+1}, i^* annihilates z_r, and i^*c_r is a cocycle in $F(A_r^*)$. We define ϖ_r to be the cohomology class of i^*c_r in $H^{2^r,3\cdot 2^r}(A_r)$.

LEMMA 4·3. *$H^{4,12}(A_1) \cong Z_2$, generated by the image of ϖ_2.*

Proof. Since A_1 is so small, we may easily make an explicit resolution and check that $H^{4,12}(A_1) \cong Z_2$. Let B be the subalgebra of A generated by $Sq^{0,1}$; thus B is an exterior algebra, and $H^{**}(B)$ is a polynomial algebra on one generator which lies in $H^{1,3}(B)$. In (3) this generator is called $h_{2,0}$. By direct calculation again, we check that the injection $B \to A_1$ induces an isomorphism

$$H^{4,12}(A_1) \to H^{4,12}(B).$$

It is now sufficient to check that the image of ϖ_2 in $H^{4,12}(B)$ is $(h_{2,0})^4$. Next recall that in (3) one obtains information about $H^{**}(A)$ by using a family of spectral sequences (see especially (3), p. 45); here we shall require the first spectral sequence of this family, namely that with $n = 2$. We propose to check that in this spectral sequence, the transgression τ is defined on $(h_{2,}0)^4$ and takes the value $h_3h_0^4$; this will give exactly what is wanted. We calculate as follows:

By ((3), p. 45, Lemma 2·5·2 (i)) we have

$$\tau h_{2,0} = h_1 h_0.$$

Hence

$$\begin{aligned}\tau(h_{2,0})^2 &= \tau Sq^1 h_{2,0}\\ &= Sq^1\tau h_{2,0}\\ &= Sq^1(h_1h_0)\\ &= Sq^1h_1.Sq^0h_0+Sq^0h_1.Sq^1h_0\\ &\qquad\text{(by the Cartan formula)}\\ &= h_1^3+h_2h_0^2.\end{aligned}$$

(This formula is also given by (3), p. 45, Lemma 2·5·2 (i).) Hence

$$\begin{aligned}\tau(h_{2,0})^4 &= \tau Sq^2(h_{2,0})^2\\ &= Sq^2\tau(h_{2,0})^2\\ &= Sq^2(h_2h_0^2+h_1^3)\\ &= Sq^0h_2.(Sq^1h_0)^2+Sq^0h_1.(Sq^1h_1)^2\\ &\quad\text{(by the Cartan formula)}\\ &= h_3h_0^4+h_2h_1^4\\ &= h_3h_0^4\end{aligned}$$

(since $h_2h_1^4$ is a boundary under d_2). This completes the proof.

Let us now consider the injection $j: A_r \to A_{r+1}$.

LEMMA 4·4. *$j^*\varpi_{r+1} = (\varpi_r)^2$ for $r \geq 2$.*

Proof. This follows immediately from (4·2), since z_r and all the cochains γ map to zero in $F(A_r^*)$.

In order to introduce a 'periodicity map' in $H^{**}(A)$, we shall need to consider Massey products. Let L be a left A-module. In order to avoid discussing the dependence of our constructions on the choice of resolution, we use a standard resolution, namely that given by the bar construction; this is defined in terms of symbols

$$a_0[a_1|a_2|\dots|a_s|\,l],$$

with $a_i\in A$, $l\in L$. This allows us to obtain $\mathrm{Ext}_A^{**}(L, Z_2)$ as the cohomology of a standard cochain complex, namely that given by the cobar construction; this is defined in terms of symbols

$$[\alpha_1|\alpha_2|\dots|\alpha_s|\,\lambda]$$

with $\alpha_i\in A^*$, $\lambda\in L^*$. This cochain complex is a left module over $F(A^*)$. It is now clear how to define Massey products. In particular, let

$$\mathrm{Ker}\,(h_0^{2^r}) \subset \mathrm{Ext}_A^{s,t}(L, Z_2)$$

be the subgroup of elements e such that $h_0^{2^r}e = 0$. For any such e, take a representative cocycle x; then we have

$$[\xi_1|\dots|\xi_1]\,x = \delta y;$$

let $\pi_r e$ be the class of the cocycle

$$[\xi_1^{2^{r+1}}]\,y+c_rx.$$

We have defined a homomorphism

$$\pi_r\colon \mathrm{Ker}\,(h_0^{2^r}) \to \frac{\mathrm{Ext}_A^{s+2^r,\,t+3.2^r}(L, Z_2)}{h_{r+1}\,\mathrm{Ext}_A^{s+2^r-1,\,t+2^r}(L, Z_2)},$$

and $\pi_r e$ is a representative for the Massey product $\langle h_{r+1}, h_0^{2^r}, e\rangle$.

Remark. As soon as we have proved that $H^{s,t}(A) = 0$ for $s = 2^r$, $t = 3.2^r$ (Corollary 5·6) we shall know that $\pi_r e$ coincides exactly with this Massey product.

LEMMA 4·5. *The following diagram is commutative:*

$$\begin{array}{ccc}\mathrm{Ker}\,(h_0^{2^r}) & \xrightarrow{\pi_r} & \dfrac{\mathrm{Ext}_A^{s+2^r,\,t+3.2^r}(L, Z_2)}{h_{r+1}\,\mathrm{Ext}_A^{s+2^r-1,\,t+2^r}(L, Z_2)}\\ {\scriptstyle i^*}\downarrow & & \downarrow{\scriptstyle i^*}\\ \mathrm{Ext}_{A_r}^{s,t}(L, Z_2) & \xrightarrow{\varpi_r} & \mathrm{Ext}_{A_r}^{s+2^r,\,t+3.2^r}(L, Z_2)\end{array}$$

Note. The arrow marked 'ϖ_r' is defined by multiplication on the left with ϖ_r.

Proof. $i^*([\xi_1^{2^{r+1}}]y + c_r x) = 0 + (i^*c_r)(i^*x)$.

5. *The periodicity theorems.* In this section we will state and prove the periodicity theorems. The order of proof is as follows. We begin with the result for modules over A_1 (Theorem 5·1); we deduce the result for modules over A_r $(2 \leqslant r < \infty)$ (Theorem 5·3); from this we deduce the result for modules over A (Theorem 5·4 plus Corollary 5·7); finally we deduce the result for $H^{**}(A)$.

Theorem 5·1. *Let L be a left A_1-module which is free over A_0. Then the homomorphism*

$$\varpi_2 \colon \mathrm{Ext}_{A_1}^{s,t}(L, Z_2) \to \mathrm{Ext}_{A_1}^{s+4,t+12}(L, Z_2)$$

is an epimorphism for $s = 0$ and an isomorphism for $s > 0$.

Note. The homomorphism ϖ_2 is defined by left multiplication with the image of $\varpi_2 \in H^{4,12}(A_2)$ in $H^{4,12}(A_1)$ (see Lemma 4·3).

LEMMA 5·2. *Theorem* 5·1 *is true in the special case $L = A_0$.*

The proof is by direct calculation, which is fairly light since the algebra A_1 is so small. Note that the minimal resolution of A_0 over A_1 is periodic with period 4.

Proof of Theorem 5·1. Let $0 \to L \to M \to N \to 0$ be an extension of A_1-modules all free over A_0; then we have the following commutative diagram.

$$\begin{array}{ccccccccc}
\mathrm{Ext}_{A_1}^{s-1,t}(L,Z_2) & \xrightarrow{\delta} & \mathrm{Ext}_{A_1}^{s,t}(N,Z_2) & \to & \mathrm{Ext}_{A_1}^{s,t}(M,Z_2) & \to & \mathrm{Ext}_{A_1}^{s,t}(L,Z_2) & \xrightarrow{\delta} & \mathrm{Ext}_{A_1}^{s+1,t}(N,Z_2) \\
\downarrow \varpi_2 & & \downarrow \varpi_2 & & \downarrow \varpi_2 & & \downarrow \varpi_2 & & \downarrow \varpi_2 \\
\mathrm{Ext}_{A_1}^{s+3,t+12}(L,Z_2) & \xrightarrow{\delta} & \mathrm{Ext}_{A_1}^{s+4,t+12}(N,Z_2) & \to & \mathrm{Ext}_{A_1}^{s+4,t+12}(M,Z_2) & \to & \mathrm{Ext}_{A_1}^{s+4,t+12}(L,Z_2) & \xrightarrow{\delta} & \mathrm{Ext}_{A_1}^{s+5,t+12}(N,Z_2)
\end{array}$$

(The squares involving δ are commutative because δ is right multiplication by the class of the extension $0 \to L \to M \to N \to 0$ in $\mathrm{Ext}_A^{1,0}(N, L)$.) By the Five Lemma, if Theorem 5·1 is true for L and N, it is true for M. By induction, Theorem 5·1 is true for any finite extension of modules isomorphic to A_0. Hence (arguing as for Lemma 2·3) it is true for all L.

For the next result, we assume that L is a left module over A_r $(2 \leqslant r < \infty)$, that L is free over A_0, and that $L_t = 0$ for $t < l$.

THEOREM 5·3. *The homorphism*

$$\varpi_r \colon \mathrm{Ext}_{A_r}^{s,t}(L, Z_2) \to \mathrm{Ext}_{A_r}^{s+2^r,t+3\cdot 2^r}(L, Z_2)$$

is an isomorphism for $s \geqslant 0$, $t < l + 4s$.

Note. The homomorphism ϖ_r is defined by left multiplication with the element $\varpi_r \in \mathrm{Ext}_A^{2^r,3\cdot 2^r}(Z_2, Z_2)$ (see section 4).

Proof. For $s = 0$ both Ext groups are zero (using Theorem 2·1), so the result is trivially true. We may therefore restrict attention to the case $s > 0$.

For $s > 0, t - l < 0$ both Ext groups are zero (using Theorem 2·1 again), so the result is true in this case. We may now proceed by induction over $t - l$; as an inductive hypothesis, we assume the result known for smaller values of $t - l$.

Let K be the kernel of the obvious map

$$A_r \otimes_{A_1} L \to L,$$

so that we have an exact sequence

$$0 \to K \to A_r \otimes_{A_1} L \to L \to 0.$$

The module $A_r \otimes_{A_1} L$ is A_0-free by Proposition 2·6. Hence K is A_0-free, by Lemma 2·4; also $K_t = 0$ for $t < l+4$. Now we can consider the following commutative diagram.

$$\begin{array}{ccccc}
\mathrm{Ext}_{A_r}^{s-1,t}(A_r \otimes_{A_1} L, Z_2) & \to & \mathrm{Ext}_{A_r}^{s-1,t}(K, Z_2) & \to & \mathrm{Ext}_{A_r}^{s,t}(L, Z_2) \to \\
\downarrow \varpi_r & & \downarrow \varpi_r & & \downarrow \varpi_r \\
\mathrm{Ext}_{A_r}^{s+2^r-1,t+3.2^r}(A_r \otimes_{A_1} L, Z_2) & \to & \mathrm{Ext}_{A_r}^{s+2^r-1,t+3.2^r}(K, Z_2) & \to & \mathrm{Ext}_{A_r}^{s+2^r,t+3.2^r}(L, Z_2) \to
\end{array}$$

$$\begin{array}{ccc}
\to \mathrm{Ext}_{A_r}^{s,t}(A_r \otimes_{A_1} L, Z_2) & \to & \mathrm{Ext}_{A_r}^{s,t}(K, Z_2) \\
\downarrow \varpi_r & & \downarrow \varpi_r \\
\to \mathrm{Ext}_{A_r}^{s+2^r,t+3.2^r}(A_r \otimes_{A_1} L, Z_2) & \to & \mathrm{Ext}_{A_r}^{s+2^r,t+3.2^r}(K, Z_2)
\end{array}$$

Since we are assuming $s > 0$, we have $s-1 \geqslant 0$. The inductive hypothesis applies to K, and shows that the second vertical arrow is an isomorphism for $t < l+4s$, while the fifth vertical arrow is an isomorphism for $t < l+4s+4$. As for the fourth vertical arrow, we have

$$\mathrm{Ext}_{A_r}^{s,t}(A_r \otimes_{A_1} L, Z_2) \cong \mathrm{Ext}_{A_1}^{s,t}(L, Z_2),$$
$$\mathrm{Ext}_{A_r}^{s+2^r,t+3.2^r}(A_r \otimes_{A_1} L, Z_2) \cong \mathrm{Ext}_{A_1}^{s+2^r,t+3.2^r}(L, Z_2).$$

These isomorphisms carry the homomorphism ϖ_r into $(\varpi_2)^{2^{r-2}}$ (using Lemma 4·4); and this is an isomorphism for $s > 0$, by Theorem 5·1. Similarly for the first vertical arrow, which is an epimorphism. The required conclusion now follows by the Five Lemma. This completes the induction, and proves the theorem.

For the next result we assume that L is a left A-module, that L is free over A_0, and that $L_t = 0$ for $t < l$.

THEOREM 5·4. *For each* r *(in the range* $2 \leqslant r < \infty$*) the map* π_r *of* section 4 *gives an isomorphism*

$$\pi_r : \mathrm{Ext}_A^{s,t}(L, Z_2) \xrightarrow{\cong} \mathrm{Ext}_A^{s+2^r,t+3.2^r}(L, Z_2)$$

valid for $s > 0$, $t < l + \min(4s, 2^{r+1} + T(s-1))$.

Proof. Since $t < l + 2^{r+1} + T(s-1)$, it follows (using Theorem 2·1) that

$$\mathrm{Ker}(h_0^{2^r}) = \mathrm{Ext}_A^{s,t}(L, Z_2),$$
$$h_{r+1}\, \mathrm{Ext}_A^{s+2^r-1,t+2^r}(L, Z_2) = 0.$$

Lemma 4·5 now provides the following commutative diagram.

$$\begin{array}{ccc}
\mathrm{Ext}_A^{s,t}(L, Z_2) & \xrightarrow{\pi_r} & \mathrm{Ext}_A^{s+2^r,t+3.2^r}(L, Z_2) \\
i^* \downarrow & & \downarrow i^* \\
\mathrm{Ext}_{A_r}^{s,t}(L, Z_2) & \xrightarrow{\varpi_r} & \mathrm{Ext}_{A_r}^{s+2^r,t+3.2^r}(L, Z_2)
\end{array}$$

The maps i^* are isomorphisms for $t < l + 2^{r+1} + T(s-1)$, by Theorem 3·1; the map ϖ_r is iso for $t < l+4s$, by Theorem 5·3. This proves the theorem.

Corollary 5·5. *For each r (in the range $2 \leqslant r < \infty$) there is an isomorphism*

$$\pi_r\colon \operatorname{Ext}_A^{s,t}(Z_2, Z_2) \to \operatorname{Ext}_A^{s+2^r, t+3\cdot 2^r}(Z_2, Z_2)$$

valid for $1 < s < t < \min(4s-2, 2+2^{r+1}+T(s-2))$.

Proof. Let $I(A) = \sum_{t>0} A_t$ and let $L = I(A)/A\,Sq^1$, so that L is free over A_0 and $l = 2$. Then we have an isomorphism

$$\operatorname{Ext}_A^{s,t}(Z_2, Z_2) \cong \operatorname{Ext}_A^{s-1,t}(L, Z_2)$$

valid for $0 < s < t$. So Corollary 5·5 follows immediately from Theorem 5·4.

Proof of Theorem 1·1. Let $L = I(A)/A\,Sq^1$; we have to prove that $\operatorname{Ext}_A^{s-1,t}(L, Z_2) = 0$ for $0 < s$, $t < U(s)$. By applying Theorem 2·1 we would only obtain the result for $t < V(s)$, where

$$V(4k) = 12k-3, \quad V(4k+1) = 12k+2, \quad V(4k+2) = 12k+4, \quad V(4k+3) = 12k+6.$$

However, we may assume the result known for $s = 4$ by direct calculation; for larger values of s, of the form $s = 4k$, it follows by periodicity (Theorem 5·4 with $r = 2$).

Corollary 5·6. *We have* $H^{s,t}(A) = 0$ *for* $s = 4k$, $t = 12k$ $(k > 0)$.

Proof. For $k = 1$ we may assume the result known by direct calculation. For larger values of k it follows by periodicity (Corollary 5·5 with $r = 2$).

Corollary 5·7. *The map* π_r *of section* 4 *is defined by*

$$\pi_r x = \langle h_{r+1}, h_0^{2^r}, x\rangle.$$

Proof. This follows immediately from Corollary 5·6, as remarked in section 4.

Corollary 5·8. *The isomorphism* π_r *of Corollary* 5·5 *is defined by*

$$\pi_r x = \langle h_{r+1}, h_0^{2^r}, x\rangle.$$

Proof. Let $L = I(A)/A\,Sq^1$. Then the isomorphism

$$\operatorname{Ext}_A^{s-1,t}(L, Z_2) \cong \operatorname{Ext}_A^{s,t}(Z_2, Z_2)$$

(for $0 < s < t$) is defined by multiplication on the right with a fixed element of $\operatorname{Ext}_A^{1,0}(Z_2, L)$, namely the class of the extension

$$0 \to I(A)/A\,Sq^1 \to A/A\,Sq^1 \to Z_2 \to 0.$$

Such multiplication commutes with Massey products on the left.

Theorem 1·2 follows by combining Corollaries 5·5 and 5·8.

Remark. It follows from our constructions, using Lemma 4·4, that our periodicity isomorphisms satisfy

$$\pi_{r+1} = (\pi_r)^2$$

in the region where π_r is valid. This may also be checked by the following manipulation.

$$\begin{aligned}
&\langle h_{r+1}, h_0^{2^r}, \langle h_{r+1}, h_0^{2^r}, x\rangle\rangle \\
&\quad = \langle h_{r+1}, \langle h_0^{2^r}, h_{r+1}, h_0^{2^r}\rangle, x\rangle + \langle\langle h_{r+1}, h_0^{2^r}, h_{r+1}\rangle, h_0^{2^r}, x\rangle \\
&\quad = \langle h_{r+1}, 0, x\rangle + \langle h_{r+2} h_0^{2^r}, h_0^{2^r}, x\rangle \\
&\quad = \langle h_{r+2}, h_0^{2^{r+1}}, x\rangle.
\end{aligned}$$

REFERENCES

(1) ADAMS, J. F. On the structure and applications of the Steenrod algebra. *Comment. Math. Helv.* **32** (1958), 180–214.
(2) ADAMS, J. F. Théorie de l'homotopie stable. *Bull. Soc. Math. France*, **87** (1959), 277–280.
(3) ADAMS, J. F. On the non-existence of elements of Hopf invariant one. *Ann. of Math.* **72** (1960), 20–104.
(4) ADAMS, J. F. A finiteness theorem in homological algebra. *Proc. Cambridge Philos. Soc.* **57** (1961), 31–36.
(5) ADAMS, J. F. *Stable homotopy theory* (Springer-Verlag, 1964).
(6) CARTAN SEMINAR NOTES, 1958/59.
(7) CARTAN H. and EILENBERG, S. *Homological algebra* (Princeton, 1956).
(8) LIULEVICIUS, A. The factorisation of cyclic reduced powers by secondary cohomology operations. *Mem. Amer. Math. Soc.* **42** (1962).
(9) LIULEVICIUS, A. Zeroes of the cohomology of the Steenrod algebra. *Proc. Amer. Math. Soc.* **14** (1963), 972–976.
(10) MAY, J. P. Thesis (Princeton, 1964).
(11) MILNOR, J. The Steenrod algebra and its dual. *Ann. of Math.* **67** (1958), 150–171.
(12) SHIMADA, N. and YAMANOSHITA, T. On triviality of the mod p Hopf invariant. *Japanese Journal of Math.* **31** (1961), 1–25.

Topology Vol. 10, pp. 271–282. Pergamon Press, 1971.

MODULES OVER THE STEENROD ALGEBRA

J. F. ADAMS and H. R. MARGOLIS

(*Received* 4 *November* 1970; *revised* 26 *March* 1971)

§0. INTRODUCTION

IN THIS paper we will investigate certain aspects of the structure of the mod 2 Steenrod algebra, A, and of modules over it. Because of the central role taken by the cohomology groups of spaces considered as A-modules in much of recent algebraic topology (e.g. [1], [2], [7]), it is not unreasonable to suppose that the algebraic structure that we elucidate will be of benefit in the study of topological problems.

The main result of this paper, Theorem 3.1, gives a criterion for a module M to be free over the Steenrod algebra, or over some subHopf algebra B of A. To describe this criterion consider a ring R and a left module over R, M; for $e \in R$ we define the map $e: M \to M$ by $e(m) = em$. If the element e satisfies the condition that $e^2 = 0$ then $\operatorname{im} e \subset \ker e$, therefore we can define the homology group $H(M, e) = \ker e/\operatorname{im} e$. Then, for example, $H(R, e) = 0$ implies that $H(F, e) = 0$ for any free R-module F. We can now describe the criterion referred to above: for any subHopf algebra B of A there are elements e_i—with $e_i{}^2 = 0$ and $H(B, e_i) = 0$—such that a connected B-module M is free over B if and only if $H(M, e_i) = 0$ for all i. This criterion we attribute to C.T.C. Wall (unpublished); he studied the cases $B = A_1, A_2$ proving our Theorem 3.1 for $B = A_1$. Here A_n is the subalgebra of A generated by $Sq^1, \ldots, Sq^{2^n}$. It should also be noted that some of these homology groups have already been found useful in algebraic topology, for example in the work of Anderson, Brown and Peterson [2].

This paper is organized as follows. Section 1 is devoted to a brief description of the Steenrod algebra from the point of view taken by Milnor [6]. The section also quotes a characterization, due to the second author, of subHopf algebras of A. In Section 2 we develop the algebraic tools needed to prove the main theorem by considering the particular situation of exterior algebras. Section 3 is primarily devoted to the proof of the main theorem, Theorem 3.1, which was described above. The section also includes some of the more immediate corollaries, for example, if we are given a short exact sequence of A-modules and any two are free then so is the third. In Section 4 we begin a more detailed study of A-modules using the homology groups introduced in Section 3. The results are of two types, "global" and "local". We prove, for example, that free A-Modules are injective and use this to answer positively the question: is there a nice relation between A-modules M and N if there is an A-map between them that induces isomorphisms of the homologies?

A brief word about the genesis of this paper: all the results were originally proven by the second author [3]; however this paper incorporates a proof of the main theorem due to the first author that is substantially easier to follow.

§1. THE STRUCTURE OF THE Mod 2 STEENROD ALGEBRA

In this section we recall some basic properties of the Steenrod algebra primarily from the point of view of the Milnor basis. The mod 2 Steenrod algebra, A, is the algebra of stable operations in cohomology with Z_2-coefficients. We recall the description due to John Milnor of this algebra:

THEOREM 1.1. *The Steenrod algebra A is a locally finite Hopf algebra such that the dual Hopf algebra A^* satisfies*

(1) $A^* = Z_2[\xi_1, \xi_2, \ldots]$ $\deg \xi_i = 2^i - 1$, *and*

(2) $\psi\xi_i = \sum_{k=0}^{i} \xi_{i-k}^{2^k} \otimes \xi_k$ $(\xi_0 = 1)$.

Therefore A has a Z_2-basis dual to the monomial basis of A^* which we denote $\{Sq(r_1, r_2, \ldots)\}$ with $Sq(r_1, r_2, \ldots)$ dual to $\xi_1^{r_1}\xi_2^{r_2}$ .... Further with respect to this basis the Hopf algebra structure of A is given by

(1) $\psi Sq(r_1, \ldots) = \sum_{s_i + t_i = r_i} Sq(s_1, \ldots) \otimes Sq(t_1, \ldots)$, and

(2) $Sq(r_1, \ldots) \cdot Sq(s_1, \ldots) = \sum_X \beta(X) Sq(t_1, \ldots)$, the summation being over all matrices

$$X = \begin{vmatrix} * & x_{01} & x_{02} & \cdots \\ x_{10} & x_{11} & \cdots & \\ x_{20} & \vdots & & \\ \vdots & & & \end{vmatrix}$$ that satisfy $s_j = \sum x_{ij}$, $r_i = \sum 2^j x_{ij}$ and with $t_k = \sum_{i+j=k} x_{ij}$

and $\beta(X) = \prod_k \dfrac{t_k!}{x_{k0}! \cdots x_{0k}!} \in Z_2$.

Proof. See [6] for details.

Definition 1.1. We will be interested in certain particular elements of A to which we give special notation. Let $P_t(r) = Sq(0, \ldots, r)$ with the r in the tth position and $P_t^s = P_t(2^s)$. This notation agrees with May's [5].

Notation. We will say that $2^k \in r$ if 2^k appears in the dyadic expansion of r.

The following lemma is the reason that the multiplication described in Theorem 1.1 is not as bad as it looks.

LEMMA 1.2. *The coefficient $\beta(X) = 0$ if and only if for some $x_{i_1j_1}, x_{i_2j_2}$ with $i_1 + j_1 = i_2 + j_2$ and some k, $2^k \in x_{i_1j_1}, x_{i_2j_2}$.*

Proof. It suffices to show that the multinomial coefficient $(x_1, \ldots, x_n) = \dfrac{(x_1 + \cdots + x_n)!}{x_1! \ldots x_n!}$ is zero mod 2 if and only if for some i, j, k, $2^k \in x_i, x_j$. We will do this by induction on n.

If $n = 2$ we have the standard result that $\dfrac{(x_1 + x_2)!}{x_1!\, x_2!} \equiv \prod \binom{a_i}{b_i}$ mod 2 where $x_1 + x_2 = \sum a_i 2^i$ and $x_1 = \sum b_i 2^i$ (see [9]). Since $\binom{a_i}{b_i} \equiv 0$ mod 2 if and only if $a_i = 0$ and $b_i = 1$, the coefficient is zero mod 2 if and only if $2^k \in x_1, x_2$ for some k. Assuming the result for $n - 1$ we note that $(x_1, \ldots, x_n) = (x_1 + \cdots + x_{n-1}, x_n) \cdot (x_1, \ldots, x_{n-1})$. Therefore $(x_1, \ldots, x_n) \equiv 0$ mod 2 if and only if $(x_1 + \cdots + x_{n-1}, x_n) \equiv 0$ mod 2 or $(x_1, \ldots, x_{n-1}) \equiv 0$ mod 2. If $(x_1, \ldots, x_{n-1}) \equiv 0$ mod 2, then for some $i, j < n$ and some k, $2^k \in x_i, x_j$ and if not then $(x_1 + \cdots + x_{n-1}, x_n) \equiv 0$ mod 2 implies for some $i < n$ and some k, $2^k \in x_i, x_n$,

By way of illustration we record the computations that arise later in this work.

LEMMA 1.3. (1) *If* $r, s < 2^t$ *then* $[P_t(r), P_t(s)] = 0$.

(2) *If* $r < 2^t$ *then* $P_t(r)P_t(s) = \dfrac{(r + s)!}{r!\, s!} P_t(r + s)$, *in particular* $P_t(r)^2 = 0$.

(3) $P_t^t \cdot P_t^t = P_t(2^t - 1)P_{2t}^0$.

(4) *If* $1 \leq r < 2^t$ *then* $[P_t^t, P_t(r)] = P_t(r - 1)P_{2t}^0$.

(5) $[P_t^t, P_{2t}^0] = 0$.

Proof. As a sample consider (3). Because of the restrictions on the rows of the matrices that we must consider, the only ones to look at are

$$\begin{vmatrix} * & 0 & \cdots & 2^t \\ 1 & & & \\ \vdots & & \cdot^{\cdot^{\cdot}} & \\ 2^t & & & \end{vmatrix} \quad \text{and} \quad \begin{vmatrix} * & \cdots & 2^t - 1 \\ 0 & & \\ \vdots & & \vdots \\ 0 & \cdots & 1 \end{vmatrix}.$$

From Lemma 1.2 it is obvious that the multinomial coefficient associated with the first matrix is 0 and with the second matrix is 1. This gives (3).

In Section 3 we will be interested in the structure of the subHopf algebras of A. Therefore we quote the following proposition of [4] which gives a characterization of the subHopf algebras suitable for our purposes.

PROPOSITION 1.4. *Let* $B \subset A$ *be a subHopf algebra, let*

$$r_B(t) = \max\{s \mid 2^s \in b_t \text{ for } Sq(b_1, \ldots, b_t, \ldots) \in B\}$$

or $r_B(t) = \infty$ *if there is no maximum or*

$$r_B(t) = -1 \text{ if } b_t = 0 \text{ for all } Sq(b_1, \ldots, b_t, \ldots) \in B.$$

(1) *B has a Z_2-basis* $\{Sq(r_1, \ldots) \mid r_t < 2^{r_B(t)+1}\}$.

(2) *As an algebra B is generated by* $\{P_t^s \mid s \leq r_B(t)\}$.

Note. As the proposition makes clear, a subHopf algebra B is determined by its function $r_B(t)$. However not all such functions are so realizable—for example $r(1) = 1$, $r(2) = -1$ is not.

Examples. (1) Let A_n denote the subHopf algebra determined by the function

$$r(t) = \max\{n+1-t, -1\} \text{ then } \bigcup_n A_n = A.$$

(2) Let $A(0)$ denote the subHopf algebra of A determined by the function $r(t) = 0$ all t, then $A(0)$ is an exterior algebra on P_t^0 all t.

(3) Let B be a subHopf algebra of A, let $B(0)$ denote the subHopf algebra $A(0) \cap B$—it is determined by $r(t) = 0$ if $r_B(t) \geq 0$ and $r(t) = -1$ otherwise. Note that $B(0)$ is a normal subalgebra of B—a subalgebra C of B is normal if $B \cdot I(c)$ is a two-sided ideal of B.

Definition 1.2. Let B be a subHopf algebra of A, then htB is the maximum of $r_B(t)$ for all t—we say $htB = \infty$ if either $r_B(t) = \infty$ for some t or there is no maximum. So for example $htA_n = n$ and $htB(0) = 0$ (unless B is trivial).

The main result of this paper will be proved by induction on htB and in that induction the following proposition will play a key role.

PROPOSITION 1.5. *Let B be a subHopf algebra of A with $htB = n < \infty$. Then $B/\!/B(0)$ is isomorphic to a subHopf algebra of A, B', with $htB' = n-1$. For instance $A_n/\!/A_n(0) \approx A_{n-1}$.*

Recall that for C normal in B, $B/\!/C$ is the algebra $B/BI(C) = Z_2 \otimes_C B$.

Proof. Let $\theta: A^* \to A^*$ be given by $\theta x = x^2$, then it is easy to check that θ is a map of Hopf algebras. Therefore there is a dual map $\theta^*: A \to A$ of Hopf algebras, θ^* halving degree and $\theta^* P_t^s = P_t^{s-1}$, $s > 0$, $\theta^* P_t^0 = 0$. Since $\ker \theta^* = A \cdot I(A(0))$, θ^* induces an isomorphism $\theta': A/\!/A(0) \to A$, so let $B' = \theta'(B/\!/B(0))$ (θ' being an isomorphism of Hopf algebras and $B/\!/B(0)$ a subHopf algebra of $A/\!/A(0)$ imply that B' is a subHopf algebra of A). Further since $\theta' P_t^s = P_t^{s-1}$ it follows from Proposition 1.4 that $r_{B'}(t) = r_B(t) - 1$ and so $htB' = htB - 1$ if $htB < \infty$.

§2. MODULES OVER EXTERIOR ALGEBRAS

For the remainder of this paper we will be working with categories of *connected* (left) modules over *connected* algebras—a graded algebra R over K is connected if $R_i = 0$ for $i < 0$ and $R_0 = K$ and a graded R-module M is connected if $M_i = 0$ for $i < r$ for some integer r.

Definition 2.1. Let M be such a module over the algebra R. For $e \in R$ we define the map $e: M \to M$ by $e(m) = em$. We say that e *is exact on* M if $M \xrightarrow{e} M \xrightarrow{e} M$ is exact. For $e \in R$ satisfying $e^2 = 0$ we also have $H(M; e) = \ker e|M / \operatorname{im} e|M$ and then e is exact on M if and only if $H(M; e) = 0$.

It is immediate that if e is exact on R and M is free over R then e is exact on M. This leads us to the following definition.

Definition 2.2. An algebra R is *pseudo-exterior* on $\{e_i\}$ with respect to a category of R-modules, $\mathscr{C}$, if there are elements $e_i \in R$ which are exact on R such that the following condition is satisfied:

an R-module $M \in \mathscr{C}$ is R-free if and only if $H(M, e_i) = 0$ for all i.

The name we have chosen is reasonable because we shall show that exterior algebras on generators of distinct degree are pseudo-exterior. However the subject of pseudo-exterior algebras only becomes truly non-trivial when we add that we will prove that every subHopf algebra of the mod 2 Steenrod algebra—including A itself—is pseudo-exterior.

THEOREM 2.1. *Let $R = E[x_1, \ldots, x_n]$ the exterior algebra on generators $x_1, \ldots, x_n$. We further assume that the dimensions of the generators are distinct. Then a connected R module M is R-free if and only if $H(M, x_i) = 0$ for all i; that is R is pseudo-exterior with respect to the category of connected R-modules.*

Note. (1) The condition on the degrees of the generators is essential as the following example shows. Let $R = E[x_1, x_2]$ where $\deg x_1 = \deg x_2$; we will exhibit an R-module M such that $H(M, x_1) = H(M, x_2) = H(M, x_3) = 0$ where $x_3 = x_1 + x_2$ (these are all the elements of R which are exact on R itself) but M is not free over R. One such M is given by generators a, b and relations $x_1a + x_3b$, $x_2a + x_1b$, $x_3a + x_2b$.

(2) Since the fact that the generators have distinct degrees is crucial, we will need the following in Section 3: $\deg P_t^s = \deg P_{t_1}^{s_1}$ if and only if $P_t^s = P_{t_1}^{s_1}$. That is $\deg P_t^s = 2^s(2^t - 1)$ and the t and s can be recovered from the dyadic expansion of this number.

In order to prove Theorem 2.1 we prove two propositions that will also be of use in §3. For an R-module M let $Q_R(M) = M/I(R)M = K \otimes_R M$ (then for $B \subset R$, $R/\!/B = Q_B(R)$ and $Q_B(M)$ is a $Q_B(R)$-module). Let $\{m_i\}$ be a set of elements in M whose images in $Q_R(M)$ form a K-base for $Q_R(M)$. Then the following facts are well known and easily proved.

(i) The elements m_i generate M as an R-module.
(ii) If M is free over R on any base, then the elements m_i form an R-base for M.

PROPOSITION 2.2. *Let B be normal in R and $C = R/\!/B$. If M is free over B and $Q_B(M)$ is free over C then M is free over R.*

Proof. Take elements b_i forming a K-base for B, and elements c_j in A whose images in C form a K-base for C. Also take elements m_k in M whose images in $Q_R(M) = Q_C(Q_B(M))$ form a K-base. Since $Q_B(M)$ is C-free, the elements $c_j m_k$ give a K-base there. Since M is B-free, the elements $b_i c_j m_k$ form a K-base in M. Either there are no elements m_k, in which case the result is trivial, or else the elements $b_i c_j$ are linearly independent in R; so we assume the latter. Since the elements $b_i c_j$ certainly span R they form a K-base there. So the elements m_k form an R-base for M.

Let $R = E[x, y]$ where $\deg x = d$, $\deg y = e$ with $d \neq e$. Let M be a connected R-module. Then both yM and M/yM are modules over $E[x]$.

LEMMA 2.3. *If x and y are exact on M then x is exact on $M/(yM)$.*

Proof. Since y is exact on M we have an isomorphism $M/(yM) \xrightarrow{y} yM$ commuting with x. Therefore $H_n(M/(yM); x) = H_{n+e}(yM; x)$. But also we have the following exact sequence of $E[x]$-modules: $0 \to yM \to M \to M/(yM) \to 0$. This yields an exact homology sequence and since $H(M; x) = 0$ we have an isomorphism $H_n(M/(yM); x) = H_{n+d}(yM; x)$. So $H_n(M/(yM); x) = H_{n+d-e}(M/(yM); x)$. But $M/(yM)$ is zero in degrees less than some degree, so $H_r(M/(yM); x) = 0$ for r small enough. Since $d - e \neq 0$ we can use this isomorphism to show that $H_r(M/(yM); x) = 0$ for all r.

Proof of Theorem 2.1. The proof is by induction on the number n of generators. The result is trivially true for $n = 1$; suppose it is true for $E[x_1, \ldots, x_{n-1}]$. Let M be a module over R such that $H(M, x_i) = 0$ for each x_i. We apply Proposition 2.2 taking $B = E[x_n]$, so that $C = E[x_1, x_2, \ldots, x_{n-1}]$. Then M is free over B by the trivial case $n = 1$. We have $Q_B(M) = M/(x_n M)$; the elements x_i are exact on $Q_B(M)$ for $1 \leq i \leq n - 1$ by Lemma 2.3. By induction $Q_B(M)$ is C-free and so by Proposition 2.2, M is free over R.

We end this section by proving one final result that will be needed in the next section. Let R be an exterior algebra $E[x_1, \ldots, x_n]$ in which the dimensions of the generators are all distinct. Let M be an R-module. Then the quotient module

$N = M/(x_1 M + \cdots + x_{n-1}M) = Q_B(M) \quad (B = E[x_1, \ldots, x_{n-1}])$ is a module over $E[x_n]$.

LEMMA 2.4. *If each x_i is exact on M then x_n is exact on N.*

Of course this follows immediately from Theorem 2.1; but the obvious direct proof is by induction over n. The result is true for $n = 2$ by Lemma 2.3, so suppose it true for $n - 1$. Let R and M be as above. Then $P = M/(x_1 M + \cdots + x_{n-2}M)$ is a module over $E[x_{n-1}, x_n]$. By the inductive hypothesis, x_{n-1} and x_n are exact on P. By Lemma 2.3 x_n is exact on $P/(x_{n-1}P) = N$.

§3. THE MAIN THEOREM AND COROLLARIES

The primary aim of this paper is to prove that a wide class of interesting algebras are pseudo-exterior. Recall that for an arbitrary subHopf algebra $B \subset A$ (as usual A the mod 2 Steenrod algebra), B is generated as an algebra by the P_t^s's in B.

THEOREM 3.1. *Let B be a subHopf algebra of A then B is pseudo-exterior* (*with respect to the category of connected B-modules*) *on the P_t^s's in B with $s < t$. In particular A is pseudo-exterior on $\{P_t^s \mid s < t\}$.*

The brunt of our work will be to prove the following theorem of which Theorem 3.1 is an easy corollary.

THEOREM 3.2. *Let B be a finite subHopf algebra of A, then B is pseudo-exterior on the P_t^s's in B with $s < t$.*

We observe that any particular application of Theorem 3.2 can be made independent of proposition 1.4. To do so, one simply has to check that the given subalgebra B, and the subalgebra B', B'' etc. which arise from it by using Proposition 1.5, are as described in Proposition 1.4. For example the reader who does not wish to check the proof of Proposition 1.4 can still be sure that Theorem 3.2 is true for A_n and Theorem 3.1 is true for A.

Proof of Theorem 3.1 *from Theorem* 3.2. There are two things that must be proven. First that for any B and $P_t^s \in B$ with $s < t$, P_t^s is exact on B. Second if a connected B-module satisfies $H(M, P_t^s) = 0$ for all $P_t^s \in B$ with $s < t$ then M is free over B.

To prove the first let C be the subalgebra of B generated by $P_t^{s'}$ with $s' \leq s$. Since $s < t$, C is an exterior on the $P_t^{s'}$'s with $s' \leq t$ (see for example Lemma 1.3) and since C

has a Z_2-basis given by all $P_t(r)$ with $r < 2^t$, it is in fact a subHopf algebra of B. Therefore by an oft-quoted result of Milnor and Moore [8], B is free over C. So P_t^s is exact on B if it is exact on C and C being an exterior algebra with P_t^s as one of the generators, this is obvious.

To prove the second statement assuming Theorem 3.2 we proceed as follows: let $B_n = B \cap A_n$, then the B_n's satisfy

(1) B_n is a finite subHopf algebra of B

(2) $B_n \subset B_{n+1}$ and $\cup B_n = B$.

Let M be a connected B-module satisfying $H(M, P_t^s) = 0$ for all $P_t^s \in B$ with $s < t$. In particular $H(M, P_t^s) = 0$ for all $P_t^s \in B_n$ for any n and so by Theorem 3.2 M is B_n-free for any n. So it suffices to show that if M is B_n-free for all n then M is B-free. Let $\{m_i\}$ be a set of elements in M whose images in $Q_B(M)$ form a Z_2-base for $Q_B(M)$. We will prove that M is B-free on $\{m_i\}$ so assume that there are elements $b_i \in B$ (with only a finite number non-zero) such that $\Sigma b_i m_i = 0$. For n large enough $b_i \in B_n$ for all i. Also since $M \to Q_{B_n}(M) \to Q_B(M)$, the images of the m_i's in $Q_{B_n}(M)$ are linearly independent so can be expanded to a base for $Q_{B_n}(M)$ over Z_2 coming from $\{m_j'\} \supset \{m_i\}$. Then since M is B_n-free we have that M has a B_n-base of $\{m_j'\}$. So in particular we must have that $\Sigma b_i m_i = 0$ implies $b_i = 0$ for all i.

Before proving Theorem 3.2 we state some of the more immediate corollaries.

COROLLARY 3.3. *Let B be a subHopf algebra of A. If we are given an exact sequence of B-modules $0 \to M_1 \to M_2 \to M_3 \to 0$ such that any two of the modules are free over B then so is the third.*

Proof. For any $P_t^s \in B$ with $s < t$ we can regard M_i as a complex with differential P_t^s and maps of B-modules induce maps of complexes. Therefore the exact sequence $0 \to M_1 \to M_2 \to M_3 \to 0$ induces long exact sequences in the P_t^s homologies. Then the freeness of two of the M_i's implies that their P_t^s homologies are both zero, so the same is true of the third which is therefore free by Theorem 3.1.

We recall the following definition from homological algebra.

Definition 3.1. An R-module M has *infinite homological degree* if there are no projective resolutions of M over R of finite length.

COROLLARY 3.4. *Let M be a connected B-module (B a subHopf algebra of A), then either M is B-free or M has infinite homological degree.*

Proof. If M_0 has a finite projective resolution over B then there is a sequence of exact sequences $0 \to M_i \to F_{i-1} \to M_{i-1} \to 0$ with F_{i-1} B-free and for i large enough M_i is B-free. Therefore iterated application of Corollary 3.3 gives us that M_0 is B-free.

Note. The results of Corollaries 3.3 and 3.4 as well as many of the results of the next section are true for an arbitrary pseudo-exterior algebra, as the proofs make evident.

We now commence with the proof of Theorem 3.2. The proof will be by induction on htB (see Definition 1.2), the inductive step based on Proposition 2.2. In order to apply the inductive hypothesis we need the following lemma, whose proof we postpone until the end of this section.

LEMMA 3.5. *Let B be a finite subHopf algebra of A and let M be a connected B-module. If P_t^s is exact on M for $P_t^s \in B$ with $s < t$ then P_t^s is exact on $M/\sum_{P_t^0 \in B} P_t^0 M = Q_{B(0)}(M)$ for $P_t^s \in B$ with $s \leq t$.*

Proof of Theorem 3.2. In the light of what has already been shown in proving Theorem 3.1 it suffices to show that if M is a connected B-module such that $H(M, P_t^s) = 0$ for all $P_t^s \in B$ with $s < t$, then M is B-free. As stated above the proof will be by induction on htB. If $htB = 0$ then B is an exterior algebra on a subset of the P_t^0's—this follows from Proposition 1.4—and so the theorem is true having been proven under the guise of Theorem 2.1. So assume the result for connected modules over B' with $htB' < n$ and let $htB = n$.

So consider M a module over B such that P_t^s is exact on M for $P_t^s \in B$, $s < t$. As in Section 1, let $B(0)$ be the exterior algebra generated by the P_t^0's in B and let $B' = B/\!/B(0)$. Then, as we have shown in Proposition 1.5, B' is a finite subHopf algebra of A with $htB' = n - 1$. Since the P_t^s's with $s \leq t$ in B project to the P_t^s's with $s < t$ in B' it follows from Lemma 3.5 that P_t^s is exact on $Q_{B(0)}(M)$ for all $P_t^s \in B'$ with $s<t$. Therefore by induction $Q_{B(0)}(M)$ is free over B'. In addition M is free over $B(0)$ as noted above. Therefore by Proposition 2.2 M is free over B, which completes the proof of the theorem modulo the proof of Lemma 3.5.

If we look at Lemma 3.5, it is clear that a special role is played by the operations P_t^s with $s = t$ because they have to be proven exact on the quotient although they are not given to be exact on M. We therefore study the elements $P_t^t \in B$. Since $P_t^t \in B$ and B is a subHopf algebra of A it follows that $P_t^0, P_t^1, \ldots, P_t^{t-1} \in B$ and since $[P_t^t, P_t^0] = P_{2t}^0$ we also have $P_{2t}^0 \in B$. As in Lemma 1.3 it is clear that $P_t^0, P_t^1, \ldots, P_t^{t-1}, P_{2t}$ generate a sub-exterior algebra of B. Let $D(t)$ denote the subalgebra of B generated by that exterior algebra and P_t^t. The structure of $D(t)$ is given by the following formulae of Lemma 1.3:

$$(P_t^t)^2 = P_t(2^t - 1)P_{2t}^0,$$
$$[P_t^t, P_t(r)] = P_t(r-1)P_{2t}^0 \text{ for } 1 \leq r < 2^t,$$
$$[P_t^t, P_{2t}^0] = 0.$$

It follows that $D(t)/\!/E[P_{2t}^0] = E[P_t^0, P_t^1, \ldots, P_t^{t-1}, P_t^t]$ and if M is a module over $D(t)$ then $Q_E(M)$ $(E = E[P_{2t}^0])$ is a module over $E[P_t^0, P_t^1, \ldots, P_t^{t-1}, P_t^t]$.

LEMMA 3.6. *Let M be a connected $D(t)$-module such that P_{2t}^0 is exact on M, then P_t^t is exact on $Q_E(M)$.*

Proof. The result to be proved may be expressed as follows. Suppose given an element x of M of degree n such that $P_t^t x = P_{2t}^0 y$ for some y in M. We have to show that we can write x in the form $x = P_t^t u + P_{2t}^0 v$ for some u, v in M.

The proof is by a double induction and the main induction is over n, so we assume as our main inductive hypothesis that the result is true for elements x' of degree less than n. Suppose given an element x of degree n such that $P_t^t x = P_{2t}^0 y$. Applying P_t^t we get

$P_t(2^t - 1)P^0_{2t}x = P_t^t P^0_{2t}y$, that is $P^0_{2t}[P_t(2^t - 1)x + P_t^t y] = 0$. Since P^0_{2t} is exact on M we have $P_t(2^t - 1)x = P_t^t y + P^0_{2t}z$ for some z in M.

Now observe that

$$\begin{aligned} P_t(2^t - 2)P^0_{2t}x &= P_t^t P_t(2^t - 1)x + P_t(2^t - 1)P_t^t x \\ &= P_t^t P_t^t y + P_t^t P^0_{2t}z + P_t(2^t - 1)P^0_{2t}y \\ &= P_t^t P^0_{2t}z. \end{aligned}$$

Again by the exactness of P^0_{2t} we see that $P_t(2^t - 2)x = P_t^t z + P^0_{2t}w$.

If $t = 1$ this completes the proof since $P_t(0) = 1$. Otherwise we continue by induction. More precisely we will prove by downwards induction over r that $P_t(r)x = P_t^t c + P^0_{2t}d$ for $0 \leq r \leq 2^t - 1$. Then for $r = 0$ this will express x in the required form. We have obtained this result for $r = 2^t - 1$ and $r = 2^t - 2$ above, so we assume as our subsidiary inductive hypothesis that $P_t(r)x$ has the desired form for some r in the range $1 \leq r \leq 2^t - 2$. Let $s = 2^t - 1 - r$ $(s > 0)$. We have $P_t(r)P_t(s) = P_t(s)P_t(r) = P_t(2^t - 1)$.

So

$$\begin{aligned} P_t^t y + P^0_{2t}z &= P_t(2^t - 1)x \\ &= P_t(s)P_t(r)x \\ &= P_t(s)P_t^t c + P_t(s)P^0_{2t}d \\ &= P_t^t P_t(s)c + P_t(s - 1)P^0_{2t}c + P_t(s)P^0_{2t}(d). \end{aligned}$$

That is, $P_t^t(y + P_t(s)c) = P^0_{2t}(z + P_t(s - 1)c + P_t(s)d)$. But the dimension of y is less than n, so the main inductive hypothesis shows that $y + P_t(s)c = P_t^t e + P^0_{2t}f$. Now observe that

$$\begin{aligned} P_t(r - 1)P^0_{2t}x &= P_t(r)P_t^t x + P_t^t P_t(r)x \\ &= P_t(r)P^0_{2t}y + P_t^t P_t^t c + P_t^t P^0_{2t}d \\ &= P_t(r)P^0_{2t}P_t(s)c + P_t(r)P^0_{2t}P_t^t e + P_t^t P_t^t c + P_t^t P^0_{2t}d \\ &= P^0_{2t}P_t^t P_t(r)e + P^0_{2t}P_t^t d. \end{aligned}$$

That is, $P^0_{2t}(P_t(r - 1)x + P_t^t h) = 0$ where $h = d + P_t(r)e$. Since P^0_{2t} is exact we have $P_t(r - 1)x = P_t^t h + P^0_{2t}k$. This completes the subsidiary induction over r, which completes the main induction and hence the lemma.

Proof of Lemma 3.5. Now let M be a module over B such that P_t^s is exact on M for $P_t^s \in B$ with $s < t$. Consider $P_t^s \in B$ with $s \leq t$. We form first the quotient $M/P^0_{2t}M$. Then P_t^s is exact on $M/P^0_{2t}M$ by Lemma 3.6 if $s = t$ and by Lemma 2.4 if $s < t$. Also for $P_i^0 \in B$ with $i \neq 2t$, P_i^0 is exact on $M/P^0_{2t}M$ by Lemma 2.3.

We now consider the quotient $N = Q_{B_s}(M)$ where B_s is the subexterior algebra of B generated by all P_i^0 with $i > s$. Then N is a module over an exterior algebra C on generators $P_i^0 \in B$ with $i \leq s$ and P_t^s; this being a subalgebra of $Q_{B_s}(B)$ as can be seen from the following relations:

$$\begin{aligned} [P_i^0, P_t^s] &= P^0_{i+t}P_t(2^s - 2^i) \quad \text{if } i \leq s, \\ P_t^s P_t^s &= 0 \quad \text{if } s < t, \\ P_t^t P_t^t &= P^0_{2t}P_t(2^s - 1). \end{aligned}$$

Further P_i^0 is exact on N for $1 \leq i \leq s$, as we see by applying Lemma 2.4 to M considered as a module over the exterior algebra generated by B_s and P_i^0. Again P_t^s is exact on N, as we see by applying Lemma 2.4 to $M/P_{2t}^0 M$ considered as a module over the exterior algebra generated by $P_i^0 \in B$ with $i > s$ and $i \neq 2t$ and P_t^s. These results allow us to apply Lemma 2.3 to N with respect to the exterior algebra C. We conclude that P_t^s is exact on $Q_{B(0)}(M)$.

§4. PROPERTIES OF THE HOMOLOGY GROUPS

This section is the beginning of a more detailed study of the homology groups introduced in Sections 2 and 3. The results are of two primary types.

On the one hand we would like to know something about the "global" nature of the homology groups. In particular we wish to answer the following question: let M and N be connected A-modules, and suppose that we are given an A-map $M \to N$ which induces isomorphisms of all the homology groups, is there a new relationship between M and N? Further, in the process of answering this question we prove that connected free A-modules are injective.

On the other hand we wish to consider properties that can perhaps be described as "local". In this work we consider the case in which B is a subHopf algebra of A. The localization results are of two types. If an A-module M is an extended B-module ($M = A \otimes_B N$ for some B-module N) then only the homology groups in B arise, i.e. $H(M, P_t^s) = 0$ if $P_t^s \notin B$. We can also localize with respect to degree and get that if B is finite and $H_i(M, P_t^s) = 0$ for $i < I$ and $P_t^s \in B$ then M is B-free through deg $I - \alpha$ (α independent of M or I).

THEOREM 4.1. *If F is a connected free A-module then F is injective in the category of arbitrary A-modules.**

Proof. (a) We first reduce the problem to one of showing that A itself is an injective A-module in the stated category. Suppose that A is injective then the arbitrary product of copies of A, $\Pi A x_\alpha$, is injective (in general the product of injectives is injective). Let $\Sigma A x_\alpha$ be a connected free A-module; then $\Pi A x_\alpha$ is connected. We have the short exact sequence $0 \to \Sigma A x_\alpha \to \Pi A x_\alpha \to M \to 0$ and since $\Pi A x_\alpha$ is injective it will suffice to show that this sequence splits (in general the direct summand of an injective is injective).

SUBLEMMA. *For all* P_t^s, $H(\Pi A x_\alpha, P_t^s) = 0$ *(therefore since* $\Pi A x_\alpha$ *is connected we have also that* $\Pi A x_\alpha$ *is A-free).*

Proof. Let $(a_\alpha) \in \ker P_t^s | \Pi A x_\alpha$ and let $\Pi_\alpha : \Pi A x_\alpha \to A x_\alpha$ be the projection. Then $P_t^s a_\alpha = P_t^s \Pi_\alpha(a_\alpha) = \Pi_\alpha P_t^s(a_\alpha) = 0$ and therefore $a_\alpha = P_t^s b_\alpha$. Thus $(a_\alpha) = P_t^s(b_\alpha)$.

From the sublemma and the sequence above we conclude that $H(M, P_t^s) = 0$ and since M is connected it is a free A-module. And so as we desired, we have the splitting of

$$0 \to \Sigma A x_\alpha \to \Pi A x_\alpha \to M \to 0.$$

(b) We must now show that A is injective in the category of arbitrary A-modules.

---

* We would like to thank J. C. Moore and F. Peterson for pointing out an error in the original proof of this result. Also the proof of part (b) is a modification due to F. Peterson of a proof of ours.

Consider $0 \to A \xrightarrow{\iota} M$ with M an arbitrary A-module. To show that ι splits consider all pairs (N, p) with $A \subset N \subset M$ and $p:N \to A$ satisfying $p\iota = 1$. Partially order these pairs by setting $(N, p) < (N', p')$ if $N \subset N'$ and $p = p' \mid N$. Every linearly ordered subset has an upper bound so we may extract a maximal element (N, p). Assume that $N \neq M$ and consider $m \in M - N$. We show how to extend p to $N + Am$.

As an A_n-module A is locally finite and free, and since A_n is a Poincare algebra we conclude that A is an injective A_n-module. So p extends to an A_n-map $p_n:M \to A$. Let $p_n(m) = a_n$, the sequence $\{a_n\}$ is a subset of A_k (where $k = \deg(m)$) which is a finitely generated vector space over Z_2. Therefore for some n, a_n occurs infinitely often and it is easy to check that $p':N + Am \to A$ defined by $p' \mid N = p$ and $p'(m) = a_n$ is an A-map (any relation in A occurs within some A_n).

We are now in a position to answer the question posed at the beginning of this section.

Definition 4.1. Two B-modules M and N are *stably equivalent* if there are free B-modules F_1 and F_2 such that $F_1 \oplus M$ and $F_2 \oplus N$ are isomorphic.

THEOREM 4.2. *Let $f:M \to N$ be a map of connected A-modules and suppose that for all P_t^s with $s < t$ $f_*:H(M, P_t^s) \to H(N, P_t^s)$ is an isomorphism. Then M and N are stably equivalent.*

Proof. First assume that f is epic, i.e. $0 \to K \to M \to N \to 0$. Then $H(K, P_t^s) = 0$ for all P_t^s and therefore by Theorem 3.1 K is A-free. By Theorem 4.1 we conclude that the sequence splits and $M \oplus K \approx N$.

Now consider the general case $0 \to K \to M \xrightarrow{f} N \to L \to 0$. Let $F \xrightarrow{g} N \to 0$ be exact with F a free A-module. Then $M \oplus F \xrightarrow{f+g} N$ is onto giving $0 \to K' \to M \oplus F \to N \to 0$. But since $H(F, P_t^s) = 0$ and $f_*:H(M, P_t^s) \to H(N, P_t^s)$ is an isomorphism, we get that

$$(f + g)_*:H(M \oplus F, P_t^s) \to H(N, P_t^s)$$

is an isomorphism, which completes the proof.

We come now to the "local" results described at the beginning of this section.

Definition 4.2. An A-module M is an *extended B-module* (B a subHopf algebra of A) if M is isomorphic to $A \otimes_B N$ for some B-module N.

THEOREM 4.3. *Let M be an extended B-module, then for $P_t^s \notin B$ (with $s < t$) $H(M, P_t^s) = 0$.*

The proof of this result can be found in [4].

The other type of localization that we wish to consider is that with respect to degree. In order to do this we make the following obvious definition.

Definition 4.3. Let M be an R-module (both R and M being graded) then M is *free through degree r* if there is a free R-module F and map $f:F \to M$ such that f is an isomorphism in degree $\leq r$.

THEOREM 4.4. *Let B be a finite subHopf algebra of A and let M be a connected B-module. If $H_i(M, P_t^s) = 0$ for $P_t^s \in B$, $s < t$, and for $i \leq r$ then M is B-free through degree $r - c$ where c is a constant depending on B.*

Since we are emphasizing the qualitative nature of the result, no attempt will be made to determine a best possible such c. Also we will only sketch the proof of this result since it is a straightforward modification of the proof of Theorem 3.2 (and the proofs of Proposition 2.2 and Lemmas 2.3, 2.4, 3.5, 3.6).

We first note that in the case of Lemmas 2.3, 2.4, 3.6 the desired modification can easily be made since each is essentially proved by induction on degree starting with $M_r = 0$ for r sufficiently small. In the case of Proposition 2.2 our proof involved an explicit construction of a basis of M over R. Since we are working with connected modules over connected algebras the same construction can be performed to give a basis through a range (in the sense of Definition 4.3).

With these results the modified version of Lemma 3.5 is easily provable. From this the proof of Theorem 4.4 follows.

REFERENCES

1. J. F. ADAMS: On the Structure and Applications of the Steenrod Algebra, *Comm. Math. Helv.* **32** (1958).
2. D. W. ANDERSON, E. H. BROWN and F. PETERSON: The Structure of the Spin Cobordism Ring, *Ann. Math.* **86** (1967).
3. H. R. MARGOLIS: *Modules over the Steenrod Algebra I* (unpublished).
4. H. R. MARGOLIS: *Coalgebras over the Steenrod Algebra* (to appear).
5. J. P. MAY: Ph.D. Thesis, Princeton University (1964).
6. J. MILNOR: The Steenrod Algebra and its Dual, *Ann Math.* **67** (1958).
7. J. MILNOR: The Cobordism Ring and a Complex Analogue, *Am. J. Math.* **87** (1960).
8. J. MILNOR and J. MOORE: On the Structure of Hopf Algebras, *Ann. Math.* **81** (1965).
9. N. STEENROD and D. B. A. EPSTEIN: Cohomology Operations, *Ann. Math. Stud.* No. 50

The University
Manchester

Proc. Camb. Phil. Soc. (1974), **76**, 45
PCPS 76–5

Sub-Hopf-algebras of the Steenrod algebra

By J. F. ADAMS* and H. R. MARGOLIS†‡

* *Cambridge University, Cambridge, England*
† *Boston College, Chestnut Hill, Massachusetts*

(*Received* 24 *August* 1973)

1. *Introduction.* In (3) the second author has shown that all sub-Hopf-algebras of the mod 2 Steenrod algebra have a certain form; this result has been used in (1) and (2). The analogous result for the mod p Steenrod algebra, where p is an odd prime, is contained in (6). The object of this paper is to give a shorter proof of a slightly sharper result; we construct all the subalgebras in question. In the mod 2 case this sharper result has also been obtained by Anderson and Davis (2).

In most of what follows, we will give the details for the case in which p is an odd prime; we will merely indicate the modifications necessary for the simpler case $p = 2$.

The rest of this paper is arranged as follows. In section 2 we state our main result. In section 3 we give a general lemma about sub-Hopf-algebras of an extension, and indicate how we propose to apply it to the Steenrod algebra. The proof of the main result is carried out in sections 4, 5.

2. *Notation and results.* We write A^* for the mod p Steenrod algebra and A for its dual. According to Milnor (4), if p is odd, then A is the tensor product of a polynomial algebra on generators ξ_i of degree $2(p^i-1)$, $i \geqslant 1$, and an exterior algebra on generators τ_j of degree $2p^j-1$, $j \geqslant 0$. The coproduct is given by

$$\psi\xi_n = \sum_{i+j=n} \xi_i^{p^j} \otimes \xi_j,$$
$$\psi\tau_n = \tau_n \otimes 1 + \sum_{i+j=n} \xi_i^{p^j} \otimes \tau_j.$$

(Here ξ_0 is interpreted as 1.) In the case $p = 2$, A is a polynomial algebra on generators ξ_i of degree 2^i-1, and the coproduct is given by

$$\psi\xi_n = \sum_{i+j=n} \xi_i^{2^j} \otimes \xi_j.$$

We proposed to construct all the sub-Hopf-algebras of A^*, and we begin by constructing certain quotient algebras of A. Let h be a function from the set $\{1, 2, 3, \ldots\}$ to the set $\{0, 1, 2, \ldots, \infty\}$; let k be a function from the set $\{0, 1, 2, \ldots\}$ to the set $\{1, 2\}$. Then we construct the algebra $B(h, k)$ as a quotient of A by imposing the relations

$$\xi^{p^{h(i)}} = 0 \quad (i = 1, 2, 3, \ldots),$$
$$\tau^{k(j)} = 0 \quad (j = 0, 1, 2, \ldots).$$

(If for some i we have $h(i) = \infty$, then we impose no relation on ξ_i.) The dual of $B(h, k)$ is a subcoalgebra $B^*(h, k)$ of A^*.

‡ The second author was partially supported by NSF grant no. GP 28565.

If $p = 2$ we need only one function $h:\{1, 2, 3, \ldots\} \to \{0, 1, 2, \ldots, \infty\}$; a similar construction gives the quotient algebra $B(h)$ of A and the subcoalgebra $B^*(h)$ of A^*.

We can make $B^*(h, k)$ a subalgebra of A^* by imposing conditions on the functions h and k.

Condition 2·1. For all $i, j \geqslant 1$ either $h(i) \leqslant j + h(i+j)$ or $h(j) \leqslant h(i+j)$.

Condition 2·2. For all $i \geqslant 1, j \geqslant 0$ such that $k(i+j) = 1$, either $h(i) \leqslant j$ or $k(j) = 1$.

If $p = 2$ we impose merely Condition 2·1.

PROPOSITION 2·3. (*a*) *If p is odd and the functions h, k satisfy* (2·1), (2·2) *then $B(h, k)$ inherits from A the structure of a Hopf algebra, and $B^*(h, k)$ is a sub-Hopf-algebra of A^*.*

(*b*) *If $p = 2$ and h satisfies* (2·1), *then the corresponding conclusions hold for $B(h)$ and $B^*(h)$.*

THEOREM 2·4. (*a*) *If p is odd every sub-Hopf-algebra of the mod p Steenrod algebra A^* is of the form $B^*(h, k)$ for some h and k satisfying* (2·1) *and* (2·2).

(*b*) *If $p = 2$ every sub-Hopf-algebra of the mod 2 Steenrod algebra A^* is of the form $B^*(h)$ for some h satisfying* (2·1).

The proof of Proposition 2·3 is easy. We must check that the coproduct $\psi: A \to A \otimes A$ passes to the quotient to give a map $\psi: B \to B \otimes B$, where $B = B(h, k)$. We have

$$\psi \xi_n^{p^{h(n)}} = \sum_{i+j=n} \xi_i^{p^{j+h(n)}} \otimes \xi_j^{p^{h(n)}}.$$

Assuming Condition 2·1, either

$$\xi_i^{p^{j+h(n)}} = 0 \quad \text{in } B,$$

or

$$\xi_j^{p^{h(n)}} = 0 \quad \text{in } B.$$

So $\psi \xi_n^{p^{h(n)}}$ gives the zero element of $B \otimes B$. Similarly, we have

$$\psi \tau_n = \tau_n \otimes 1 + \sum_{i+j=n} \xi_i^{p^j} \otimes \tau_j.$$

If we assume $k(n) = 1$ and Condition 2·2, then either

$$\xi_i^{p^j} = 0 \quad \text{in } B,$$

or

$$\tau_j = 0 \quad \text{in } B.$$

So $\psi \tau_n$ gives the zero element of $B \otimes B$. Thus ψ passes to the quotient as required. The statement about $B^*(h, k)$ follows by duality. The case $p = 2$ is similar.

This completes the proof of Proposition 2·3.

3. *Subalgebras of extensions*. Before we start on the main proof, it seems best to isolate a general result which allows us to deal with sub-Hopf-algebras of extensions. All our Hopf algebras will be connected and locally finite. We refer to (5) for all basic material on Hopf algebras.

Let

$$0 \to A \xrightarrow{i} B \xrightarrow{j} C \to 0$$

be a diagram in the category of Hopf algebras. We call it an *extension* if i is the injection of a normal subalgebra, and j is the projection of A on the quotient algebra $C = B//A$. The dual of such an extension is again an extension, say

$$0 \leftarrow A^* \overset{i^*}{\leftarrow} B^* \overset{j^*}{\leftarrow} C^* \leftarrow 0.$$

PROPOSITION 3·1. (*a*) *Suppose given an extension*

$$0 \rightarrow A \overset{i}{\rightarrow} B \overset{j}{\rightarrow} C \rightarrow 0$$

such that A is central in B, and an epimorphism $\pi: B \rightarrow E$. Then we can complete the diagram

$$\begin{array}{ccccccccc} 0 \rightarrow & A & \overset{i}{\rightarrow} & B & \overset{j}{\rightarrow} & C & \rightarrow 0 \\ & \downarrow & & \downarrow \pi & & \downarrow & \\ 0 \rightarrow & D & \leftarrow & E & \rightarrow & F & \rightarrow 0 \end{array}$$

so that the lower row is an extension and all the vertical arrows are epimorphisms.

(*b*) *Suppose given an extension*

$$0 \leftarrow A^* \overset{i^*}{\leftarrow} B^* \overset{j^*}{\leftarrow} C^* \leftarrow 0$$

and a monomorphism $\pi^*: E^* \rightarrow B^*$. Suppose also that for the dual algebras A is central in B. Then we can complete the diagram

$$\begin{array}{ccccccccc} 0 \leftarrow & A^* & \overset{i^*}{\leftarrow} & B^* & \overset{j^*}{\leftarrow} & C^* & \leftarrow 0 \\ & \uparrow & & \uparrow \pi^* & & \uparrow & \\ 0 \leftarrow & D^* & \leftarrow & E^* & \leftarrow & F^* & \leftarrow 0 \end{array}$$

so that the lower row is an extension and all the vertical arrows are monomorphisms.

Proof. (*a*) Set $D = \operatorname{Im}(\pi i)$. Since A is central in B, D is central in E. Set $F = E//D$. Then π passes to the quotient to give the required epimorphism $C \rightarrow F$.

(*b*) Immediate from (*a*) by duality.

In order to apply this lemma, we build up the Steenrod algebra A^* by an infinite sequence of extensions. Let P_n be the polynomial subalgebra of A generated by $\xi_1, \xi_2, \ldots, \xi_n$; for $n = \infty$ we interpret P_∞ as the polynomial subalgebra of A generated by $\xi_1, \xi_2, \ldots$. If p is odd, let R_n be the subalgebra of A generated by $\xi_1, \xi_2, \ldots$ and $\tau_0, \tau_1, \ldots, \tau_n$. For $n = -1$ we interpret R_{-1} as P_∞ and for $n = \infty$ we interpret R_∞ as A. For $p = 2$ we interpret P_∞ as A and we do not need any algebras R_n.

We aim to prove by induction over n the results about P_n and R_n which correspond to Theorem 2·4 for A. In order to state these results, we proceed exactly as in section 2. First, let h be a function from the set $\{1, 2, 3, \ldots, n\}$ to the set $\{0, 1, 2, \ldots, \infty\}$; then we can construct the algebra $C_n(h)$ as a quotient of P_n by imposing the relations

$$\xi_i^{p^{h(i)}} = 0 \quad (i = 1, 2, \ldots, n).$$

Secondly, if p is odd, let h be a function from the set $\{1, 2, 3, \ldots\}$ to the set $\{0, 1, 2, \ldots, \infty\}$ as in section 2, and let k be a function from the set $\{0, 1, 2, \ldots, n\}$ to the set $\{1, 2\}$; then we can construct the algebra $D_n(h, k)$ as a quotient of R_n by imposing the relations

$$\xi_i^{p^{h(i)}} = 0 \quad (i = 1, 2, \ldots),$$

$$\tau_j^{k(j)} = 0 \quad (j = 0, 1, 2, \ldots, n).$$

If $p = 2$ we omit this step.

The duals of $C_n(h)$ and $D_n(h,k)$ are subcoalgebras $C_n^*(h)$ and $D_n^*(h,k)$ of P_n^* and R_n^*. We can make $C_n^*(h)$ or $D_n^*(h,k)$ a subalgebra of P_n^* or R_n^* by imposing suitable conditions on h and k; that is, if we are dealing with $C_n(h)$ we impose on h the condition (2·1) for $i+j \leqslant n$; if we are dealing with $D_n(h,k)$ we impose on h the condition (2·1), and on k the condition (2·2) for $i+j \leqslant n$.

PROPOSITION 3·2. (*a*) *If the function h satisfies* (2·1) *for* $i+j \leqslant n$, *then* $C_n^*(h)$ *is a sub-Hopf-algebra of* P_n^*.

(*b*) *If p is odd, if the function h satisfies* (2·1), *and if the function k satisfies* (2·2) *for* $i+j \leqslant n$, *then* $D_n^*(h,k)$ *is a sub-Hopf-algebra of* R_n^*.

The proof is exactly the same as the proof of Proposition 2·3.

THEOREM 3·3. (*a*) *Every sub-Hopf-algebra of* P_n^* *is of the form* $C_n^*(h)$, *where h satisfies* (2·1) *for* $i+j \leqslant n$.

(*b*) *If p is odd, every sub-Hopf-algebra of* R_n^* *is of the form* $D_n^*(h,k)$, *where h satisfies* (2·1) *and k satisfies* (2·2) *for* $i+j \leqslant n$.

THEOREM 2·4(*a*) is to be regarded as the case $n = \infty$ of Theorem 3·3(*b*); Theorem 2·4(*b*) is to be regarded as the case $p = 2$, $n = \infty$ of Theorem 3·3(*a*).

4. *Proof of Theorem* 3·3(*a*). Before proceeding, we need further information about the Hopf algebra $C_n^*(h)$. In order to state it, recall that P_n has a base consisting of the monomials

$$\xi_1^{r_1}\xi_2^{r_2}\dots\xi_n^{r_n};$$

and P_n^* has a dual base, whose elements may be written $P(r_1, r_2, \dots, r_n)$. In particular, we write P_t^s for the element of the dual base corresponding to $\xi_t^{p^s}$. Similarly, $C_n(h)$ has a base consisting of the monomials

$$\xi_1^{r_1}\xi_2^{r_2}\dots\xi_n^{r_n}$$

with $0 \leqslant r_1 < p^{h(i)}$ for each i; all other monomials are zero in $C_n(h)$. Therefore $C_n^*(h)$ has a base consisting of the elements $P(r_i, r_2, \dots, r_n)$ with $0 \leqslant r_1 < p^{h(i)}$ for each i. In particular, P_t^s lies in $C_n^*(h)$ if $t \leqslant n$ and $s < h(t)$.

LEMMA 4·1. (*a*) $C_n^*(h)$ *is generated as an algebra by the elements* P_t^s *with* $t \leqslant n$, $s < h(t)$.

(*b*) *A primitive element in* $C_n(h)$ *must be zero unless its degree is of the form* $2p^s(p^t-1)$ *with* $t \leqslant n$, $s < h(t)$.

Proof. Order the sequences $(r_1, r_2, \dots, r_n)$ lexicographically from the left. By (4) we have

$$P(r_1, r_2, \dots, r_n)\,P(s_1, s_2, \dots, s_n) = \lambda P(r_1+s_1, r_2+s_2, \dots, r_n+s_n) \text{ modulo lower terms,}$$

where

$$\lambda = \prod_i \frac{(r_i+s_i)!}{(r_i)!\,(s_i)!}.$$

A simple induction now proves part (*a*). Part (*b*) follows by duality, because the primitive subspace of $C_n(h)$ is dual to the indecomposable quotient of $C_n^*(h)$.

The proof of Theorem 3·3(*a*) for $n < \infty$ is now by induction over n. For $n = 0$ the result is trivially true (if suitably interpreted), so we assume it true for $n-1$.

We have dual extensions

$$0 \to P_{n-1} \to P_n \to Z_p[\xi_n] \to 0$$
$$0 \leftarrow P^*_{n-1} \leftarrow P^*_n \leftarrow Z_p[\xi_n]^* \leftarrow 0.$$

Let C^* be an arbitrary sub-Hopf-algebra of P^*_n. Then Proposition 3·1 applies, and gives us the following diagrams of extensions.

$$\begin{array}{ccccccccc} 0 \to & P_{n-1} & \to & P_n & \to & Z_p[\xi_n] & \to 0 \\ & \downarrow & & \downarrow & & \downarrow & \\ 0 \to & B & \to & C & \to & D & \to 0 \end{array}$$

$$\begin{array}{ccccccccc} 0 \leftarrow & P^*_{n-1} & \leftarrow & P^*_n & \leftarrow & Z_p[\xi_n]^* & \leftarrow 0 \\ & \uparrow & & \uparrow & & \uparrow & \\ 0 \leftarrow & B^* & \leftarrow & C^* & \leftarrow & D^* & \leftarrow 0 \end{array}$$

By the inductive hypothesis, $B = C_{n-1}(h)$ and $B^* = C^*_{n-1}(h)$, where h is a function defined on $\{1, 2, \ldots, n-1\}$ and h satisfies (2·1) for $i+j \leqslant n-1$. The algebra D is generated by one element ξ_n which is primitive in D. By a classical theorem of Borel ((5), Proposition 7·8), D is either a truncated algebra given by one relation

$$\xi_n^{p^m} = 0,$$

or else the polynomial algebra $Z_p[\xi_n]$ (in which case we interpret m as ∞). We extend h to a function defined on $\{1, 2, \ldots, n\}$ by setting $h(n) = m$. We will show that h satisfies (2·1) for $i+j \leqslant n$ and that $C^* = C^*_n(h)$.

By a standard result ((5), Theorem 4·4), C is free over B with a base consisting of the ξ_n^i for $0 \leqslant i < p^m$. If $m = h(n) = \infty$, then this already shows that $C = C_n(h)$, and (2·1) is trivially satisfied for $i+j = n$. So we assume $m = h(n) < \infty$. Then the structure of C is entirely given by the relation

$$\xi_n^{p^m} = \sum_i b_i \xi_n^i \quad \text{in} \quad C, \tag{4·2}$$

where $b_i \in B$ and i runs over the range $0 \leqslant i < p^m$.

LEMMA 4·3. *In degrees less than* $2p^m(p^n - 1)$, C^* *has a base consisting of the elements* $P(r_1, r_2, \ldots, r_n)$ *such that* $0 \leqslant r_i < p^{h(i)}$ *for each* i.

This follows immediately from the remarks above.

LEMMA 4·4. *In* (4·2), *we have* $b_i = 0$ *for* $i > 0$.

Proof. The proof is by downward induction over i. Suppose that $b_i = 0$ for $i > d$, where $0 < d < p^m$. Let $\alpha \in C^*$ be any element of the same degree as b_d, and let

$$\beta = P(0, 0, \ldots, 0, d)$$

(where the d is in the nth place). Then β lies in C^* by (4·3); so $\alpha\beta$ lies in C^*. We have

$$\langle \alpha\beta, \ \xi_n^{p^m} \rangle = 0,$$

and

$$\langle \alpha\beta, \ b_d \xi_d^n \rangle = \langle \alpha, b_d \rangle,$$

and

$$\langle \alpha\beta, \ b_i \xi_n^i \rangle = 0 \quad \text{if} \quad i < d.$$

Using (4·2), we see that $\langle \alpha, b_d \rangle = 0$ for all α; this shows that $b_d = 0$. This completes the induction, and proves Lemma 4·4.

LEMMA 4·5. *h satisfies Condition* (2·1) *for* $i+j = n$.

Proof. If not then there exist integers $i \geq 1, j \geq 1$ such that $i+j = n, h(i) > j+h(n)$ and $h(j) > h(n)$. By Lemma 4·3, the elements

$$\alpha = P_i^{j+m} \quad \text{and} \quad \beta = P_j^m$$

lie in C^*. Then the commutator $\alpha\beta - \beta\alpha$ lies in C^*, and

$$\langle \alpha\beta - \beta\alpha, \quad \xi_n^{p^m} \rangle = 1$$

$$\langle \alpha\beta - \beta\alpha, \quad b_0 \rangle = 0.$$

This contradicts (4·2) and (4·4) and thus proves Lemma 4·5.

LEMMA 4·6. *In* (4·2) *we have* $b_0 = 0$.

Proof. It follows from Lemma 4·5 that $\xi_n^{p^m}$ is a primitive element in C. Thus b_0 is a primitive element of degree $2p^m(p^n - 1)$ in $B = C_{n-1}(h)$. But now $b_0 = 0$ by Lemma 4·1 (b).

This completes the inductive step in the proof of Theorem 3·3 (a), and proves that theorem for $n < \infty$.

It remains to deal with the case $n = \infty$. Let C^* be a sub-Hopf-algebra of P_∞^*. Then for $n < \infty$ the image of P_n in C is a quotient Hopf algebra of P_n, which by the work just done has the form $C_n(h_n)$ for some function h_n defined on $\{1, 2, \ldots, n\}$ and satisfying (2·1) for $i+j \leq n$. Clearly the restriction of h_{n+1} to $\{1, 2, \ldots, n\}$ must coincide with h_n; therefore the functions h_n give a function h_∞ defined on $\{1, 2, 3, \ldots\}$, and satisfying (2·1). It is now clear that $C = C_\infty(h_\infty)$. This completes the proof of Theorem 3·3 (a).

5. *Proof of Theorem* 3·3 (b). We proceed exactly as in section 4. First we need further information about the Hopf algebras $D_n^*(h, k)$. Recall that R_n has a base consisting of the monomials

$$\tau_0^{s_0}\tau_1^{s_1} \ldots \tau_n^{s_n} \xi_1^{r_1} \xi_2^{r_2} \ldots;$$

and R_n^* has a dual base, whose elements may be written $P(s_0, s_1, \ldots, s_n; r_1, r_2, \ldots)$. In particular, we write P_t^s for the element of the dual base corresponding to $\xi_t^{p^s}$, and Q_i for the element of the dual base corresponding to τ_i. Similarly, $D_n(h, k)$ has a base consisting of the monomials

$$\tau_0^{s_0}\tau_1^{s_1} \ldots \tau_n^{s_n} \xi_1^{r_1} \xi_2^{r_2} \ldots$$

with $0 \leq r_i < p^{h(i)}$, $0 \leq s_j < k(j)$ for each i, j; all other monomials are zero in $D_n(h, k)$. Therefore $D_n^*(h, k)$ has a base consisting of the elements $P(s_0, s_1, \ldots, s_n; r_1, r_2, \ldots)$ with $0 \leq r_i < p^{h(i)}$, $0 \leq s_j < k(j)$ for each i, j. In particular, P_t^s lies in $D_n^*(h, k)$ if $s < h(t)$; Q_j lies in $D_n^*(h, k)$ if $j \leq n, k(j) = 2$.

LEMMA 5·1 (a). *$D_n^*(h, k)$ is generated as an algebra by the elements P_t^s with $s < h(t)$ and the elements Q_j with $j \leq n$, $k(j) = 2$.*

(b) *A primitive element in $D_n(h, k)$ is zero unless its degree is of the form $2p^s(p^t - 1)$ with $s < h(t)$ or $2p^j - 1$ with $j \leq n, k(j) = 2$.*

Part (a) is comparable with Lemma 4·1 (a), and part (b) follows by duality.

We now prove Theorem 3·3 (b) for finite n by induction over n. For $n = -1$ the result is true by the case $n = \infty$ of Theorem 3·3 (a); so we assume the result true for $n - 1$.

We have dual extensions

$$0 \to R_{n-1} \to R_n \to \Lambda[\tau_n] \to 0$$
$$0 \leftarrow R^*_{n-1} \leftarrow R^*_n \leftarrow \Lambda[\tau_n]^* \leftarrow 0.$$

(Here $\Lambda[\tau_n]$ means the exterior algebra on one generator τ_n.) Let D^* be an arbitrary sub-Hopf-algebra of R^*_n. Then Proposition 3·1 applies, and gives us the following diagrams of extensions.

$$\begin{array}{ccccccccc} 0 & \to & R_{n-1} & \to & R_n & \to & \Lambda[\tau_n] & \to & 0 \\ & & \downarrow & & \downarrow & & \downarrow & & \\ 0 & \to & C & \to & D & \to & E & \to & 0 \end{array}$$

$$\begin{array}{ccccccccc} 0 & \leftarrow & R^*_{n-1} & \leftarrow & R^*_n & \leftarrow & \Lambda[\tau_n]^* & \leftarrow & 0 \\ & & \uparrow & & \uparrow & & \uparrow & & \\ 0 & \leftarrow & C^* & \leftarrow & D^* & \leftarrow & E^* & \leftarrow & 0 \end{array}$$

By the inductive hypothesis, $C = D_{n-1}(h, k)$ and $C^* = D^*_{n-1}(h, k)$, where k is defined on $\{0, 1, 2, \ldots, n-1\}$ and h, k satisfy (2·1), (2·2) so far as they apply. The quotient E of $\Lambda[\tau_n]$ is generated by one generator τ_n which is primitive in E; its structure must be given by one relation

$$\tau_n^m = 0 \quad \text{in} \quad E$$

where $m = 1$ or 2. We extend k to a function defined on $\{0, 1, 2, \ldots, n\}$ by setting $k(n) = m$. As before, D is free over C with a base consisting of the powers τ_n^j for $0 \leqslant j < m$. If $m = k(n) = 2$, then this already shows that $D = D_n(h, k)$, and (2·2) is trivially satisfied for $i + j = n$. So we assume $m = k(n) = 1$. Then the structure of D is entirely given by one relation

$$\tau_n = c_0 \quad \text{in} \quad D, \tag{5·2}$$

where $c_0 \in C$.

Lemma 5·3. *In degress less than* $2p^n - 1$, D^* *has a base consisting of the elements* $P(s_0, s_1, \ldots, s_n; r_1, r_2, \ldots)$ *such that* $0 \leqslant r_i < p^{h(i)}$ *and* $0 \leqslant s_j < k(j)$ *for each* i, j.

This follows immediately from the remarks above.

The analogue of Lemma 4·4 is vacuous.

Lemma 5·5. *k satisfies Condition* (2·2) *for* $i + j = n$.

Proof. If not then there exist integers $i \geqslant 1, j \geqslant 0$ such that $i + j = n$, $h(i) > j$ and $k(j) = 2$. By Lemma 5·3, the elements

$$\alpha = P_i^j \quad \text{and} \quad \beta = Q_j$$

lie in D^*. Then the commutator $\alpha\beta - \beta\alpha$ lies in D^*, and

$$\langle \alpha\beta - \beta\alpha, \quad \tau_n \rangle = 1$$
$$\langle \alpha\beta - \beta\alpha, \quad c_0 \rangle = 0.$$

This contradiction proves Lemma 5·5.

Lemma 5·6. *In* (5·2) *we have* $c_0 = 0$.

Proof. It follows from Lemma 5·5 that τ_n is a primitive element in D. Thus c_0 is a primitive element of degree $2p^n - 1$ in $C = D_{n-1}(h, k)$. But now $c_0 = 0$ by Lemma 5·1 (*b*).

This completes the inductive step in the proof of Theorem 3·3 (*b*), and proves that theorem for $n < \infty$. The step to the case $n = \infty$ is done exactly as in section 4. This proves all the results claimed.

REFERENCES

(1) ADAMS, J. F. and MARGOLIS, H. R. Modules over the Steenrod Algebra, *Topology* **10** (1971), 271–282.

(2) ANDERSON, D. W. and DAVIS, D. M. A Vanishing Theorem in Homological Algebra, to appear.

(3) MARGOLIS, H. R. Coalgebras over the Steenrod Algebra, to appear.

(4) MILNOR, J. The Steenrod Algebra and its Dual. *Ann. of Math.* (2) **67** (1958), 150–171.

(5) MILNOR, J. and MOORE, J. C. On the structure of Hopf Algebras. *Ann. of Math.* (2) **81** (1965), 211–264.

(6) ROSEN, S. S. On Torsion in Connective Complex Cobordism. Thesis, Northwestern University, 1972.

II. What we don't know about RP^{∞}

Here I would like to advertise an unsolved problem, in the hope of throwing it open to wider participation.

When we consider the iterated suspension homomorphism in unstable homotopy groups of spheres we need information about the homotopy groups of truncated real projective spaces RP^{n+r}/RP^n, and in particular about the stable homotopy groups of these spaces. Now, for r fixed the stable homotopy type of RP^{n+r}/RP^n is a periodic function of n; this allows us to speak of the stable homotopy type of RP^{n+r}/RP^n even when n is negative, interpreting it by periodicity; that is, if n is negative we interpret the stable homotopy type of RP^{n+r}/RP^n as that of RP^{n+r+2^m}/RP^{n+2^m} (but shifted down by 2^m dimensions), where m is appropriately large. This periodicity allows us to speak in certain ways which are pleasantly simple and dramatic. For example, it is more memorable to talk about dimension -1, rather than about a positive dimension congruent to -1 modulo 2^m for m appropriately large.

If we filter the spaces RP^{n+r}/RP^n by their subspaces RP^{n+s}/RP^n, and apply stable homotopy, we obtain a spectral sequence. The E^2 term of this spectral sequence has the following form:

$$E^2_{p,q} = \begin{cases} Z_2 \otimes \pi^S_q(S^0) & (p \text{ odd}) \\ \mathrm{Tor}^Z_1(Z_2, \pi^S_q(S^0)) & (p \text{ even}). \end{cases}$$

Here $\pi^S_q(S^0)$ is the qth stable homotopy group of the sphere. I emphasise that this equation is intended to be equally valid for $p > 0$ and for $p \leq 0$. For fixed finite q and r, $E^r_{p,q}$ is periodic in p with period 2^m for some m depending on q and r. The groups $E^\infty_{p,q}$ may be defined, but they are not periodic.

There is some evidence for the following conjecture.

Conjecture (after Mahowald). This spectral sequence converges to $\pi^S_*(S^{-1})$.

The sphere S^{-1} of stable dimension -1 appears because we have

$$RP^{-1}/RP^{-n} \simeq S^{-1} \vee (RP^{-2}/RP^{-n}).$$

This equation, of course, has to be interpreted as a statement about the stable homotopy type of RP^{2^m-1}/RP^{2^m-n} for m sufficiently large, as explained above.

It is tempting to talk about this conjecture in picturesque language, and speak as if there were a spectrum which is like the suspension spectrum of RP^∞, but has one cell in each dimension p whether p is positive, negative or zero. Just as the cohomology ring $H^*(RP^\infty; Z_2)$ is the polynomial ring $Z_2[x]$, where $x \in H^1(RP^\infty; Z_2)$, so the cohomology of this hypothetical spectrum should be the ring of finite Laurent series $Z_2[x, x^{-1}]$. It is now obvious what the cohomology of this hypothetical spectrum should be as a module over the Steenrod algebra. In fact, for $n \geq 0$ we have

$$Sq^i x^n = c(n,i)\, x^{n+i},$$

where the coefficient $c(n, i) \in Z_2$ is a periodic fraction of n; we define the coefficient $c(n, i)$ for negative values of n by periodicity and define $Sq^i x^n$ accordingly. This construction does make $Z_2[x, x^{-1}]$ into a module over the mod 2 Steenrod algebra; in fact, the Adem relations are satisfied, because it is sufficient to observe that they hold on x^n when n is positive. One can now make simple calculations. For example, let $Sq = \sum_{i\geq 0} Sq^i$; then $Sq\, x = x(1+x)$ and Sq is multiplicative, so $Sq\, x^{-1} = x^{-1}(1+x)^{-1} = x^{-1} + 1 + x + x^2 \ldots$, that is $Sq^i x^{-1} = x^{i-1}$.

It is necessary to point out firmly that there is no spectrum, in Boardman's category or in any other sensible category, whose cohomology is this A-module. In fact, if there were, then the generator for the 0-dimensional homology group would come from a finite subspectrum, say X. Then we could choose i so large that x^{-i} would restrict to zero in $H^{-i}(X; Z_2)$. But then $(\chi\, Sq^i)x^{-i} = 1$ would restrict to zero in $H^0(X; Z_2)$, a contradiction.

The 'hypothetical spectrum' is therefore a mythical beast. The statements we want to make do not refer to it: they are all to be interpreted in terms of finite complexes RP^{n+r}/RP^n and limits.

One line of thought which might lead one towards the conjecture stated is the consideration of

$$\mathrm{Ext}_A^{**}(Z_2[x, x^{-1}], Z_2).$$

First we consider Ext_{A_n}, where A_n is the subalgebra of A generated by $Sq^1, Sq^2, \ldots, Sq^{2^n}$. As a module over A_n, $Z_2[x, x^{-1}]$ is generated by the powers x^i with $i \equiv -1 \bmod 2^{n+1}$. We may filter $Z_2[x, x^{-1}]$ by the A_n-submodule generated by these generators x^i with $i = 2^{n+1}r - 1$, $r \leq p$. The successive subquotients have the form $A_n \otimes_{A_{n-1}} Z_2$, and their Ext groups are given by

$$\mathrm{Ext}_{A_n}^{**}(A_n \otimes_{A_{n-1}} Z_2, Z_2) \cong \mathrm{Ext}_{A_{n-1}}^{**}(Z_2, Z_2).$$

If we have to conjecture the structure of

$$\mathrm{Ext}_{A_n}^{**}(Z_2[x, x^{-1}], Z_2),$$

the simplest conjecture is

$$\sum_{r \in Z} \operatorname{Ext}^{**}_{A_{n-1}} (Z_2, Z_2),$$

on generators of dimension $2^{n+1}r - 1$. This conjecture is true for $n = 2, 3$. We then have

$$\operatorname{Ext}^{**}_{A} (Z_2[x, x^{-1}], Z_2) = \lim_{\substack{\leftarrow \\ n}} \operatorname{Ext}^{**}_{A_n} (Z_2[x, x^{-1}], Z_2);$$

if we make the simplest conjecture for the maps of the inverse system, the inverse limit would be $\operatorname{Ext}^{**}_{A} (Z_2, Z_2)$, on one generator of dimension -1..

Alternatively, we might proceed as follows. Let M be the sub-A-module of $Z_2[x, x^{-1}]$ generated by the x^i with $i \neq -1$. If M were connected, we would count its generators by computing $Z_2 \otimes_A M = 0$. That is, M is a module 'with no generators'. Similarly, $\operatorname{Tor}_1^A(Z_2, M) = 0$; that is, M has 'no relations' between its 'no generators'. If M were connected it would follow that $\operatorname{Ext}^{**}_{A} (M, Z_2) = 0$; as M is not connected this does not follow, but we might still conjecture it.

Unfortunately, such statements are unlikely to lead anywhere, because even if there is an Adams spectral sequence starting from $\operatorname{Ext}^{**}_{A}(Z_2[x, x^{-1}], Z_2)$, it is likely to be very hard to prove anything useful about its convergence. We therefore turn to other evidence. In what follows, the groups $E^{\infty}_{p,q}$ are those of the spectral sequence in the conjecture.

Proposition 1. <u>For</u> $q = 0$, <u>the groups</u> $E^{\infty}_{p,0}$ <u>are zero except for</u> $p = -1$; $E^{\infty}_{-1,0} = Z_2$.

We can even state the differentials which lead to this state of affairs. Let us write i_n for the homology generator in dimension n (with coefficients Z or Z_2 as may be needed). Then we have

$$d_2 i_p = \eta i_{p-2} \quad \text{for } p \equiv 1 \mod 4$$
$$d_4 i_p = \nu i_{p-4} \quad \text{for } p \equiv 3 \mod 8$$
$$d_8 i_p = \sigma i_{p-8} \quad \text{for } p \equiv 7 \mod 16$$
$$d_9 i_p = \sigma\eta i_{p-9} \quad \text{for } p \equiv 15 \mod 32$$

$$d_{10}i_p = \sigma\eta^2 i_{p-10} \quad \text{for } p \equiv 31 \mod 64$$

$$d_{12}i_p = \zeta i_{p-12} \quad \text{for } p \equiv 63 \mod 128$$

$$d_{16}i_p = \rho i_{p-16} \quad \text{for } p \equiv 127 \mod 256 \quad \text{etc.}$$

Proposition 2. For $q = 1$, the groups $E^{\infty}_{p,1}$ are zero except for $p = -2$; $E^{\infty}_{-2,1} = Z_2$.

The group for $p = -2$ arises as follows. The complex RP^0/RP^{-3} has the form $(S^{-2} \vee S^{-1}) \cup e^0$, where the components of the attaching map for e^0 are η and 2. So in RP^0/RP^{-n}, the element $2 \in \pi^S_{-1}(S^{-1})$ compresses to RP^{-2}/RP^{-n}, and gives a generator for the group $E^{\infty}_{-2,1}$.

For some distance we can inspect the differentials which lead to this state of affairs. For $p \equiv 3 \mod 4$, $E^2_{p,1}$ consists of boundaries. We have

$$d_2\eta i_p = \eta^2 i_{p-2} \quad \text{for } p \equiv 0, 1 \mod 4$$

$$d_6\eta i_p = \nu^2 i_{p-6} \quad \text{for } p \equiv 2 \mod 8.$$

At this point one guesses that $d_{14}\eta i_p = \sigma^2 i_{p-14}$ for $p \equiv 6 \mod 16$, but this is wrong. We have

$$d_8\eta i_p = \bar{\nu} i_{p-8} \quad \text{for } p \equiv 6 \mod 16$$

$$d_9\eta i_p = \sigma\eta^2 i_{p-9} \quad \text{for } p \equiv 14 \mod 32.$$

It may be as well to point out the first place where we get an element of $\pi_{i-1}(S^{-1})$ with $i > 0$.

Proposition 3. For $q = 3$ and p odd, the groups $E^{\infty}_{p,3}$ are zero except for $p = -3$; $E^{\infty}_{-3,3} = Z_2$; the element $\eta \in \pi_0(S^{-1})$ compresses to RP^{-3}/RP^{-n} and gives a generator for $E^{\infty}_{-3,3}$.

We would be on firmer ground, of course, if we could prove that $E^{\infty}_{p,q} = 0$ for $p + q < -1$ - for the conjecture implies this. Applying S-duality to the finite complexes and then passing to the limit, we would like to prove something like this:

Conjecture. The group of maps in Boardman's category from the suspension spectrum of RP^∞ to the suspension spectrum of S^n is zero if $n > 0$.

This would seem to be a stable analogue of one of the following two related conjectures, which are due to Sullivan. For the first, let X be a finite simplicial complex on which Z_2 acts simplicially; let Z_2 act on S^∞ by the antipodal map.

Conjecture. The evident map, from the fixed-point set of Z_2 in X, to the function space of equivariant maps from S^∞ to X, induces an isomorphism of mod 2 cohomology.

Conjecture. If Y is a finite complex, the function-space of base-point-preserving maps from RP^∞ to Y is contractible.

The conjectures as stated seem to be inaccessible. Even if we replace them by suitable stable analogues, there is reason to think that no reformulation will be simultaneously convenient to prove and useful for the present purpose.

Finally, the Kahn-Priddy theorem is relevant to the present conjecture; it proves that for $p = -1$, $E^\infty_{-1,q}$ is zero for $q > 0$. It is just possible that comparable methods, based on the consideration of infinite loop-spaces, would be helpful in studying the present conjecture. The plan would be to take the suspension-spectrum $(RP^N/RP^{-n})/S^{-1}$, replace it by an Ω-spectrum, obtain information on the homology of the loop-spaces in the Ω-spectrum, and pass to limits.

Department of Pure Mathematics and Mathematical Statistics
16 Mill Lane
Cambridge CB2 1SB

Math. Proc. Camb. Phil. Soc. (1980), **87**, 459 459

Calculation of Lin's Ext groups

BY W. H. LIN, D. M. DAVIS, M. E. MAHOWALD AND J. F. ADAMS

(*Received* 29 *October* 1979)

1. The first-named author has proved interesting results about the stable homotopy and cohomotopy of spaces related to real projective space RP^∞; these are presented in an accompanying paper (6). His proof is by the Adams spectral sequence, and so depends on the calculation of certain Ext groups. The object of this paper is to prove the required result about Ext groups. The proof to be given is not Lin's original proof, which involved substantial calculation; it follows an idea of the second and third authors. The version to be given incorporates modifications suggested later by the fourth author.

We begin by introducing the modules whose Ext is needed. Let $P = Z_2[x, x^{-1}]$ be the ring of Laurent polynomials in one variable x of degree 1; we make it into a module over the mod 2 Steenrod algebra A by setting

$$Sq^i x^j = \frac{j(j-1)\dots(j-i+1)}{1.2\dots i} x^{i+j}.$$

The topological background for this may be found in (1) or (6).

Z_2, as a graded module, will mean Z_2 in degree 0; then we have a monomorphism

$$\phi: Z_2 \to P$$

defined by $\phi(\lambda) = \lambda x^0$. The suspension $\Sigma^j M$ of a graded module M is defined by $(\Sigma^j M)^{i+j} = M^i$; then we have an epimorphism

$$\gamma: P \to \Sigma^{-1} Z_2$$

defined by $\gamma(\sum_i \lambda_i x^i) = \lambda_{-1}$.

THEOREM 1·1. (Lin). *The induced maps*

$$\gamma_*: \mathrm{Tor}^A_{s,t}(Z_2, P) \to \mathrm{Tor}^A_{s,t}(Z_2, \Sigma^{-1} Z_2),$$
$$\gamma^*: \mathrm{Ext}^{s,t}_A(\Sigma^{-1} Z_2, Z_2) \to \mathrm{Ext}^{s,t}_A(P, Z_2),$$
$$\phi_*: \mathrm{Ext}^{s,t}_A(Z_2, Z_2) \to \mathrm{Ext}^{s,t}_A(Z_2, P)$$

are all iso.

The result that γ^* is mono was published in (5).

Our proof of Theorem 1·1, which is given in §4, is based on a study of the structure of P. In fact, we will now sketch a series of reductions of (1·1). Let A_r be the subalgebra of A generated by the Sq^{2^i} with $i \leqslant r$; then we wish to compute (for example) $\mathrm{Tor}^A_{s,t}(Z_2, P)$; but it is sufficient to compute $\mathrm{Tor}^{A_r}_{s,t}(Z_2, P)$, because we can pass to a direct limit over r.

LEMMA 1·2. *As a module over A_r, P is generated by the powers x^j with $j \equiv -1 \bmod 2^{r+1}$.*

Proof. If $j \equiv -1 \bmod 2^{r+1}$ and $0 \leqslant i < 2^{r+1}$, then $Sq^i \in A_r$ and $Sq^i x^j = x^{i+j}$.

Let $F_{l,r}$ be the A_r-submodule of P generated by the x^j with $j < l$. By Lemma 1·2 it is sufficient to consider the $F_{l,r}$ with $l \equiv -1 \bmod 2^{r+1}$; but for (1·5) it will be convenient to index the $F_{l,r}$ as we have just done.

We wish to compute (say) $\mathrm{Tor}^A_{s,t}(Z_2, P)$, but it is sufficient to compute

$$\mathrm{Tor}^{A_r}_{s,t}(Z_2, P/F_{l,r}),$$

because we can pass to an (attained) limit as $l \to -\infty$. It is now sufficient to compute $\mathrm{Tor}^A_{s,t}(Z_2, A \otimes_{A_r}(P/F_{l,r}))$, by a change-of-rings theorem.

Lemma 1·3 (Davis & Mahowald). *There is an isomorphism of A-modules*

$$A \otimes_{A_r}(P/F_{l,r}) \cong \bigoplus_j \Sigma^j(A \otimes_{A_{r-1}} Z_2),$$

where j runs over the set $j \equiv -1 \bmod 2^{r+1}$, $j \geqslant l$.

This lemma answers the purpose of computing $\mathrm{Tor}^A_{s,t}(Z_2, A \otimes_{A_r}(P/F_{l,r}))$.

We remark that $P/F_{l,r}$ does not split as a sum of cyclic modules over A_r; it is essential to pass to $A \otimes_{A_r}(P/F_{l,r})$. (For all homological purposes, to look at $A \otimes_{A_r}(P/F_{l,r})$ over A is the same as looking at $P/F_{l,r}$ over A_r; but structure-theory is not part of homology.) Moreover, it is no use to go to a limit over l and try to state a similar result for $A \otimes_{A_r} P$; it is essential to pass to Tor before taking the limit over l.

In order to make good the steps of the reduction, one must consider the behaviour of the isomorphism in (1·3) as l and r vary, so that we can pass to the limits in question; and to prove (1·1), we must consider the relation of the map γ which occurs there to the isomorphism in (1·3). In the former direction, we must consider the following diagrams (1·4), (1·5).

$$(1\cdot4)\qquad \begin{array}{ccc} A\otimes_{A_r}P/F_{l,r} & \overset{\cong}{\longleftrightarrow} & \bigoplus_j \Sigma^j(A\otimes_{A_{r-1}}Z_2) \\ \downarrow & & \downarrow \theta \\ A\otimes_{A_r}P/F_{m,r} & \overset{\cong}{\longleftrightarrow} & \bigoplus_k \Sigma^k(A\otimes_{A_{r-1}}Z_2) \end{array}$$

Here the left-hand vertical arrow is the obvious quotient map, which exists when $l \leqslant m$; and the horizontal arrows are as in Lemma 1·3. The index j runs over the set $j \equiv -1 \bmod 2^{r+1}$, $j \geqslant l$, and the index k runs over the set $k \equiv -1 \bmod 2^{r+1}$, $k \geqslant m$. The map θ has the obvious components, namely the zero map of $\Sigma^j(A\otimes_{A_{r-1}}Z_2)$ if $j < m$, and the identity map

$$\Sigma^j(A \otimes_{A_{r-1}} Z_2) \to \Sigma^k(A \otimes_{A_{r-1}} Z_2) \quad \text{if} \quad j = k \geqslant m.$$

$$(1\cdot5)\qquad \begin{array}{ccc} A\otimes_{A_r}P/F_{l,r} & \overset{\cong}{\longleftrightarrow} & \bigoplus_j \Sigma^j(A\otimes_{A_{r-1}}Z_2) \\ \downarrow & & \downarrow \psi \\ A\otimes_{A_{r+1}}P/F_{l,r+1} & \overset{\cong}{\longleftrightarrow} & \bigoplus_k \Sigma^k(A\otimes_{A_r}Z_2) \end{array}$$

Here the left-hand vertical arrow is the obvious quotient map, and the horizontal arrows are as in Lemma 1·3. The index j runs over the set $j \equiv -1 \bmod 2^{r+1}$, $j \geqslant l$, and the index k runs over the set $k \equiv -1 \bmod 2^{r+2}$, $k \geqslant l$; that is, just half of the values of j correspond to values of k. The map ψ has the obvious components: if

$$j = k \equiv -1 \bmod 2^{r+2}$$

we take the obvious quotient map $\Sigma^j(A \otimes_{A_{r-1}} Z_2) \to \Sigma^k(A \otimes_{A_r} Z_2)$, and if

$$j \equiv 2^{r+1} - 1 \bmod 2^{r+2}$$

we take the zero map of $\Sigma^j(A \otimes_{A_{r-1}} Z_2)$.

LEMMA 1·6 (Davis and Mahowald). *The isomorphisms of Lemma* 1·3 *can be chosen so that Diagrams* 1·4 *and* 1·5 *commute, and so that for* $l \leqslant -1$ *the composite*

$$\Sigma^{-1}(A \otimes_{A_{r-1}} Z_2) \to \bigoplus_j \Sigma^j(A \otimes_{A_{r-1}} Z_2) \cong A \otimes_{A_r} P/F_{l,r} \xrightarrow{1\otimes\gamma} A \otimes_{A_r} \Sigma^{-1} Z_2$$

is the obvious quotient map.

The proof of Lemma 1·3 will be given in § 2, and the proof of Lemma 1·6 will be given in § 3.

2. In this section we will prove Lemma 1·3.

The modules $P/F_{l,r}$ for different values of l become isomorphic after regrading; so it is sufficient to consider one value of l, say $l = -1$. And as we only have to consider one value of r in this section, there is no need to display r either; so for brevity let us write

$$F = F_{-1,r}, \quad F' = F_{2^{r+1}-1,r}, \quad F'' = F_{2\cdot 2^{r+1}-1,r}.$$

LEMMA 2·1. *In P we have* $\mathrm{Sq}^{2^i} x^{-1} \in F = F_{-1,r}$ *if* $i < r$.

Proof. It is sufficient to display the equation

$$\mathrm{Sq}^{2^i} x^{-1} = \mathrm{Sq}^{2^r} x^{2^i - 1 - 2^r}.$$

LEMMA 2·2. *We have the following exact sequence of* A_r*-modules.*

$$0 \to \Sigma^{-1} A_r \otimes_{A_{r-1}} Z_2 \to P/F \to P/F' \to 0.$$

Proof. It is clear that we have an exact sequence

$$0 \to F'/F \to P/F \to P/F' \to 0;$$

moreover, Lemma 2·1 shows that we can define a map

$$\Sigma^{-1} A_r \otimes_{A_{r-1}} Z_2 \to F'/F$$

by sending $a \otimes 1$ to ax^{-1}. This map is onto, by (1.2); to show that it is iso, it is sufficient to show that both sides have rank 2^{r+1} over Z_2. This is known for $\Sigma^{-1} A_r \otimes_{A_{r-1}} Z_2$, and we prove it for F'/F.

In fact, choose a residue class $\rho \bmod 2^{r+1}$. Then the values of j in ρ for which x^j lies in F' form a descending segment, say $(j_0, j_0 - 2^{r+1}, j_0 - 2\cdot 2^{r+1}, \ldots)$; and the values of j in ρ for which x^j lies in F form the subsegment $(j_0 - 2^{r+1}, j_0 - 2\cdot 2^{r+1}, \ldots)$. So F'/F has a Z_2-base consisting of one power x^{j_0} for each residue class $\bmod 2^{r+1}$. This proves Lemma 2·2.

LEMMA 2·3. *We have the following exact sequence of A-modules.*

$$0 \to \Sigma^{-1} A \otimes_{A_{r-1}} Z_2 \xrightarrow{\alpha} A \otimes_{A_r} P/F \to A \otimes_{A_r} P/F' \to 0.$$

Proof. This follows by taking (2·2) and applying the functor $A \otimes_{A_r}$, which preserves exactness since A is free as a right module over A_r.

We will next show that the exact sequence in (2·3) splits. For this purpose we recall Milnor's work on the dual of the Steenrod algebra (7). We write A^* for the dual of A; it is a polynomial algebra $Z_2[\xi_1, \xi_2, \ldots, \xi_k, \ldots]$ on generators ξ_k of degree $2^k - 1$; and we have

$$\psi \xi_k = \sum_{i+j=k} \xi_i^{2^j} \otimes \xi_j,$$

where ξ_0 is interpreted as 1. The dual of the subalgebra A_r is the quotient

$$A_* / (\xi_1^{2^{r+1}}, \xi_2^{2^r}, \ldots, \xi_r^{2^2}, \xi_{r+1}^2, \xi_{r+2}, \xi_{r+3}, \ldots).$$

We now introduce the quotient of A_* by a smaller ideal, namely

$$B_* = A_* / (\xi_2^{2^r}, \ldots, \xi_r^{2^2}, \xi_{r+1}^2, \xi_{r+2}, \xi_{r+3}, \ldots).$$

It is easy to verify that B_* is a left comodule with respect to A_{r*} and a right comodule with respect to A_{r-1*}. Let $B = B^*$ be the dual of B_*; it is a subspace of $A^* = A$; it is a left module over A_r and a right module over A_{r-1}.

LEMMA 2·4. *There is an isomorphism of A_r-modules*

$$\beta \colon \Sigma^{-1} B \otimes_{A_{r-1}} Z_2 \to P/F$$

which sends $b \otimes 1$ to bx^{-1}.

Proof. The prescription $\beta(b \otimes 1) = bx^{-1}$ gives a well-defined map of $\Sigma^{-1} B \otimes_{A_{r-1}} Z_2$, by (2·1); and it is a map of A_r-modules. It is onto, for $\mathrm{Sq}^i \in B$ (each $i \geqslant 0$), $\mathrm{Sq}^i x^{-1} = x^{i-1}$ and the elements x^{i-1} span P/F. In order to prove that β is iso, it is sufficient to note that $\Sigma^{-1} B \otimes_{A_{r-1}} Z_2$ and P/F have the same Poincaré series. In fact, since we know the structure of B, and B is free as a right module over A_{r-1} by Theorem 4·4 of Milnor-Moore (8) we find that the Poincaré series for $B \otimes_{A_{r-1}} Z_2$ is

$$\frac{1}{1 - t^{2^r}} (1 + t^{3 \cdot 2^{r-1}})(1 + t^{7 \cdot 2^{r-2}}) \ldots (1 - t^{(2^r - 1)2})(1 - t^{2^{r+1} - 1}).$$

On the other hand, using Lemma 2·2 we can filter P/F so as to obtain a subquotient $A_r \otimes_{A_{r-1}} Z_2$ every 2^{r+1} dimensions, and we find that the Poincaré series for $\Sigma(P/F)$ is

$$\frac{1}{1 - t^{2^{r+1}}} (1 + t^{2^r})(1 + t^{3 \cdot 2^{r-1}}) \ldots (1 - t^{(2^r - 1)2})(1 - t^{2^{r+1}} - 1).$$

This proves Lemma 2·4.

Proof of Lemma 1·3. Consider the following diagram.

$$\begin{array}{ccc}
\Sigma^{-1} A \otimes_{A_{r-1}} Z_2 & \xrightarrow{\alpha} & A \otimes_{A_r} P/F \\
{\scriptstyle 1} \big\uparrow & & \big\uparrow {\scriptstyle 1 \otimes \beta} \; \cong \\
\Sigma^{-1} A \otimes_{A_{r-1}} Z_2 & \xleftarrow{\mu \otimes 1} & \Sigma^{-1} A \otimes_{A_r} B \otimes_{A_{r-1}} Z_2
\end{array}$$

Here α and β are as in (2·3), (2·4), while μ is given by the product map for A, that is, $\mu(a \otimes b) = ab$. It is easy to check that the diagram is commutative. Thus the exact sequence in (2·3) splits and gives

$$A \otimes_{A_r} P/F \cong (\Sigma^{-1} A \otimes_{A_{r-1}} Z_2) \oplus (A \otimes_{A_r} P/F').$$

But the same conclusion applies to P/F', so that

$$A \otimes_{A_r} P/F \cong (\Sigma^{-1} A \otimes_{A_{r-1}} Z_2) \oplus (\Sigma^{2^{r+1}-1} A \otimes_{A_{r-1}} Z_2) \oplus (A \otimes_{A_r} P/F'').$$

Continuing by induction, we obtain Lemma 1·3.

3. In this section we will prove Lemma 1·6. To this end, we begin by giving explicit formulae for the splitting which was obtained by induction at the end of § 2.

We first introduce the element

$$y_k = \sum_{i+j=k} \chi(\mathrm{Sq}^i) \otimes x^j \in A \otimes_{A_r} /F_{l,r}.$$

Here χ is the canonical anti-automorphism of A; and the sum is finite, since we only have to consider the range $i \geq 0, j \geq l$. Then we have the following more precise form of Lemma 1·3.

LEMMA 3·1. *The A-module $A \otimes_{A_r} P/F_{l,r}$ is a direct sum (over k such that*

$$k \equiv -1 \bmod 2^{r+1}, \; k \geq l)$$

of cyclic summands $\Sigma^k(A \otimes_{A_{r-1}} Z_2)$ with generators y_k.

Proof. Consider the explicit splitting used in proving Lemma 1·3; it displays P/F as the direct sum of the cyclic submodule $\Sigma^{-1} A \otimes_{A_{r-1}} Z_2$, on the generator $x^{-1} = y_{-1}$, and a complementary summand, namely the kernel of the splitting map

$$(\mu \otimes 1)(1 \otimes \beta)^{-1}.$$

We claim that this kernel contains the remaining elements y_k, that is, those with $k \geq 0$. In fact, we have

$$\beta \, \mathrm{Sq}^{j+1} = x^j$$

(where j runs over the range $j \geq -1$, so that $j+1$ runs over the range $j+1 \geq 0$). Thus

$$\begin{aligned}(\mu \otimes 1)(1 \otimes \beta)^{-1}\Big(\sum_{i+j=k} \chi(\mathrm{Sq}^i) \otimes x^j\Big) &= \sum_{i+(j+1)=(k+1)} \chi(\mathrm{Sq}^i)\,\mathrm{Sq}^{j+1} \\ &= 0 \quad \text{if} \quad k+1 \geq 1.\end{aligned}$$

On the other hand, the periodicity isomorphisms

$$P/F' \cong \Sigma^{2^{r+1}} P/F$$
$$P/F'' \cong \Sigma^{2^{r+1}} P/F', \text{etc.},$$

clearly carry elements y_k to other elements $y_{k'}$. It is now clear that the inductive process used in proving Lemma 1·3 displays P/F as a direct sum of cyclic submodules $\Sigma^k(A \otimes_{A_{r-1}} Z_2)$ on generators y_k. This proves Lemma 3·1.

It is now clear that Diagram (1·4) commutes; in fact, this has been clear since we constructed the splitting by induction. Moreover, the composite

$$\Sigma^{-1}(A \otimes_{A_{r-1}} Z_2) \to \bigoplus_j \Sigma^j (A \otimes_{A_{r-1}} Z_2) \cong A \otimes_{A_r} P/F_{l,r} \xrightarrow{1\otimes\gamma} A \otimes_{A_r} \Sigma^{-1} Z_2$$

carries the generator 1, via y_{-1}, to 1. To complete the proof of Lemma 1·6, we have to show that Diagram 1·5 commutes; and this will follow from the first half of the following lemma.

LEMMA 3·2. *The element* $y_k \in A \otimes_{A_r} P/F_{l,r}$ *is zero unless* $k \equiv -1 \bmod 2^{r+1}$; *and then it is equal to the sum*

$$\sum_{i+j=k} \chi(\mathrm{Sq}^i) \otimes x^j,$$

where i and j are restricted to the residue classes

$$i \equiv 0 \bmod 2^{r+1}, \quad j \equiv -1 \bmod 2^{r+1}.$$

The proof requires identities in A and P.

LEMMA 3·3. *There exist a finite number of elements* $a_i = a_{i,r} \in A_r$, *of degree* $2^{r+1}i + 2^r$, *such that*

(i) $\mathrm{Sq}^{2^{r+1}k+2^r} = \sum_{i+j=k} a_i \mathrm{Sq}^{2^{r+1}j}$,

(ii) $\sum_{i+j=k} \chi(a_i)\, x^{2^{r+1}j-1} = x^{2^{r+1}k+2^r-1}$,

(iii) $\sum_{i+j=k} \chi(a_i)\, x^{2^{r+1}j+2^r-1} = 0$.

The prototype of these identities may be seen for $r = 0$; we have one element Sq^1, and

$$\mathrm{Sq}^{2k+1} = \mathrm{Sq}^1\mathrm{Sq}^{2k},$$
$$\mathrm{Sq}^1 x^{2k-1} = x^{2k},$$
$$\mathrm{Sq}^1 x^{2k} = 0.$$

Otherwise, the best way to justify these identities is to use them.

Proof of Lemma 3·2, *assuming Lemma* 3·3. We proceed by induction over r. The result is trivially true for $r = -1$ provided we interpret A_{-1} as Z_2, so we assume it true for $r-1$. Then y_k is zero in $A \otimes_{A_{r-1}} P/F_{l,r-1}$ unless $k \equiv -1 \bmod 2^r$, so we have to consider only two cases, $k \equiv -1 \bmod 2^{r+1}$ and $k \equiv 2^r - 1 \bmod 2^{r+1}$.

In the first case, we write $k = 2^{r+1}m - 1$, and the inductive hypothesis gives

$$y_k = \sum_{i+j=m} \chi(\mathrm{Sq}^{2^{r+1}i}) \otimes x^{2^{r+1}j-1} + \sum_{i+j=m-1} \chi(\mathrm{Sq}^{2^{r+1}i+2^r}) \otimes x^{2^{r+1}j+2^r-1}.$$

We can rewrite the second sum using (3·3) (i), and we obtain

$$\sum_{e+h+j=m-1} \chi(\mathrm{Sq}^{2^{r+1}e})\,\chi(a_h) \otimes x^{2^{r+1}j+2^r-1} = \sum_{e+h+j=m-1} \chi(\mathrm{Sq}^{2^{r+1}e}) \otimes \chi(a_h)\, x^{2^{r+1}j+2^r-1}$$

(since the tensor product is taken over A_r, and $\chi(a_h) \in A_r$). But this gives zero, by (3·3) (iii).

In the second case, we write $k = 2^{r+1}m + 2^r - 1$, and the inductive hypothesis gives

$$y_k = \sum_{i+j=m} \chi(\mathrm{Sq}^{2^{r+1}i+2^r}) \otimes x^{2^{r+1}j-1} + \sum_{i+j=m} \chi(\mathrm{Sq}^{2^{r+1}i}) \otimes x^{2^{r+1}j+2^r-1}.$$

We can rewrite the first sum using (3·3) (i) as above, and we get

$$\sum_{e+h+j=m} \chi(\mathrm{Sq}^{2^{r+1}e})\,\chi(a_h) \otimes x^{2^{r+1}j-1} = \sum_{e+h+j=m} \chi(\mathrm{Sq}^{2^{r+1}e}) \otimes \chi(a_h)\,x^{2^{r+1}j-1}$$
$$= \sum_{e+n=m} \chi(\mathrm{Sq}^{2^{r+1}e}) \otimes x^{2^{r+1}n+2^r-1}$$

(using (3·3) (ii)). So we see that $y_k = 0$ in this case. This proves Lemma 3·2, assuming Lemma 3·3.

Proof of Lemma 3·3. With the notation of § 2, B is free as a left module over A_r, by Theorem 4·4 of Milnor–Moore (8); and in fact we can take the elements $\mathrm{Sq}^{2^{r+1}j}$ as an A_r-base. Therefore, we have for each k a unique formula

$$\mathrm{Sq}^{2^{r+1}k+2^r} = \sum_{i+j=k} a_i(k)\,\mathrm{Sq}^{2^{r+1}j}$$

with coefficients $a_i(k) \in A_r$. Since A_r is a finite algebra and $a_i(k)$ is of degree $2^{r+1}i + 2^r$, the sum can be taken over a finite range of i which does not depend on k. Moreover, the coefficient $a_i(k)$ does not depend on k provided that k is sufficiently large; for in the dual, multiplying by $\xi_1^{2^{r+1}}$ gives an isomorphism of everything at issue. Let us write a_i for the common value of $a_i(k)$; then the formula

$$\mathrm{Sq}^{2^{r+1}k+2^r} = \sum_{i+j=k} a_i\,\mathrm{Sq}^{2^{r+1}j}$$

remains true for small values of k, provided we interpret $\mathrm{Sq}^{2^{r+1}j}$ as zero for $j < 0$. This proves part (i); we turn to parts (ii) and (iii).

First we must recall that the operation $a: P^i \to P^j$ is zero or not according as $\chi a: P^{-1-j} \to P^{-1-i}$ is zero or not. To a topologist, the obvious proof is by S-duality, since any finite segment of P may be realized as the cohomology of a truncated projective space RP^{k+l}/RP^{k-1}, and we know the S-dual of that by [2]. To an algebraist, the obvious way is to use the map

$$P \otimes P \xrightarrow{\ \mu\ } P \xrightarrow{\ \gamma\ } \Sigma^{-1}Z_2;$$

this establishes P^k as the dual of P^l whenever $k + l = -1$, in a way compatible with operations $a \in A$.

We now want to determine the sum of the operations $\chi(a_i)$ in P which map into a degree congruent mod 2^{r+1} to -1 or $2^r - 1$ according to the case (where this sum is counted as 0 or 1 in the obvious way). In part (iii) the sum is obviously zero, for every operation $a \in I(A)$ in P is zero out of degree 0 or (dually) into degree -1, and every operation $a \in I(A_r)$ is zero into a degree congruent to $-1 \bmod 2^{r+1}$. In part (ii), it is equivalent to determine the sum of the operations a_i mapping out of a degree congruent to 2^r mod 2^{r+1}. But we can take the original formula

$$\mathrm{Sq}^{2^{r+1}k+2^r} = \sum_{i+j=k} a_i\,\mathrm{Sq}^{2^{r+1}j}$$

and apply it to the class x^{-2^r}; on this class all the operations $\mathrm{Sq}^{2^{r+1}k+2^r}$ and $\mathrm{Sq}^{2^{r+1}j}$ are 1, so we find that the required sum of the operations a_i is 1. This proves Lemma 3·3, which completes the proof of Lemmas 3·2 and 1·6.

4. In this section we will prove Theorem 1·1. First we prepare some general lemmas on homological algebra.

Let A_r be an algebra over Z_2 which is zero in degrees less than 0 and greater than d, and let M be a left module over A_r.

LEMMA 4·1. *If $M^t = 0$ for $t < l$, then $\mathrm{Tor}^{A_r}_{s,t}(Z_2, M) = 0$ for $t < l$; if $M^t = 0$ for $t > u$, then $\mathrm{Tor}^{A_r}_{s,t}(Z_2, M) = 0$ for $t > u+(s+1)d$.*

These bounds are of course extremely crude, but they will serve.

Proof. It is easy to construct a resolution of M by free modules C_s with the property that $C_{s\,t} = 0$ if $t < l$ or $t > u+(s+1)d$ according to the case.

LEMMA 4·1. *Let L be a left module and M a right module over an algebra A which comes as the union of an ascending sequence of subalgebras A_r. Then the map*

$$\underset{\overrightarrow{r}}{\mathrm{Lim}}\, \mathrm{Tor}^{A_r}_{s,t}(M, L) \to \mathrm{Tor}^{A}_{s,t}(M, L)$$

is iso.

This may be found in (4), Theorem 2·4 (*a*); or the reader can quickly supply his own proof.

If V is a graded vector space over Z_2, V^* will mean its graded dual. Let A be a graded algebra over Z_2. If V is an A-module (on either side), then V^* becomes an A-module (on the other side).

LEMMA 4·3. *Let L be a left A-module and M a right A-module. Then we have a natural isomorphism*

$$\mathrm{Ext}^{s,t}_A(L, M^*) \cong [\mathrm{Tor}^A_{s,t}(M, L)]^*.$$

Proof. Choose a resolution of L by projective modules C_s. We have a natural (1–1) correspondence between bilinear maps $M \otimes_{Z_2} C_s \to Z_2$ and linear maps

$$C_s \to \mathrm{Hom}_{Z_2}(M, Z_2) = M^*.$$

Under this $(1-1)$ correspondence, the subset of bilinear maps $M \otimes_A C_s \to Z_2$ corresponds to the subset of A-linear maps $C_s \to M^*$; that is, we have a natural isomorphism

$$\mathrm{Hom}_A(C_s, M^*) \leftrightarrow (M \otimes_A C_s)^*.$$

The homology groups of the complex $M \otimes_A C_s$ are $\mathrm{Tor}^A_{s,t}(M, L)$; since dualizing preserves exactness, the homology groups of $(M \otimes_A C_s)^*$ are $[\mathrm{Tor}^A_{s,t}(M, L)]^*$. But the homology groups of $\mathrm{Hom}_A(C_s, M^*)$ are $\mathrm{Ext}^{s,t}_A(L, M^*)$. This proves the lemma.

Alternatively, the reader may prefer to invoke (the graded analogue of) (3), p. 120, Proposition 5·1.

Proof of Theorem 1·1. We transcribe the information in Lemmas 1·3, 1·6 by applying the functor $\mathrm{Tor}^A_{s,t}(Z_2, \quad)$ to the modules, maps and diagrams. We recall that for any A_r-module M we have a natural change-of-rings isomorphism

$$\mathrm{Tor}^A_{s,t}(Z_2, A \otimes_{A_r} M) \cong \mathrm{Tor}^{A_r}_{s,t}(Z_2, M).$$

We see that we have an isomorphism

$$\mathrm{Tor}^{A_r}_{s,t}(Z_2, P/F_{l,r}) \cong \bigoplus_j \mathrm{Tor}^{A_{r-1}}_{s,t}(Z_2, \Sigma^j Z_2)$$

where j runs over the set $j \equiv -1 \bmod 2^{r+1}, j \geqslant l$. This isomorphism makes the following diagrams commute.

$$\begin{array}{ccc}
Tor^{A_r}_{s,t}(Z_2, P/F_{l,r}) & \xleftrightarrow{\cong} & \bigoplus_j Tor^{A_{r-1}}_{s,t}(Z_2, \Sigma^j Z_2) \\
\downarrow & & \downarrow \theta_* \\
Tor^{A_r}_{s,t}(Z_2, P/F_{m,r}) & \xleftrightarrow{\cong} & \bigoplus_k Tor^{A_{r-1}}_{s,t}(Z_2, \Sigma^k Z_2)
\end{array}$$

$$\begin{array}{ccc}
Tor^{A_r}_{s,t}(Z_2, P/F_{l,r}) & \xleftrightarrow{\cong} & \bigoplus_j Tor^{A_{r-1}}_{s,t}(Z_2, \Sigma^j Z_2) \\
\downarrow & & \downarrow \psi_* \\
Tor^{A_{r+1}}_{s,t}(Z_2, P/F_{l,r+1}) & \xleftrightarrow{\cong} & \bigoplus_k Tor^{A_r}_{s,t}(Z_2, \Sigma^k Z_2)
\end{array}$$

The details of the maps are of course exactly as in Lemma 1·6. Moreover, for $l \leqslant -1$ the composite

$$\mathrm{Tor}^{A_{r-1}}_{s,t}(Z_2, \Sigma^{-1} Z_2) \to \bigoplus_j \mathrm{Tor}^{A_{r-1}}_{s,t}(Z_2, \Sigma^j Z_2) \cong$$

$$\cong \mathrm{Tor}^{A_r}_{s,t}(Z_2, P/F_{l,r}) \xrightarrow{\gamma_*} \mathrm{Tor}^{A_r}_{s,t}(Z_2, \Sigma^{-1} Z_2)$$

is the obvious projection.

For a fixed r, the A_r-module P is unfortunately the inverse limit of the modules $P/F_{l,r}$ (as $l \to -\infty$). However, on applying $\mathrm{Tor}^{A_r}_{s,t}(Z_2, \quad)$ we get an attained limit. More precisely, Lemma 4·1 shows that

$$\mathrm{Tor}^{A_r}_{s,t}(Z_2, F_{l,r}) = 0$$

for l sufficiently negative (depending on s and t). It follows that the induced map

$$\mathrm{Tor}^{A_r}_{s,t}(Z_2, P) \to \mathrm{Tor}^{A_r}_{s,t}(Z_2, P/F_{l,r})$$

is iso for l sufficiently negative (depending on s and t). Similarly, Lemma 4·1 applies to $\mathrm{Tor}^{A_{r-1}}_{s,t}(Z_2, \Sigma^j Z_2)$; if we form the direct sum

$$\bigoplus_j \mathrm{Tor}^{A_{r-1}}_{s,t}(Z_2, \Sigma^j Z_2),$$

where j runs over the whole set $j \equiv -1 \bmod 2^{r+1}$, then for a given s and t this sum contains only a finite number of non-zero terms. Therefore by taking l sufficiently negative depending on s and t, we obtain an isomorphism

$$\mathrm{Tor}^{A_r}_{s,t}(Z_2, P) \xleftrightarrow{\cong} \bigoplus_j \mathrm{Tor}^{A_{r-1}}_{s,t}(Z_2, \Sigma^j Z_2)$$

where j runs over the set $j \equiv -1 \bmod 2^{r+1}$. This isomorphism was conjected in [1]. It makes the following diagram commute.

$$\begin{array}{ccc} Tor^{A_r}_{s,t}(Z_2, P) & \overset{\cong}{\longleftrightarrow} & \bigoplus_j Tor^{A_{r-1}}_{s,t}(Z_2, \Sigma^j Z_2) \\ \downarrow & & \downarrow \psi_* \\ Tor^{A_{r+1}}_{s,t}(Z_2, P) & \overset{\cong}{\longleftrightarrow} & \bigoplus_k Tor^{A_r}_{s,t}(Z_2, \Sigma^k Z_2) \end{array}$$

This also was suggested in [1] as the simplest possibility. Moreover, the composite

$$\mathrm{Tor}^{A_{r-1}}_{s,t}(Z_2, \Sigma^{-1}Z_2) \to \bigoplus_j \mathrm{Tor}^{A_{r-1}}_{s,t}(Z_2, \Sigma^j Z_2) \cong$$

$$\cong \mathrm{Tor}^{A_r}_{s,t}(Z_2, P) \xrightarrow{\gamma_*} \mathrm{Tor}^{A_r}_{s,t}(Z_2, \Sigma^{-1}Z_2)$$

is the obvious projection.

We can now pass to direct limits, using Lemma 4·2. Owing to the nature of the maps ψ_*, the only summands in the direct system $\bigoplus_j \mathrm{Tor}^{A_{r-1}}_{s,t}(Z_2, \Sigma^j Z_2)$ which survive are those with $j = -1$. We obtain an isomorphism

$$\mathrm{Tor}^{A}_{s,t}(Z_2, P) \overset{\cong}{\longleftrightarrow} \mathrm{Tor}^{A}_{s,t}(Z_2, \Sigma^{-1}Z_2);$$

and we also find that the map

$$\mathrm{Tor}^{A}_{s,t}(Z_2, \Sigma^{-1}Z_2) \overset{\cong}{\longleftrightarrow} \mathrm{Tor}^{A}_{s,t}(Z_2, P) \xrightarrow{\gamma_*} \mathrm{Tor}^{A}_{s,t}(Z_2, \Sigma^{-1}Z_2)$$

is the identity. This proves that γ_* is iso.

Dualizing this result with the help of Lemma 4·3, we see that

$$\gamma^*\colon \mathrm{Ext}^{s,t}_A(\Sigma^{-1}Z_2, Z_2) \to \mathrm{Ext}^{s,t}_A(P, Z_2)$$

is iso.

On the other hand, we can use the canonical anti-automorphism $\chi\colon A \to A$ to change any left A-module into a right A-module and vice versa. If we allow this, then we have the symmetry of Tor in the form of a natural isomorphism

$$\mathrm{Tor}^{A}_{s,t}(L, M) \cong \mathrm{Tor}^{A}_{s,t}(M, L).$$

So the result on Tor also shows that the map

$$\gamma_*\colon \mathrm{Tor}^{A}_{s,t}(P, Z_2) \to \mathrm{Tor}^{A}_{s,t}(\Sigma^{-1}Z_2, Z_2)$$

is iso. But now we can again dualize with the help of Lemma 4·3. According to the discussion in the proof of Lemma 3·3, the dual of P is ΣP, and the dual of $\gamma\colon P \to \Sigma^{-1}Z_2$ is $\Sigma\phi\colon \Sigma Z_2 \to \Sigma P$. Removing some Σ's, we see that

$$\phi_*\colon \mathrm{Ext}^{s,t}_A(Z_2, Z_2) \to \mathrm{Ext}^{s,t}_A(Z_2, P)$$

is iso. This completes the proof of Theorem 1·1.

W. H. Lin acknowledges financial support from the National Council of Sciences of the Republic of China. D. M. Davis and M. E. Mahowald were partially supported by grants from the National Science Foundation.

REFERENCES

(1) Adams, J. F. Operations of the nth kind in K-theory, and what we don't know about RP^{∞}. *London Math. Soc. Lecture Notes* no. 11, pp. 1–9. (Cambridge University Press, 1974).

(2) Atiyah, M. F. Thom complexes. *Proc. London Math. Soc.* (3) **11** (1961), 291–310.

(3) Cartan, H. and Eilenberg, S. *Homological Algebra* (Princeton University Press, 1956).

(4) Lin, T. Y. and Margolis, H. R. Homological aspects of modules over the Steenrod algebra. *J. Pure and Applied Algebra* **9** (1977), 121–129.

(5) Lin, W. H. The Adams–Mahowald conjecture on real projective spaces. *Math. Proc. Cambridge Philos. Soc.* **86** (1979), 237–241.

(6) Lin, W. H. On conjectures of Mahowald, Segal and Sullivan. *Math. Proc. Cambridge Philos. Soc.* **87** (1980), 449–458.

(7) Milnor, J. The Steenrod algebra and its dual. *Ann. of Math.* (2) **67** (1958), 150–171.

(8) Milnor, J. and Moore, J. C. On the structure of Hopf algebras. *Ann. of Math.* (2) **81** (1965), 211–264.

Lin, National Chengchi University, Taipei, Taiwan.
Davis, Lehigh University, Bethlehem, Pa. 18015, USA.
Mahowald, Northwestern University, Evanston, Ill. 60201, USA.
Adams, DPMMS, 16 Mill Lane, Cambridge CB2 1SB.

Topology Vol. 24, No. 4, pp. 435–460, 1985.

THE SEGAL CONJECTURE FOR ELEMENTARY ABELIAN p-GROUPS

J. F. ADAMS, J. H. GUNAWARDENA and H. MILLER

(*Received* 20 *November* 1984)

§1. INTRODUCTION

CARLSSON's proof of the Segal conjecture [2, 3] depends on an input from calculation; the object of this paper is to provide the input needed.

More precisely, we originally confirmed by calculation that a non-equivariant form of Segal's conjecture, describing the cohomotopy of the classifying space BG, is true when G is an elementary abelian p-group. Our approach was to calculate the cohomotopy groups $\pi^r(BG)$ by an Adams spectral sequence, and so most of the work lay in computing the requisite Ext groups over the Steenrod algebra A.

Carlsson [2, 3] invented an inductive argument, which proves the Segal conjecture in general, provided one can assume as input that an equivariant form of the Segal conjecture is true when G is an elementary abelian p-group. This he deduced from our non-equivariant result, by quoting work of Lewis, May and McClure [6].

Carlsson [2, 3] also observed that while his inductive argument by itself does not suffice to prove the case of an elementary abelian p-group, it does enable one to reduce the input from calculation. Instead of calculating the cohomotopy group $\pi^r(BG)$, it is sufficient to calculate the relevant group in Carlsson's "fundamental exact sequence", and to prove that the boundary map in this exact sequence is an isomorphism. For this we refer the reader to May and Priddy [10].

With this reduction, there is still work to be done in calculating an Ext group, but the work is less. Once this Ext group is calculated, there are two ways to calculate the relevant group in Carlsson's exact sequence. (i) One may follow Carlsson and reduce problems of equivariant homotopy theory to problems in non-equivariant homotopy theory; one then resolves the latter by using the classical Adams spectral sequence. (ii) Alternatively, one may set up an equivariant analogue of the Adams spectral sequence, capable of answering at least some of the problems of equivariant homotopy theory; one then uses an equivariant spectral sequence directly to calculate the relevant group in Carlsson's "fundamental exact sequence". The Ext group needed is the same either way.

In this paper we will not supply details for either of the arguments (i), (ii) above. We prefer not to write out (i) because it involves no essential novelty or serious difficulty, and because (ii) may well be preferable in the long run. We prefer not to write out (ii) because this method needs time to mature. We will therefore take as our object the calculation of the relevant Ext group, and we regard this as the theorem whose proof we have a duty to publish.

We also have results about other Ext groups which arise in studying the Segal conjecture, including those needed for our original calculation of $\pi^r(BG)$, but for these a statement and sketch proof will suffice.

The minimum which will serve Carlsson's purpose is provided by parts (a), (b) of the following result. We will explain it after stating it—but we assume that p is a fixed prime and V is an elementary abelian p-group of rank n.

THEOREM 1.1. (a) *The quotient map*

$$H^*(V)_{loc} \to F_p \otimes_A H^*(V)_{loc}$$

is a Tor-equivalence.

(b) *$F_p \otimes_A H^*(V)_{loc}$ is zero except in degree $-n$, where its rank is $p^{n(n-1)/2}$.*

(c) *The representation of $Aut(V) = GL(V) = GL(n, F_p)$ afforded by $F_p \otimes_A H^*(V)_{loc}$ is the Steinberg representation* [16].

(d) *A base for $F_p \otimes_A H^*(V)_{loc}$ is provided by the elements*

$$g(e_1 x_1^{-1} e_2 x_2^{-1} \dots e_n x_n^{-1}), \quad g \in Syl(V).$$

Here we write $H^*(G)$ for $H^*(BG; F_p)$. We use the letters U, V, W . . . for elementary abelian p-groups because they often have to be regarded as vector spaces over F_p. We define $H^*(V)_{loc}$ by localizing $H^*(V)$ so as to invert $\beta h \in H^2(V)$ for every non-zero $h \in H^1(V)$. The ring $H^*(V)_{loc}$ is an algebra over the mod p Steenrod algebra A. If M is an A-module, $F_p \otimes_A M$ is regarded as a quotient A-module on which A acts trivially. We say that a map $\theta: L \to M$ of A-modules is a "Tor-equivalence" if the induced map

$$\theta_*: \mathrm{Tor}^A_{**}(F_p, L) \to \mathrm{Tor}^A_{**}(F_p, M)$$

is iso. The point of this emerges from the following result.

PROPOSITION 1.2. *If $\theta: L \to M$ is a Tor-equivalence, then the induced map*

$$\theta_*: Tor^A_{**}(K, L) \to Tor^A_{**}(K, M)$$

is iso for any (right) A-module K which is bounded above; the induced map

$$\theta^*: Ext_A^{**}(L, N) \leftarrow Ext_A^{**}(M, N)$$

is iso for any (left) A-module N which is bounded below and finite-dimensional over F_p in each degree.

The hypotheses of boundedness are essential. The proof will be omitted on the grounds that it is sufficiently obvious.

In (1.1) (d), the elements $e_1, e_2, \dots, e_n$ are a base chosen in $H^1(V)$, which may be identified with V^*, the dual of V. We then set $x_r = \beta e_r$, so that $x_1, x_2, \dots x_n$ are a corresponding base in $\beta V^* \subset H^2(V)$. Thus, whether $p > 2$ or $p = 2$, $H^*(V)$ contains a symmetric or polynomial algebra $S[\beta V^*]$ on generators $\{x_r\}$. Syl(V) is the subgroup of $GL(V)$ consisting of upper unitriangular matrices (with respect to these bases); it is a Sylow subgroup of $GL(V)$. We keep all this notation as standard, except that in the case $n = 1$, $V = Z_p$ we write e, x for the generators e_1, x_1.

The case $n = 1$, $p = 2$ of (1.1) is due to [8], while the case $n = 1$, $p > 2$ is due to [4]. Thus (1..1)) generalizes results previously known to be relevant to the Segal conjecture.

Our proof of Theorem 1.1 is based upon the "Singer construction" [14, 15, 7]. For the moment we need only explain three points about this. First, the Singer construction gives a functor $T(M)$ from A-modules to A-modules, which comes provided with a natural transformation $\varepsilon: T(M) \to M$. Secondly, the Singer construction allows one to reduce the calculation of Ext groups for a larger module, namely $T(M)$, to the calculation of Ext for a smaller module, namely M.

THEOREM 1.3. *The map $\varepsilon: T(M) \to M$ of Singer's construction is a Tor-equivalence.*

This reduction theorem was originally found by the second and third authors independently.

Thirdly, there is a relation between $H^*(V)_{loc}$ and the iterated Singer construction

$$T^n(F_p) = T(T(\dots T(F_p) \dots)).$$

THEOREM 1.4. *There is an isomorphism of A-algebras*

$$T^n(F_p) \cong H^*(V)_{loc}^{Syl(V)}.$$

Here M^G means the subobject of elements in M fixed under G, as usual. The localization required may be done by inverting

$$\Pi\beta h \mid h \in V^*, h \neq 0;$$

this element is fixed under $GL(V)$; it makes no difference whether we localize before or after passing to a subalgebra of fixed elements.

The case $p = 2$ of (1.4) is due to Singer [15], while the case $p > 2$ is modelled on a result of Li and Singer [7]. More precisely, Li and Singer prove the corresponding result for the subalgebra of invariants $H^*(V)_{\mathrm{loc}}^{\mathrm{Bor}(V)}$, where Bor$(V)$ is the Borel subgroup of upper-triangular matrices in $GL(V)$. At this point we should explain that for $p > 2$ our version of the "Singer construction" is not quite the same as that of Li and Singer [7]. Theorem 1.3 is true for both versions; but for the purposes of our proof, a reduction theorem like (1.3) grows more useful as $T(M)$ grows larger. Our version of $T(M)$ is (roughly speaking) (p-1) times as large as that of Li and Singer [7], and our subalgebra of invariants is (roughly speaking) $(p\text{-}1)^n$ times as large as theirs; this allows us to get closer to $H^*(V)_{\mathrm{loc}}$. (See §2).

Priddy and Wilkerson [13] have shown how the deduction of (1.1) (a), (b) from (1.3) and (1.4) may be illuminated by their observation that $H^*(V)_{\mathrm{loc}}$ is projective as a module over $F_p[GL(V)]$. However, we will indicate our original argument, which is elementary.

If we localize $H^*(V)$ less than in (1.1) then it becomes harder to prove homological results about it, but we can still do so. Let S be a subset of $\beta V^* \subset H^2(V)$. We form $H^*(V)_S$ by localizing $H^*(V)$ so as to invert all the non-zero elements of S. The ring $H^*(V)_S$ is an algebra over A. We assume $S > \{0\}$ and suppose given a non-zero element $x = x_1 \in S$.

THEOREM 1.5. *The map*

$$H^*(V)_S \xrightarrow{\{res_W\}} \bigoplus_W H^*(W)_{S\cap\beta W^*}$$

is a Tor-equivalence.

Here W runs over certain quotients of V, so that W^* runs over certain subspaces of V^*. More precisely, βW^* runs over complements in βV^* for the subspace $\langle x \rangle$ generated by x; that is, we require $\beta V^* = \langle x \rangle \oplus \beta W^*$. There are p^{n-1} choices for W. The A-maps

$$H^*(V)_S \xrightarrow{\mathrm{res}_W} H^*(W)_{S\cap\beta W^*}$$

will be explained in §8; they raise degree by 1.

(1.5) enables one to reduce the calculation of Ext groups for any localized algebra $H^*(V)_S$ to the unlocalized case. In fact, if on the right we have an algebra $H^*(W)_{S\cap\beta W^*}$ with $S\cap\beta W^*$ non-zero, then we may choose a non-zero element $x_2 \in S\cap\beta W^*$ and apply the theorem again, and so on by induction.

As our work was originally conceived, we needed to compute Ext groups for the unlocalized case (at least in terms of more familiar Ext groups). The best version of the result is conceptual, and we will give this version in §9; but in this introduction we avoid explaining it, stating instead a form which is more explicit. We assume that U and V are elementary abelian p-groups and that M is an A-module, bounded below and finite-dimensional over F_p in each degree.

THEOREM 1.6. *Then the map*

$$\bigoplus_X Ext_A^{s-s(X),t-s(X)}(H^*(W(X)), M) \xrightarrow{\omega} Ext_A^{s,t}(H^*(V), M \otimes H^*(U))$$

is iso.

Here we explain that in §9 we shall associate to U and V a finite set of indices X. We shall also associate to each index X an integer $s(X)$ and an elementary abelian p-group $W(X)$. Finally we shall introduce the map ω.

For $s = 0, t = 0, M = F_p$ we can do without the indexing apparatus: our result reduces to the statement that the obvious map

$$F_p[\mathrm{Hom}_{F_p}(U, V)] \to \mathrm{Hom}_A^0(H^*(V), H^*(U))$$

is an isomorphism. However, for $s > 0$ we need more indices than are provided by the F_p-maps from U to V.

The level of generality in (1.6) is such as to compute the E_2-term of the Adams spectral sequence

$$\mathrm{Ext}_A^{**}(H^*(\mathbf{B}V), H^*(\mathbf{T}) \otimes H^*(\mathbf{B}U) \Rightarrow [\mathbf{T} \wedge \mathbf{B}U, \mathbf{B}V]_*$$

for any suitable choice of the test-object $\mathbf{T}$. (The bold-face letters stand for spectra, and $\mathbf{B}G = \Sigma^\infty(BG \sqcup P)$.)

We can justify this level of generality by considering the proof of (1.6). This proof flows by a simple and inevitable induction over the rank of U; we sketch the step from "rank 1" to "rank 2". Obviously, if you can compute $\mathrm{Ext}_A^{**}(H^*(V), M \otimes H^*(Z_p))$ for general M, then you can substitute $M = L \otimes H^*(Z_p)$; since $H^*(Z_p) \otimes H^*(Z_p) = H^*(Z_p \times Z_p)$, you can compute $\mathrm{Ext}_A^{**}(H^*(V), L \otimes H^*(Z_p \times Z_p))$ in terms of groups

$$\mathrm{Ext}_A^{**}(H^*(W(X)), L \otimes H^*(Z_p)),$$

which you can compute by the same token. Notice that if you want to compute the cohomotopy groups $\pi^*(BU)$, so that you want the final result only for $V = 0$ and $M = F_p$, you still need the inductive hypothesis in the generality given.

The body of this paper is arranged as follows. The proofs of (1.3), (1.4) and (1.1) (a), (b), (d) will be given in §2, §3 and §4 respectively. These proofs involve a certain amount of forward reference. In particular, we proceed by stating and using any fact about Singer's functor T which we know to be true; in §5 we sketch an approach to T which allows one to prove all these facts. Similarly, in §4 we use a proof by induction, which involves algebras of invariants $H^*(V)_{\mathrm{loc}}^G$ for various subgroups $G \subset \mathrm{Syl}(V)$. In §6 we explain the subgroups G concerned, and obtain information about these algebras of invariants $H^*(V)_{\mathrm{loc}}^G$. As a corollary, we justify an explicit description of the algebra of invariants $H^*(V)_{\mathrm{loc}}^{\mathrm{Syl}(V)}$, which is stated at (3.3) and used in §3.

§7 deals with the Steinberg representation and proves (1.1)(c). §8 proves (1.5).

The final sections, §9–§12, are devoted to sketching the proof of (1.6). We wish to draw the reader's attention to the categorical considerations involved in giving a conceptual statement of (1.6), and in particular to a construct we call the "Burnside category" $\mathscr{A}$; we hope it may be of wider use. We therefore urge the reader to study §9.

We are grateful to W. M. Singer for keeping us informed of his work, and similarly to G. Carlsson and to Priddy and Wilkerson. We are grateful to the Sloan Foundation, to the University of Aarhus, and to the University of Chicago for enabling us to meet in spite of our usual geographical separation. Finally, we thank the editors and referees of *Topology* (ably seconded by J. P. May) for lifting from our consciences the duty of publishing our proofs in such complete detail as may be more appropriate to private archives.

§2. PROOF OF (1.3)

To prove (1.3), we shall need some facts about the Singer construction.

Additively, the Singer construction $T(M)$ is isomorphic to the tensor product $L \otimes M$ of M with a fixed object L. (However, the A-module structure on $T(M)$ is not given by the usual "diagonal" formula.) Just as one assigns to each cohomology theory K^* the "coefficient groups" $K^*(P)$, so to each functor T from A-modules to A-modules one assigns the "coefficient module" $T(F_p)$. In our case $T(F_p)$ is $L \otimes F_p$, that is, L; thus L becomes an A-module, and plays the role of a "coefficient module" for the Singer construction. It is usual to write T for this coefficient module, and to write $T(M) = T \otimes M$. In fact T is an A-algebra;

and the obvious action of T on $T(M) = T \otimes M$ is an A-action, that is, it satisfies the Cartan formula.

We can now explain the relationship between our version of the Singer construction (say $T(M)$) and the version of Li and Singer (say $T'(M)$). In our version the coefficient algebra is $T = H^*(Z_p)_{loc}$, the case $n = 1$ of the algebra considered in (1.1). In the version of Li and Singer the coefficient algebra is $T' = H^*(\Sigma_p)_{loc}$. Here Σ_p is the symmetric group on p letters; the inclusion $Z_p \to \Sigma_p$ induces an inclusion

$$H^*(\Sigma_p) \to H^*(Z_p).$$

We localize $H^*(\Sigma_p)$ by inverting x^{p-1}. If the version $T'(M)$ of the Singer construction is taken as known, one can define the version $T(M)$ by

$$T(M) = T \otimes_{T'} T'(M).$$

We shall need some facts about $T'(M)$. The instance $T'(A)$ must be a bimodule over A (for any $b \in A$, the map $a \mapsto ab$: $A \to A$ is a map of left A-modules, and $T'(-)$ is a functor). For homological purposes, the most convenient way to construct $T'(M)$ is to give an explicit description of the bimodule $T'(A)$, and then set

$$T'(M) = T'(A) \otimes_A M.$$

We shall give a description of this sort.

As in [8] we use the dual A_* of the mod p Steenrod algebra A [11]. This dual has exterior generators $\tau_0, \tau_1, \ldots$ and polynomial generators $\xi_1, \xi_2, \ldots$. (We omit the modifications necessary in the case $p = 2$, which are standard.) We have to use the usual finite subalgebras of the Steenrod algebra. We write $A_*(n)$ for the quotient $A_*/I(n)$, where the ideal $I(n)$ is generated by the τ_r with $r > n$ and the $\xi_r^{p^s}$ with $r + s \geq n+1$. The quotient $A_*(n)$ is dual to a sub-Hopf-algebra $A(n)$ of A. The subalgebra $A(-1)$ is F_p; the subalgebra $A(0)$ is the exterior algebra generated by β.

We also introduce a localized quotient

$$B_*(n) = (A_*/J(n))\,[\xi_1^{-1}] \quad (n \geq 0)$$

where the ideal $J(n)$ is generated by the τ_r with $r > n$ and the $\xi_r^{p^s}$ with $r \geq 2, r+s \geq n+1$. The object $A_*/J(n)$ is a left comodule over $A_*(n)$ and a right comodule over $A_*(n-1)$. Multiplication by $\xi_1^{p^n}$ preserves both comodule structures. Since $B_*(n)$ may be regarded as the direct limit of $A_*/J(n)$ under multiplication by $\xi_1^{p^n}$, it becomes a left comodule over $A_*(n)$ and a right comodule over $A_*(n-1)$. It is also an algebra, and is finite-dimensional over F_p in each degree.

We define $B(n)$ to be the dual of $B_*(n)$. This object is a bimodule; it is a left module over $A(n)$ and a right module over $A(n-1)$.

For example, $B_*(0)$ has a base consisting of the elements ξ_1^k and $\tau_0 \xi_1^k$ for $k \in Z$. We take the dual base in $B(0)$ and call its elements P^k and βP^k for $k \in Z$.

Since we have canonical maps $A_* \to B_*(n+1) \to B_*(n)$, wwe have canonical maps $B(n) \to B(n+1) \to A$ preserving all the relevant structure. The element written P^k in $B(0)$ maps to P^k in A if $k \geq 0$, to 0 if $k < 0$; similarly for βP^k.

LEMMA 2.1. (i) *$B(n)$ is free as a left module over $A(n)$; the elements P^k with $k \equiv 0$ mod p^n may be taken as a base; the left-primitive submodule of $B_*(n)$ is $F_p[\xi_1^{p^n}, \xi_1^{-p^n}]$.*

(ii) *$B(n)$ is free as a right module over $A(n-1)$; the elements $P^k, \beta P^k$ with $k \in Z$ may be taken as a base. Equivalently, the map $B(0) \otimes A(n-1) \to B(n)$ is iso.*

We defer the proof in order to complete our description of $T'(M)$.

If M is an $A(n-1)$-module, we may now construct

$$B(n) \otimes_{A(n-1)} M;$$

this is an $A(n)$-module. If M is an $A(n)$-module, then the canonical map

$$B(n) \otimes_{A(n-1)} M \to B(n+1) \otimes_{A(n)} M$$

is iso, since both groups are isomorphic to $B(0) \otimes M$ by (2.1) (ii). Thus the construction is essentially independent of n. If M is an A-module, we may now construct the attained limit

$$\varinjlim_{n} (B(n) \otimes_{A(n-1)} M),$$

and this is an A-module.

Unfortunately this is not yet exactly what we want for $T'(M)$. The relevant isomorphism between $B(n) \otimes_{A(n-1)} F_p$ and $H_*(\Sigma_p)_{\text{loc}}$ is of degree -1, while for any purpose which involves products we want the isomorphism between T' and $H^*(\Sigma_p)_{\text{loc}}$ to have degree 0. While we defer any more detail to §5, we indicate that it is better to define a preliminary version $T''(M)$ of the Singer construction to be $B(n) \otimes_{A(n-1)} M$ or $\varinjlim_{n} B(n) \otimes_{A(n-1)} M$ according to the case, and to define $T'(M)$ as isomorphic to $T''(M)$ under an isomorphism which changes degrees by 1.

If M is an A-module, the map

$$B(n) \otimes_{A(n-1)} M \to A \otimes_{A(n-1)} M \to M$$

passes to the limit, and gives a map of A-modules

$$\varepsilon'' \colon T''(M) = \varinjlim_{n} B(n) \otimes_{A(n-1)} M \to M.$$

Replacing $T''(M)$ by the isomorphic A-module $T'(M)$, we get

$$\varepsilon' \colon T'(M) \to M$$

which is now an A-map of degree $+1$. To construct

$$\varepsilon \colon T(M) = T \otimes_{T'} T'(M) \to M,$$

one first projects $T = H^*(Z_p)_{\text{loc}}$ onto the direct summand $T' = H^*(\Sigma_p)_{\text{loc}}$ and then applies ε'.

This completes all we need explain about the Singer construction in order to prove (1.3).

Proof of (2.1). (i) It is clear that as a left comodule over $A_*(n)$, $B_*(n)$ is a direct sum of copies of $A_*(n)$ shifted by multiplication with the powers $\xi_1^{rp^n}$, $r \in Z$.

(ii) It is easy to show that in A, the elements P^k, βP^k with k sufficiently large (say $k \geq k_0$) are linearly independent under right multiplication by $A(n-1)$. Using the canonical map $B(n) \to A$, we see that the same result holds also in $B(n)$.

We can deduce that all the elements P^k, βP^k in $B(n)$ are linearly independent under right multiplication by $A(n-1)$. In fact, multiplication by $\xi_1^{-rp^n}$ gives a linear map $B_*(n) \to B_*(n)$ which is a map of bicomodules. Its dual is a linear map $B(n) \to B(n)$ which is a map of bimodules. Suppose we had any linear relation over $A(n-1)$ between the elements P^k, βP^k in $B(n)$; by applying this map for a suitable value of r, we could shift the relation up until it involved only elements P^k, βP^k with $k \geq k_0$.

Thus we see that the map

$$B(0) \otimes A(n-1) \to B(n)$$

is mono. On the other hand, the objects $B(0) \otimes A(n-1)$ and $B(n)$ have the same (finite) dimension over F_p in each degree; so the map is iso. This proves (2.1).

LEMMA 2.2. (i) *If M is A-free then $T(M)$ is A-flat.*

(ii) *If M is A-free then the map*

$$F_p \otimes_A T(M) \xrightarrow{1 \otimes \varepsilon} F_p \otimes_A M$$

is iso.

Proof. (i) If M is A-free then it is $A(n-1)$-free. If M is free over $A(n-1)$ then $B(n)\otimes_{A(n-1)} M$ is a direct sum of copies of $B(n)$, so it is free over $A(n)$ by (2.1) (i). This shows that $T'(M)$ is free over $A(n)$. Over $A(n), T(M) = T\otimes_{T'} T'(M)$ is a direct sum of $(p-1)$ copies of $T'(M)$, because multiplication by x^{p^n} gives a shift map commuting with $A(n)$. Therefore $T(M)$ is free over $A(n)$; this holds for all n. So

$$\begin{aligned}\mathrm{Tor}^A_{s,t}(K, T(M)) &= \varinjlim_n \mathrm{Tor}^{A(n)}_{s,t}(K, T(M))\\ &= 0 \text{ for } s > 0.\end{aligned}$$

Thus $T(M)$ is A-flat.

(ii) It is sufficient to prove the special case $M = A$, for the general case follows by passing to direct sums. By (2.1) (i), $B(n)\otimes_{A(n-1)} A(n-1)$ is $A(n)$-free on generators P^k, $k \equiv 0 \bmod p^n$. Thus $F_p\otimes_{A(n)} T''(A(n-1))$ is F_p-free on generators P^k, $k \equiv 0 \bmod p^n$. Passing to the limit over n, we see that $F_p\otimes_A T''(A)$ is F_p-free on one generator P^0. Thus the map

$$F_p\otimes_A T''(A) \xrightarrow{1\otimes\varepsilon''} F_p\otimes_A A = F_p$$

is iso. Therefore the corresponding result holds for T'. For T, we can use x^{p^n} as a shift map, as above; we see that $F_p\otimes_{A(n)} T(A(n-1))$ is F_p-free on generators in degrees congruent to -1 mod $2p^n$. Passing to the limit over n, we see that $F_p\otimes_A T(A)$ is zero except in degree -1. This proves (2.2).

The deduction of (1.3) from (2.2) may be omitted as routine.

§3. PROOF OF (1.4)

Let us take $V = Z_p \times W$, so that

$$H^*(V) \cong H^*(Z_p)\otimes H^*(W).$$

THEOREM 3.1. *There is an isomorphism of A-algebras*

$$H^*(V)^{Syl(V)}_{loc} \cong T(H^*(W)^{Syl(W)}_{loc}).$$

Here Syl (V) and Syl (W) are groups of upper unitriangular matrices with respect to bases chosen so that e_1 is a base in Z_p^*, $e_2, e_3, \ldots, e_n$ are a base in W^*, and $x_r = \beta e_r$ as in §1.

If (3.1) is granted, (1.4) will follow immediately by induction over the rank n of V.

In order to construct the isomorphism in (3.1), we need to know more about the Singer construction. It comes provided with a structure map

$$T(M) = T\otimes M \xrightarrow{f} T\hat{\otimes} M.$$

Here $T\hat{\otimes} M$ is a completed tensor product; we get it by completing $T\otimes M$ with respect to a topology in which a typical neighbourhood of zero is

$$\left(\sum_{r\leq -N} T^r\right)\otimes M.$$

A typical element of $T\hat{\otimes} M$ is a "downward-going formal Laurent series"

$$\sum_{r\leq R} x^r\otimes m'_r + \sum_{r\leq R} ex^r\otimes m''_r$$

where e, x are the generators in $T = H^*(Z_p)_{\mathrm{loc}}$. To make A act on $T\hat{\otimes} M$, we take the usual (diagonal) action on $T\otimes M = H^*(Z_p)_{\mathrm{loc}}\otimes M$ and pass to the completion.

The map f is an A-map and a map of T-modules; it is always mono. If M is an A-algebra, then the obvious product on $T\otimes M$ makes $T(M)$ and $T\hat{\otimes} M$ into algebras, and f becomes a map of algebras.

We apply this with $M = H^*(W)_{loc}$. Since $V = Z_p \times W$ we have an embedding

$$H^*(V) \subset T \otimes H^*(W)_{loc} \subset T \hat{\otimes} H^*(W)_{loc}.$$

This embedding extends to the localization $H^*(V)_{loc}$. (For any element c of degree 2 in $H^*(W)_{loc}$ the element $x+c$ is invertible in $T \hat{\otimes} M^*(W)_{loc}$, with inverse

$$x^{-1} - x^{-2}c + x^{-3}c^2 - \dots .)$$

A more precise version of (3.1) is now as follows.

THEOREM 3.2. *The image of* $H^*(V)_{loc}^{Syl(V)}$ *under the embedding*

$$H^*(V)_{loc} \to T \hat{\otimes} H^*(W)_{loc}$$

is the same as the image of $T(H^*(W)_{loc}^{Syl(W)})$ *under the embedding* f.

To prove (3.2), we need to know the algebras of invariants and to calculate f. We refer to §6 for the following.

PROPOSITION 3.3. $H^*(V)_{loc}^{Syl(V)}$ *is a free module on the* 2^n *generators*

$$f_1^{i_1} f_2^{i_2} \dots f_n^{i_n} \quad (\text{where each } i_r \text{ is } 0 \text{ or } 1)$$

over the algebra of finite Laurent series

$$F_p[y_1, y_1^{-1}] \otimes F_p[y_2, y_2^{-1}] \otimes \dots \otimes F_p[y_n, y_n^{-1}].$$

Here the generators f_r, y_r are defined as follows.

$$f_r = \begin{vmatrix} x_1^{p^r-2} & \dots & x_r^{p^r-2} \\ \vdots & & \\ x_1^p & \dots & x_r^p \\ x_1 & \dots & x_r \\ e_1 & \dots & e_r \end{vmatrix}$$

$$y_r = \begin{vmatrix} x_1^{p^r-1} & \dots & x_r^{p^r-1} \\ \vdots & & \\ x_1^{p^2} & \dots & x_r^{p^2} \\ x_1^p & \dots & x_r^p \\ x_1 & \dots & x_r \end{vmatrix}$$

These elements and their constructions go back to Mui [12]. The elements f_r and y_r are easily seen to be invariant under Syl(V). The determinant y_r is a product of factors which are non-zero elements of βV^*; thus y_r is invertible in $H^*(V)_{loc}$.

We write g_r, z_r for the generators in $H^*(W)^{Syl(W)}$ constructed in the same way as the generators f_r, y_r in $H^*(V)^{Syl(V)}$.

PROPOSITION 3.4. *The map*

$$T \otimes H^*(W)_{loc} \xrightarrow{f} T \hat{\otimes} H^*(W)_{loc}$$

has

$$f(x_1 \otimes 1) = y_1, \quad f(e_1 \otimes 1) = f_1$$

and

$$f(x_1^{p^r} \otimes z_r) = y_{r+1}, \quad f(x_1^{p^{r-1}} \otimes g_r) = f_{r+1} \quad \textit{for } r \geq 1.$$

If (3.3) and (3.4) are granted, then (3.2) follows at once.

In order to calculate the map f, we need to know the Steenrod operations on the generators g_r, z_r for $H^*(W)_{\text{loc}}^{\text{Syl}(W)}$.

LEMMA 3.5. (i) *We have* $P^k z_r = 0$ *unless* $k = (p^r - p^j)/(p-1)$ *for some* j *such that* $0 \leq j \leq r$. *In this case*

$$P^k z_r = \begin{vmatrix} x_2^{p^r} & \cdots & x_{r+1}^{p^r} \\ \vdots & & \\ x_2^{p^{j+1}} & \cdots & x_{r+1}^{p^{j+1}} \\ x_2^{p^{j-1}} & \cdots & x_{r+1}^{p^{j-1}} \\ \vdots & & \\ x_2 & \cdots & x_{r+1} \end{vmatrix}.$$

(ii) *We have* $\beta P^k z_r = 0$.

(iii) *We have* $P^k g_r = 0$ *unless* $k = (p^{r-1} - p^j)/(p-1)$ *for some* j *such that* $0 \leq j \leq r-1$. *In this case*

$$P^k g_r = \begin{vmatrix} x_2^{p^{r-1}} & \cdots & x_{r+1}^{p^{r-1}} \\ \vdots & & \\ x_2^{p^{j+1}} & \cdots & x_{r+1}^{p^{j+1}} \\ x_2^{p^{j-1}} & \cdots & x_{r+1}^{p^{j-1}} \\ \vdots & & \\ x_2 & \cdots & x_{r+1} \\ e_2 & \cdots & e_{r+1} \end{vmatrix}.$$

(iv) *We have* $\beta P^k g_r = 0$ *unless* $k = (p^{r-1} - 1)/(p-1)$. *In this case*

$$\beta P^k g_r = z_r.$$

We calculate (i) by applying the total Steenrod power $p = \sum_{k=0}^{\infty} P^k$ to the determinant for z_r and evaluating the resulting determinant; similarly for (iii). Parts (ii) and (iv) follow.

Proof of (3.4). We need to known that f is a map of T-modules and satisfies the following explicit formula.

$$f(1 \otimes m) = \sum_{k \geq 0} (-1|^k x^{-k(p-1)} \otimes P^k m + \sum_{k \geq 0} (-1)^{k+1} e x^{-k(p-1)-1} \otimes \beta P^k m. \quad (3.6)$$

Using the Steenrod operations given by (3.5), we calculate as follows.

$$f(x_1^{p^r} \otimes z_r) = \sum_{0 \leq j \leq r} (-1)^{r-j} x_1^{p^j} \begin{vmatrix} x_2^{p^r} & \cdots & x_{r+1}^{p^r} \\ \vdots & & \\ x_2^{p^{j+1}} & \cdots & x_{r+1}^{p^{j+1}} \\ x_2^{p^{j-1}} & \cdots & x_{r+1}^{p^{j-1}} \\ \vdots & & \\ x_2 & \cdots & x_{r+1} \end{vmatrix}$$

$$= y_{r+1}.$$

Similarly for $f(x_1^{p^{r-1}} \otimes g_r)$.

This proves (3.4), which completes the proof of (3.1) and (1.4), modulo the facts used.

§4. PROOF OF (1.1) (a), (b), (d)

As this proof is by induction, we must formulate the inductive hypothesis. For suitable subgroups G normal in Syl(V), we prove the following.

THEOREM 4.1. (a) *The quotient map*

$$H^*(V)^G_{\text{loc}} \xrightarrow{q} F_p \otimes_A H^*(V)^G_{\text{loc}}$$

is a Tor-equivalence.

(b) $F_p \otimes_A H^*(V)^G_{\text{loc}}$ *is zero except in degree* $-n$.

(c) *In degree* $-n$ *it is of rank* $|Syl(V):G|$.

(d) *More precisely, a base for* $F_p \otimes_A H^*(V)^G_{\text{loc}}$ *is provided by the sums*

$$\sum_{\gamma \in \Gamma} \gamma(e_1 x_1^{-1} e_2 x_2^{-2} \ldots e_n x_n^{-1})$$

where Γ *runs over the cosets of* G *in* $Syl(V)$.

The special case $G = 1$ of (4.1) proves (1.1) (a), (b), (d). For more detail on the other subgroups G considered, see §6. In part (d), the notation is as in (1.1). It is clear that the sum

$$\sum_{\gamma \in \Gamma} \gamma(e_1 x_1^{-1} e_2 x_2^{-1} \ldots e_n x_n^{-1})$$

is invariant under G.

The first step in proving (4.1) is to prove the special case $G = \text{Syl}(V)$. In this case we have

$$H^*(V)^{\text{Syl}(V)}_{\text{loc}} \cong T^n(F_p)$$

by (1.4). By (1.3) we have n Tor-equivalences

$$T^n F_p \xrightarrow{\varepsilon} T^{n-1} F_p \longrightarrow \cdots \longrightarrow TF_p \xrightarrow{\varepsilon} F_p$$

each of degree $+1$. Thus we have a Tor-equivalence (of degree n)

$$H^*(V)^{\text{Syl}(V)}_{\text{loc}} \xrightarrow{\phi} F_p.$$

It is now easy to deduce (4.1) (a), (b), (c) for $G = \text{Syl}(V)$ by commuting q with ϕ.

We now proceed by downwards induction over G. For the subgroups G with which we work (see §6), the inductive step presents itself as follows. We have a subgroup F normal in G, with quotient $G/F \cong Z_p$ generated by g. We suppose as our inductive hypothesis that (4.1) (a), (b) are true for G, and we wish to deduce them for F. We have a filtration

$$0 = M_0 \subset M_1 \subset M_2 \subset \ldots \subset M_p = H^*(V)^F_{\text{loc}}$$

of $H^*(V)^F_{\text{loc}}$ by A-submodules M_j, in which each subquotient M_j/M_{j-1} is isomorphic to $H^*(V)^G_{\text{loc}}$. The definition of the filtration is

$$M_j = \text{Ker}\,(g-1)^j : H^*(V)^F_{\text{loc}} \longrightarrow H^*(V)^F_{\text{loc}}$$

and the isomorphism $M_j/M_{j-1} \to M_1/M_0 = H^*(V)^G_{\text{loc}}$ is given by $(g-1)^{j-1}$.

We suppose, as the hypothesis of a subsidiary induction over j, that the quotient map

$$M_j \xrightarrow{q} F_p \otimes_A M_j$$

is a Tor-equivalence and that $F_p \otimes_A M_j$ is zero except in degree $-n$. Consider the following diagram.

$$\begin{array}{ccccccccc}
 & & 0 & \longrightarrow & M_j & \longrightarrow & M_{j+1} & \longrightarrow & H^*(V)^G_{\mathrm{loc}} & \longrightarrow & 0 \\
 & & & & \downarrow{\scriptstyle q_j} & & \downarrow{\scriptstyle q_{j+1}} & & \downarrow{\scriptstyle q} & & \\
\mathrm{Tor}^A_{1*}(F_p, H^*(V)^G_{\mathrm{loc}}) & \longrightarrow & & & F_p \otimes_A M_j & \longrightarrow & F_p \otimes_A M_{j+1} & \longrightarrow & F_p \otimes_A H^*(V)^G_{\mathrm{loc}} & \longrightarrow & 0
\end{array}$$

By hypothesis, $F_p \otimes_A M_j$ is zero except in degree $-n$. By the main inductive hypothesis,

$$\begin{aligned}\mathrm{Tor}^A_{1,-n}(F_p, H^*(V)^G_{\mathrm{loc}}) &\cong \bigoplus \mathrm{Torr}^A_{1,0}(F_p, F_p) \\ &= 0.\end{aligned}$$

So the lower sequence is short exact. We see that $F_p \otimes_A M_{j+1}$ is zero except in degree $-n$. Now we use the Five Lemma; by our hypotheses, q_j and q are Tor-equivalences, and therefore q_{j+1} is a Tor-equivalence. This completes the subsidiary induction, which runs up to $j = p$ and proves (4.1) (a), (b) for F.

It is easy to carry (4.1) (c) through this induction, and it is not hard to carry through (4.1) (d) provided we have the necessary starting-point, as follows.

LEMMA 4.2. *The sum*

$$\sum_{\gamma \in Syl(V)} \gamma(e_1 x_1^{-1} e_2 x_2^{-1} \ldots e_n x_n^{-1})$$

gives a non-zero element of $F_p \otimes_A H^*(V)_{\mathrm{loc}}$.

The proof of (4.2) is best approached by further remarks about the Singer construction. The map ε used to state and prove (1.3) can be factored through the map f used to prove (1.4), to give the following diagram.

$$\begin{array}{ccc}
T(M) & \xrightarrow{\ f\ } & T \hat{\otimes} M \\
 & {\scriptstyle \varepsilon}\searrow & \downarrow{\scriptstyle \mathrm{res}} \\
 & & M
\end{array}$$

More precisely, the map "res" is defined by

$$\mathrm{res}\left(\sum_{r \le R} x^r \otimes m'_r + \sum_{r \le R} e x^r \otimes m''_r\right) = m''_{-1}.$$

It is reasonable to think of this map as a "residue", since it takes the coefficient of the term of degree -1 in a Laurent series. The map res is an A-map of degree $+1$.

We can restrict the map res to parts of $T \hat{\otimes} M$ constructed by localization. These remarks, taken with §3, suggest the following. Take $M = H^*(W)_{\mathrm{loc}}$, as in §3. Restrict the map

$$T \hat{\otimes} H^*(W)_{\mathrm{loc}} \xrightarrow{\mathrm{res}} H^*(W)_{\mathrm{loc}}$$

to the subalgebra $H^*(V)_{\mathrm{loc}}$, embedded in $T \hat{\otimes} H^*(W)_{\mathrm{loc}}$ as in §3. We get an A-map

$$H^*(V)_{\mathrm{loc}} \xrightarrow{\mathrm{res}} H^*(W)_{\mathrm{loc}}.$$

LEMMA 4.3. *The map res carries the sum*

$$\sum_{\gamma \in Syl(V)} \gamma(e_1 x_1^{-1} e_2 x_2^{-1} \ldots e_n x_n^{-1})$$

in $H^*(V)_{loc}$ *to the sum*

$$\sum_{\delta \in Syl(W)} \delta(e_2 x_2^{-1} \ldots e_n x_n^{-1})$$

in $H^*(W)_{loc}$.

(4.3) follows from simple properties of the residue, and (4.2) follows easily from (4.3) by induction over the rank n of V.

The same ideas allow an alternative proof of (1.1) (d) (or (4.1) (d)). For this we use $p^{n(n-1)/2}$ different iterated residues to show that the elements

$$\gamma(e_1 x_1^{-1} e_2 x_2^{-1} \ldots e_n x_n^{-1}) \quad (\gamma \in \mathrm{Syl}(V))$$

give linearly independent elements of $F_p \otimes_A H^*(V)_{loc}$ to the full number allowed by (1.1) (b).

§5. THE SINGER CONSTRUCTION

To a topologist, the way to understand Singer's functor T is that it computes the limiting cohomology of a certain construction on spectra. For this we refer the reader to [1].

To a conceptual algebraist, the way to understand Singer's functor T is via the derived functors of the functor when takes any A-module and assigns to it the quotient where the "unstable" axiom is satisfied. For this we refer the reader to [5].

Although we realize the value and interest of these viewpoints, we neglect them for brevity. In this paper we need to treat Singer's functor T as a matter of computational algebra.

Indeed, our approach in §2, §3, §4 has been to prove the theorems at issue, stating as we go any necessary facts about T which we know to be true. However, the approach in §2 already provides a self-contained account of $T'(M)$ so far as its A-module structure goes; it is natural to ask if it can be elaborated into a self-contained account of $T(M)$ which proves all the results we have used (most notably (3.6)). The answer is that such an approach is possible; indeed, we have carried it out in detail; but in this section we will merely sketch the ideas needed.

The approach in §2 is sufficiently detailed up to the point where we define a preliminary version $T''(M)$ of the Singer construction to be $B(n) \otimes_{A(n-1)} M$ or $\varinjlim_n B(n) \otimes_{A(n-1)} M$, according as M is given as a module over $A(n-1)$ or over A. This gives $T''(M)$ as a module over $A(n)$ or A according to the case. It also gives the map $\varepsilon'' : T''(M) \to M$. However, we need to see a map f'' which we can later process to give the structure map f used in §3. For this purpose we first introduce a diagonal map

$$T''(M \otimes N) \xrightarrow{\Psi} T''(M) \hat{\otimes} T''(N)$$

by dualizing the product map in $B_*(n)$. We then form the composite

$$\begin{array}{ccc} T''(M) = T''(F_p \otimes M) & \xrightarrow{\Psi} & T''(F_p) \hat{\otimes} T''(M) \\ & & \Big\downarrow {\scriptstyle 1 \otimes \varepsilon''} \\ & & T''(F_p) \otimes M \end{array}$$

and take it for our map

$$T''(M) \xrightarrow{f''} T''(F_p) \hat{\otimes} M.$$

The map f'' has good properties; one can give explicit formulae for it, and show that it is mono.

Next we need to see a ring of coefficients acting on $T''(M)$. We will introduce such a ring, and later reconcile it with the account in §2.

Let $T'(n)$ be the right-primitive subobject of $B_*(n)$. The maps

$$\cdots \to B_*(n+1) \to B_*(n) \to \cdots \to B_*(0)$$

induce

$$\cdots \to T'(n+1) \to T'(n) \to \cdots \to T'(0)$$

and all these maps are iso, since each $T'(n)$ is dual to a quotient $B(n) \otimes_{A(n-1)} F_p$ with base P^k, βP^k for $k \in Z$. Let us write T' for the (attained) limit $\underleftarrow{\mathrm{Lim}}_n\, T'(n)$; this is an algebra. It contains an algebra $F_p[\xi, \xi^{-1}]$ of finite Laurent series on one generator ξ, which maps to ξ_1 in $B_*(0)$ and (for example) to $\xi_1 - \xi_1^{-p}\xi_2$ in $B_*(2)$. As a module over $F_p[\xi, \xi^{-1}]$, T' is free on two generators $1, \tau$ which map to $1, \tau_0$ in $B_*(0)$. If we use cohomological degrees, we must give ξ degree $-2(p-1)$ and τ degree -1.

Multiplication by $t' \in T'$ gives a linear map $B_*(n) \to B_*(n)$ which is a map of right comodules. Its dual is a map of right modules

$$B(n) \xrightarrow{t'} B(n).$$

This defines

$$B(n) \otimes_{A(n-1)} M \xrightarrow{t'} B(n) \otimes_{A(n-1)} M$$

and passes to the limit to give

$$T''(M) \xrightarrow{t'} T''(M).$$

In this way T' comes to act on $T''(M)$.

We now need an explicit isomorphism between $T''(M)$ and $T' \otimes M$. With $M = F_p$, for example, $T''(F_p)$ is a free T'-module on one generator β, but it is not a free T'-module on the generator 1 (at least if $p > 2$). For this or other reasons, we define an isomorphism of T'-modules

$$T' \otimes M \xrightarrow{\theta} T''(M)$$

by

$$\theta(t' \otimes m) = (-1)^{\deg t'} t'(\beta \otimes m).$$

We now define $T'(M) = T' \otimes M$ and give it an action of $A(n)$ or A, as the case may be, by using θ to pull back the action of $A(n)$ or A on $T''(M)$. Of course, since θ is of degree 1, this introduces the usual signs.

By rewriting the source and target of f'', we obtain a map

$$T'(M) = T' \otimes M \xrightarrow{f'} T' \hat{\otimes} M.$$

This map has good properties, and it is still mono. Similarly we obtain

$$T'(M) \xrightarrow{\varepsilon'} M.$$

Finally, we need to identify T' with $H^*(\Sigma_p)_{loc}$.

LEMMA 5.1. *There is an isomorphism of algebras*

$$T' \xrightarrow{\phi} H^*(\Sigma_p)_{loc}$$

which is also an isomorphism of A-modules (provided that the A-action on T' is that which it gets as $T'(F_p)$). Explicitly,

$$\phi(\xi) = -x^{-(p-1)}, \quad \phi(\tau) = ex^{-1}.$$

The idea is as follows. We can define an $A(n)$-map of degree -1 by

$$B(n) \to A \xrightarrow{\gamma} H^*(\Sigma_p)_{\mathrm{loc}}$$

where γ is defined by

$$\gamma(a) = (-1)^{1+\deg(a)} a(ex^{-1}).$$

We use the unstable axiom to show that in sufficiently high degrees, this map factors through $B(n) \otimes_{A(n-1)} F_p$. Thus in sufficiently high degrees we get a composite

$$T'(F_p) \xrightarrow{\theta} T''(F_p) \to H^*(\Sigma_p)_{\mathrm{loc}}$$

which is an $A(n)$-map and coincides with the map ϕ of algebras given by the formulae in the enunciation. Now we use periodicity to show that the map ϕ is an A-map in all degrees.

Lemma 5.1 provides an A-map

$$H^*(\Sigma_p)_{\mathrm{loc}} \xrightarrow{\phi^{-1}} T'(F_p) \xrightarrow{\varepsilon'} F_p;$$

this shows that the map "res" of §4, §8 is an A-map (if the reader does not already have his preferred proof).

We mention briefly two more ideas. First, since the map

$$T'(M) = T' \otimes M \xrightarrow{f'} T' \hat{\otimes} M$$

is mono, suitable properties (such as the Cartan formula) can be checked after applying f'. Secondly, suppose we start with a structure map, such as f'' or ε'', which is given conceptually and by transparent explicit formulae. If we replace source and target using explicitly-given isomorphisms, we shall still have good explicit formulae; but if we iterate the process, the formulae may not stay so transparent. This leads to results such as (3.6).

This completes our sketch of a self-contained approach to the Singer construction.

§6. ALGEBRAS OF INVARIANTS

In this section our first object is to indicate the subgroups G which can be used in the argument of §4; we also indicate results about their algebras of invariants, including (3.3).

Our subgroups G can be considered as matrix groups, defined by restricting the matrix A to agree with the identity below a certain stepwise boundary line. More precisely, with notation as in (1.1), let U_r be the subspace of $U = \beta V^*$ generated by $x_1, x_2, \ldots, x_r$. Let q: $\{1, 2, \ldots, n\} \to \{0, 1, 2, \ldots, n-1\}$ be a function which has $q(r) \le r-1$ and is non-decreasing, so that $r \le s$ implies $q(r) \le q(s)$. Let $G \subset \mathrm{Syl}(V)$ be the subgroup of matrices A which induce the identity map of $U_r/U_{q(r)}$ for each r; equivalently,

$$a_{ij} = \delta_{ij} \quad \text{for} \quad i > q(j).$$

Next we explain the lemma which we use to prove that pairs $F \subset G$ of such subgroups have the property needed in §4. We suppose given a group Z_p (such as G/F) with generator g, acting on a polynomial algebra $R[x]$ of characteristic p so that g fixes R and $g(x) = x + c$ where c is some invertible constant in R. We define

$$M_j = \mathrm{Ker}((g-1)^j : R[x] \to R[x])$$

as in §3.

LEMMA 6.1. *Then the map*

$$(g-1)^{j-1} : M_j/M_{j-1} \to M_1/M_0 = R[x]^{Z_p}$$

is iso for $1 \le j \le p$. *Moreover,* $R[x]^{Z_p}$ *is a polynomial algebra* $R[y]$, *where*

$$y = \prod_{0 \le i < p} (g^i x) = x^p - c^{p-1}x.$$

The proof is elementary.

We must now explain why the action of $G/F = Z_p$ on the subalgebra $H^*(V)^F_{\mathrm{loc}}$ is such that we can apply this lemma.

First, since we can in fact apply (6.1), we can determine the subalgebras $H^*(V)^F_{\mathrm{loc}}$ by induction upwards over F, starting from $F = 1$ and computing $H^*(V)^G_{\mathrm{loc}}$ as $H^*(V)^F_{\mathrm{loc}})^{G/F}$. (There are enough subgroups F of the sort we consider to reach any one by an induction, either downwards from $\mathrm{Syl}(V)$ as in §4, or upwards from 1 as here.) To give the answer, we construct generators in $S[\beta V^*] \subset H^*(V)$ as follows. For each $r \in \{1, 2, \ldots, n\}$, we choose an F-orbit C_r in U_r which is not in U_{r-1}. Let $\pi(C_r)$ be the product of the elements in this orbit.

PROPOSITION 6.2. (a) *For each such choice, the algebra of invariant elements* $S[\beta V^*]^F_{\mathrm{loc}}$ *is*

$$F_p[\pi(C_1), \pi(C_2), \ldots, \pi(C_n)]_{\mathrm{loc}}.$$

(b) $H^*(V)^F_{\mathrm{loc}}$ *is a free module over* $S[\beta V^*]^F_{\mathrm{loc}}$ *on the* 2^n *generators*

$$f_1^{i_1} f_2^{i_2} \ldots f_n^{i_n}$$

described in §3.

Secondly, we have good control over the pair $F \subset G$. If $F \subset G$ are subgroups such as we consider with $G/F \cong Z_p$, then F differs from G only by the imposition of one extra condition $a_{ij} = 0$ for some pair (i, j) with $i < j$. A generator g for $G/F \cong Z_p$ is given by the elementary matrix which agrees with the identity matrix except for $a_{ij} = 1$. From this we see that G fixes all but one of the generators $\pi(C_r)$ in (6.2)(a). We can thus take

$$R = F_p[\pi(C_1), \ldots, \pi(C_{j-1})]_{\mathrm{loc}}[\pi(C_{j+1}), \ldots, \pi(C_n)].$$

Moreover, we can take $x = \pi(C_j)$, because g moves this generator in the required way. (This point does take some elementary algebra.)

If we want information about algebras of invariants, there is never any trouble in passing from less localized objects, such as $R[x]$, to information about more localized objects, such as $S[\beta V^*]^F_{\mathrm{loc}}$. We can thus prove (6.2) (a) by induction upwards over F.

As for (6.2) (b), there is never any trouble in throwing in the 2^n passive generators.

Both the last paragraphs apply also to proving $(g-1)^j$ iso, as is asserted for $R[x]$ by (6.1) and needed for $H^*(V)^F_{\mathrm{loc}}$ in §4.

Finally, the case $F = \mathrm{Syl}(V)$ of (6.2) can be rewritten to give (3.3).

§7. THE STEINBERG REPRESENTATION

(1.1) (c) states that $F_p \otimes_A H^*(V)_{\mathrm{loc}}$ affords the (mod p) Steinberg representation of $GL(V)$. We will sketch a proof of this by conceptual algebra, avoiding any explicit formula for a Steinberg idempotent. We subdivide the proof into two parts, (7.1) and (7.2) below, by introducing an alternative construction of the Steinberg module M. From our definition of M we prove the following.

PROPOSITION 7.1. *There is a canonical map from* $F_p \otimes_Z M$ *to* $H^*(V)_{\mathrm{loc}}$ *such that the composite*

$$F_p \otimes_Z M \to H^*(V)_{\mathrm{loc}} \xrightarrow{q} F_p \otimes_A H^*(V)_{\mathrm{loc}}$$

is iso.

If we assume the result of Priddy and Wilkerson [13] that $H^*(V)_{\mathrm{loc}}$ is projective over $F_p[GL(V)]$, then the splitting in (7.1) shows that $F_p \otimes_A H^*(V)_{\mathrm{loc}}$ is projective over

$F_p[GL(V)]$. By (1.1) (d), $F_p \otimes_A H^*(V)_{\text{loc}}$ restricts to the regular representation of Syl(V) (over F_p). These two points show that $F_p \otimes_A H^*(V)_{\text{loc}}$ satisfies one characterization of the mod p Steinberg representation; but we do not need to argue in this way.

PROPOSITION 7.2. *M is canonically isomorphic to $\tilde{H}_{n-2}(TB)$, the homology of the Tits building.*

We present the Z-module $M = M(V^*)$ by generators and relations, as follows. We take one generator $m(x_1, x_2, \ldots, x_n)$ for each base $(x_1, x_2, \ldots, x_n)$ of V^*. We prescribe the following relations.

(i) m is antisymmetric in its arguments, that is,

$$m(x_{\rho 1}, x_{\rho 2}, \ldots, x_{\rho n}) = \varepsilon(\rho) m(x_1, x_2, \ldots, x_n)$$

for each permutation ρ.

(ii) If λ is a non-zero scalar then

$$m(\lambda x_1, x_2, \ldots, x_n) = m(x_1, x_2, \ldots, x_n).$$

(iii) Suppose that V^* comes as the direct sum $V^* = X^* \oplus Y^*$ of a subspace X^* of dimension 2 and a subspace Y^* of dimension $n-2$. Suppose that any two of x_1, x_2, x_3 form a base for X^*, while $y_3, y_4, \ldots, y_n$ form a base for Y^*. Then

$$\begin{aligned} & m(x_1, x_2, y_3, y_4, \ldots, y_n) \\ & + m(x_2, x_3, y_3, y_4, \ldots, y_n) \\ & + m(x_3, x_1, y_3, y_4, \ldots, y_n) = 0. \end{aligned}$$

It is clear how $GL(V)$ acts on M.

We give the map

$$F_p \otimes_Z M \to H^*(V)_{\text{loc}}$$

of (7.1) by giving it on the generators. For present purposes the vector-spaces V^* and βV^* can be identified under β; let $e_1, e_2, \ldots, e_n$ and $x_1, x_2, \ldots, x_n$ be corresponding bases in them. Then we send the generator

$$m(x_1, x_2, \ldots, x_n)$$

to

$$e_1 x_1^{-1} e_2 x_2^{-1} \ldots e_n x_n^{-1} \in H^*(V)_{\text{loc}}.$$

We leave to the reader the exercise of checking that this map preserves the relations; the fact that it does so explains the construction of M.

We can analyse the structure of M. Let $(x_1, x_2, \ldots, x_n)$ be one base for V^*, and let g run over the corresponding group of upper uni-triangular matrices Syl(V).

PROPOSITION 7.3. *Then the generators $m(gx_1, gx_2, \ldots gx_n)$ form a Z-base for M.*

If this is granted, (7.1) follows: the map in (7.1) takes the base of $F_p \otimes_Z M$ given by (7.3), and sends it to the base for $F_p \otimes_A H^*(V)_{\text{loc}}$ given by (1.1) (d).

We sketch the proof that the generators in (7.3) span M. This is done by induction over n, using the following lemma. Let W^* be a subspace of dimension $(n-1)$ in V^*.

LEMMA 7.4. *M is spanned by generators $m(y_1, y_2, \ldots y_n)$ in which all but one of the y_r lie in W^*.*

This is proved from the given relations by induction over the number of y_r which do not lie in W^*.

We sketch the proof that the generators in (7.3) are linearly independent over Z. This is done by setting up suitable homomorphisms

$$\theta : M = M(V^*) \to Z.$$

For each base $x_1, x_2, \ldots, x_n$ in V^* and each maximal flag F in V^* there is at most one permutation ρ such that $x_{\rho 1}, x_{\rho 2}, \ldots, x_{\rho n}$ is a base adapted to the flag F. We let θ_F carry $m(x_1, x_2, \ldots, x_n)$ to $\varepsilon(\rho) = \pm 1$ if there is such a permutation ρ, to 0 otherwise.

We move on to (7.2). We recall that the Tits building TB (for V^*) is a certain finite simplicial complex. It has a vertex for each subspace of V^* other than the trivial subspaces 0 and V^*. Its top-dimensional simplexes are in $(1-1)$ correspondence with the maximal flags F in V^*. So by using as components the maps θ_F just introduced, we obtain a map

$$M = M(V^*) \xrightarrow{\{\theta_F\}} \bigoplus_F Z = C_{n-2}(TB);$$

from the proof of (7.3) we know that this map is mono. Since the subgroup of boundaries is zero, the reduced homology group $\tilde{H}_{n-2}(TB)$ is the subgroup of cycles $\tilde{Z}_{n-2}(TB)$. One checks directly that $\{\theta_F\}$ maps M into $\tilde{Z}_{n-2}(TB)$. It remains to show that $\{\theta_F\}$ maps M onto $\tilde{Z}_{n-2}(TB)$; the proof is combinatorial, and we omit it for brevity.

§8. DELOCALIZATION

In this section we sketch the proof of (1.5). The theme of our argument is that we take information about objects which are more localised, and deduce information about object which are less localized.

We first explain the map in (1.5). In §4 we said that a direct-sum splitting $V \cong Z_p \times W$ leads to an A-map

$$H^*(V)_{\mathrm{loc}} \xrightarrow{\mathrm{res}} H^*(W)_{\mathrm{loc}}.$$

We now write it res_w to indicate its dependence on W. If we localize less, this map carries $H^*(V)_S$ into $H^*(W)_{S \cap \beta W^*}$. We simplify the notation by dropping the symbol β, identifying the subspace $\beta V^* \subset H^2(V)$ with V^*, as in §7.

LEMMA 8.1. (1.5) *is true for* $S = V^*$.

The proof is based on the following diagram.

$$\begin{array}{ccc} H^*(V)_{V^*} & \xrightarrow{\{\mathrm{res}_W\}} & \bigoplus_W H^*(W)_{W^*} \\ \Big\downarrow{\scriptstyle q} & & \Big\downarrow{\scriptstyle \{q_W\}} \\ F_p \otimes_A H^*(V)_{V^*} & \xrightarrow{\{1 \otimes \mathrm{res}_W\}} & \bigoplus_W F_p \otimes_A H^*(W)_{W^*} \end{array}$$

The two vertical arrows are Tor-equivalences by (1.1) (a). We see that the lower horizontal arrow is iso by calculating its effect on the base given by (1.1) (d).

We prove (1.5) by induction over n. For $n = 1$ there is only one way to localize, and the result is true by (8.1). We therefore assume the result is true in dimension $(n-1)$. We now proceed by downwards induction over S. Lemma 8.1 shows that the result is true for $S = V^*$; for the induction step, we must assume that T contains just one more line than S, say $T = S \cup \langle y \rangle$, $y \notin S$, and assume that the result holds for T. We now have the following commutative diagram, in which ρ_W is defined by passing to the quotient from res_W.

$$\begin{array}{ccccccccc} 0 & \longrightarrow & H^*(V)_S & \longrightarrow & H^*(V)_T & \longrightarrow & \dfrac{H^*(V)_T}{H^*(V)_S} & \longrightarrow & 0 \\ & & \Big\downarrow{\scriptstyle \{\mathrm{res}_W\}} & & \Big\downarrow{\scriptstyle \{\mathrm{res}_W\}} & & \Big\downarrow{\scriptstyle \{\rho_W\}} & & \\ 0 & \longrightarrow & \bigoplus_W H^*(W)_{S \cap W^*} & \longrightarrow & \bigoplus_W H^*(W)_{T \cap W^*} & \longrightarrow & \bigoplus_W \dfrac{H^*(W)_{T \cap W^*}}{H^*(W)_{S \cap W^*}} & \longrightarrow & 0 \end{array}$$

Here the middle verticle arrow is a Tor-equivalence by the inductive hypothesis; we can prove that the left-hand vertical arrow is a Tor-equivalence by the Five Lemma, provided we prove the following.

LEMMA 8.2. *The map*

$$\frac{H^*(V)_T}{H^*(V)_S} \xrightarrow{\{\rho_W\}} \bigoplus_W \frac{H^*(W)_{T\cap W^\bullet}}{H^*(W)_{S\cap W^\bullet}}$$

is a Tor-equivalence.

Here we can restrict W^* to run over the p^{n-2} complements for $\langle x\rangle$ which contain $\langle y\rangle$, because $H^*(W)_{T\cap W^\bullet}/H^*(W)_{S\cap W^\bullet}$ is zero for the other choices of W^*.

The proof of (8.2) depends on a lemma.

LEMMA 8.3. *In* (8.2), *the truth or falsity of the conclusion depends only on the image of S in the quotient space $V^*/\langle y\rangle$.*

We sketch the proof of (8.3). It is sufficient to study the effect of replacing S by S', where $S\subset S'$ and both have the same image in $V^*/\langle y\rangle$. Of course we take $T' = S'\cup\langle y\rangle$. We now check that the map

$$\frac{H^*(V)_T}{H^*(V)_S} \longrightarrow \frac{H^*(V)_{T'}}{H^*(V)_{S'}}$$

is iso. The reason is that S' already acts invertibly on $H^*(V)_T/H^*(V)_S$. In fact, the series

$$s^{-1}-\lambda y s^{-2}+\lambda^2y^2s^3- \ldots$$

provides an inverse for any element $s+\lambda y$ in S'; on any particular element of $H^*(V)_T/H^*(V)_S$ this series converges after a finite number of terms, because for any $z\in H^*(V)_T$ there is a power y^m of y such that $y^m z\in H^*(V)_S$. The same considerations show that the map

$$\frac{H^*(W)_{T\cap W^\bullet}}{H^*(W)_{S\cap W^\bullet}} \longrightarrow \frac{H^*(W)_{T'\cap W^\bullet}}{H^*(W)_{S'\cap W^\bullet}}$$

is iso. Therefore $\{\rho_W\}$ is a Tor-equivalence for S if and only if it is so for S'.

We sketch the proof of (8.2). Here we use (8.3) to clean up the position of S. Choose a complement $\bar V^*$ for $\langle y\rangle$ in V^* such that $\langle x\rangle\subset\bar V^*$. Then $\bar V^*$ provides one representative for each coset in $V^*/\langle y\rangle$, and so (8.7) allows us to suppose that $S\subset\bar V^*$.

The p^{n-2} complements W^* for $\langle x\rangle$ in V^* which contain $\langle y\rangle$ are now in $(1-1)$ correspondence with the p^{n-2} complements $\bar W^*$ for $\langle x\rangle$ in $\bar V^*$. By using our special choice of S, and properties of the residue, the map ρ_W can be thrown by isomorphisms of the source and target onto

$$H^*(\bar V)_S\otimes\frac{H^*(Z_p)_{\mathrm{loc}}}{H^*(Z_p)} \xrightarrow{\mathrm{res}_{\bar W}\otimes 1} H^*(\bar W)_{S\cap\bar W^\bullet}\otimes\frac{H^*(Z_p)_{\mathrm{loc}}}{H^*(Z_p)}.$$

(Here $\langle y\rangle = (Z_p)^*$.) Since $\bar V$ is of rank $(n-1)$, the hypothesis of our main induction over n says that the map

$$H^*(\bar V)_S \xrightarrow{\{\mathrm{res}_{\bar W}\}} \bigoplus_{\bar W} H^*(\bar W)_{S\cap\bar W^\bullet}$$

is a Tor-equivalence. An easy lemma, comparable with (1.2), says that Tor-equivalences remain Tor-equivalences if you tensor with an A-module bounded above, such as $H^*(Z_p)_{\mathrm{loc}}/H^*(Z_p)$. So the map

$$H^*(V)_S\otimes\frac{H^*(Z_p)_{\mathrm{loc}}}{H^*(Z_p)} \xrightarrow{\{\mathrm{res}_{\bar W}\otimes 1\}} \bigoplus_{\bar W} H^*(\bar W)_{S\cap\bar W^\bullet}\otimes\frac{H^*(Z_p)_{\mathrm{loc}}}{H^*(Z_p)}$$

is a Tor-equivalence. This proves (8.2).

§9. FUNCTION-OBJECTS

In this section we will explain the correct, conceptual version of (1.6) promised in §1.

We recall that the level of generality in (1.6) is such that the algebra corresponds to the topological problem of computing $[\mathbf{T} \wedge \mathbf{B}U, \mathbf{B}V]$, where the bold-face letters mean spectra. Here the group $[\mathbf{T} \wedge \mathbf{B}U, \mathbf{B}V]$ is a representable functor of the test-object $\mathbf{T}$; the representing object is the function-spectrum $\mathbf{F}$ "of maps from $\mathbf{B}U$ to $\mathbf{B}V$", and information about the functor is equivalent to information about $\mathbf{F}$. We need to formulate the algebraic analogue of a function-spectrum.

For this we need to work in a category $\mathscr{C}$ which is preadditive and monoidal [9], and we explain our first example.

In our "Ext category" $\mathscr{E}$ the objects are A-modules $L, M, N, \ldots$ which are bounded below and finite-dimensional over F_p in each degree. The hom-set $E(L, M)$ from L to M in $\mathscr{E}$ is the bigraded Ext group $\mathrm{Ext}_A^{**}(M, L)$. Thus $\mathscr{E}$ is the opposite of the usual Ext category; this makes some formulae look better; in particular, cohomology is a covariant functor with values in $\mathscr{E}$. Composition in $\mathscr{E}$ is the usual Yoneda product. The monoidal operation on objects is the usual tensor product $L \otimes M$; on morphisms it is the usual tensor product in Ext.

We will explain the notion of a "function-object" in such a category $\mathscr{C}$. Let L and M be given objects in $\mathscr{C}$; we plan to consider "functions from L to M". Suppose given further a finite number of objects W_i in $\mathscr{C}$ and morphisms

$$W_i \otimes L \xrightarrow{w_i} M.$$

If the hom-sets of $\mathscr{C}$ are bigraded then the morphisms w_i may be of any bidegrees (s_i, t_i). For each "test object" T in $\mathscr{C}$ we get a map

$$C^{s-s_i, t-t_i}(T, W_i) \xrightarrow{\omega_i} C^{s,t}(T \otimes L, M)$$

which carries $T \xrightarrow{f} W_i$ to

$$T \otimes L \xrightarrow{f \otimes 1} W_i \otimes L \xrightarrow{w_i} M.$$

With these maps as components we get a map

$$\bigoplus_i C^{s-s_i, t-t_i}(T, W_i) \xrightarrow{\omega} C^{s,t}(T \otimes L, M).$$

If this map ω is an isomorphism for all objects T in $\mathscr{C}$, we will say that the data $\{W_i, w_i\}$ are a "function-object" from L to M. In this case the data $\{W_i, w_i\}$ allow us to express the group $C(T \otimes L, M)$ in terms of representable functors of T.

Of course, if there were in $\mathscr{C}$ a categorical product of the objects W_i suitably regraded, then this object (with a suitable map) would be a function-object in the usual sense; but we do not assume that any such object exists in $\mathscr{C}$.

The content of (1.6) is that certain data constitute a function-object from $H^*(U)$ to $H^*(V)$ in $\mathscr{E}$, and thereby allow us to compute $\mathrm{Ext}_A^{**}(H^*(V), -\otimes H^*(U))$.

The task of saying what data constitute this function-object is usually called "book-keeping". The art of book-keeping is to establish a correspondence between entries in a ledger, where the information is easy to find, and certain aspects of the real world, where things may be harder. The analogue of the real world, for us, is the category $\mathscr{E}$ where we keep our unknown Ext groups. The analogue of the ledger is a category $\mathscr{A}^{gr}$ where things are easy. The objects of $\mathscr{A}^{gr}$ are the elementary abelian p-groups $U, V, W, \ldots$; the monoidal operation on objects is the Cartesian product $U \times V$; we explain the morphisms of $\mathscr{A}^{gr}$ later. As for the correspondence between the ledger and the world, its analogue is a certain functor β from $\mathscr{A}^{gr}$ to $\mathscr{E}$. On objects the functor β is given by $\beta(V) = H^*(V)$; since $H^*(U \times V) = H^*(U) \otimes H^*(V)$, β preserves the monoidal operation on objects; we explain the effect of β on morphisms later. We now explain how β allows us to transfer constructs from $\mathscr{A}^{gr}$ to $\mathscr{E}$.

Suppose given a suitable functor β from one preadditive monoidal category to another. Suppose given a function-object $\{W_i, w_i\}$ from U to V in the source category. We will say that β "preserves this function-object" if $\{\beta W_i, \beta w_i\}$ is a function-object from βU to βV in the target category. That is, in our applications, the appropriate induced map

$$\bigoplus_i \mathrm{Ext}_A^{s-s_i, t-t_i}(H^*(W_i), M)) \xrightarrow{\omega} \mathrm{Ext}_A^{s,t}(H^*(V), M \otimes H^*(U))$$

is to be iso for every A-module M which is bounded below and finite-dimensional in each degree.

We will sketch the proof of the following results.

PROPOSITION 9.1. *For each U and V there is a function-object $\{W(X), w(X)\}$ from U to V in $\mathscr{A}^{gr}$.*

THEOREM 9.2. *The functor β: $\mathscr{A}^{gr} \to \mathscr{E}$ of §10 preserves all function-objects.*

When we explain the function-object $\{W(X), w(X)\}$ in (9.1) and the functor β in (9.2), that will complete our explanation of (1.6).

We owe the reader details about $\mathscr{A}^{gr}$, and first we must explain a category $\mathscr{A}$ from which we construct $\mathscr{A}^{gr}$ by passing to an associated graded category.

If we consider the topological problem of computing $[\mathbf{T} \wedge \mathbf{B}G_1, \mathbf{B}G_2]$, it is natural to begin with the special case $\mathbf{T} = \mathbf{S}^0$ and study $[\mathbf{B}G_1, \mathbf{B}G_2]$. The most reasonable approach is to follow the ideas which Segal proposed for the special case $G_2 = 1$. The first step should be to define an algebraic construct $A(G_1, G_2)$ and a homomorphism

$$A(G_1, G_2) \xrightarrow{\alpha} [\mathbf{B}G_1, \mathbf{B}G_2].$$

Here the construct $A(G_1, G_2)$ should play the same role that the usual Burnside ring does in the special case $G_2 = 1$; it should be the closest approximation to $[\mathbf{B}G_1, \mathbf{B}G_2]$ that can be constructed by algebraic means (without using analytic methods such as completion).

In the special case $G_1 = G_2$, our construct $A(G, G)$ has already appeared in the work of C. M. Witten [17], for the same reason and purpose.

In general, these groups A (G_1, G_2) should become the hom-sets of a category, under a product corresponding to the composition of maps of spectra. This category should be monoidal, with the monodial operation corresponding to the smash product in the category of spectra.

We therefore set up the "Burnside category" $\mathscr{A}$ as follows.

The objects of the Burnside category $\mathscr{A}$ will be the finite groups $G, H, \ldots$. We wish to describe the hom-set of morphisms from G to H in $\mathscr{A}$. We consider finite sets X which come provided with an action of G on the left of X and an action of H on the right of X, so that these two actions commute and the action of H on the right of X is free. Such sets X we call "(G, H)-sets". We take the (G, H)-sets and classify them into isomorphism classes. The operation of disjoint union passes to isomorphism classes, and turns the set of isomorphism classes into a commutative monoid. This monoid is a free cummutative monoid; we obtain a base by considering the isomorphism classes of (G, H)-sets X which are irreducible under disjoint union. (It is equivalent to say that the action of G on X/H is transitive.) We define $A(G, H)$ to be the Grothendieck group or universal group associated to this monoid. This is a free abelian group; we obtain a base by considering the same irreducibles as before.

For example, if $H = 1$, then a $(G, 1)$-set is essentially just a G-set, and so $A(G, 1)$ reduces to the usual group $A(G)$.

We define the set of morphisms in $\mathscr{A}$ from G to H to be $A(G, H)$. We have to define the composition product

$$A(G, H) \otimes A(H, K) \to A(G, K)$$

(where the notation reveals that for this purpose we shall compose morphisms from left to right). Let X be a (G, H)-set and Y an (H, K)-set; then $X \times_H Y$ is a (G, K)-set. This operation passes to isomorphism classes and is biadditive with respect to disjoint union; so it defines a product as stated. This product is associative and has units; $1_G \in A(G, G)$ is the class of G, considered as a (G, G)-set with the obvious left and right actions. This makes $\mathscr{A}$ into a category.

We now make $\mathscr{A}$ into a monoidal category. The product on objects in the Cartesian product $G \times H$ of groups. (In the ordinary category of groups and homomorphisms this is a categorical product; it is no longer a categorical product in $\mathscr{A}$.) The product on morphisms is defined as follows. Let X_1 be a (G_1, H_1)-set and let X_2 be a (G_2, H_2)-set; then $X_1 \times X_2$ is a $(G_1 \times G_2, H_1 \times H_2)$-set. This consutruction passes to isomorphism classes and is biadditive with respect to disjoint unions; so it defines a product

$$A(G_1, H_1) \otimes A(G_2, H_2) \rightarrow A(G_1 \times G_2, H_1 \times H_2)$$

as required.

We omit discussion of the good formal properties of the constructs just introduced.

For guidance it is useful to note that one can define a functor α from $\mathscr{A}$ to the category of spectra, that is, a set of homomorphisms

$$A(G_1, G_2) \xrightarrow{\alpha} [\mathbf{B}G_1, \mathbf{B}G_2]$$

which preserve the structure. This is done using transfer. It is not needed for the algebraic purposes of the present paper, and so we omit it.

We also define particular morphisms in $\mathscr{A}$. For each homomorphism $\theta: G \rightarrow H$ we introduce an element $\theta_* \in A(G, H)$; this is the class of H, with G acting on its left via θ and H acting on its right. For each monomorphism $\phi: H \rightarrow G$ we introduce an element $\phi^* \in A(G, H)$; this is the class of G, with G acting on its left and H acting on its right via ϕ. This action of H is free because we assume that ϕ is mono.

We can now give more motivation for the category $\mathscr{A}$. A functor T defined on $\mathscr{A}$ provides a functor defined on the usual category of finite groups: on objects G we take $T(G)$ and on homomorphisms $\theta: G \rightarrow H$ we take $T(\theta_*)$. But beyond this we get homomorphisms $T(\phi^*)$, which correspond to the possibility of "induction". (For example, the "homology of groups" is such a functor T, essentially because it factors as a composite of two functors: the functor α from $\mathscr{A}$ to spectra, and the homology-functor from spectra to graded groups.) If T is a functor defined on $\mathscr{A}$, then the homomorphisms $T(\theta_*)$ and $T(\phi^*)$ satisfy all the usual axioms for "induction" and "restriction", including the double coset formula. However, we do not have to state these axioms explicitly; they are implicit in the structure of the category $\mathscr{A}$. We regard the category $\mathscr{A}$ as the place where one can do "universal" calculations with induction and restriction subject to the usual axioms.

We will now move towards our associated graded category $\mathscr{A}^{gr}$. First we take the full subcategory of the Burnside category in which the objects are elementary abelian p-groups. Next we shall define a filtration on its hom-sets $A(U, V)$.

For guidance it is useful to note that the algebraic filtration of a morphism $f \in A(U, V)$ is in fact the Adams filtration of the resulting map of spectra

$$\mathbf{B}U \xrightarrow{\alpha f} \mathbf{B}V.$$

Indeed, for the purposes which we originally had in mind, it was important to know that our algebra had the correct relation to the topological world; if one uses geometrical means to set up a comparison map between two spectral sequences, then it is important to know that the geometrically-induced map of E_2-terms agrees with the map of Ext_A proved to be iso in the algebraic work. For the purposes of the present paper we need not worry.

If X is an irreducible (U, V)-set, we define $s(X)$ by

$$p^{s(X)} = |X/V|.$$

Clearly this depends only on the isomorphism class of X.

We define the filtration subgroup

$$F_sA(U, V) \subset A(U, V)$$

to be the subgroup generated by the elements $p^\lambda[X]$, where X runs over the irreducible (U, V)-sets and λ, X satisfy

$$\lambda + s(X) \geq s.$$

LEMMA 9.3. *Composition and cross product preserve this filtration. More precisely, if X is an irreducible (U, V)-set and Y is an irreducible (V, W)-set then $[X][Y] = p^\lambda[Z]$ where Z is an irreducible (U, W)-set with*

$$\lambda + s(Z) = s(X) + s(Y);$$

similarly for the cross product, with $\lambda = 0$.

We omit the proof.

We can now define the associated graded category $\mathscr{A}^{gr}$. The objects of $\mathscr{A}^{gr}$ are to be the elementary abelian p-groups $U, V, W, \ldots$. The hom-set $\mathscr{A}^{gr}(U, V)$ from U to V is to be a graded vector-space over F_p, whose sth component is

$$F_sA(U, V)/F_{s+1}A(U, V).$$

Lemma 9.3 shows that composition and cross product pass to the quotient and give operations on $\mathscr{A}^{gr}$.

Finally, we return to (9.1).

Let U, V be any two objects of $\mathscr{A}^{gr}$, that is, any two elemetary abelian p-groups. Let X run over a set of representatives for the isomorphism classes of irreducible (U, V)-sets. For each X, let $W(X)$ be the automorphism group of X; of course, we mean "automorphisms of X" to preserve the left action of $G = U$ and the right action of $H = V$. We can consider X as a $(W(X) \times U, V)$-set; let

$$w(X) \in \mathscr{A}^{gr}(W(X) \times U, V)$$

be the class of X. Then the data $\{W(X), w(X)\}$ constitute a function-object from U to V in $\mathscr{A}^{gr}$. The proof is an essentially straightforward exericse about sets with groups acting on them, and for brevity we omit it.

§10. THE FUNCTOR β

In this section we will describe the functor β promised in §9.

We suppose given an element

$$E \in \mathrm{Ext}_A^{1,1}(H^*(1), H^*(Z_p))$$

with the following properties.

(10.1) If $\theta: Z_p \to Z_p$ is an automorphism, then

$$\theta_* E = E.$$

(This reveals that we have reverted to the usual order of composition.)

(10.2) Let $\theta: Z_p \times Z_p \to Z_p \times Z_p$ be the homomorphism

$$\theta(x, y) = (x + \lambda y, y)$$

for some fixed $\lambda \in F_p$. Then

$$\theta_*(1 \times E) = (1 \times E).$$

PROPOSITION 10.3. *For each such element E there is a unique functor β: $\mathscr{A}^{gr} \to \mathscr{E}$ with the following properties.*

(a) *β is given on objects by $\beta(V) = H^*(V)$.*

(b) *β is additive and preserves the monoidal structure.*

(c) *For each morphism $\theta: U \to V$ we have*

$$\beta(\theta_*) = H^*(\theta): H^*(V) \to H^*(U).$$

(d) *For the injection* $i: 1 \to Z_p$ *we have*

$$\beta(i^*) = E \in Ext_A^{1,1}(H^*(1), H^*(Z_p)).$$

The idea is that with a suitable choice of E, we can secure the following: if $f \in A(U, V)$, then

$$\beta f \in \mathrm{Ext}_A^{**}(H^*(V), H^*(U))$$

gives the position of

$$\alpha f: \mathbf{B}U \to \mathbf{B}V$$

in the Adams spectral sequence for computing $[\mathbf{B}U, \mathbf{B}V]$. More formally, if f is of filtration s in $A(U, V)$, then

$$\beta f \in \mathrm{Ext}_A^{s,s}(H^*(V), H^*(U))$$

is a permanent cycle in the Adams spectral sequence

$$\mathrm{Ext}_A^{**}(H^*(V), H^*(U)) \Rightarrow [\mathbf{B}U, \mathbf{B}V],$$

and αf, βf have the same image in E_∞.

Unfortunately, in the absence of special information about our Adams spectral sequences, we cannot claim that a statement of this form defines βf uniquely except for $s = 0, 1$; *a priori*, differentials might cause some permanent cycles in E_2 to map to zero in E_∞. Moreover, one can hardly expect to obtain special information about our Adams spectral sequences until we have calculated their E_2 terms, and this would risk a circular argument.

However, we have a purely algebraic proof of (10.3), and its idea is as follows.

Given the objects of a preadditive monoidal category, one can present its morphisms by generators and relations, just as one presents a group by generators and relations. (For a group one builds up words from generators by multiplication, but here one builds up words from generators by the operations allowed in a preadditive monoidal category.) The category $\mathscr{A}^{gr}$ can be presented by generators and relations. A suitable presentation has as its generators all morphisms θ_*, and a single morphism i^* corresponding to the map $i: 1 \to Z_p$. (Each θ_* is of filtration 0, while i^* is of filtration 1.) As relations we have various formal relations which involve the generators θ_* but not i^*, and two special relations: if $\theta: Z_p \to Z_p$ is an automorphism (as in (10.1)) then $\theta_* i^* = i^*$, and if $\theta: Z_p \times Z_p \to Z_p \times Z_p$ is as in (10.2), then $\theta_*(1 \times i^*) = (1 \times i^*)$. Now, (10.3) (a)–(d) specify β on generators and ensure that β preserves the relations, so there is a unique β as asserted.

To carry out this proof would involve two things: to make precise the notion of a "presentation", and to show that $\mathscr{A}^{gr}$ can be presented by the presentation in question. In fact one need not do the first; the mathematics involved in the second can be rewritten as a proof that the requirements lead to a unique choice of β, with all the required properties, on successively larger parts of $\mathscr{A}^{gr}$. The work is somewhat long; it requires linear algebra and no new ideas, and we omit it.

We explain the choice of E we propose. In fact there is only one choice up to a scalar factor; and if we replace E by λE, we multiply β by λ^s in degree s, so as long as we don't take $\lambda = 0$ we don't alter the truth of (9.2) or (1.6). Evidently we should take E to be the position in the appropriate Adams spectral sequence of the map $\alpha(i^*): \mathbf{B}Z_p \to \mathbf{B}1$. This choice agrees (up to sign) with the purely algebraic choice we will give.

Let M be the submodule of $H^*(Z_p)_{\mathrm{loc}}$ which consists of the groups in degrees ≥ -1. It takes part in the following short exact sequence.

$$0 \to H^*(Z_p) \to M \xrightarrow{\mathrm{res}} F_p \to 0.$$

We take the class of this extension as our element

$$E \in \mathrm{Ext}_A^{1,1}(F_p, H^*(Z_p)).$$

This settles the functor β.

§11. THE CASE $U = Z_p$

We sketch the proof of the following.

THEOREM 11.1. (9.2) *and* (1.6) *are true for* $U = Z_p$.

The proof is based upon the following diagram.

$$\begin{array}{ccc}
\uparrow & & \\
\mathrm{Ext}^s_A(H^*(V)\otimes H^*(Z_p)_{\mathrm{loc}}, M) & \xleftarrow[\cong]{\{res_k\}} & \bigoplus_{k=1}^{p^n} \mathrm{Ext}^s_A(H^*(W_k), M) \\
\uparrow{\scriptstyle 1\otimes j} & & \downarrow{\scriptstyle \{X_k\}} \\
\mathrm{Ext}^s_A\left(H^*(V)\otimes \dfrac{H^*(Z_p)_{\mathrm{loc}}}{H^*(Z_p)}, M\right) & \xleftarrow[\cong]{D} & \mathrm{Ext}^s_A(H^*(V), M\otimes H^*(Z_p)) \\
\uparrow{\scriptstyle \Delta} & \nearrow{\scriptstyle X_0} & \\
\mathrm{Ext}^{s-1}_A(H^*(V)\otimes H^*(Z_p), M) & & \\
\uparrow & &
\end{array}$$

Diagram 11.2

In this diagram, the vertical exact sequence on the left is obtained from the short exact sequence

$$0 \to H^*(Z_p) \to H^*(Z_p)_{\mathrm{loc}} \xrightarrow{j} \frac{H^*(Z_p)_{\mathrm{loc}}}{H^*(Z_p)} \to 0$$

by applying $H^*(V)\otimes\_\_$ and then applying Ext_A. The indices X are those which arise in (1.6); they are as described in §9. If $U = Z_p$, then p^n of the indices X correspond to the homomorphisms $\theta_k : Z_p \to V$; we assign them the numbers $k = 1, 2, \ldots, p^n$. There is one more index, namely the (Z_p, V)-set $V \times Z_p$; we assign it the number $k = 0$. The arrows labelled X_k, X_0 are induced as described in §9.

The map

$$\mathrm{res}_k : H^*(V)\otimes H^*(Z_p)_{\mathrm{loc}} \to H^*(W_k)$$

is a residue of the sort used in (1.5). In fact, the group W_k corresponding to X_k, although isomorphic to V, should be considered as a quotient of $V \times Z_p$ via the map $V \times Z_p \to W_k$ which carries (v, z) to $v + \theta_k z$; therefore $H^*(W_k)$ is a subalgebra of $H^*(V)\otimes H^*(Z_p)$, and we can take residues of formal Laurent series with coefficients in $H^*(W_k)$. The map $\{\mathrm{res}_k\}$ in (11.2) is iso by (1.5) plus (1.2).

The map marked D (for "duality") is provided as follows.

Let L, M, P be left A-modules; let P^* be the dual of P, made into a left A-module in the usual way so that the evaluation map $P^* \otimes P \xrightarrow{ev} F_p$ is A-linear. We assume that M is bounded below, and that P is bounded above and finite-dimensional over F_p in each degree.

LEMMA 11.3. *Then the natural transformation*

$$Ext^{**}_A(L, M\otimes P^*) \to Ext^{**}_A(L\otimes P, M)$$

is iso.

The natural transformation carries an element $L \xrightarrow{f} M\otimes P^*$ of the Ext category to the composite

$$L\otimes P \xrightarrow{f\otimes 1} M\otimes P^*\otimes P \xrightarrow{1\otimes ev} M.$$

The proof of (11.3) is easy, and we omit it.

To apply (11.3), we replace the map $P^* \otimes P \xrightarrow{ev} F_p$ by a map

$$H^*(Z_p) \otimes \frac{H^*(Z_p)_{\mathrm{loc}}}{H^*(Z_p)} \to F_p$$

obtained as follows. Consider

$$H^*(Z_p)_{\mathrm{loc}} \otimes H^*(Z_p)_{\mathrm{loc}} \xrightarrow{\bar{\Delta}^*} H^*(Z_p)_{\mathrm{loc}} \xrightarrow{\mathrm{res}} F_p;$$

this is an A-map (of degree $+1$) which is also a dual pairing, and in which $H^*(Z_p)$ annihilates $H^*(Z_p)$. Here we define $\bar{\Delta}: Z_p \to Z_p \times Z_p$ by $\bar{\Delta}(z) = (-z, z)$. The sign serves to get some details correct when we check that diagram (11.2) commutes (up to a fixed sign for the triangle). It takes several lemmas to prove that (11.2) commutes; we omit them for brevity. The result (11.1) then follows by diagram-chasing.

§12. PROOF OF (1.6)

We will sketch the proof of (9.2).

First we remark that if β preserves one function-object from U to V, then it preserves all function-objects from U to V. (If we have two function-objects $\{W'_i, w'_i\}$, $\{{}''_j \cdot w''_j\}$ for the same source and target, then one can be thrown onto the other by an invertible matrix of maps $W'_i \to W''_j$ of suitable degrees; β carries an invertible matrix to an invertible matrix.)

Secondly we show how to make new function-objects from old. Suppose given a monoidal category $\mathscr{C}$ and three objects F, G, H in $\mathscr{C}$. Suppose given a function-object

$$\{W_j, W_j \otimes G \xrightarrow{w_j} H\}$$

from G to H, and suppose that for each W_j we have a function-object

$$\{V_{ij}, V_{ij} \otimes F \xrightarrow{v_{ij}} W_j\}$$

from F to W_j. Then we can form the morphism

$$V_{ij} \otimes F \otimes G \xrightarrow{v_{ij} \otimes 1} W_j \otimes G \xrightarrow{w_j} H.$$

LEMMA 12.1. *$\{V_{ij}, (v_{ij} \otimes 1)w_j\}$ is a function-object from $F \otimes G$ to H.*

The proof is easy.

We will call this construction of a function-object from $F \otimes G$ to H the "product construction".

LEMMA 12.2. *Suppose that a functor β preserves the function-object $\{W_j, w_j\}$ from G to H and also preserves the function-object $\{V_{ij}, v_{ij}\}$ from F to W_j for all j. Then it preserves the function-object $\{V_{ij}, (v_{ij} \otimes 1)w_j\}$ from $F \otimes G$ to H given by the product construction.*

The proof is easy.

Proof of (9.2). Consider a function-object in $\mathscr{A}^{gr}$ from U to V. If U is of rank 0 or 1 the result is trivial or true by (11.1); so we proceed by induction over the rank of U. Suppose $U = U' \times U''$ where U' and U'' are of less rank. Then by (9.1) there is a function-object $\{W_j, w_j\}$ from U'' to V and there is also a function-object $\{V_{ij}, v_{ij}\}$ from U' to W_j. By the inductive hypothesis β preserves these function-objects; so by (12.2) it preserves the function-object from $U' \times U''$ to V given by the product construction. Therefore β preserves any other function-object from $U' \times U''$ to V. This completes the induction and proves (9.2), which finishes the sketch proof of (1.6).

REFERENCES

1. R. R. BRUNER, J. P. MAY, J. E. MCCLURE and M. STEINBERGER: H_∞ ring spectra and applications, to appear in *Lecture Notes in Mathematics*, Springer-Verlag. Especially Chap. 2.
2. G. CARLSSON: Equivariant stable homotopy and Segal's Burnside ring conjecture, preprint, University of California at San Diego, 1982.
3. G. CARLSSON: Equivariant stable homotopy and Segal's Burnside ring conjecture, to appear in the *Ann. Math.*
4. J. H. GUNAWARDENA: Cohomotopy of some classifying spaces, thesis, Cambridge, 1981.
5. J. LANNES and S. ZARATI: Dérivés de la déstabilisation, invariants de Hopf d'ordre supérieur, et suite spectrale d'Adams, preprint, Ecole Polytechnique/Université Paris-Sud.
6. L. G. LEWIS, J. P. MAY and J. E. MCCLURE: Classifying G-spaces and the Segal conjecture, in *Canadian Math. Soc. Conference Proceedings* Vol. 2 part 2, Amer. Math. Soc. 1982, pp. 165–179.
7. H. H. LI and W. M. SINGER: Resolutions of modules over the Steenrod algebra and the classical theory of invariants, *Math. Zeit.* **181** (1982), 269–286.
8. W. H. LIN, D. M. DAVIS, M. E. MAHOWALD and J. F. ADAMS: Calculation of Lin's Ext groups, *Math. Proc. Camb. Phil. Soc.* **87** (1980), 459–469, especially p. 462.
9. S. MACLANE: *Categories for the Working Mathematician*, Springer-Verlag 1971, Chap. 7, especially pp. 157–158.
10. J. P. MAY and S. B. PRIDDY: The Segal conjecture for elementary abelian p-groups, II, to appear.
11. J. W. MILNOR: The Steenrod algebra and its dual, *Ann. Math.* **67** (1958), 150–171.
12. H. MUI: Modular invariant theory and the cohomology algebras of symmetric groups, *J. Fac. Sci. Univ. Tokyo* **22** (1975), 319–369.
13. S. B. PRIDDY and C. WILKERSON: Hilbert's Theorem 90 and the Segal conjecture for elementary abelian p-groups, to appear in *Am. J. Math.*
14. W. M. SINGER: On the localisation of modules over the Steenrod algebra, *J. Pure Appl. Algebra* **16** (1980), 75–84.
15. W. M. SINGER: A new chain complex for the homology of the Steenrod algebra, *Math. Proc. Camb. Phil. Soc.* **90** (1981), 279–282.
16. R. STEINBERG: Prime power representations of finite linear groups I, II, *Can. J. Math.* **8** (1956), 580–591; **9** (1957), 347–351.
17. C. M. WITTEN: thesis, Stanford, 1978.

Department of Pure Mathematics,
University of Cambridge,
16 Mill Lane,
Cambridge CB2 1SB, UK

THE SPHERE, CONSIDERED AS AN H-SPACE MOD p

By J. F. ADAMS (*Cambridge*)

[Received 28 May 1960]

I. M. JAMES (4) has suggested a classification of H-spaces into ten classes and has given examples to show that five of these classes are non-empty. It is one of the objects of this paper to give examples showing that all ten classes are non-empty. I begin by summarizing James' classification.

By an *H-space* we shall understand a space X provided with a base-point $e \in X$ and a product map $\mu\colon X \times X \to X$ such that $\mu(x,e) = x$, $\mu(e,x) = x$. We shall say that two H-spaces (X,e,μ), (X',e',μ') are *equivalent* if there is a homotopy-equivalence $f\colon X, e \to X', e'$ such that the following diagram is homotopy-commutative.

$$\begin{array}{ccc} X \times X & \xrightarrow{\mu} & X \\ {\scriptstyle f\times f}\downarrow & & \downarrow{\scriptstyle f} \\ X' \times X' & \xrightarrow{\mu'} & X' \end{array}$$

Let A denote an adjective applicable to H-spaces, as for example 'associative' or 'commutative'; then we shall say that an H-space (X,e,μ) is *equivalent-A* if there is an equivalent H-space (X',e',μ') which is A.

An H-space may be

(*a*) not homotopy-associative, or

(*b*) homotopy-associative but not equivalent-associative, or

(*c*) equivalent-associative.

Similarly, it may be

(*d*) not homotopy-commutative, or

(*e*) homotopy-commutative, but not equivalent-commutative, or

(*f*) equivalent-commutative.

In this way we obtain nine classes of H-spaces. One of these classes, however, can be divided into two, for, if an H-space is equivalent-associative and also equivalent-commutative, it may be

(*g*) equivalent-(associative and commutative), or

(*h*) not so.

In this way H-spaces can be divided into ten classes.

Quart. J. Math. Oxford (2), 12 (1961), 52–60.

I remark that our definition of 'equivalent-associative' (chosen for its obvious homotopy-invariance) is different from James' definition and not obviously equivalent to it. It is therefore possible that the ten classes defined above differ from James' ten classes. However, the ten examples I shall give serve equally well in either context.

These ten examples are all spheres, mutilated in one fashion or another. To prove that they are not equivalent-associative, or not homotopy-associative, we rely on the following two lemmas:

LEMMA 1. *For each odd integer $n \geqslant 5$ there is an infinity of primes p such that the exterior algebra $E(n; Z_p)$ is not a possible cohomology ring $H^*(X; Z_p)$ for an equivalent-associative H-space.*

LEMMA 2. *For suitable odd n (e.g. $n = 7$) the exterior algebra $E(n; Z_3)$ is not a possible cohomology ring $H^*(X; Z_3)$ for a homotopy-associative H-space.*

Proof of Lemma 1. Suppose that X is an equivalent-associative H-space whose cohomology ring is $E(n; Z_p)$. Then there is an equivalent H-space X' which is associative and has the same cohomology ring. By (**2**) it has a classifying space B; and $H^*(B; Z_p)$ is a polynomial ring $P(n+1; Z_p)$. Let x be a generator of this ring; then $x^p \neq 0$: that is, $P_p^f(x) \neq 0$, where $f = \frac{1}{2}(n+1)$. By the Adem relations (**1**), P_p^f can be factorized in terms of operations P_p^g, where $g = p^h$. Therefore at least one such operation P_p^g is non-zero in $P(n+1; Z_p)$. This implies that

$$n+1 \mid 2(p-1)p^h.$$

This is impossible if

$$p > n+1, \qquad p \not\equiv 1 \bmod \tfrac{1}{2}(n+1).$$

If $n \geqslant 5$, we can find an infinity of primes p satisfying these conditions.

The proof of Lemma 2 is closely similar. If X is a homotopy-associative H-space, we can construct a 'projective 3-space over X', say B. The cohomology ring $H^*(B; Z_3)$ is then a truncated polynomial ring, in which the cube of the generator is non-zero. We now apply the Adem relations, as above, with $p = 3$.

It is natural to study H-spaces X such that $H^*(X; Z_p) = E(n; Z_p)$, because this is one of the simplest cohomology rings possible for an H-space. Our next theorem will provide examples of such H-spaces.

Let n be an odd integer greater than one; let us divide the primes p into two classes, P and Q. Let Z be the additive group of integers, and F the additive group of fractions t/u such that all the prime factors of u lie in Q.

It is easy to see that one can construct a countable CW-complex X

containing the n-sphere S^n as a subcomplex, so that X is a Moore space of type $Y(F, n)$, and so that the induced homomorphism

$$i_*: H_n(S^n) \to H_n(X)$$

realizes the embedding of Z in F. (Further details will be given below.) We have the theorem:

Theorem. *Such a space X satisfies the following conditions.*

(i) $H^*(X; Z_p) = E(n; Z_p)$ *if* $p \in P$.

(ii) *If* $r \neq n$, *then* $\pi_r(X) \cong \sum_{p \in P} {}_p\pi_r(S^n)$, *where* ${}_p\pi_r(S^n)$ *denotes the p-component of the torsion group* $\pi_r(S^n)$.

(iii) *If* $2 \in Q$, *then X is a commutative H-space; if (in addition)* $3 \in Q$, *then X is homotopy-associative.*

We may summarize this theorem by the 'slogan' that 'an odd sphere is an H-space mod p'. The author originally made this remark in 1956, in order to answer a rather technical question concerned with the p-components of the homotopy groups of spheres. It was not published at that time. Taken in conjunction with Lemmas 1 and 2, it evidently yields examples of H-spaces of the following classes:

(i) commutative and not homotopy-associative;
(ii) commutative and homotopy-associative, but not equivalent-associative.

It also yields examples in connexion with the work of Stasheff and the author on A_n-spaces (7); however, no details of this application will be given in the present note.

Proof of the Theorem. We first make explicit the construction of X. Let d_m ($m = 1, 2, \ldots$) be a sequence of positive integers, to be chosen later; let e be a vertex in S^n and let $f_m: S^n, e \to S^n, e$ be a map of degree d_m. We may construct a CW-complex X by taking the product

$$\left(\bigcup_m [2m-1, 2m]\right) \times S^n$$

and identifying each point $(2m, x)$ with $(2m+1, f_m(x))$. (X is thus a 'telescope'.) The sphere S^n is embedded in X as $1 \times S^n$. We easily see that

$$H_n(X) = F',$$

where F' is the additive group of fractions $t/(d_1 d_2 \ldots d_m)$; the induced homomorphism

$$i_*: H_n(S^n) \to H_n(X)$$

realizes the embedding of Z in F'. We can ensure that $F' = F$ by a suitable choice of the integers d_m. By our construction, X is $(n-1)$-connected and has $H_r(X) = 0$ for $r \neq 0, n$; X is therefore a Moore space, of the sort required.

We establish part (ii) of the theorem by two applications of Serre's C-theory (6). Let C_Q be the class of torsion groups in which the order of each element is a product of primes in Q; similarly for C_P. Then the homomorphism

$$i_*: H_r(S^n) \to H_r(X)$$

is an isomorphism mod C_Q for each r. By the main theorems of (6) we see that

$$i_*: \pi_r(S^n) \to \pi_r(X)$$

is an isomorphism mod C_Q for all r.

Next, let R be the group of rationals; then the Eilenberg–Maclane groups $H_r(R, n)$ may be described as follows. $H^*(R, n; R)$ is an algebra over R on one generator of dimension n; it is a polynomial algebra or an exterior algebra according as n is even or odd. For $r > 0$, $H_r(R, n; Z)$ is the vector-space dual (over R) of $H^r(R, n; R)$. (These facts are easily established by induction over n.) Let Y be a space of type $K(R, n)$; then it is possible to find a map $f: X \to Y$ such that the homomorphism

$$f_*: H_n(X) \to H_n(Y)$$

realizes the embedding of F in R. Since n is odd,

$$f_*: H_r(X) \to H_r(Y)$$

is an isomorphism mod C_P for all r. Hence

$$f_*: \pi_r(X) \to \pi_r(Y)$$

is an isomorphism mod C_P for all r. These conclusions establish part (ii) of the theorem.

We begin work on part (iii) of the theorem by considering the symmetric square ΣX of X; a commutative product on X is equivalent to a retraction of ΣX onto X. We shall show that, if $2 \in Q$, then X is a deformation retract of ΣX, so that X admits one and only one homotopy class of commutative products.

For this purpose only the homotopy-type of the pair ΣX, X is relevant; and this is determined by the homotopy-type of X, which is determined by the data. We can therefore work in terms of any chosen X. Let us take X to be a 'simplicial telescope', in the obvious sense. Then X is an enumerable CW-complex, and X^2 admits a triangulation which is preserved when the two coordinates are permuted. In this way ΣX becomes a simplicial CW-complex. Moreover, X is the union of an increasing sequence of subcomplexes, each equivalent to

S^n; therefore ΣX is the union of an increasing sequence of subcomplexes, each equivalent to ΣS^n. Since n is odd, we have

$$H_r(\Sigma S^n) = \begin{cases} Z & (r = 0, n), \\ Z_2 & (r = n+2, n+4, \dots, 2n-1), \\ 0 & \text{otherwise.} \end{cases}$$

In order to calculate $H_r(\Sigma X)$ as a direct limit we need only know the homomorphism

$$(\Sigma f_m)_*\colon H_r(\Sigma S^n) \to H_r(\Sigma S^n)$$

induced by a map $f_m\colon S^n \to S^n$ of degree d_m. This homomorphism is zero if $r > n$ and d_m is even, as one sees by the method of [(**8**) § 11]. We conclude that, if $2 \in Q$, then

$$H_r(\Sigma X, X) = 0$$

for each r. It follows that X is a deformation retract of ΣX.

We must now insert a slight digression. Since the homotopy-type of X is well-determined, we can prove part (i) of the theorem by calculating $H^*(X; Z_p)$ for some convenient X, for example, a telescope. More generally, let A be an abelian group and let G be a group in C_P; then we have

$$F \otimes_Z F = F, \qquad \mathrm{Tor}_Z(F, A) = 0,$$
$$\mathrm{Hom}_Z(F, G) = G, \qquad \mathrm{Ext}_Z(F, G) = 0.$$

For example, the group $\mathrm{Ext}_Z(F, G)$ can be found by calculating $H^{n+1}(X; G)$ when X is a telescope; or, equivalently, we can remark that the chain groups of such a telescope provide a resolution of F over Z.

We complete the proof of part (iii) of the theorem by comparing X with a loop-space. Let S^2X be the (reduced) double suspension of X, taking e as the base-point; S^2X is thus a Moore space of type $Y(F, n+2)$. We can embed $S(\Sigma X)$ in $\Sigma(SX)$ by the rule

$$i(t, (x, y)) = ((t, x), (t, y)) \quad (0 \leqslant t \leqslant 1).$$

This is consistent with the embedding of SX in each. Therefore we can embed $S^2(\Sigma X)$ in $\Sigma(S^2X)$ consistently with the embedding of S^2X in each; we see that

$$H_r(\Sigma S^2X, S^2\Sigma X) = 0.$$

The retraction from ΣX to X gives (by double suspension) a retraction from $S^2\Sigma X$ to S^2X; we can now extend this to a retraction on ΣS^2X.

Let us write

$$r\colon \Sigma X \to X, \qquad R\colon \Sigma S^2X \to S^2X$$

for the retractions we have constructed, and let Ω^2S^2X be the double loop-space on S^2X. The product R on S^2X induces a product μ_R on

$\Omega^2 S^2 X$. We have an embedding

$$j\colon X \to \Omega^2 S^2 X;$$

we now show that j is a homomorphism, with respect to the products r, μ_R. The verification is trivial. For

$$(jr(x_1, x_2))(t,u) = (t,u,r(x_1,x_2)) \quad \text{(by definition of } j)$$

$$\begin{aligned}(\mu_R(jx_1, jx_2))(t,u) &= R((jx_1)(t,u), (jx_2)(t,u)) \quad \text{(by definition of } \mu_R) \\ &= R((t,u,x_1), (t,u,x_2)) \\ &= Ri(t,u,(x_1,x_2)) \\ &\qquad \text{(where } i \text{ is the embedding of } S^2\Sigma X \text{ in } \Sigma S^2 X) \\ &= (t,u,r(x_1,x_2)) \quad \text{(by construction of } R).\end{aligned}$$

We have thus shown that j is a homomorphism provided that the product used in $\Omega^2 S^2 X$ is μ_R. On the other hand, this product in $\Omega^2 S^2 X$ is homotopic to the ordinary product of loops, as is well known; and the latter is homotopy-associative. We conclude that there is a homotopy

$$h\colon I \times X^3 \to \Omega^2 S^2 X$$

in $\Omega^2 S^2 X$, between the maps

$$\mu(\mu \times 1),\ \mu(1 \times \mu)\colon X^3 \to X.$$

It remains only to compress this homotopy into X.

Since the groups $\pi_r(X)$, $\pi_{r+2}(S^2 X)$ are in C_P for $r \neq n$, the groups $\pi_r(\Omega^2 S^2 X, X)$ are also in C_P. It is easy to calculate $H_r(\Omega^2 S^2 X)$, $H^r(\Omega^2 S^2 X;\, R)$ in a limited range of dimensions; since n is odd, we find that

$$H_r(\Omega^2 S^2 X, X) = \begin{cases} 0 & \text{if } 0 \leqslant r < p(n+1)-2, \\ Z_p & \text{if } r = p(n+1)-2, \end{cases}$$

where p is the smallest prime in P. Therefore $\pi_r(\Omega^2 S^2 X, X) = 0$ if $r < p(n+1)-2$. On the other hand, if G is a group in C_P, we have

$$H^r(I \times X^3, (0 \cup 1) \times X^3;\, G) = 0$$

if $r > 3n+1$. If $p > 3$, we have $3n+1 < p(n+1)-2$, and the theory of obstructions shows that the required compression is possible. This completes the proof.

To give our next example, we retain some of the notations used above, so that p is the smallest prime in the set P; we suppose that $p > 2$. Let Y be a space of type $Y(F, n)$, as above, and let W be a space obtained from Y by the method of 'killing homotopy groups', retaining its first two non-zero groups (which lie in dimensions n, $n+2p-3$) and killing the remainder. Then W is not equivalent to any H-space W'

which is both associative and commutative; for by a theorem of J. C. Moore, A. Dold, and R. Thom (**3**) such a space W' is weakly equivalent to a Cartesian product of Eilenberg–Maclane spaces, which W is not. (In fact, W has a non-zero k-invariant, given by the Steenrod operation P^1.)

However, W is weakly equivalent to a loop-space because the Postnikov invariant of W is stable; we therefore dispose of various devices for showing that W is equivalent to an associative H-space. We shall also show that W admits a commutative product. In what follows, we may suppose (as before) that our spaces are simplicial CW-complexes. We have a retraction from $W \cup \Sigma Y$ to W; we shall show that there is no obstruction to extending this retraction over ΣW. In fact, W may be obtained from Y by adjoining cells of dimension $n+2p+2$ and higher; therefore

$$H^r(\Sigma W,\ W \cup \Sigma Y;\ G) = 0$$

for $r < n+2p+2$. On the other hand, $\pi_r(W) = 0$ for $r > n+2p-3$. The extension is therefore possible.

Our further constructions depend on a modification of the 'method of killing homotopy groups'. As above, we suppose the primes p divided into two classes P and Q.

LEMMA 3. *Suppose given an integer n and a CW-complex X such that $\pi_r(X)$ is a torsion group for $r \geqslant n$. Then there is a CW-complex Y containing X such that*

$$i_*\colon \pi_r(X) \to \pi_r(Y)$$

is an isomorphism mod C_Q *for all r, while*

$$\pi_r(Y) = 0 \bmod C_P \quad \textit{for } r \geqslant n.$$

Proof. Suppose that we form a space X' by attaching to X the cone on a $Y(G,n)$, where $G \in C_Q$. Then we have

$$\pi_r(X', X) = 0 \bmod C_Q$$

for all r. If we take

$$G = \sum_{p \in Q} {}_p\pi_n(X),$$

we can ensure that

$$\pi_n(X') = \sum_{p \in P} {}_p\pi_n(X).$$

This gives the first step of an obvious induction, proving the lemma.

It is easily shown (by obstruction-theory) that, if we start with a map $f\colon X \to X'$ between two spaces such as X, we can extend it to a map $g\colon Y \to Y'$.

Lemma 3 will provide us with all our remaining examples in a very simple fashion. Let us begin from the 7-sphere S^7; it admits various products, but they are neither homotopy-associative nor homotopy-commutative. For each product μ, the separation element which shows that μ is not homotopy-associative lies in $\pi_{21}(S^7)$. Let Q be the set of primes p such that ${}_p\pi_{21}(S^7) \neq 0$; then Q is finite. Using Lemma 3, we may kill the corresponding p-components from dimension 21 upwards, obtaining a space X. Any product $\mu\colon S^7 \times S^7 \to S^7$ can be extended to a product $\mu'\colon X \times X \to X$; choose one such product μ'. By construction, we have an associating homotopy defined over $S^7 \times S^7 \times S^7$; and the further obstructions to extending this homotopy over $X \times X \times X$ lie in cohomology groups which vanish.

The H-space X is therefore homotopy-associative. It is not equivalent-associative because Lemma 1 applies for an infinity of primes p. It is not homotopy-commutative because we have done nothing to alter the separation elements in $\pi_{14}(S^7)$.

Similarly, let us begin with a product

$$\mu\colon S^7 \times S^7 \to S^7$$

such that the separation element to homotopy-commutativity is of order 2^k in $\pi_{14}(S^7)$. [It is easy to see that such a product exists; cf. (3.1) of (5)]. Using Lemma 3, we may kill the 2-components from dimension 14 upwards, obtaining a space X. Arguing as before, X is an H-space and is homotopy-commutative. It is not equivalent-commutative because ΣX cannot be retracted onto X, owing to the non-zero operation

$$Sq^2\colon H^7(\Sigma X;\, Z_2) \to H^9(\Sigma X;\, Z_2).$$

It is not homotopy-associative, by Lemma 2.

Similarly, we may employ both the above devices. Let X be a space obtained from S^7 by killing ${}_2\pi_r(S^7)$ for $r \geqslant 14$ and ${}_p\pi_r(S^7)$ for $r \geqslant 21$ and a suitable finite set of p; then X admits a product which is homotopy-commutative and homotopy-associative, but neither equivalent-commutative nor equivalent-associative.

Lastly, let X be a space obtained from S^7 by killing $\pi_r(S^7)$ for $r \geqslant 13$. Then the Postnikov system of X is stable, so X is weakly equivalent to an iterated loop-space $\Omega^m Y$ for as large an m as we please. In particular, X is equivalent-associative and homotopy-commutative. It is not equivalent-commutative, for the same reason as before:

$$Sq^2\colon H^7(\Sigma X;\, Z_2) \to H^9(\Sigma X;\, Z_2)$$

is non-zero.

If we add to these examples the standard H-space structures on S^1, S^3, and S^7, we obtain the ten examples required.

We have in fact done more than was needed. At various points we were committed to prove propositions of the following form: 'a certain H-space (X, e, μ) does not have the property P'. On each occasion, we have proved in addition that (X, e, μ') does not have the property P for any other product μ'.

REFERENCES

1. J. Adem, 'Relations on iterated reduced powers', *Proc. Nat. Acad. Sci. U.S.A.* 39 (1953) 636–8.
2. A. Dold and R. Lashof, 'Principal quasi-fibrations and fibre homotopy equivalence of bundles', *Illinois J. of Math.* 3 (1959) 285–305.
3. A. Dold and R. Thom, 'Quasifaserungen und unendliche symmetrische Produkte', *Annals of Math.* 67 (1958) 239–81.
4. I. M. James, *The ten types of H-spaces* (mimeographed).
5. —— 'Products on spheres', *Mathematika* 6 (1959) 1–13.
6. J.-P. Serre, 'Groupes d'homotopie et classes de groupes abéliens', *Annals of Math.* 58 (1953) 258–94.
7. J. Stasheff [to appear].
8. S. D. Liao, 'On the topology of cyclic products of spheres', *Trans. American Math. Soc.* 77 (1954) 520–51.

Topology Vol. 1, pp. 67–72. Pergamon Press, 1961.

H-SPACES WITH FEW CELLS

J. F. ADAMS

(*Received* 1 *April* 1960)

I. M. JAMES has recently studied the question: when can a q-sphere bundle over the n-sphere be an H-space? The possibilities are much reduced by the theorem below. We shall suppose that G is an H-space such that $H_*(G; Z_2)$ is an exterior algebra on two generators of (positive) dimensions q, n. (Without loss of generality, we shall suppose that $q \leq n$).

THEOREM. *Such an H-space G is only possible if the dimensions* (q, n) *satisfy one of the following conditions:*

(i) *q and n both belong to the set* (1, 3, 7);
(ii) $(q, n) = (1, 2), (3, 5)$ *or* $(7, 11)$;
(iii) $(q, n) = (7, 15)$.

It is clear that the products $S^q \times S^n$ provide examples of H-spaces for which condition (i) holds. Again, the groups $SO(3)$ and $SU(3)$ provide examples in which alternative (ii) holds, with $(q, n) = (1, 2)$ and $(3, 5)$. I do not know whether the remaining two pairs of dimensions are possible or not.

The theorem is proved by studying cohomology operations in a "projective plane" P corresponding to G. Let TG be the cone on G; then the projective plane P is formed by attaching $(TG)^2$ to the suspension SG of G by a "Hopf" map; this map is constructed as follows. The boundary B of $(TG)^2$ consists of $(TG \times G) \cup (G \times TG)$. We decompose the suspension SG into the two cones $T_- G$, $T_+ G$. We extend the product map

$$\mu : G \times G \to G$$

to give

$$\mu_+ : TG \times G \to T_+G$$

$$\mu_- : G \times TG \to T_-G.$$

We thus obtain a "Hopf" map

$$J(\mu) : B \to SG$$

and so construct P.

We must now discuss the cohomology of P. (From this point on, we agree that Z_2 coefficients are to be understood in all our cohomology groups). The additive structure of $H^*(P)$ is easily determined from the exact cohomology sequence of the pair (P, SG). The coboundary

$$\delta : H^*(SG) \to H^*(P, SG)$$

is determined by the following diagram.

$$\begin{array}{ccccc} H^*(G) & \xrightarrow{\Delta_{SG}} & H^*(SG) & \xrightarrow{\delta} & H^*(P, SG) \\ \downarrow{\mu^*} & & \downarrow{J(\mu)^*} & & \downarrow{\gamma^*} \cong \\ H^*(G \times G) & \xrightarrow{\Delta_B} & H^*(B) & \xrightarrow{\delta} & H^*((TG)^2, B) \end{array}$$

Here, of course, the homomorphisms Δ_B, Δ_{SG} belong to the Mayer-Vietoris sequences for $B = (TG \times G) \cup (G \times TG)$, $SG = (T_+G) \cup (T_-G)$. In positive dimensions, Δ_{SG} is an isomorphism, while $\delta\Delta_B$ is an epimorphism, with kernel $H^*(G \vee G)$. The map $\gamma : (TG)^2 \to P$ is the quotient map.

Let us write $\phi : H^*(P) \to H^*(G)$ for the composite

$$H^*(P) \xrightarrow{i^*} H^*(SG) \overset{\Delta_{SG}}{\cong} H^*(G),$$

and

$$\theta : H^*(G \times G) \to H^*(P)$$

for the composite

$$H^*(G \times G) \xrightarrow{\Delta_B} H^*(B) \xrightarrow{\delta} H^*((TG)^2, B) \underset{\cong}{\xleftarrow{\gamma^*}} H^*(P, SG) \xrightarrow{j^*} H^*(P).$$

We obtain the following result.

LEMMA 1. *Let x, y be primitive generators in $H^q(G)$, $H^r(G)$. Then there exist elements ξ, η in $H^*(P)$ such that $\phi\xi = x$, $\phi\eta = y$. The three elements x, y, xy in $H^*(G)$ yield nine external products in $H^*(G \times G)$, and the images of these by θ satisfy only one relation, namely $\theta(x \otimes y) = \theta(y \otimes x)$.*

We thus obtain eight linearly independent elements in $H^*(P)$; we may form a base of $H^*(P)$ by adjoining 1, ξ and η.

LEMMA 2. *We have $\xi^2 = \theta(x \otimes x)$, $\xi\eta = \theta(x \otimes y)$, $\eta^2 = \theta(y \otimes y)$. All other non-trivial cup-products in $H^*(P)$ are zero.*

Proof. This follows immediately from the following commutative diagram.

$$\begin{array}{ccccc} H^*(P) & \otimes & H^*(P) & \longrightarrow & H^*(P) \\ \downarrow{\phi} & & \downarrow{\phi} & & \uparrow \\ H^*(G) & \otimes & H^*(G) & & \theta \\ \downarrow{\pi_1^*} & & \downarrow{\pi_2^*} & & \\ H^*(G \times G) & \otimes & H^*(G \times G) & \longrightarrow & H^*(G \times G) \end{array}$$

Here the horizontal arrows indicate cup-products, while the maps $\pi_1, \pi_2 : G \times G \to G$ are the projections onto the first and second factors. This diagram is deduced from the following one.

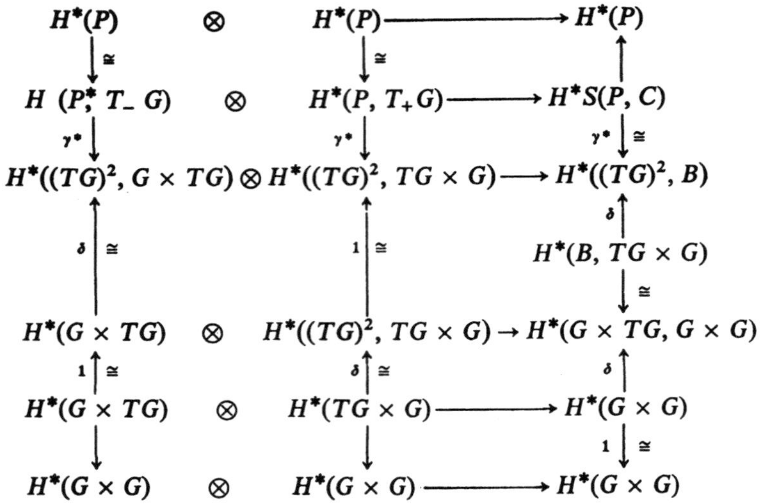
H*(P) ⊗ H*(P) → H*(P)
H (P; T₋ G) ⊗ H*(P, T₊G) → H*S(P, C)
γ* γ* γ*
H*((TG)², G × TG) ⊗ H*((TG)², TG × G) → H*((TG)², B)
δ 1 δ
H*(B, TG × G)
H*(G × TG) ⊗ H*((TG)², TG × G) → H*(G × TG, G × G)
1 δ δ
H*(G × TG) ⊗ H*(TG × G) → H*(G × G)
1
H*(G × G) ⊗ H*(G × G) → H*(G × G)

In this diagram, the columns may be reduced to the form given above, by considering subsidiary diagrams, such as the following one, in which e denotes a homotopy-unit in G.

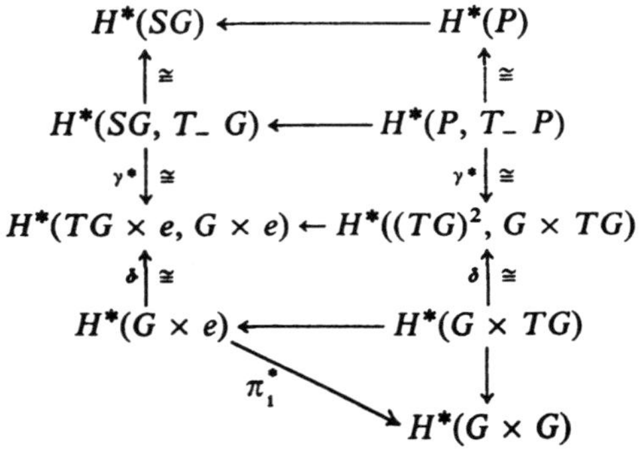
H*(SG) ← H*(P)
H*(SG, T₋ G) ← H*(P, T₋ P)
γ* γ*
H*(TG × e, G × e) ← H*((TG)², G × TG)
δ δ
H*(G × e) ← H*(G × TG)
π₁*
H*(G × G)

This completes the proof of Lemma 2.

We now investigate the Steenrod squares in $H^*(P)$. Let us write Sq for $\sum_{i>0} Sq^i$, and express $Sq\xi$ in terms of our base in $H^*(P)$; assume that its component in η is $\lambda\eta$. Then, applying φ, we have

$$Sqx = x + \lambda y$$

$$Sqy = \quad y.$$

From this, we can compute the Steenrod squares in $H^*(G \times G)$, by using the Cartan formula. Applying θ, we obtain the following results.

LEMMA 3.

$$(Sq - 1)\xi^2 = \lambda\eta^2$$

$$(Sq - 1)\xi\eta = \lambda\eta^2$$

$$(Sq - 1)\theta(x \otimes xy) = \lambda\theta(y \otimes xy)$$

$$(Sq - 1)\theta(xy \otimes x) = \lambda\theta(xy \otimes y).$$

We note that these values are consistent with the Cartan formula for products in $H^*(P)$, and that the case $\lambda \neq 0$ can arise.

For reference, we now collect two further lemmas.

LEMMA 4. *Assume that* $H^r(X) = 0$ *for* $m < r < m + a$ *and that* $Sq^a : H^m(X) \to H^{m+a}(X)$ *is non-zero. Then* $a = 1, 2, 4$ *or* 8.

This follows immediately from the decomposition formulae for Sq^a given in [2, 1].

LEMMA 5. *Assume that* $H^r(X) = 0$ *for* $m < r < m + a$ *and for* $m + a < r < m + a + b$ *that* $H^{m+a}(X) = Z_2$ *and that* $Sq^{a+b} : H^m(X) \to H^{m+a+b}(X)$ *is non-zero. Then the dimensions* (a, b) *and the operations*

$$Sq^a : H^m(X) \to H^{m+a}(X), \qquad Sq^b : H^{m+a}(X) \to H^{m+a+b}(X)$$

satitfy one of the following conditions:

(i) $a + b = 1, 2, 4$ *or* 8;
(ii) $(a, b) = (2,1), (4, 2)$ *or* $(8, 4)$; Sq^a *and* Sq^b *are both non-zero;*
(iii a) $(a, b) = (8, 8)$; Sq^a *is zero,* Sq^b *non-zero;*
(iii b) $(a, b) = (8, 8)$; Sq^a *is non-zero,* Sq^b *zero;*
(iv a) $(a, b) = (2^k - 1, 1)$, $k \geqslant 4$; Sq^a *is zero,* Sq^b *non-zero;*
(iv b) $(a, b) = (1, 2^k - 1)$, $k \geqslant 4$; Sq^a *is non-zero,* Sq^b *zero.*

We note that all the cases mentioned under (i), (ii), (iii) are possible, while (iv a) and (iv b) are possible for $k = 4$ at least.

Proof. Consider first the case when $a + b$ is not a power of 2. Then Sq^{a+b} decomposes in terms of products Sq^cSq^d $(c + d = a + b, c > 0, d > 0.)$ Hence

$$Sq^a : H^m(X) \to H^{m+a}(X)$$
$$Sq^b : H^{m+a}(X) \to H^{m+a+b}(X)$$

are both non-zero. By Lemma 4, we have $a = 2^i$, $b = 2^j$. Since $a + b$ is not a power of 2, we have $i \neq j$. If $|i - j| \neq 1$, then Sq^{a+b} admits two decompositions, of which one contains $Sq^{2^i}Sq^{2^j}$ but not $Sq^{2^j}Sq^{2^i}$ while the other contains $Sq^{2^j}Sq^{2^i}$ but not $Sq^{2^i}Sq^{2^j}$. The former gives a contradiction. If $|i - j| = 1$, then $a + b = 3.2^k$, and the decomposition of $Sq^{3.2^k}$ contains $Sq^{2^k}Sq^{2^{k+1}}$ but not $Sq^{2^{k+1}}Sq^{2^k}$. So in this case we have $a = 2^{k+1}, b = 2^k$; this gives the cases of alternative (ii).

In what follows, then, we may assume that $a + b = 2^k$, $k \geqslant 4$. Suppose first that Sq^a and Sq^b are both non-zero. By Proposition 1 we have $a = 2^i, b = 2^j$, whence $i = j = k - 1$. The relation

$$Sq^{2^{k-1}}Sq^{2^{k-1}} = \sum_{0 \leqslant l \leqslant 2^{k-2}} \lambda_l Sq^{2^k - l}Sq^l$$

now yields a contradiction.

Suppose next that Sq^a is zero. Then we may apply the decomposition of Sq^{2^k} given in [1]; we find that some pair of operations

$$\Phi_{i,j} : H^m(X) \to H^{m+a}(X)$$
$$a_{i,j,k} : H^{m+a}(X) \to H^{m+a+b}(X)$$

must both be non-zero. This shows that Sq^b is non-zero, and that $b = 2^k - 2^i - 2^j + 1$, where $0 \leqslant i \leqslant j \leqslant k-1$, $i \neq j-1$. Using Lemma 4 again, the only cases that arise are $i = 0$, $j = 3$, $k = 4$ and $i = j = k-1$, giving cases (iii a) and (iv a).

Lastly, suppose that Sq^b is zero. Then we may pass to the Spanier–Whitehead dual DX of X, and we are in the situation of the preceding case. This leads to cases (iii b) and (iv b). (One should perhaps explain how the Steenrod squares in X and DX are related. The Steenrod square Sq^n in $H^*(X)$ is the vector-space dual of the operation $c(Sq^n)$ in $H^*(DX)$, where c is the canonical anti-automorphism of the Steenrod algebra A [3]. The element $c(Sq^n) + Sq^n$ is decomposable in A.)

This completes the proof of Lemma 5.

We now turn to the proof of the theorem. We have to show that the proposed cohomology ring $H^*(P)$ is impossible for most values of q and n. Let us set $Q = q + 1$, $N = n + 1$, and begin by considering the case in which $2Q \leqslant N$. In this case our base elements fall into the following order of increasing dimensions:

$$1, \xi, \xi^2, \eta, \xi\eta, \theta(x \otimes xy) \text{ and } \theta(xy \otimes x), \eta^2, \ldots$$

Since $Sq^Q\xi = \xi^2 \neq 0$ and $q \geqslant 1$, Lemma 4 shows immediately that $Q = 2$, 4 or 8. With the notation of Lemma 3, we have $\lambda = 0$, because $Sq^i x = 0$ for $i > q$. Therefore $Sq(\xi\eta) = \xi\eta$, and similarly for $Sq\theta(x \otimes xy)$ and $Sq\theta(xy \otimes x)$. However, $Sq^N\eta = \eta^2 \neq 0$; hence Sq^N is not decomposable in terms of Steenrod squares, and so N is a power of 2. The cases $N = 2, 4, 8$ lead to case (i) of the theorem. If $N = 2^k$, $k \geqslant 4$ we may pass to the Spanier–Whitehead dual DP of P and study the decomposition of Sq^N in terms of secondary operations. This clearly leads to the following alternatives.

(a) $$2^k - 2^i - 2^j + 1 = 2Q - 1;$$

moreover, there is a non-zero operation Sq^{2Q-1} in DP, going from the dimension corresponding to $\theta(x \otimes xy)$ to the dimension corresponding to η.

This is contradicted by the relation

$$Sq^{2Q-1} = Sq^1 Sq^{2Q-2}.$$

(b) $$2^k - 2^i - 2^j + 1 = Q.$$

The only possibility is $k = 4$, $i = 0$, $j = 3$, leading to $(Q, N) = (8, 16)$, that is, to case (iii) of the theorem.

We now consider the case $Q \leq N < 2Q$. In this case our base elements fall into the following order of increasing dimensions:

$$1, \xi, \eta, \xi^2, \xi\eta, \eta^2, \ldots$$

Since $Sq^Q\xi = \xi^2 \neq 0$, we can apply Lemma 5, and we find the following cases.

(i) $Q = 2$, 4 or 8. In this case we have to examine the equation $Sq^N\eta = \eta^2 \neq 0$. Since we have assumed $Q \leqslant N < 2Q$, the only case in which N can be a power of 2 is the case $N = Q$; this leads to case (i) of the theorem. There remain a finite number of cases (eleven, in fact) in which Sq^N is decomposable in terms of Steenrod squares. Such a decomposition leads easily to a contradiction in each of the cases except $(Q, N) = (2, 3), (4, 6), (8, 12)$. These give case (ii) of the theorem.

(ii) $$a = 2^{k+1}, b = 2^k \ (k = 0, 1, 2).$$

Then $Q = 3.2^k$, $N = 5.2^k$. We have a contradiction between the equation $Sq^N\eta = \eta^2 \neq 0$ and the relation

$$Sq^N = Sq^{2^k}Sq^{2^{k+2}} + \sum_{0<i\leqslant 2^{k-1}} \lambda_i Sq^{N-i}Sq^i.$$

(iii) $a = 8$, $b = 8$. Then $Q = 16$, $N = 24$. The relation

$$Sq^{24} = Sq^8Sq^{16} + \sum_{0<i\leqslant 4} \lambda_i Sq^{24-i}Sq^i$$

combined with the fact that $Sq^{24}\eta = \eta^2 \neq 0$, shows that $Sq^{16}\eta = \xi\eta$, $Sq^8\xi\eta = \eta^2$. It follows that in Lemma 3 we have $\lambda = 1$; in particular, we have $Sq^8\xi = \eta$, $Sq^8\xi^2 = 0$. The equation $Sq^8\xi = \eta$ implies $Sq^8\eta = 0$. The equations

$$Sq^{16}\eta = \xi\eta \neq 0, \qquad Sq^8\eta = 0, \qquad Sq^8\xi^2 = 0$$

contradict Lemma 5.

(iv a) $$a = 2^{kn} - 1, \qquad b = 1, \qquad k \geqslant 4.$$

Then $Q = 2^k$, $N = 2^{k+1} - 1$. We have a contradiction between the equation

$$Sq^N\eta = \eta^2 \neq 0$$

and the relation

$$Sq^N = Sq^1Sq^{N-1}.$$

(iv b) $a = 1$, $b = 2^k - 1$, $k \geqslant 4$. Then $Q = 2^k$, $N = 2^k + 1$.

In this case the equation $Sq^N\eta = \eta^2 \neq 0$ and the relation

$$Sq^N = Sq^1Sq^{N-1}$$

show that $$Sq^{2^k}\eta = \xi\eta, \qquad Sq^1\xi\eta = \eta^2.$$

The equations

$$Sq^{2^k} = \xi\eta \neq 0, \qquad Sa^1\xi^2 = 0$$

contradict Lemma 5.

This completes the proof.

REFERENCES

1. J. F. Adams: On the non-existence of elements of Hopf invariant one, *Ann. Math., Princeton*, **72** (1960), 20–104.
2. J. Adem: The iteration of the Steenrod squares in algebraic topology, *Proc. Nat. Acad. Sci., Wash.* **38** (1952), 720–726.
3. J. Milnor: The Steenrod algebra and its dual, *Ann. Math., Princeton*, **67** (1958), 150–171.

Trinity Hall,
Cambridge;
The University,
Manchester 13.

Annals of Mathematics, **111** (1980), 95–143

Finite *H*-spaces and algebras over the Steenrod algebra

By J. F. Adams and C. W. Wilkerson*

1. Introduction and statement of results

A well-known problem asks: what polynomial algebras can arise as cohomology rings of spaces? More precisely, let p be a fixed prime, and let $\mathbf{F}_p[x_1, x_2, \cdots, x_l]$ be a polynomial algebra on generators $x_1, x_2, \cdots, x_l$ of degrees $2d_1, 2d_2, \cdots, 2d_l$; then is there or is there not a space X such that

$$H^*(X; \mathbf{F}_p) \cong \mathbf{F}_p[x_1, x_2, \cdots, x_l]?$$

This problem is related to the study of "finite H-spaces". More precisely, let X be a 1-connected space such that ΩX is homotopy-equivalent to a finite complex; then $H^*(X; \mathbf{F}_p)$ has the form considered above for all but a finite number of primes p; and one would like to infer restrictions on the "type" $(2d_1, 2d_2, \cdots, 2d_l)$.

We will complete the solution of this problem when the prime p is sufficiently large, in the sense that p does not divide $d_1 d_2 \cdots d_l$. In fact, it turns out that sufficient spaces X have already been constructed by Clark and Ewing [8]; the task of this paper is to prove that no more polynomial algebras can arise as the cohomology rings of spaces.

The topological information we use for this purpose is restricted to primary cohomology operations. In fact, we shall obtain a complete classification of all unstable algebras H^* over the mod p Steenrod algebra which are polynomial on generators of degrees $2d_1, 2d_2, \cdots$ with $d_1, d_2, \cdots$ all prime to p. Roughly speaking, we show that as a matter of pure algebra, all proceeds as in the "classical case". Here, by the "classical case", we mean the range of examples in which we start from a compact connected Lie group G, take X to be the classifying space BG, and take $H^* = H^*(BG; \mathbf{F}_p)$. In this case we have a maximal torus T, and if p is sufficiently large the induced map

* The second author was partially supported by National Science Foundation grants MCS 77-01813 and MCS 78-02284 during this research, and is an Alfred P. Sloan fellow.

$$H^*(BG; \mathbf{F}_p) \longrightarrow H^*(BT; \mathbf{F}_p)$$

is mono. As an analogous result in pure algebra, we offer the following.

THEOREM 1.1. *In order that an (anticommutative) algebra H^* over the mod p Steenrod algebra should admit an embedding in $H^*(BT; \mathbf{F}_p)$, where T is a torus of some suitable dimension, the following four conditions are necessary and sufficient.*

(1.1.1) *H^* is zero in odd degrees.*

(1.1.2) *H^* is an integral domain.*

(1.1.3) *H^* satisfies the "unstable" condition.*

(1.1.4) *There is an upper bound to the number of elements in H^* which can be algebraically independent over $\mathbf{F}_p$.*

Here we recall that if T^n is a torus of dimension n, then $H^*(BT^n; \mathbf{F}_p)$ is a polynomial algebra $\mathbf{F}_p[x_1, x_2, \cdots, x_n]$ on n generators of degree 2. From this it is clear that the four conditions are necessary; we have to work to prove them sufficient. We make a standing convention (and we will not repeat it) that results stated without proof are proved later.

If $p > 2$, condition (1.1.1) follows from condition (1.1.2). For $p = 2$, there is an alternative form of the theorem in which one omits condition (1.1.1) and replaces $BT = \mathbf{C}P^\infty \times \cdots \times \mathbf{C}P^\infty$ by $RP^\infty \times \cdots \times RP^\infty$, so that $H^*(BT^n; \mathbf{F}_p)$ is replaced by a polynomical algebra on n generators of degree 1. This form may be deduced from that given by a dimension-doubling trick.

In the classical case, we have not only a maximal torus T, but also a Weyl group W; and if p is sufficiently large, the induced map gives an isomorphism

$$H^*(BG; \mathbf{F}_p) \cong H^*(BT; \mathbf{F}_p)^W .$$

Here the right-hand side means the subalgebra of elements in $H^*(BT; \mathbf{F}_p)$ which are invariant under W. As an analogous result in pure algebra, we offer the next theorem. The statement uses certain Steenrod operations Q^r which will be defined in Section 2.

THEOREM 1.2. *In order that an (anticommutative) algebra H^* over the mod p Steenrod algebra should admit an isomorphism*

$$H^* \cong H^*(BT^n; \mathbf{F}_p)^W$$

for some n and some suitable group W of automorphisms of $H^*(BT^n; \mathbf{F}_p)$, the following two conditions (in addition to those in Theorem 1.1) are necessary and sufficient.

(1.2.1) *The integral domain H^* is integrally closed in its field of fractions.*

(1.2.2) *If $y \in H^{2dp}$ and $Q^r y = 0$ for each $r \geqq 1$, then $y = x^p$ for some $x \in H^{2d}$.*

It is sufficient to assume (1.2.1) *and replace* (1.2.2) *by the following condition* (1.2.3); *and this also ensures that the order $|W|$ of W is prime to p.*

(1.2.3) *H^* is generated as an $\mathbf{F}_p$-algebra by a finite number of generators $x_1, x_2, \cdots$ of degrees $2d_1, 2d_2, \cdots$ where each d_i is prime to p.*

If H^* is a polynomial algebra $\mathbf{F}_p[x_1, x_2, \cdots, x_l]$ on generators $x_1, x_2, \cdots, x_l$ of degrees $2d_1, 2d_2, \cdots, 2d_l$ where each d_i is prime to p, we obtain the further conclusion that W is a p-adic generalized reflection group in the sense of Clark and Ewing [8].

We pause to comment. Condition (1.2.1) holds automatically whenever H^* is a polynomial algebra. In applications where we do not know enough about the Steenrod operations to verify (1.2.2), we fall back on (1.2.3). The conclusion in the last sentence of the theorem can be stated more completely as follows: the dimension n of the torus T^n is equal to the number l of polynomial generators; the action of W on $H^2(BT^l; \mathbf{F}_p)$ lifts to an action on $H^2(BT^l; \mathbf{Z}_p^\wedge)$, where $\mathbf{Z}_p^\wedge$ is the p-adic integers; and then W acts on $H^2(BT^l; \mathbf{Z}_p^\wedge)$ as a p-adic generalized reflection group. The point of this conclusion is that the possible reflection groups W are known, and therefore we get an explicit classification. In particular, the possible "types" $(2d_1, 2d_2, \cdots, 2d_l)$ are such as may be inferred from the tables of Clark and Ewing [8]. (One has to note that the product of two reflection groups is again a reflection group, and the tables list only the irreducible factors.)

Although we have presented Theorems 1.1 and 1.2 as results of pure algebra, the fact is that all the algebras over the Steenrod algebra which we have just classified do arise as cohomology rings of spaces.

COROLLARY 1.3. *Let H^* be an (anticommutative) algebra over the* mod p *Steenrod algebra which satisfies* (1.1.1)-(1.1.4), (1.2.1) *and* (1.2.3). *Then there is a space X such that $H^* \cong H^*(X; \mathbf{F}_p)$.*

The reader may wish to know whether Theorem 1.2 settles the old conjecture that "every finite H-space with a classifying space has the type of a Lie group". The answer is that Theorem 1.2 does not settle this conjecture.

Example 1.4. The type

$$\{4, 4, 4, 8, 8, 8, 12, 12, 16, 16, 20, 24, 24, 28\}$$

is not the type of a Lie group; but for every prime $p > 3$ there is a space X_p such that $H^*(X_p; \mathbf{F}_p)$ is polynomial of this type.

It seems likely that one can construct a single space X which will serve for all primes $p > 3$, but we do not need to insist on this. The moral is that we need further work at the small primes.

To prove our main results we shall follow the programme laid down by the second author in [14]. The crucial insight is the following: in a suitable category C of algebras over the mod p Steenrod algebra, the algebra $H^*(BT^n; \mathbf{F}_p)$ has properties analogous to those of an algebraically closed field in the category of fields. In fact, one can give a precise treatment of "algebraic closure in C"; we will do so in Section 4, but it seems better not to encumber this introduction with detailed definitions. Terms in quotation marks, like "algebraic closure in C", will be defined properly in Section 4; till then we ask the reader to imagine their meaning by analogy with the familiar case of fields.

PROPOSITION 1.5. *Let C be the category of those algebras H^* over the mod p Steenrod algebra which satisfy* (1.1.1), (1.1.2) *and* (1.1.3). *Then every object H^* in C has an "algebraic closure" $H^* \subset K^*$ in C. If H^* satisfies* (1.1.4), *then so does K^*.*

Here the definition of an "algebraic closure" $H^* \subset K^*$ will demand two things: the map $H^* \to K^*$ has to be an "algebraic extension", and the object K^* has to be "algebraically closed".

THEOREM 1.6. *The objects K^* in the above category C which are "algebraically closed" and satisfy* (1.1.4) *are precisely the algebras $H^*(BT^n; \mathbf{F}_p)$, where n runs over the range $n \geqq 0$.*

This result carries out the programme of [14], and confirms the insights of that paper.

Let H^* be an algebra which satisfies (1.1.1) to (1.1.4). Then results (1.5) and (1.6) yield a particular embedding of H^* in $H^*(BT; \mathbf{F}_p)$.

Definition 1.7. The *canonical embedding* of H^* in $H^*(BT; \mathbf{F}_p)$ is that obtained by composing the embedding $H^* \subset K^*$ of (1.5) with the isomorphism $K^* \cong H^*(BT^n; \mathbf{F}_p)$ of (1.6).

At the price of defining the terms involved, one can add that the dimension n of the torus T^n is equal to the transcendency degree over $\mathbf{F}_p$ of the field of fractions of H^*.

The existence of the canonical embedding proves Theorem 1.1; but to prove Theorem 1.2, we need to ask when this embedding has good properties.

In the classical case, $H^*(BT; \mathbf{F}_p)$ is finitely-generated as a module over $H^*(BG; \mathbf{F}_p)$. As an analogous result in pure algebra, we offer the following.

THEOREM 1.8. *Let $H^* \subset K^* \cong H^*(BT^n; \mathbf{F}_p)$ be as in* (1.7). *Then K^* is finitely-generated as a module over H^* if and only if H^* is finitely-generated as an algebra over $\mathbf{F}_p$.*

Of course the condition is satisfied in our applications.

We would also like to prove, under suitable assumptions, that $H^* \subset K^* \cong H^*(BT^n; \mathbf{F}_p)$ is a Galois extension, whose Galois group may be used as the group W in Theorem 1.2. The "normality" of the extension results from the proof of (1.5), but its "separability" does not.

THEOREM 1.9. *For the embedding $H^* \subset K^* \cong H^*(BT^n; \mathbf{F}_p)$ of* (1.7) *to be a "separable extension", condition* (1.2.2) *is necessary and sufficient; condition* (1.2.3) *is sufficient.*

We have two more supplementary results. First, it is natural to ask for information about maps comparable to that which we have about objects. Now, when we set up the theory of "algebraic closure", we have to legislate that the only morphisms to be considered are monomorphisms (because we are trying to carry over the usual theory for fields). So it takes a little extra work to answer the present question, which concerns maps that need not be mono.

PROPOSITION 1.10. *Suppose given a diagram of the following form, in which H^* is a finitely-generated $\mathbf{F}_p$-algebra as in* (1.8), *and the rows are as in* (1.7), *and θ is a homomorphism of algebras over the* mod p *Steenrod algebra.*

$$\begin{array}{ccc} H^* \subset K^* & \cong & H^*(BT^n; \mathbf{F}_p) \\ \theta\downarrow & & \downarrow\phi \\ H'^* \subset K'^* & \cong & H^*(BT^{n'}; \mathbf{F}_p) \end{array}$$

Then we can find a homomorphism ϕ to complete the commutative diagram.

This generalizes certain results of [1] and [14].

Finally, we can reformulate (1.1) as a result about "invariant prime ideals". Let H^* be an (anticommutative) algebra over the mod p Steenrod algebra, and let $I^* \subset H^*$ be an ideal closed under Steenrod operations.

COROLLARY 1.11. *If the quotient algebra H^*/I^* satisfies conditions* (1.1.1) *to* (1.1.4), *then I^* is the kernel of a map $\theta\colon H^* \to H^*(BT^n; \mathbf{F}_p)$.*

This follows immediately, by using Theorem 1.1 to embed H^*/I^* in $H^*(BT^n; \mathbf{F}_p)$. In practice H^* usually satisfies conditions (1.1.3) and (1.1.4), and then H^*/I^* inherits these conditions. Of course, H^*/I^* satisfies condition (1.1.2) if and only if the ideal I^* is prime. As for condition (1.1.1), the reader may consult the remarks about it which follow Theorem 1.1.

The special case of Corollary 1.11 in which $H^* = H^*(BT^m; \mathbf{F}_p)$ is due to Serre [12], and was the starting-point for the work of the second author in [14]. Quillen [11] has studied cases in which $H^* = H^*(EG \times_G X; \mathbf{F}_p)$ and the map θ arises as an induced homomorphism.

This completes our statement of results. The remainder of this paper is organised as follows. Sections 2 and 3 are service sections; Section 2 records all the lemmas we need about Steenrod operations, and Section 3 deals with derivations. Section 4 sets up the theory of "algebraic closure", and so proves Proposition 1.5. This section also proves certain properties of "algebraically closed objects" which are not stated in Proposition 1.5, but are needed later; however, these properties should be regarded as formal rather than substantial. The crucial part of our proof is contained in Sections 5 and 6. In proving Theorem 1.6, we have to start from an object K^* which is "algebraically closed" and satisfies (1.1.4), and we have to prove that it is isomorphic to $H^*(BT^n; \mathbf{F}_p)$ for some n. The first step is to locate in K^* a subalgebra isomorphic to $H^*(BT^n; \mathbf{F}_p)$, and this is done in Section 5. The second step is to show that the inclusion $H^*(BT^n; \mathbf{F}_p) \subset K^*$ is an isomorphism, and this is done in Section 6. However, Section 5 does more than we have just stated, for it is also necessary to lay the foundations for proving the other results. Finally, Section 7 completes the proof of the remaining results, notably (1.8), (1.9), (1.2) and (1.3).

We thank John Ewing for arousing our interest in the problem and for helpful conversations and correspondence.

2. Steenrod operations

In this section we will record all the lemmas about Steenrod operations which are needed in what follows. We have to cover three topics. First, we have to discuss the "unstable" condition. Secondly, we have to discuss the particular operations Q^r. Thirdly, we have to discuss the Cartan formula on p^{th} powers.

At this point we had better insist that all our objects H^* are to be graded. In Section 1 we systematically omitted the word "graded", leaving it to be understood; any reader who overlooked this point didn't miss much, but for accuracy it has to borne in mind. When we have a graded object H^* and write "x lies in H^*" or "$x \in H^*$" we mean that $x \in H^d$ for some d. We prefer not to form $\bigoplus_d H^d$ or $\coprod_d H^d$, and we prefer not to consider elements in them; so inhomogeneous elements are disenfranchised from now on.

In view of (1.1.1), we agree that all our graded objects H^* are to be zero in odd degrees. It follows that if the Bockstein coboundary β acts on any of our objects H^*, then it will act as zero. We therefore define A^* to be the quotient of the usual mod p Steenrod algebra by the two-sided ideal generated by β. We lay down that (for the purposes of this section) an "A^*-algebra" is to be a commutative graded algebra over $\mathbf{F}_p$, zero in odd degrees, which is a graded module over A^*, and in which the Steenrod operations on products satisfy the Cartan formula. (Later on we shall add (1.1.2) as a standing assumption, but in this section we will mention (1.1.2) if we use it.)

Let H^* be such an A^*-algebra. We say that an element $x \in H^{2d}$ is "unstable" if

$$P^k x = \begin{cases} x^p & \text{for } k = d \\ 0 & \text{for } k > d . \end{cases}$$

Here P^k must be interpreted as Sq^{2k} if $p = 2$. Let U^* be the set of unstable elements in H^*.

LEMMA 2.1. (i) *U^* is an A^*-subalgebra of H^*, and zero in negative degrees.*

(ii) *Assume further that H^* satisfies* (1.1.2). *Then $U^0 = \mathbf{F}_p$; moreover $x \in H^{2d}$ is unstable if x^p is unstable.*

We say that an A^*-algebra H^* is "unstable" if every element of H^* is unstable. Of course, the cohomology algebra of any space is unstable.

We now pass on to discuss particular elements in A^*. For this purpose we use the dual A_* of A^* [10]. Of course, both A_* and A^* are Hopf algebras. First assume $p > 2$; in this case the dual A_* of A^* is the subalgebra of the Milnor algebra generated by the polynomial generators $\xi_1, \xi_2, \xi_3, \cdots$ without using the exterior generators $\tau_0, \tau_1, \tau_2, \cdots$. Thus we have

$$A_* = \mathbf{F}_p[\xi_1, \xi_2, \xi_3, \cdots] .$$

In the case $p = 2$ we have to interpret ξ_r as the square ζ_r^2 of the generator ζ_r in the Milnor algebra.

The algebra A_* has a base consisting of the monomials

$$\xi^I = \xi_1^{i_1}\xi_2^{i_2}\xi_3^{i_3} \cdots ,$$

where $I = \{i_1, i_2, i_3, \cdots\}$ runs over sequences in which all but a finite number of the i's are zero. We have a corresponding dual base in A^*, and its elements are the Milnor operations

$$P^I = P^{(i_1, i_2, i_3, \cdots)} .$$

For each $r \geqq 1$ we define Q^r to be the operation P^I corresponding to the sequence

$$I = \{0, 0, \cdots, 0, 1, 0, \cdots\}$$

where the 1 comes in the rth place. In other words, Q^r is the element of the Milnor base for A^* corresponding to the monomial ξ_r in the monomial base for A_*. This operation is not the same as the one which Milnor writes Q_r; the latter is zero in all our graded algebras H^*. The Cartan formula for Q^r is

$$Q^r(xy) = (Q^r x)y + x(Q^r y) ;$$

that is, Q_r is primitive.

We also define an operation Q^0 on all our graded algebras H^* by setting

$$Q^0 x = dx \text{ for all } x \in H^{2d} .$$

Of course Q^0 does not lie in A^*, but it does satisfy

$$Q^0(xy) = (Q^0 x)y + x(Q^0 y) .$$

The degree of Q^r is $2(p^r - 1)$ for each $r \geqq 0$.

We need the following properties of the operations Q^r.

LEMMA 2.2. *We have*

$$P^k Q^i = \begin{cases} Q^i P^k & \text{if } k < p^i \\ Q^i P^k + Q^{i+1} P^{k-p^i} & \text{if } k \geqq p^i . \end{cases}$$

By taking $k = p^i$, we obtain (if we wish) the following inductive definition of the Q's in terms of the P^k.

$$Q^1 = P^1 ,$$
$$Q^{i+1} = P^{p^i} Q^i - Q^i P^{p^i} .$$

LEMMA 2.3. *Given d and $r \geqq 0$, there exists an operation $a \in A^*$ such that*

$$Q^{r+s}(ax) = (Q^s x)^{p^r}$$

for each $s \geqq 0$ and each unstable x of degree $2d$.

We turn to the Cartan formula on p^{th} powers. In any A^*-algebra we have

$$P^I(x^p) = \begin{cases} (P^J x)^p & \text{if } I = pJ \\ 0 & \text{if } I \text{ is not of the form } pJ . \end{cases}$$

We have another way to express this. There is a Frobenius map $\phi: A_* \to A_*$ given by $\phi(z) = z^p$, as usual. Of course it does not preserve degrees; it carries A_{2d} to A_{2dp}. There is a dual map $\phi^*: A^* \to A^*$, which does not preserve degrees either; and we have

$$\phi^* P^I = \begin{cases} P^J & \text{if } I = pJ \\ 0 & \text{if } I \text{ is not of the form } pJ . \end{cases}$$

Thus the formula above yields

$$a(x^p) = ((\phi^* a)x)^p \tag{2.4}$$

for any $a \in A^*$ and any x in any A^*-algebra.

LEMMA 2.5. *$\operatorname{Ker}\phi^*$ is the left ideal generated by the Q^r with $r \geqq 1$; it is also the right ideal generated by the Q^r with $r \geqq 1$.*

This completes the statement of the results which will be quoted later; before proving them, we offer the reader the chance to skip to Section 3.

We begin work on Lemma 2.1. In order to show that U^* is closed under Steenrod operations, we introduce the following lemma.

LEMMA 2.6. *If $k \geqq (p+1)j$ then there is a relation in A^* of the form*

$$P^{k-j}P^j = \sum_{0 \leq r \leq j} \lambda_r P^{pr} P^{k-pr}$$

with $\lambda_r \in \mathbf{F}_p$, $\lambda_j = 1$.

Proof. We use the dual A_* of A^* [10]. The product in A^* is dual to the coproduct ψ in A_*, which has the following form:

$$\psi \xi_k = \sum_{i+j=k} \xi_i^{p^j} \otimes \xi_j .$$

(Here ξ_0 counts as 1.) Consider the monomial

$$m = \xi_1^{e_1} \xi_2^{e_2} \xi_3^{e_3} \cdots$$

in A_*, and assume first that $e_s > 0$ for some $s \geqq 3$. Then ψm can be written as a sum of terms $m' \otimes m''$, in each of which either m' or m'' contains a factor ξ_t with $t \geqq 2$. In either case, $m' \otimes m''$ annihilates $P^u \otimes P^v$ for any u, v; thus m annihilates the product $P^u\, P^v$. In particular, the monomial m annihilates all the products $P^{k-j}\, P^j$ and $P^{pr} P^{k-pr}$ in the proposed relation. We can argue similarly with ξ_2. We see that ψm can be written as a sum of terms $m' \otimes m''$ of the sort described above, plus one further term of the form

$$a \xi_1^{pe_2} \otimes b \xi_1^{e_2} .$$

Thus m annihilates all the products in the proposed relation if $e_2 > j$. Consider then the monomial

$$m = \xi_1^{k-(p+1)r} \xi_2^r$$

for $0 \leqq r \leqq j$. Arguing as above, we see that the Kronecker product of m

with $P^{pr} \otimes P^{k-pr}$ is 1; also its Kronecker product with $P^{pt}\, P^{k-pt}$ is 0 for $t < r$. Therefore we can determine the scalars λ_r by downwards induction over r so that every monomial m has the same Kronecker product with $\sum_{0 \leqq r \leqq j} \lambda_r P^{pr} P^{k-pr}$ as with $P^{k-j} P^j$. Since the Kroncker product of $\xi_1^{k-(p+1)j} \xi_2^j$ and $P^{k-j}P^j$ is 1, the first step of the induction gives $\lambda_j = 1$. This proves Lemma 2.6.

Proof of Lemma 2.1. First we prove that U^* is zero in negative degrees. In fact, if $x \in U^{2d}$ with $d < 0$ then we get $P^0 x = 0$; since $P^0 x = x$ we get $x = 0$.

It is trivial that U^* is closed under $\mathbf{F}_p$-linear combinations. As for products, let $x \in U^{2d}$, $y \in U^{2e}$ where we may assume $d \geqq 0$, $e \geqq 0$; then by the Cartan formula,

$$\begin{aligned} P^k(xy) &= \sum_{i+j=k} (P^i x)(P^j y) \\ &= \begin{cases} x^p y^p & \text{if } k = d + e \\ 0 & \text{if } k > d + e \,. \end{cases} \end{aligned}$$

So $xy \in U^*$ and U^* is an $\mathbf{F}_p$-subalgebra.

Suppose now that $x \in U^{2d}$; we wish to show that $P^j x \in U^*$. If $j > d$ then $P^j x = 0$, which lies in U^*. So we may assume $j \leqq d$. In this case it is sufficient to consider $P^i P^j x$ where $i \geqq d + j(p-1)$. We may apply Lemma 2.6, for we have

$$k = i + j \geqq d + pj \geqq (p+1)j \,.$$

We get $P^i P^j x = \sum_{0 \leqq r \leqq j} \lambda_r P^{pr} P^{i+j-pr} x$.

Now if $i > d + j(p-1)$ we have

$$i + j - pr > d + p(j - r) \geqq d \,,$$

so since x is unstable we get $P^{i+j-pr} x = 0$ and $P^i P^j x = 0$. If $i = d + j(p-1)$ then the same argument applies to all the terms with $r < j$; retaining the term with $r = j$, we get

$$\begin{aligned} P^i P^j x &= P^{pj} P^d x \\ &= P^{pj}(x^p) \\ &= (P^j x)^p \end{aligned}$$

(by the Cartan formula). This proves that $P^j x$ is unstable. So the set U^* is closed under P^j for all j, and hence under all the elements of A^*. This proves part (i).

We turn to part (ii). If $x \in U^0$ then we get $x^p = P^0 x = x$. If H^* is a (graded) integral domain then the equation $x^p = x$ has in H^* only the p solutions which lie in $\mathbf{F}_p$.

Finally assume that $x \in H^{2d}$ and x^p is unstable. The Cartan formula gives

$$P^{pk}x^p = (P^k x)^p ;$$

since x^p is unstable (of degree $2pd$) we get

$$(P^k x)^p = \begin{cases} (x^p)^p & \text{if } pk = pd \\ 0 & \text{if } pk > pd . \end{cases}$$

If H^* is a (graded) integral domain of characteristic p, then p^{th} roots in H^* are unique, and we get

$$P^k x = \begin{cases} x^p & \text{if } k = d \\ 0 & \text{if } k > d . \end{cases}$$

Thus x is unstable. This proves Lemma 2.1.

The proof of Lemma 2.2 is a simple calculation in the dual A_* of A^*, rather easier than the proof of (2.6); we omit the details.

As for the proof of Lemma 2.3, we shall have to work up to it gradually. First we remark that it is convenient to test our operations in the space BT^d. By choosing a base $x_1, x_2, \cdots, x_d$ in $H^2(BT^d; \mathbf{F}_p)$ we can write

$$H^*(BT^d; \mathbf{F}_p) = \mathbf{F}_p[x_1, x_2, \cdots, x_d] .$$

Thus we have the product

$$x_1 x_2 \cdots x_d \in H^{2d}(BT^d; \mathbf{F}_p) .$$

LEMMA 2.7. *If an element $a \in A^*$ annihilates $x_1 x_2 \cdots x_d$, then it annihilates every unstable element of degree $2d$.*

Proof. First we need to recall that A^* has an alternative base consisting of the composites

$$P^{k_1}P^{k_2}P^{k_3} \cdots$$

where $K = \{k_1, k_2, k_3, \cdots\}$ runs over sequences in which all but a finite number of the k's are zero and $k_r \geqq pk_{r+1}$ for each r. The "excess" of such a composite or sequence is defined by

$$\varepsilon(K) = \sum_r (k_r - pk_{r+1}) .$$

If we have $\varepsilon(K) > d$, then the composite $P^{k_1}P^{k_2}P^{k_3} \cdots$ annihilates every unstable element of degree $2d$; for $y = P^{k_2}P^{k_3} \cdots x$ is of degree $2e$ where

$$e = d + (p-1)\sum_{r \geqq 2} k_r ,$$

and y is unstable by Lemma 2.1; since $k_1 > e$, we get $P^{k_1}y = 0$.

On the other hand, the composites $P^{k_1}P^{k_2}P^{k_3} \cdots$ with $\varepsilon(K) \leqq d$ take linearly independent values on $x_1 x_2 \cdots x_d$, as one proves by calculation, the argument being essentically due to Cartan [6].

So if we are given $a\,(x_1x_2\cdots x_d)=0$, it follows that a is a linear combination of composites $P^{k_1}P^{k_2}P^{k_3}\cdots$ with $\varepsilon(K)>d$, and hence that a annihilates every unstable element of degree $2d$. This proves the lemma.

We now turn back to the Milnor operations P^I. For these we define "excess" by

$$e(I)=\sum_r i_r\ ;$$

we also define a "shift operator" on sequences I, as follows. If

$$I=\{i_1,\,i_2,\,i_3,\,\cdots\}\ ,$$

then we set

$$I'=\{i_2,\,i_3,\,i_4,\,\cdots\}$$

so that $i'_r=i_{r+1}$.

LEMMA 2.8. *We have*

$$P^I x=(P^{I'}x)^p$$

for every sequence I such that $e(I)=d$ and every unstable element x of degree $2d$.

Proof. Let $x_1x_2\cdots x_d$ be as in Lemma 2.7; the operations on this class are given by the following formula:

$$a(x_1x_2\cdots x_d)=\sum_{r_1,r_2,\cdots,r_d}\langle a,\,\xi_{r_1}\xi_{r_2}\cdots\xi_{r_d}\rangle x_1^{p^{r_1}}x_2^{p^{r_2}}\cdots x_d^{p^{r_d}}\ .$$

(Here ξ_0 counts as 1.) From this formula we check by calculation that for the class $u=x_1x_2\cdots x_d$ we have

$$P^I u=(P^{I'}u)^p$$

when $e(I)=d$. If x is any unstable element of degree $2d$, then $P^{I'}x$ is unstable by Lemma 2.1; let its degree be $2k$, so that we get

$$P^kP^{I'}x=(P^{I'}x)^p\ .$$

Let $a=P^I-P^kP^{I'}$; we have just shown that the operation a annihilates $u=x_1x_2\cdots x_d$; by Lemma 2.7, it must annihilate every unstable element x of degree $2d$. That is, we have

$$P^Ix=P^kP^{I'}x=(P^{I'}x)^p\ .$$

This proves the lemma.

Proof of Lemma 2.3. The result is trivially true for $r=0$ (take $a=1$) and for $d=0$ (take $a=0$). So we may assume $r\geqq 1$, $d\geqq 1$. Now take

$$a=P^{(0,\cdots,0,d-1,0,\cdots)}$$

where the entry $(d-1)$ comes in the rth place. For $s>0$ we get

$$\begin{aligned} Q^{r+s}(ax) &= Q^{r+s}P^{(0,\cdots,0,d-1,0,\cdots)}x \\ &= P^{(0,\cdots,0,d-1,0,\cdots,0,1,0,\cdots)}x \end{aligned}$$

where the entry 1 comes in the $(r+s)^{\text{th}}$ place. By using (2.8) r times, we get

$$Q^{r+s}(ax) = (Q^s x)^{p^r} .$$

Similarly, for $s = 0$ we get

$$\begin{aligned} Q^r(ax) &= Q^r P^{(0,\cdots,0,d-1,0,\cdots)}x \\ &= dP^{(0,\cdots 0,d,0,\cdots)}x . \end{aligned}$$

By using (2.8) r times, we get

$$Q^r(ax) = dx^{p^r} .$$

On the other hand, we have

$$(Q^0 x)^{p^r} = (dx)^{p^r} = dx^{p^r} .$$

This proves Lemma 2.3.

Proof of Lemma 2.5. Let I^* be the right ideal generated by the Q^r with $r \geqq 1$, that is, the set of sums $\sum_r Q^r a_r$. We have

$$Q^r P^I = (i_r + 1)P^{(i_1,\cdots,i_{r-1},i_r+1,i_{r+1},\cdots)}$$

and so I^* has a base consisting of the P^J such that $j_r \not\equiv 0 \bmod p$ for some r. But these P^J form a base for $\operatorname{Ker}\phi^*$, so $I^* = \operatorname{Ker}\phi^*$.

Now, since $\phi^*: A^* \to A^*$ preserves products, it is clear that $\operatorname{Ker}\phi^*$ is a two-sided ideal. Alternatively, Lemma 2.2 shows that I^* is closed under left multiplication by P^k for each k, and so is a two-sided ideal.

Let us now apply the canonical anti-automorphism $\chi: A^* \to A^*$. Since $\chi(Q^r) = -Q^r$, we see that the left ideal generated by the elements $-Q^r$ is the same as the two-sided ideal generated by the elements $-Q^r$. That is, the left ideal generated by the Q^r is $\operatorname{Ker}\phi^*$. This proves Lemma 2.5.

3. Derivations

In this section we will record what we need about derivations; this includes a little background, and just one result.

For brevity, it seems best to agree that throughout the rest of this paper the word "algebra" will mean a graded algebra H^* over $\mathbf{F}_p$ which satisfies (1.1.1) (so that $H^d = 0$ when d is odd), and is commutative, and satisfies (1.1.2) (so that when $x, y \in H^*$ and $xy = 0$ we have either $x = 0$ or $y = 0$). Such an algebra is a (graded) "field" if for each d, every non-zero $x \in H^{2d}$ has an inverse $x^{-1} \in H^{-2d}$. Every algebra has a (graded) field of fractions, obtained by localizing so as to provide the required inverses.

Let $H^* \subset K^*$ be a pair of algebras. Then a derivation $\partial\colon H^* \to K^*$ (of degree d) is a function $\partial\colon H^* \to K^*$ (of degree d) such that

$$\partial(x + y) = (\partial x) + (\partial y)$$
$$\partial(xy) = (\partial x)y + x(\partial y) .$$

(Since our ground field is $\mathbf{F}_p$, derivations are automatically linear.) For example, the operations Q^r of Section 2 are derivations. For background on derivations, we refer the reader to [3, Chapter IV, pp. 41-47, Chapter V, pp. 136-145], [9, pp. 167-189] or [16, pp. 120-131].

We can form linear combinations of derivations. Suppose $\partial_1, \partial_2, \cdots, \partial_n\colon H^* \to K^*$ are derivations and $\lambda_1, \lambda_2, \cdots, \lambda_n$ are scalars in K^* and of appropriate degrees; then $\sum_i \lambda_i\partial_i\colon H^* \to K^*$ is defined by

$$(\textstyle\sum_i \lambda_i\partial_i)x = \sum_i \lambda_i(\partial_i x) ,$$

and it is again a derivation.

The definitions of linear dependence, linear independence, and so on, are phrased so that they work when K^* is any algebra, and reduce to the graded analogues of the usual ones when K^* is a (graded) field. For example, derivations $\partial_1, \partial_2, \cdots, \partial_n\colon H^* \to K^*$ are linearly dependent if there exist scalars $\lambda_1, \lambda_2, \cdots, \lambda_n$ in K^*, of suitable degrees and not all zero, such that $\sum_i \lambda_i\partial_i = 0$; otherwise they are linearly independent. A derivation ∂_0 is linearly dependent on $\partial_1, \partial_2, \cdots, \partial_n$ if we have $\lambda_0\partial_0 = \sum_i \lambda_i\partial_i$ for some $\lambda_0 \neq 0$; and so on.

These notions from linear algebra are invariant under extension of the algebra K^* where we keep our values ∂x and our coefficients λ. For example, although we are happy for the reader to use his own preferred arguments about the "extension of scalars", we point out that the linear independence of $\partial_1, \partial_2, \cdots, \partial_n$ can be expressed by the non-vanishing of a determinant

$$\det(\partial_i x_j)$$

where $x_1, x_2, \cdots, x_n$ are suitable elements of H^*. This determinant can be defined as the usual polynomial in the entries of the matrix, even though we are working in an algebra graded over the even integers.

LEMMA 3.1. *Suppose that $\partial_1, \partial_2, \cdots, \partial_n\colon H^* \to K^*$ are derivations, $x_1, x_2, \cdots, x_n \in H^*$ and*

$$\det(\partial_i x_j) \neq 0 ;$$

in other words, $\partial_1, \partial_2, \cdots, \partial_n$ take linearly independent values on $x_1, x_2, \cdots, x_n$. Then $x_1, x_2, \cdots, x_n$ are algebraically independent over $\mathbf{F}_p$.

Sketch proof. We may suppose (as the hypothesis of an induction over n) that $x_1, x_2, \cdots, x_{n-1}$ are algebraically independent over $\mathbf{F}_p$, and also (as the hypothesis of a subsidiary induction over m) that zero is the only polynomial

$$g(X) = a_{m-1}X^{m-1} + \cdots + a_1X + a_0$$

in $\mathbf{F}_p[x_1, x_2, \cdots, x_{n-1}, X]$ such that $g(x_n) = 0$. Take then a polynomial

$$g(X) = a_mX^m + \cdots + a_1X + a_0$$

in $\mathbf{F}_p[x_1, x_2, \cdots, x_{n-1}, X]$ and assume $g(x_n) = 0$. We may assume $a_m \neq 0$, and without loss of generality that a_m is a pth power (otherwise multiply by a_m^{p-1}). Take the equation $g(x_n) = 0$, apply ∂_i and use the fact that $\det(\partial_i x_j) \neq 0$; we find that

$$(\partial a_{m-1}/\partial x_j)x_n^{m-1} + \cdots + (\partial a_1/\partial x_j)x_n + (\partial a_0/\partial x_j) = 0$$

for $1 \leqq j \leqq n - 1$, and

$$(\partial g/\partial X)(x_n) = 0 .$$

Using the inductive hypothesis, we see that $(\partial a_i/\partial x_j) = 0$ for $1 \leqq j \leqq n - 1$ and $(\partial g/\partial X) = 0$. The former result rather easily gives $a_i = (b_i)^p$ for each i, and then the latter gives $g(X) = \big(h(X)\big)^p$. From $g(x_n) = 0$ we deduce $h(x_n) = 0$; now the inductive hypothesis gives $h(X) = 0$, so $g(X) = 0$. Compare [3, pp. 141–142].

4. Algebraic closure

In this section we shall develop the theory of "algebraic closure", and so prove Proposition 1.5. We shall also record a few results we need later.

Our plan is to erect scaffolding by treating algebraic closure in two simpler categories before we reach the category that really interests us, namely the category of "unstable A^*-algebras". More precisely, in the category of "algebras" the objects will be algebras as defined in Section 3—so that they satisfy both (1.1.1) and (1.1.2). In the category of "A^*-algebras", the objects will be A^*-algebras in the sense of Section 2 which are algebras in the sense above—so we are now imposing condition (1.1.2). In the category of "unstable A^*-algebras", the objects will be A^*-algebras (in the sense above) which are unstable. In each case, the morphisms are to be the injections preserving the structure. As we remarked in Section 1, we specify "injections" because we wish to carry over the usual theory for fields. For variety we may refer to the injections as "monomorphisms", "embeddings", "extensions" and so on.

In the category of algebras, everything goes exactly as in the familiar case of fields, except that one has to be willing to carry everything over from the ungraded case to the graded case; also the textbooks usually talk about algebraic closure for fields, and we want it for integral domains, but one can easily pass from an integral domain to its field of fractions.

As an example, let us explain when an embedding $H^* \subset K^*$ of algebras is an "algebraic extension". Suppose given an embedding $H^* \subset K^*$ and an element $x \in K^{2d}$. Let $H^*[X; 2d]^*$ be the algebra of polynomials over H^* in one variable X of degree $2d$. We say that x is "algebraic over H^*" if it satisfies a non-trival homogeneous polynomial equation $f(x) = 0$, where $f(X)$ lies in $H^*[X; 2d]^*$ and is non-zero. We say that x is "separable over H^*" if we can choose the polynomial $f(X)$ so that it has distinct roots (in some extension of K^*). We say that x is "purely inseparable over H^*" if we can choose the polynomial to have the form $f(X) = X^{p^e} - c$. We say that the extension $H^* \subset K^*$ is "algebraic", or that K^* is algebraic over H^*, if every element of K^* is algebraic over H^*; similarly with "algebraic" replaced by "separable" or "purely inseparable". If we have $H^* \subset K^* \subset L^*$ and L^* is algebraic over H^*, then L^* is algebraic over K^* and K^* is algebraic over H^*.

While the preceding definitions have been given for algebras, we can apply them to A^*-algebras by neglecting the extra structure given by the Steenrod operations; similarly for unstable A^*-algebras. Now we go on to definitions which do depend on the category in an essential way.

Let C be one of the three categories defined above. We will say that an object L^* is "algebraically closed in C" if every diagram

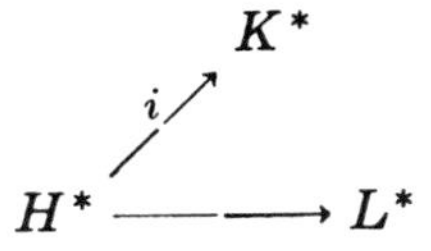

in C, in which i is an algebraic extension, can be completed to a commutative triangle in C of the following form.

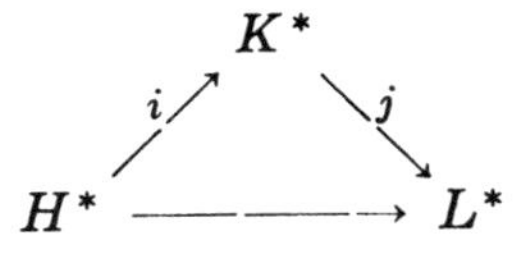

As one would expect from the example of fields, there is no need for j to be unique. This definition asks that L^* should be "relatively injective" with respect to the class of algebraic extensions. One can proceed similarly with the word "algebraic" replaced by "separable" or "purely inseparable".

If L^* is algebraically closed in C and $i: L^* \to M^*$ is an extension in C which is algebraic, then i is iso.

We say that an embedding $i: H^* \to K^*$ in C is "an algebraic closure of

H^* in C" if i is an algebraic extension and K^* is algebraically closed in C. If such an algebraic closure of H^* exists, then it is unique up to isomorphism (not necessarily canonical).

We may sometimes omit to mention the category C when it is clear which category is meant. For example, from Section 5 onwards it will be the category of unstable A^*-algebras.

PROPOSITION 4.1. *Any algebra H^* admits an embedding $H^* \subset M^*$ (in the category of algebras) with the following properties.*

(i) *$H^* \subset M^*$ is an algebraic closure (in the category of algebras).*

(ii) *Any polynomial $f(X)$ in $M^*[X; 2d]^*$ factorizes as a product of linear factors*

$$f(X) = c(X - m_1)(X - m_2) \cdots (X - m_n) .$$

One can prove this proposition by carrying over to the graded case one's preferred textbook treatment for the ungraded case. We have written it out, but we would not expect an editor to print it.

Our plan is now to construct the other "algebraic closures" we need inside the algebraic closure M^* provided by (4.1). This has the effect of relegating all "choice" to the construction of M^*, where of course the situation is well understood.

PROPOSITION 4.2. *Let H^* be an A^*-algebra, and let $H^* \subset M^*$ be as in* (4.1). *Then we can find an algebra L^* such that $H^* \subset L^* \subset M^*$ and an A^*-algebra structure on L^* extending that on H^* with the following properties.*

(i) *$H^* \subset L^*$ is an algebraic closure in the category of A^*-algebras.*

(ii) *L^* is a graded field.*

(iii) *Any element of M^* which is separable over L^* actually lies in L^*.*

(iv) *Let I^* be the subalgebra of elements in L^* which are purely inseparable over H^*; then L^* is separable over I^*.*

(v) *L^* satisfies* (1.2.2).

We do not need to prove the next remark, but the significance of (1.2.2) here is as follows: whether we work in the category of A^*-algebras or the category of unstable A^*-algebras, (1.2.2) is a necessary and sufficient condition for an object to be closed with respect to purely inseparable extensions.

PROPOSITION 4.3. *Let H^* be an unstable A^*-algebra, let $H^* \subset L^* \subset M^*$ be as in* (4.2), *and let K^* be the set of unstable elements in L^*. Then we have the following properties.*

(i) *$H^* \subset K^*$ is an algebraic closure in the category of unstable*

A^*-algebras.

(ii) *K^* satisfies* (1.2.2).

(iii) *The extension $H^* \subset K^*$ is normal, in the sense that every automorphism $\alpha: M^* \to M^*$ which fixes H^* maps K^* into K^*.*

(iv) *The subalgebra I^* of* (4.2) (iv) *is contained in K^*.*

Let us continue the notation of this proposition. When we come to prove Theorem 1.8, we shall need to make a controlled approach to I^*. Let I_r^* be the subalgebra of elements $x \in K^*$ such that $x^{p^r} \in H^*$; thus $I^* = \bigcup_r I_r^*$, by (4.3) (iv).

LEMMA 4.4. *If H^* satisfies* (1.1.4), *then there is an integer r with the following property: if a derivation ∂ on K^* annihilates I_r^*, then ∂ is zero on K^*.*

This completes the statement of the results which will be quoted later; before proving them, we offer the reader the chance to skip to Section 5.

We now start work on Proposition 4.2; we find it convenient to begin with the separable part of the work. We need an existence statement which allows us to extend Steenrod operations from a subalgebra H^* to a separable extension S^*. We also need a corresponding uniqueness statement; but for the applications of the uniqueness statement, it is important to allow the Steenrod operations in question to take values in an algebra T^* which might be larger than S^*.

LEMMA 4.5. *Suppose given an A^*-algebra H^* and embeddings of algebras $H^* \subset S^* \subset T^*$, where S^* is a (graded) field separable over H^*. Then the Steenrod operations $P^k: H^* \to H^*$ admit a unique extension to functions $P^k: S^* \to T^*$ such that*

$$P^0x = x,\ P^k(x + y) = (P^kx) + (P^ky) \text{ and } P^k(xy) = \sum_{i+j=k}(P^ix)(P^jy)\ .$$

These functions map S^* into S^* and make S^* an A^*-algebra.

This result is due to [14]. The proof is essentially forced, although its presentation can be varied slightly. After assuring the reader that we have written it out, we omit the details.

We now pass on to the purely inseparable part of our work. For this purpose we need a construction. Let M^* be as in (4.1), and let J^* be a subalgebra of M^* with an A^*-algebra structure on it. Let $R(J^*)^*$ be the set of those elements $x \in M^*$ such that $x^p \in J^*$ and $Q^r(x^p) = 0$ for all $r \geqq 1$. (The letter R here stands for "roots".)

LEMMA 4.6. *$R(J^*)^*$ is an algebra containing J^*, and there is a unique way to give it an A^*-algebra structure extending that on J^*.*

Proof. Let $G^* \subset J^*$ be the set of elements $y \in J^*$ such that $Q^r y = 0$ for all $r \geqq 1$; then $R(J^*)^*$ is the set of all elements $x \in M^*$ such that $x^p \in G^*$. Since each Q^r is a derivation, G^* is a subalgebra of J^*; it follows that $R(J^*)^*$ is an algebra. If $x \in J^*$ then $Q^r(x^p) = 0$ (because Q^r is a derivation defined on x); thus $J^* \subset R(J^*)^*$.

We will show that G^* is an A^*-algebra. Suppose given $y \in G^*$ and $a \in A^*$; by Lemma 2.5 we can write $Q^r a = \sum_s b_s Q^s$. Thus

$$Q^r(ay) = \sum_s b_s(Q^s y) = 0 ,$$

so $ay \in G^*$.

If there are any Steenrod operations on $R(J^*)^*$, then by (2.4) they must satisfy

$$\big((\phi^* a)x\big)^p = a(x^p) .$$

But this determines them uniquely, because every element $b \in A^{2d}$ can be written in the form $\phi^* a$ for some $a \in A^{2pd}$, and if $a(x^p)$ has a pth root in $R(J^*)^*$ then that p^{th} root is unique.

But conversely, we claim that this formula gives a valid definition of the Steenrod operations on $R(J^*)^*$. In fact, if we alter a by an element of $\operatorname{Ker} \phi^*$, this does not alter $a(x^p)$ by Lemma 2.5, since $x^p \in G^*$. Since we must have $a \in A^{2pd}$, the element $a(x^p)$ has degree divisible by $2p$ and so has a unique p^{th} root y in M^*; we have $a(x^p) \in G^*$ by the remarks above, and so $y \in R(J^*)^*$. This shows that the formula gives well-defined Steenrod operations. It is easy to check that they have the required formal properties. This proves the lemma.

We can now construct the algebra L^* needed in (4.2). Let H^* be an A^*-algebra, and let $H^* \subset M^*$ be as in (4.1). Let $S^* \subset M^*$ be the subset of elements $x \in M^*$ which are separable over H^*; then S^* is a subalgebra; this takes a little work, but it goes exactly as in the ungraded case. Suppose $m \in M^*$ is non-zero and separable over H; then m^{-1} exists in M^* (because we can solve the equation $mX - 1 = 0$ in M^*) and m^{-1} is separable over H^*; so S^* is a (graded) field. Thus Lemma 4.5 yields an A^*-algebra structure on S^*. Now we define $S(n)^*$ inductively as follows. Start from $S(0)^* = S^*$; suppose that $S(n-1)^*$ is defined and is an A^*-algebra contained in M^*. Set $S(n)^* = R\big(S(n-1)^*\big)^*$ and use Lemma 4.6 to give it an A^*-algebra structure extending that on $S(n-1)^*$. Let $L^* = \bigcup_n S(n)^*$; then we have $H \subset L^* \subset M^*$ and L^* is an A^*-algebra.

It is clear that L^* has property (1.2.2). In fact, suppose that $y \in L^{2dp}$ and $Q^r y = 0$ for all $r \geqq 1$; then $y \in S(n)^*$ for some n and $y = x^p$ for some $x \in M^{2d}$. So $x \in R\big(S(n)^*\big)^*$; that is, $x \in S(n+1)^*$, and $x \in L^*$.

Next we note that L^* is maximal among subalgebras of M^* which carry an A^*-algebra structure. For suppose that the A^*-algebra structure on L^* could be extended to T^*, where $L^* \subset T^* \subset M^*$; and let $x \in T^*$; then x is algebraic over H^*. By the same argument that one uses in the ungraded case, there is a power p^e of p such that x^{p^e} is separable over H^*; that is, $x^{p^e} \in S^*$. Since Q^r is defined and is a derivation, we have $Q^r(x^{p^f}) = 0$ for $f \geqq 1$. So we see by induction over n that $x^{p^{e-n}} \in S(n)^*$ for $n \leqq e$; in particular, $x \in L^*$. Thus $T^* = L^*$.

Now we can deduce that every element of M^* separable over L^* is actually in L^*. For let T^* be the set of elements in M^* which are separable over L^*; then (as above) T^* is a subalgebra, and even a (graded) field; so Lemma 4.5 extends the A^*-algebra structure from L^* to T^*; by the last paragraph, we have $T^* = L^*$.

It follows that L^* is a graded field. In fact, if $l \in L^*$ and $l \neq 0$, then the equation $lX - 1 = 0$ has a solution in M^* by (4.1) (ii); and since this solution is separable over L^*, it lies in L^*.

Let I^* be as in (4.2) (iv). We need a lemma.

LEMMA 4.7. *If $x \in L^{2d}$ and x^p is separable over I^*, then x is separable over I^*.*

Proof. Let $y = x^p$, and let

$$f(Y) = i_m Y^m + i_{m-1} Y^{m-1} + \cdots + i_1 Y + i_0 \in I^*[Y; 2dp]^*$$

be a non-zero homogeneous polynomial such that $f(y) = 0$ and with m as small as possible. If we work over the (graded) field of fractions of I^*, this polynomial divides any other which has y as a root; since y is separable the roots of f are distinct. We must have $i_m \neq 0$, and we can assume without loss of generality that i_m is a pth power, for otherwise we can multiply f by i_m^{p-1}. Take the equation $f(y) = 0$ and apply Q^r; since

$$Q^r(y) = Q^r(x^p) = 0 ,$$

we find

$$(Q^r i_{m-1}) y^{m-1} + \cdots + (Q^r i_1) y + (Q^r i_0) = 0 .$$

By our choice of f, this shows that $Q^r i_s = 0$ for each s, and this holds for all r. Since L^* satisfies (1.2.2), there exists $j_s \in L^*$ such that $i_s = (j_s)^p$; and here of course $j_s \in I^*$. Set

$$g(X) = j_m X^m + j_{m-1} X^{m-1} + \cdots + j_1 X + j_0 \in I^*[X; 2d]^* ;$$

then we get

$$(g(x))^p = f(x^p) = 0 ,$$

and so $g(x) = 0$. Also the roots of g are distinct, for their p^{th} powers are the roots of f. Thus x is separable over I^*. This proves the lemma.

We can now prove (4.2) (iv). Let $x \in L^*$; then x is algebraic over H^*; so (arguing as in the ungraded case) we find a power p^e of p such that x^{p^e} is separable over H^*. A fortiori, x^{p^e} is separable over I^*. Now Lemma 4.7 shows successively that

$$x^{p^{e-1}},\ x^{p^{e-2}},\ \cdots,\ x^{p^2},\ x^p \quad \text{and} \quad x$$

are separable over I^*.

Since the extension $H^* \subset L^*$ is algebraic, it only remains to show that L^* is algebraically closed (in the category of A^*-algebras). Here we record a lemma for use later.

LEMMA 4.8. *Suppose given a diagram*

$$\begin{array}{ccccc} & & Q^* & & \\ & {}^{i}\nearrow & & \searrow^{\theta} & \\ P^* & & & & M^* \\ & \searrow & & \nearrow & \\ & & L^* & & \end{array}$$

in the category of algebras, in which the part

$$\begin{array}{ccc} & & Q^* \\ & {}^{i}\nearrow & \\ P^* & & \\ & \searrow & \\ & & L^* \end{array}$$

lies in the category of A^*-algebras, i is an algebraic extension, and $L^* \to M^*$ is as above. Then θ maps into L^* and is a map of A^*-algebras.

Proof. First we claim that there is no loss of generality in assuming that Q^* is a (graded) field. For if not, let R^* be the (graded) field of fractions of Q^*; since M^* is a (graded) field, we can easily extend the map $\theta: Q^* \to M^*$ to a map (of algebras) from R^* to M^*; Lemma 4.5 allows us to extend the A^*-algebra structure from Q^* to R^*; and R^* is still algebraic over P^*. So we can assume that Q^* is a (graded) field.

By using the diagram

$$\begin{array}{ccccc} & & Q^* & & \\ & {}^{i}\nearrow & & \searrow^{\theta} & \\ P^* & & & & M^* \\ & \searrow & & \nearrow & \\ & & L^* & & \end{array}$$

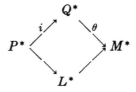

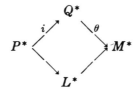

we can consider all the algebras in this diagram as subalgebras of M^*.

Let T^* be the set of those elements in Q^* which are separable over P^*. These elements are separable over L^*, and hence lie in L^*; thus $T^* \subset L^*$. By an argument we have used before, T^* is a (graded) field. Lemma 4.5 shows that for any element $t \in T^*$ the operations $P^k(t)$ defined in Q^* agree with the operations $P^k(t)$ defined in L^*; moreover these operations make T^* an A^*-subalgebra. We now define $T(n)^*$ to be the subalgebra of elements $x \in Q^*$ such that $x^{p^n} \in T^*$. We make the following inductive hypotheses: we have $T(n-1)^* \subset L^*$; and for any element $x \in T(n-1)^*$, the operations ax defined in Q^* agree with the operations ax defined in L^*. If $x \in T(n)^*$, we have $x^p \in T(n-1)^* \subset L^*$ and $Q^r(x^p) = 0$; since L^* satisfies (1.2.2) we have $x^p = y^p$ for some $y \in L^*$; since p^{th} roots are unique we have $x = y$, that is, $x \in L^*$. Thus $T(n)^* \subset L^*$. The Steenrod operations in Q^* and L^* both satisfy

$$a(x^p) = \big((\phi^* a)x\big)^p$$

by (2.4); here $a(x^p)$ is the same whether we use the Steenrod operations from Q^* or L^*, and p^{th} roots are unique; so $(\phi^* a)x$ is the same whether we use the Steenrod operations from Q^* or L^*. This completes the induction.

But now we have $Q^* = \bigcup_n T(n)^*$ for the usual reason: any element x of Q^* is algebraic over P^*, and so there is a power p^e of p such that x^{p^e} is separable over P^* and lies in T^*. We conclude that θ maps Q^* into L^* and is a map of A^*-algebras. This proves the lemma.

It is now easy to deduce that L^* is algebraically closed in the category of A^*-algebras. For this purpose we have to assume given a diagram

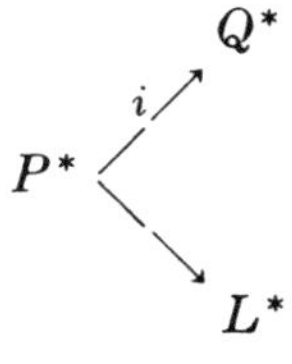

in the category of A^*-algebras, in which i is an algebraic extension. But then Proposition 4.1 allows us to complete the diagram

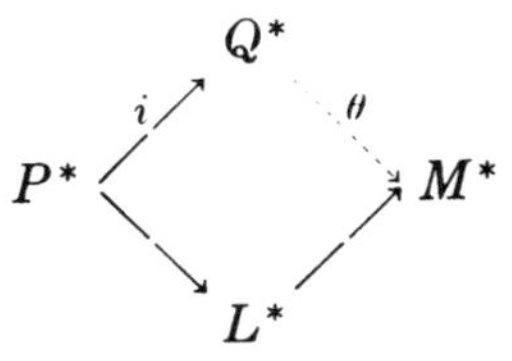

in the category of algebras, and Lemma 4.8 guarantees that θ maps into L^* and is a map of A^*-algebras. This completes the required commutative triangle, so L^* is algebraically closed in the category of A^*-algebras. This proves Proposition 4.2.

We turn to the proof of Proposition 4.3. Let H^* be an unstable A^*-algebra, let $H^* \subset L^* \subset M^*$ be as in (4.2), and let K^* be the set of unstable elements in L^*. Then K^* is an A^*-algebra, by Lemma 2.1. The extension $H^* \subset K^*$ is algebraic; we show that K^* is algebraically closed in the category of unstable A^*-algebras. Suppose given a diagram

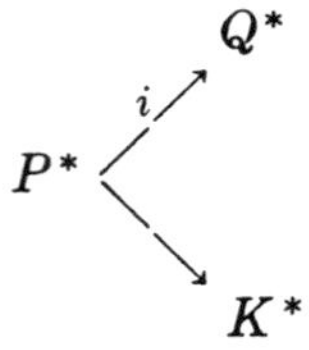

in the category of unstable A^*-algebras, where i is an algebraic extension. By (4.2) we can complete the following diagram in the category of A^*-algebras.

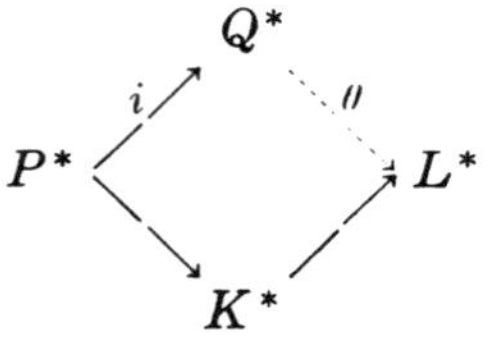

Let $q \in Q^*$; since q is unstable, it follows that θq is unstable; thus θq lies in K^*, by definition. So θ maps Q^* into K^*. This completes the required commutative triangle, and shows that K^* is algebraically closed in the category of unstable A^*-algebras.

We prove that the extension $H^* \subset K^*$ is normal. Suppose given an automorphism $\alpha\colon M^* \to M^*$ which fixes H^*. Consider the following commutative diagram.

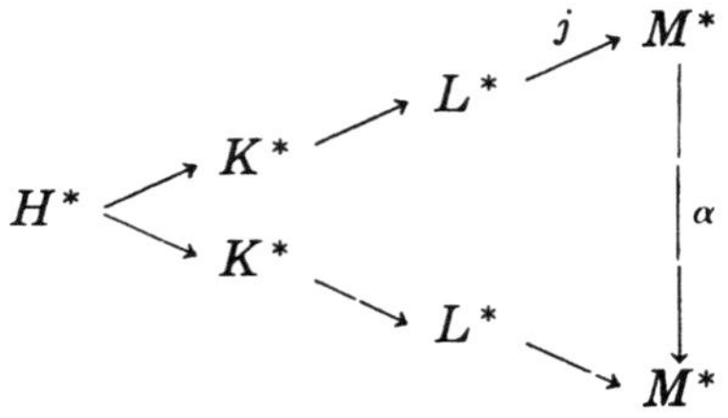

By Lemma 4.8, αj maps L^* into L^* and is a map of A^*-algebras. But then it follows that α maps the set K^* of unstable elements into K^*.

We prove that K^* satisfies (1.2.2). Suppose $y \in K^{2dp}$ and $Q^r y = 0$ for all $r \geqq 1$. Since L^* satisfies (1.2.2), there exists $x \in L^{2d}$ such that $x^p = y$. Since x^p is unstable, x is unstable by Lemma 2.1; thus $x \in K^*$.

Similarly we prove (4.3) (iv). Suppose that x lies in the subalgebra I^* of (4.2) (iv); that is, $x \in L^*$ and $x^{p^e} \in H^*$ for some e. Then Lemma 2.1 shows successively that the elements

$$x^{p^{e-1}}, x^{p^{e-2}}, \cdots, x^{p^2}, x^p \quad \text{and} \quad x$$

are unstable; in particular, $x \in K^*$. This completes the proof of Proposition 4.3.

In order to prove Proposition 1.5, it remains to prove its last clause: if H^* satisfies (1.1.4) then so does K^*. Suppose H^* satisfies (1.1.4); then its (graded) field of fractions has some finite transcendency degree n over $\mathbf{F}_p$. Since K^* is algebraic over H^*, its (graded) field of fractions has the same transcendency degree n over $\mathbf{F}_p$. But then K^* satisfies (1.1.4). This completes the proof of (1.5).

Proof of Lemma 4.4. If H^* satisfies (1.1.4), then so does K^*, as we have just remarked. If there were m derivations $\partial_1, \partial_2, \cdots, \partial_m$ linearly independent on K^*, then Lemma 3.1 would provide m elements of K^* algebraically independent over $\mathbf{F}_p$; so there is an upper bound to the m for which this can happen. Thus we can choose a finite set of derivations $\partial_1, \partial_2, \cdots, \partial_n$ on K^* which are linearly independent, but such that every other derivation ∂ on K^* is linearly dependent on them. (We recall from §3 that for this purpose it does not much matter where our derivations take their values.) We claim that $\partial_1, \partial_2, \cdots, \partial_n$ remain linearly independent on I^*; for K^* is separable over I^* by (4.2) (iv), (4.3) (iv), and so a derivation $\lambda_1\partial_1 + \lambda_2\partial_2 + \cdots + \lambda_n\partial_n$ which was zero on I^* would be zero on K^* (compare [3, p. 139], [16, p. 124]). It follows that there is a finite set of elements $i_1, i_2, \cdots, i_n$ in I^* such that $\partial_1, \partial_2, \cdots, \partial_n$ take linearly independent values on $i_1, i_2, \cdots, i_n$. But then $i_1, i_2, \cdots, i_n$ must all lie in I^*_r for some r, and so $\partial_1, \partial_2, \cdots, \partial_n$ remain linearly independent on I^*_r. This proves Lemma 4.4.

5. Location of a subalgebra $H^*(BT^n; \mathbf{F}_p)$

One of our objectives is to prove that every unstable A^*-algebra K^* which is algebraically closed and satisfies (1.1.4) is isomorphic to $H^*(BT^n; \mathbf{F}_p)$ for some n. This will be done in two main steps. In this section we shall locate inside K^* a suitable subalgebra $H^*(BT^n; \mathbf{F}_p)$; in the next, Section 6, we shall prove that the injection $H^*(BT^n; \mathbf{F}_p) \to K^*$ is iso.

Let us sketch the method of this section. Suppose given an unstable A^*-algebra K^* which is algebraically closed and satisfies (1.1.4). Then our first step is to prove that the derivations Q^r on K^* work just the same way that they do on $H^*(BT^n; \mathbf{F}_p)$ for some n; after all, they must do so if the result is correct. More precisely, we show that the derivations Q^r on K^* satisfy a certain linear relation

$$c_0Q^0 + c_1Q^1 + c_2Q^2 + \cdots + c_nQ^n = 0$$

with coefficients $c_r \in K^*$. After that, if there is any element of degree 2 in the unstable A^*-algebra K^*, it must satisfy

$$c_0Q^0x + c_1Q^1x + c_2Q^2x + \cdots + c_nQ^nx = 0 ;$$

that is,

$$c_0x + c_1x^p + c_2x^{p^2} + \cdots + c_nx^{p^n} = 0 .$$

We therefore write down the equation

$$c_0X + c_1X^p + c_2X^{p^2} + \cdots + c_nX^{p^n} = 0 ;$$

we prove that its roots lie in K^2, and we prove that they generate a subalgebra isomorphic to $H^*(BT^n; \mathbf{F}_p)$.

The last paragraph may serve as a sketch of the proof needed for Theorem 1.1. However, our plans go further: we wish to re-use the results of this section to prove Theorem 1.2. For this purpose we must economize on our data, and begin by studying the derivations Q^r on an unstable A^*-algebra which is not necessarily algebraically closed.

THEOREM 5.1. *Let H^* be an unstable A^*-algebra which satisfies* (1.1.4), *and let $Q^r: H^* \to H^*$ be as in Section* 2 *for $r \geqq 0$. Then there exists $n \geqq 0$ with the following property: the derivations Q^r for any n distinct values of r are linearly independent on H^*, but the derivations Q^r for any $n+1$ values of r are linearly dependent. Let*

$$c_0Q^0 + c_1Q^1 + c_2Q^2 + \cdots + c_nQ^n = 0$$

be any non-trivial linear relation between $Q^0, Q^1, Q^2, \cdots, Q^n$; then all the coefficients c_r are non-zero; and if L^* is any (graded) field containing H^* and the coefficients c_r, then a finite L^*-linear combination $\sum_{0\leq r} l_rQ^r$ annihilates H^* if and only if the polynomial $\sum_{0\leq r} l_rX^{p^r}$ is divisible in the polynomial algebra $L^*[X; 2]^*$ by the polynomial $f(x) = \sum_{0\leq r\leq n} c_rX^{p^r}$.

After this, the second step of our programme gives the following precise result.

THEOREM 5.2. (i) *The roots of the equation $f(X) = 0$ are all separable over H^* and lie in K^2, where K^* is the algebraic closure of H^* in the category of unstable A^*-algebras.*

(ii) *The subalgebra of K^* generated by these roots is, up to isomorphism, $H^*(BT^n; \mathbf{F}^p)$, where n is the same as in* (5.1).

(iii) *A finite linear combination of derivations* Q^r *vanishes on* H^* *if and only if it vanishes on* $H^*(BT^n; \mathbf{F}_p)$.

(iv) *If* $y \in H^{2d}$ *where* $d \not\equiv 0 \bmod p$, *then there is a power* y^e *of* y *which lies in the ideal of* K^* *generated by the augmentation ideal of* $H^*(BT^n; \mathbf{F}_p)$.

We begin work on the proof of Theorem 5.1.

LEMMA 5.3. *If*

$$Q^{r+r_0}, Q^{r+r_1}, Q^{r+r_2}, \cdots, Q^{r+r_n}: H^* \longrightarrow H^*$$

are linearly dependent, then so are $Q^{r_0}, Q^{r_1}, Q^{r_2}, \cdots, Q^{r_n}$.

Proof. Let $y_0, y_1, \cdots, y_n$ be any $n+1$ elements in H^*; say y_j is of degree $2d_j$. Since H^* is unstable, Lemma 2.3 delivers an operation a_j such that

$$Q^{r+s}(a_j y_j) = (Q^s y_j)^{p^r}$$

for each $s \geqq 0$. Consider the determinant

$$\det\left(Q^{r+r_i}(a_j y_j)\right)$$

where i, j run over the range $0 \leqq i \leqq n$, $0 \leqq j \leqq n$. Since $Q^{r+r_0}, Q^{r+r_1}, \cdots, Q^{r+r_n}$ are linearly dependent, this determinant is zero. But since

$$Q^{r+r_i}(a_j y_j) = (Q^{r_i} y_j)^{p^r}$$

and $z \mapsto z^{p^r}$ is a homomorphism of algebras over $\mathbf{F}_p$, we have

$$\det\left(Q^{r+r_i}(a_j y_j)\right) = \left(\det\left(Q^{r_i} y_j\right)\right)^{p^r} .$$

Since H^* is a (graded) integral domain, we deduce

$$\det\left(Q^{r_i} y_j\right) = 0 .$$

Since this holds for all choices of $y_0, y_1, \cdots, y_n$, we conclude that $Q^{r_0}, Q^{r_1}, \cdots, Q^{r_n}$ are linearly dependent. This proves the lemma.

LEMMA 5.4. *Suppose* $Q^0: H^* \to H^*$ *is linearly dependent on* $Q^{r_1}, Q^{r_2}, \cdots, Q^{r_n}$, *where* $r_1, r_2, \cdots, r_n$ *are positive. Then every* $Q^r: H^* \to H^*$ *is linearly dependent on* $Q^{r_1}, Q^{r_2}, \cdots, Q^{r_n}$.

Proof. Suppose given a linear relation

$$h_0 Q^0 = h_1 Q^{r_1} + h_2 Q^{r_2} + \cdots + h_n Q^{r_n}$$

with $h_0 \neq 0$. Without loss of generality we can suppose that $h_i \in H^*$, and also that h_0 is a pth power, for if not we can multiply the relation by h_0^{p-1}. Take the relation

$$h_0(Q^0 x) = h_1(Q^{r_1} x) + h_2(Q^{r_2} x) + \cdots + h_n(Q^{r_n} x)$$

and apply Q^r, where $r > 0$; since $Q^r h_0 = 0$, we get

$$h_0(Q^rQ^0x) = (Q^rh_1)(Q^{r_1}x) + (Q^rh_2)(Q^{r_2}x) + \cdots + (Q^rh_n)(Q^{r_n}x) + h_1(Q^rQ^{r_1}x) + h_2(Q^rQ^{r_2}x) + \cdots + h_n(Q^rQ^{r_n}x) .$$

But applying the same relation to Q^rx, we get

$$h_0(Q^0Q^rx) = h_1(Q^{r_1}Q^rx) + h_2(Q^{r_2}Q^rx) + \cdots + h_n(Q^{r_n}Q^rx) .$$

Here we have

$$Q^rQ^s - Q^sQ^r = 0 \quad \text{for} \quad s > 0 ,$$

but

$$Q^rQ^0 - Q^0Q^r = Q^r$$

because Q^r has degree $2(p^r - 1)$. So subtracting, we get

$$h_0(Q^rx) = (Q^rh_1)(Q^{r_1}x) + (Q^rh_2)(Q^{r_2}x) + \cdots + (Q^rh_n)(Q^{r_n}x) .$$

Since this holds for each x in H^*, we get

$$h_0Q^r = (Q^rh_1)Q^{r_1} + (Q^rh_2)Q^{r_2} + \cdots + (Q^rh_n)Q^{r_n} .$$

This proves the lemma.

We can now prove the first statement in Theorem 5.1. First suppose that $Q^{r_1}, Q^{r_2}, \cdots, Q^{r_m}$ are linearly independent on H^*; then Lemma 3.1 yields m elements of H^* algebraically independent over $\mathbf{F}_p$; and since we assume that H^* satisfies (1.1.4), there is an upper bound to the m for which this can happen. In particular, we can determine $n \geqq 0$ so that $Q^{r_1}, Q^{r_2}, \cdots, Q^{r_n}$ are linearly independent whenever $r_1 < r_2 < \cdots < r_n$, but $Q^{r_0}, Q^{r_1}, Q^{r_2}, \cdots, Q^{r_n}$ are linearly dependent for some choice of $r_0 < r_1 < r_2 < \cdots < r_n$. By Lemma 5.3 we can assume $r_0 = 0$; since $Q^{r_1}, Q^{r_2}, \cdots, Q^{r_n}$ are linearly independent by our choice of n, Q^0 must be linearly dependent on $Q^{r_1}, Q^{r_2}, \cdots, Q^{r_n}$. So Lemma 5.4 shows that every Q^r is linearly dependent on $Q^{r_1}, Q^{r_2}, \cdots, Q^{r_n}$; it follows that any $n + 1$ of the Q's are linearly dependent. This proves our claim.

Now let

$$c_0Q^0 + c_1Q^1 + c_2Q^2 + \cdots + c_nQ^n = 0$$

be any non-trivial linear relation between $Q^0, Q^1, Q^2, \cdots, Q^n$. Since any n of $Q^0, Q^1, Q^2, \cdots, Q^n$ are linearly independent, none of $c_0, c_1, c_2, \cdots, c_n$ can be zero; and in fact these coefficients are unique up to a non-zero scalar factor from some suitable (graded) field.

LEMMA 5.5. *The relation*

$$c_0^{p^r}Q^r + c_1^{p^r}Q^{r+1} + c_2^{p^r}Q^{r+2} + \cdots + c_n^{p^r}Q^{r+n} = 0$$

holds on H^* *for each* $r \geqq 0$.

Proof. We copy the proof of (5.3). Since $Q^1, Q^2, \cdots, Q^n \colon H^* \to H^*$ are linearly independent, we can choose elements $y_1, y_2, \cdots, y_n$ in H^* where they take linearly independent values; that is, we have

$$\det(Q^i y^j) \neq 0 .$$

Suppose y_j is of degree $2d_j$. Since H^* is unstable, Lemma 2.3 delivers an operation a_j such that

$$Q^{r+i}(a_j y_j) = (Q^i y_j)^{p^r} \qquad \text{for each} \quad i.$$

Then

$$\begin{aligned} \det\left(Q^{r+i}(a_j y_j)\right) &= \det\left((Q^i y_j)^{p^r}\right) \\ &= \left(\det(Q^i y_j)\right)^{p^r} \\ &\neq 0 . \end{aligned}$$

That is, the derivations $Q^{r+1}, Q^{r+2}, \cdots, Q^{r+n}$ take linearly independent values on the elements $a_j y_j$.

On the other hand, by what we have proved, $Q^r, Q^{r+1}, Q^{r+2}, \cdots, Q^{r+n}$ are linearly dependent, although any n of them are linearly independent. So there is a linear relation

$$d_0 Q^r + d_1 Q^{r+1} + d_2 Q^{r+2} + \cdots + d_n Q^{r+n} = 0$$

in which the d_i are not all zero, and in fact none of the d_i are zero, and the d_i are unique up to a non-zero scalar factor. Indeed, the coefficients $d_0, d_1, d_2, \cdots, d_n$ may be taken to be the cofactors in the matrix

$$[Q^{r+i}(a_j y_j)]$$

where i, j run over $0 \leqq i \leqq n$, $1 \leqq j \leqq n$. But these cofactors are the $p^{r\text{th}}$ powers of those in the matrix

$$[Q^i y_j] ;$$

that is, they are

$$c_0^{p^r}, c_1^{p^r}, c_2^{p^r}, \cdots, c_n^{p^r} .$$

This proves the lemma.

We can now prove the second statement in Theorem 5.1. By Lemma 5.5, we do not affect the action of $\sum_{0 \leq r} l_r Q^r$ on H^* if we subtract an L^*-linear combination of operations

$$c_0^{p^s} Q^s + c_1^{p^s} Q^{s+1} + c_2^{p^s} Q^{s+2} + \cdots + c_n^{p^s} Q^{s+n} .$$

This does not affect the divisibility of $\sum_{0 \leq r} l_r X^{p^r}$ by $f(X)$ either, for it changes $\sum_{0 \leq r} l_r X^{p^r}$ by subtracting an L^*-linear combination of the polynomials $(f(X))^{p^s}$. In this way we can reduce the question to the case of a sum

$$\sum_{0 \leq r < n} l_r Q^r .$$

But such an operation vanishes on H^* if and only if $l_r = 0$ for each r, by what we have already proved; and the divisibility criterion gives the same condition. This completes the proof of Theorem 5.1.

We turn to discuss Theorem 5.2. Let $H^* \subset K^* \subset L^* \subset M^*$ be as in Section 4, and let

$$f(X) = c_0 X + c_1 X^p + c_2 X^{p^2} + \cdots + c_n X^{p^n}$$

be as above. The equation $f(X) = 0$ has roots in M^2 by (4.1), since we can assume for this purpose that the coefficients c_r lie in H^*. These roots are all distinct, for we have

$$f'(X) = c_0 \neq 0 .$$

Thus the roots are separable over H^*, and lie in L^2 by (4.2) (ii). In order to prove that they lie in K^2, we have to prove that they are unstable.

For this purpose, we make the polynomial algebra $L^*[X; 2]^*$ into an A^*-algebra by regarding it as

$$L^* \otimes H^*(BT^1; \mathbf{F}_p) ;$$

that is, we define

$$P^k X = \begin{cases} X & \text{for } k = 0 \\ X^p & \text{for } k = 1 \\ 0 & \text{for } k > 1 . \end{cases}$$

We hope that the following lemmas now support and motivate each other.

LEMMA 5.6. *If a polynomial* $g(X) = \sum_{0 \leq r} l_r X^{p^r}$ *corresponds to a derivation* $\partial = \sum_{0 \leq r} l_r Q^r$, *then the polynomial* $P^k g(X)$ *corresponds to the derivation* $\partial_k = \sum_{i+j=k} P^i \partial(\chi P^j)$.

Here χ is the canonical anti-automorphism of A^*, as usual. We think of the series of derivations ∂_k as being obtained from ∂ by conjugation with the "automorphism" $\sum_{k=0}^{\infty} P^k$.

LEMMA 5.7. *Suppose* $f(X)$ *is as in* (5.1) *and is chosen to have its coefficients* c_r *in* L^*; *then* $P^k f(X)$ *is divisible by* $f(X)$ *in* $L^*[X; 2]^*$ *for each* k.

LEMMA 5.8. *If a polynomial* $f(X) \in L^*[X; 2]^*$ *has roots* $\lambda_1, \lambda_2, \cdots, \lambda_m$ *in* L^2, *and if* $P^k f(X)$ *is divisible by* $f(X)$ *for each* k, *then the roots* $\lambda_1, \lambda_2, \cdots, \lambda_m$ *are unstable.*

Proof of Lemma 5.6. On the one hand, we have

$$\begin{aligned} P^k g(X) &= P^k(\textstyle\sum_{0 \leq r} l_r X^{p^r}) \\ &= \textstyle\sum_{0 \leq r} (P^k l_r) X^{p^r} + \sum_{0 \leq r} (P^{k-p^r} l_r) X^{r+1} \end{aligned}$$

where P^m is interpreted as 0 if $m < 0$. On the other hand, we have

$$\sum_{i+j=k} P^i \partial(\chi P^j) = \sum_{i+j=k} P^i \left(\sum_{0 \leq r} l_r Q^r\right)(\chi P^j) = \sum_{a+b+j=k,r} (P^a l_r)(P^b Q^r(\chi P^j)) .$$

By use of Lemma 2.2, this gives

$$\sum_{a+b+j=k,r} (P^a l_r)(Q^r P^b(\chi P^j)) + \sum_{a+b+j=k,r} (P^a l_r)(Q^{r+1} P^{b-p^r}(\chi P^j)) .$$

By the properties of χ, this gives

$$\sum_r (P^k l_r) Q^r + \sum_r (P^{k-p^r} l_r) Q^{r+1} ,$$

which agrees with the result for $P^k g(X)$. This proves the lemma.

Proof of Lemma 5.7. Let $f(X)$ be as above; then the corresponding derivation ∂ annihilates H^*. Since χP^j maps H^* into H^*, $\partial_k = \sum_{i+j=k} P^i \partial(\chi P^j)$ also annihilates H^*. By Lemma 5.6, ∂_k corresponds to $P^k f(X)$; by the final clause of Theorem 5.1, $P^k f(X)$ is divisible by $f(X)$ in $L^*[X;2]^*$. This proves the lemma.

The reader may perhaps wonder whether one can prove at this stage an explicit formula for $P^k f(X)$ which makes it clear that it is divisible by $f(X)$. The answer is that one can. Since we do not need it, we suppress it.

Proof of Lemma 5.8. Again we define the total Steenrod operation on $L^*[X;2]^*$ by

$$P_T(x) = \sum_k (P^k x) T^k ;$$

it takes values in the ring of formal power-series

$$(L^*[X;2]^*)[[T]]^*$$

(where we omit the notation which should show that T is of degree $-2(p-1)$). By evaluating each polynomial in X at $X = \lambda_i$ we obtain homomorphisms

$$\varphi_i \colon L^*[X;2]^* \longrightarrow L^* ,$$
$$\psi_i \colon (L^*[X;2]^*)[[T]]^* \longrightarrow L^*[[T]]^* .$$

Since we suppose that $P^k f(X)$ is divisible by $f(X)$ for each k, and $\varphi_i f(X) = 0$, we have

$$\varphi_i P^k f(X) = 0$$

and

$$\psi_i P_T f(X) = 0 .$$

But since we may write

$$f(X) = c(X - \lambda_1)(X - \lambda_2) \cdots (X - \lambda_m)$$

with $c \neq 0$, we have

$$\psi_i P_T f(X) = (\psi_i P_T c)(\psi_i P_T(X - \lambda_1)) \cdots (\psi_i P_T(X - \lambda_m)) .$$

This takes place in $L^*[[T]]^*$, which is a (graded) integral domain, and $\psi_i P_T c = P_T c \neq 0$, so we must have

$$\psi_i P_T(X - \lambda_j) = 0 \quad \text{for some } j .$$

Taking the coefficient of T^0, we have

$$\lambda_i - \lambda_j = 0$$

so that $\lambda_j = \lambda_i$; thus we get

$$\psi_i P_T(X - \lambda_i) = 0 .$$

This gives

$$\psi_i(X + X^p T - P_T\lambda_i) = 0 ,$$

that is,

$$\lambda_i + \lambda_i^p T - P_T\lambda_i = 0$$

or

$$P_T\lambda_i = \lambda_i + \lambda_i^p T .$$

Thus λ_i is unstable. This holds for each i, which proves the lemma.

We have seen that the roots of $f(X) = 0$ lie in L^2; Lemmas 5.7 and 5.8 now show that these roots are unstable, and so lie in K^2. This proves part (i) of Theorem 5.2.

We have already seen that the roots of $f(X) = 0$ are all distinct; so there are p^n of them. They form a vector space over $\mathbf{F}_p$, for the function $x \mapsto x^{p^r}$ is $\mathbf{F}_p$-linear. Thus the roots form a vector space of dimension n over $\mathbf{F}_p$. Let us choose a base in this vector space; then there is a unique map

$$H^*(BT^n; \mathbf{F}_p) = \mathbf{F}_p[x_1, x_2, \cdots, x_n] \xrightarrow{\theta} K^*$$

such that $\theta x_1, \theta x_2, \cdots, \theta x_n$ are the chosen base vectors. Here, in the first instance, θ commutes with addition and multiplication; but it also commutes with Steenrod operations, for we have just proved that the Steenrod operations on $\theta x_1, \theta x_2, \cdots, \theta x_n$ are given by the usual formulae. Next we wish to prove that θ is mono; and we shall do this by applying Lemma 3.1. For this purpose we have to check that some derivations are linearly independent. In $H^*(BT^n; \mathbf{F}_p)$, the derivations Q^s are given by

$$Q^s = \textstyle\sum_r A_{rs}\partial/\partial x_r$$

where

$$A_{rs} = Q^s(x_r) = (x_r)^{p^s} ,$$

so that

$$A = \begin{bmatrix} x_1 & x_1^p & x_1^{p^2} & \cdots & x_1^{p^{n-1}} \\ x_2 & x_2^p & x_2^{p^2} & \cdots & x_2^{p^{n-1}} \\ \vdots & \vdots & \vdots & & \vdots \\ x_n & x_n^p & x_n^{p^2} & \cdots & x_n^{p^{n-1}} \end{bmatrix}.$$

(The reason for using this matrix rather than its transpose will appear in § 6.)

LEMMA 5.9. *The determinant Δ of A is given by*

$$\Delta = \det A = \prod (\mu_1 x_1 + \mu_2 x_2 + \cdots + \mu_n x_n)$$

where the product runs over n-tuples $(\mu_1, \mu_2, \cdots, \mu_n)$ of elements of $\mathbf{F}_p$ in which not all the μ_i are zero and the last non-zero μ_i is 1.

Proof. Take the matrix A and add to the ith row λ_1 times the first row, λ_2 times the second row, and so on, up to λ_{i-1} times the $(i-1)^{\text{th}}$ row. Since $z \mapsto z^{p^r}$ is $\mathbf{F}_p$-linear, we see that the ith row becomes divisible by

$$\lambda_1 x_1 + \lambda_2 x_2 + \cdots + \lambda_{i-1} x_{i-1} + x_i \; ;$$

so Δ is divisible by this linear factor. Since the polynomial algebra $\mathbf{F}_p[x_1, x_2, \cdots, x_n]$ is a unique factorization domain, we see that Δ is divisible by the product

$$\pi = \prod (\mu_1 x_1 + \mu_2 x_2 + \cdots + \mu_n x_n)$$

of all these linear factors. Since Δ and π have the same degree, namely

$$2(1 + p + p^2 + \cdots + p^{n-1}) ,$$

the remaining factor is a scalar. Since the monomial

$$x_1 x_2^p x_3^{p^2} \cdots x_n^{p^{n-1}}$$

occurs with coefficient 1 both in Δ and in π, the remaining factor is 1; that is, $\Delta = \pi$. This proves the lemma.

We can now return to the proof of Theorem 5.2. Since we choose $\theta x_1, \theta x_2, \cdots, \theta x_n$ to be linearly independent, we have

$$\theta(\mu_1 x_1 + \mu_2 x_2 + \cdots + \mu_n x_n) \neq 0$$

for each factor of $\det A$. Since K^* is a (graded) integral domain, we deduce $\theta(\det A) \neq 0$; that is, the derivations $Q^0, Q^1, \cdots, Q^{n-1}$ take linearly independent values on $\theta x_1, \theta x_2, \cdots, \theta x_n$. Now Lemma 3.1 shows that $\theta x_1, \theta x_2, \cdots, \theta x_n$ are algebraically independent; that is, θ is mono. This proves (5.2) (ii).

Next we claim that Theorem 5.1 applies to $H^*(BT^n; \mathbf{F}_p)$ with the same value of n and the same polynomial $f(X)$ that served for H^*. In fact, we have just seen that $Q^0, Q^1, Q^2, \cdots, Q^{n-1}$ are linearly independent on

$H^*(BT^n; \mathbf{F}_p)$; on the other hand, every $n+1$ of the derivations Q^r are linearly dependent on $H^*(BT^n; \mathbf{F}_p)$, for they are all linearly dependent on $\partial/\partial x_1, \partial/\partial x_2, \cdots, \partial/\partial x_n$. Thus the value of n is the same. Moreover, the non-trivial linear relation

$$c_0Q^0 + c_1Q^1 + c_2Q^2 + \cdots + c_nQ^n = 0$$

holds on $H^2(BT^n; \mathbf{F}_p)$ (by construction) and hence on $H^*(BT^n; \mathbf{F}_p)$. So a finite linear combination $\sum_{0 \le r} l_r Q^r$ annihilates $H^*(BT^n; \mathbf{F}_p)$ if and only if $\sum_{0 \le r} l_r X^{p^r}$ is divisible by $f(X) = \sum_{0 \le r \le n} c_r X^{p^r}$, which happens if and only if $\sum_{0 \le r} l_r Q^r$ annihilates H^*. This proves (5.2) (iii).

We turn to part (iv). We have seen that the equation $f(X) = 0$ has as its roots the elements of $H^2(BT^n; \mathbf{F}_p)$. Therefore we can normalize the coefficients c_r so that $c_n = 1$ and still keep the remaining coefficients c_r in K^*, and even in the augmentation ideal of $H^*(BT^n; \mathbf{F}_p)$; for with this normalization the coefficients of $f(X)$ are just the elementary symmetric functions of the elements of $H^2(BT^n; \mathbf{F}_p)$. Suppose given $y \in H^{2d}$; since y is unstable, Lemma 2.3 yields an operation a such that

$$Q^n a y = (Q^0 y)^{p^n} .$$

Here $Q^0 y = dy$ and we assume $d \neq 0 \bmod p$, so we may absorb the factor $d^{p^n} = d$ into a and write

$$y^{p^n} = Q^n b y .$$

But now since $by \in H^*$, our basic relation between the operations Q^r on H^* yields

$$\begin{aligned} y^{p^n} &= Q^n b y \\ &= -c_0(Q^0 b y) - c_1(Q^1 b y) - \cdots - c_{n-1}(Q^{n-1} b y) ; \end{aligned}$$

and this lies in the ideal of K^* generated by the augmentation ideal of $H^*(BT^n; \mathbf{F}_p)$. This completes the proof of Theorem 5.2.

6. Proof of Theorem 1.6

In Section 5 we started from an unstable A^*-algebra which satisfied (1.1.4), took an extension $H^* \subset K^*$ as in Section 4, and located a subalgebra $H^*(BT^n; \mathbf{F}_p)$ embedded in K^*. If we add the assumption that H^* is algebraically closed, then the embedding $H^* \subset K^*$ must be iso. The main object of this section is to prove the following result.

THEOREM 6.1. *If H^* is algebraically closed, then the embedding $H^*(BT^n; \mathbf{F}_p) \subset K^*$ is iso.*

This will prove half of Theorem 1.6; the other half will then follow fairly easily.

In an earlier version of this paper we had a different proof for the corresponding result; and the strategy of this proof was as follows. Let $y \in K^{2d}$. Then one can establish a large number of algebraic equations in a large number of "unknowns". The "unknowns" include y, and the remaining "unknowns" are elements ay for various Steenrod operations $a \in A^*$. The coefficients in these equations are polynomials (over $\mathbf{F}_p$) in the elements c_r considered in Section 5. The equations depend only on n and d. By showing that the appropriate Jacobian does not vanish, one can prove that these algebraic equations define an algebraic variety of dimension zero. By applying Bezout's Theorem, one can then show that this variety has at most a certain number N of points. But by running y over $H^{2d}(BT^n; \mathbf{F}_p)$ one can obtain the full number N of distinct solutions; therefore, we conclude that every element $y \in K^{2d}$ lies in $H^{2d}(BT^n; \mathbf{F}_p)$.

We hope that the reader finds something appealing in this strategy; we shall not risk spoiling this impression by giving the details. The proof we shall present is our second, and the guiding idea is as follows.

We have $H^*(BT^n; \mathbf{F}_p) = \mathbf{F}_p[x_1, x_2, \cdots, x_n]$. We shall work up to the following final lemma.

LEMMA 6.2. *With the assumptions of* (6.1), *each element* $y \in K^{2d}$ *admits an expression*

$$y = \sum_I (y_I)^p x_1^{i_1} x_2^{i_2} \cdots x_n^{i_n}$$

where $y_I \in K^*$ *and the sum runs over indices* $I = (i_1, i_2, \cdots, i_n)$ *such that* $0 \leqq i_r < p$ *for each* r *and* $\sum_r i_r \equiv d \bmod p$.

Certainly, if we have

$$K^* = H^*(BT^n; \mathbf{F}_p) = \mathbf{F}_p[x_1, x_2, \cdots, x_n]$$

then this lemma must be true; but conversely, if this lemma is true, then one can infer $K^{2d} = H^{2d}(BT^n; \mathbf{F}_p)$ by induction over d, for clearly y_I is of less degree than y except in trivial cases.

By applying a theorem of field-theory due to Jacobson, [9, p. 186], one can prove a formula like that in (6.2), but with elements y_I in the (graded) field of fractions of K^*. Our proof will therefore be in two stages. As our first stage, we replace the idea of quoting Jacobson's result by the idea of working through his proof, exercising the utmost economy over denominators. The outcome is Lemma 6.5, which gives us a formula like that in (6.2) with a good bound on the denominators occurring in the coefficients $(y_I)^p$. As a second stage, we use a divisibility argument to show that the elements y_I must lie in K^*.

We now start to discuss Jacobson's method. In order to determine the coefficients $(y_I)^p$ in the formula

$$y = \sum_I (y_I)^p x_1^{i_1} x_1^{i_2} \cdots x_n^{i_n}$$

the obvious method is to assume that the formula holds and apply the differential operators

$$\partial^J/\partial x^J = (\partial/\partial x_1)^{j_1}(\partial/\partial x_2)^{j_2} \cdots (\partial/\partial x_n)^{j_n}$$

where J runs over the indices $(j_1, j_2, \cdots, j_n)$ such that $0 \leqq j_r < p$ for each r; we thus arrive at formulae for the coefficients $(y_I)^p$ in terms of the derivatives $\partial^J y/\partial x^J$. The only snag in this is that, on the face of it, the differential operators $\partial^J/\partial x^J$ are not defined on K^*, but only on $H^*(BT^n; \mathbf{F}_p)$. However, we can replace the derivations $\partial/\partial x_r$ by the derivations Q^s, which are defined on K^*.

We must now discuss the composites of $Q^0, Q^1, \cdots, Q^{n-1}$ which replace the operators $\partial^J/\partial x^J$. Let $K^* \subset L^*$ be as in Section 4; we adopt L^* as a handy place on which the Q^s are defined. Let B^* be the algebra of functions $f: L^* \to L^*$ generated under composition and addition by the following functions.

(i) The derivations Q^r for $0 \leqq r < n$.

(ii) The multiplication maps $x \mapsto lx$, where the constant l runs over L^*.

In this algebra we need the following particular functions. Let

$$E(i) = Q^0(Q^0 - 1)(Q^0 - 2) \cdots (Q^0 - i + 1) ,$$

so that $E(i)$ carries an element x of degree $2d$ to

$$d(d - 1)(d - 2) \cdots (d - i + 1)x .$$

For each index $I = (i_0, i_1, i_2, \cdots, i_{n-1})$ we introduce the operation

$$Q^I = E(i_0)(Q^1)^{i_1}(Q^2)^{i_2} \cdots (Q^{n-1})^{i_{n-1}} .$$

LEMMA 6.3. *B^* is generated as a left L^*-module by the functions Q^I, where I runs over indices $(i_0, i_1, \cdots, i_{n-1})$ such that $0 \leqq i_r < p$ for each r.*

Proof. Since

$$Q^r(lx) = (Q^r l)x + l(Q^r x)$$

we can move all the multiplication maps $x \mapsto lx$ to the left of the composites in which they occur. Since

$$Q^0 Q^r - Q^r Q^0 = -Q^r \qquad \text{for} \quad 0 < r ,$$
$$Q^r Q^s - Q^s Q^r = 0 \qquad \text{for} \quad 0 < r < s ,$$

it is sufficient to consider composites

$$(Q^0)^{i_0}(Q^1)^{i_1}(Q^2)^{i_2}\cdots(Q^{n-1})^{i_{n-1}}$$

in which the Q^r occur in the correct order. Since

$$(Q^0)^p = Q^0 ,$$
$$(Q^r)^p = 0 \qquad \text{for} \quad r > 0 ,$$

it is sufficient to consider composites

$$(Q^0)^{i_0}(Q^1)^{i_1}(Q^2)^{i_2}\cdots(Q^{n-1})^{i_{n-1}}$$

with $0 \leqq i_r < p$ for each r. This proves the lemma.

We must now relate the operations Q^I on $H^*(BT^n; \mathbf{F}_p)$ to the operations $\partial^I/\partial x^I$. This must be possible, since we know how to write each Q^i in terms of the $\partial/\partial x_j$; but to do it in a useful way, we need a construction on matrices that takes a little explanation.

If we were working over an ungraded (commutative) ring R, then a matrix $[A_{ij}]$ with $1 \leqq i \leqq m$, $1 \leqq j \leqq n$ would correspond to a linear map with target R^m and source R^n. But since we work over a graded ring R^*, there is more than one free module of a given rank. We must allow for a target M^* which is a free graded R^*-module on m given basis elements b_i of degrees $2d_1, 2d_2, \cdots, 2d_m$; and we must allow for a source N^* which is a free graded R^*-module on n given basis elements c_j of degrees $2e_1, 2e_2, \cdots, 2e_n$. Then a linear map α (of degree zero) corresponds to a matrix A in which the entry A_{ij} is of degree $2(e_j - d_i)$; and we have

$$\alpha(c_j) = \textstyle\sum_i A_{ij} b_i .$$

(We assure the reader of the correctness of this formula, though it may look peculiar.) And of course, we can form a product AB of matrices if and only if the source of A has the same description $(n; e_1, e_2, \cdots, e_n)$ as the target of B.

Let M^* be a free graded R^*-module on m given basis elements $b_1, b_2, \cdots, b_m$. Let $S(M^*)$ be the symmetric algebra (over R^*) on M^*; and let $T(M^*)$ be the corresponding algebra "truncated at height p", that is, the quotient of $S(M^*)$ by the ideal generated by all p^{th} powers x^p of elements x in M^*. Then $T(M^*)$ has an R^*-base consisting of the monomials

$$b_1^{i_1} b_2^{i_2} \cdots b_m^{i_m}$$

where $0 \leqq i_r < p$ for each r. We thus obtain a functor T defined on the category of such modules-with-bases; in particular, if A is a matrix, then $T(A)$ is a matrix.

LEMMA 6.4. (i) *The formula for writing the derivations Q^i in terms of the derivations $\partial/\partial x_j$ on $H^*(BT^n; \mathbf{F}_p)$ is*

$$Q^i = \sum_j A_{ji} \partial/\partial x_j$$

where $A_{ji} = Q^i(x_j) = (x_j)^{p^i}$, *as in* (5.9).

(ii) *The formula for writing the operations* Q^I *in terms of the operations* $\partial^J/\partial x^J$ *on* $H^*(BT^n; \mathbf{F}_p)$ *is*

$$Q^I = \sum_J \big(T(A)\big)_{JI} \partial^J/\partial x^J$$

where T *is as above and* A *is as in* (i).

(iii) *The matrix* A *in* (i), (ii) *can be written, over* $R^* = H^*(BT^n; \mathbf{F}_p)$, *as the product of diagonal matrices and invertible matrices.*

(iv) *If a matrix* A *with determinant* Δ *can be written* (*over* R^*) *as a product of diagonal matrices and invertible matrices, then there is a matrix* B (*over the same graded ring* R^*) *such that* $B \cdot T(A)$ *and* $T(A) \cdot B$ *are each* Δ^{p-1} *times the identity matrix.*

Proof. Since part (i) is immediate, we begin work with part (ii). First we consider the operations

$$Q^I = (Q^1)^{i_1}(Q^2)^{i_2} \cdots (Q^{n-1})^{i_{n-1}}$$

with $i_0 = 0$. Initially we have

$$Q^r = x_1^{p^r}(\partial/\partial x_1) + x_2^{p^r}(\partial/\partial x_2) + \cdots + x_n^{p^r}(\partial/\partial x_n) .$$

Suppose then that for some I, K with $i_0 = 0$, $k_0 = 0$ we have

$$Q^I = \sum_J \lambda_J^p (\partial^J/\partial x^J) ,$$
$$Q^K = \sum_L \mu_L^p (\partial^L/\partial x^L)$$

for some scalars λ_J, μ_L in $R^* = H^*(BT^n; \mathbf{F}_p)$. Since the derivations $\partial/\partial x_s$ annihilate the pth powers μ_L^p, we find

$$Q^{I+K} = Q^I Q^K = \sum_{J,L} \lambda_J^p \mu_L^p (\partial^{J+L}/\partial x^{J+L}) .$$

The coefficient of $\partial^M/\partial x^M$ is a pth power, and this completes the induction. We obtain the same coefficients that appear in $T(A)$.

We must discuss separately the operations $E(i_0)$. Suppose as an inductive hypothesis that

$$E(r) = \sum_{j_1, j_2 \cdots j_r} x_{j_1} x_{j_2} \cdots x_{j_r} (\partial/\partial x_{j_1})(\partial/\partial x_{j_2}) \cdots (\partial/\partial x_{j_r}) ,$$

where each j runs over the range $1 \leq j \leq n$. Then the definition of $E(r+1)$ is

$$E(r+1) = Q^0 E(r) - rE(r) .$$

But by definition of Q^0 we have

$$Q^0(x_{j_1} x_{j_2} \cdots x_{j_r}) = r x_{j_1} x_{j_2} \cdots x_{j_r} ,$$

so that

$$\sum_{j_1, j_2, \cdots, j_r} (Q^0(x_{j_1} x_{j_2} \cdots x_{j_r}))(\partial/\partial x_{j_1})(\partial/\partial x_{j_2}) \cdots (\partial/\partial x_{j_r}) = rE(r) ,$$

and we are left with

$$\begin{aligned} E(r+1) &= \sum_{j_1, \cdots, j_r} x_{j_1} \cdots x_{j_r}(Q^0(\partial/\partial x_{j_1}) \cdots (\partial/\partial x_{j_r})) \\ &= \sum_{j_0, j_1, \cdots, j_r} x_{j_0} x_{j_1} \cdots x_{j_r}((\partial/\partial x_{j_0})(\partial/\partial x_{j_1}) \cdots (\partial/\partial x_{j_r})) . \end{aligned}$$

This completes the induction.

So now suppose that

$$E(i_0) = \sum_J \lambda_J (\partial^J/\partial x^J) ,$$
$$(Q^1)^{i_1}(Q^2)^{i_2} \cdots (Q^{n-1})^{i_{n-1}} = \sum_K (\mu_K)^p (\partial^K/\partial x^K) .$$

Since the derivations $\partial/\partial x_s$ annihilate the pth powers $(\mu_K)^p$, we find

$$\begin{aligned} Q^I &= E(i_0)(Q^1)^{i_1}(Q^2)^{i_2} \cdots (Q^{n-1})^{i_{n-1}} \\ &= \sum_{J,K} \lambda_J (\mu_K)^p (\partial^{J+K}/\partial x^{J+K}) . \end{aligned}$$

We obtain the same coefficients that appear in $T(A)$. This proves part (ii).

We turn to part (iii). We have two R^*-modules of derivations on $H^*(BT^n; \mathbf{F}_p)$, one with base $Q^0, Q^1, \cdots, Q^{n-1}$ and one with base $\partial/\partial x_1, \partial/\partial x_2, \cdots, \partial/\partial x_n$; the injection of the former in the latter is the linear map which corresponds to the matrix A; we wish to factorize this injection.

It is natural to consider the R^*-module of derivations of $H^*(BT^n; \mathbf{F}_p)$ with the base

$$Q^0, Q^1, \cdots, Q^r, \partial/\partial x_{r+2}, \partial/\partial x_{r+3}, \cdots, \partial/\partial x_n$$

(where $r \leqq n-1$). Let us identify $H^*(BT^r; \mathbf{F}_p)$ with

$$\mathbf{F}_p[x_1, x_2, \cdots, x_r] \subset \mathbf{F}_p[x_1, x_2, \cdots, x_n] .$$

On $H^*(BT^r; \mathbf{F}_p)$ we have a relation

$$c_0 Q^0 + c_1 Q^1 + \cdots + c_{r-1} Q^{r-1} + Q^r = 0$$

where the c's are the elementary symmetric functions of the elements of $H^2(BT^r; \mathbf{F}_p)$, as Section 5. Working with derivations on $H^*(BT^n; \mathbf{F}_p)$, we first form

$$\bar{Q}^r = c_0 Q^0 + c_1 Q^1 + \cdots + c_{r-1} Q^{r-1} + Q^r$$

and then construct

$$\partial = \bar{Q}^r - \sum_{r+2 \leqq i \leqq n} (\bar{Q}^r x_i) \partial/\partial x_i .$$

Thus the module with the base

$$Q^0, Q^1, \cdots, Q^{r-1}, Q^r, \partial/\partial x_{r+2}, \cdots, \partial/\partial x_n$$

also admits the base

$$Q^0, Q^1, \cdots, Q^{r-1}, \partial, \partial/\partial x_{r+2}, \cdots, \partial/\partial x_n ;$$

the transformation from either base to the other is an invertible matrix. On the other hand, δ vanishes on $x_1, x_2, \cdots, x_r$ and also on $x_{r+2}, x_{r+3}, \cdots, x_n$, by construction; so we have

$$\delta = \lambda(\partial/\partial x_{r+1})$$

for a suitable scalar

$$\lambda = \delta(x_{r+1}) = \bar{Q}^r(x_{r+1}) \in H^*(BT^{r+1}; \mathbf{F}_p) .$$

So the injection of the module with base

$$Q^0, Q^1, \cdots, Q^{r-1}, \delta, \partial/\partial x_{r+2}, \cdots, \partial/\partial x_n$$

into the module with base

$$Q^0, Q^1, \cdots, Q^{r-1}, \partial/\partial x_{r+1}, \partial/\partial x_{r+2}, \cdots, \partial/\partial x_n$$

is given by a diagonal matrix. This factorizes the matrix A of (i) as the product of $2n$ factors which are alternately invertible and diagonal; and this proves part (iii).

We turn to part (iv). Evidently it is sufficient to consider separately the case of a diagonal matrix and the case of an invertible matrix. If A is invertible the result is trivial, so we suppose that A is diagonal; say it carries the base element $c_j \in N^*$ to $\lambda_j b_j \in M^*$. Then evidently $T(A)$ is diagonal, carrying the base element $c_1^{j_1} c_2^{j_2} \cdots c_n^{j_n}$ to

$$\lambda_1^{j_1}\lambda_2^{j_2} \cdots \lambda_n^{j_n} b_1^{j_1} b_2^{j_2} \cdots b_n^{j_n} ;$$

so B will have to carry $b_1^{j_1} b_2^{j_2} \cdots b_n^{j_n}$ to

$$\lambda_1^{p-1-j_1}\lambda_2^{p-1-j_2} \cdots \lambda_n^{p-1-j_n} c_1^{j_1} c_2^{j_2} \cdots c_n^{j_n} ;$$

and the entries needed in B do lie in R^*, because we have $j_s \leqq p-1$ for each s. This completes the proof of Lemma 6.4.

LEMMA 6.5. (i) *Suppose given $y \in L^{2d}$. Then there exist coefficients $l_I \in L^*$ such that*

$$by = \sum_I l_I(b(x_1^{i_1} x_2^{i_2} \cdots x_n^{i_n}))$$

for each $b \in B^*$, where the sum runs over indices $I = (i_1, i_2, \cdots, i_n)$ such that $0 \leqq i_r < p$ for each r.

(ii) *Moreover, we have $Q^r l_I = 0$ for $0 \leqq r < n$.*

(iii) *If H^* is algebraically closed then we have $Q^r l_I = 0$ for all r, so that $l_I = 0$ when $\sum_r i_r \not\equiv d \bmod p$, and l_I is a pth power in L^* when $\sum_r i_r \equiv d \bmod p$.*

(iv) *If $y \in K^*$ then $\Delta^{p-1} l_I \in K^*$, where Δ is as in (5.9).*

We obtain a formula of the sort considered in (6.2) by taking $b = 1$.

Proof. In order to establish the equation in part (i), it is sufficient (by Lemma 6.3) to show that

$$Q^J y = \sum_I l_I (Q^J(x_1^{i_1} x_2^{i_2} \cdots x_n^{i_n}))$$

where J runs over indices such that $0 \leqq j_r < p$ for each r. This gives a system of p^n linear equations for the p^n unknowns l_I, and we must consider the matrix of this system, say for brevity

$$Q^J(x^I) = Q^J(x_1^{i_1} x_2^{i_2} \cdots x_n^{i_n}) .$$

We can factorize this matrix as the product of two others: the matrix $T(A)$ for writing each Q^J as a linear combination of the operations $\partial^K/\partial x^K$, and a second matrix

$$(\partial^K/\partial x^K)(x^I) .$$

The latter is triangular, with diagonal entries

$$i_1! \, i_2! \cdots i_n! \neq 0 ;$$

thus it is invertible over $R^* = H^*(BT^n; \mathbf{F}_p)$. The matrix $T(A)$ can be inverted over R^* at the price of introducing a factor Δ^{p-1}, by (6.4); and Δ is non-zero by (5.9). We conclude that part (i) holds for a unique choice of the l_I; and this solution satisfies (iv) also (since the elements $Q^J y$ lie in K^* whenever y does).

We now take the equation in (i) and apply Q^r, where $0 \leqq r < n$. We get

$$Q^r by = \sum_I (Q^r l_I)(bx^I) + \sum_I l_I (Q^r bx^I) .$$

But since $Q^r b \in B^*$, part (i) also gives

$$(Q^r b) y = \sum_I l_I ((Q^r b) x^I) .$$

Subtracting, we get

$$0 = \sum_I (Q^r l_I)(bx^I) .$$

That is, the coefficients $Q^r l_I$ solve the problem of part (i) for $y = 0$; so by the work above, we have $Q^r l_I = 0$ for $0 \leqq r < n$. This proves part (ii).

We turn to part (iii). For each r (not necessarily less than n) we have a relation

$$Q^r = \sum_{0 \leqq s < n} \lambda_s Q^s$$

valid on $H^*(BT^n; \mathbf{F}_p)$; here the scalars λ_s actually lie in $R^* = H^*(BT^n; \mathbf{F}_p)$, as we see from Section 5, but it wouldn't hurt us if they lay anywhere else. By (5.2) (iii) this relation also holds on H^*. We are now assuming that H^* is algebraically closed, so that the embedding $H^* \subset K^*$ is iso, and this relation holds on K^*. By (4.2) (iv), (4.3) (iv), L^* is separable over K^*, so this relation holds on L^*. This relation allows us to infer that $Q^r l_I = 0$ for all r.

In particular, we have $Q^0 l_I = 0$, and so $l_I = 0$ unless l_I is of degree $2ep$

for some e. When l_I is of degree $2ep$, it must be a pth power, since L^* satisfies (1.2.2) by (4.2) (v). This proves Lemma 6.5.

We now prepare for the divisibility argument.

LEMMA 6.6. *Let x be an element of degree 2 in K^*; then x is prime in K^*. Two such primes x, x' are associated if and only if their ratio is in $\mathbf{F}_p$.*

Proof. Let $x \in K^2$; we wish to show that x is prime; we may assume $x \neq 0$. Suppose that x divides yz in K^*; say $wx = yz$ with $y \in K^{2d}$, $z \in K^{2e}$ and therefore $w \in K^{2(d+e-1)}$. Let us define the total Steenrod power by

$$P_T(u) = \sum_{i \geq 0} (P^i u) T^i$$

as usual, where T is a new variable; then we have

$$(P_T w)(P_T x) = (P_T y)(P_T z) .$$

Since w, x, y and z are all unstable, it follows that $P_T w$, $P_T x$, $P_T y$ and $P_T z$ are polynomials in T; we interpret them as elements of $L^*[T]$. Since L^* is a (graded) field, $L^*[T]$ is a unique factorization domain; and the element

$$P_T x = x + x^p T$$

is prime. So $P_T x$ divides either $P_T y$ or $P_T z$; suppose it divides $P_T y$. From

$$P_T y = y + \cdots + y^p T^d ,$$
$$P_T x = x + x^p T$$

we deduce

$$P_T(y/x) = y/x + \cdots + (y/x)^p T^{d-1} ;$$

that is, the element y/x is unstable. So y/x lies in K^*, by the definition of K^*; and so x divides y in K^*. Thus x is prime.

As for the final sentence of the enunciation, any unit in K^* is in $\mathbf{F}_p$ by (2.1). This proves the lemma.

Proof of Lemma 6.2. Suppose $y \in K^{2d}$. By Lemma 6.5, taking $b = 1$, we get an expression

$$y = \sum_I l_I x_1^{i_1} x_2^{i_2} \cdots x_n^{i_n} ;$$

after omitting the terms with $l_I = 0$, each l_I is a pth power in L^*; and we have $\Delta^{p-1} l_I \in K^*$. A fortiori $\Delta^p l_I$ lies in K^*, and since K^* satisfies (1.2.2) by (4.3) (ii), $\Delta^p l_I$ is a pth power in K^*. Thus we can write $l_I = (u/v)^p$ with u, v in K^*, for example by taking $v = \Delta$; but we choose such an expression in which u, v are of minimal degree (this being possible by (2.1)). Set $w = \Delta^{p-1} l_I \in K^*$; then we have

$$\Delta^{p-1} u^p = v^p w \quad \text{in} \quad K^* .$$

By Lemma 5.9, Δ is the product of factors of degree 2, which are prime in

K^* and not associated in K^* by Lemma 6.6. Let l be such a "linear" factor; then l divides $v^p w$. If l divides v, then l^p divides $v^p w$, and since l divides Δ^{p-1} only $p-1$ times, l must divide u. Then we can replace u/v by ul^{-1}/vl^{-1}, contradicting the assumption that u, v are of minimum degree. If l does not divide v, then l^{p-1} must divide w. Since we can repeat this argument for each linear factor l of Δ, we see that Δ^{p-1} divides w, and $l_I = (u/v)^p$ lies in K^*. Since K^* satisfies (1.2.2) by (4.3) (ii), we see that $l_I = (y_I)^p$ for some $y_I \in K^*$. (Of course this gives $u/v = y_I$, and so $v = 1$, but that doesn't matter.) This proves Lemma 6.2.

Proof of Theorem 6.1. We proceed by induction over the degree. Suppose, as an inductive hypothesis, that the embedding $H^*(BT^n; \mathbf{F}_p) \subset K^*$ is iso in degrees less than $2d$; this is true for $d = 1$ by (2.1), so we assume $d \geqq 1$. Let $y \in K^{2d}$. By Lemma 6.2 we have an equation

$$y = \sum\nolimits_I (y_I)^p x_1^{i_1} x_2^{i_2} \cdots x_n^{i_n}$$

where each y_I lies in K^* and is of less degree than y, and so lies in $H^*(BT^n; \mathbf{F}_p)$ by the inductive hypothesis. So $y \in H^{2d}(BT^n; \mathbf{F}_p)$. This completes the induction and proves Theorem 6.1.

At this stage we have proved half of Theorem 1.6; if H^* is an unstable A^*-algebra which satisfies (1.1.4) and is algebraically closed, then H^* is isomorphic to $H^*(BT^n; \mathbf{F}_p)$ for some $n \geqq 0$.

Proof of Corollary 1.7. Let H^* be an unstable A^*-algebra which satisfies (1.1.4); in particular, its (graded) field of fractions has finite transcendency degree over $\mathbf{F}_p$, say transcendency degree m. By (1.5), H^* admits an algebraic closure $H^* \subset K^*$; here K^* is algebraic over H^*, and so its (graded) field of fractions has the same transcendency degree m over $\mathbf{F}_p$. In particular, K^* satisfies (1.1.4), as we already remarked in proving (1.5). By the half of Theorem 1.6 which we have just proved, we must have $K^* \cong H^*(BT^n; \mathbf{F}_p)$ for some n; and then its (graded) field of fractions has transcendency degree n over $\mathbf{F}_p$, so we must have $n = m$. This proves Corollary 1.7, which of course proves Theorem 1.1.

From Corollary 1.7 we will deduce the other half of Theorem 1.6. In fact, for any n we can choose to substitute in Corollary 1.7 an object H^* such that the transcendency degree of its (graded) field of fractions is exactly n; for example, we can take $H^* = H^*(BT^n; \mathbf{F}_p)$. Then Corollary 1.7 shows that its algebraic closure has the form

$$H^* \subset K^* \cong H^*(BT^n; \mathbf{F}_p)$$

for that value of n; so $H^*(BT^n; \mathbf{F}_p)$ must be algebraically closed. This proves Theorem 1.6.

7. Integrality, separability and tidying-up

In this section we clear up the proofs of the results which remain to be proved, beginning with (1.8) and (1.9).

LEMMA 7.1. *Let $R^* \supset S^*$ be (graded) $\mathbf{F}_p$-algebras with R^* finitely-generated (as an $\mathbf{F}_p$-algebra) and integral over S^*. Then S^* is finitely-generated (as an $\mathbf{F}_p$-algebra).*

The reader who wants to see this easy Noetherian exercise written out is referred to [2, p. 81] or [4, Chapter 5, p. 33].

Proof of Theorem 1.8. In the easy direction, suppose that $K^* = H^*(BT^n; \mathbf{F}_p)$ is finitely-generated as a module over H^*. Then K^* is integral over H^*, and so H^* is finitely-generated by Lemma 7.1.

In the harder direction, suppose that H^* is finitely-generated (as an $\mathbf{F}_p$-algebra). As in Section 4, let I_r^* be the set of those elements $x \in K^*$ such that $x^{p^r} \in H^*$. Then I_r^* is an A^*-algebra, for the usual reason: if $x^{p^r} \in H^*$, then $(P^k x)^{p^r} = P^{kp^r} x^{p^r} \in H^*$. We claim that I_r^* is finitely-generated as an $\mathbf{F}_p$-algebra. In fact, I_r^* is isomorphic under $z \mapsto z^{p^r}$ to a subalgebra $S^* \subset H^*$; and here H^* is integral over S^*, for if $x \in H^*$, then $x \in I_r^*$ and $x^{p^r} \in S^*$; so S^* is finitely-generated by (7.1).

Of course we choose r as in Lemma 4.4, so that if a derivation ∂ on K^* annihilates I_r^*, then ∂ is zero on K^*. Let D be the set of derivations

$$\partial \colon I_r^* \longrightarrow I_r^*$$

which are given by finite I_r^*-linear combinations of the Q^s. We claim D is a finitely-generated module over I_r^*. In fact, by the paragraph above, I_r^* has a finite set of generators $z_1, z_2, \cdots, z_q$; let us define a map

$$D \longrightarrow \bigoplus_{1 \le i \le q} I_r^*$$

by assigning to each derivation ∂ the q-tuple $\{\partial z_i\} \in \bigoplus_{1 \le i \le q} I_r^*$; this map is mono, since a derivation ∂ is determined by its effect on the generators z_i; so D is isomorphic to a submodule of a finitely-generated module, and hence finitely-generated since I_r^* is Noetherian.

Take a finite set of generators for D, and express each as a linear combination of the Q^s; we see that D is generated by a finite number of the Q^s, say by $Q^0, Q^1, Q^2, \cdots, Q^{m-1}$ for some m. Then Q^m is a linear combination of $Q^0, Q^1, Q^2, \cdots, Q^{m-1}$; say we have a linear relation

$$d_0 Q^0 + d_1 Q^1 + d_2 Q^2 + \cdots + d_m Q^m = 0$$

with coefficients $d_i \in I_r^*$ and with $d_m = 1$. This relation holds on I_r^*; but then it holds on K^* by our choice of r. It now follows that for each element

$x \in K^2 = H^2(BT^n; \mathbf{F}_p)$ we have the relation

$$d_0 x + d_1 x^p + d_2 x^{p^2} + \cdots + d_m x^{p^m} = 0$$

with coefficients $d_i \in I_r^*$ and with $d_m = 1$. Take the $p^{r\text{th}}$ power; we get a relation

$$e_0 x^{p^r} + e_1 x^{p^{r+1}} + e_2 x^{p^{r+2}} + \cdots + e_m x^{p^{r+m}} = 0$$

with coefficients $e_i = (d_i)^{p^r}$ in H^* and with $e_m = 1$. That is, x is integral over H^*. Now form the monomials

$$x_1^{\nu_1} x_2^{\nu_2} \cdots x_n^{\nu_n}$$

with $0 \leqq \nu_i < p^{r+m}$ for each i; they give a finite set of generators for K^* as a module over H^*. This proves Theorem 1.8.

Proof of Proposition 1.10. Let $H^* \subset K^* \cong H^*(BT^n; \mathbf{F}_p)$ and θ be as assumed in (1.10). Then $\operatorname{Ker}\theta$ is a prime ideal in H^*. Since $\operatorname{Ker}\theta$ is prime and K^* is integral over H^*, there is a prime ideal $J^* \subset K^*$ such that $J^* \cap H^* = \operatorname{Ker}\theta$; for this standard result, the reader may consult [2, p. 62] or [4, Chapter 5, p.38]. Let I^* be the subset of elements $x \in K^*$ such that $P^k x \in J^*$ for all $k \geqq 0$. Since $k = 0$ is allowed, we have $I^* \subset J^*$.

First we claim that I^* is an ideal. It is clear that it is an additive subgroup. Suppose $x \in I^*$ and $y \in K^*$; then

$$P^k(xy) = \textstyle\sum_{i+j=k} (P^i x)(P^j y) \; ;$$

in each term $P^i x$ lies in J^*, and J^* is an ideal, so $P^k(xy)$ lies in J^*; this holds for all k, so xy lies in I^*. Thus I^* is an ideal.

We claim that I^* is a prime ideal. Suppose $x \notin I^*$ and $y \notin I^*$. Then there exists i such that $P^i x \notin J^*$; but $P^i x \in J^*$ for all sufficiently large i, because x is unstable; so we can find i such that $P^i x \notin J^*$ but $P^r x \in J^*$ for $r > i$. Similarly, we can find j such that $P^j y \notin J^*$ but $P^s y \in J^*$ for $s > j$. Take $k = i + j$; then

$$P^k(xy) = (P^i x)(P^j y) + \textstyle\sum_{r+s=k\; r>i} (P^r x)(P^s y) + \sum_{r+s=k\; s>j} (P^r x)(P^s y) \; .$$

The two sums lie in J^*, but the term $(P^i x)\,(P^j y)$ does not lie in J^* because J^* is a prime ideal. So $P^k(xy) \notin J^*$ and $xy \notin I^*$. Thus I^* is prime.

Since $J^* \cap H^* = \operatorname{Ker}\theta$, which is closed under Steenrod operations, we see that

$$I^* \cap H^* = J^* \cap H^* \; .$$

Since I^* and J^* are both prime, $I^* \subset J^*$, $I^* \cap H^* = J^* \cap H^*$ and K^* is integral over H^*, it follows that $I^* = J^*$; for this standard result, the reader may consult [2, p. 61] or [4, Chapter 5, p. 36]. This shows that each operation P^k maps J^* into J^*; in other words, J^* is closed under Steenrod operations.

We now have an algebraic extension

$$H^*/\mathrm{Ker}\,\theta \subset K^*/J^*$$

in the category of unstable A^*-algebras. Because K'^* is algebraically closed, we can complete the following commutative diagram.

$$\begin{array}{ccc} H^* & \xrightarrow{\subset} & K^* \\ \downarrow & & \downarrow \\ H^*/\mathrm{Ker}\,\theta & \xrightarrow{\subset} & K^*/J^* \\ \bar{\theta}\downarrow & & \vdots \\ H'^* & \xrightarrow{\subset} & K'^* \end{array}$$

This proves Proposition 1.10.

Proof of Theorem 1.9. Let $H^* \subset I^* \subset K^*$ be as in Section 4. First suppose that H^* satisfies (1.2.2); then we have $H^* = I^*$. Since K^* is always separable over I^* by (4.2) (iv), it follows that K^* is separable over H^*.

Conversely, suppose that K^* is separable over H^*. Then a fortiori I^* is separable over H^*; and since it is also purely inseparable over H^* by construction, this is only possible if $H^* = I^*$. Now suppose $y \in H^{2dp}$ satisfies $Q^r y = 0$ for all $r \geqq 1$. Since K^* satisfies (1.2.2), there exists $x \in K^*$ such that $x^p = y$; but then $x \in I^*$, and hence $x \in H^*$. Thus H^* satisfies (1.2.2).

To tackle the harder part of the theorem, we suppose that H^* is generated by a finite number of elements y_i of degrees $2d_i$ with d_i prime to p for each i. Applying the work in Section 5 directly to H^*, we obtain an embedding

$$H^*(BT^n; \mathbf{F}_p) \subset K^*$$

with the properties stated in Theorem 5.2. In particular, every element of $H^2(BT^n; \mathbf{F}_p)$ is separable over H^*, and hence every element of $H^*(BT; \mathbf{F}_p)$ is separable over H^*. Of course all this is without prejudice to the fact that we also have

$$K^* \cong H^*(BT^m; \mathbf{F}_p)$$

for some m, possibly larger than n. The quotient of K^* by the ideal in K^* generated by the augmentation ideal of $H^*(BT^n; \mathbf{F}_p)$ is clearly isomorphic to $H^*(BT^{m-n}; \mathbf{F}_p)$. According to (5.2) (iv), for each generator y_i of H^* there is a power y_i^q which maps to zero in $H^*(BT^{m-n}; \mathbf{F}_p)$; since $H^*(BT^{m-n}; \mathbf{F}_p)$ is a (graded) integral domain, it follows that y_i maps to zero in $H^*(BT^{m-n}; \mathbf{F}_p)$. This shows that the image of H^* in $H^*(BT^{m-n}; \mathbf{F}_p)$ is precisely $\mathbf{F}_p$. By

Theorem 1.8, K^* is finitely-generated as a module over H^*; passing to their images, $H^*(BT^{m-n}; \mathbf{F}_p)$ is finitely-generated as a module over $\mathbf{F}_p$. But this is only possible if $m - n = 0$. Therefore the embedding

$$H^*(BT^n; \mathbf{F}_p) \subset K^* \cong H^*(BT^m; \mathbf{F}_p)$$

is iso, and K^* is separable over H^*. This proves Theorem 1.9.

Proof of Theorem 1.2. In the easy direction, suppose given an isomorphism

$$H^* \cong H^*(BT^n; \mathbf{F}_p)^W ;$$

we wish to prove (1.2.1) and (1.2.2).

Suppose then that x lies in the (graded) field of fractions of H^* and is integral over H^*. A fortiori it lies in the (graded) field of fractions of $K^* = H^*(BT^n; \mathbf{F}_p)$ and is integral over K^*. But K^* (being a polynomial algebra) is integrally closed in its field of fractions, so $x \in K^*$. Since x is fixed under W, we have $x \in H^*$. This proves (1.2.1).

Similarly, suppose $y \in H^{2dp}$ and $Q^r y = 0$ for all $r \geqq 1$. Since K^* satisfies (1.2.2), there exists $x \in K^*$ such that $x^p = y$. But then x must be fixed under W, so $x \in H^*$. This proves (1.2.2).

To prove the converse, let H^* be an unstable A^*-algebra which satisfies (1.1.1)-(1.1.4), (1.2.1), and either (1.2.2) or (1.2.3). Then the effect of our work so far is that there is an extension

$$H^* \subset K^* \cong H^*(BT^n; \mathbf{F}_p)$$

such that K^* is normal, integral and separable over H^*. More precisely, let M^* be as in Section 4; then every automorphism of M^* which fixes H^* maps K^* to K^*; and in this way we get a Galois group W of automorphisms of K^*. Of course we have $H^* \subset (K^*)^W$. By carrying over the usual arguments from the classical theory of ungraded fields, we see that every element of K^* which is fixed under W can be written as the quotient u/v of two elements u, v in H^* with $v \neq 0$. But then since K^* is integral over H^*, u/v must be integral over H^*; and since we assume (1.2.1), we must have $u/v \in H^*$. Thus $H^* = (K^*)^W$.

Next we need to infer from (1.2.3) that $|W|$ is prime to p. The argument is by Poincaré series. Since we assume that H^* is generated as an $\mathbf{F}_p$-algebra by a finite number of generators of degrees $2d_1, 2d_2, \cdots, 2d_l$, the Poincaré series of H^* has the form

$$f(T) = \frac{g(T)}{(1 - T^{d_1})(1 - T^{d_2}) \cdots (1 - T^{d_l})} \in \mathbf{Z}[[T]]$$

where $g(T)$ is a polynomial. Since H^* is embedded in $K^* = H^*(BT^n; \mathbf{F}_p)$,

$f(T)$ has a pole of order $\leqq n$ at $T = 1$; thus we can write

$$f(T) = \frac{h(T)}{(1-T)^n} \frac{1-T}{1-T^{d_1}} \frac{1-T}{1-T^{d_2}} \frac{1-T}{1-T^{d_l}}$$

where $h(T)$ is a polynomial. On the other hand, let $\chi^i(w)$ be the trace of the action of $w \in W$ on $K^{2i} = H^{2i}(BT^n; \mathbf{F}_p)$. Form the Poincaré series

$$\chi_T(w) = \sum_{i \geq 0} \chi^i(w) T^i \in \mathbf{F}_p[[T]] ;$$

then if the eigenvalues of w on $H^2(BT^n; \mathbf{F}_p)$ are $\lambda_1(w), \lambda_2(w), \cdots, \lambda_n(w)$, we have

$$\chi_T(w) = \frac{1}{(1-\lambda_1(w)T)(1-\lambda_2(w)T) \cdots (1-\lambda_n(w)T)} .$$

This allows us to compute the trace of $\sum_W w$; but $\sum_W w$ maps K^* into $H^* = (K^*)^W$, and on H^* is multiplication by $|W|$, the order of W. Therefore we find

$$\frac{|W| h(T)}{(1-T)^n} \frac{1-T}{1-T^{d_1}} \frac{1-T}{1-T^{d_2}} \cdots \frac{1-T}{1-T^{d_l}}$$
$$\equiv \sum_W \frac{1}{(1-\lambda_1(w)T)(1-\lambda_2(w)T) \cdots (1-\lambda_n(w)T)} \bmod p .$$

On the right the term with $w = 1$ is

$$\frac{1}{(1-T)^n} ,$$

and all the other terms have denominators which contain $(1 - T)$ to a power less than n. Multiply up by

$$(1-T)^n \frac{1-T^{d_1}}{1-T} \frac{1-T^{d_2}}{1-T} \cdots \frac{1-T^{d_l}}{1-T}$$

and substitute $T = 1$; we find

$$|W| h(1) \equiv d_1 d_2 \cdots d_l \bmod p .$$

Thus $|W|$ is prime to p.

It is now easy to lift W to a group of automorphisms of $H^*(BT^n; \mathbf{Z}_p^\wedge)$. To see this, let G_r be the group of automorphisms of $H^2(BT^n; \mathbf{Z}_p^\wedge/p^r\mathbf{Z}_p^\wedge)$. The map $G_{r+1} \to G_r$ is an epimorphism with kernel of order p^{n^2}. Since the order of W is prime to p, we can lift the map $W \to G_1$ inductively and get maps $W \to G_r$ which by passing to the limit define an action of W on $H^2(BT^n; \mathbf{Z}_p^\wedge)$.

We claim that the resulting action of W on $H^2(BT^n; \mathbf{Z}_p^\wedge)$ is essentially unique. In fact, suppose given representations M, N of W over the p-adic integers which become equivalent when reduced mod p. Then the equivalence $\alpha: M/pM \to N/pN$ can be lifted to a $\mathbf{Z}_p^\wedge$-linear map $\theta: M \to N$; by averaging θ

over W we obtain a $\mathbf{Z}_p^\wedge[W]$-map $\phi: M \to N$ which also covers α, and is therefore an isomorphism.

For the final sentence of (1.2), we have to assume that H^* is a polynomial algebra. We wish to infer that the subalgebra $H^*(BT^n; \mathbf{Z}_p^\wedge)^W$ of elements invariant under W is again polynomial (over $\mathbf{Z}_p^\wedge$). First we have to prove that the map

$$H^*(BT^n; \mathbf{Z}_p^\wedge)^W \otimes \mathbf{F}_p \longrightarrow H^*(BT^n; \mathbf{F}_p)^W$$

is iso. In fact, any element of $H^*(BT^n; \mathbf{F}_p)^W$ can be lifted to $H^*(BT^n; \mathbf{Z}_p^\wedge)^W$, for we first lift to an element of $H^*(BT^n; \mathbf{Z}_p^\wedge)$ and then average over W; and if $v \in H^*(BT^n; \mathbf{Z}_p^\wedge)^W$ maps to 0 in $H^*(BT^n; \mathbf{F}_p)$, we must have $v = pu$ for some $u \in H^*(BT^n; \mathbf{Z}_p^\wedge)$, and here u must be fixed under W because v is.

If we now take elements $v_i \in H^*(BT^n; \mathbf{Z}_p^\wedge)^W$ covering a set of polynomial generators in $H^*(BT^n; \mathbf{F}_p)^W$ we can form the map

$$\mathbf{Z}_p^\wedge[v_1, v_2, \cdots] \longrightarrow H^*(BT^n; \mathbf{Z}_p^\wedge)^W$$

and observe that it is iso.

It now follows that W is a generalized reflection group. The first proof (written for the complex case) was given by Shephard and Todd [13, (5.1), p. 282]. Their method was as follows. First they determined the number of generalized reflections in their group G; then they allowed these generalized reflections to generate a subgroup H; then they proved $H = G$. For this purpose they needed to apply to H results about generalized reflection groups which they had proved case-by-case. For the case-by-case proof one may substitute that due to Chevalley [7]. For a modern treatment, the reader may refer to Bourbaki [5, Chapter 5, p. 115]. This completes the proof of Theorem 1.2.

Proof of Corollary 1.3. This follows exactly the pattern given in Clark and Ewing [8]. From Theorem 1.2 one obtains an isomorphism

$$H^* \cong H^*(BT^n; \mathbf{F}_p)^W .$$

The space X is then constructed as a fibering

$$F \longrightarrow X \longrightarrow B$$

in which $H^*(F; \mathbf{F}_p) \cong H^*(BT^n; \mathbf{F}_p)$ and B is an Eilenberg-MacLane space of type $(W, 1)$, with W acting on F so as to act in the required way on $H^*(F; \mathbf{F}_p)$. One may take F to be a suitable Eilenberg-MacLane space, either of type $(\times_1^n \mathbf{Z}_p^\wedge, 2)$, or of type $(\times_1^n \mathbf{Z}_{p^\infty}, 1)$, where $\mathbf{Z}_{p^\infty}$ is the p-torsion subgroup of S^1.

It remains to discuss Example 1.4. One can settle whether or not the given type is that of a Lie group by a finite search, and the answer is that it is not. Nevertheless, it has the form

$$\{B_7 \times F_4 \times B_2\} \cup \{8\} ,$$

so it is realizable by [8, case 3] for $p > 2$, $p \equiv 1 \bmod 4$, that is, $p \equiv 1, 5 \bmod 12$. It also has the form

$$\{D_8 \times B_2 \times B_2\} \cup \{12, 24\} ,$$

so it is realizable by [8, case 5] for $p > 3$, $p \equiv 1 \bmod 3$, that is, $p \equiv 1, 7 \bmod 12$. (Alternatively, one could use case 2a.) Again, it has the form

$$\{D_8 \times D_4\} \cup \{4, 24\} ,$$

so it is realizable by [8, case 2b] for $p > 3$, $p \equiv \pm 1 \bmod 12$, that is, $p \equiv 1, 11 \bmod 12$.

This completes the proof of all the results stated.

University of Cambridge, England
Wayne State University, Detroit, Michigan

References

[1] J. F. Adams and Z. Mahmud, Maps between classifying spaces, Inventiones Math. **35** (1976), 1-41.
[2] M. F. Atiyah and I. G. Macdonald, *Commutative Algebra*, Addison-Wesley, 1969.
[3] N. Bourbaki, *Algèbre*, Chapters 4, 5, Hermann, Paris, 1950.
[4] ———, *Algèbre Commutative*, Chapters 5, 6, Hermann, Paris, 1964.
[5] ———, *Groupes et Algèbres de Lie*, Chapters 4, 5, 6, Hermann, Paris, 1968.
[6] H. Cartan, Sur l'itération des opérations de Steenrod, Comment. Math. Helv. **29** (1955), 40-58.
[7] C. Chevalley, Invariants of finite groups generated by reflections, Amer. J. Math. **77** (1955), 778-782.
[8] A. Clark and J. Ewing, The realization of polynomial algebras as cohomology rings, Pacific J. Math. **50** (1974), 425-434.
[9] N. Jacobson, *Lectures in Abstract Algebra*, Vol. 3, Van Nostrand 1964.
[10] J. Milnor, The Steenrod algebra and its dual, Ann. of Math. **67** (1958), 150-171.
[11] D. Quillen, The spectrum of an equivariant cohomology ring, I, II, Ann. of Math. **94** (1971), 549-572, 573-602.
[12] J-P. Serre, Sur la dimension cohomologique des groupes profinis, Topology **3** (1965), 413-420.
[13] G. C. Shephard and J. A. Todd, Finite unitary reflection groups, Canadian J. Math. **6** (1954), 274-304.
[14] C. Wilkerson, Classifying spaces, Steenrod operations and algebraic closure, Topology **16** (1977), 227-237.
[15] ———, Some polynomial algebras over the Steenrod algebra A_p, Bull. A. M. S. **79** (1973), 1274-1276.
[16] O. Zariski and P. Samuel, *Commutative Algebra*, Vol. I, Van Nostrand 1958.

(Received October 23, 1978)

Annals of Mathematics, **113** (1981), 621–622

Finite H-spaces and algebras over the Steenrod algebra; a correction

By J. F. ADAMS and C. W. WILKERSON

We are grateful to K. Ishiguro for pointing out a regrettable fallacy which vitiates a passage of our paper [1] starting with p. 141, line 3, "On the other hand", and finishing with p. 141, line 21, "Thus $|W|$ is prime to p". We wish to replace this passage with the following argument.

Let $\bar{H}^*$ and $\bar{K}^*$ be the (graded) fields of fractions of H^* and K^* respectively. We recall from (1.8) that K^* is integral over H^*. It follows that for each $x \neq 0$ in K^* we can find $y \neq 0$ in K^* such that $xy \in H^*$. It is now easy to choose elements $u_1, u_2, \cdots, u_m$ in K^* which form a base for $\bar{K}^*$ over $\bar{H}^*$. In particular, the elements $u_1, u_2, \cdots, u_m$ generate over H^* a free module M^*, whose Poincaré series has the form $f(T)k(T)$, where $k(T)$ is a polynomial with $k(1) = m$. By (1.8) again, K^* is finitely generated as a module over H^*, and so choosing a common denominator $\Delta \neq 0$ in H^* we can arrange that K^* is contained in $\Delta^{-1}M^*$, or equivalently $\Delta K^* \subset M^*$. From the inclusions

$$\Delta K^* \subset M^* \subset K^*$$

we get

$$\frac{T^\delta}{(1-T)^n} \leqq f(T)k(T) \leqq \frac{1}{(1-T)^n} \qquad (0 < T < 1)$$

where δ is the degree of Δ; or equivalently,

$$T^\delta \leqq h(T)k(T)\frac{1-T}{1-T^{d_1}}\,\frac{1-T}{1-T^{d_2}}\cdots\frac{1-T}{1-T^{d_l}} \leqq 1\,.$$

Letting T tend to 1, we find that

$$h(1)m = d_1 d_2 \cdots d_l\,.$$

Thus the degree m of the field extension $\bar{H}^* \subset \bar{K}^*$ is prime to p; we note that this method gives a new proof that the extension is separable, independent of the proof of Theorem 1.9 above. Since $|W| = m$, $|W|$ is prime to p.

With this argument, all the results in our paper survive.

UNIVERSITY OF CAMBRIDGE, ENGLAND
WAYNE STATE UNIVERSITY, DETROIT, MICHIGAN

REFERENCE

[1] J. F. ADAMS and C. W. WILKERSON, Finite H-spaces and algebras over the Steenrod algebra, Annals of Math. **111** (1980), 95–143.

(Received November 24, 1980)

Journal of Pure and Applied Algebra 19 (1980) 1–8

FINITE *H*-SPACES AND LIE GROUPS

J.F. ADAMS

University of Cambridge, 16 Mill Lane, Cambridge CB2 1SB, England

The definition of an H-space goes back to Serre, in his thesis [7]. The given structure of an H-space comprises three things: a topological space X, a base-point $e \in X$, and a "product map" $\mu : X \times X \to X$. These, of course, may be required to satisfy various axioms. For present purposes we don't need to know the axioms; we do need to know some examples.

In the first class of examples, the space X is a topological group G; the point e is the unit element in G; and the map μ is given by the product in the group, $\mu(g, h) = gh$. The topological groups of most important to us here are the Lie groups.

In the second class of examples, X is a loop-space ΩY. That is, one starts from a space Y with base-point y_0; and one forms the space ΩY of continuous functions ω: $[0, 1], 0, 1 \to Y, y_0, y_0$. These functions are called loops, and one gives the set of loops the compact-open topology. The base-point e is the loop constant at y_0; and one defines the product $\mu(\omega, \omega')$ of two loops in the usual way, that is,

$$\mu(\omega, \omega')(t) = \begin{cases} \omega(2t) & (0 \le t \le \frac{1}{2}), \\ \omega'(2t-1) & (\frac{1}{2} \le t \le 1). \end{cases}$$

Loop-spaces are of course the sort of function-space which Serre exploited with such success; and one may say that at this point he was proving some basic lemmas about their topology, by analogy with the known case of topological groups.

It was realised quite early that the second class of examples essentially contains the first. More precisely, let G be a topological group; then under mild restrictions we can form its classifying space BG, and the loop-space ΩBG gives us back G up to equivalence.

There are certainly H-spaces which are not loop-spaces; for example, let X be the unit sphere in the space of Cayley numbers, with μ defined by the multiplication of Cayley numbers. However, we understand such phenomena well enough; in this lecture I mostly want to talk about H-spaces which are loop-spaces ΩY, or as we usually say, H-spaces with classifying spaces Y.

Among such, it has always seemed that the Lie groups are distinguished by their finiteness properties. For example, a compact Lie group G is a finite complex, and a general loop-space ΩY is not even equivalent to a finite complex. In the theory of finite H-spaces, one tends to assume that X is an H-space which is equivalent to some finite complex.

In this subject, like any other, people prove theorems and construct counter-examples. The theorems tend to say that finite H-spaces behave in some way like Lie groups; the counter-examples show that there are finite H-spaces which are not Lie groups.

In 1963 [11] I.M. James suggested that one should look for such counterexamples among sphere bundles over spheres. This suggestion was apparently forgotten for a time. However, the celebrated counterexample of Hilton and Roitberg [3] is of this nature. We now have a good understanding of the range of counterexamples which have been constructed, owing to the advent of the method of localisation [9].

As a representative theorem, I quote the fine result of Hubbuck [4] that a finite H-space which is homotopy-commutative is actually equivalent to a torus T^n. I certainly do not want to disparage this result in any way when I say that unfortunately, as in other parts of group-theory, the general case is more difficult than the abelian case.

We should therefore ask, what are the fundamental statements about the topology of Lie groups which we should try to carry over to finite H-spaces?

To begin with, the Borel theorem carries over. If we exclude a finite number of primes p, then the mod p cohomology of the classifying space Y is a polynomial algebra, on generators whose number and degrees do not depend on p. More formally, there is an integer l, the *rank*, and integers $(2d_1, 2d_2, \ldots, 2d_l)$, the *type*, so that for $p \geq p_0(Y)$ we have

$$H^*(Y;\ F_p) \cong F_p[y_1, y_2, \ldots, y_l]$$

with y_i of degree $2d_i$.

In the classical case, when we start from a compact connected Lie group G, we have a maximal torus T and a Weyl group W. Then a classical result says that

$$H^*(BG;\ F_p) \to H^*(BT;\ F_p)^W$$

is iso for $p \geq p_1(G)$. Here $H^*(BT;\ F_p)^W$ means the subalgebra of elements in $H^*(BT;\ F_p)$ which are invariant under W.

A result of Adams and Wilkerson [2] carries this over to finite H-spaces, in the following way. Suppose that for a particular prime p we have an isomorphism

$$H^*(Y;\ F_p) \cong F_p[y_1, y_2, \ldots, y_l],$$

with y_i of degree $2d_i$, as above; and suppose also that p does not divide $d_1 d_2 \cdots d_l$. It is not assumed that there is any particular geometric relation between ΩY and a torus. Nevertheless, the proof constructs something like a Weyl group, namely a finite subgroup W_p of $\mathrm{GL}(l, Z_p^\wedge)$ which is generated by generalised reflections; and since $\mathrm{GL}(l, Z_p^\wedge)$ acts on $H^*(BT^l;\ F_p)$, of course the subgroup W_p does so. Then the result gives an isomorphism

$$H^*(Y;\ F_p) \xrightarrow{\cong} H^*(BT^l;\ F_p)^{W_p},$$

and this isomorphism preserves Steenrod operations. The point of all this is that the p-adic reflection groups are classified by known results of algebra; so one comes down to a list of 37 cases, some of which are infinite families.

We may accept this as reasonably satisfactory for the primes p which are sufficiently large, in the sense that $p \geq p_0(Y)$ and p does not divide $d_1 d_2 \cdots d_l$. The next problem is, what further information can one get using the small primes? One can convince oneself that one cannot get a good answer using Steenrod operations, and one had better turn to K-theory. I will not offer reasons, since I think that you will believe it. In any case, we glimpse the prospect of running the Adams–Wilkerson programme over again, but using K-theory instead of ordinary cohomology $H^*(\ ; F_p)$. The counterexamples still force us to work one prime at a time; you might think of using $K^*(\ ; F_p)$, but that seems to be a bad idea; it is probably best to use $K^*(\ ; Z_p^\wedge)$. So when I write $K^*(\ )$, I mean $K^*(\ ; Z_p^\wedge)$.

In the classical case we have a result

$$K(BG) \xrightarrow{\cong} K(BT)^W$$

without any restriction on the prime p. So we glimpse the possibility of doing away with the condition $p \geq p_0(Y)$ which was essential to the Borel theorem.

We can now envisage a programme in two steps. The first step should belong to topology: we should assume given a finite H-space in the sense that $X \simeq \Omega Y$ and X is equivalent to a finite complex; and we should deduce that $K(Y)$ has good algebraic properties. The second step should belong to algebra: we should assume given an algebraic object R with the same structure that $K(Y)$ has, and with the good properties proved for $K(Y)$ in the first step; and we should deduce an isomorphism

$$R \xrightarrow{\cong} K(BT)^{W_p}.$$

There is something known about the first step. Namely, if $p > 2$ and $\pi_1(X) = 0$, then

$$K(Y) \cong Z_p^\wedge[[y_1, y_2, \ldots, y_l]],$$

just as in the classical case. This follows from work of J.P. Lin [5, 6]. I have the impression that there is a great deal of information implicit in this result; unfortunately, we don't yet know how to get it out.

Next we must face the question: how much structure must we consider on $K(Y)$, and how many good properties of it must we prove and use? I take it as read that our algebraic objects R will be algebras over $Z_p^\wedge$, with operations λ^i, and complete for the topology defined by powers of the augmentation ideal. We have to deal with things a bit less obvious.

When we were using $H^*(Y; F_p)$, we made essential use of the grading and the "unstable axiom" on Steenrod operations. It is orthodox belief that when you use K-theory, you substitute the filtration on $K(Y)$ for the grading on $H^*(Y; F_p)$. We believe that we have to use a filtration on $K(Y)$; the only question is, which filtration? A priori one can think of two which make sense to an algebraist: Atiyah's γ-filtration, and the rational filtration. Here I take Atiyah's γ-filtration as known, but I should say something about the rational filtration. To a topologist, I say that $y \in K(Y)$ has rational filtration $\geq 2n$ if $\mathrm{ch}_r\, y = 0$ in $H^{2r}(Y;\, Q_{\hat{p}})$ for $r < n$; and then I tell you that you can make sense of it for an algebraist too.

You might conjecture that the two filtrations are the same for good spaces like BG. They are not. Take $p = 3$ and take G to be the exceptional Lie group F_4. If you look in Tits' tables [10] you will find that the first two non-trivial irreducible representations are (say) α of degree 26 and β of degree 52. β is the adjoint representation, and α is the one whose highest weight is a short root. Form

$$y = (\beta - 52) - 3(\alpha - 26) = \beta - 3\alpha + 26.$$

Then y has rational filtration 8, and $3y$ has γ-filtration 8, but y has γ-filtration 4.

So we have to choose. To guide our choice, we recall that we are interested in embedding $R = K(Y)$ in $K(BT)$; so evidently the filtration we want is the one that comes by pulling back the unique filtration on $K(BT)$, if such an embedding is possible. This describes the rational filtration; so we are committed to using the rational filtration.

Next it should be reasonable to study the relation between the operations λ^i or Ψ^k and the rational filtration. After all, we have one precedent to go on, and this is an unpublished proof by me that if

$$H^*(Y;\, Z_{\hat{p}}) = Z_{\hat{p}}[y], \quad y \in H^4,$$

then

$$K(Y) \cong K(BSU(2))$$

where the isomorphism preserves the operations. The nature of this proof is as follows. Let E_0 mean the associated graded, using the rational filtration. Then the hypothesis shows that there is an isomorphism

$$E_0 K(Y) \cong E_0 K(BSU(2)).$$

But if there is one isomorphism then there is more than one: at least one which commutes with some operations and at least one which doesn't. It is essential to pick an isomorphism which does commute with the operations before one tries to lift it to an isomorphism $K(Y) \cong K(BSU(2))$.

Therefore, it should be reasonable to consider associated operations on the associated graded – which comes back to the question of studying the relation between the operations λ^i or Ψ^k and the rational filtration. Investigation reveals the following situation.

Suppose $y \in K(Y)$ and y is of rational filtration $\geq 2n$, so that $\mathrm{ch}_m\, y = 0$ for $m \leq n$. Set $s = [r/(p-1)]$. Then an old theorem of mine [1] says that $p^s \mathrm{ch}_{n+r}\,(y)$ is integral, in the sense that it lies in the image of

$$H^{2n+2r}(Y;\, Z_{\hat{p}}) \to H^{2n+2r}(Y;\, Q_{\hat{p}}).$$

Now suppose that $K(Y)$ is torsion free, as happens in our applications; and imagine that we don't know about $H^*(Y; Z_p)$, but we make a new definition: an element $h \in H^{2n+2r}(Y; Q_{\hat{p}})$ is integral if there exists $y \in K(Y)$ such that ch $y = h +$ (higher terms). This will agree with the usual definition if $H^*(Y; Z_{\hat{p}})$ is torsion-free – and we have to assume Y is innocent until it is proved guilty. With this new definition, we get a different theorem: if y is of rational filtration $\geq 2n$, then $s!\, p^s\, \mathrm{ch}_{n+r}(y)$ is integral in the new sense. This is different from the old result if $s \geq p$, and apparently it cannot be improved without further assumptions. Let us therefore make a further definition, and say that $K(Y)$ "has good integrality" if $p^s\, \mathrm{ch}_{n+r}(y)$ is integral in the new sense (whenever y has rational filtration $\geq 2n$). If this holds, then the relation between operations and rational filtration will be such as we are used to for torsion-free spaces. All torsion-free spaces have good integrality, but maybe some other spaces do also.

For example, consider the exceptional Lie group G_2; then BG_2 has good integrality (at the prime 2), although it has 2-torsion. Similarly, $BPSU(p)$ has good integrality at the prime p, although it has p-torsion. I also checked on $R(T)^{W_p}$ for the first p-adic reflection group which is not a Weyl group and seemed interesting, namely type 12 of the list [8], with $p = 3$. Finally, while I was thinking of these things I received a letter to which I shall return at the end.

Well, one might be ready to frame a conjecture that if G is a compact connected Lie group, then BG has good integrality. No such luck. Take $p = 2$ and consider the group $G = \mathrm{Spin}\ (8)$. The rational filtration on $K(B\ \mathrm{Spin}\ 8)$ coincides with the usual filtration of the Atiyah–Hirzebruch spectral sequence. Let $y = \Delta^+ - \Delta^-$; calculation shows that

$$\mathrm{ch}\ y = \chi + \frac{1}{24} P_1 \chi + \cdots$$

where χ is the Euler class. This is just consistent with my old theorem, because $p^s = 4$ and $\frac{1}{2} P_1 \chi$ lies in the image of

$$H^{12}(B\ \mathrm{Spin}\ 8;\ Z_{\hat{2}}) \to H^{12}(B\ \mathrm{Spin}\ 8;\ Q_{\hat{2}}),$$

with mod 2 reduction $w_4 w_8$. However, it doesn't survive the Atiyah–Hirzebruch spectral sequence (because $\mathrm{Sq}^3\, w_4 w_8 = w_7 w_8$); so it is not integral in the new sense.

The only way forward, alas, is to change the definition of integrality again for our limited purpose. We appeal to the same argument as for the filtration. That is, we seek an embedding in $K(BT)$; so we pull back the definition of "integrality" from $K(BT)$ under the putative embedding, although we don't yet know it exists. We see that we have to add the following to our list of good properties of R.

"There exists a valuation v on $E_0 R$ such that $v(p) = 1$ and $v(y) \geq n$ if and only if there exist $z \in E_0 R$ and m such that $x^m = p^{mn} z$."

Clearly the second clause determines v uniquely if it exists at all, so this is indeed a property of the ring $E_0 R$. Moreover, it is a necessary condition for the existence of an embedding; to see this, you pull back the usual p-adic valuation on $E_0 K(BT) = H^*(BT; Z_{\hat{p}})$.

Similarly, we see that we have to add the following to our list of good properties of R.

"If $y \in R$ is of rational filtration $\geq 2n$, then $\nu(\mathrm{ch}_{n+r}\, y) \geq -s$ where $s = [r/(p-1)]$."
This also is a necessary condition for the existence of an embedding; and so we cannot succeed unless we prove it for $K(Y)$. But both conditions look inconvenient to verify.

To begin with, let us try to define a valuation ν on E_0R by

$$\nu(y) = \mathrm{Sup}\ \{q/m \mid y^m \in p^q E_0 R\}.$$

Then, setting aside the question of whether ν has the properties of a valuation, it looks like a struggle to prove that $\nu(y)$ is an integer. In fact, come to think of it, it looks like a struggle to prove that $\nu(y)$ is even finite. Meditating upon this, we arrive at one more necessary condition. We have to add the following to our list of good properties of R.

"E_0R is finitely generated as an algebra over $\hat{Z_p}$."

In fact, at this point it hits one that there are maybe half-a-dozen finiteness properties that one knows for the classifying space of a compact Lie group, but which one does not know for the classifying space of a finite H-space. This situation seems to call for further work.

After this unsatisfactory report, I imagine that you might like some entertainment. I recall a famous reference [12]. This purports to be a letter from a mathematician long since dead, saying how glad he is to see his results rediscovered independently after such a lapse of time, and giving some explicit formulae which his successors had not found. I have received a letter of a similar nature; let it speak for itself.

"Gentlemen,

Mathematicians may be divided into two classes; those who know and love Lie groups, and those who do not. Among the latter, one may observe and regret the prevalence of the following opinions concerning the compact exceptional simple Lie group of rank 8 and dimension 248, commonly called E_8.

(1) That he is remote and unapproachable, so that those who desire to make his acquaintance are well advised to undertake an arduous course of preparation with E_6 and E_7.

(2) That he is secretive; so that any useful fact about him is to be found, if at all, only at the end of a long, dark tunnel.

(3) That he holds world records for torsion.

Point (1) deserves the following comment. Any right-thinking mathematician who wishes to construct the root-system of E_6 does so as follows: first he constructs the root-system of E_8, and then inside it he locates the root-system of E_6. In this way he benefits from the great symmetry of the root-system of E_8, and its perspicuous nature. If this good precedent is not followed in other researches, one should consider whether to infer a lack of boldness in the investigator rather than a lack of cooperation from the subject-matter.

Since point (2) is equivalent to point (1), we may pass to point (3). And here we should first reject the defences offered by some who might otherwise pass as well-informed. For they appear to regard it as a venial blemish on an otherwise worthy character, comparable to holding world records for the drinking of beer. This will not do. Let us first consider the riotous profusion of torsion displayed by such groups as $PSU(n)$. It then becomes clear that one can award a title to E_8 only by restricting the competition to simply-connected groups. This is as if one were to award a title for drinking beer, having first fixed the rules so as to exclude all citizens of Heidelberg, Munich, Burton-on-Trent, and any other place where they actually brew or drink much of the stuff. In other words, it is contrary to natural justice.

In the second place, to consider the question at all reveals a certain preoccupation with ordinary cohomology. Any impartial observer must marvel at your obsession with this obscure and unhelpful invariant. The author, like all respectable Lie groups, is much concerned to present a decorous and seemly appearance to the eyes of K-theory; and taken in conjunction with other general theorems, this forces him to have a modest amount of torsion in ordinary cohomology. I shall seek some suitable person to inform you in an Appendix.

As a further argument against points (1) and (2), it is natural to release some small scrap of information which you would not otherwise possess. And this may also serve to guarantee the authenticity of this letter; for you must at least believe that it comes via some mathematician who would not mislead you about my views. You may then be expecting me to reveal, for example, $H^{*}(BE_8; F_5)$. I shall not oblige you. That could only encourage you in the low tastes that I have already condemned. Instead, I shall note the following possibility. It may happen that a space Y has its K-theory $K^{*}(Y)$ torsion-free and zero in odd degrees, but nevertheless a careful study of $K^{*}(Y)$ will reveal that Y must have torsion in its ordinary cohomology. Again, I shall seek some suitable person to inform you in an Appendix.

Be it therefore known and proclaimed among you, that my K-theory $K(E_8)$ and that of my classifying space $K(BE_8)$ cannot be criticised in this respect, at least at the prime 5. Their conduct is such as would be blameless and above reproach in the K-theory of a space without 5-torsion in its ordinary cohomology.

Given at our palace, etc, etc,

and signed

E_8."

Appendix 1. We have $\pi_3(E_8) = Z$. The generator is represented by a homomorphism $\theta: S^3 \to E_8$. Using the Hurewicz theorem, and so on, the induced map

$$(B\theta)^{*}: H^4(BS^3; Z) \leftarrow H^4(BE_8; Z)$$

must be iso. On the other hand, the representation rings $R(S^3)$ and $R(E_8)$ restrict so differently on their respective tori that the induced map of their associated graded objects can't be iso in degree 4; its image can be identified with $60\, H^4(BS^3; Z)$. Therefore, in the Atiyah–Hirzebruch spectral sequence for BE_8, the permanent cycles in degree 4 are $60\, H^4(BE_8; Z)$. So this Atiyah–Hirzebruch spectral sequence has non-zero differentials, and this can only happen if BE_8 has 2, 3 and 5-torsion in its ordinary cohomology.

Appendix 2. The discussion of "good integrality" in the body of the paper covers this point.

References

[1] J.F. Adams, On Chern characters and the structure of the unitary group, Proc. Camb. Phil. Soc. 57 (1961) 189–199.

[2] J.F. Adams and C.W. Wilkerson, Finite *H*-spaces and algebras over the Steenrod algebra, Annals of Math. 111 (1980) 95–143.

[3] P. Hilton and J. Roitberg, On principal S^3-bundles over spheres, Annals of Math. (2) 90 (1969) 91–107.

[4] J.R. Hubbuck, On homotopy commutative *H*-spaces, Topology 8 (1969) 119–126.

[5] J.P. Lin, Torsion in *H*-spaces, I, Annals of Math. (2) 103 (1976) 457–487.

[6] J.P. Lin, Torsion in *H*-spaces, II, Annals of Math. (2) 107 (1978) 41–88.

[7] J.-P. Serre, Homologie singulière des espaces fibrés, Annals of Math. 54 (1951) 425–505.

[8] G.C. Shephard and J.A. Todd, Finite unitary reflection groups, Canad. J. Math. 6 (1954) 274–304.

[9] D. Sullivan, Genetics of homotopy theory and the Adams conjecture, Annals of Math. (2) 100 (1974) 1–79.

[10] J. Tits, Tabellen zu den einfachen Lie Gruppen und ihren Darstellungen, Lecture Notes in Mathematics, Vol. 40 (Springer-Verlag, Berlin, 1967).

[11] Problems in Differential and Algebraic Topology, A.M.S. Summer Topology Institute, Seattle 1963, duplicated notes; see pp. 12–13.

[12] "Correspondence", Annals of Math. 69 (1959) 247–251.

SPIN(8), TRIALITY, F_4 AND ALL THAT

J. F. Adams

UNIVERSITY OF CAMBRIDGE

1. INTRODUCTION

This piece will be written from the viewpoint of a pure mathematician, because that is what I am. However, if I make an effort I can probably leave out a lot of the sort of detail that applied mathematicians would like to neglect. Let me sketch the material to be covered.

Both pure and applied mathematicians need groups, because symmetry is important to both. Now, a group G can itself show symmetry. More precisely, we can first form the group Aut G of all automorphisms of G. Inside this we can distinguish the subgroup Inn G of inner automorphisms of G, that is, automorphisms α_h given by conjugation in the sense $\alpha_h(g) = ghg^{-1}$ $(g,h \in G)$. We wish to discard the inner automorphisms as trivial, so we check that Inn G is a normal subgroup of Aut G and form the quotient Out G = Aut G/Inn G. This is called the group of outer automorphisms; thus, an "outer automorphism" is a coset of ordinary automorphisms - it is not a particular kind of automorphism.

When we apply this process to the simple Lie groups, we find that one (namely D_4) has more symmetry than all the rest, because its outer automorphism group is Σ_3, the symmetric group on three letters, rather than Σ_2 or the trivial group Σ_1 as we find in all other cases.

In this statement, D_4 means the compact, connected and simply-connected group Spin(8). (At least, this explanation is good enough for present purposes, even if it involves the judicious neglect of detail which I promised you.)

Here I should pause to remind just a few readers that Spin(n) is the double covering group of the rotation group SO(n). That is, there is a homomorphism of Lie groups from Spin(n) onto SO(n); the kernel of this homomorphism has exactly two elements; and Spin(n) is connected (for $n \geq 2$). Readers whose tastes run to topology will realise that these properties characterise Spin(n), and may indeed be used to construct Spin(n) starting from SO(n). Readers whose tastes run to algebra will probably prefer a direct construction of the spinor groups using Clifford algebras; I advise them to consult Atiyah, Bott & Shapiro (1964), or Porteous (1969), but to take the representation-theory of Clifford algebras from Eckmann (1942).

The word "triality" is applied to the algebraic and geometric aspects of the Σ_3 symmetry which Spin(8) has. The word was chosen by analogy with the word "duality"; the reader already knows about duality, and if one examines the algebraic and geometric aspects of the Σ_2 symmetry which SU(n+1) has, one finds that they agree with what the reader knows already. I gather that applied mathematicians are not much interested in the geometric aspects.

All this goes to show that the group D_4 is not quite as the other groups D_n are. The group D_4 is almost exceptional; there is a close relationship between it and the exceptional group F_4. In fact, we can embed Spin(8) in F_4, and then the Σ_3 symmetry of Spin(8) is worked by conjugation with suitable elements of F_4.

In order to understand this, we shall need to know more about F_4. Two concrete presentations of F_4 are helpful: (i) F_4 is the group of automorphisms of the exceptional Jordan algebra J; (ii) F_4 is the groups of isometries of the Cayley projective plane Π (if Π is provided with the appropriate Riemannian metric). These two presentations are closely related, since Π may be embedded in J so that Π inherits its F_4-action from J (and similarly inherits its Riemannian metric from that on J). All these matters deserve explanation.

This completes my sketch of my programme.

2. THE Σ_3 SYMMETRY

In this setion I will be more specific about the Σ_3 symmetry. We need to know what three objects the symmetric group Σ_3 permutes; and the answer is that it permutes representations.

More precisely, let V be a vector space (over R or C), and let Aut(V) be the general linear group of automorphisms of V. Then a representation in which G acts on V may be viewed as a homomorphism $\theta: G \to \text{Aut}V$. If we are given also an automorphism $\alpha: G \to G$, we can form the composite

$$G \xrightarrow{\alpha} G \xrightarrow{\theta} \text{Aut } V.$$

If α is an inner automorphism, then the representation $\theta\alpha$ is equivalent to θ. Thus the outer automorphism group Out G = Aut G/Inn G permutes the (equivalence classes of) representations of G (of a given degree).

Next we need to recall how many representations there are to be permuted. Let G be a group which comes with a given action on R^n or C^n (as Spin(n) does); then I write λ^i for the representation of G of degree $n!/(i!(n-i)!)$ on the ith exterior power of R^n or C^n, that is, the ith component of the exterior algebra or Grassmann algebra. In particular, λ^1 means the originally-given representation on R^n or C^n. The group Spin(2r) has two basic half-spinor representations Δ^+, Δ^- of degree 2^{r-1}; for the group Spin(8s), these are real representations (of degree 2^{4s-1}). In particular, the group Spin(8) has just three irreducible representations of degree 8, all real, namely λ^1, Δ^+ and Δ^-. Since the outer automorphism group of Spin(8) permutes λ^1, Δ^+ and Δ^-, we get a homomorphism from it to Σ_3.

Theorem 2.1. This homomorphism is an isomorphism from the outer automorphism group of Spin(8) to Σ_3.

This result should be compared with the following one for the case of "duality".

Theorem 2.2. The outer automorphism group of SU(n+1) (for $n > 1$) is Σ_2, acting by permuting λ^1 and λ^n.

If you interpret λ^1 as the space of column vectors, with matrices A acting on the left in the usual way, then λ^n may be interpreted as the space of row vectors v, with action by forming vA^{-1}. However you interpret it, λ^n is the dual space of λ^1.

I pause to illustrate Theorem 2.1. The group SO(8) has an outer automorphism $\bar{\alpha}$ given by conjugation with a matrix A of determinant -1 in O(8). This automorphism $\bar{\alpha}$ lifts to an automorphism α: Spin(8) $\to$ Spin(8). The automorphism α fixes λ^1 and interchanges Δ^+, Δ^-; this is easily checked using characters.

Given the preceding remarks, the proof of Theorem 2.1 is easy, and the idea is this. A map G $\to$ Spin(8) can be interpreted as giving an 8-dimensional representation of G; but you can also argue in the converse direction if you take care.

The approach sketched in this section is due to Élie Cartan (1938). He had found the Σ_3 symmetry rather earlier, as soon as he set himself the task of determining the automorphism groups of all the simple Lie groups (Cartan, 1925).

3. MOTIVATION

In this section I want to motivate the constructions I will give below.

If one wishes to construct the Cayley projective plane, it is natural to ask how one would construct the projective plane over the quaternions H if one were forbidden to use their associativity. To construct this projective plane one normally begins with the space of column vectors H^3 (considered as a right module over H); one then takes the one-dimensional subspaces. With each such subspace S one can associate a linear map $H^3 \to H^3$, namely orthogonal projection on S. In this way the subspaces S correspond to those 3 x 3 matrices over H which are Hermitian, idempotent, and of trace 1. These conditions on matrices never involve multiplying more than two quaternions together, so they do not involve associativity. We can now generalise. The exceptional Jordan algebra J is normally defined as the set of 3 x 3 matrices over the Cayley numbers which are Hermitian. Therefore, we may hope to find the Cayley projective plane Π embedded inside J as the set of idempotents of trace 1; and this is exactly what happens.

However, if we adopt this usual construction of J, then the symmetry group which is visible from the beginning is only $\Sigma_3 \times G_2$; Σ_3 because you can permute the rows of a matrix in any way provided you permute the columns the same way, and G_2 because that is the group of automorphisms of the Cayley numbers.[2] It would be nice to have the Spin(8) symmetry visible from the beginning, so I will do it that way. To do so I need a little algebra, and I devote the next section to it.

4. THE ALGEBRA OF TRIALITY

In the context of Theorem 2.2 (duality), we have a bilinear map $\lambda^n \otimes \lambda^1 \to C$ which is invariant under SU(n+1); with the usual interpretations, the product of a row vector and a column vector is a scalar. In this section we will look at the corresponding phenomena in the case of triality.

Let V^1, V^2, V^3, be three vector-spaces of dimension 8 over R which afford the three inequivalent representations of G = Spin(8). (I put it this way to emphasise the symmetry of my assumptions.) I assume that V^1, V^2, V^3 come with given inner products (,) invariant under G = Spin(8).

Lemma 4.1. There is a trilinear map

$$f: V^1 \otimes V^2 \otimes V^3 \to R$$

which is invariant under G = Spin(8) and not identically zero; it is unique up to a non-zero scalar factor. It may be normalised so that for $x^i \in V^i$ with $(x^i, x^i) = 1$ we have

$$-1 \leq f(x^1 \otimes x^2 \otimes x^3) \leq +1,$$

and these bounds are attained. It is then unique up to sign.

Sketch proof. An easy calculation with characters gives

$$\Delta^+ \otimes \Delta^- \cong \lambda^1 + \lambda^3.$$

Therefore

$$\lambda^1 \otimes \Delta^+ \otimes \Delta^- \cong (\lambda^1 \otimes \lambda^1) + (\lambda^1 \otimes \lambda^3).$$

Now λ^1 is dual to λ^1 but λ^3 is not, so that the right-hand side contains the trivial representation R with multiplicity 1. Now, if you do things properly, the defined meaning of that phrase "with multiplicity 1" is

$$\dim_R \operatorname{Hom}_{RG}(\lambda^1 \otimes \Delta^+ \otimes \Delta^-, R) = 1.$$

The rest is easy.

The map f has various interesting aspects. Let us define the associated product

$$\mu: V^2 \otimes V^3 \to V^1$$

by

$$(x^1, \mu(x^2 \otimes x^3)) = f(x^1 \otimes x^2 \otimes x^3);$$

for example, μ might be the projection of $\Delta^+ \otimes \Delta^-$ on λ^1, suitability normalised.

Lemma 4.2. Then

$$||\mu(x^2 \otimes x^3)|| = ||x^2|| \; ||x^3|| \, .$$

Sketch proof. Consider arguments with $||x^2|| = 1$, $||x^3|| = 1$. According to our normalisation of f we have

$$||\mu(x^2 \otimes x^3)|| \leq 1,$$

and this bound is attained for at least one pair (x^2, x^3). It can be shown that the group $G = \mathrm{Spin}(8)$ acts transitively on pairs (x^2, x^3) with $||x^2|| = 1$, $||x^3|| = 1$, so we have

$$||\mu(x^2 \otimes x^3)|| = 1$$

for all such pairs.

Lemma 4.2 shows us again that there is an 8-dimensional normed algebra over R. In fact, we have only to choose base-points e^2, e^3 of norm 1 in V^2, V^3, and identify V^2, V^3 with V^1 under multiplication by e^3, e^2 respectively; we get on V^1 a normed product with unit. However, our strategy will be precisely to avoid breaking the symmetry by choosing these base-points.

The next lemma tells us that the map f has no more symmetry than we know already.

Lemma 4.3. Suppose given a triple of maps $(\alpha^1, \alpha^2, \alpha^3)$ with $\alpha^i \in SO(V^i)$ and such that

$$f(\alpha^1 x^1 \otimes \alpha^2 x^2 \otimes \alpha^3 x^3) = f(x^1 \otimes x^2 \otimes x^3).$$

Then there exists $g \in G = Spin(8)$ such that $\alpha^i x^i = g x^i$ for each i.

If one is given only the map α^1, there are two solutions for $g \in G = Spin(8)$; going from one solution to the other changes the signs of both α^2 and α^3.

Lemma 4.3 is usually stated after identifying V^1, V^2 and V^3 with the Cayley numbers (as indicated above). It is then regarded as one of the standard algebraic expressions of triality.

Sketch proof of (4.3). Without loss of generality we can assume $\alpha^2 = 1$. Choose a base-point e^2 in V^2. Pass from f to μ as in (4.2), and define an action of V^2 on V^1 as follows: given $x^1 \in V^1$, $x^2 \in V^2$, first solve for x^3 in

$$\mu(e^2 \otimes x^3) = x^1$$

and then take $\mu(x^2 \otimes x^3) \in V^1$. Proceeding in the standard way, we get a Clifford algebra acting on V^1. Granted the dimensions, V^1 has to be an irreducible representation of this Clifford algebra, and such that any map $\theta: V^1 \to V^1$ of Clifford modules is multiplication by a real scalar. But α^1 is such a map, so α^1 is multiplication by a real scalar, necessarily ± 1. Similarly for α^3.

5. CONSTRUCTION OF THE EXCEPTIONAL JORDAN ALGEBRA J

In this section I will construct J.

I begin with the vector space $R^3 \oplus V^1 \oplus V^2 \oplus V^3$, where V^1, V^2, V^3 are as in Section 4 and the standard basis vectors e^1, e^2, e^3 in R^3 are in (1-1) correspondence with V^1, V^2, V^3. This gives me a vector-space of dimension 27, which is correct for J; but of course it needs more structure yet.

I had better give some explanation. The vector $\lambda_1 e^1 + \lambda_2 e^2 + \lambda_3 e^3 \in R^3$ in my construction will correspond to the diagonal matrix

$$\begin{bmatrix} \lambda_1 & 0 & 0 \\ 0 & \lambda_2 & 0 \\ 0 & 0 & \lambda_3 \end{bmatrix}$$

in the usual construction. After I have finished explaining the structure of J, I shall be able to state that any element of J can be thrown by an automorphism of J onto an element $\lambda_1 e^1 + \lambda_2 e^2 + \lambda_3 e^3 \in R^3$. Two such elements $\Sigma\lambda_i e^i$, $\Sigma\mu_i e^i$ can be thrown onto one another by automorphisms of J if and only if $\{\mu_1, \mu_2, \mu_3\}$ is a permutation of $\{\lambda_1, \lambda_2, \lambda_3\}$. It follows that the polynomial functions on J which are invariant under automorphisms of J can be generated by three, which I may take to be the power-sums

$$\lambda_1 + \lambda_2 + \lambda_3, \qquad \lambda_1^2 + \lambda_2^2 + \lambda_3^2, \qquad \lambda_1^3 + \lambda_2^3 + \lambda_3^3.$$

I propose to give a direct construction of these crucial invariant functions, or rather the associated symmetric multilinear functions. So I must define a linear function ℓ, a bilinear function b and a trilinear function t.

First I define ℓ: $J \to R$. I set

$$\ell(\sum_i \lambda_i e^i) = \sum_i \lambda_i$$
$$\ell | V^i = 0 \quad \text{for } i = 1, 2, 3.$$

In the usual construction, this corresponds to the <u>trace</u> of a 3 x 3 Hermitian matrix over the Cayley numbers.

Secondly I define the inner product b: $J \otimes J \to R$. The summands R^3, V^1, V^2, V^3 are to be orthogonal. On R^3 you are to take the usual inner product, so that

$$b(e^i \otimes e^i) = \delta_{ij}.$$

On each V^i you are to take twice the originally-given inner product:

$$b(x \otimes y) = 2(x,y).$$

This factor 2 has to come in somewhere, and this is where it causes least trouble.

Thirdly I define the trilinear function t: $J \otimes J \otimes J \to R$. On any summand $V^i \otimes V^j \otimes V^k$ with i, j, k, all distinct I permute the factors to get to $V^1 \otimes V^2 \otimes V^3$ and use the map f of lemma 4.1. On $e^i \otimes y \otimes z$ with $y \in V^j$, $z \in V^j$ I take

$$\begin{cases} 0 & \text{if } i=j \\ (y,z) & \text{if } i \neq j, \end{cases}$$

and similarly if e^i comes in the second or third place. On $e^i \otimes e^j \otimes e^k$ I take

$$\begin{cases} 1 & \text{if } i = j = k \\ 0 & \text{otherwise.} \end{cases}$$

On all summands of $J \otimes J \otimes J$ not yet mentioned I take zero. This completes the construction of the trilinear map t.

Finally I will construct the Jordan product, and so make J into an algebra. The function $t(x\otimes y\otimes z)$ is evidently a linear function of x if I fix y and z, and the inner product b is non-degenerate, so I can write

$$t(x\otimes y\otimes z) = b(x, y\circ z)$$

for a unique $y\circ z$ in J, which clearly depends linearly on y and z. This gives the Jordan product $y\circ z$; since t is symmetric in y and z, so is $y\circ z$. However, we have clearly lost on symmetry by going from t to the Jordan product.

The Jordan product has $e^1 + e^2 + e^3$ as a unit.

6. THE GROUP F_4

The group F_4 can now be constructed as the set of R-linear maps $J \to J$ which preserve ℓ, b and t, or equivalently, which preserve the Jordan product. It has been visible from the beginnning that all our constructions are invariant under Spin(8), so we get

$$\mathrm{Spin}(8) \subset F_4.$$

Proposition 6.1. (i) The subgroup of F_4 which fixed e^1, e^2 and e^3 is exactly Spin(8). (ii) The subgroup of F_4 which fixes e^1 is isomorphic to Spin(9); more precisely, it is the Spin group of the subspace $V^1\oplus(\lambda(e^2-e^3))$, acting so as to fix e^1, e^2+e^3 and by a 16-dimensional spin representaion on $V^2\oplus V^3$. Similarly with e^1 replaced by e^2 or e^3.

Sketch proof. Part (i) comes easily from lemma 4.3. From this it is as easy to see that the subgroup of F_4 which fixes e^1 is no larger than claimed in part (ii). The essential point is to show that J does have the automorphisms asserted in part (ii). Here I must explain that when I constructed t, I used the components of some functions which I already knew to be invariant under Spin(9). More precisely, Spin(9) has a 16-dimensional real spin-representation Δ, and I can construct invariant maps which involve it just as in the sketch proof of lemma 4.1. When I restrict from Spin(9) to Spin(8), Δ splits as $\Delta^+ + \Delta^-$, and I can identify the components of my invariant maps. These are the components I already used in constructing t, so t is invariant under Spin(9). Since ℓ and b are clearly invariant under Spin(9), I get $\mathrm{Spin}(9) \subset F_4$.

Corollary 6.2. The subgroup of F_4 which preserves the set $\{e^1, e^2, e^3\}$ maps onto Σ_3 with kernel Spin(8), and the quotient Σ_3 acts on Spin(8) by its outer automorphism group.

Sketch proof. Consider the subgroup Spin(9) of F_4 which fixes e^1, as in (6.1). In this subgroup we find at least one element which reverses the sign of e^2-e^3, so that it interchanges e^2 and e^3, and therefore interchanges V^2 and V^3. Similarly with e^1 replaced by e^2 or e^3.

With this I have carried out my brief and shown you that the Σ_3 symmetry of Spin(8) is worked by conjugation in F_4. It is not my brief to say as much as I would like about the geometry of the Cayley projective plane Π, but I will define it for you. It is the F_4-orbit of the point $e^1 \in J$; equivalently, it is the set of points $j \in J$ where

$$\ell(j) = 1, \qquad b(j \otimes j) = 1, \qquad t(j \otimes j \otimes j) = 1;$$

equivalently, it is the set of points $j \in J$ where

$$\ell(j) = 1, \quad j \circ j = j.$$

The last description reconciles it with the "idempotents of trace 1" in Section 3.

7. TRIALITY AND COMPLEX QUADRICS

The oldest phenomenon associated with triality goes back before Élie Cartan. To see it, it will be well to complexify our groups. The group SO(8) is a group of real 8 x 8 matrices which preserve the real quadratic form

$$\sum_1^8 x_i^2;$$

it is embedded in the corresponding group of complex 8 x 8 matrices of determinant 1, which preserve the complex quadratic form

$$\sum_1^8 z_i^2.$$

The injection is a homotopy equivalence, so the complex group has a double cover G which contains the group Spin(8). Let us complexify the vector spaces V^1, V^2 and V^3 of Section 4; our complex group G acts on $V^1 \otimes C$, $V^2 \otimes C$ and $V^3 \otimes C$ so as to preserve a non-singular complex quadratic form on each. It therefore acts on the three projective spaces $P(V^1 \otimes C)$, $P(V^2 \otimes C)$ and $P(V^3 \otimes C)$, which are of dimension 7. It also acts on the quadrics Q^1, Q^2 and Q^3 in these projective spaces which are determined by the vanishing of the quadratic forms. These quadrics are of dimension 6. Now just as a complex quadric of dimension 2 has on it two systems of lines, so a complex quadric of dimension 6 has on it two systems of linear subspaces of dimension 3.

I now introduce an idea of Tits (1956); each complex group G of a suitable kind can be interrogated and made to yield "the" geometric system on which it consents to act as a group of automorphisms. For brevity I will not go into the way in which this is done. However, in our case we get three equally valid realisations of Tit's geometrical system, in terms of the geometries of Q^1, Q^2 and Q^3.

Proposition 7.1. These three geometries correspond as follows.

(Points on Q^1) ↔ (3-spaces of one system on Q^2) ↔ (3-spaces of one system on Q^3)

(3-spaces of one system on Q^1) ↔ (Points on Q^2) ↔ (3-spaces of the other system on Q^3)

(3-spaces of the other system on Q^1) ↔ (3-spaces of the other system on Q^2) ↔ (Points on Q^3)

(Lines on Q^1) ↔ (Lines on Q^2) ↔ (Lines on Q^3).

Moreover, these (1-1) correspondences preserve incidence.

Here two linear subspaces on Q^i are "incident" if one is contained in the other, or, when both are of dimension 3, if their intersection is of dimension 2.

I will briefly indicate the first correspondence in the proposition. Take a point of Q^1, represented by a non-zero vector $x \in V^1 \otimes C$. Consider the set of vectors $y \in V^2 \otimes C$ such that

$$f(x \otimes y \otimes z) = 0 \text{ for all } z \in V^3 \otimes C.$$

Such vectors y form a vector subspace of dimension 4 in $V^2 \otimes C$, yielding a 3-space on Q^2.

SPIN(8), TRIALITY, F_4 AND ALL THAT

REFERENCES

M. F. Atiyah, R. Bott & A. Shapiro, (1964), Clifford Modules, Topology 3, Suppl. 1, pp. 1-38.

E. Cartan, (1925), Le principe de dualité et la théorie des groupes simples et semisimples, Bull. des Sciences Mathématiques, 49 pp. 367-374.

E. Cartan, (1938), Leçons sur la théorie des spineurs II Hermann et Cie, Act. Sci.et Ind., 701, pp. 53-54.

B. Eckmann, (1942), Gruppentheoretische Beweis des Satzes von Hurwitz-Radon über die Komposition quadratischer Formen, Comment. Math. Helv., 15 pp. 358-366.

I. Porteous, (1969) Topological Geometry, Van Nostrand Reinhold , especially Chapter 13.

J. Tits, (1956), Les groupes de Lie exceptionnels et leur interprétation géometrique, Bull. Soc. Math. Belgique, 8, pp. 48-81.

(1959)"Correspondence", Ann. Math., 69, pp. 247-251.

Contemporary Mathematics
Volume 37, 1985

THE FUNDAMENTAL REPRESENTATIONS OF E_8

J. F. Adams

This paper is a snippet of a book I have in preparation about the exceptional Lie groups. This snippet argues that in order to understand the representations of E_8, one should begin by understanding just three of them; one is the adjoint representation, and it goes on to construct the other two.

To begin with, one might reason that eight fundamental representations would suffice. In fact, let G be a compact simply-connected Lie group (such as E_8). Then a celebrated theorem of Hermann Weyl says that the representation ring $R(G)$ is a polynomial algebra

$$Z[\rho_1, \rho_2, \ldots, \rho_\ell],$$

where ℓ is the rank of G. So if we understood the representations $\rho_1, \rho_2, \ldots, \rho_8$ of E_8, then we could construct all virtual representations of E_8 by taking tensor products and Z-linear combinations.

The proof of Weyl's theorem tells us more about the fundamental representations ρ_i. Each one is irreducible, and has a highest weight which lies on an edge of the fundamental dual Weyl chamber. Each edge of this Weyl chamber corresponds to the opposite face, which corresponds to a simple root, which in turn corresponds to a node of the Dynkin diagram. Thus the fundamental representations ρ_i in Weyl's theorem correspond to the nodes of the Dynkin diagram.

To cut down on the number 8, we can allow ourselves to construct new representations from old by taking exterior powers, as well as tensor products and Z-linear combinations. We will see how to calculate the exterior powers we need "modulo less significant terms." To formalize this, let "top" be the function which takes any G-module M and assigns to it the smallest G-submodule $\mathrm{top}(M)$ which has the same extreme weights with the same multiplicities. We think of $\mathrm{top}(M)$ as "the most significant part of M."

Let us take our compact simply-connected Lie group G, and suppose that its Dynkin diagram has "an arm of length i", in the following sense. We suppose given nodes $v_1, v_2, \ldots, v_i$ of the Dynkin diagram so that

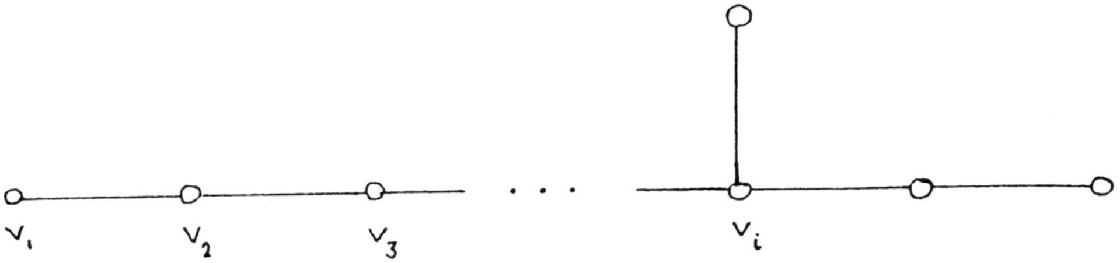

(i) v_j is joined to v_{j+1} by a single bond for $1 \leq j < i$, and (ii) no other edge runs to any of $v_i, v_2, \ldots, v_{i-1}$. There is no limitation on the number or type of edges which may run to v_i, but v_1 is "a free end."

Let ρ_j be the irreducible representation which corresponds, as in Weyl's theorem, to the node v_j.

PROPOSITION 1. For $j \leq i$ we have

$$\rho_j = \operatorname{top}(\lambda^j(\rho_1)).$$

The application of this result is clear.

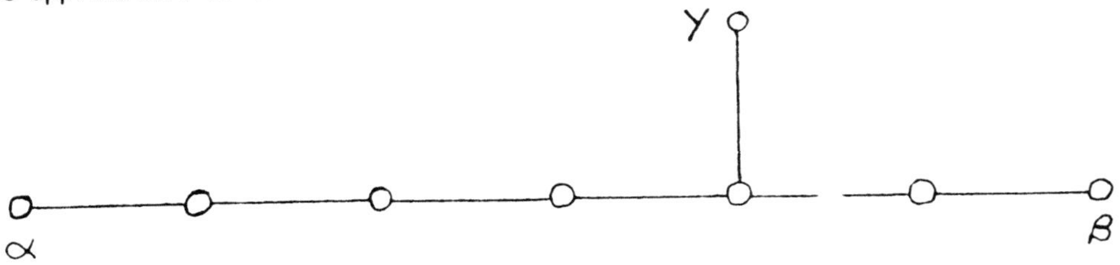

The Dynkin diagram for E_8 has three arms; let α, β and γ be the irreducible representations which correspond to their free ends, as shown in the diagram. In particular, α is the adjoint representation in which E_8 acts on its own Lie algebra $L(E_8)$; I take this representation as known. (Here and later we ignore the difference between $L(E_8)$ and its complexification; we treat representations as complex, even though Lie algebras may come over R.)

COROLLARY 2. Seven of the eight generators for the polynomial ring $R(E_8)$ may be taken as

$$\alpha,\ \lambda^2\alpha,\ \lambda^3\alpha,\ \lambda^4\alpha,\ \beta,\ \lambda^2\beta,\ \gamma\ .$$

The eighth may be taken either as $\lambda^5\alpha$, or as $\lambda^3\beta$, or as $\lambda^2\gamma$.

The corresponding argument for D_n would say that one should begin by understanding three representations of $D_n = \mathrm{Spin}(2n)$, namely the usual representation on R^{2n} and the two half-spinor representations Δ^+, Δ^-. This we believe, so perhaps we can accept the analogue for E_8.

I owe the statement of Proposition 1 to J. H. Conway. He said he "thought he saw (vaguely) how one might prove it." I wrote out a proof, so the result is sound. Before publishing a proof I would want to check whether anybody else proved it before.

The proof of Proposition 1 yields a bit more. For example, one can obtain the corresponding result about $\lambda^2(\rho_j)$ for any node v_j. (This saves us saying anything about triple bonds in what follows next; it is so embarassing to have a general result which applies to only one special case.) One can also obtain the corresponding result about $\lambda^{i+1}(\rho_1)$, at the price of dividing cases. For example, suppose our arm continues by forking, like this.

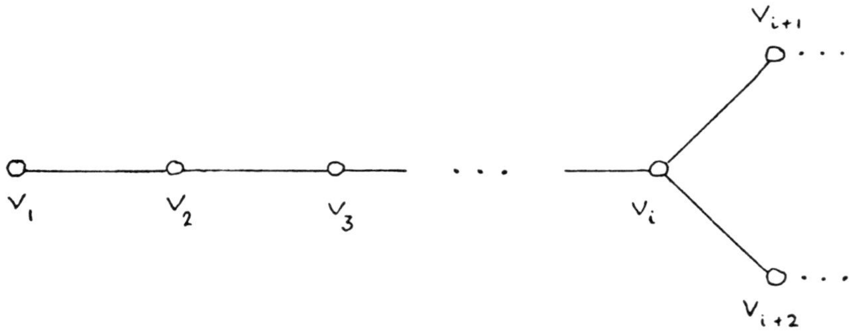

Then we find

$$\text{top}(\lambda^{i+1}(\rho_1)) = \text{top}(\rho_{i+1} \otimes \rho_{i+2}).$$

Similarly if our arm continues along a double bond; but then one must divide cases, according as the double bond goes from a short root to a long one or vice versa. The reader can guess the result by considering the cases B_n and C_n.

Well, all this was fun while it lasted, but we must get back to business and construct β and γ. Here any author faces a dilemma. If he gives a conceptual method then all his readers (including me) will long for concrete details and explicit formulae so that they can get their hands on things. If he gives concrete details and explicit formulae then all his readers (including me) will yearn for conceptual enlightenment and understanding.

The reader willing to persevere to the end may be reassured; there follows a method which is conceptual and leads to explicit formulae. But for purposes of writing, I choose the second horn of the dilemma. Our business is to get our hands on things; so I shall treat the explicit formulae as the results to be recorded, and consign the conceptual methods to the proof.

To give explicit formulae we must agree on a coordinate system. The group E_8 contains a subgroup of type A_8. We may arrange details to the following effect. The subgroup $A_8 = (SU(9)/Z_3)$ inherits its root $x_1 - x_2$ (with the usual notation) from the lowest negative root of E_8; it inherits its other simple roots as shown by the following diagram.

Diagram 3

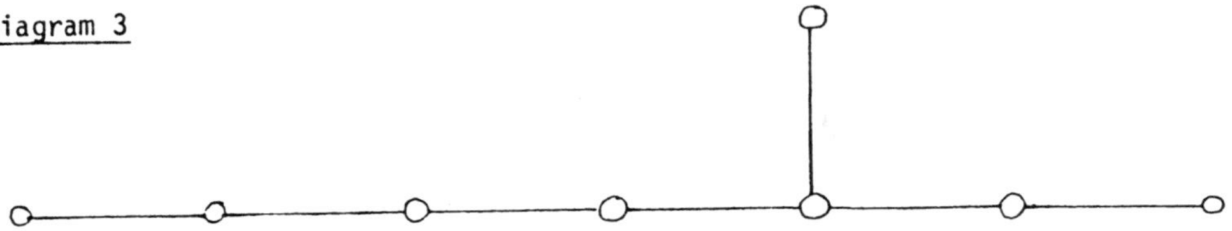

As a representation of A_8, the Lie algebra $L(E_8)$ splits to give

$$L(E_8) = L(A_8) + \lambda^3 + \lambda^{3*}.$$

Let $e_1, e_2, \ldots, e_9$ be the standard base in the vector space $V = C^9$ on which A_8 acts; in other words, e_i is an eigenvector for the weight x_i. Let $e_1^*, e_2^*, \ldots, e_9^*$ be the dual base in V^*. We regard $L(A_8)$ as embedded in $\mathrm{Hom}(V,V) = V \otimes V^*$, and use the elements $e_i \otimes e_j^*$ as a base in $V \otimes V^*$. In other words, $e_i \otimes e_j^*$ is an eigenvector for the weight $x_i - x_j$ if $i \neq j$. Similarly, the product $e_i^* \wedge e_j^* \wedge e_k^*$ indicates an element of λ^{3*}. (Of course the isomorphism between λ^{3*} and the corresponding part of $L(E_8)$ is only determined up to a scalar factor, but this scalar factor will not hurt us.) Usually we abbreviate $e_i \otimes e_j^*$ to $e_i e_j^*$ and $e_i^* \wedge e_j^* \wedge e_k^*$ to $e_i^* e_j^* e_k^*$.

We now introduce the element

$$v_{ik} = \sum_j (e_i^* e_j^* e_k^* \otimes e_j e_k^* + e_j e_k^* \otimes e_i^* e_j^* e_k^*)$$

in the symmetric square $\sigma^2(\alpha) \subset \alpha \otimes \alpha$, where $\alpha = L(E_8)$. In view of the anti-symmetry of $e_i^* e_j^* e_k^* \in \lambda^{3*}$, the element v_{ik} is zero for $i = k$, and it is sufficient to run the sum over j distinct from i and k.

THEOREM 4. (a) The representation $\sigma^2(\alpha)$ of E_8 contains a unique copy of β. (b) This copy of β contains the elements v_{ik}. (c) For $i \neq k$ the elements v_{ik} are eigenvectors corresponding to extreme weights of β. (d) In particular (with our choice of details) v_{91} is an eigenvector corresponding to the highest weight of β.

The use of this is as follows. An irreducible G-module M can be generated by any non-zero element $m \in M$; therefore, Theorem 4 allows us to realize β as the E_8-submodule of $\sigma^2(\alpha)$ generated by v_{91} (or any other v_{ik} with $i \neq k$.) Eigenvectors corresponding to extreme weights are particularly useful for further calculations.

In part (a), the uniqueness is easy. In fact, $\sigma^2(\alpha)$ contains a trivial summand 1, and also an irreducible summand of highest weight $\mathrm{top}(\sigma^2(\alpha))$, whose dimension can be found from the formula of Hermann Weyl. It turns out that the remaining summand has dimension 3,875, which is precisely the dimension of β.

We now introduce the element

$$w_k = \sum_i e_i e_k^* \otimes v_{ik}$$

$$= \sum_{i,j} e_i e_k^* \otimes (e_i^* e_j^* e_k^* \otimes e_j e_k^* + e_j e_k^* \otimes e_i^* e_j^* e_k^*)$$

in $\alpha \otimes \beta \subset \alpha \otimes \sigma^2(\alpha) \subset \alpha \otimes \alpha \otimes \alpha$. It is sufficient to run the first sum over i distinct from k, and the second over distinct i and j which are both distinct from k.

THEOREM 5. (a) The representation $\alpha \otimes \beta$ of E_8 contains a unique copy of γ. (b) This copy of γ contains the elements w_k. (c) The elements w_k are eigenvectors corresponding to extreme weights of γ. (d) In particular (with our choice of details) w_1 is an eigenvector corresponding to the highest weight of γ.

The use of this is as for Theorem 4; we may realize γ as the E_8-submodule of $\alpha \otimes \beta$ generated by w_1 (or any other w_k).

In part (a), the uniqueness is easy. In fact, $\alpha \otimes \beta$ contains an irreducible summand of highest weight $\mathrm{top}(\alpha\beta)$, and the remaining summand has too small a dimension to contain two copies of γ.

To prove Theorems 4 and 5 we need a general lemma of the sort which takes longer to state than to prove. Let ω be a weight of a compact connected Lie group G.

LEMMA 6. (a) Let M be an irreducible representation of G which has ω as an extreme weight, and let $m \in M$ be an eigenvector for the weight ω. Then

$$[\ell_\theta, m] = 0$$

whenever $\ell_\theta \in L(G)$ is an eigenvector corresponding to a root θ with $(\theta,\omega) \geq 0$.

(b) Conversely, let M be any representation of G, and let $m \in M$ be an eigenvector for the weight ω such that

$$[\ell_\theta, m] = 0$$

whenever $\ell_\theta \in L(G)$ is an eigenvector corresponding to a root θ with $(\theta,\omega) \geq 0$. Then the G-submodule of M generated by m is an irreducible representation of G which has ω as an extreme weight.

Sketch proof. Take the obvious steps yourself - it's quicker than going to the library.

PROOF OF THEOREM 4. We first examine the relevant weights. For definiteness, let us normalize the inner product of weights so that roots θ have

$(\theta,\theta) = 2$. Then a weight ω of E_8 qualifies as an extreme weight of β if and only if $(\omega,\omega) = 4$. In $\alpha \otimes \alpha$, the eigenspace for such a weight ω has a base consisting of the vectors $\ell_\theta \otimes \ell_\phi$, where ℓ_θ, ℓ_ϕ are eigenvectors in $L(E_8)$ corresponding to the roots θ, ϕ, and where $\theta + \phi = \omega$. The solutions of this equation are as follows. In any solution, both the roots θ and ϕ must have $(\theta,\omega) = 2$; there are just 14 roots θ with this property; any one of the 14 may be chosen for θ, and then the formula $\phi = \omega - \theta$ gives another root which is one of the 14. For example, with the notation of Theorem 4, we have

$$\omega_{ik} = (-x_i - x_j - x_k) + (x_j - x_k)$$

where j runs over the 7 possibilities distinct from i and k. We seek the linear combination of vectors $\ell_\theta \otimes \ell_\phi$ to which Lemma 6 applies, and we will find it by symmetry.

At this point we specialize and take ω to be the highest weight of β.

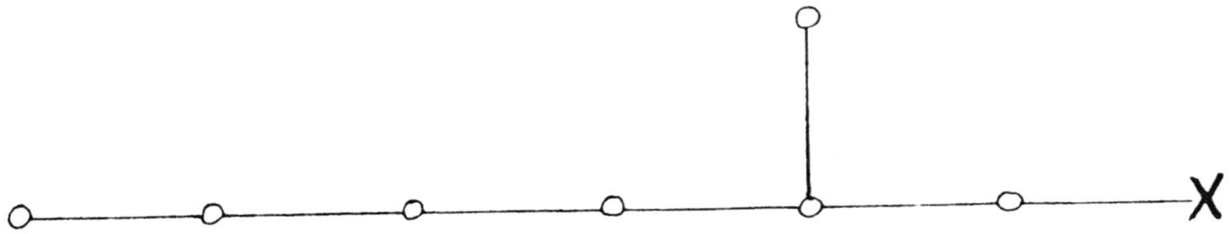

The marked node of the Dynkin diagram corresponds to an edge of the fundamental Weyl chamber and to a subgroup S^1 of the maximal torus T. If we take the subgroup $Z_2 \subset S^1$, then its centralizer $Z(Z_2)$ is a known subgroup of type D_8. We may arrange details to the following effect. The subgroup $D_8 = Spin(16)/Z_2$ inherits its root $(x_1 - x_2)$ (with the usual notation) from the lowest negative root of E_8, and inherits its other simple roots as shown by the following diagram.

Diagram 7

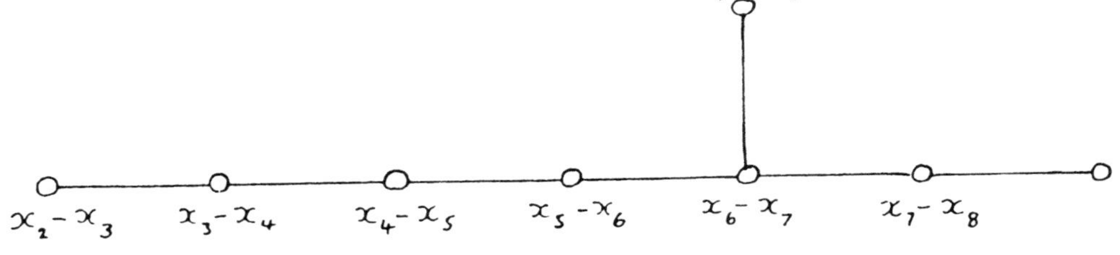

As a representation of D_8, the Lie algebra $L(E_8)$ splits to give

$$L(E_8) = L(D_8) + \Delta^-.$$

(The usual choice of details gives Δ^+ here; the present unusual choice is taken with an eye to what follows.)

Our circle S^1 lies inside $T \subset D_8$ as

$$(-2x, 0, 0, 0, 0, 0, 0, 0).$$

We have $Z(S^1) \subset Z(Z_2) = D_8$; in fact $Z(S^1)$ is a subgroup of local type $S^1 \times D_7$, where the D_7 inherits its simple roots as shown in Diagram 7 above. As a representation of $S^1 \times D_7$ we get

$$L(E_8) = \xi^2 \otimes \lambda^1_{14} + \xi \otimes \Delta^+ + L(S^1 \times D_7) + \xi^{-1} \otimes \Delta^- + \xi^{-2} \otimes \lambda^1_{14}\ .$$

Here ξ is the fundamental 1-dimensional representation of S^1 and λ^1_{14} is the obvious representation of $D_7 = \mathrm{Spin}(14)$ on R^{14}. The roots of the part $\xi^2 \otimes \lambda^1_{14}$ are the 14 we have already mentioned, which have $(\theta, \omega) = 2$ (where ω is the highest weight of β).

It follows that $\alpha^2 = L(E_8) \otimes L(E_8)$ contains a summand

$$\xi^4 \otimes \lambda^1_{14} \otimes \lambda^1_{14}.$$

As a representation of D_7, $\lambda^1_{14} \otimes \lambda^1_{14}$ contains a unique copy of the trivial representation 1. The summand

$$\xi^4 \otimes 1 \subset \xi^4 \otimes \lambda^1_{14} \otimes \lambda^1_{14}$$

lies inside $\sigma^2(\alpha)$; it is annihilated by $L(D_7)$, and (clearly) by $\xi \otimes \Delta^+$ and $\xi^2 \otimes \lambda^1_{14}$. Lemma 6 applies; so this summand $\xi^4 \otimes 1$ generates a copy of β inside $\sigma^2(\alpha)$, and provides an eigenvector for the highest weight of β.

The generator for β which we have just found can be characterized by using a subgroup smaller than D_7. Take $SU(7)$ embedded in $\mathrm{Spin}(14)$ in the usual way. This is a subgroup of type A_6 in E_8 which inherits its simple roots as shown in the following diagram.

Diagram 8

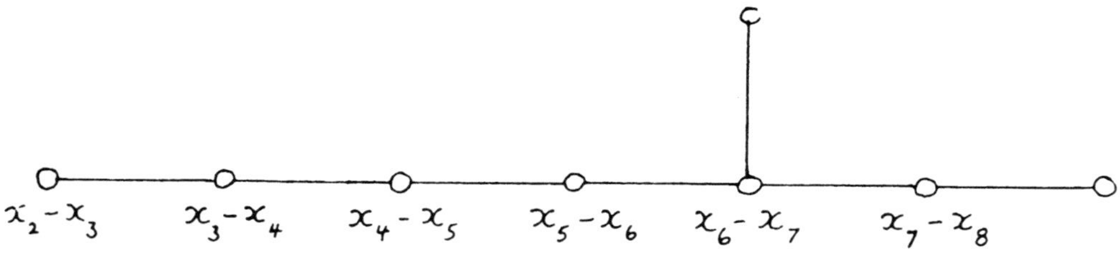

The virtue of this subgroup A_6 is that it also lies inside our subgroup A_8 (as the subgroup fixing e_1 and e_9). The restriction of λ^1_{14} to SU(7) is $\lambda^1_7 + \lambda^{1*}_7$; the representation

$$\sigma^2(\lambda^1_7 + \lambda^{1*}_7) = \sigma^2(\lambda^1_7) + \lambda^1_7 \otimes \lambda^{1*}_7 + \sigma^2(\lambda^{1*}_7).$$

contains a unique summand 1 invariant under SU(7). Therefore the vector we want can be characterized (up to a scalar multiple) as the unique vector in $\sigma^2(\alpha)$ which has the correct weight ω and is invariant under A_6.

The vector

$$v_{91} = \sum_{2 \le j \le 8} (e^*_9 e^*_j e^*_1 \otimes e_j e^*_1 + e_j e^*_1 \otimes e^*_9 e^*_j e^*_1)$$

satisfies this characterization. This completes the proof of parts (a) and (d) of Theorem 4.

Parts (b) and (c) (about the other vectors v_{ik}) follow, since the action of A_8 allows us to work any even permutation of $e_1, e_2, \ldots, e_9$ while preserving T.

PROOF OF THEOREM 5. Again we begin by examining the relevant weights. A weight ω of E_8 qualifies as an extreme weight of γ if and only if $(\omega,\omega) = 8$ and $\frac{1}{2}\omega$ is not a weight. In $\alpha \otimes \alpha \otimes \alpha$, the eigenspace for such a weight ω has a base consisting of the vectors $\ell_\theta \otimes \ell_\phi \otimes \ell_\psi$, where ℓ_θ, ℓ_ϕ, ℓ_ψ are eigenvectors in $L(E_8)$ corresponding to the roots θ, ϕ, ψ, and where $\theta + \psi + \psi = \omega$. The solutions of this equation are as follows. In any solution, just two of the roots (say θ and ϕ) must have $(\theta,\omega) = 3$; the third has $(\psi,\omega) = 2$. There are precisely 8 roots θ with the property that $(\theta,\omega) = 3$; any two distinct roots of these 8 may be taken for θ and ϕ, and the formula $\psi = \omega - \theta - \phi$ then gives a root to complete the solution. For example, with the notation of Theorem 5, we have

$$\omega_k = \theta + \phi + \psi$$

where

$$\theta = (x_i - x_k),\ \phi = (x_j - x_k),\ \psi = (-x_i - x_j - x_k);$$

here i and j run over the 8 values distinct from k, and i,j must be distinct.

We now proceed much as for β. We specialize and take ω to be the highest weight of γ. The node of the Dynkin diagram , marked in the figure on the next page, corresponds to an edge of the fundamental Weyl chamber or to a subgroup S^1 of the maximal torus T. If we take the subgroup $Z_3 \subset S^1$, its centralizer $Z(Z_3)$ is the subgroup A_8 with which we started. Our circle S^1 lies inside $T \subset A_8$ as

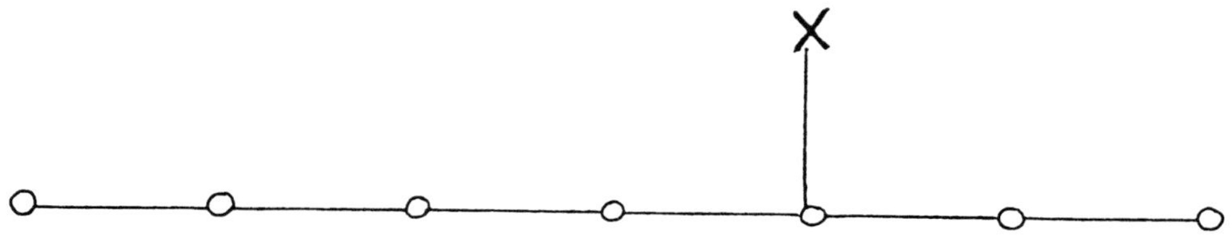

$$\tfrac{1}{3}(-8y, y, y, y, y, y, y, y, y, y).$$

We have $Z(S^1) \subset Z(Z_3)$; in fact $Z(S^1)$ is a subgroup of local type $S^1 \times A_7$, where A_7 is the subgroup of A_8 fixing e_1, and inherits its simple roots as shown in Diagram 3. As a representation of $S^1 \times A_7$ we get

$$L(E_8) = \eta^3 \otimes \lambda^1 + \eta^2 \otimes \lambda^{2*} + \eta \otimes \lambda^3 + L(S^1 \times A_7) + \eta^{-1} \otimes \lambda^{3*} + \eta^{-2} \otimes \lambda^2 + \eta^{-3} \otimes \lambda^{1*}.$$

Here η is the fundamental 1-dimensional representation of S^1. The roots of the part $\eta^3 \otimes \lambda^1$ are the 8 possibilities for θ, ϕ we mentioned, and the roots of the part $\eta^2 \otimes \lambda^{2*}$ are the 28 possibilities for ψ.

It follows that $\alpha^3 = L(E_8) \otimes L(E_8) \otimes L(E_8)$ contains a summand

$$\eta^8 \otimes (\lambda^{2*} \otimes \lambda^1 \otimes \lambda^1 + \lambda^1 \otimes \lambda^{2*} \otimes \lambda^1 + \lambda^1 \otimes \lambda^1 \otimes \lambda^{2*}).$$

Here $\lambda^{2*} \otimes \lambda^1 \otimes \lambda^1$, for example, contains a unique copy of the trivial representation 1 of A_7; let v be a generator. Let ρ be the cyclic permutation defined by

$$\rho(x \otimes y \otimes z) = (z \otimes x \otimes y);$$

then ρv and $\rho^2 v$ will serve as generators in the other two summands. The vector v is annihilated by $L(A_7)$ and (clearly) by $\eta^3 \otimes \lambda^1$ and $\eta^2 \otimes \lambda^{2*}$, but not necessarily by $\eta \otimes \lambda^3$. However, if we act on the fixed vector v with a variable vector $\ell \in \eta \otimes \lambda^3$, we must obtain an A_7-invariant map from λ^3 to $\lambda^1 \otimes \lambda^1 \otimes \lambda^1$, and there is only one of those up to a scalar factor; in particular, the image is contained in the alternating tensors. We can now apply ρ and ρ^2. We see that a linear combination

$$cv + c'\rho v + c''\rho^2 v$$

is annihilated by $\eta \otimes \lambda^3$ provided

$$c + c' + c'' = 0.$$

That is, it is sufficient to take an A_7-invariant vector annihilated by $1 + \rho + \rho^2$; if we do so, Lemma 6 applies and we have a vector which generates a submodule γ.

The obvious expression for a vector v is

$$\sum_{i,j=2}^{9} e_i^* e_j^* e_1^* \otimes e_i e_1^* \otimes e_j e_1^* .$$

To obtain an element annihilated by $1 + \rho + \rho^2$, we apply $\rho^2 - \rho$ and get

$$\sum_{i=2}^{9} e_i e_1^* \otimes \sum_{j=2}^{9} (e_i^* e_j^* e_1^* \otimes e_j e_1^* + e_j e_1^* \otimes e_i^* e_j^* e_1^*) .$$

This is the vector w_1, and this completes the proof of parts (a) and (d) of Theorem 5.

Parts (b) and (c) (about the other vectors w_k) follow, since the action of A_8 allows us to work any even permutation of $e_1, e_2, \ldots, e_9$ while preserving T.

Well, you know what they say: when you have a good thing going, don't stop. We wish to see characterizations of the submodules $\beta \subset \sigma^2(\alpha)$ and $\gamma \subset \alpha \otimes \beta$. We wish to see calculations done which exploit the grasp we have gained. In a book one could not stop here; one would take profit. This argument does not apply to short trailers for forthcoming attractions.

Math. Ann. 278, 29–39 (1987)

Mathematische Annalen

2-Tori in E_8

J. F. Adams

Department of Pure Mathematics and Mathematical Statistics, University, Cambridge CB2 1SB, UK

Dedicated to Friedrich Hirzebruch

1. Introduction

A p-torus in a compact Lie group G is a subgroup $A \subset G$ which is an elementary abelian p-group, $A \cong Z_p \times Z_p \times \ldots \times Z_p$. Many authors, from Borel [1] to Quillen [2], have shown that they influence the behaviour of $H^*(BG; F_p)$; this is only one of the reasons for studying them. In this note I will show that one can handle the worst case.

Theorem 1.1. *Maximal 2-tori in the compact exceptional group E_8 fall into just two conjugacy classes, namely the classes of the examples $D(T^8)$ (of* rank 9*) and EC^8 (of* rank 8*) which will be given explicitly in Sect. 2.*

The proof does not involve any preliminary study of E_7; on the contrary, (1.1) enables one to study 2-tori in E_7 afterwards, by remarking that every 2-torus in E_7 is a 2-torus in E_8.

The proof does rely on the fact that E_8 has a subgroup of local type D_8. Here we recall that the centre of Spin(16) is $Z_2 \times Z_2$. Therefore Spin(16) has three quotients Spin(16)/Z_2: the obvious quotient $SO(16) = \mathrm{Spin}(16)/\{1, z\}$, and the two semi-spin groups

$$Ss^+(16) = \mathrm{Spin}(16)/\{1, z^+\},$$
$$Ss^-(16) = \mathrm{Spin}(16)/\{1, z^-\}.$$

The last two are isomorphic (under the outer automorphism of Spin(16)); but it is usual to choose the map $\mathrm{Spin}(16) \to E_8$ so that its image is $Ss^+(16)$.

In Sect. 3 we will determine the maximal 2-tori in $Ss^+(n)$, after a preliminary study of the projective groups $PO(n)$ and $PSO(n)$. Since we need only the case $n = 16$, we assume for convenience $n \equiv 0 \bmod 8$ and $n > 8$; this excludes the cases $n = 4$, 8 which are special, and the case $n \equiv 4 \bmod 8$, $n > 4$ which is of less interest. The answer for $Ss^+(n)$ is given modulo the answer for Spin(n) – a problem usually regarded as a nuisance which grows exponentially with n. However, we need only the case $n = 16$: this is treated in Sect. 4, and the result for E_8 follows in Sects. 5.

2. Construction of Examples

Any compact connected Lie group G contains a maximal torus T. The solutions of $t^2=1$ in T form a 2-torus $A(T)$ of the same rank as T, say l. Of course, it is not necessarily maximal.

Suppose now that the Weyl group $W=N(T)/T$ contains the element -1; that is, there exists $x\in N(T)$ such that $xtx^{-1}=t^{-1}$ for all $t\in T$. Then x commutes with $A(T)$. We propose to form the subgroup generated by x and $A(T)$, (cf. [3, p. 139]); but before doing so, we need a few remarks about x.

The element x is unique up to conjugacy in G fixing T. For any other choice has the form xu for some $u\in T$, and in T we can solve $v^{-2}=u$, whence $vxv^{-1}=xv^{-2}=xu$.

The element x has order 4 at most. In fact $x^2tx^{-2}=t$, so $x^2\in T$; thus $x\cdot x^2\cdot x^{-1}=x^{-2}$, and $x^4=1$.

The square x^2 is independent of the choice of x and lies in the centre of G. For the first point, $(xu)(xu)=x^2u^{-1}u=x$. For the second, it follows that x^2 is invariant under all automorphisms of G preserving T, in particular, under W; hence x^2 lies on every plane of the diagram.

It is certainly possible for x to have order 4 [as in the case $G=Sp(n)$]. However, we assume that x has order 2. For example, it is so in the following cases.

(a) $SO(4m+\varepsilon)$ and $\mathrm{Spin}(8m+\varepsilon)$ where $\varepsilon=-1, 0, 1$.

(b) G_2, F_4, $\mathrm{Ad}\,E_7$, E_8.

In such cases, x generates with $A(T)$ a 2-torus of rank $l+1$, which we call $D(T)$ (for "double 2-torus"). If we need to display l or G we write $D(T^l)$ or $D(T)\subset G$. This explains the notation $D(T^8)$ in (1.1).

In particular, the construction gives a 2-torus $D(T^4)\subset\mathrm{Spin}(8)$. The maps

$$\mathrm{Spin}(8)\times\mathrm{Spin}(8)\to\mathrm{Spin}(16)$$
$$\mathrm{Spin}(8)\times\mathrm{Spin}(8)\to Ss^+(16)\subset E_8$$

have kernels Z_2 and $Z_2\times Z_2$ respectively; these kernels lie in the centre of $\mathrm{Spin}(8)\times\mathrm{Spin}(8)$, so they lie in $D(T^4)\times D(T^4)$. Thus we obtain the following quotients of $D(T^4)\times D(T^4)$. First, a 2-torus of rank 9, $EC^9\subset\mathrm{Spin}(16)$. Secondly, a 2-torus of rank 8, $EC^8\subset Ss^+(16)\subset E_8$. This explains the notation EC^8 in (1.1).

The following construction is less convenient for our proof, but does more to justify the letters EC for "exotic candidate". We have a 2-torus of rank 8

$$D(T^4)\times D(T^2)\subset F_4\times G_2\subset E_8;$$

this is conjugate to EC^8, either by direct inspection, or by (1.1) [since the methods of Sect. 5 show that it is not subconjugate to $D(T^8)\subset E_8$].

3. 2-Tori in $PO(n)$, $PSO(n)$, and $Ss^+(n)$

First we must explain the invariants we use in describing these 2-tori.

Typically we have to consider a 2-torus A in a compact Lie group G which comes with a double cover

$$Z_2\to\tilde{G}\to G\,.$$

Then by pullback we get an extension

$$Z_2 \to \tilde{A} \to A,$$

and this is classified by an element $q \in H^2(BA; Z_2)$; we write it q because we often consider it as a quadratic form on A.

In particular, we have to consider the following three extensions $Z_2 \to \tilde{G} \to G$.

$$\begin{array}{ll} Z_2 \to O(n) \to PO(n) & (n \text{ even}) \\ \left.\begin{array}{l} Z_2 \to Ss^+(n) \to PSO(n) \\ Z_2 \to Ss^-(n) \to PSO(n) \end{array}\right\} & (n \equiv 0 \bmod 4). \end{array}$$

We write v, v^+, v^- for the corresponding quadratic forms q on $A \subset G$.

We note that if A lies in $PSO(n)$, then the quadratic form v [to which it is entitled as a subtorus of $PO(n)$] becomes $v^+ + v^-$.

Proposition 3.1. *Assume n even. Then maximal 2-tori in $PO(n) = O(n)/\{\pm I\}$ fall into conjugacy classes corresponding to the solutions (r, s) of $r \cdot 2^s = n$ with $r \neq 2$. The 2-torus $A(r, s)$ corresponding to (r, s) is of rank $r - 1 + 2s$; its quadratic form v is of rank $2s$ and plus type, that is, equivalent to*

$$x_1x_2 + x_3x_4 + \ldots + x_{2s-1}x_{2s}.$$

Of course, $O(n)$ has a maximal 2-torus unique up to conjugacy, namely the subgroup Δ of diagonal matrices with diagonal entries ± 1. Its image in $PO(n)$, that is $\Delta/\{\pm I\}$, appears in (3.1) as the unique case with $v = 0$, that is $A(n, 0)$.

We turn to $PSO(n)$, and here we expect to find $S\Delta/\{\pm I\}$, where $S\Delta = \Delta \cap SO(n)$.

Proposition 3.2. *Assume $n \equiv 0 \bmod 8$ and $n > 8$. Then maximal 2-tori in $PSO(n)$ fall into the following conjugacy classes.*

(Case $s = 0$). The class of $S\Delta/\{\pm I\}$.

(Case $s = 1$). The class of $D(T) \subset PSO(n)$ (see Sect. 2).

(General case). For each solution (r, s) of $r \cdot 2^s = n$ with $r \neq 2$ and $s \geqq 2$, two classes conjugate in $PO(n)$ to $A(r, s)$. Of these, $A(r, s)^+$ has

$$\begin{cases} v^+ = 0 & \\ v^- & \textit{of rank } 2s \textit{ and plus type,} \end{cases}$$

while $A(r, s)^-$ has

$$\begin{cases} v^- = 0 & \\ v^+ & \textit{of rank } 2s \textit{ and plus type.} \end{cases}$$

We turn to $Ss^+(n)$.

Proposition 3.3. *Assume $n \equiv 0 \bmod 8$ and $n > 8$. Then maximal 2-tori in $Ss^+(n)$ fall into the following conjugacy classes.*

(General case). For each solution (r, s) of $r \cdot 2^s = n$ with $r \neq 2$ and $s \geqq 2$, the class of $\tilde{A}(r, s)^+$, the counterimage in $Ss^+(n)$ of the 2-torus $A(r, s)^+$ in (3.2).

(Case $s = 1$). The class of $D(T) \subset Ss^+(n)$ (see Sect. 2).

(Case $s = 0$). The image in $Ss^+(n)$ of each conjugacy class of maximal 2-tori in Spin(n) *other than $D(T) \subset$* Spin(n).

Proof of (3.1). Suppose given a 2-torus $A \subset PO(n)$. By pullback from

$$Z_2 = \{\pm I\} \to O(n) \to PO(n)$$

we get an extension

$$Z_2 \to \tilde{A} \to A.$$

The classification of such "special 2-groups" is known, and is of course the same as the classification of the quadratic forms q to which they correspond. The classification shows that $\tilde{A}$ is a central product

$$F_1 * F_2 * \ldots * F_t.$$

Here the "central product" $G * H$ involves amalgamating the given central subgroups Z_2 in G and H; each factor F_i may be one of the extensions

$$\begin{array}{l} Z_2 \to Z_2 \times Z_2 \to \quad Z_2 \\ Z_2 \to \quad \delta_8 \quad \to Z_2 \times Z_2 \end{array}$$

where δ_8 is the ordinary dihedral group of order 8; and we are allowed one further factor, which may be either

$$Z_2 \to Z_4 \to \quad Z_2$$

or

$$Z_2 \to Q_8 \to Z_2 \times Z_2$$

where Q_8 is the ordinary quaternion group of order 8.

It is easy to determine the real representations of such a central product in which the given central subgroup Z_2 acts as $\{\pm 1\}$. This rests on the following considerations. (i) A representation of $G * H$ gives a representation of $G \times H$. (ii) The (complex) representation-theory of $G \times H$ is known in terms of the (complex) representation-theory of G and H. (iii) The representation-theory of δ_8, Z_4, and Q_8 is both easy and favourable.

The outcome is as follows. We write $\tilde{A}$ as $G * H$, where G is the central product of all the trivial extensions

$$Z_2 \to Z_2 \times Z_2 \to Z_2$$

and H is the central product of the remaining, non-trivial extensions. Then any real representation of $\tilde{A}$, in which the given central subgroup Z_2 acts as $\{\pm 1\}$, may be written as a tensor product $U \otimes_{\mathbb{R}} V$, in which U, V are real representations of G, H in which Z_2 acts as $\{\pm 1\}$, and V is the unique such representation of H which is irreducible over $\mathbb{R}$. This V may be constructed as the following tensor product over $\mathbb{R}$. For each dihedral factor δ_8 we take one copy of the usual real 2-dimensional representation of δ_8, i.e. the representation by the matrices

$$\begin{bmatrix} \pm 1 & 0 \\ 0 & \pm 1 \end{bmatrix}, \quad \begin{bmatrix} 0 & \pm 1 \\ \pm 1 & 0 \end{bmatrix}.$$

If there is a copy of Z_4 we take one copy of the usual action of Z_4 on $\mathbb{C}$; alternatively, if there is a copy of Q_8 we take one copy of the usual action of Q_8 on $\mathbb{H}$.

In particular, all this applies to our assumed embedding $\tilde{A}\subset O(n)$. Indeed, what we have done describes all the 2-tori A in $PO(n)$, up to conjugacy. It remains to determine which of them are maximal.

Since G is a 2-torus we can split U into real eigenspaces for the action of G; we can choose a base in U adapted to this decomposition, and then enlarge G to the 2-torus G' of diagonal matrices with respect to this base. We still obtain a faithful representation of our enlarged group $\tilde{A}'$; so if A was originally maximal, we must have $G=G'$.

If H contains a factor Z_4 acting on $\mathbb{C}$, then we can replace it by δ_8 acting on its usual 2-dimensional representation, and so obtain a faithful representation of a larger group $\tilde{A}'$. So this case is excluded.

Similarly, if H contains a factor Q_8 acting on $\mathbb{H}$, we can replace it by $Q_8 * Q_8$ acting on $\mathbb{H}$ from left and right; of course $Q_8 * Q_8 \cong \delta_8 * \delta_8$, and this construction has to give the tensor product of 2 copies of the usual 2-dimensional representation of δ_8. Anyway, this case is also excluded.

This leads to the description of $A(r,s)$. Here our group $\tilde{A}$ comes as a quotient of $(Z_2)^r\times(\delta_8)^s$, with $(Z_2)^r$ acting on $U=\mathbb{R}^r$ as the group of diagonal matrices, and $(\delta_8)^s$ acting on $V=\bigotimes_1^s \mathbb{R}^2$ by s copies of the usual 2-dimensional representation.

The case $r=2$ can be excluded as follows. If $G=(Z_2)^2$, acting on $\mathbb{R}^2$ as

$$\begin{bmatrix} \pm 1 & 0 \\ 0 & \pm 1 \end{bmatrix},$$

then we can replace it by a copy of δ_8 acting as usual, and so enlarge $\tilde{A}$.

It remains to show that $A=A(r,s)$ cannot be subjugate to $A'=A(r',s')$ unless $r=r'$, $s=s'$. For this we consider the quadratic form v. On A it has rank $2s$ and isotropic subspaces of dimension $r+s-1$, and similarly for A'. We can have $A\subset A'$ only if $s\leqq s'$ and $r+s\leqq r'+s'$. Since $r\cdot 2^s=n=r'\cdot 2^{s'}$, this leads to

$$r'(2^{s'-s}-1)\leqq s'-s\,.$$

This is possible only for $s'-s=1$, $r'=1$, leading to the case $r=2$ which has been excluded. This proves (3.1).

The proof of (3.2) needs some preparation. First, the task of describing v^+ and v^- is not as bad as one might expect from the problem of simultaneously reducing two general quadratic forms over F_2.

Lemma 3.4. *Let A be a 2-torus in $PSO(n)$, where $n\equiv 0 \bmod 8$. Then either* (i) $v^+=0$, *or* (ii) $v^-=0$, *or* (iii) $v^+=v^-$, *or* (iv)

$$v^+=a(a+l)\,,\qquad v^-=b(b+l)\,,\qquad v^++v^-=c(c+l)$$

for some a, b, c, l of degree 1 in $H^*(BA;\,F_2)$.

The reason is that our quadratic forms are not general; they come from special elements of $H^2(BPSO(n);\,F_2)$.

In fact, from this point on our extensions

$$Z_2\rightarrow\tilde{G}\rightarrow G$$

all have G connected. In this case the covering $\tilde{G}\to G$ corresponds to a homomorphism $\pi_1(G)\to Z_2$, or equivalently $\pi_2(BG)\to Z_2$, yielding a unique element $q\in H^2(BG;\ Z_2)$. On any 2-torus $A\subset G$ this restricts to the element $q\in H^2(BA;\ Z_2)$ we considered before; we omit the proof, which is easy. All this parallels what one usually does for the covering

$$Z_2\to \mathrm{Spin}(n)\to SO(n)$$

and the characteristic class w_2.

In fact, the characteristic classes v^+, $v^-\in H^2(BPSO(n);\ Z_2)$ both map to $w_2\in H^2(BSO(n);\ Z_2)$; and if we omit the symbols for the induced maps, the characteristic classes for the extensions

$$Z_2\to \mathrm{Spin}(n)\to Ss^+(n)$$
$$Z_2\to \mathrm{Spin}(n)\to Ss^-(n)$$

are $v=v^-$ and $v=v^+$.

Lemma 3.5. *The classes v^+, v^- in $H^2(BPSO(n);\ Z_2)$ satisfy*

$$v^+\cdot Sq'v^- + v^-\cdot Sq^1v^+ = 0 \qquad \textit{if} \quad n\equiv 0 \bmod 8$$
$$Sq^2Sq^1v^+ + v^+\cdot Sq^1v^+ + Sq^2Sq^1v^- + v^-\cdot Sq^1v^- = 0 \quad \textit{if} \quad n\equiv 4 \bmod 8 .$$

This pinpoints a crucial difference between the cases $n\equiv 0,\ 4 \bmod 8$.

Proof of (3.5). First we dismiss the case $n=4$, in which we have $PSO(4)\cong SO(3)\times SO(3)$. Then $H^*(BPSO(4);\ Z_2)$ is a polynomial algebra on generators v^+, Sq^1v^+, v^-, Sq^1v^-, and the relations

$$Sq^2Sq^1v^+ + v^+\cdot Sq^1v^+ = 0$$
$$Sq^2Sq^1v^- + v^-\cdot Sq^1v^- = 0$$

hold separately.

Consider now the fibering

$$F = B\,\mathrm{Spin}(n)\to E = BPSO(n)\to B .$$

The base B is an Eilenberg-MacLane space of type $(Z_2\times Z_2, 2)$, with fundamental classes v^+, $v^-\in H^2(B;\ Z_2)$. We assume $n>4$, so the fibre F has a fundamental class $f\in H^4(F;\ Z_2)$. A priori we have

$$\begin{aligned}\tau f = \alpha^+Sq^2Sq^1v^+ &\quad +\alpha^-Sq^2Sq^1v^-\\ +\beta^+v^+\cdot Sq^1v^+ &\quad +\beta^-v^-\cdot Sq^1v^-\\ +\gamma^+v^+\cdot Sq^1v^- &\quad +\gamma^-v^-\cdot Sq^1v^+\end{aligned}$$

for some coefficients $\alpha^+,\ldots,\gamma^-$ which remain to be determined. By symmetry under the outer automorphism of $PSO(n)$ we have

$$\alpha^+ = \alpha^-, \qquad \beta^+ = \beta^-, \qquad \gamma^+ = \gamma^- .$$

Since $Sq^1f = 0$ we have

$$\alpha^+ = \beta^+, \qquad \alpha^- = \beta^-, \qquad \gamma^+ = \gamma^- .$$

Thus

$$\tau f = \alpha \begin{pmatrix} Sq^2 Sq^1 v^+ & + Sq^2 Sq^1 v^- \\ + v^+ \cdot Sq^1 v^+ & + v^- \cdot Sq^1 v^- \end{pmatrix} + \gamma (v^+ \cdot Sq^1 v^- \quad + v^- \cdot Sq^1 v^+).$$

Next we consider the map

$$SO(4) \xrightarrow{\Delta} SO(4) \times \ldots \times SO(4) \xrightarrow{\oplus} SO(4n).$$

This passes to the quotient to give

$$PSO(4) \xrightarrow{\theta} PSO(4n).$$

For n odd, $(B\theta)^*$ carries v^+ to v^+, v^- to v^-; this gives $\gamma = 0$. For n even, $(B\theta)^*$ carries v^+ to 0, v^- to $v^+ + v^-$; this gives $\alpha = 0$.

Finally we consider the fibering

$$F = BZ_2 \to E = BSO(n) \to B = BPSO(n).$$

I claim that $w_4 \in H^4(E; Z_2)$ cannot come from $H^4(B; Z_2)$. In fact, if $n \equiv 4 \bmod 8$ then the restriction of w_4 to F is non-zero; thus w_4 is of filtration 0. This does not happen if $n \equiv 0 \bmod 8$; but then w_4 restricts to zero on the Z_2 subgroup generated by $(-I_2) \oplus I_{n-2}$, and restricts non-zero on the Z_2 subgroup generated by $I_2 \oplus (-I_{n-2})$ – although these subgroups become the same in $PSO(n)$. So in this case w_4 is of filtration 2.

If we had $\alpha = 0$ and $\gamma = 0$ then this spectral sequence would prove $H^4(B; Z_2) \to H^4(E; Z_2)$ epi. This is a contradiction; so $\alpha = 1$ or $\gamma = 1$ according to the case. This proves (3.5).

Proof of (3.4). By (3.5), the equation

$$v^+ \cdot Sq^1 v^- = v^- \cdot Sq^1 v^+$$

holds in $H^*(BA; Z_2)$, which is a polynomial algebra and therefore a unique factorisation domain. This leads to the following possibilities only.

(i) $v^+ = 0$.

(ii) $v^- = 0$.

(iii) $v^+ = v^-$.

(iv) There is an element l of degree 1 in $H^*(BA; F_2)$ such that

$$Sq^1 v^+ = l v^+, \quad Sq^1 v^- = l v^-.$$

(v) There are elements l, m, n of degree 1 and q of degree 2 such that

$$v^+ = lm, \quad Sq^1 v^- = nq$$
$$v^- = ln, \quad Sq^1 v^+ = mq.$$

In case (v) we work out that if l, m, n are all non-zero then

$$q = l(l+m), \quad q = l(l+n)$$

and $m = n$. So case (v) is covered by cases (i), (ii), (iii).

By running through the classification of quadratic forms q, we see that the equation

$$Sq^1 q = lq$$

can hold only if $q = mn$ for some elements m, n of degree 1 (which may be equal or zero) and then $l = m + n$. So case (iv) leads to the result stated in case (iv) of (3.4).

Proof of (3.2) It is clear from (3.1) that any 2-torus in $PSO(n)$ is subconjugate in $PO(n)$ to $A(r,s) \cap PSO(n)$ for some (r,s) with $r \cdot 2^s = n$, $r \neq 2$. The case $s=0$ gives

$$A(n,0) \cap PSO(n) = S\Delta/\{\pm I\};$$

for $s > 0$ we have $A(r,s) \subset PSO(n)$.

We show that $S\Delta/\{\pm I\}$ cannot be subconjugate in $PO(n)$ to any $A(r',s')$ with $s' > 0$. For this we repeat the last part of the proof of (3.1). The inequality from the rank of the isotropic subspaces becomes $n - 1 \leqq r' + s'$, that is

$$r'(2^{s'} - 1) \leqq s' + 1 .$$

This is possible only for $r' = 1$, $s' = 1$ or 2, and these cases are excluded since $n > 4$.

It remains to see whether the class of $A(r,s)$ under conjugacy in $PO(n)$ falls into one class or two under conjugacy in $PSO(n)$. The subgroup $S\Delta/\{\pm I\}$ maps to itself under conjugation by any diagonal matrix of determinant -1. For $A(\frac{1}{2}n, 1) = D(T)$, the notion of a "double 2-torus" $D(T)$ is invariant even under outer automorphisms. For $s \geqq 2$ the rank of $v = v^+ + v^-$ is at least 4; Lemma 3.4 shows that either $v^+ = 0$ or $v^- = 0$, and this settles the question.

Proof of (3.3). Any 2-torus in $Ss^+(n)$ maps to some 2-torus in $PSO(n)$, so we must consider the possible maximal 2-tori in $PSO(n)$ and look for the maximal isotropic subspaces of v^+.

(General case). For $A(r,s)^+$ the maximal isotropic subspace is clearly $A(r,s)^+$ itself. For $A(r,s)^-$, any subspace on which v^+ vanishes lifts to a 2-torus in Spin(n), which brings us to the final case "$s=0$".

(Case $s=1$). Half the elements of $A(T) \subset PSO(n)$ lift to elements of order 2 in $Ss^+(n)$, and (since we assume $n \equiv 0 \bmod 8$) all the elements of $D(T) \subset PSO(n)$ which are not in $A(T)$ lift to elements of order 2 in $Ss^+(n)$. There are two choices for a maximal isotropic subspace, but either way, the lift is a $D(T)$ in $Ss^+(n)$.

(Case $s=0$). The 2-torus $S\Delta/\{\pm I\}$ already lifts to $SO(n)$, so any subspace which lifts to $Ss^+(n)$ must lift to Spin(n).

In this case it is clearly enough to consider maximal 2-tori in Spin(n). We can exlude $D(T) \subset \mathrm{Spin}(n)$ because it maps to a proper subgroup of $D(T) \subset Ss^+(n)$.

This shows that any maximal 2-torus in $Ss^+(n)$ is one of the 2-tori listed in (3.3). It remains to show that the 2-tori listed in (3.3) are not subconjugate to one another.

If one $\tilde{A}(r,s)^+$ contained another $\tilde{A}(r',s')^+$ (up to conjugacy), then the same would hold for their images in $PSO(n)$, contradicting (3.2).

Similarly, let write A_0 for the image in $Ss^+(n)$ of a maximal 2-torus in Spin(n) (case $s=0$). If one such A_0 contained another A_0' (up to conjugacy), then the same would hold for their counterimages in Spin(n). Also A_0 cannot contain $D(T)$ or an $\tilde{A}(r,s)^+$, because $v=v^-$ is zero on A_0 and non-zero on $D(T)$, $\tilde{A}(r,s)^+$.

Similarly, $D(T)$ cannot contain any $\tilde{A}(r,s)^+$, becuase $v=v^-$ has greater rank on the latter. If $D(T)$ contains an A_0, then (by counting ranks) A_0 must be a maximal isotropic subspace of $D(T)$, in which case A_0 comes from a $D(T)\subset \mathrm{Spin}(n)$; and that case is excluded.

An $\tilde{A}(r,s)^+$ cannot contain any A_0, because A_0 has rank $\frac{1}{2}n$, and in $\tilde{A}(r,s)^+$ the rank of a maximal isotropic subspace is $r+s$; the inequality $\frac{1}{2}n \leqq r+s$ cannot be satisfied for $n>8$. A fortiori, an $\tilde{A}(r,s)^+$ cannot contain $D(T)$, because that contains an A_0 of the sort just excluded.

This proves (3.3).

4. 2-Tori in Spin(16) and $Ss^+(16)$

Proposition 4.1. *Maximal 2-tori in* Spin(16) *fall into just two conjugacy classes, namely the classes of the examples* $D(T^8)\subset \mathrm{Spin}(16)$ *and* EC^9 *described in Sect. 2.*

Corollary 4.2. *Maximal 2-tori in* $Ss^+(16)$ *fall into the following four conjugacy classes.*

(0) EC^8 *of* rank 8 *(see Sect. 2)*.

(i) $D(T)$ *of* rank 9.

(ii) $\tilde{A}(4,2)^+$ *of* rank 8.

(iv) $\tilde{A}(1,4)^+$ *of* rank 9.

This will follow immediately from (3.3), (4.1).

Proof of (4.1). First we show that the two examples are not conjugate.

Let "class $2r$" be the conjugacy class in Spin(n) of elements whose images in $SO(n)$ have $2r$ eigenvalues -1 and $n-2r$ eigenvalues $+1$ (where $0<r<2n$). Then the 256 elements which are in $D(T^8)\subset \mathrm{Spin}(16)$ but not in $A(T^8)$ are all of class 8. So all the elements of class 4 lie in a proper subgroup, namely $A(T)$.

By contrast, in $D(T)\subset \mathrm{Spin}(8)$ the 28 elements of class 4 generate the remaining 4 elements. It follows that in $EC^9\subset \mathrm{Spin}(16)$, the elements of class 4 generate the whole group.

We now start the main proof. Any 2-torus A in Spin(16) must project to some 2-torus $\bar{A}$ in $SO(16)$, and after conjugation we may suppose $\bar{A}\subset S\Delta$. The first problem is then to locate the maximal isotropic subspaces of w_2, considered as a quadratic form on the vector-space $S\Delta$ of dimension 15. For this we need to classify w_2 according to the theory for classifying quadratic forms over F_2; the answer is that w_2 is of rank 14 and plus type, equivalent to

$$x_2x_3+x_4x_5+\ldots+x_{14}x_{15}$$

(independent of x_1). Thus the maximal isotropic subspaces are all of rank 8; so the maximal 2-tori in Spin(16) must all be of rank 9, and the 2-tori of rank 9 named in (4.1) must be maximal.

The number of maximal isotropic subspaces is

$$\begin{aligned}&(2^0+1)(2^1+1)(2^2+1)(2^3+1)(2^4+1)(2^5+1)(2^6+1)\\&=2\cdot 3\cdot 5\cdot 9\cdot 17\cdot 33\cdot 65\\&=2\cdot 3^4\cdot 5^2\cdot 11\cdot 13\cdot 17.\end{aligned}$$

We wish to classify them under conjugacy in $SO(16)$, or equivalently, under the action of the symmetric group Σ_{16} permuting the coordinates of R^{16}.

With sufficient care one can determine the stabiliser in Σ_{16} of our two examples. The number of isotropic subspaces in the Σ_{16}-orbit of $D(T)$ turns out to be

$$\frac{16!}{8!2^7}=2\cdot 3^4\cdot 5^2\cdot 7\cdot 11\cdot 13.$$

For EC^9 we get

$$\frac{16!}{2(14\cdot 12\cdot 8)^2}=2^2\cdot 3^4\cdot 5^3\cdot 11\cdot 13.$$

Since $7+2\cdot 5=17$, the total for these two orbits is

$$2\cdot 3^4\cdot 5^2\cdot 11\cdot 13\cdot 17;$$

that is, any maximal isotropic subspace belongs to one of these two orbits. This proves (4.1).

5. 2-Tori in E_8

Lemma 5.1. *Any 2-torus in E_8 is conjugate to one in $Ss^+(16)$.*

Proof. First suppose that a 2-torus A acts on a compact Lie group G. If T is maximal among tori preserved by A, then NT/T is 0-dimensional; in fact, the usual proof for $A=1$ carries over, because if NT/T is positive-dimensional, the tangent vector which one uses at T/T may as well be an eigenvector for A. It follows that such a T is maximal in the usual sense; this shows that A preserves at least one maximal torus T in G. (This gives a simple proof for 2-tori of a result true for more general A [3].)

It follows that if A is a 2-torus in G, then A is conjugate to a subgroup of $N(T)$. Then the image of $\bar{A}$ of A in $W=N(T)/T$ is again a 2-torus, and after conjugation we can suppose that $\bar{A}$ lies in our favourite Sylow 2-subgroup of W. For example, our favourite Sylow 2-subgroup of $W(E_8)$ is a Sylow 2-subgroup of $W(Ss^+(16))$. This conjugates A into the normaliser $N(T)$ in $Ss^+(16)$.

Lemma 5.2. *Any 2-torus maximal in $Ss^+(16)$ remains maximal in E_8.*

Proof. Such a 2-torus must contain the non-trivial central element z of $Ss^+(16)$; so any larger 2-torus must lie in the centraliser of z, i.e. in $Ss^+(16)$.

To prove (1.1), it remains only to show that of the 2-tori listed in (4.2), the two of rank 9 become conjugate in E_8, and so do the two of rank 8.

For this, we need to know that elements of order 2 in E_8 fall into just two conjugacy classes. Elements of class "a" have centraliser of local type $A_1 \times E_7$, and their shortest representatives in $L(T)$ have norm $\frac{1}{2}$. For example, "class 4" in Spin (16) is represented by $(\frac{1}{2}, \frac{1}{2}, 0, 0, 0, 0, 0, 0)$ with respect to the usual coordinates in $L(T)$, and maps to class "a". Elements of class "b" have centraliser $Ss^+(16)$, and their shortest representatives in $L(T)$ have norm 1. For example, "class 8" in Spin(16) is represented by $(\frac{1}{2}, \frac{1}{2}, \frac{1}{2}, \frac{1}{2}, 0, 0, 0, 0)$, and maps to class "$b$". Again, the nontrivial element z in the centre of $Ss^+(16)$ maps to class "b", either because it is represented by $(1, 0, 0, 0, 0, 0, 0, 0)$ or because its centraliser is $Ss^+(16)$.

Next let A be one of the 2-tori $D(T)$, EC^8 in $Ss^+(16)$. I will show that in either case, the non-trivial element z in the centre of $Ss^+(16)$ lies in a proper subgroup $B \subset A$ which can be distinguished even in E_8.

In $D(T) \subset Ss^+(16)$, z lies in $A(T)$, which can be distinguished even in E_8 as the subgroup generated by the 120 elements of class "a" in $D(T)$. [The elements in $D(T)$ which are not in $A(T)$ are all of class "b".]

The case of EC^8 is a little more subtle.

In E_8 a 2-torus of rank 3, which consists of the identity and 7 elements of class "a", can be distinguished from one which contains at least one element of class "b". We call the former "2-tori of type G_2", thinking of the example $D(T^2) \subset G_2 \subset E_8$.

Both the original copies of $D(T^4) \subset \text{Spin}(8)$ inject into EC^8. In EC^8 we have 56 elements of class "a", 28 in one copy of $D(T^4)$ and 28 in the other; let x be one of them. Of the remaining 55 elements of class "a" in EC^8, we can distinguish between the 24 which lie with x in some subgroup of type G_2, and the 31 which do not. The former 24 generate the copy of $D(T^4)$ containing x. So we can distinguish a pair of rank-5 subspaces, namely the two copies of $D(T^4)$. So we can distinguish their intersection, that is, the common image of the centres of the two copies of Spin(8). This gives a distinguishable rank-2 subspace B containing z.

Given the subgroup B, we choose an element $x \in A$ of class "b" which is not in B. (There are 256 choices for x when $A = DT$, 196 when $A = EC^8$.) In E_8, x is conjugate to z; choose a conjugation α throwing x on z. Since A centralises x, αA centralises z, so it lies in $Ss^+(16)$. Thus αA is a maximal 2-torus in $Ss^+(16)$, of the same rank as A, and the element z does not lie in its distinguished subgroup $\alpha(B)$. By (4.2), αA is of type $\tilde{A}(1,4)^+$ or $\tilde{A}(4,2)^+$ according to its rank.

This completes the proof of (1.1).

References

1. Borel, A.: Sous-groupes commutatifs et torsion des groupes de Lie compacts connexes. Tokoku Math. J. **13**, 216–240 (1961)
2. Quillen, D.: The spectrum of an equivariant cohomology ring. I, II. Ann. Math. **94**, 549–572, 573–602 (1971)
3. Borel, A., Serre, J.-P.: Sur certains sous-groupes des groupes de Lie compacts. Comment. Math. Helv. **27**, 128–139 (1953)

Received July 26, 1986

Inventiones math. 35, 1 – 41 (1976)

Inventiones mathematicae

Maps between Classifying Spaces

J.F. Adams (Cambridge) and Z. Mahmud (Cambridge)

To Jean-Pierre Serre

§ 1. Introduction and Statement of Main Theorems

Let G and G' be compact connected Lie groups. (The assumption that G and G' are compact and connected will remain in force throughout this paper, and will not be repeated.) Let BG and BG' be the classifying spaces of G and G'. The object of this paper is to study maps $f\colon BG \to BG'$, since this seems to be a case of the homotopy classification problem which is both particularly interesting and particularly favourable. We also generalise this problem by studying maps f which are defined only after localisation in the sense of Sullivan [19]. However, it is convenient to insist that one localises so as to invert only a finite number of primes; for brevity, the phrase "finite localisation" will mean "localisation so as to invert only a finite number of primes." Corollary 1.11 below determines exactly which homomorphisms $\theta\colon H^*(BG;\mathbf{Q}) \leftarrow H^*(BG';\mathbf{Q})$ are induced by maps f defined after finite localisation. More precisely, we succeed in reducing this problem to a problem in classical Lie theory.

In this section, § 1, we will explain our results so far as they reduce this problem in algebraic topology to one in Lie theory. In the next section, § 2, we will consider the problem in Lie theory; so many examples will come only in § 2. In order to avoid interrupting the exposition, the proofs of several results in § 1 are postponed to later sections. § 3 contains the proof of Theorem 1.7; § 4 and § 5 go to proving Theorem 1.5; § 6 contains the proof of Theorem 1.10; and § 7 proves Corollary 1.12.

We proceed to explain our results. The most optimistic conjecture [17] would be that every map $f\colon BG \to BG'$ is of the form Bh, where $h\colon G \to G'$ is a homomorphism of Lie groups. This is certainly false, as has been shown by Sullivan [18]. Nevertheless, we will show that any map $f\colon BG \to BG'$ has "weights" just as if it were a representation. More precisely, suppose given a homomorphism $h\colon G \to G'$; and let T, T' be maximal tori in G, G'. Then there is a homomorphism $k\colon T \to T'$ such that the following diagram is homotopy-commutative.

$$\begin{array}{ccc} BG & \xrightarrow{Bh} & BG' \\ \uparrow & & \uparrow \\ BT & \xrightarrow{Bk} & BT' \end{array}$$

(The vertical arrows are, of course, induced by the inclusions $T \subset G$, $T' \subset G'$. The homomorphism h will not necessarily carry T into T', but we can move the torus hT into T' by an inner automorphism of G'; and such an inner automorphism of G' induces a map of BG' homotopic to the identity.) We would like a similar conclusion for a map f: $BG \to BG'$ which is not necessarily of the form Bh. One can find a suitable homomorphism k: $T \to T'$, but it seems unreasonable to hope for a proof that the diagram

$$\begin{array}{ccc} BG & \xrightarrow{f} & BG' \\ \uparrow & & \uparrow \\ BT & \xrightarrow{Bk} & BT' \end{array}$$

is homotopy-commutative; we have to be content with a homological conclusion.

Theorem 1.1. *Let f: $BG \to BG'$ be a map. Then there is a homomorphism k: $T \to T'$ such that the following diagram is commutative.*

$$\begin{array}{ccc} H^*(BG; \mathbf{Q}) & \xleftarrow{f^*} & H^*(BG'; \mathbf{Q}) \\ \downarrow & & \downarrow \\ H^*(BT; \mathbf{Q}) & \xleftarrow{(Bk)^*} & H^*(BT'; \mathbf{Q}) \end{array}$$

We can obtain substantially the same result with less data; it is sufficient to assume that the map f is given after finite localisation. The advantage is that in this form the theorem has a converse; any homomorphism k: $T \to T'$, satisfying certain necessary conditions we give below, arises from a map f defined after finite localisation (although it might not arise from any map f defined without localisation). This is stated precisely as Theorem 1.10 below. However, there is a price to be paid for using less data about f; we have to use rational "weights" rather than integral ones, and we now explain how to formulate this.

Let $\tilde{T}$ and $\tilde{T}'$ be the universal covers of T and T'; they may also be described as the Lie algebras of T and T', or as the Stiefel diagrams of G and G'; they are vector-spaces over $\mathbf{R}$. Let I and I' be the kernels of the projections $\tilde{T} \to T$ and $\tilde{T}' \to T'$; we regard them as "integer lattices" in $\tilde{T}$ and $\tilde{T}'$. In Lie theory it is usual to pass from a homomorphism k: $T \to T'$ to its universal cover $\tilde{k}$: $\tilde{T} \to \tilde{T}'$; this is an $\mathbf{R}$-linear map carrying I into I'. We have now to consider $\mathbf{R}$-linear maps ϕ: $\tilde{T} \to \tilde{T}'$ which carry $I \otimes \mathbf{Q}$ into $I' \otimes \mathbf{Q}$.

Lemma 1.2. *There is a (1-1) correspondence between $\mathbf{R}$-linear maps ϕ: $\tilde{T} \to \tilde{T}'$ which carry $I \otimes \mathbf{Q}$ into $I' \otimes \mathbf{Q}$ and homomorphisms*

$$\phi^*: \ H^*(BT; \mathbf{Q}) \leftarrow H^*(BT'; \mathbf{Q})$$

of $\mathbf{Q}$-algebras. If ϕ should carry I into I' and so induce $\bar{\phi}$: $T = \tilde{T}/I \to \tilde{T}'/I' = T'$, then $\phi^* = (B\bar{\phi})^*$.

The proof is immediate. We may identify I with $\pi_1(T)$ or $H_1(T)$, and this identifies $I \otimes \mathbf{Q}$ with $H_1(T; \mathbf{Q})$; similarly for I'. So maps such as ϕ correspond to

Q-linear maps

$$H_1(T;\mathbf{Q}) \to H_1(T';\mathbf{Q})$$

or equivalently to **Q**-linear maps

$$H_2(BT;\mathbf{Q}) \to H_2(BT';\mathbf{Q}).$$

By duality they correspond to **Q**-linear maps

$$H^2(BT;\mathbf{Q}) \leftarrow H^2(BT';\mathbf{Q}).$$

Since $H^*(BT';\mathbf{Q})$ is a polynomial algebra on generators of degree 2, such maps correspond to **Q**-algebra maps

$$H^*(BT;\mathbf{Q}) \leftarrow H^*(BT';\mathbf{Q}).$$

At this stage, then, we can envisage a form of Theorem 1.1 in which the homomorphism

$$H^*(BT;\mathbf{Q}) \xleftarrow{(Bk)^*} H^*(BT';\mathbf{Q})$$

is replaced by a homomorphism ϕ^* induced by a map $\phi\colon \tilde{T} \to \tilde{T}'$ as in Lemma 1.2. However we can economise on our data further than we have explained yet. It is sufficient to suppose given a homomorphism

$$\theta\colon\ H^*(BG;\mathbf{Q}) \leftarrow H^*(BG';\mathbf{Q})$$

(of algebras over **Q**) which "commutes with Steenrod operations for all sufficiently large primes p". Now of course on the face of it this clause makes no sense, for the Steenrod operations are not defined on rational cohomology. We proceed to give it sense. Let $\mathbf{Z}_{(p)}$ be the ring of integers localised at p, that is, the ring of rational numbers m/n with n prime to p. Let $\mathbf{F}_p$ be the field of integers modulo p.

Lemma 1.3. *Suppose given a homomorphism*

$$\theta\colon\ H^*(BG;\mathbf{Q}) \leftarrow H^*(BG';\mathbf{Q})$$

(*of algebras over* **Q**). *Then for all but a finite number of primes p there are unique homomorphisms* θ_p, $\bar{\theta}_p$ *which make the following diagram commute.*

$$\begin{array}{ccc}
H^*(BG;\mathbf{Q}) & \xleftarrow{\ \theta\ } & H^*(BG';\mathbf{Q}) \\
\uparrow & & \uparrow \\
H^*(BG;\mathbf{Z}_{(p)}) & \xleftarrow{\ \theta_p\ } & H^*(BG';\mathbf{Z}_{(p)}) \\
\downarrow & & \downarrow \\
H^*(BG;\mathbf{F}_p) & \xleftarrow{\ \bar{\theta}_p\ } & H^*(BG';\mathbf{F}_p)
\end{array}$$

(The vertical arrows are induced by the obvious maps

$$\mathbf{Z}_{(p)} \to \mathbf{Q}, \qquad \mathbf{Z}_{(p)} \to \mathbf{F}_p.)$$

With the notation of Lemma 1.3, we say that "θ commutes with Steenrod operations for all sufficiently large primes p" if $\bar{\theta}_p$ commutes with Steenrod operations except for a finite number of primes p.

It is clear that if we start from a map $f\colon BG \to BG'$, or from a map f defined after finite localisation, then the homomorphism

$$f^*\colon H^*(BG;\mathbf{Q}) \leftarrow H^*(BG';\mathbf{Q})$$

commutes with Steenrod operations for all sufficiently large primes p. Conversely, our theorems imply the following result.

Corollary 1.4. *Any homomorphism*

$$\theta\colon H^*(BG;\mathbf{Q}) \leftarrow H^*(BG';\mathbf{Q})$$

of **Q**-*algebras, which commutes with Steenrod operations for all sufficiently large primes p, is induced by some map f defined after finite localisation.*

This seems to be a remarkable fact about classifying spaces, for the statement would cease to be true if BG and BG' were replaced by more general spaces.

The generalised form of Theorem 1.1 is now as follows.

Theorem 1.5. (a) *Let*

$$\theta\colon H^*(BG;\mathbf{Q}) \leftarrow H^*(BG';\mathbf{Q})$$

be a homomorphism of **Q**-*algebras which commutes with Steenrod operations for all sufficiently large primes p. Then there is an* **R**-*linear map* $\phi\colon \tilde{T} \to \tilde{T}'$ *carrying* $I \otimes \mathbf{Q}$ *into* $I' \otimes \mathbf{Q}$ *such that the following diagram is commutative.*

$$\begin{array}{ccc} H^*(BG;\mathbf{Q}) & \xleftarrow{\theta} & H^*(BG';\mathbf{Q}) \\ \downarrow & & \downarrow \\ H^*(BT;\mathbf{Q}) & \xleftarrow{\phi^*} & H^*(BT';\mathbf{Q}) \end{array} \tag{1.6}$$

(b) *Moreover, suppose further that θ carries*

$$\operatorname{Im}\{H^*(BG';\mathbf{Z}[1/m]) \to H^*(BG';\mathbf{Q})\}$$

into

$$\operatorname{Im}\{H^*(BG;\mathbf{Z}[1/m]) \to H^*(BG;\mathbf{Q})\}$$

for some positive integer m. Then ϕ may be chosen so as to carry $I \otimes \mathbf{Z}[1/m]$ *into* $I' \otimes \mathbf{Z}[1/m]$.

Clause (b) serves to set a bound on the amount of "localisation" needed in constructing ϕ. For example, Theorem 1.1 follows from Theorem 1.5 by taking $m = 1$.

To go with Theorem 1.5 on the existence of ϕ, we have a result on its uniqueness. To state our next two results, let W and W' be the Weyl groups of G and G'; they act on $\tilde{T}$ and $\tilde{T}'$.

Theorem 1.7. *Let ϕ, ψ: $\tilde{T} \to \tilde{T}'$ be* **R**-*linear maps carrying $I \otimes \mathbf{Q}$ into $I' \otimes \mathbf{Q}$ and such that the composites*

$$H^*(BT; \mathbf{Q}) \xleftarrow{\phi^*} H^*(BT'; \mathbf{Q}) \longleftarrow H^*(BG'; \mathbf{Q})$$

$$H^*(BT; \mathbf{Q}) \xleftarrow{\psi^*} H^*(BT'; \mathbf{Q}) \longleftarrow H^*(BG'; \mathbf{Q})$$

are equal. Then there is an element w' in W' such that $\psi = w'\phi$.

Here $w'\phi$ means the composite

$$\tilde{T} \xrightarrow{\phi} \tilde{T}' \xrightarrow{w'} \tilde{T}'.$$

Corollary 1.8. *Let ϕ be as in Theorem 1.5; then it has the following property.*

(1.9) *For each w in W there exists w' in W' such that $\phi w = w'\phi$.*

This follows immediately by applying Theorem 1.7 to the map $\psi = \phi w$. (Here ϕw means the composite

$$\tilde{T} \xrightarrow{w} \tilde{T} \xrightarrow{\phi} \tilde{T}',$$

which is as good as ϕ.)

In order to avoid repeated references to (1.9), we make a definition: an "admissible map" will be an **R**-linear map ϕ: $\tilde{T} \to \tilde{T}'$ which carries $I \otimes \mathbf{Q}$ into $I' \otimes \mathbf{Q}$ and satisfies (1.9). For an example of an admissible map which we consider particularly instructive, we refer the reader to Example 2.5. We can now state the converse of Theorem 1.5.

Theorem 1.10. (a) *Let ϕ: $\tilde{T} \to \tilde{T}'$ be an admissible map. Then there exists a unique homomorphism θ which makes diagram* (1.6) *commute; and θ can be induced by a map f defined after finite localisation.*

(b) *Moreover, suppose further that ϕ carries $I \otimes \mathbf{Z}[1/m]$ into $I' \otimes \mathbf{Z}[1/m]$. Then the map f may be defined after localisation so as to invert $m|W|$, where $|W|$ is the order of the Weyl group W.*

We add a few words of explanation. In clause (a) the uniqueness of θ is immediate, because

$$H^*(BG; \mathbf{Q}) \to H^*(BT; \mathbf{Q})$$

is monomorphic. The existence of θ is easily seen, as follows. By a theorem of Borel [5, p. 411, 7], the map

$$H^*(BG; \mathbf{Q}) \to H^*(BT; \mathbf{Q})$$

identifies $H^*(BG; \mathbf{Q})$ with the subalgebra $H^*(BT; \mathbf{Q})^W$ of elements invariant under the Weyl group W; and similarly for G'. Using (1.9), we check that ϕ^* carries $H^*(BT'; \mathbf{Q})^{W'}$ into $H^*(BT; \mathbf{Q})^W$.

The essential content of clause (a), then, is that θ can be induced by a map f defined after finite localisation. Clause (b) serves to set a bound on the amount of localisation needed in constructing f; however the estimate $m|W|$ is often not the best possible. For example, if f is of the form Bh, it exists without inverting anything.

Corollary 1.11. *The commutative diagram* (1.6) *gives a* (1-1) *correspondence between*
(a) *on the one hand, those homomorphisms*

$$\theta\colon H^*(BG;\mathbf{Q}) \leftarrow H^*(BG';\mathbf{Q})$$

which can be induced by maps f defined after finite localisation, and
(b) *on the other hand, W'-equivalence classes of admissible maps $\phi\colon \tilde{T} \to \tilde{T}'$.*

This follows immediately from the results (1.5) to (1.10). It makes good our claim that we reduce a problem in algebraic topology to one in Lie theory. More explicitly, the problem of determining how many homomorphisms θ arise at (a) is, on the face of it, a problem in algebraic topology; we reduce it to the problem of determining how many maps ϕ arise at (b), and this is a problem in classical Lie theory (which we will consider in § 2).

The results (1.5) to (1.10) also imply Corollary 1.4.

We have further corollaries. We recall that the (complex) representation ring $R(G)$ of G is embedded in the (complex) K-cohomology ring $K(BG)$ [2, 3].

Corollary 1.12. *The subring $R(G) \subset K(BG)$ can be characterised topologically as the set of finite **Z**-linear combinations of finite-dimensional vector bundles over BG.*

Corollary 1.13. *If $f\colon BG \to BG'$ is a map, then $f^*\colon K(BG) \leftarrow K(BG')$ carries $R(G')$ into $R(G)$.*

These two results were first obtained by Atiyah (unpublished). In Corollary 1.13, one should note that f^* need not carry honest (positive) representations of G' into honest (positive) representations of G; counterexamples may be found among Sullivan's maps from $BSU(2)$ to $BSU(2)$ [18].

It is clear that Corollary 1.13 follows immediately from Corollary 1.12, because f^* carries finite-dimensional bundles over BG' to finite-dimensional bundles over BG.

For a third corollary, we assume as in (1.5) that

$$\theta\colon H^*(BG;\mathbf{Q}) \leftarrow H^*(BG';\mathbf{Q})$$

is a homomorphism of **Q**-algebras which commutes with Steenrod operations for all sufficiently large primes p (for example, an induced map $\theta = f^*$).

Corollary 1.14. (a) *If such a homomorphism*

$$\theta\colon H^*(BG;\mathbf{Q}) \leftarrow H^*(BG';\mathbf{Q})$$

is zero in degree 4, *then it is zero in all degrees.*
(b) *If (further) G is simple and $G' = G$, then*

$$\theta\colon H^4(BG;\mathbf{Q}) \leftarrow H^4(BG;\mathbf{Q})$$

is multiplication by a non-negative scalar.

Proof. (a) We can find a positive definite rational quadratic form q on $I' \otimes \mathbf{Q}$ which is invariant under W'; this corresponds to an element $x \in H^4(BG'; \mathbf{Q})$. Suppose $\theta x = 0$; then with ϕ as in (1.5), $\phi^* q$ is the zero quadratic form on $I \otimes \mathbf{Q}$; since q is positive definite, this implies $\phi = 0$ and $\theta = 0$.

The proof of (b) is similar – see (2.12).

To conclude this introduction, we wish to acknowledge our indebtedness to Dr. J. Hubbuck; his pioneering work on this problem led to our interest in it, and he directed the research of the second author. The first author also wishes to acknowledge that the prior investigation of many special cases by Dr. J. Hubbuck and by the second author gave some valuable leads. We are grateful to M.F. Atiyah for correspondence about (1.12) and (1.13), and to C. Wilkerson and E.M. Friedlander for correspondence and discussion on a topic which will be explained in § 2.

§ 2. Lie Theory and Examples

In the previous section, we reduced a problem in algebraic topology to a problem in Lie theory. In this section we will consider the problem in Lie theory. We may state it as follows: given any pair of Lie groups G and G', we would like to know the set of admissible maps $\phi: \tilde{T} \to \tilde{T}'$.

Our first aim will be to present sufficient examples, and this we do from (2.1) to (2.11) inclusive. More precisely, from (2.4) to (2.11) we give examples of admissible maps, while (2.1) to (2.3) are examples of cases in which there are no admissible maps of interest. We regard Example 2.5 as particularly instructive.

Our second aim will be to build up enough theory to give a conceptual understanding of the examples. This theory comes in three parts. The first part centres around the case in which rank G = rank G' and $|W| = |W'|$; this covers Examples 2.4 to 2.7, and occupies (2.12) to (2.15) inclusive. In particular, we mention some recent work, using étale homotopy theory, which leads to topological maps $f: BG \to BG'$ corresponding to the admissible maps we consider. The second part of the theory deals with the case in which G' is classical. This covers Examples 2.8 to 2.10, and occupies (2.16) to (2.20) inclusive. The fact that Example 2.11 is not covered by the first two parts of the theory shows the need for the third part, which considers the general case; this occupies (2.21) to (2.29) inclusive. In particular, Theorem 2.21 proves that if ϕ is admissible, then in the equation

$$\phi w = w' \phi$$

we may choose w' to depend homomorphically on w.

It is a basic observation that our topic is "local" in the sense of Lie theory. In fact, if we pass from a group G to a finite cover of it, we do not alter $\tilde{T}$ or W; we may alter I, but we do not alter $I \otimes \mathbf{Q}$. Similarly for G'; therefore, we do not alter anything which is used in the definition of an "admissible map". It is equally easy to make the corresponding observation about the topological problem considered in § 1 (because in § 1 we considered maps f defined after finite localisation.)

Since our topic is local, we can use the classification theorem of Lie theory, and we use the notation of that theory for the simple groups $A_n, B_n, \ldots, E_8$. We can now proceed to the examples.

As "example 0", we see that for any pair (G, G') the zero map $0: \tilde{T} \to \tilde{T}'$ is admissible; it corresponds to the constant map $BG \to BG'$. One may then ask: for which pairs (G, G') is the zero map the only admissible map, and for which pairs (G, G') are there non-zero admissible maps? If the only admissible map from $\tilde{T}$ to $\tilde{T}'$ is zero, we can infer that every map $f: BG \to BG'$ induces the zero map of rational cohomology; and in this way we can cover a number of results previously obtained ad hoc. A result of some generality is given as (2.12). For the moment we give examples; their proofs are deferred.

Example 2.1. Suppose $G = \mathrm{Spin}(2m)$ *and* $G' = U(n)$. *If* $m \geqq 4$ *and* $2m > n$, *then every admissible map is zero.*

The result is sharp; if $m = 3$ then the classical isomorphism $\mathrm{Spin}(6) \cong SU(4)$ gives a non-zero admissible map, and if $2m \leqq n$ we obtain a non-zero admissible map from the inclusion

$$SO(2m) \subset SU(2m) \subset U(n).$$

Example 2.2. Suppose $G = B_5 = \mathrm{Spin}(11)$ *or* $G = A_6 = SU(7)$, *and* $G' = E_6$. *Then every admissible map is zero.*

The result is sharp, because with $G = D_5 = \mathrm{Spin}(10)$ or $G = A_5 = SU(6)$ and $G' = E_6$ there are non-trivial homomorphisms $G \to G'$ [8].

Example 2.3. Suppose $G = B_6 = \mathrm{Spin}(13)$ *or* $G = D_7 = \mathrm{Spin}(14)$, *and* $G' = E_7$. *Then every admissible map is zero.*

The result is sharp; with $G = D_6 = \mathrm{Spin}(12)$ and $G' = E_7$ there is a non-trivial homomorphism $G \to G'$ [8].

So far our examples of non-zero admissible maps have come from homomorphisms of Lie groups. However not all admissible maps do so.

Example 2.4. Take $G = G'$. For any rational number λ, define $\psi^\lambda: \tilde{T} \to \tilde{T}$ to be λ times the identity map, so that $\psi^\lambda(\mathbf{v}) = \lambda \mathbf{v}$. The associated induced map of $H^{2q}(BG; \mathbf{Q})$ is multiplication by λ^q. Any corresponding map $f: BG \to BG$ may be called a "map of type Ψ^λ"; the maps Ψ^p constructed by Sullivan [18] are of this form, with $G = U(n)$ or $SU(n)$.

It can be shown that the admissible map ψ^λ is induced by a homomorphism of Lie groups exactly when $\lambda = 1, 0$ or -1.

Example 2.5. Take $G = G' = G_2$. We may identify the Stiefel diagram $\tilde{T}$ with the complex plane $\mathbf{C}$ so that the integer lattice I becomes the lattice $a + b\omega$ (where $a, b, \in \mathbf{Z}$ and $\omega = \exp(2\pi i/3)$) and the Weyl group W becomes the dihedral group of order 12, acting in the usual way on the hexagon ± 1, $\pm\omega$, $\pm\omega^2$. We can define a map $\phi: \mathbf{C} \to \mathbf{C}$ by

$$\phi(z) = (\omega - \omega^2)\, z$$

or equivalently

$$\phi(r\exp(i\theta))=3^{\frac{1}{2}}r\exp i(\theta+\pi/2).$$

(The reader is urged to draw a picture.) The formula

$$\alpha(w)=\phi w\phi^{-1}$$

defines an outer automorphism of W, so we have

$$\phi w=w'\phi$$

where

$$w'=\alpha(w).$$

The map ϕ carries the integer lattice into itself and induces an endomorphism of the maximal torus T.

Example 2.6. We may treat in parallel the cases $G=G'=Sp(2)$ and $G=G'=F_4$. It becomes convenient to write $T(G)$, $W(G)$ etc. for the maximal torus, Weyl group etc. of a named group G. We introduce coordinates (x_1, x_2) into the Stiefel diagram $\tilde{T}(Sp(2))$ so that the long roots are $\pm 2x_1$, $\pm 2x_2$ (as usual); then we may identify $\tilde{T}(Sp(2))$ with $\mathbf{C}$ by throwing (x_1, x_2) on x_1+ix_2. We introduce coordinates (x_1, x_2, x_3, x_4) into the Stiefel diagram $\tilde{T}(\mathrm{Spin}(9))$ so that the short roots are $\pm x_1$, $\pm x_2$, $\pm x_3$, $\pm x_4$ (as usual); since Spin(9) is a subgroup of maximal rank in F_4 we may use the same coordinates in $\tilde{T}(F_4)$; then we may identify $\tilde{T}(F_4)$ with $\mathbf{C}^2$ by throwing (x_1, x_2, x_3, x_4) on (x_1+ix_2, x_3+ix_4). In either case, we may define a map $\phi\colon \mathbf{C}\to\mathbf{C}$ or $\phi\colon \mathbf{C}^2\to\mathbf{C}^2$ by

$$\phi(\mathbf{v})=(1+i)\,\mathbf{v}.$$

Again, the formula

$$\alpha(w)=\phi w\phi^{-1}$$

defines an outer automorphism of W, and we have

$$\phi w=w'\phi$$

where

$$w'=\alpha(w).$$

The map ϕ carries the integer lattice I to itself, and induces an endomorphism of the maximal torus T.

Example 2.7. Take $G=B_n$, $G'=C_n$ or vice versa. It is an old observation of Serre [5, foot of p. 428] that both from the viewpoint of homology theory and from the viewpoint of Lie theory, the groups $SO(2n+1)$ and $Sp(n)$ look alike except at the prime 2. It is easy to display an admissible map which is iso. Let us take in $U(n)$ the usual maximal torus consisting of the diagonal matrices; let us take the maximal tori in $SO(2n+1)$, $Sp(n)$ to be the images of $T(U(n))$ under the usual embeddings

$$U(n) \subset SO(2n) \subset SO(2n+1),$$

$$U(n) \subset Sp(n);$$

then the isomorphism

$$T(SO(2n+1)) \xleftarrow{\cong} T(U(n)) \xrightarrow{\cong} T(Sp(n))$$

is compatible with an isomorphism between the Weyl groups $W(SO(2n+1))$ and $W(Sp(n))$; its universal cover is an admissible map ϕ which is iso.

It will be convenient to use the letter ε to indicate one of the four admissible maps defined in Examples 2.5, 2.6, 2.7. (The context will specify G and G' and so determine which map is meant.) The reader may regard ε as standing for "exotic equivalence".

So far our examples of admissible maps of interest have been iso (as maps of vector spaces over **R**). However there are admissible maps of interest which are not iso.

Example 2.8. Take $G = A_2 = SU(3)$, $G' = A_5 = SU(6)$. We have a "Whitney sum map"

$$SU(3) \times SU(3) \xrightarrow{\oplus} SU(6)$$

which carries (A, B) to

$$\begin{bmatrix} A & 0 \\ 0 & B \end{bmatrix};$$

this induces

$$\tilde{T} \times \tilde{T} \xrightarrow{\oplus} \tilde{T}'.$$

If λ and μ are rational we have the map

$$\tilde{T} \xrightarrow{(\psi^\lambda, \psi^\mu)} \tilde{T} \times \tilde{T}$$

which carries $\mathbf{v} \in \tilde{T}$ to $(\lambda\mathbf{v}, \mu\mathbf{v}) \in \tilde{T} \times \tilde{T}$. The composite

$$\tilde{T} \xrightarrow{(\psi^\lambda, \psi^\mu)} \tilde{T} \times \tilde{T} \xrightarrow{\oplus} \tilde{T}'$$

is an admissible map, which we may write $\psi^\lambda \oplus \psi^\mu$; it corresponds to a "map of type $\Psi^\lambda \oplus \Psi^\mu$" in topology.

Example 2.9. Again take $G = A_2 = SU(3)$, $G' = A_5 = SU(6)$. Take the maximal torus in $U(n)$ to consist, as usual, of the diagonal matrices with diagonal entries

$$(\exp(2\pi i x_1), \exp(2\pi i x_2), \ldots, \exp(2\pi i x_n));$$

then $x_1, x_2, \ldots, x_n$ serve as coordinates in the Stiefel diagram of $U(n)$, and the Stiefel diagram of $SU(n)$ is the subspace

$$x_1 + x_2 + \cdots + x_n = 0.$$

Take rational numbers $\lambda \geqq \mu \geqq 0$, and define a map

$$\phi = \phi_{\lambda,\mu}: \tilde{T}(SU(3)) \to \tilde{T}(SU(6))$$

by

$$\phi(x_1, x_2, x_3)$$
$$= (\lambda x_1 + \mu x_2, \lambda x_2 + \mu x_3, \lambda x_3 + \mu x_1, \lambda x_1 + \mu x_3, \lambda x_2 + \mu x_1, \lambda x_3 + \mu x_2).$$

Then ϕ is admissible; for given any permutation w of (x_1, x_2, x_3), we can find a permutation w' of the coordinates in $\tilde{T}(SU(6))$ so that $\phi w = w' \phi$.

Examples 2.8 and 2.9 overlap; when $\mu = 0$ we have $\phi_{\lambda,0} = \psi^\lambda \oplus \psi^\lambda$, and when $\lambda = \mu$ we have $\phi_{\lambda,\lambda} = \psi^{-\lambda} \oplus \psi^{-\lambda}$ (up to the action of W').

Example 2.10. Take $G = C_3 = Sp(3)$, $G' = C_4 = Sp(4)$. Introduce coordinates $(x_1, x_2, \ldots, x_n)$ into $\tilde{T}(Sp(n))$ by using the inclusion $U(n) \subset Sp(n)$ as in (2.7). Define

$$\phi: \tilde{T}(Sp(3)) \to \tilde{T}(Sp(4))$$

by

$$\phi(x_1, x_2, x_3) = (x_1 + x_2 + x_3, -x_1 + x_2 + x_3, x_1 - x_2 + x_3, x_1 + x_2 - x_3).$$

This example can in fact be recovered from our previous ones plus homomorphisms of Lie groups. First, let us classify admissible maps $\phi: \tilde{T} \to \tilde{T}'$ into equivalence classes, regarding ϕ and ψ as "equal" if $\psi = w'\phi$ for some w' in W'. The equivalence classes still form a category, because composition passes to equivalence classes. As the objects of this category it is useful to display the groups concerned rather than their tori; we therefore write

$$G \overset{\phi}{\dashrightarrow} G'$$

to indicate an admissible map or its equivalence class; we write

$$G \xrightarrow{h} G'$$

when we actually have a homomorphism of Lie groups (which of course gives us an equivalence class of admissible maps.) We can now write down the following composite.

$$Sp(3) \overset{\psi^2}{\dashrightarrow} Sp(3) \overset{\varepsilon}{\dashrightarrow} \mathrm{Spin}(7) \xrightarrow{\Delta} \mathrm{Spin}(8) \xrightarrow{i} \mathrm{Spin}(9) \overset{\varepsilon}{\dashrightarrow} Sp(4).$$

Here ψ^2 is as in (2.4) (and we could put the factor ψ^2 in any place since maps ψ^λ commute with all other admissible maps). The admissible maps ε are as in (2.7), according to our standing convention; Δ is the findamental spin-representation of Spin(7), and i: Spin(8) $\to$ Spin(9) is the inclusion. This composite gives the admissible map of Example 2.10.

The final example of this set is perhaps the most amusing.

Example 2.11. Take $G = G_2$, $G' = F_4$. In F_4 there is a subgroup of maximal rank $A_2 \times A_2$ [8]. Here the two factors A_2 do not enter in a symmetrical way, because half the roots of $A_2 \times A_2$ are short roots of F_4 and half the roots of $A_2 \times A_2$ are long roots of F_4. We specify that it is the left-hand factor A_2 whose roots come by restriction from roots of $A_2 \times A_2$ which are long roots of F_4; equivalently, the injection of the left-hand factor A_2 in F_4 extends to an embedding of G_2 in F_4 but the injection of the right-hand factor does not.

At this point it becomes convenient to agree that for the rest of this section, the letter i will mean the injection of a subgroup of maximal rank. Such an injection $i: G \to G'$ can always be chosen to carry T into T', giving an isomorphism $i: \tilde{T} \to \tilde{T}'$; this identifies W with a subgroup of W'.

For example, the injection $i: A_2 \to G_2$ identifies $W(G_2)$ with $W(A_2) \times \{\pm 1\}$, and the injection $i: A_2 \times A_2 \to F_4$ embeds $W(A_2 \times A_2)$ in $W(F_4)$. But $W(F_4)$ also contains $\{\pm 1\}$. Take then the composite

$$\tilde{T}(G_2) \xleftarrow[\cong]{i} \tilde{T}(A_2) \xrightarrow{(\psi^\lambda, \psi^\mu)} \tilde{T}(A_2) \times \tilde{T}(A_2) \xrightarrow[\cong]{i} \tilde{T}(F_4).$$

(Here the map (ψ^λ, ψ^μ) carries $\mathbf{v} \in \tilde{T}(A_2)$ to $(\lambda \mathbf{v}, \mu \mathbf{v})$, as in (2.8).) We obtain an admissible map

$$\phi_{\lambda, \mu}: G_2 \dashrightarrow F_4.$$

We have $\phi_{-\lambda, -\mu} = \phi_{\lambda, \mu}$ (up to the action of W', as usual). The map $\phi_{1,0}$ is induced by the classical embedding of Lie groups $G_2 \to F_4$. With our usual convention about ε, the composite

$$G_2 \overset{\phi_{\lambda,\mu}}{\dashrightarrow} F_4 \overset{\varepsilon}{\dashrightarrow} F_4$$

is $\phi_{2\mu, -\lambda}$, and the composite

$$G_2 \overset{\varepsilon}{\dashrightarrow} G_2 \overset{\phi_{\lambda,\mu}}{\dashrightarrow} F_4$$

is $\phi_{\lambda', \mu'}$, where $\lambda' = -\lambda + 2\mu$, $\mu' = \lambda + \mu$. (We do not claim that these two assertions are obvious; indeed the reader who can verify the latter quickly can be satisfied with his grasp of F_4.)

We now start on the theory. The following proposition helps to explain why in Examples 2.4 to 2.7 we have treated the case in which rank G = rank G' and $|W| = |W'|$ as the first case of interest.

Proposition 2.12. *Assume that G is simple.*

(i) *If the admissible map $\phi: \tilde{T} \to \tilde{T}'$ is non-zero, then it is mono, and any rational W'-invariant inner product on $\tilde{T}'$ restricts (under ϕ) to a positive rational multiple of the rational W-invariant inner product on $\tilde{T}$.*

(ii) *If there is a non-zero admissible map ϕ, then* rank $G \leqq$ rank G' *and* $|W|$ *divides* $|W'|$.

Proof. Let K be the kernel of $\phi: \tilde{T} \to \tilde{T}'$; then by (1.9), W acts on K. Since G is simple, $\tilde{T}$ is irreducible. If ϕ is non-zero, then $K \neq \tilde{T}$; so $K = 0$. Thus ϕ is mono and rank $G \leqq$ rank G'.

Using (1.9) again, we see that a rational W'-invariant inner product on $\tilde{T}'$ must restrict (under ϕ) to a rational W-invariant inner product on $\tilde{T}$. But the latter is unique up to a rational scalar factor, which must be positive since our inner products are positive definite.

Let S be the subset of those pairs (w, w') in $W \times W'$ such that $\phi w = w' \phi$; then S is a subgroup. The projection map $S \to W$ is epi by (1.9); if ϕ is mono then the projection map $S \to W'$ is mono. Thus $|W|$ divides $|S|$ and $|S|$ divides $|W'|$. This proves the result.

The conditions given in (2.12)(ii) are necessary for the existence of a non-zero admissible map, and they have the virtue that they are simple and easy to apply. For example, (2.2) follows at once, because in (2.2) $|W|$ is either $2^8 \cdot 3.5$ or $7!$, and $|W|$ does not divide $|W'| = 2^7 \cdot 3^4 \cdot 5$. However it would be too much to hope that the conditions are sufficient in general. For example, some cases of (2.1) do not follow from (2.12) (consider the cases in which m is odd and $n = 2m - 1$); and (2.3) does not follow at all, because $|W|$ is either $2^{10} \cdot 3^2 \cdot 5$ or $2^{10} \cdot 3^2 \cdot 5 \cdot 7$ and $|W'|$ is $2^{10} \cdot 3^4 \cdot 5 \cdot 7$.

For the next result, we consider simple groups $G, G', \ldots$ with a given rank and with Weyl groups of a given order.

Proposition 2.13. *There is a contravariant functor which assigns to each non-zero admissible map $\phi: G \dashrightarrow G'$ between such groups an isomorphism ϕ^* from the Dynkin diagram* [11] *of G' to the Dynkin diagram of G. Two such maps ϕ, ψ give the same induced isomorphism of Dynkin diagrams if and only if $\psi = \lambda w' \phi$ for some positive rational λ and some $w' \in W'$. Every isomorphism of Dynkin diagrams is induced by some admissible map.*

We explain that for the purposes of this result, a Dynkin diagram consists of a set of nodes, joined by edges which are marked as single, double or triple bonds; the nodes are not to be marked to show whether they correspond to short or long roots. In particular, the Dynkin diagrams of B_n and C_n are isomorphic for present purposes.

It will be sufficient to sketch the proof of Proposition 2.13. First we set up the functor. Suppose given a non-zero admissible map $\phi: \tilde{T} \to \tilde{T}'$. By Proposition 2.12, ϕ is mono; since we assume rank $G =$ rank G', ϕ is iso. Next recall that the "infinitesimal Stiefel diagram of G" consists of the vectorspace $\tilde{T}$ together with a set of hyperplanes through the origin in $\tilde{T}$, namely the rootplanes of G. Using (1.9), it is easy to check that every rootplane of G corresponds under ϕ to a rootplane of G'; and using the assumption $|W| = |W'|$, we see that every rootplane of G' arises in this way. Replacing ϕ by $w'\phi$ if necessary, we may assume that ϕ carries the fundamental Weyl chamber of G to that of G'. Now the nodes of the Dynkin diagram for G' correspond to the simple roots θ'_i of G' or to the walls H'_i of the fundamental Weyl chamber for G'. We may associate to each wall H'_i (of G') the wall $\phi^{-1} H'_i$ (of G); equivalently, we associate to each simple root θ'_i of G' that simple root of G which is a scalar multiple of $\theta'_i \phi$.

The competent reader can easily check (i) that this procedure defines an isomorphism ϕ^* from the Dynkin diagram of G' to that of G (because ϕ preserves the angle between roots) (ii) that $\phi \mapsto \phi^*$ is a functor (that is, it preserves composition and identity maps) and (iii) that $\phi^* = \psi^*$ if and only if $\psi = \lambda w' \phi$ for some positive rational λ and some $w' \in W'$.

There remains the statement that every isomorphism of Dynkin diagrams is induced by some admissible map. It is not necessary to go through a general proof (although such a proof is not hard); the statement can be verified by case-by-case checking, using the classification theorem. We proceed to survey the cases.

The only cases in which isomorphic Dynkin diagrams come from simple groups which are not locally isomorphic is the case $G = B_n$, $G' = C_n$ or vice versa;

and this case is covered by Example 2.7. So in what follows we may assume $G = G'$.

First we assume that all the bonds of the Dynkin diagram are single. In this case it is a classical result of Lie theory [12, or for those who prefer a reference in English, 15, 20] that each automorphism of the Dynkin diagram arises from an outer automorphism $h: G \to G$. We therefore find maps f of the form $Bh: BG \to BG$, and their corresponding maps ϕ. When the admissible map $\psi^{-1}: \tilde{T} \to \tilde{T}$ does not lie in W, a two-element subgroup of the group of outer automorphisms goes into providing the admissible maps $\psi^{\pm 1}: \tilde{T} \to \tilde{T}$; the corresponding maps Bh are the identity and a map of type Ψ^{-1}. This happens exactly for $G = A_n$, D_{2n+1} and E_6.

Secondly we assume that the Dynkin diagram contains a double or a triple bond. Then there are exactly three cases in which the Dynkin diagram has a non-trivial group of automorphisms, and they are the cases $G = G_2$, $Sp(2)$ and F_4 considered in Examples 2.5, 2.6. In each case there is just one non-trivial automorphism, which turns the Dynkin diagram end for end.

Examining the map ϕ given in Example 2.5, we see that it carries each short root of G_2 to a long root, and each long root to 3 times a short root. Similarly, examining the maps ϕ given in Example 2.6, we see that they carry each short root of $Sp(2)$ or F_4 to a long root, and each long root to twice a short root. Therefore these admissible maps induce the non-trivial automorphisms of the appropriate Dynkin diagrams. This completes the discussion of Proposition 2.13.

The reader may perhaps expect us to record something about the homomorphisms of rational cohomology corresponding to the admissible maps we have just considered. We omit Example 2.7 as trivial. The composite

$$G \xrightarrow{\varepsilon} G \xrightarrow{\varepsilon} G$$

is ψ^2 in Example 2.6 and ψ^3 in Example 2.5 (up to the action of W', as usual). So the corresponding composite

$$H^{4d}(BG; \mathbf{Q}) \xleftarrow{\theta} H^{4d}(BG; \mathbf{Q}) \xleftarrow{\theta} H^{4d}(BG; \mathbf{Q})$$

is multiplication by 2^{2d} or 3^{2d}. Therefore θ acts on $H^{4d}(BG; \mathbf{Q})$ with eigenvalues $\pm 2^d$ or $\pm 3^d$ according to the case. To give the eigenvectors would involve fixing the choice of cohomology generators; we omit this and record only the eigenvalues. It is sufficient to give the eigenvalues when θ acts on the indecomposable quotient of $H^*(BG; \mathbf{Q})$.

Table 2.14. If $G = Sp(2)$ then θ acts on the indecomposable quotient in degree 4, 8 *by multiplication with* 2, -2^2.

If $G = G_2$ then θ acts on the indecomposable quotient in degree 4, 12 *by multiplication with* 3, -3^3.

If $G = F_4$ then θ acts on the indecomposable quotient in degree 4, 12, 16, 24 *by multiplication with* 2, -2^3, 2^4, -2^6.

The calculations which lead to these results are omitted.

It may also be of interest to discuss how much localisation is necessary before we can construct maps $f: BG \to BG'$ corresponding to the admissible maps we have considered. The following results are due mainly to Friedlander [13].

Proposition 2.15. (i) *For any simple G and any non-zero rational $\lambda = m/n$ a map $f: BG \to BG$ of type Ψ^λ exists if you localise so as to invert mn* [13, 22].

(ii) *A map f: $BSp(2) \to BSp(2)$ corresponding to the admissible map of Example 2.6 exists if you invert the prime 2 (even by Theorem 1.10) and without localisation there is no map f: $BSp(2) \to BSp(2)$ which induces a non-trivial automorphism of the Dynkin diagram* [16].

(iii) *A map f: $BG_2 \to BG_2$ corresponding to the admissible map of Example 2.5 exists if you invert the prime 3* [13].

(iv) *A map f: $BF_4 \to BF_4$ corresponding to the admissible map of Example 2.6 exists if you invert the prime 2* [13].

(v) *A map f: $BSO(2n+1) \to BSp(n)$ corresponding to the admissible map of Example 2.7 exists if you invert the prime 2, and is then an equivalence* [13].

This information comes from the study of algebraic groups via étale homotopy theory. The pattern is as follows. Each Lie group can be "lifted" to an algebraic group defined over **Z**. Suppose that one knows a morphism between the corresponding algebraic groups over some field of characteristic p; then étale homotopy theory delivers a map f: $BG \to BG'$ between the classifying spaces of the honest Lie groups, but at the price of localising so as to invert the prime p. Although the admissible maps of Examples 2.5 and 2.6 do not correspond to maps of Lie groups, they are well known in the theory of algebraic groups, where they correspond to "exceptional isogenies" defined when one works over a field of characteristic 3 (in Example 2.5) or 2 (in Example 2.6).

Unfortunately, those who study algebraic groups seem (so far as we know) to have studied morphisms other than isogenies in less detail than perhaps the subject would repay. Examples such as (2.8) and (2.9) clearly call for a version of representation-theory which will deliver exotic morphisms of algebraic groups from SL_n to SL_m in characteristic p; and similarly for the other classical groups. It seems possible to discern the direction such a theory should take. For the exceptional groups we have no information. In view of Example 2.11, perhaps it would be worth-while to look for exotic morphisms of algebraic groups from G_2 to F_4 in characteristic p?

We turn now to the second part of the theory, which covers the case in which G' is classical. First we consider the case $G' = U(n)$. In this case we have coordinates $(x_1, x_2, \ldots, x_n)$ in $\tilde{T}'$, so an **R**-linear map $\phi: \tilde{T} \to \tilde{T}'$ has components

$$(\phi_1, \phi_2, \ldots, \phi_n),$$

where each ϕ_r is an **R**-linear map from $\tilde{T}$ to **R**. In fact ϕ_r is the composite $x_r\phi$, where x_r means the coordinate function x_r: $\tilde{T}' \to \mathbf{R}$, considered as a weight of $U(n)$. The map ϕ carries $I \otimes \mathbf{Q}$ into $I' \otimes \mathbf{Q}$ if and only if each ϕ_r is a rational weight of G, and we assume this in what follows. We want to make explicit condition (1.9) on ϕ.

Proposition 2.16. *Assume $G' = U(n)$. Then ϕ is admissible if and only if the following condition is satisfied for each l: $\tilde{T} \to \mathbf{R}$ and each $w \in W$: if l appears exactly d times among $\phi_1, \phi_2, \ldots, \phi_n$, then lw also appears exactly d times among $\phi_1, \phi_2, \ldots, \phi_n$.*

For an example, see (2.9).

Proof. The map ϕw has components $(\phi_1 w, \phi_2 w, \ldots, \phi_n w)$. If we assume the condition stated in the proposition, then these components differ from $(\phi_1, \phi_2, \ldots, \phi_n)$ by a permutation; and since the Weyl group $W(U(n))$ is the symmetric group which permutes $(x_1, x_2, \ldots, x_n)$, this gives an element w' such that $w'\phi = \phi w$. Similarly for the converse.

Let us provisionally call an admissible map $\phi\colon G \dashrightarrow U(n)$ "irreducible" if its components $\phi_1, \phi_2, \ldots, \phi_n$ form a single orbit order W. For example, the map $\phi_{\lambda,\mu}$ of (2.9) is irreducible if $\lambda > \mu > 0$.

Corollary 2.17. *An admissible map $\phi\colon G \dashrightarrow U(n)$ may be factored in the form*

$$G \overset{\psi}{\dashrightarrow} U(n_1) \times U(n_2) \times \cdots \times U(n_r) \xrightarrow{\oplus} U(n)$$

where $n_1 + n_2 + \cdots + n_r = n$, "$\oplus$" is the Whitney sum map, and ψ is determined by admissible maps

$$\psi(i)\colon\ G \dashrightarrow U(n_i)$$

which are irreducible.

For an example, see (2.8).

Proof. Partition the components of ϕ into orbits under W. (These orbits need not be distinct; one orbit may be repeated, as happens in Example 2.9 when $\mu = 0$ or $\lambda = \mu$.)

This corollary emphasises the strong analogy between the theory of admissible maps $G \dashrightarrow U(n)$ and representation-theory. However, the "irreducibles" are different in the two theories. For example, the adjoint representation $SU(3) \to SU(8)$ is an irreducible representation; but as an admissible map it is the Whitney sum of $\phi_{2,1}\colon SU(3) \dashrightarrow SU(6)$ and the zero admissible map $SU(3) \dashrightarrow SU(2)$. Similarly we have an irreducible representation $G_2 \to SU(7)$ (the fundamental 7-dimensional complex representation of G_2); but as an admissible map it is the Whitney sum of an admissible map $G_2 \dashrightarrow SU(6)$ and a zero map. The admissible maps into $U(n)$ correspond to the "symmetric sums" in representation-theory, that is, the representations of T which are invariant under W.

Proof of Example 2.1. Suppose $G = \mathrm{Spin}(2m)$ and $G' = U(n)$. According to the work we have just done, it will be sufficient to examine the orbits under W of the non-zero weights $l\colon \tilde{T} \to \mathbf{R}$, and show that each orbit has at least $2m$ elements.

Using the usual coordinates, consider the weight

$$l = c_1 x_1 + c_2 x_2 + \cdots + c_m x_m$$

where $c_i \in \mathbf{Q}$. If none of the coefficients c_i are zero, then by changing the signs of an even number of them we obtain 2^{m-1} distinct weights lw, and $2^{m-1} \geqq 2m$ since we assume $m \geqq 4$. If r of the coefficients c_i are zero with $0 < r < m$, then by permuting the coefficients and changing their signs we obtain at least $2^{m-r} m!/r!(m-r)!$ distinct weights lw, and again this is at least $2m$. This completes the proof.

Proposition 2.16 has analogues for the other classical groups. To obtain the analogue for $SU(n)$, we have only to impose the extra condition

$$\phi_1+\phi_2+\cdots+\phi_n=0$$

(which is automatically satisfied when G is semisimple). In the cases $G'=B_n$, C_n and D_n we also have coordinates $(x_1, x_2, \ldots, x_n)$ in $\tilde{T}$, so that maps $\phi\colon \tilde{T}\to\tilde{T}'$ again have components $(\phi_1, \phi_2, \ldots, \phi_n)$.

Proposition 2.18. *Assume $G'=B_n=SO(2n+1)$ or $G'=C_n=Sp(n)$. Then ϕ is admissible if and only if the following condition is satisfied for each $l\colon \tilde{T}\to\mathbf{R}$ and $w\in W$: if the pair $\pm l$ appears exactly d times among the pairs $\pm\phi_1, \pm\phi_2, \ldots, \pm\phi_n$, then the pair $\pm lw$ also appears exactly d times.*

For an example, see (2.10).

The proof is exactly as for Proposition 2.16. The map ϕw has components $(\phi_1 w, \phi_2 w, \ldots, \phi_n w)$; if we assume the condition stated, then these components differ from $(\phi_1, \phi_2, \ldots, \phi_n)$ by a permutation plus a number of sign-changes, that is, by an element of $W(SO(2n+1))=W(Sp(n))$. Similarly for the converse.

Corollary 2.19. *The injection $Sp(n)\subset U(2n)$ induces a (1-1) correspondence between the admissible maps considered in (2.18) and those admissible maps $\phi\colon G \dashrightarrow U(2n)$ which are self-conjugate in the sense that $\psi^{-1}\phi=\phi$.*

In fact, an admissible map ϕ into $U(2n)$ with components $(\phi_1, \phi_2, \ldots, \phi_{2n})$ is self-conjugate if and only if the components $(-\phi_1, -\phi_2, \ldots, -\phi_{2n})$ differ from $(\phi_1, \phi_2, \ldots, \phi_{2n})$ by a permutation, and this ensures that the components can be permuted into the form $(\pm\phi'_1, \pm\phi'_2, \ldots, \pm\phi'_n)$.

Proposition 2.20. *Assume $G'=D_n=SO(2n)$. Then ϕ is admissible if and only if its components satisfy the condition stated in Proposition 2.18, and further, the element*

$$\phi_1\phi_2\ldots\phi_n\in H^{2n}(BT;\mathbf{Q})$$

is invariant under W.

The condition on $\phi_1\phi_2\ldots\phi_n$ is obviously necessary, as one sees by chasing the Euler class round the following diagram.

$$\begin{array}{ccc} H^*(BG;\mathbf{Q}) & \longleftarrow & H^*(BG';\mathbf{Q}) \\ \downarrow & & \downarrow \\ H^*(BT;\mathbf{Q}) & \longleftarrow & H^*(BT';\mathbf{Q}) \end{array}$$

Otherwise the proof is as for (2.16) and (2.18); the given conditions ensure that the components $(\phi_1 w, \phi_2 w, \ldots, \phi_n w)$ differ from $(\phi_1, \phi_2, \ldots, \phi_n)$ by a permutation plus an even number of sign changes.

In Propositions 2.16, 2.18 and 2.20 we have admissible maps which are completely open to inspection, and the reader can see that in the equation $\phi w=w'\phi$, it is possible to choose w' to depend homomorphically on w. We will show how to prove this in general.

Theorem 2.21. *Let $\phi\colon \tilde{T} \to \tilde{T}'$ be an admissible map. Then there exists a homomorphism $\alpha\colon W \to W'$ such that*

$$\phi w = (\alpha w)\phi.$$

It is convenient to begin with a definition. We will say that a linear map $\phi\colon \tilde{T} \to \tilde{T}'$ is "regular" if its image $\phi(\tilde{T})$ contains a vector $\phi(\mathbf{x})$ which is regular in the usual sense; that is, $\phi(\mathbf{x})$ lies on no plane of the Stiefel diagram of G'. (It is sufficient to find a vector $\mathbf{y}$ which lies on no plane through the origin, for by taking a scalar multiple $\phi(\lambda\mathbf{y})$ we can ensure that $\phi(\lambda\mathbf{y})$ lies on no other plane.)

For example, the map $\psi^\lambda \oplus \psi^\mu$ of Example 2.8 is regular if and only if $\lambda \neq \mu$.

Lemma 2.22. *If an admissible map $\phi\colon \tilde{T} \to \tilde{T}'$ is regular, then for each $w \in W$ the equation $\phi w = w'\phi$ has a unique solution for w'; and this solution depends homomorphically on w.*

Proof. If $\phi\mathbf{x}$ is regular, then the equation

$$w'\phi\mathbf{x} = \phi w\mathbf{x}$$

is sufficient to determine w' uniquely. And if the solution w' is unique, then it is elementary to check that it depends homomorphically on w.

Proof of Theorem 2.21. If ϕ is regular the result follows from Lemma 2.22. Suppose then that ϕ is not regular; then the rootplanes of G' cover $\phi(\tilde{T})$. By Lemma 3.1 below, at least one of these rootplanes contains $\phi(\tilde{T})$. Let the rootplanes of G' which contain $\phi(\tilde{T})$ be $H_1, H_2, \ldots, H_p$. They partition $\tilde{T}'$ into open convex regions, which we will call "prisms". Let S be the subgroup of W' consisting of elements which fix $\phi(\tilde{T})$. We will prove that S permutes the prisms in a way which is simply-transitive. Let $K_1, K_2, \ldots, K_q$ be the root planes of G' which do not contain $\phi(\tilde{T})$; each intersects $\phi(\tilde{T})$ in a proper subspace; so by Lemma 3.1 we can find a vector $\phi(\mathbf{x})$ in $\phi(\tilde{T})$ which lies on no rootplane K_j. Let S' be the subgroup of W' consisting of elements which fix $\phi(\mathbf{x})$. Then clearly $S \subset S'$; we will prove that $S = S'$. In fact, by [1] S' is generated by the reflections in the hyperplanes H_i; therefore each element of S' fixes the whole of $\phi(\tilde{T})$, and $S' = S$. Now consider the Weyl chambers $C_1, C_2, \ldots, C_r$ of G' whose closures contain $\phi(\mathbf{x})$. Then each prism contains at least one chamber C_k, because each prism contains points arbitrarily close to $\phi(\mathbf{x})$. Also each prism contains at most one chamber C_k; for two chambers C_k, C_l are separated by some wall of C_k, and this must be a wall containing $\phi(\mathbf{x})$, and hence one of the planes H_i. But the action of S' on the chambers C_k is simply-transitive, by [1]. Therefore the action of S on the prisms is simply-transitive.

For each $w \in W$, consider the set of w' such that $\phi w = w'\phi$. This set is a coset $w'S$ of S. Any element in it maps $\phi(\tilde{T})$ to $\phi(\tilde{T})$, permutes the rootplanes H_i, and therefore permutes the prisms. Choose once for all a fixed prism P; then the coset $w'S$ contains a unique element $\alpha(w)$ which carries P to P. It is now easy to check that α is a homomorphism. This proves Theorem 2.21.

We note that the only arbitrary element in this construction is the choice of P; this can be changed by an element $s \in S$, and therefore α can be changed by conjugation with s to get the homomorphism

$$w \mapsto s(\alpha w)s^{-1}.$$

Proof of Example 2.3. Take $G = B_6 = \text{Spin}(13)$ and $G' = E_7$. Suppose given a non-zero admissible map $\phi: G \dashrightarrow G'$. Use Theorem 2.21 to choose a homomorphism $\alpha: W \to W'$ such that $\phi w = (\alpha w)\phi$. By Proposition 2.12 ϕ is mono, and it follows that α is mono. The Sylow 2-subgroup of W is of order 2^{10}, and α must map it to a subgroup of W' of order 2^{10}, which will then be a Sylow 2-subgroup of W'. But Sylow 2-subgroups of W' are all conjugate; replacing ϕ by $\omega\phi$ and α by $w \mapsto \omega(\alpha w)\omega^{-1}$ for some suitable $\omega \in W'$, we may assume that α maps the Sylow 2-subgroup of W into any Sylow 2-subgroup of W' we prefer, say S'. Now E_7 has a subgroup of maximal rank $A_1 \times D_6$ [8], so we get an embedding of $W(A_1 \times D_6)$ in $W(E_7)$; the Sylow 2-subgroup of $W(A_1 \times D_6)$ has order 2^{10}, so we choose it for S'. When S' acts on $\tilde{T}'$ it has irreducible subspaces of dimensions 1, 2 and 4; these are therefore inequivalent irreducible representations of S'. By hypothesis the subspace $\phi(\tilde{T})$ is a subspace of dimension 6 invariant under S'; it must be the sum of the subspaces of dimensions 2 and 4; that is, it must be the Stiefel diagram of D_6. But now the non-trivial element of $W(A_1)$ lies in S' and acts trivially on $\phi(\tilde{T})$; it must come from a non-trivial element of W which acts trivially on $\tilde{T}$. This contradiction proves the result.

Given Theorem 2.21, we can view our problem as follows. Suppose given two groups G, G'. Then in principle we can enumerate the homomorphisms $\alpha: W \to W'$; there are only finitely many. For each homomorphism $\alpha: W \to W'$ the linear maps

$$\phi: \tilde{T} \to \tilde{T}'$$

such that

$$\phi w = (\alpha w)\phi$$

(for all $w \in W$) form a vector space, and in principle we can determine it. This vector space will have some dimension d; thus in Example 2.4 we have $d=1$, while in Examples 2.8, 2.9 and 2.11 we have $d=2$. We may say that we have found a "d-parameter family" of admissible maps, and all admissible maps fall into a finite number of such families. Indeed, as we only want to find the admissible maps up to the action of W', it is sufficient to consider one homomorphism α out of each equivalence class under conjugation in W'.

For example, take $G = A_2 = SU(3)$ and $G' = A_5 = SU(6)$; then by Proposition 2.16 we have precisely the two classes of admissible maps given in Examples 2.8 and 2.9.

Proposition 2.23. *Assume G simple. Then with the notation above, we have*

$$d \leqq \frac{\text{rank } G'}{\text{rank } G}.$$

For example, if $\text{rank } G' < 2 \text{ rank } G$, then $d = 0$ or 1, and the non-zero admissible maps fall into a finite number of classes under the equivalence relation used in Proposition 2.13 ($\phi \simeq \psi$ if and only if $\psi = \lambda w' \phi$ for some positive rational λ and some $w' \in W'$).

Proof. The space $\tilde{T}$ gives a representation of W which is irreducible over $\mathbf{R}$ and stays irreducible over $\mathbf{C}$ (for otherwise the action of W on $\tilde{T}$ would preserve the orientation, and it certainly does not). Consider $\tilde{T}'$ as a representation of W via α; then $\tilde{T}'$ contains the representation $\tilde{T}$ exactly d times, where

$$d = \dim_{\mathbf{R}} \operatorname{Hom}_W(\tilde{T}, \tilde{T}'),$$

agreeing with the definition above. But then

$$\dim \tilde{T}' \geqq d \dim \tilde{T}.$$

This proves the proposition.

The next task which faces us is to extend the "representation-theoretic" viewpoint of (2.16) and (2.17) to Lie groups G' which are not necessarily classical. We suggest that the appropriate definition is as follows. We say that an admissible map $\phi\colon G \dashrightarrow G'$ is "reducible" if it factors in the form

$$G \overset{\psi}{\dashrightarrow} G'' \xrightarrow{\ i\ } G',$$

where ψ is admissible, and i is the injection of a proper subgroup G'' of maximum rank in G'. Otherwise we say that ϕ is "irreducible".

To support this definition, we note that the subgroups of maximum rank in $U(n)$ are precisely the subgroups

$$U(n_1) \times U(n_2) \times \cdots \times U(n_r)$$

considered in Corollary 2.17; it follows that in the special case $G' = U(n)$, the present definition of "irreducible" agrees with the provisional definition we gave before. It is this analogy with representation-theory which leads us to use the word "irreducible".

Example 2.24. *The identity map* $1\colon G \to G$ *is irreducible. The admissible maps* ε *of Examples* 2.5, 2.6 *and* 2.7 *are irreducible. However, the following composites are reducible.*

$$A_1 \times C_3 \xrightarrow{\ i\ } F_4 \overset{\varepsilon}{\dashrightarrow} F_4$$

$$A_2 \times A_2 \xrightarrow{\ i\ } F_4 \overset{\varepsilon}{\dashrightarrow} F_4.$$

(We recall that we use the letter i to indicate the injection of any required subgroup of maximal rank.)

Proof. The first two sentences are easily proved; if we suppose given a factorisation through $i\colon G'' \to G'$, we find we must have $W'' = W'$ and so $G'' = G'$. To prove the assertion about $A_1 \times C_3$, we use the following commutative diagram.

$$\begin{array}{ccccc} C_4 & \overset{\varepsilon}{\dashrightarrow} & B_4 & \xrightarrow{\ i\ } & F_4 \\ \uparrow{\scriptstyle \oplus} & & & & \downarrow{\scriptstyle \varepsilon} \\ A_1 \times C_3 & & \xrightarrow{\qquad i \qquad} & & F_4 \end{array}$$

Since the composite

$$F_4 \overset{\varepsilon}{\dashrightarrow} F_4 \overset{\varepsilon}{\dashrightarrow} F_4$$

is ψ^2, the composite εi: $A_1 \times C_3 \to F_4$ factors through B_4. To prove the assertion about $A_2 \times A_2$, we use the following commutative diagram.

$$\begin{array}{ccc} A_2 \times A_2 & \xrightarrow{i} & F_4 \\ {\scriptstyle\tau}\,\vdots & & \vdots\,{\scriptstyle\varepsilon} \\ \downarrow & & \downarrow \\ A_2 \times A_2 & \xrightarrow{i} & F_4 \end{array}$$

Here (assuming i is as specified in Example 2.11) the map

$$\tau\colon\ \tilde{T}(A_2) \times \tilde{T}(A_2) \to \tilde{T}(A_2) \times \tilde{T}(A_2)$$

is given by

$$\tau(\mathbf{u}, \mathbf{v}) = (2\mathbf{v},\ -\mathbf{u}).$$

As further motivation for our definition of "irreducible", we note that when one compiles lists of admissible maps, it seems wasteful and redundant to list maps of the form

$$G \overset{\psi}{\dashrightarrow} G'' \xrightarrow{i} G'$$

where ψ occurs in an earlier list. More precisely, we have the following result.

Proposition 2.25. *In order to determine all admissible maps, it is sufficient to determine those for which the source G is semi-simple, the target G′ is simple and the admissible map* $\phi\colon \tilde{T} \to \tilde{T}'$ *is both irreducible and mono.*

The proof is deferred.

This proposition allows one to use comparatively short lists, as in the following examples.

Example 2.26. Suppose $G' = G_2$. *Then the admissible maps which are both irreducible and mono, with semi-simple source and target* G_2, *are the non-zero scalar multiples of the following.*

(i) *The identity map* $1\colon G_2 \to G_2$.

(ii) *The map* $\varepsilon\colon G_2 \dashrightarrow G_2$ *of Example* 2.5.

(iii) *The composite* $A_2 \xrightarrow{i} G_2 \overset{\varepsilon}{\dashrightarrow} G_2$.

Example 2.27. Suppose $G' = F_4$. *The admissible maps which are both irreducible and mono, with semi-simple source and target* F_4, *are the following.*

(i) *The non-zero scalar multiples of*

$$F_4 \xrightarrow{1} F_4,$$

$$F_4 \overset{\varepsilon}{\dashrightarrow} F_4,$$

$$B_4 \xrightarrow{i} F_4 \overset{\varepsilon}{\dashrightarrow} F_4,$$

$$C_4 \overset{\varepsilon}{\dashrightarrow} B_4 \xrightarrow{i} F_4 \overset{\varepsilon}{\dashrightarrow} F_4, \quad \textit{and}$$

$$D_4 \xrightarrow{i} F_4 \overset{\varepsilon}{\dashrightarrow} F_4.$$

(ii) *The maps* $\phi_{\lambda,\mu}$: $G_2 \dashrightarrow F_4$ *of Example* 2.11, *provided we exclude the following ratios* $\lambda:\mu$ *which give reducible maps:*

$$\lambda:\ \mu = 1:0, 0:1, -1:1, 2:1.$$

We will omit the calculations which lead to these results. However we believe the main ideas will emerge from the theory which follows.

Let $Z' \subset \tilde{T}'$ be the counterimage of the centre of G'; this counterimage can be described in the usual way as the set of points $\mathbf{v} \in \tilde{T}'$ such that each family of parallel planes of the diagram contains a plane through $\mathbf{v}$.

Let Γ' be the extended Weyl group of G' (generated by the reflections in all the planes of the Stiefel diagram, not just those which pass through the origin). Let Γ_0' be the subgroup of translations in Γ'. For any subgroup $S' \subset W'$, let $Z'(S')$ be the set of vectors $\mathbf{v} \in \tilde{T}'$ such that

$$s'\,\mathbf{v} \equiv \mathbf{v} \mod \Gamma_0'$$

for all $s' \in S' \subset W'$. The notation is justified by the following result, which will be proved in due course.

Proposition 2.28. *If* $S' = W'$, *then* $Z'(W') = Z'$.

Theorem 2.29. *An admissible map* $\phi: G \dashrightarrow G'$ *is irreducible if and only if it satisfies the following two conditions.*

(i) *It is regular. (If so it defines a unique homomorphism* α: $W \to W'$, *by Lemma* 2.22.)

(ii) $Z'(\alpha(W)) = Z'$.

Proof of (2.28) and (2.29). We begin by showing that $Z' \subset Z'(W')$. Suppose $\mathbf{v} \in Z'$ and w' is the reflection in a rootplane through the origin. Let ω' be the reflection in a parallel plane through $\mathbf{v}$; then

$$w'\,\mathbf{v} = w'\,\omega'\,\mathbf{v}$$

and $w'\omega' \in \Gamma_0'$. Since W' is generated by such reflections, it follows that

$$w'\,\mathbf{v} \equiv \mathbf{v} \mod \Gamma_0'$$

for all $w' \in W'$. Hence $Z' \subset Z'(W')$.

The proof of (2.28) will be complete when we complete the proof of (2.29); for the identity map $1: G' \to G'$ is irreducible, so (2.29) asserts that $Z'(W') = Z'$.

Next we show that if ϕ satisfies the two conditions given in (2.29), then it is irreducible. For suppose that ϕ factors in the form

$$G \overset{\psi}{\dashrightarrow} G'' \xrightarrow{\ i\ } G$$

where i is the injection of a subgroup G'' of maximum rank. If ϕ is regular, then ψ must also be regular. We have $\Gamma_0'' \subset \Gamma_0'$ (they might not be equal) and since the elements αw are the same for both, we have

$$Z''(\alpha(W)) \subset Z'(\alpha W)).$$

Thus

$$Z' \subset Z'' \subset Z''(W'') \subset Z''(\alpha W) \subset Z'(\alpha W).$$

If $Z'(\alpha W) = Z'$, then we infer $Z' = Z''$; but subgroups of maximum rank are characterised by their centres [8], so $G'' = G'$ and ϕ is irreducible.

Conversely, we wish to show that the two conditions are necessary. Let us argue for a contradiction and suppose that $\phi: G \dashrightarrow G'$ is not regular. Then we can go through the proof of Theorem 2.21 and find planes $H_1, H_2, \ldots, H_p$ and a prism P. All elements $\alpha(w)$ carry P to P; by choosing a vector in the open convex set P and averaging it over the subgroup $\alpha(W)$, we can find a vector $\mathbf{y}$ in P such that all elements $\alpha(w)$ fix $\mathbf{y}$. Let Σ be the subgroup of elements in W' which fix $\mathbf{y}$; then α maps W into Σ. We will show that Σ is the Weyl group of a subgroup G'' of maximum rank.

In fact, let $G'' \subset G'$ be the identity-component of the centraliser of the 1-parameter subgroup corresponding to $\mathbf{y}$. Then clearly $G'' \supset T'$, so G'' is a subgroup of maximum rank in G'; we choose T' as the maximal torus T'' in G''. Since all inner automorphisms in G'' fix $\mathbf{y}$, it is clear that W'' fixes $\mathbf{y}$ and $W'' \subset \Sigma$. Conversely, it is easy to see that the rootplanes of G'' are exactly those rootplanes of G' which contain $\mathbf{y}$ (compare [8]). So W'' contains the reflections in those rootplanes; but such reflections suffice to generate Σ [1], so $W'' \supset \Sigma$.

By construction, the rootplanes $H_1, H_2, \ldots, H_p$ of G' do not contain $\mathbf{y}$, so they are not rootplanes of G'', and G'' is a proper subgroup of G'.

Let us now reintepret ϕ as a map $\psi: \tilde{T} \to \tilde{T}''$. Then we have factored ϕ in the form

$$T \xrightarrow{\psi} \tilde{T}'' = \tilde{T}'.$$

Here ψ is admissible because α maps W into Σ, that is, into W''. So ϕ is reducible.

To complete the proof of Theorem 2.29, we argue by contradiction again. Suppose that $\phi: G \dashrightarrow G'$ is regular, but $Z'(\alpha(W)) \neq Z'$. Choose then an element $\mathbf{y} \in Z'(\alpha(W))$ which is not in Z'; let Σ be the subgroup of elements $w' \in W'$ such that

$$w'\mathbf{y} \equiv \mathbf{y} \mod \Gamma_0';$$

then α maps W into Σ. We will show that Σ is contained in the Weyl group of a proper subgroup of maximum rank.

In fact, let $G'' \subset G'$ be the identity-component of the centraliser of the image of $\mathbf{y}$ in G'. Then clearly $G'' \supset T'$, so G'' is a subgroup of maximum rank in G'; we choose T' as the maximal torus T'' in G''. It is easy to see that the rootplanes of G'' are exactly those rootplanes of G' for which some parallel plane of the same family contains $\mathbf{y}$ (compare [8]). In particular, since we chose $\mathbf{y} \notin Z'$, not all rootplanes of G' qualify as rootplanes of G'', and G'' is a proper subgroup of G'. We will show $\Sigma \subset W''$.

It is a known result [9] that the action of Γ' on the set of alcoves in $\tilde{T}'$ is simply-transitive; therefore we get a simply-transitive action of $W' = \Gamma'/\Gamma_0'$ on the set of Γ_0'-equivalence classes of alcoves. Consider the alcoves whose closures contain $\mathbf{y}$, and take their Γ_0'-equivalence classes; we show that W'' acts transitively on the set of such classes. This can be seen by a standard argument – one can get from one alcove containing $\mathbf{y}$ to any adjacent alcove containing $\mathbf{y}$ by reflection in the appropriate wall, which is an element of W'' since the wall in question must contain $\mathbf{y}$. Choose then a Γ_0'-equivalence class of alcoves, say A_0; and take $\sigma\in\Sigma$. Since

$$\sigma\mathbf{y}\equiv\mathbf{y} \mod \Gamma_0'.$$

σA_0 is another class of the sort we consider. Since W'' acts transitively on such classes, we have

$$\sigma A_0 = w'' A_0$$

for some $w''\in W''$. Since the action of W' is simply-transitive, we have $\sigma = w''$. Thus $\Sigma\subset W''$.

Let us now reintepret ϕ as a map ψ: $\tilde{T}\to\tilde{T}''$. Then we have factored ϕ in the form

$$\tilde{T}\xrightarrow{\psi}\tilde{T}''=\tilde{T}',$$

and ψ is admissible because α maps W into Σ and $\Sigma\subset W''$. Thus ϕ is reducible.

This completes the proof of (2.29) and (2.28).

Proof of Proposition 2.25. Suppose given a general admissible map ϕ: $G\dashrightarrow G'$. We can factor ϕ in the form

$$G\overset{\psi}{\dashrightarrow}G''\xrightarrow{i}G',$$

where ψ is irreducible and i is the injection of a subgroup G'' of maximum rank in G'. (Subgroups of maximum rank clearly satisfy the descending chain condition.) Lists of subgroups of maximal rank are easy to compile and use, and are available in classical references [8, 11]; so it is sufficient to determine the possibilities for ψ, given G''.

Up to local isomorphism G'' is a product of circle groups S^1 and simple groups. Now a product of groups is a product in our category; to determine the admissible maps ψ: $G\dashrightarrow H_1\times H_2$, it is sufficient to determine their components ψ_1: $G\dashrightarrow H_1$ and ψ_2: $G\dashrightarrow H_2$. Moreover if ψ is irreducible then the components ψ_1 and ψ_2 are irreducible, for if ψ_1 factored through H_1'' then ψ would factor through $H_1''\times H_2$. It is trivial to determine the admissible maps into a circle S^1, so we have reduced to the case in which the target G' is simple and the map ϕ is irreducible.

Now express G as the product $G_1\times G_2$ of a torus G_1 and a semisimple group G_2. An irreducible map ϕ of $G_1\times G_2$ into a simple group G' must annihilate $\tilde{T}(G_1)$; for W acts trivially on $\tilde{T}(G_1)$, and so the image $\phi(\tilde{T}(G_1))$ is contained in $Z'(\alpha(W))=Z'$ and must be zero. Thus ϕ factors in the form

$$G_1\times G_2\xrightarrow{j}G_2\overset{\psi}{\dashrightarrow}G'$$

where j is the obvious projection. Here ψ is still irreducible, so we have reduced to the case in which G is semisimple, G' simple and ϕ irreducible.

Now express G as a product $G_1\times G_2\times\cdots\times G_n$ of simple groups. Then $W=W_1\times W_2\times\cdots\times W_n$; and when W acts on $\tilde{T}$, the Stiefel diagrams $\tilde{T}(G_i)$ of the factors give irreducible representations of W, which are inequivalent because W

acts on $\tilde{T}(G_i)$ via the projection $W \to W_i$ onto the i-th factor. Suppose given an admissible map $\phi\colon \tilde{T} \to \tilde{T}'$, and consider $\operatorname{Ker}\phi$; as before, $\operatorname{Ker}\phi$ is a subspace invariant under W; therefore it is the product of some subset of the $\tilde{T}(G_i)$. Define G'' to be the quotient of G by the corresponding G_i (equivalently, it is the product of the remaining G_i). Then we can factor ϕ in the form

$$G \xrightarrow{\ j\ } G'' \overset{\psi}{\dashrightarrow} G'$$

where j is the quotient map; and here ψ is both irreducible and mono. This completes the proof.

§ 3. Uniqueness Results

In this section our main aim is to prove Theorem 1.7; but we will also prove another uniqueness result for use in § 4. Our method relies on the following lemma.

Lemma 3.1. *Let V be a finite-dimensional vector space (over an infinite field k) which is the union of finitely many vector spaces V_α. Then there is at least one α for which $V_\alpha = V$.*

Proof. Without loss of generality we may assume that V is coordinate n-space k^n. Consider the vectors

$$\mathbf{v}_x = (1, x, x^2, \ldots, x^{n-1})$$

as x runs over k. There are infinitely many of them, so at least one of the subspaces V_α must contain n of the $\mathbf{v}_x$. But any n of the vectors $\mathbf{v}_x$ form a base for V, because their determinant is a Vandermonde determinant which is non-zero. This proves Lemma 3.1.

Proof of Theorem 1.7. Our notation and assumptions will be as in Theorem 1.7, so that $\phi, \psi\colon \tilde{T} \to \tilde{T}'$ are $\mathbf{R}$-linear maps carrying $I \otimes \mathbf{Q}$ into $I' \otimes \mathbf{Q}$. We apply Lemma 3.1, taking V to be the space $H^2(BT'; \mathbf{Q})$. For each w' in W', let $V(w')$ be the subspace of elements v in $H^2(BT'; \mathbf{Q})$ such that

$$\psi^* v = \phi^* (Bw')^* v \qquad \text{in } H^2(BT; \mathbf{Q}).$$

We will prove that the subspaces $V(w')$ cover V.

Take then an element $v \in V = H^2(BT'; \mathbf{Q})$. Form the product

$$\pi = \prod_{w' \in W'} (Bw')^* (1 + v) \qquad \text{in } H^*(BT'; \mathbf{Q}).$$

Since π is invariant under W', it is the restriction of an element $\tilde{\pi}$ in $H^*(BG'; \mathbf{Q})$. We are assuming that the composites

$$H^*(BT; \mathbf{Q}) \xleftarrow{\psi^*} H^*(BT'; \mathbf{Q}) \longleftarrow H^*(BG'; \mathbf{Q})$$
$$H^*(BT; \mathbf{Q}) \xleftarrow{\phi^*} H^*(BT'; \mathbf{Q}) \longleftarrow H^*(BG'; \mathbf{Q})$$

are equal; applying them both to $\tilde{\pi}$, we see that

$$\psi^* \pi = \phi^* \pi$$

or equivalently

$$\prod_{w' \in W'} (1+\psi^*(Bw')^* v) = \prod_{w' \in W'} (1+\phi^*(Bw')^* v).$$

Now the left-hand side contains a term with $w'=1$. Since $H^*(BT; \mathbf{Q})$ is a unique factorisation domain, we conclude that there exists w' in W' and a unit λ in $\mathbf{Q}$ such that

$$(1+\psi^* v) = \lambda(1+\phi^*(Bw')^* v).$$

By considering the component of degree 0 we see that $\lambda = 1$; thus

$$\psi^* v = \phi^*(Bw')^* v$$

and v lies in $V(w')$. This proves that the spaces $V(w')$ cover V.

Lemma 3.1 now shows that there is a w' for which $V(w') = V$. That is,

$$\psi^* = \phi^*(Bw')^* \colon H^2(BT; \mathbf{Q}) \leftarrow H^2(BT'; \mathbf{Q}).$$

As explained in the proof of Lemma 1.2, this proves that $\psi = w'\phi$. This completes the proof of Theorem 1.7.

We now set up an elementary result which is needed in §4. In §4 we take G to be a torus T of dimension d; we may take $\tilde{T}$ to be $\mathbf{R}^d$ and I to be $\mathbf{Z}^d$. We take G' to be the unitary group $U(n)$; by taking coordinates as in §2, we may take $\tilde{T}'$ to be $\mathbf{R}^n$ and I' to be $\mathbf{Z}^n$. Any matrix A with n rows, d columns and rational entries defines a map $\phi\colon \tilde{T} \to \tilde{T}'$ by $\phi(\mathbf{v}) = A\mathbf{v}$; at this point we are interested in matrices A with integral entries, which map I to I' and define homomorphisms $\bar{A}\colon T \to T'$.

Lemma 3.2. *Let A, M be integral matrices such that the composites*

$$H^*(BT; \mathbf{Z}) \xleftarrow{(B\bar{A})^*} H^*(BT'; \mathbf{Z}) \longleftarrow H^*(BU(n); \mathbf{Z})$$
$$H^*(BT; \mathbf{Z}) \xleftarrow{(B\bar{M})^*} H^*(BT'; \mathbf{Z}) \longleftarrow H^*(BU(n); \mathbf{Z})$$

are equal modulo p (where p is prime). Then there is an $n \times n$ permutation matrix P (corresponding to an element in W') such that $PM \equiv A \bmod p$.

Proof. We may take BT^d to be $CP^\infty \times CP^\infty \times \cdots \times CP^\infty$ (with d factors); let $x_1, x_2, \ldots, x_d$ be the usual generators in $H^2(BT^d; \mathbf{Z})$. Take the two composites mentioned in the lemma and apply them both to the total Chern class $\sum_{r=0}^{n} c_r$ in $H^*(BU(n); Z)$. We find

$$\prod_{i=1}^{n}\left(1+\sum_{j=1}^{d} a_{ij}x_j\right) = \prod_{i=1}^{n}\left(1+\sum_{j=1}^{d} m_{ij}x_j\right)$$

in $H^*(BT^d; \mathbf{F}_p)$. But $H^*(BT^d; \mathbf{F}_p)$ is a unique factorisation domain; so the two factorisations agree up to the order of the factors. This proves the result stated.

Of course the same method provides a more elementary proof for the special case $G' = U(n)$ of Theorem 1.7, by working over $\mathbf{Q}$ instead of over $\mathbf{F}_p$.

§ 4. Proof of Theorem 1.5; Special Case

In this section we will prove the special case of Theorem 1.5 in which $G' = U(n)$. On the way we will prove Lemma 1.3.

First we show that in proving Theorem 1.5(a), we may assume the data of Theorem 1.5(b) for some m.

Lemma 4.1. *For any homomorphism*

$$\theta\colon H^*(BG;\mathbf{Q}) \leftarrow H^*(BG';\mathbf{Q})$$

of $\mathbf{Q}$*-algebras there exists* m *such that* θ *carries*

$$\operatorname{Im}\{H^*(BG';\mathbf{Z}[1/m]) \rightarrow H^*(BG';\mathbf{Q})\}$$

into

$$\operatorname{Im}\{H^*(BG;\mathbf{Z}[1/m]) \rightarrow H^*(BG;\mathbf{Q})\}.$$

Proof. We may choose a finite set of elements h_α in $H^*(BG';\mathbf{Z})$ which generate it as an algebra over $\mathbf{Z}$; this may be proved by the method of Venkov [21], since Noetherian arguments work as well over $\mathbf{Z}$ as over $\mathbf{F}_p$. Consider the images of the h_α in $H^*(BG;\mathbf{Q})$, using θ. By a suitable choice of m we can ensure that all these images lie in

$$\operatorname{Im}\{H^*(BG;\mathbf{Z}[1/m]) \rightarrow H^*(BG;\mathbf{Q})\}.$$

Since the h_α suffice to generate $H^*(BG';\mathbf{Z}[1/m])$ over $\mathbf{Z}[1/m]$, this proves the result stated.

Proof of Lemma 1.3. Consider the following diagram.

$$\begin{array}{ccc}
H^*(BG;\mathbf{Q}) & \xleftarrow{\ \theta\ } & H^*(BG';\mathbf{Q}) \\
\uparrow{\scriptstyle i} & & \uparrow{\scriptstyle i'} \\
H^*(BG;\mathbf{Z}_{(p)}) & \xleftarrow{\ \theta_p\ } & H^*(BG';\mathbf{Z}_{(p)}) \\
\downarrow{\scriptstyle j} & & \downarrow{\scriptstyle j'} \\
H^*(BG;\mathbf{F}_p) & \xleftarrow{\ \bar{\theta}_p\ } & H^*(BG';\mathbf{F}_p)
\end{array}$$

By a result of Borel [6], $H^*(BG;\mathbf{Z})$ and $H^*(BG';\mathbf{Z})$ will have no p-torsion if we exclude a finite set of primes p; then i will be mono and j will identify $H^*(BG;\mathbf{F}_p)$ with $H^*(BG;\mathbf{Z}_{(p)})\otimes\mathbf{F}_p$, and similarly for i', j'. Let us also exclude all primes p dividing the integer m obtained in Lemma 4.1; for all other primes p, θ carries $\operatorname{Im} i'$ into $\operatorname{Im} i'$, so that there exists a unique map θ_p. The existence and uniqueness of $\bar{\theta}_p$ follow immediately. This proves Lemma 1.3.

Lemma 4.2. *In proving Theorem* 1.5, *it is sufficient to consider the special case* $G = T$, $m = 1$.

Proof. First we consider the restriction on G, which is almost trivial. Suppose given a homomorphism

$$\theta\colon H^*(BG;\mathbf{Q})\leftarrow H^*(BG';\mathbf{Q})$$

with the properties assumed in Theorem 1.5. Then the composite

$$H^*(BT;\mathbf{Q})\longleftarrow H^*(BG;\mathbf{Q})\xleftarrow{\ \theta\ } H^*(BG';\mathbf{Q})$$

has the same properties. If we know the case $G=T$ of Theorem 1.5, we deduce the existence of a map $\phi\colon \tilde{T}\to\tilde{T}'$ with the required properties.

Secondly we consider the restriction on m. Let

$$\theta\colon H^*(BG;\mathbf{Q})\leftarrow H^*(BG';\mathbf{Q})$$

be a homomorphism of $\mathbf{Q}$-algebras which commutes with Steenrod operations for all sufficiently large primes p, and assume (using Lemma 4.1 if necessary) that θ carries

$$\mathrm{Im}\,\{H^*(BG';\mathbf{Z}[1/m])\to H^*(BG';\mathbf{Q})\}$$

into

$$\mathrm{Im}\,\{H^*(BG;\mathbf{Z}[1/m])\to H^*(BG;\mathbf{Q})\}.$$

Define a new homomorphism

$$\theta'\colon H^*(BG;\mathbf{Q})\leftarrow H^*(BG';\mathbf{Q})$$

by

$$\theta'=m^{dn}\theta\colon H^{2n}(BG;\mathbf{Q})\leftarrow H^{2n}(BG';\mathbf{Q}),$$

where d is to be chosen later. Using the fact that $m^p\equiv m \bmod p$, we check that θ' commutes with Steenrod operations whenever θ does so. By choosing a finite set of generators h_α as in the proof of Lemma 4.1 and choosing d suitably, we may arrange that θ' carries

$$\mathrm{Im}\,\{H^*(BG';\mathbf{Z})\to H^*(BG';\mathbf{Q})\}$$

into

$$\mathrm{Im}\,\{H^*(BG;\mathbf{Z})\to H^*(BG';\mathbf{Q})\}.$$

If we know the case $m=1$ of Theorem 1.5, we deduce the existence of a map $\phi'\colon \tilde{T}\to\tilde{T}'$, carrying I into I', and related to θ' as in diagram (1.6). Then we take

$$\phi=m^{-d}\phi',$$

and it has the required properties. This proves Lemma 4.2.

Lemma 4.3. *Theorem* 1.5 *is true in the special case in which G is a torus T of dimension* 1 *and $G'=U(n)$.*

The first result in this direction is due to Schwarzenberger [14]. The proof to be given is due to the second author (in his thesis).

We recall from § 3 that we take BT to be CP^∞, and that we are using matrices A (in this case with n rows and 1 column) to describe linear maps from $\tilde{T}=\mathbf{R}^1$

to $\tilde{T}' = \mathbf{R}^n$. For later use we record the connection between a column vector

$$A = \begin{bmatrix} a_1 \\ a_2 \\ \vdots \\ a_n \end{bmatrix}$$

and the corresponding homomorphism

$$\theta \colon H^*(CP^\infty; \mathbf{Q}) \leftarrow H^*(BU(n); \mathbf{Q}).$$

Lemma 4.4. *The homomorphism θ corresponding to A is given by $\theta c_r = b_r x^r$, where c_r is the r-th Chern class, b_r is the r-th elementary symmetric function of $a_1, a_2, \ldots, a_n$, and $x \in H^2(CP^\infty; \mathbf{Z})$ is the generator. Conversely, starting from $\theta c_r = b_r x^r$, the numbers $a_1, a_2, \ldots, a_n$ are the roots of the equation*

$$z^n - b_1 z^{n-1} + b_2 z^{n-2} - \cdots + (-1)^n b_n = 0.$$

This is an elementary calculation.

In proving Lemma 4.3, we may assume (by Lemma 4.2) that $m = 1$, and therefore (since CP^∞ and $BU(n)$ are torsion-free) that we are given a homomorphism

$$\theta \colon H^*(CP^\infty; \mathbf{Z}) \leftarrow H^*(BU(n); \mathbf{Z})$$

(commuting with Steenrod operations for all sufficiently large primes p). We write

$$\theta c_r = b_r x^r \quad (\text{where } b_r \in \mathbf{Z}),$$

and form the equation

$$z^n - b_1 z^{n-1} + b_2 z^{n-2} - \cdots + (-1)^n b_n = 0.$$

If we prove that all its roots are integral, the proof of Lemma 4.3 will be complete. This will be done by the next two lemmas.

Lemma 4.5. *Suppose a polynomial*

$$z^n - b_1 z^{n-1} + b_2 z^{n-2} - \cdots + (-1)^n b_n,$$

with integer coefficients, splits into linear factors over $\mathbf{F}_p$ for all but a finite number of primes p. Then it splits into linear factors over $\mathbf{Z}$.

This is a known result of number-theory [10].

Lemma 4.6. *In our case, the polynomial*

$$z^n - b_1 z^{n-1} + b_2 z^{n-2} - \cdots + (-1)^n b_n$$

does split into linear factors over $\mathbf{F}_p$ for all but a finite number of primes p.

Proof. Let $\mathbf{F}_q$ be a finite extension of $\mathbf{F}_p$ in which the polynomial does split into linear factors

$$(z - z_1)(z - z_2) \ldots (z - z_n).$$

We wish to prove that $z_1, z_2, \ldots, z_n$ lie in $\mathbf{F}_p$. First observe that

$$\theta c_r = b_r x^r = \sigma_r(z_1, z_2, \ldots, z_n) x^r \mod p,$$

where σ_r is the r-th elementary symmetric function. Let $r!\,ch_r$ be, as usual, the element in $H^{2r}(BU(n); \mathbf{Z})$ which restricts to $x_1^r + x_2^r + \cdots + x_n^r$ in $H^{2r}(BT'; \mathbf{Z})$; then we have

$$\theta(r!\,ch_r) = (z_1^r + z_2^r + \cdots + z_n^r) x^r \mod p.$$

Consider now the Steenrod operation

$$P^1 \colon H^{2r}(CP^\infty; \mathbf{F}_p) \to H^{2s}(CP^\infty; \mathbf{F}_p)$$

where $s = r + (p-1)$. We have

$$P^1 x^r = r x^s.$$

Thus in $H^*(BU(n); \mathbf{F}_p)$ we have

$$P^1(r!\,ch_r) = r(s!\,ch_s),$$

for it is sufficient to calculate in $H^*(BT'; \mathbf{F}_p)$. Since θ is supposed to commute with P^1 (if p is sufficiently large), we have

$$P^1 \theta(r!\,ch_r) = \theta P^1(r!\,ch_r) = r\theta(s!\,ch_s),$$

that is

$$(z_1^r + z_2^r + \cdots + z_n^r) r x^s = r(z_1^s + z_2^s + \cdots + z_n^s) x^s \mod p.$$

Thus

$$z_1^r + z_2^r + \cdots + z_n^r = z_1^s + z_2^s + \cdots + z_n^s$$

provided $1 \leqq r < p$.

Let us now restrict attention to primes $p > n$, and suppose that the distinct roots $z_1, z_2, \ldots, z_n$ are $w_1, w_2, \ldots, w_t$, where w_i occurs v_i times. Thus we have $\sum_i v_i = n < p$, so $1 \leqq v_i < p$. We have shown that

$$v_1 w_1^r + v_2 w_2^r + \cdots + v_t w_t^r = v_1 w_1^s + v_2 w_2^s + \cdots + v_t w_t^s,$$

or equivalently

$$\sum_i v_i (w_i^p - w_i) w_i^{r-1} = 0;$$

since $t \leqq n < p$ this holds for $1 \leqq r \leqq t$. We can treat these equations as t linear equations for the t unknowns $v_i(w_i^p - w_i)$; the determinant is a Vandermonde determinant, and is non-zero since the w_i are distinct; thus $v_i(w_i^p - w_i) = 0$ for each i. Since $1 \leqq v_i < p$, we deduce that $w_i^p = w_i$ for each i. By elementary Galois theory, this shows that $w_i \in \mathbf{F}_p$. This proves Lemma 4.6, which completes the proof of Lemma 4.3.

Lemma 4.7. *Theorem* 1.5 *is true in the special case in which G is a torus T of dimension d and $G' = U(n)$.*

Previous results in this direction are due to Atiyah (unpublished, but see the second footnote of [4]) and Berstein [4].

We prove Lemma 4.7 by induction over d. The result is true for $d=1$, by Lemma 4.3; so we assume the result true for tori of dimension $(d-1)$. A torus of dimension $d>1$ contains so many subtori that we can get a lot of mileage out of the inductive hypothesis.

In particular, for each integer q let us define

$$J_q = \begin{bmatrix} 1 & 0 & \dots & 0 & 0 \\ 0 & 1 & \dots & 0 & 0 \\ \vdots & \vdots & & \vdots & \vdots \\ 0 & 0 & \dots & 1 & 0 \\ 0 & 0 & \dots & 0 & 1 \\ 0 & 0 & \dots & 0 & q \end{bmatrix} : \mathbf{R}^{d-1} \to \mathbf{R}^d.$$

This yields a map

$$\bar{J}_q : T^{d-1} = \mathbf{R}^{d-1}/\mathbf{Z}^{d-1} \to \mathbf{R}^d/\mathbf{Z}^d = T^d$$

as explained in § 3. The induced map

$$J_q^* = (B\bar{J}_q)^* : H^*(BT^{d-1}; \mathbf{Z}) \leftarrow H^*(BT^d; \mathbf{Z})$$

considered in (1.2) is given by

$$J_q^* x_r = x_r \quad \text{for } r<d$$
$$J_q^* x_d = q x_{d-1},$$

where the x_i are the usual cohomology generators.

Lemma 4.8. *Let*

$$\theta, \theta' : H^*(BT^d; \mathbf{Z}) \leftarrow H^*(BU(n); \mathbf{Z})$$

be two homomorphisms of $\mathbf{Z}$*-algebras such that*

$$J_q^* \theta = J_q^* \theta'$$

for $(n+1)$ *distinct values of* q*. Then* $\theta = \theta'$.

Proof. Suppose

$$\theta c_r = f(x_1, x_2, \dots, x_d)$$
$$\theta' c_r = f'(x_1, x_2, \dots, x_d)$$

where f and f' are two homogeneous polynomials of degree $r \leqq n$. Then according to our assumption,

$$f(x_1, \dots, x_{d-1}, q x_{d-1}) = f'(x_1, \dots, x_{d-1}, q x_{d-1})$$

for $(n+1)$ distinct values of q. We equate the coefficients of $x_1^{\nu_1} x_2^{\nu_2} \dots x_{d-1}^{\nu_{d-1}}$ and apply the principle that two polynomials in q of degree $\leqq n$ which agree for $(n+1)$ distinct values of q must be identically equal. We find that $f=f'$ and $\theta c_r = \theta' c_r$. Since this holds for each r, we find $\theta = \theta'$. This proves Lemma 4.8.

We can now explain the pattern of our inductive proof. In proving Lemma 4.7, we may assume (by 4.2) that $m=1$, and therefore (since BT^d and $BU(n)$) are torsion-free) that we are given a homomorphism

$$\theta\colon H^*(BT^d;\mathbf{Z}) \leftarrow H^*(BU(n);\mathbf{Z})$$

(commuting with Steenrod operations for all sufficiently large primes p). Let us agree to omit coefficients $\mathbf{Z}$ in our cohomology groups. Then the composite

$$H^*(BT^{d-1}) \xleftarrow{J_0^*} H^*(BT^d) \xleftarrow{\theta} H^*(BU(n))$$

is one to which we can apply our inductive hypothesis, and we obtain a matrix of integers A (with n rows and $(d-1)$ columns) such that the composite

$$H^*(BT^{d-1}) \xleftarrow{A^*} H^*(BT') \xleftarrow{i^*} H^*(BU(n))$$

is $J_0^*\theta$.

Similarly we introduce

$$K = \begin{bmatrix} 0 \\ 0 \\ \vdots \\ 0 \\ 0 \\ 1 \end{bmatrix} : \mathbf{R}^1 \to \mathbf{R}^d;$$

then the composite

$$H^*(BT^1) \xleftarrow{K^*} H^*(BT^d) \xleftarrow{\theta} H^*(BU(n))$$

is one to which we can apply Lemma 4.3, and we obtain a matrix of integers $\mathbf{v}$ (with n rows and 1 column) such that

$$\mathbf{v}^* i^* = K^*\theta.$$

We would now like to form a partitioned matrix such as

$$[A \mid \mathbf{v}] : \mathbf{R}^d \to \mathbf{R}^n;$$

our trouble is that A and $\mathbf{v}$ are only determined up to a permutation of their rows, and we do not know how to permute their rows before matching them up. It is for this purpose that we use the maps J_q with $q \neq 0$.

Lemma 4.9. *Assume Lemma 4.7 true for T^{d-1}; let θ, A and $\mathbf{v}$ be as above. Then for each sufficiently large prime p there is a vector $\mathbf{w}$ obtained by permuting the rows of $\mathbf{v}$ (in a way which may depend on p) such that the maps*

$$H^*(BT^{d-1}) \xleftarrow{J_p^*} H^*(BT^d) \xleftarrow{\qquad\theta\qquad} H^*(BU(n))$$

$$H^*(BT^{d-1}) \xleftarrow{J_p^*} H^*(BT^d) \xleftarrow{[A|\mathbf{w}]^*} H^*(BT') \xleftarrow{i^*} H^*(BU(n))$$

coincide.

The proof is postponed.

Assuming Lemma 4.9, we can easily complete the proof of Lemma 4.7. For in Lemma 4.9 there are only $n!$ possible permutations of the rows of $\mathbf{v}$; so if we apply Lemma 4.9 using $(n+1)!$ sufficiently large primes p, we must find a matrix $[A|\mathbf{w}]$ which works for at least $(n+1)$ such primes. Now we apply Lemma 4.8, taking θ' to be the composite

$$H^*(BT^d) \xleftarrow{[A|\mathbf{w}]^*} H^*(BT') \xleftarrow{i^*} H^*(BU(n)).$$

We see that $\theta = [A|\mathbf{w}]^* i^*$. This completes the induction and proves Lemma 4.7.

It remains to prove Lemma 4.9. In proving this lemma, we assume that Lemma 4.7 is true for T^{d-1}, so we may apply it to the composite

$$H^*(BT^{d-1}) \xleftarrow{J_p^*} H^*(BT^d) \xleftarrow{\theta} H^*(BU(n)).$$

We infer that there is a matrix of integers M (with n rows and $(d-1)$ columns) such that $J_p^* \theta$ coincides with

$$H^*(BT^{d-1}) \xleftarrow{M^*} H^*(BT') \xleftarrow{i^*} H^*(BU(n)).$$

(Of course M will depend on p, though we do not display p in the notation.) Our plan is to compare A with M and M with $\mathbf{v}$, by arguments using both p-adic approximation and approximation in the usual absolute value.

First we remark that the induced maps

$$J_0^*, J_p^*: H^*(BT^{d-1}) \leftarrow H^*(BT^d)$$

are equal mod p. Composing with θ, we have

$$A^* i^* = M^* i^* \quad \bmod p.$$

Applying Lemma 3.2, we see that by permuting the rows of M, we can ensure that

$$A \equiv M \quad \bmod p.$$

Next we claim that by choosing p sufficiently large (in a way depending only on the initial data) we can ensure that the first $(d-2)$ columns of $M-A$ are zero. We take p so large that

$$|a_{ij}| < \tfrac{1}{2} p \quad \text{for each } i, j. \tag{4.10}$$

Let

$$L = \begin{bmatrix} 1 & 0 & \dots & 0 \\ 0 & 1 & \dots & 0 \\ \vdots & \vdots & & \\ 0 & 0 & \dots & 1 \\ 0 & 0 & \dots & 0 \end{bmatrix} : R^{d-2} \to R^{d-1};$$

then $J_0 L = J_p L$. Therefore

$$L^* J_0^* \theta = L^* J_p^* \theta,$$

that is

$$L^* A^* i^* = L^* M^* i^*.$$

By Theorem 1.7 (proved in § 3) AL, ML agree up to a permutation of their rows; that is, the first $(d-2)$ columns of A agree with the first $(d-2)$ columns of M up to a permutation of their rows. So we have

$$|m_{ij}|<\tfrac{1}{2}p \qquad \text{for } j\leqq d-2,$$

whence

$$|m_{ij}-a_{ij}|<p,$$

and since

$$m_{ij}-a_{ij}\equiv 0 \mod p,$$

we have

$$m_{ij}=a_{ij} \qquad \text{for } j\leqq d-2.$$

It follows that we can write M in the form

$$M=[A\,|\,\mathbf{w}]J_p$$

where $\mathbf{w}$ is a vector which remains to be determined. As soon as we can show that $\mathbf{w}$ is obtained from $\mathbf{v}$ by permuting rows, the proof of Lemma 4.9 will be complete.

Before proceeding we recall the following material. Consider two polynomials

$$f(z)=z^n-b_1z^{n-1}+b_2z^{n-2}-\cdots+(-1)^nb_n$$
$$g(z)=z^n-d_1z^{n-1}+d_2z^{n-2}-\cdots+(-1)^nd_n$$

with complex coefficients b_r, d_r. Given f, we can ensure that the complex roots of $g(z)=0$ lie close to those of $f(z)=0$ by taking each coefficient d_r close to the corresponding coefficient b_r. This is stated more precisely by the following lemma.

Lemma 4.11. *Suppose given $b_1, b_2, \ldots, b_n$ and $\varepsilon>0$. Then there exists $\delta>0$ with the following property: if $|d_r-b_r|<\delta$ for each r, then there is a (1-1) correspondence between the roots $z_1, z_2, \ldots, z_n$ of $f(z)=0$ and the roots $\zeta_1, \zeta_2, \ldots, \zeta_n$ of $g(z)=0$ such that*

$$|\zeta_i-z_j|<\varepsilon$$

for corresponding roots ζ_i, z_j.

This lemma is standard.

When we apply this lemma, the polynomial $f(z)$ will be the one whose roots are $v_1, v_2, \ldots, v_n$, and the polynomial $g(z)$ will be one whose roots are close to $w_1, w_2, \ldots, w_n$. Recalling the definition of $\mathbf{v}$ and using Lemma 4.4, we see that the coefficients of $f(z)$ are given by

$$b_rx^r=\begin{bmatrix}0\\0\\\vdots\\0\\0\\1\end{bmatrix}^* \theta c_r.$$

Similarly, an equation whose roots are

$$m_{i\,d-1} = a_{i\,d-1} + p w_i$$

has coefficients e_r given by

$$e_r x^r = \begin{bmatrix} 0 \\ 0 \\ \vdots \\ 0 \\ 1 \\ p \end{bmatrix}^* \theta c_r.$$

Therefore an equation $g(z) = 0$ whose roots are

$$p^{-1} m_{i\,d-1} = p^{-1} a_{i\,d-1} + w_i$$

has coefficients given by $d_r = p^{-r} e_r$, that is by

$$d_r x^r = \begin{bmatrix} 0 \\ 0 \\ \vdots \\ 0 \\ p^{-1} \\ 1 \end{bmatrix}^* \theta c_r.$$

(Here, of course, the column vector induces a homomorphism

$$H^*(BT^1; \mathbf{Q}) \leftarrow H^*(BT^d; \mathbf{Z}).)$$

Now as $p \to \infty$,

$$\begin{bmatrix} 0 \\ 0 \\ \vdots \\ 0 \\ p^{-1} \\ 1 \end{bmatrix} \quad \text{tends to} \quad \begin{bmatrix} 0 \\ 0 \\ \vdots \\ 0 \\ 0 \\ 1 \end{bmatrix}$$

and d_r tends to b_r for each r. According to Lemma 4.11, by taking p sufficiently large we can find a (1-1) correspondence between the w_i and the v_j such that

$$|w_i + p^{-1} a_{i\,d-1} - v_j| < \tfrac{1}{2}$$

for corresponding roots. Here we have

$$|p^{-1} a_{i\,d-1}| < \tfrac{1}{2} \quad \text{by (4.10)},$$

so

$$|w_i - v_j| < 1;$$

since w_i and v_j are integers we have $w_i = v_j$ for corresponding roots. This completes the proof of Lemma 4.9, and so proves Lemma 4.7.

At this point we have proved that Theorem 1.5 is true when the group G' is a unitary group $U(n)$.

§ 5. Proof of Theorem 1.5; General Case

In this section we will prove Theorem 1.5 in the general case. We will use a variant of the method of "unitary embedding"; it involves three steps. (i) First we note that the theorem is true when G' is a product of unitary groups. (ii) Secondly we show that for any G' there is a suitable embedding of G' in a product of unitary groups, say G''. (iii) Thirdly, we use the result for G'' to infer the result for G'.

Throughout we will assume $G = T$, as we may by Lemma 4.2.

Lemma 5.1. *Theorem* 1.5 *is true if G' is a product of unitary groups*

$$U(n_1) \times U(n_2) \times \cdots \times U(n_r).$$

Proof. Suppose given a homomorphism

$$\theta\colon H^*(BT; \mathbf{Q}) \leftarrow H^*(BG'; \mathbf{Q})$$

which has the properties assumed in (1.5)(a) and (b). Consider the composite

$$H^*(BT; \mathbf{Q}) \xleftarrow{\ \theta\ } H^*(BG'; \mathbf{Q}) \xleftarrow{(B\pi_i)^*} H^*(BU(n_i); \mathbf{Q})$$

where $\pi_i\colon G' \to U(n_i)$ is the projection onto the i-th factor. This composite has the same properties, so by Lemma 4.7 it is induced by a map $\phi_i\colon \tilde{T} \to \tilde{T}'_i$, where T'_i is the maximal torus in $U(n_i)$. Using these maps as components we obtain a map $\phi\colon \tilde{T} \to \tilde{T}' = \bigtimes_i \tilde{T}'_i$. It is now easy to check that ϕ has the required properties.

Lemma 5.2. *For any G', there is an embedding $i\colon G' \to G''$ such that*

(a) *G'' is a product of unitary groups, and*

(b) $(Bi)^*\colon H^*(BG'; \mathbf{Q}) \leftarrow H^*(BG''; \mathbf{Q})$

is epimorphic.

Proof. Using the Peter-Weyl theorem we can find an embedding $i_1\colon G' \to U(n_1)$. The $\mathbf{Q}$-algebra $H^*(BG'; \mathbf{Q})$ is finitely-generated; using the results of Atiyah [2, 3], we can find a finite number of unitary representations $i_2\colon G' \to U(n_2), \ldots, i_r\colon G' \to U(n_r)$ so that the components of their Chern characters generate $H^*(BG'; \mathbf{Q})$. Let

$$i\colon G' \to U(n_1) \times U(n_2) \times \cdots \times U(n_r)$$

be the homomorphism with components $i_1, i_2, \ldots, i_r$. This proves Lemma 5.2.

Proof of Theorem 1.5. Let $i\colon G' \to G''$ be as in Lemma 5.2. Since the maximal torus T' of G' is contained in some maximal torus of G'', we can assume without loss of generality that T'' contains T'. Suppose given a map

$$\theta\colon H^*(BT; \mathbf{Q}) \leftarrow H^*(BG'; \mathbf{Q})$$

with the properties assumed in (1.5)(a) and (b). Then the composite

$$H^*(BT;\mathbf{Q}) \xleftarrow{\theta} H^*(BG';\mathbf{Q}) \xleftarrow{(Bi)^*} H^*(BG'';\mathbf{Q})$$

has the same properties; so by Lemma 5.1 there is a map $\psi\colon \tilde{T} \to \tilde{T}''$ with the property stated in (1.5)(b) and such that the following diagram is commutative.

$$\begin{array}{ccc} H^*(BG';\mathbf{Q}) & \xleftarrow{(Bi)^*} & H^*(BG'';\mathbf{Q}) \\ \downarrow{\scriptstyle\theta} & & \downarrow \\ H^*(BT;\mathbf{Q}) & \xleftarrow{\psi^*} & H^*(BT'';\mathbf{Q}). \end{array} \tag{5.3}$$

Lemma 5.4. *There is then an element w'' in W'' such that $w''\psi$ factors through $\tilde{T}'$.*

This will complete the proof of Theorem 1.5; for if $w''\psi$ factors in the form

$$\tilde{T} \xrightarrow{\phi} \tilde{T}' \longrightarrow \tilde{T}'',$$

then it is easy to check that ϕ has the properties asserted in Theorem 1.5.

Proof of Lemma 5.4. We apply Lemma 3.1, taking V to be the kernel of the homomorphism

$$H^2(BT';\mathbf{Q}) \leftarrow H^2(BT'';\mathbf{Q}).$$

For each w'' in W'', let $V(w'')$ be the subspace of elements v in V such that

$$\psi^*(Bw'')^* v = 0 \quad \text{in } H^2(BT;\mathbf{Q}).$$

We will prove that the subspaces $V(w'')$ cover V.

Take then an element $v \in V \subset H^2(BT'';\mathbf{Q})$. Form the product

$$\pi = \prod_{w'' \in W''} (Bw'')^* v \quad \text{in } H^*(BT'';\mathbf{Q}).$$

Since π is invariant under W'', it is the restriction of an element $\tilde{\pi}$ in $H^*(BG'';\mathbf{Q})$.

Consider the following commutative diagram.

$$\begin{array}{ccc} H^*(BG';\mathbf{Q}) & \xleftarrow{(Bi)^*} & H^*(BG'';\mathbf{Q}) \\ \downarrow & & \downarrow \\ H^*(BT';\mathbf{Q}) & \longleftarrow & H^*(BT'';\mathbf{Q}). \end{array}$$

The product π maps to zero in $H^*(BT';\mathbf{Q})$, because it contains a factor with $w''=1$, and v maps to zero in $H^2(BT';\mathbf{Q})$. Therefore the element $\tilde{\pi}$ maps to zero in $H^*(BG';\mathbf{Q})$.

Now chase the element $\tilde{\pi}$ round diagram (5.3). We find

$$\psi^*\pi = 0,$$

that is,

$$\prod_{w'' \in W''} \psi^*(Bw'')^* v = 0.$$

This equation takes place in $H^*(BT; \mathbf{Q})$, which is an integral domain; so there exists w'' for which

$$\psi^*(Bw'')^* v = 0,$$

and v lies in $V(w'')$.

Lemma 3.1 now shows that there is an element w'' such that $V(w'') = V$. That is, the kernel of

$$H^2(BT; \mathbf{Q}) \xleftarrow{\psi^*} H^2(BT''; \mathbf{Q}) \xleftarrow{(Bw'')^*} H^2(BT''; \mathbf{Q})$$

contains the kernel of

$$H^2(BT'; \mathbf{Q}) \leftarrow H^2(BT''; \mathbf{Q}).$$

Dualising back, the image of

$$H_1(T; \mathbf{Q}) \xrightarrow{\psi} H_1(T''; \mathbf{Q}) \xrightarrow{w''} H_1(T''; \mathbf{Q})$$

is contained in the image of

$$H_1(T'; \mathbf{Q}) \to H_1(T''; \mathbf{Q}).$$

This proves Lemma 5.4, and so completes the proof of Theorem 1.5.

§ 6. Proof of Theorem 1.10

In this section we will prove Theorem 1.10. Our proof involves constructing a string of comparison maps. Let N, N' be the normalisers of T, T' in G, G'; then because f has to be constructed only after localisation, we can for our purposes replace BG by BN. The groups N, N' take part in extensions

$$T \to N \to W$$
$$T' \to N' \to W';$$

these extensions are not necessarily split, but (again because f has to be constructed only after localisation) we can for our purposes replace N by the semi-direct product of T and W. (We owe this helpful remark to G. Mislin.) Now we can get through with the aid of § 2.

Assume then that we are given an admissible map $\phi\colon \tilde{T} \to \tilde{T}'$. In proving (1.10)(a) we may assume the data of (1.10)(b) for some m; for since I is finitely-generated over $\mathbf{Z}$ and ϕ carries I into $I' \otimes \mathbf{Q}$, it is clear that ϕ carries I into $I' \otimes \mathbf{Z}[1/m]$ for some m. By writing $\psi = m^d \phi$ and choosing d suitably, we can arrange that ψ carries I into I', and so induces $\bar{\psi}\colon T \to T'$.

Using Theorem 2.21, we can choose a homomorphism $\alpha\colon W \to W'$ such that

$$\phi w = (\alpha w)\phi$$

and therefore

$$\bar{\psi} w = (\alpha w)\bar{\psi}.$$

We can now consider the extensions obtained from $T \to N \to W$, $T' \to N' \to W'$ by applying $\bar{\psi}$ to the former, α to the latter; we get extensions which take part in the following commutative diagram.

$$
\begin{array}{ccccccc}
T & \xrightarrow{\bar{\psi}} & T' & \qquad & T' & \xrightarrow{1} & T' \\
\downarrow & & \downarrow & & \downarrow & & \downarrow \\
N & \longrightarrow & E & & E' & \xrightarrow{\delta} & N' \\
\downarrow & & \downarrow & & \downarrow & & \downarrow \\
W & \xrightarrow{1} & W & & W & \xrightarrow{\alpha} & W'
\end{array}
$$

These extensions correspond to elements of the group $H^2(W; T')$ (where W acts on T' via α); and this group is annihilated by $|W|$. Therefore we can find a diagram of the following form, in which S is the semi-direct product of W and T', corresponding to the zero element of $H^2(W; T')$.

$$
\begin{array}{ccccccc}
T & \xrightarrow{|W|\bar{\psi}} & T' & \xleftarrow{m^d|W|} & T' & \xrightarrow{1} & T' \\
\downarrow & & \downarrow & & \downarrow & & \downarrow \\
N & \xrightarrow{\beta} & S & \xleftarrow{\gamma} & E' & \xrightarrow{\delta} & N' \\
\downarrow & & \downarrow & & \downarrow & & \downarrow \\
W & \xrightarrow{1} & W & \xleftarrow{1} & W & \xrightarrow{\alpha} & W'
\end{array}
$$

Let $M = m|W|$, and let $BG'\mathbf{Z}[1/M]$ be the localisation of BG' in the sense of Sullivan, where we localise so as to invert M. This localisation comes provided with a map $BG' \to BG'\mathbf{Z}[1/M]$. The injections $i\colon N \to G$, $i'\colon N' \to G'$ induce maps $Bi\colon BN \to BG$, $Bi'\colon BN' \to BG'$.

Lemma 6.1. *In the following diagram, there is a unique way to fill in the dotted arrows so that the diagram becomes commutative.*

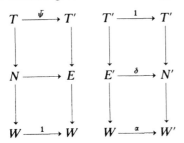

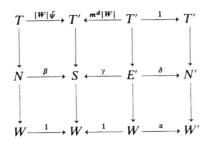

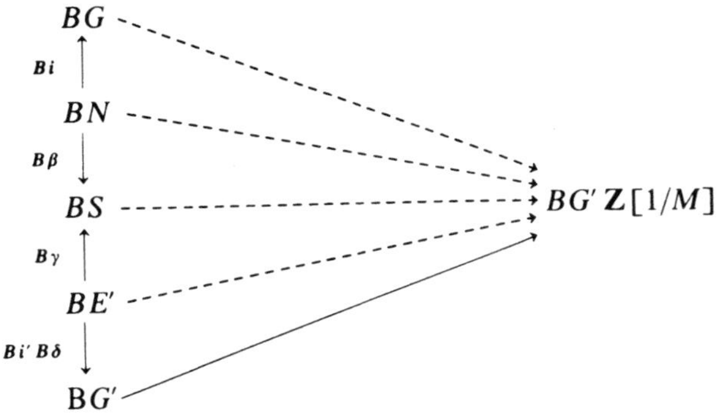

The proof is by obstruction-theory; for the maps $g = Bi$ and $g = B\gamma$ we have to prove that the cohomology groups $H^*(g; \pi_*(BG'\mathbf{Z}[1/M]))$ are zero. More generally, let C be any group of coefficients such that multiplication by $M = m|W|$

gives an isomorphism of C. Then for $g = B\gamma$ it is immediate, from the diagram in which γ first appears, that $B\gamma$ induces an isomorphism of $H^*(\ ; C)$. For Bi, both $H^*(BG; C)$ and $H^*(BN; C)$ may be identified with

$$H^*(BT; C)^W,$$

and we see that

$$(Bi)^*: H^*(BG; C) \to H^*(BN; C)$$

is iso.

This completes the proof of Lemma 6.1, and constructs our map

$$f: BG \to BG' \mathbf{Z}[1/M].$$

It remains to show that f has the required property, by showing that the diagram (1.6) commutes. This is immediate, since at each stage of our construction we have recorded the effect of our maps on the tori T, T'. Indeed we have ensured that the following diagram is homotopy-commutative.

$$\begin{array}{ccc} BT & \xrightarrow{\phi = m^{-d}\psi} & BT' \mathbf{Z}[1/M] \\ \downarrow & & \downarrow \\ BG & \xrightarrow{f} & BG' \mathbf{Z}[1/M] \end{array}$$

This completes the proof of Theorem 1.10.

§ 7. Proof of Corollary 1.12

It is clear that a representation $G \to U(n)$ induces an n-plane bundle over BG; passing to linear combinations, any element of the representation ring $R(G)$ gives a finite $\mathbf{Z}$-linear combination of vector-bundles over BG.

Conversely, suppose given an n-plane bundle over BG; then it is classified by a map $f: BG \to BU(n)$. Using Theorem 1.1 or Theorem 1.5, we obtain a homomorphism

$$T \xrightarrow{\phi} T' \longrightarrow U(n);$$

this defines an element of $R(T)$ which lies in $R(T)^W$ by (1.9), and so (by the classical theorem of H. Weyl) is the restriction of an element $\psi \in R(G)$.

Of course, Theorem 1.5 only asserts a connection between f and $B\bar{\phi}$ when we pass to rational cohomology. However, this is sufficient to show that the composite

$$BG \xrightarrow{f} BU(n) \longrightarrow BU$$

defines the same element of $K(BG)$ as ψ does. This completes the proof.

References

1. Adams, J.F.: Lectures on Lie groups. New York-Amsterdam: W.A. Benjamin 1969 (especially Theorem 5.13, pp. 110–111)
2. Atiyah, M.F., Hirzebruch, F.: Vector bundles and homogeneous spaces. Proc. Symposia in Pure Math. (AMS) **3**, 7–38 (especially p. 29 and the preceding discussion) (1961)
3. Atiyah, M.F., Segal, G.: Equivariant K-theory and completion. J. Diff. Geometry **3**, 1–18 (1969)
4. Berstein, I.: Bundles over products of infinite-dimensional projective spaces. Quart. J. Math. **19**, 275–279 (1968)
5. Borel, A.: Topology of Lie groups and characteristic classes. Bull. Amer. Math. Soc. **61**, 397–432 (1955)
6. Borel, A.: Sur la torsion des groupes de Lie. J. Math. Pures et Appl. **35**, 127–139 (1956)
7. Borel, A.: Topics in the homology theory of fibre bundles. Lecture Notes in Math. No. 36 (especially Theorem 20.3, p. 67). Berlin-Heidelberg-New York: Springer 1967
8. Borel, A., De Siebenthal, J.: Les sous-groupes fermés de rang maximum des groupes de Lie clos. Comment. Math. Helv. **23**, 200–221 (1949)
9. Bourbaki, N.: Groupes et algèbres de Lie, Chap. 4, 5, 6 (especially p. 282, paragraph 30). Paris: Hermann 1968
10. Cassels, J.W.S., Frohlich, A. (eds.): Algebraic number theory, p. 229. London-New York: Academic Press 1967
11. Dynkin, E.B.: Semisimple subalgebras of semisimple Lie algebras, and Maximal subgroups of the classical groups. In: American Mathematical Society Translations, Series 2. AMS **6**, 111–244 and 245–378 (1957)
12. Freudenthal, H., Vries, H. de: Linear Lie groups (especially § 33). London-New York: Academic Press 1969
13. Friedlander, E.M.: Exceptional isogenies and the classifying spaces of simple Lie groups. Ann. Math. **101**, 510–520 (1975)
14. Hirzebruch, F.: Topological methods in algebraic geometry (translated and with an appendix by R.L.E. Schwarzenberger) (especially p. 166). Berlin-Heidelberg-New York: Springer 1966
15. Hochschild, G.: The structure of Lie groups (especially Theorem 3.4, pp. 163–164). San Francisco: Holden-Day 1965
16. Hubbuck, J.R.: Homotopy-homomorphisms of Lie groups. London Math. Soc., Lecture Notes series No. 11, 33–41 C.U.P. 1974
17. Stasheff, J.D.: H-space problems. In: Lecture Notes in Math. 196 (especially problem no. 20, p. 127). Berlin-Heidelberg-New York: Springer 1971
18. Sullivan, D.: Geometric topology, part I. Localisation, periodicity and Galois symmetry, mimeographed notes, M.I.T. 1970 (especially Corollaries 5.10, 5.11)
19. Sullivan, D.: Genetics of homotopy theory and the Adams conjecture. Ann. Math. **100**, 1–79 (1974)
20. Tits, J.: Tabellen zu den einfachen Lie Gruppen und ihren Darstellungen. Lecture Notes in Math. 40 (especially p. 6). Berlin-Heidelberg-New York: Springer 1967
21. Venkov, B.B.: Cohomology algebras for some classifying spaces. Doklady Akad. Nauk SSSR **127**, 943–944 (1959)
22. Wilkerson, C.: Self-maps of classifying spaces. Lecture Notes in Math. 418. Berlin-Heidelberg-New York: Springer 1974

J.F. Adams
DPMMS
16, Mill Lane
Cambridge/England

Z. Mahmud
7, Crediton Close
Manchester M15 6EW/England

Received October 13, 1975

Inventiones math. 49, 1 – 65 (1978)

Inventiones mathematicae

Maps Between Classifying Spaces. II

J.F. Adams

D.P.M.M.S., 16 Mill Lane, Cambridge, England

§ 1. Introduction and Statement of Results

Let G and G' be compact Lie groups; let BG and BG' be their classifying spaces [21, 15]. In this paper I shall study the classification of maps

$$f\colon BG \to BG'.$$

What happens may be described in general terms; BG has a very rich and a very rigid structure, and the effect of this is that there are very few maps compared with what one might expect. Indeed at one time there was a conjecture [20] that any such map f is of the form $B\theta$ for some homomorphism θ of Lie groups; but as I shall explain at an appropriate point below, it is now known that this conjecture is too strong to be true.

The case in which G and G' are connected has been studied in [7]; the object of this paper is to study the case in which G and G' are not necessarily connected.

My renewed interest in this problem was stimulated by conversations with C.B. Thomas. In Thomas' applications, the group G is finite, while the group G' is a classical group. I find the following example attractive.

Example 1.1. G is the binary icosahedral group $SL(2,5)$, while G' is $SU(2)$.

This example arose in Thomas' work, and was also suggested by J. Milnor; I will give a treatment of it below – see Proposition 1.18.

In my original lecture, I gave much time to background exposition and motivation for such problems. There is now less need of this, for Thomas' work has appeared or is appearing [23, 24], and similarly for a later expository lecture by me on this subject [4, 5]. Suffice it to say that it is standard operating procedure for geometers to reduce interesting geometrical questions to problems in homotopy-theory; and when they have done so, the duty of homotopy-theorists becomes obvious.

In this case our duty is to show that the class of maps

$$f\colon BG \to BG'$$

is severely limited in some way which can be calculated. In [7], limitations on f are stated in terms of the induced map of rational cohomology

$$f^*\colon H^*(BG;Q) \leftarrow H^*(BG';Q).$$

This will not serve when G is finite, for then BG has the rational cohomology of a point. In this paper I shall work in terms of the induced map of K-theory [10, 9]

$$f^*\colon K(BG) \leftarrow K(BG').$$

Here $K(X)$ means representable K-theory, that is

$$K(X) = [X, Z \times BU]$$

where $[X, Y]$ means the set of homotopy classes of maps from X to Y. This is the appropriate definition when X is an infinite complex, and BG is usually infinite.

When G and G' are connected, classification according to the induced map

$$f^*\colon K(BG) \leftarrow K(BG')$$

comes to the same thing as the classification in [7]. When G is a finite group, results expressed in terms of $K(BG)$ issue in calculations with the character-table; such calculations tend to be both easier and more useful than calculations with the cohomology of a finite group.

In general, the connection between $K(BG)$ and representation-theory is given by Atiyah's map α [8], as I now explain.

Let G be a topological group, and let

$$\rho\colon G \to U(n)$$

be a representation; then $\alpha(\rho)$ lies in $K(BG)$, and is defined to be the following composite.

$$BG \xrightarrow{B\rho} BU(n) \subset n \times BU \subset Z \times BU$$

We have

$$\alpha(\rho \oplus \sigma) = \alpha(\rho) + \alpha(\sigma).$$

If G is a compact Lie group, we define RG to be the Grothendieck group obtained from the complex representations of G; then α passes to the Grothendieck group to give a homomorphism

$$\alpha\colon RG \to K(BG).$$

We can give RG a product by using the tensor product of representations; in fact RG is usually called the "representation ring" of G. Then α becomes a homomorphism of rings.

Theorem 1.2. *If G and G' are connected then any induced map*

$$f^*\colon K(BG) \leftarrow K(BG')$$

carries

$$\mathrm{Im}\{\alpha'\colon RG' \to K(BG')\}$$

into

$$\mathrm{Im}\{\alpha\colon RG \to K(BG)\}.$$

This is Corollary 1.13 of [7].

Theorem 1.3. *If G is finite then any induced map*

$$f^*\colon K(BG) \leftarrow K(BG')$$

carries $\mathrm{Im}\,\alpha'$ *into* $\mathrm{Im}\,\alpha$.

Results stated without proof in this introduction will be proved later.

Results such as Theorems 1.2 and 1.3 place severe limitations on the class of maps $f\colon BG \to BG'$, and indeed tend to reduce the problems to algebraic calculations. One would like to have a similar result for the general case. (The suggestion that all important groups are either finite or connected can be countered by pointing to the orthogonal group $O(n)$.) Unfortunately one cannot generalise Theorems 1.2 and 1.3 in the obvious way.

Example 1.4. *There is a compact Lie group G and a map*

$$f\colon BG \to BU(2)$$

such that $f^*\alpha'(1)$ *does not lie in* $\mathrm{Im}\,\alpha$.

Here "1" means the identity map $1: U(2) \to U(2)$, so that $\alpha'(1)$ means the element of $K(BU(2))$ given by the map

$$BU(2) \subset 2 \times BU \subset Z \times BU,$$

or equivalently by the universal bundle over $BU(2)$.

In particular, this map f is not of the form $B\theta$, for if we had $f = B\theta$ we should have $f^*\alpha'(1) = \alpha(\theta)$. Of course this is by no means the first example of a map $f\colon BG \to BG'$ which is not of the form $B\theta$, for that honour falls to the maps constructed by Sullivan [22]. However Sullivan's maps fall within the scope of Theorem 1.2; so the map f in Example 1.4 is in a sense more pathological than any of Sullivan's maps; also its construction is more elementary.

Given that we cannot generalise Theorems 1.2 and 1.3 in the form stated, the remedy is to look for related statements which do generalise. It seems that the reasonable way to prove results such as Theorems 1.2 and 1.3 is to characterise $\mathrm{Im}\,\alpha$ in some topological way which is preserved under induced maps f^*; and for this purpose it is reasonable to try the exterior powers λ^i [9]. It is usual to say that an element $x \in K(X)$ is "formally finite-dimensional" if $\lambda^i x = 0$ for i sufficiently large. The reader may wish to know how large is sufficiently large; for this we have the following result.

Lemma 1.5. *Let G be a compact Lie group and let* $x \in K(BG)$ *be an element such that* $\lambda^i x = 0$ *for i sufficiently large. Then the augmentation* εx *is a non-negative integer n,* $\lambda^n x \neq 0$, *and* $\lambda^i x = 0$ *for* $i > n$.

I shall define $FF(X) \subset K(X)$ to be the subgroup generated by the elements which are formally finite-dimensional; thus an element $x \in K(X)$ lies in $FF(X)$ if and only if it can be written as the difference $y-z$ of two elements y, z such that $\lambda^i y=0$ for i sufficiently large and $\lambda^j z=0$ for j sufficiently large. The next two results are now obvious.

Proposition 1.6. *If $f\colon X \to X'$ is a map, then*

$$f^*\colon K(X) \leftarrow K(X')$$

maps $FF(X')$ into $FF(X)$.

Proof. Suppose that $y' \in K(X')$ is formally finite-dimensional, so that $\lambda^i y'=0$ for i sufficiently large, say for $i>N$. Then

$$\lambda^i(f^* y')=f^*(\lambda^i y')=0 \quad \text{for } i>N,$$

so that $f^* y'$ is formally finite-dimensional in $K(X)$.

Of course we are interested in the case $X=BG$, $X'=BG'$.

Proposition 1.7. *In $K(BG)$ we have*

$$\operatorname{Im}\alpha \subset FF(BG).$$

Proof. RG is generated by representations, and for a representation

$$\rho\colon G \to U(n)$$

we have

$$\lambda^i(\alpha\rho)=0 \quad \text{for } i>n.$$

Example 1.4 shows that $\operatorname{Im}\alpha$ and $FF(BG)$ are not equal in general; for the element $f^*\alpha'(1)$ in that example lies in $FF(BG)$ but not in $\operatorname{Im}\alpha$. However $\operatorname{Im}\alpha$ and $FF(BG)$ are equal in important special cases, as will be shown by Theorems 1.8 and 1.10 below. To state the first, let $\pi_0 G$ be the group of components of G; I keep this as standing notation.

Theorem 1.8. *If $\pi_0 G$ is the union of its Sylow subgroups, then $FF(BG)=\operatorname{Im}\alpha$.*

Corollary 1.9. *If $\pi_0 G$ is the union of its Sylow subgroups, then any induced map*

$$f^*\colon K(BG) \leftarrow K(BG')$$

maps $\operatorname{Im}\alpha'$ into $\operatorname{Im}\alpha$.

It is clear that this corollary does follow from the results above; we have

$$\begin{aligned} \operatorname{Im}\alpha' &\subset FF(BG') && \text{by (1.7)}\\ f^*(FF(BG')) &\subset FF(BG) && \text{by (1.6)}\\ FF(BG) &\subset \operatorname{Im}\alpha && \text{by (1.8).}\end{aligned}$$

It is also clear that it gives a strong generalisation of Theorem 1.2.

Theorem 1.10. *If G is a finite group, the $FF(BG) = \operatorname{Im}\alpha$.*

It is clear that Theorem 1.3 follows from (1.7), (1.6) and (1.10), by the same argument as for (1.9).

We may therefore regard proposition 1.6 as a result analogous to Theorems 1.2 and 1.3, but of greater generality. Indeed, I hope to persuade the reader that it is the best generalisation we are likely to get, in view of the counterexamples. However, Proposition 1.6 is a trivial result; to use it, we have to gain control over $FF(BG)$, as in (1.8) and (1.10). The best general result is the next one. To state it, let $\pi_0 \colon G \to \pi_0 G$ be the quotient map; I keep this as standing notation. For each prime p, choose a Sylow p-subgroup S_p of $\pi_0 G$, and let G_p be $\pi_0^{-1} S_p$, the part of G over S_p.

Theorem 1.11. *An element $x \in K(BG)$ lies in $FF(BG)$ if and only if it satisfies the following condition: for each prime p, the image of x in $K(BG_p)$ lies in*

$$\operatorname{Im}\{\alpha_p \colon RG_p \to K(BG_p)\}.$$

We may interpret the condition as saying that "at each prime p, x lies in $\operatorname{Im}\alpha$".

However, the next theorem does more to relate $FF(BG)$ to $\operatorname{Im}\alpha$ in $K(BG)$. For background, one must first appreciate that it is often possible to divide elements of $K(BG)$ by non-zero integers. For example, consider the case in which G is a cyclic group of prime order Z_p; then we may identify $\tilde{K}(BZ)_p)$ with $Z_p^\wedge \otimes \tilde{R}Z_p$, where $Z_p^\wedge$ is the ring of p-adic integers and $\tilde{R}G$ is the augmentation ideal in RG; this is well known, and will be recalled in §2. So in this case we can divide any element of $\tilde{K}(BZ_p)$ by any integer prime to p.

Lemma 1.12. *$K(BG)$ is torsion-free.*

This shows that in $K(BG)$, division by non-zero integers is unique, if possible.

Theorem 1.13. *For each compact Lie group G there is a positive integer n such that $nFF(BG) \subset \operatorname{Im}\alpha$.*

This result contains two points. First, $FF(BG)$ is contained in the rational hull of $\operatorname{Im}\alpha$; secondly, the denominators needed to write elements of $FF(BG)$ in terms of elements of $\operatorname{Im}\alpha$ are bounded. For example, I presume that $n = 2$ will be enough for the group G in (1.4). In any case, the result gives confidence that we can keep $FF(BG)$ under control by pure algebra.

Before proceeding, I should expalin that the subgroup $FF(BG)$ defined above does not coincide with the subgroup $\bar{R}G$ defined in [4, 5]; we have $FF(BG) \subset \bar{R}G$ by Theorem 1.13, but $FF(BG)$ is usually smaller than $\bar{R}G$. The present definition is to be preferred. In particular, all the results about $\bar{R}G$ proposed in [4, 5] become true if $\bar{R}G$ is replaced by $FF(BG)$; in fact they may be found among the results of the present paper. However, Propositions 6 and 7 of [4, 5] are false if $\bar{R}G$ is defined as proposed there; a suitable counterexample is $G = Z_p$, according to the explanation above. Luckily the text of [4, 5] gives fair warning about a page earlier.

The results above complete my general results, but it is also natural to ask if we cannot strengthen some of them. I will present results which do this in particular circumstances, and explain that these results do not hold in general.

Recall that a compact Lie group G is "monogenic" if there is an element $g \in G$ whose powers are dense in G; such an element g is called a "generator". A group G

is monogenic if and only if it is the product $T \times Z_n$ of a torus T and a finite cyclic group Z_n.

Theorem 1.14. *If G is a monogenic group, $x \in K(BG)$ and $\lambda^i x = 0$ for i sufficiently large, then $x = \alpha\rho$ for some honest (positive) representation ρ of G.*

Here we need to stress the distinction between an honest (positive) representation and a general element of RG; it is convenient to refer to the latter as a "virtual representation".

Theorem 1.14 is recommended for use in practical calculations; the strength of the conclusion often outweighs the nuisance of working in $\prod_H K(BH)$, where H runs over sufficiently many monogenic subgroups of the given group G.

Unfortunately a result with this conclusion can only be true for a small class of groups G.

Example 1.15. *There is a compact Lie group G and an element $x \in K(BG)$ which have the following properties.* (i) *G is a finite p-group, so that both* (1.10) *and* (1.8) *apply at once.* (ii) *$\lambda^i x = 0$ for i sufficiently large, but x is not of the form $\alpha\rho$ for any honest (positive) representation ρ.*

Next, recall that the total exterior power λ_t is defined by

$$\lambda_t(x) = \sum_{i \geq 0} \lambda^i(x)\, t^i;$$

if $x \in K(X)$, then $\lambda_t(x)$ lies in the ring of formal power-series $K(X)[[t]]$, where t is a new variable introduced for the purpose. Suppose that $x \in FF(BG)$, so that $x = y - z$ where y and z are formally finite-dimensional; then $\lambda_t(x)$ is a rational function of t, in the sense that $\lambda_t(x) = f(t)/g(t)$ where $f(t)$ and $g(t)$ are polynomials in t and $g(t)$ is invertible in $K(BG)[[t]]$; for it is sufficient to take $f(t) = \lambda_t(y)$, $g(t) = \lambda_t(z)$. We can use this remark to characterise $FF(BG)$ when G is finite.

Theorem 1.16. *If G is finite, $x \in K(BG)$ and $\lambda_t(x)$ is a rational function of t, then $x \in \operatorname{Im}\alpha$.*

This result implies Theorem 1.10, in view of the remarks above.

Unfortunately, we cannot in general characterise $FF(BG)$ in this way.

Example 1.17. *There is a compact Lie group G and an element $x \in K(BG)$ with the following properties.* (i) *$\pi_0 G$ is a p-group, so that* (1.8) *applies.* (ii) *$\lambda_t(x)$ is a rational function of t, but we do not have $nx \in \operatorname{Im}\alpha$ for any positive integer n.*

In particular, $x \notin FF(BG)$, either by Theorem 1.8 or by Theorem 1.13.

I now turn back to Example 1.1, and so I take $G' = SU(2)$. Let us admit homomorphisms into $SU(2)$ as "representations" (by omitting notation for the inclusion $SU(2) \subset U(2)$), and so write $\alpha'(1) \in K(BSU(2))$ for the element represented by the map

$$BSU(2) \subset 2 \times BU \subset Z \times BU,$$

or equivalently, by the universal bundle over $BSU(2)$. For any map

$$f\colon BG \to BSU(2)$$

the induced homomorphism

$$f^*: K(BG) \leftarrow K(BSU(2))$$

is determined by its value on $\alpha'(1)$, and this must be an element $x \in K(BG)$ of "formal dimension 2" in the sense that $\varepsilon x = 2$ and $\lambda^i x = 0$ for $i > 2$.

I take G to be the binary icosahedral group $SL(2, 5)$. Let i: $SL(2, 5) \to SU(2)$ be a fixed choice of one of the two standard embeddings; then $\alpha(i) \in K(BSL(2, 5))$.

Proposition 1.18. (i) *The elements of formal dimension* 2 *in* $K(BSL(2, 5))$ *are precisely the elements* $\Psi^k(\alpha i)$, $k \in Z$, *where* Ψ^k *is as in* [1].

(ii) *Two elements* $\Psi^k(\alpha i)$ *and* $\Psi^l(\alpha i)$ *are equal if and only if they have the same second Chern class, that is, if and only if* $k^2 \equiv l^2 \bmod 120$.

(iii) *The elements which arise as* $f^* \alpha'(1)$ *for maps* f *are precisely the elements* $\Psi^k(\alpha i)$ *with* $k \not\equiv 2 \bmod 4$.

(iv) *The elements which arise as* $f^* \alpha'(1)$ *for maps* f *of the form* $B\theta$ *are those with invariants* $k^2 \equiv 0$, 1 *and* 49 mod 120.

I explain that there is another embedding j: $SL(2, 5) \to SU(2)$ besides the one which was chosen as i; this gives the map with invariant $k^2 \equiv 49 \bmod 120$.

To summarise, the maps f: $BSL(2, 5) \to BSU(2)$ can be classified (according to their effect on K-theory) into precisely 12 non-empty classes; of these 3 contain maps $B\theta$ and 9 do not.

This completes my statement of results. As for their proofs, I have tried to make the exposition which follows as complete, selfcontained and elementary as I can. In particular, the results for a connected group G follows from the proofs needed for the general case; I do not assume the results of [7]. The argument of this paper, unlike that of [7], does not require any major result of analytic number-theory; and unlike that of [13], it does not require the Hilbert irreducibility theorem.

In fact, all the results in this paper are proved by one method: I approximate a compact Lie group G by its finite subgroups. This may seem a very dubious expedient, but it works; Lie theorists who doubt it are referred to Proposition 7.3; topologists should be convinced by §2.

In greater detail, the rest of this paper is organised as follows. In representation-theory, the theory of "group characters" provides invariants for elements of RG; I shall provide similar invariants for elements $x \in K(BG)$; this is done in §2, and almost everything which follows is based on it. These invariants in general take values not in the complex field C, but in an extension of C; it is therefore necessary to prove that in particular cases they take values in C; this will be done in §3, and in this section we also complete the theory for finite groups, the main task being to prove (1.16). This gives us enough results to discuss the examples; accordingly, §4 is devoted to a discussion of the examples promised above, that is, (1.1)=(1.18), (1.4), (1.15) and (1.17). The reader who only wants the case in which G is finite can stop at this point. §6 is devoted to monogenic groups, and proves (1.14). This section is based on §5, in which I approximate a monogenic group by its finite subgroups; this section contains a rather brutal bare-hands argument which is the crux of the whole paper. § 7 is a sort of service section about the structure and representations of Lie groups which need not be connected. §8 completes the proof of the main results, notably (1.8), (1.11) and (1.13); this section is not too long, but it rests on almost everything that precedes it. In each section, the results which will be quoted in later sections are collected at the beginning for ease of reference.

It is a pleasure acknowledge helpful conversations and correspondence with M.F. Atiyah, C.R. Curtis, N. Katz, J. Milnor, P.M. Neumann, C.B. Thomas and J.G. Thompson.

§2. Characters

The object of this section is as stated in §1; I shall define invariants of elements $x \in K(BG)$, similar to those which the theory of "group characters" provides for elements of RG. I shall also prove such properties of these invariants as are needed later.

Let p be a prime, let $Z_p^\wedge$ be the ring of p-adic integers, and let $K(X; Z_p^\wedge)$ be K-theory with coefficients in $Z_p^\wedge$, as defined for example in [3] pp. 200–203. My invariants will actually be defined on elements of $K(BG; Z_p^\wedge)$; of course we can consider them as defined on elements of $K(BG)$, by applying the obvious map

$$K(BG) \to K(BG; Z_p^\wedge).$$

We need to carry over to $K(BG;\, Z_p^\wedge)$ some of what we said for $K(BG)$. Let G be a compact Lie group; then since $K(BG;\, Z_p^\wedge)$ is a module over $Z_p^\wedge$, there is a unique $Z_p^\wedge$-linear map

$$\alpha^\wedge : Z_p^\wedge \otimes RG \to K(BG;\, Z_p^\wedge)$$

which makes the following diagram commutative.

$$\begin{array}{ccc} RG & \xrightarrow{\alpha} & K(BG) \\ \downarrow & & \downarrow \\ Z_p^\wedge \otimes RG & \xrightarrow{\alpha^\wedge} & K(BG;\, Z_p^\wedge) \end{array}$$

Here the vertical arrows are induced by the inclusion $Z \subset Z_p^\wedge$, of course.

We shall sometimes need to topologise the groups $K(BG)$ and $K(BG;\, Z_p^\wedge)$, and of course we use the "filtration" topology, in which an element is close to zero if it restricts to zero on a skeleton $(BG)^n$.

Lemma 2.1. *Assume that G is a finite p-group. Then the map*

$$\alpha^\wedge : Z_p^\wedge \otimes RG \to K(BG;\, Z_p^\wedge)$$

is iso; the map

$$\tilde{K}(BG) \to \tilde{K}(BG;\, Z_p^\wedge)$$

is both iso and a homeomorphism; and finally, the filtration topology on $\tilde{K}(BG;\, Z_p^\wedge)$ coincides with the p-adic topology.

This is reformulation of a well-known result of Atiyah [8].

Next I explain the nature of my invariants. In a character table, the "character" appears as a complex-valued function $\chi(\rho, g)$ of two variables, a representation ρ and a point $g \in G$. One can also read each row as a function.

$$\chi_\rho \colon G \to C$$

from G to the complex field C, and this is the way one usually thinks of "the character of ρ"; but equally one can read each column as a function

$$\chi_g \colon RG \to C,$$

and this is a homomorphism of rings. I shall follow the latter pattern.

I shall say that an element $g \in G$ is "of p-power order" if $g^{p^e} = 1$ for some power p^e of p. For each such element I shall define a homomorphism of rings

$$\chi_g^\wedge \colon K(BG; Z_p^\wedge) \to Z_p^\wedge \otimes C$$

(where the tensor product is taken over Z). The existence, uniqueness and formal properties of these homomorphisms are stated in the next result.

Proposition 2.2. *There is one and only one system of $Z_p^\wedge$-algebra maps*

$$\chi_g^\wedge \colon K(BG; Z_p^\wedge) \to Z_p^\wedge \otimes C$$

(defined for all topological groups G and all elements $g \in G$ of p-power order) which have the following two properties.

(i) *For each compact Lie group G the following diagram is commutative.*

$$\begin{array}{ccc} Z_p^\wedge \otimes RG & \xrightarrow{1 \otimes \chi_g} & \\ \downarrow{\scriptstyle \alpha^\wedge} & & Z_p^\wedge \otimes C \\ K(BG; Z_p^\wedge) & \xrightarrow{\chi_g^\wedge} & \end{array}$$

(ii) *For each map $\theta \colon G \to G'$ of topological groups and each element $g \in G$ of p-power order the following diagram is commutative.*

$$\begin{array}{ccc} K(BG; Z_p^\wedge) & \xleftarrow{(B\theta)^*} & K(BG'; Z_p^\wedge) \\ & {\scriptstyle \chi_g^\wedge} \searrow \quad \swarrow {\scriptstyle \chi_{\theta g}^\wedge} & \\ & Z_p^\wedge \otimes C & \end{array}$$

Moreover, the unique system of such maps $\chi_g^\wedge$ has the following further property.

(iii) *For each g of p-power order in G and each h in G we have*

$$\chi_g^\wedge = \chi_{hgh^{-1}}^\wedge.$$

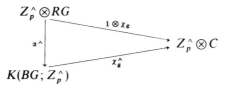

Let me summarise and comment on the differences between these invariants $\chi_g^\wedge$ and the classical group characters χ_g.

(i) χ_g is defined for all g in G, but $\chi_g^\wedge$ is only defined for elements of prime-power order.

There is no help for this. In fact, an element $\xi \in RG$ is determined by giving $\chi_g(\xi)$ for all g in G; so if we could define $\chi_g^\wedge(x)$ for all g in G, it would follow that

$$\alpha: RG \to K(BG)$$

would be mono; but we know examples in which α is not mono. Results such as (2.3) below assure us that we have defined $\chi_g^\wedge$ for enough points g.

(ii) χ_g takes values in C, but $\chi_g^\wedge$ takes values in $Z_p^\wedge \otimes C$.

There is no help for this either, even if we only want invariants defined on $K(BG)$; one can convince oneself of this by studying special cases such as $G = Z_p$, the cyclic group of order p.

(iii) χ_g is defined on a Z-module, but $\chi_g^\wedge$ is defined on the $Z_p^\wedge$-module $K(BG; Z_p^\wedge)$ rather than on $K(BG)$.

Once we accept values in $Z_p^\wedge \otimes C$, as we must, then there is no loss and some gain in getting $\chi_g^\wedge$ defined on more elements x.

The following result assures us that the invariants $\chi_g^\wedge$ are sufficient for our purposes.

Theorem 2.3. *Let G be a compact Lie group and $x \in K(BG)$. If $\chi_g^\wedge(x) = 0$ for all g of prime-power order, then $x = 0$; moreover if $\pi_0 G$ is a p-group, then it is sufficient to assume that $\chi_g^\wedge(x) = 0$ for all g of p-power order.*

There is a corresponding result if the order of $\pi_0 G$ is divisible by two or more primes; but as I shall not need it I shall omit it.

Corollary 2.4. *If an element $x \in K(BG)$ satisfies the condition given in Theorem* 1.11, *then $x \in FF(BG)$.*

This completes the statement of the results which will be quoted in later sections; but in §7 I shall also refer to the construction used in proving Lemma 2.6 below.

Proof of Lemma 2.1. First I show that $\alpha^\wedge$ is iso; this is easily proved using the method given by Atiyah in [8] §10. If $G = 1$, then BG is contractible and the result is clearly true. If $G = Z_p$, then $K(BG; Z_p^\wedge)$ may be calculated from the Atiyah-Hirzebruch spectral sequence

$$H^*(B(Z_p); Z_p^\wedge) \Rightarrow K^*(B(Z_p); Z_p^\wedge)$$

(for which see [8] §5); the behaviour of this spectral sequence can easily be calculated (as in [8] §8), and one obtains the result stated for $G = Z_p$.

To handle the general case we proceed by induction. If G is a finite p-group and $G \neq 1$ we may find a normal subgroup H such that G/H is cyclic of order p, and we may assume as an inductive hypothesis that the result is true for H. As in [8] §5 we have an obvious spectral sequence

$$H^*(B(G/H), K^*(BH; Z_p^\wedge)) \Rightarrow K(BG; Z_p^\wedge).$$

(I stress that here the cohomology of $B(G/H)$ is taken with twisted coefficients.) This spectral sequence relates $K^*(BH; Z_p^\wedge)$ to $K^*(BG; Z_p^\wedge)$; we also need to relate RH to RG.

Let the inequivalent irreducible representations of G/H be $\zeta, \zeta^2, \zeta^3, \ldots, \zeta^p = 1$; they form a cyclic group of order p (under multiplication); this group acts (by multiplication) on the irreducible representations of G, and so the irreducible representations of G fall into orbits of two types.

(i) Orbits containing one element ξ, and

(ii) orbits containing p elements $\zeta\eta, \zeta^2\eta, \zeta^3\eta, \ldots, \zeta^p\eta = \eta$.

Similarly, the cyclic group G/H acts by conjugation on H and so acts on the irreducible representations of H, and the representations of H fall into orbits of two types.

(i′) Orbits containing p elements $\xi'_1, \xi'_2, \ldots, \xi'_p$, and

(ii′) orbits containing one element η'.

The operations of restriction and induction give a (1–1) correspondence between orbits of type (i) for G and orbits of type (i′) for H; they also give a (1–1) correspondence between orbits of type (ii) for G and orbits of type (ii′) for H. Readers interested in representation-theory should be able to prove these facts for themselves if they have not seen them before; alternatively, see [8] §9.

The irreducible representations of H form a base for RH; so when we partition them into orbits we define a direct-sum splitting of $Z_p^\wedge \otimes RH$ compatible with the action of G/H, and thus we obtain a direct-sum splitting of

$$H^*(B(G/H); Z_p^\wedge \otimes RH).$$

I will show that in fact we obtain a splitting (into known parts) of the whole spectral sequence

$$H^*(B(G/H); K^*(BH; Z_p^\wedge)) \Rightarrow K^*(BG; Z_p^\wedge).$$

For each orbit of type (i′) the summand

$$H^*(B(G/H);\ Z_p^\wedge \xi'_1 \otimes Z_p^\wedge \xi'_2 \otimes \cdots \oplus Z_p^\wedge \xi'_p)$$

is $Z_p^\wedge$ in degree 0 and zero in all other degrees; the corresponding summand $Z_p^\wedge\ \xi$ in $Z_p^\wedge \otimes RG$ maps exactly to this part of our E_2-term.

For each orbit of type (ii′), let η be as above; then multiplication with η gives a homomorphism from the spectral sequence

$$H^*(B(G/H); Z_p^\wedge) \Rightarrow K^*(B(G/H); Z_p^\wedge)$$

to our spectral sequence

$$H^*(B(G/H); K(BH; Z_p^\wedge)) \Rightarrow K^*(BG; Z_p^\wedge).$$

Moreover, on the E_2-terms we get an isomorphism from the E_2-term of the former spectral sequence to the summand

$$H^*(B(G/H); Z_p^\wedge \eta')$$

of the E_2-term of the latter.

It follows from all this that the cohomology groups

$$H^*(B(G/H); K^*(BH; Z_p^\wedge))$$

are concentrated in even degrees; therefore our spectral sequence collapses. (Incidentally, this ensures that we have no trouble with its convergence.) This completes the proof that our spectral sequence splits as the sum of known parts. More precisely, the known parts are copies of

(i) $Z_p^\wedge$ concentrated in bidegree $(0, 0)$, and
(ii) the standard spectral sequence

$$H^*(B(Z_p); Z_p^\wedge) \Rightarrow Z_p^\wedge \otimes R(Z_p).$$

But by our account of the relation between RG and RH, this sum of known spectral sequences converges to $Z_p^\wedge \otimes RG$. Thus the map

$$\alpha^\wedge : Z_p^\wedge \otimes RG \to K(BG; Z_p^\wedge)$$

is iso; this competes the induction, and proves the first assertion of Lemma 2.1.

I turn to the second assertion. If G is finite and a p-group, then each homology group $\tilde{H}_i(BG)$ is finite and a p-group; it follows that the map

$$\tilde{H}^*(BG) \to \tilde{H}^*(BG; Z_p^\wedge)$$

is iso. We now see that the map $Z \to Z_p^\wedge$ induces an isomorphism from the Atiyah-Hirzebruch spectral sequence

$$\tilde{H}^*(BG) \Rightarrow \tilde{K}^*(BG)$$

to the Atiyah-Hirzebruch spectral sequence

$$\tilde{H}^*(BG; Z_p^\wedge) \Rightarrow \tilde{K}^*(BG; Z_p^\wedge);$$

so the map

$$\tilde{K}(BG) \to \tilde{K}(BG; Z_p^\wedge)$$

is both iso and a homeomorphism.

Finally I have to compare the two topologies on $\tilde{K}(BG; Z_p^\wedge)$, and I claim that the map from the p-adic topology to the filtration topology is continuous. In fact, consider the image of

$$\tilde{K}(BG; Z_p^\wedge) \to \tilde{K}((BG)^n; Z_p^\wedge);$$

the order of this image is clearly finite and a power of p, say p^ν; so by requiring that $x \in \tilde{K}(BG; Z_p^\wedge)$ is congruent to $0 \bmod p^\nu$, we can ensure that x maps to 0 in $\tilde{K}((BG)^n; Z_p^\wedge)$.

Since RG is finitely-generated the p-adic topology on $Z_p^\wedge \otimes RG$ is compact; since $\alpha^\wedge$ is iso, the p-adic topology on $\tilde{K}(BG; Z_p^\wedge)$ is compact. And since $\tilde{K}((BG)^n; Z_p^\wedge)$ is an inverse sequence of compact groups and continuous homomorphisms, its $\varprojlim^1$ is 0; that is, the filtration topology on $\tilde{K}(BG; Z_p^\wedge)$ is Hausdorff. Since the map from the p-adic topology to the filtration topology is continuous, it must be a homeomorphism. This completes the proof of Lemma 2.1.

In proving Proposition 2.2 we shall use only the first assertion of Lemma 2.1, and this only for the case in which G is a finite cyclic p-group; and of course we could obtain this by direct calculation. The rest of Lemma 2.1 is for use later.

Proof of Proposition 2.2. First consider the case in which G is a finite cyclic p-group. Then the map $\alpha^\wedge$ is iso by Lemma 2.1, and so there is a unique map $\chi_g^\wedge$ such that the diagram of (2.2) (i) is commutative.

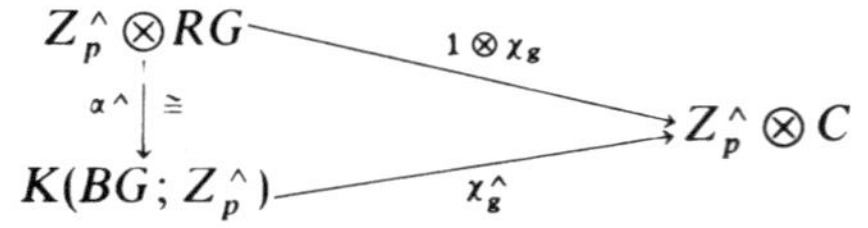

These maps satisfy (2.2) (ii) when θ is a map between finite cyclic p-groups.

Now let G' be a general topological group, and let g' be an element of order p^e in it; let $G = Z_{p^e}$ with generator g, and let $\theta\colon G \to G'$ be a map such that $\theta g = g'$. Then $\chi_g^\wedge$ is given by our first step, and there is a unique map $\chi_{\theta g}^\wedge$ such that the diagram of (2.2) (ii) is commutative.

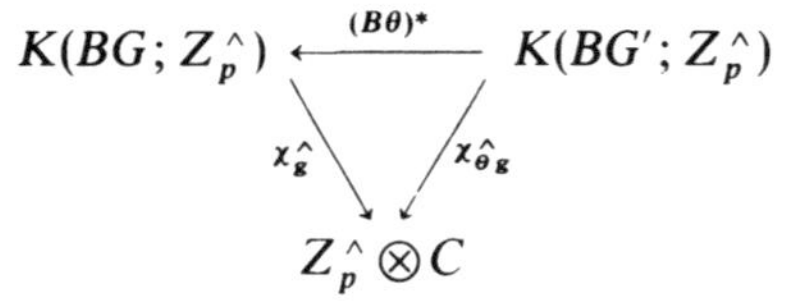
$K(BG; Z_p^\wedge)$
$(B\theta)^*$
$K(BG'; Z_p^\wedge)$
$\chi_g^\wedge$
$\chi_{\theta g}^\wedge$
$Z_p^\wedge \otimes C$

This proves the uniqueness statement in (2.2). If we adopt the last diagram as our definition of $\chi_{\theta g}^\wedge$, then it has the required properties (i), (ii); this proves the existence.

Finally, let G' be a general group, and let g', g'' be two conjugate elements of order p^e in G'; let

$$\theta', \theta''\colon Z_{p^e} \to G'$$

be maps such that $\theta' g = g'$, $\theta'' g = g''$. Then the maps $B\theta', B\theta''\colon BG \to BG'$ are homotopic, and hence $\chi_{g'}^\wedge = \chi_{g''}^\wedge$. This completes the proof of Proposition 2.2.

We turn to Theorem 2.3, but this will take several steps.

Lemma 2.5. *Let G be a finite p-group, let $x \in K(BG; Z_p^\wedge)$, and suppose $\chi_g^\wedge(x) = 0$ for all $g \in G$; then $x = 0$.*

Proof. Using the diagram of (2.2) (i), we obtain the following commutative diagram, in which the map $\alpha^\wedge$ is iso by Lemma 2.1.

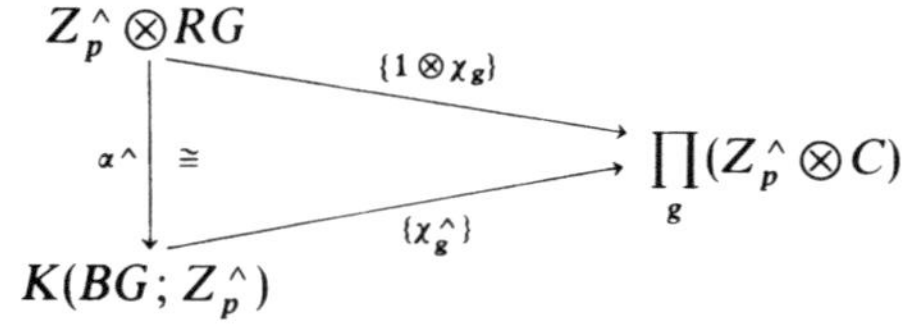

Here $\{1 \otimes \chi_g\}$ means the map whose g^{th} component is $1 \otimes \chi_g$, and similarly for $\{\chi_g^\wedge\}$. Since G is finite we have $\prod_g = \sum_g$, and we can identify $\prod_g (Z_p^\wedge \otimes C)$ with $Z_p^\wedge \otimes (\prod_g C)$;

this identifies the map $\{1\otimes\chi_g\}$ with $1\otimes\chi$, where

$$\chi\colon RG \to \prod_{g\in G} C$$

is the map with components χ_g. Now χ is mono by standard results of representation-theory, and since $Z_p^\wedge$ is torsion-free it follows that

$$1\otimes\chi\colon Z_p^\wedge\otimes RG \to Z_p^\wedge\otimes(\prod_g C)$$

is mono. This proves Lemma 2.5.

Lemma 2.6. *Suppose that the identity-component of G is a torus T, that $\pi_0 G$ is a p-group, that $x\in K(BG; Z_p^\wedge)$ and that $\chi_g^\wedge(x)=0$ for all g of p-power order in G. Then* $x=0$.

Proof. I shall approximate G by finite p-groups, and so construct a geometrical approximation to BG. I write G_1 for the identity-component of G, and keep this as standing notation. Since we assume $G_1=T$, we have an extension

$$T\to G\to\pi_0 G;$$

we may consider it as an element of $H^2(\pi_0 G; T)$; by standard transfer arguments, this element is annihilated by the order of $\pi_0 G$, say p^ν. We have a short exact sequence

$$0\to \operatorname{Ker} p^\nu \to T \xrightarrow{p^\nu} T\to 0$$

of additive groups acted on by $\pi_0 G$; it induces a long exact cohomology sequence, and we see that our extension comes from $H^2(\pi_0 G; \operatorname{Ker} p^\nu)$. We can now obtain a diagram of extensions of the following form.

$$\begin{array}{ccccccc}
\cdots\to & \operatorname{Ker} p^\mu & \longrightarrow & \operatorname{Ker} p^{\mu+1} & \to\cdots\to & T \\
 & \downarrow & & \downarrow & & \downarrow \\
\cdots\to & G_\mu & \longrightarrow & G_{\mu+1} & \to\cdots\to & G \\
 & \downarrow & & \downarrow & & \downarrow \\
\cdots\to & \pi_0 G & \xrightarrow{1} & \pi_0 G & \to\cdots\to & \pi_0 G
\end{array}$$

Here μ runs over the range $\mu\geqq\nu$, and each group G_μ is a finite p-group. Let us write G_∞ for the union of the subgroups G_μ, taken with the discrete topology; then we have the following diagram of extensions.

$$\begin{array}{ccc}
T_\infty & \longrightarrow & T \\
\downarrow & & \downarrow \\
G_\infty & \longrightarrow & G \\
\downarrow & & \downarrow \\
\pi_0 G & \xrightarrow{1} & \pi_0 G
\end{array}$$

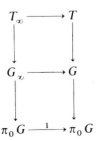

Here T_∞ is the p-torsion subgroup of T. Every element of G_∞ is of p-power order, but it does not follow that all the elements of p-power order in G lie in G_∞; indeed, it can be shown by examples that they need not form a subgroup.

It is clear that a similar construction works even if $\pi_0 G$ is not a p-group; we can replace the sequence of integers..., $p^\mu, p^{\mu+1}, \ldots$ by any sequence of integers we wish, provided that each divides the next and the first is divisible by the order of $\pi_0 G$. I shall need such a construction in § 7.

Let us return to the business in hand; we can realise BG_∞ as the union of an increasing sequence of finite complexes X_μ such that the injection $X_\mu \to BG_\infty$ factors through BG_μ. For this purpose we need only use a suitable "diagonal process"; for example, we take X_μ equivalent to the μ-skeleton of a suitable complex BG_μ.

Now assume that $x \in K(BG; Z_p^\wedge)$ and $\chi_g^\wedge(x) = 0$ for all g of p-power order in G. Then by Lemma 2.5 applied to the subgroup G_μ, x maps to zero in $K(BG_\mu; Z_p^\wedge)$; hence x maps to zero in $K(X_\mu; Z_p^\wedge)$. The inverse sequence

$$\cdots \leftarrow K^*(X_\mu; Z_p^\wedge) \leftarrow K^*(X_{\mu+1}; Z_p^\wedge) \leftarrow \cdots$$

is a sequence of compact groups and continuous homomorphisms; therefore its $\varprojlim^1$ is zero, and the map

$$\varprojlim_\mu K(X_\mu; Z_p^\wedge) \leftarrow K(BG_\infty; Z_p^\wedge)$$

is iso. So we see that x maps to zero in $K(BG_\infty; Z_p^\wedge)$.

Next I claim that the map

$$H^*(BG_\infty; Z_p^\wedge) \leftarrow H^*(BG; Z_p^\wedge)$$

is iso. If G is a circle S^1, then this can be obtained by direct calculation. If G is a torus T, one proves it by induction over the dimension d of the torus, using the Serre spectral sequences of the product fiberings in the following diagram.

$$\begin{array}{ccc} BT^1_\infty & \longrightarrow & BT^1 \\ \downarrow & & \downarrow \\ BT^d_\infty & \longrightarrow & BT^d \\ \downarrow & & \downarrow \\ BT^{d-1}_\infty & \longrightarrow & BT^{d-1} \end{array}$$

Finally one obtains the general case by the same argument applied to the fiberings in the following diagram.

$$\begin{array}{ccc} BT_\infty & \longrightarrow & BT \\ \downarrow & & \downarrow \\ BG_\infty & \longrightarrow & BG \\ \downarrow & & \downarrow \\ B\pi_0 G & \xrightarrow{1} & B\pi_0 G \end{array}$$

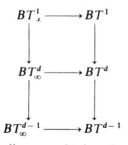

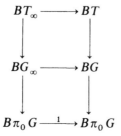

Using the Atiyah-Hirzebruch spectral sequence, we see that the map

$$K^*(BG_\infty; Z_p^\wedge) \leftarrow K^*(BG; Z_p^\wedge)$$

is iso. Therefore we have $x=0$; and this proves Lemma 2.6.

Lemma 2.7. *Suppose that G is a compact Lie group, $x \in K(BG;\ Z_p^\wedge)$ and $\chi_g^\wedge(x)=0$ for all g of p-power order in G; then $x=0$.*

Proof. Let T be a maximal torus in the identity-component G_1 of G, and let NT be the normaliser of T in G; I claim that $\pi_0 NT$ maps onto $\pi_0 G$. To prove it, take an element of $\pi_0 G$ and represent it by an element $g \in G$. Then gTg^{-1} is another maximal torus in G_1; so by the standard theory of maximal tori in connected groups, there is an element $g_1 \in G_1$ such that

$$g_1 g T g^{-1} g_1^{-1} = T;$$

then $g_1 g$ lies in NT and maps to the required element of $\pi_0 G$.

It is now clear that we have a homeomorphism

$$G_1/(G_1 \cap NT) \xrightarrow{\cong} G/NT.$$

But by the standard theory for connected groups, the Euler-Poincaré characteristic $\chi(G_1/(G_1 \cap NT))$ is 1; therefore $\chi(G/NT)=1$.

Now choose a Sylow p-subgroup S in $\pi_0 NT$, and let $H = \pi_0^{-1} S \subset NT$. Then we have a fibering

$$NT/H \to G/H \to G/NT,$$

that is, a finite covering of degree $|NT/H| = |\pi_0 NT/S|$; this is an integer q prime to p, by construction. So the Euler characteristic $\chi(G/H)$ is q and is prime to p.

Take the fibering

$$G/H \to BH \to BG$$

and apply the Becker-Gottlieb transfer [12, 6], suitably rewritten so as to apply to infinite complexes. We get the following commutative diagram.

$$\begin{array}{ccc} K^*(BG; Z_p^\wedge) & \xleftarrow{Tr} & \\ \uparrow \cong & & K^*(BH; Z_p^\wedge) \\ K^*(BG; Z_p^\wedge) & \xrightarrow{i^*} & \end{array}$$

The vertical arrow is iso because $\chi(G/H)$ is invertible in $Z_p^\wedge$. This proves that i^* is split mono.

Now let $x \in K(BG; Z_p^\wedge)$ be an element such that $\chi_g^\wedge(x)=0$ for all g of p-power order in G. Lemma 2.6 applies to H (because the identity-component of NT is T, by the standard theory for connected groups). So $i^* x = 0$ in $K(BH; Z_p^\wedge)$. Since i^* is split mono this gives $x=0$, which proves Lemma 2.7.

If the reader does not care for this application of Becker-Gottlieb transfer to infinite complexes, he is welcome to go through the usual irrelevant reduction to the case of finite complexes. This presents no essential difficulty since we can write $K^*(BG; Z_p^\wedge)$ in the form $\varprojlim_\alpha K^*(X_\alpha; Z_p^\wedge)$ with X_α finite.

Lemma 2.8. *Let G be a compact Lie group. If $x \in K(BG)$ maps to 0 in $K(BG; Z_p^\wedge)$ for all p, then $x=0$.*

Proof. Consider the exact sequence

$$0 \to Z \xrightarrow{i} \prod_p Z_p^\wedge \to \left(\prod_p Z_p^\wedge\right)/Z \to 0.$$

The quotient $(\prod_p Z_p^\wedge)/Z$ is a vector space over Q. For a compact Lie group G, $H_*(BG; Q)$ is concentrated in even dimensions; so by the Universal Coefficient Theorem, $H^*(BG; (\prod_p Z_p^\wedge)/Z)$ is concentrated in even dimensions; so by the Atiyah-Hirzebruch spectral sequence, $K^*(BG; (\prod_p Z_p^\wedge)/Z)$ is concentrated in even dimensions. Now the exact sequence

$$K^{-1}\left(BG; \left(\prod_p Z_p^\wedge\right)/Z\right) \longrightarrow K(BG; Z) \xrightarrow{i_*} K\left(BG; \prod_p Z_p^\wedge\right)$$

shows that i_* is mono. We have

$$\prod_p K(BG; Z_p^\wedge) \cong K\left(BG; \prod_p Z_p^\wedge\right)$$

(cohomology passes to products though it does not pass to sums). So the map

$$K(BG) \to \prod_p K(BG; Z_p^\wedge)$$

is mono. This proves Lemma 2.8.

Proof of Theorem 2.3. Suppose that $x \in K(BG)$ and $\chi_g^\wedge(x)=0$ for all g of prime-power order in G; then the image of x in $K(BG; Z_p^\wedge)$ is 0 by Lemma 2.7, and this holds for all p; so $x=0$ by Lemma 2.8. This proves the first assertion in (2.3).

To go further, I begin by assuming that G is connected. Let T be a maximal torus in G; I claim that the map $K(BG) \to K(BT)$ is mono. This is well known, and it also follows from the result just proved. In fact, I suppose that $x \in K(BG)$ maps to 0 in $K(BT)$; then $\chi_t^\wedge(x)=0$ for all points t of prime-power order in T; so $\chi_g^\wedge(x)=0$ for all points g of prime-power order in G, for any such point is conjugate to a point in T; so $x=0$ by the part of (2.3) just proved.

For any prime p the map $K(BT) \to K(BT; Z_p^\wedge)$ is mono, for by the Atiyah-Hirzebruch spectral sequence this map is the obvious inclusion

$$Z[[x_1, x_2, \ldots, x_d]] \to Z_p^\wedge[[x_1, x_2, \ldots, x_d]]$$

of formal power-series rings.

Suppose then that G is connected, $x \in K(BG)$ and $\chi_g^\wedge(x)=0$ for all $g \in G$ of p-power order; then x maps to 0 in $K(BT; Z_p^\wedge)$ by Lemma 2.7, and so $x=0$ by the two paragraphs above. This proves the second assertion of (2.3) when G is connected.

Suppose now that $\pi_0 G$ is a p-group, $x \in K(BG)$ and $\chi_g^\wedge(x)=0$ for all g of p-power order in G. Then the image of x in $K(BG_1)$ is 0 by the special case just proved. Let $g \in G$ be an element of q-power order for some prime $q \neq p$; then the image of g in $\pi_0 G$ is an element of q-power order in a p-group, so it is 1; that is, $g \in G_1$, and so $\chi_g^\wedge(x)=0$. This proves that $\chi_g^\wedge(x)=0$ for all g of prime-power order in G; and so $x=0$, by the first assertion of (2.3). This proves the second assertion of (2.3).

Proof of Corollary 2.4. Let the subgroups $G_p \subset G$ be as in Theorem 1.11; first I claim that the map

$$K(BG) \to \prod_p K(BG_p)$$

is mono. In fact, let g be an element of p-power order in G; then its image in $\pi_0 G$ is also of p-power order, and hence conjugate to an element of the Sylow subgroup S_p; so g is conjugate to an element $h \in G_p$. If we assume that $x \in K(BG)$ maps to 0 in $K(BG_p)$ for each p, then $\chi_h^\wedge(x)=0$ for each such h, so $\chi_g^\wedge=0$ for each such g, and so $x=0$ by Theorem 2.3.

In order for the map

$$K(BG) \to \prod_p K(BG_p)$$

to be mono, it is sufficient to run p over a finite set P of primes, because G_p reduces to the identity-component of G for all but a finite number of primes.

Now suppose that an element $x \in K(BG)$ satisfies the condition given in (1.11), so that for each prime p the image of x in $K(BG_p)$ lies in

$$\operatorname{Im}\{\alpha_p\colon RG_p \to K(BG_p)\}.$$

If we write $i_p\colon BG_p \to BG$ for the inclusion, we may suppose

$$i_p^* x = \alpha(\rho_p - \sigma_p)$$

where ρ_p and σ_p are representations of G_p. For each p we can find a representation θ_p of G such that

$$i_p^* \theta_p = \sigma_p + \phi_p$$

where ϕ_p is an honest (positive) representation; for example, since G_p is of finite index in G, we can take θ_p to be the representation of G induced from the representation σ_p of the subgroup G_p. We can even find a representation θ which works for all primes p in the finite set P; for example, we can take

$$\theta = \sum_{p \in P} \theta_p.$$

Then we have

$$i_p^* \theta = \sigma_p + \psi_p$$

for each p in P, where ψ_p is an honest (positive) representation.

Consider now the element

$$y = x + \alpha\theta \in K(BG).$$

For each p in P we get

$$i_p^* y = \alpha(\rho_p + \psi_p)$$

where $\rho_p + \psi_p$ is an honest (positive) representation. Thus

$$\lambda^k(i_p^* y) = 0$$

for k sufficiently large, in fact for $k > \varepsilon y$ (independent of p). That is,

$$i_p^*(\lambda^k y) = 0$$

for all $p \in P$, $k > \varepsilon y$. By the first two paragraphs, this shows that $\lambda^k y = 0$ for $k > \varepsilon y$; thus $x = y - \alpha\theta$ lies in $FF(BG)$. This proves Corollary 2.4.

Proof of Lemma 1.12. Suppose $x \in K(BG)$ and $nx = 0$ for some integer $n \neq 0$. For any g of p-power order in G we have

$$n\chi_g^\wedge(x) = \chi_g^\wedge(nx) = 0,$$

and hence $\chi_g^\wedge(x) = 0$ since $Z_p^\wedge \otimes C$ is torsion-free; so $x = 0$ by (2.3). This proves Lemma 1.12.

§ 3. Values of Characters; Finite Groups

This section has two main objects, and the first is as follows. In § 2 we introduced characters $\chi_g^\wedge(x)$ with values in $Z_p^\wedge \otimes C$; so now we must work to prove that under appropriate assumptions they take values in C.

Proposition 3.1. *Let G be a topological group, $g \in G$ an element of p-power order, and $x \in K(BG)$ an element such that $\lambda_t(x)$ is a rational function of t; then $\chi_g^\wedge(x)$ lies in C. In particular, this holds if $x \in FF(BG)$.*

In order to prove this, it is sufficient by (2.2)(ii) to discuss the case in which G is a finite cyclic p-group generated by g. This brings us to the second main object of this section, which is to finish dealing with the case of finite groups. The main work involved is the proof of Theorem 1.16. However, we shall also need the following result.

Proposition 3.2. *Theorem 1.14 holds when G is a finite cyclic p-group Z_{p^e}. That is, let $x \in K(B(Z_{p^e}))$ be an element such that $\lambda^i x = 0$ for i sufficiently large; then $x = \alpha\rho$ for some honest (positive) representation ρ.*

This completes the statement of results needed later.

The proof of these results requires further lemmas. Let k be a field of characteristic zero. If $a, b \in k$, we interpret

$$(1+a\,t)^b \in k[[t]]$$

as the formal binomial series

$$1+bat+\frac{b(b-1)}{1.2}a^2t^2+\cdots.$$

We consider the product

$$(1+a_1\,t)^{b_1}(1+a_2\,t)^{b_2}\dots(1+a_r\,t)^{b_r}$$

where a_i, $b_i \in k$ and the a_i are distinct and non-zero.

Lemma 3.3. *If the formal power-series*

$$(1+a_1\,t)^{b_1}(1+a_2\,t)^{b_2}\dots(1+a_r\,t)^{b_r}$$

is a rational function of t, then the exponents b_i are integers; if it is a polynomial in t, then the b_i are non-negative integers.

Proof. First assume that the formal power-series

$$(1+a_1\,t)^{b_1}(1+a_2\,t)^{b_2}\dots(1+a_r\,t)^{b_r}$$

is a rational function $f(t)/g(t)$, where $f(t)$, $g(t)$ are polynomials and $g(t)$ is invertible in $k[[t]]$ (that is, $g(0) \neq 0$). In the polynomial ring $k[t]$ we may use the remainder theorem to write

$$f(t)=(1+a_1\,t)^{d_1}(1+a_2\,t)^{d_2}\dots(1+a_r\,t)^{d_r}\,h(t)$$

where the d_i are non-negative integers and $h(t)$ is a polynomial with $h(t) \neq 0$ for

$$t=-1/a_1, -1/a_2, \dots, -1/a_r.$$

Similarly we may write

$$g(t)=(1+a_1\,t)^{e_1}(1+a_2\,t)^{e_2}\dots(1+a_r\,t)^{e_r}\,k(t).$$

Take the equation

$$\begin{aligned}(1+a_1\,t)^{b_1}&(1+a_1\,t)^{e_1}(1+a_2\,t)^{b_2}\dots(1+a_r\,t)^{e_r}\,k(t)\\ &=(1+a_1\,t)^{d_1}(1+a_2\,t)^{d_2}\dots(1+a_r\,t)^{d_r}\,h(t)\end{aligned}$$

and differentiate it; differentiation makes sense in $k[[t]]$, and can be done by the usual rules. Now divide by the original equation; for this purpose we remark that $k[[t]]$ is an integral domain, and allow ourselves to work in its field of fractions. We get

$$(3.4)\quad \frac{a_1 b_1}{1+a_1 t}+\frac{a_1 e_1}{1+a_1 t}+\frac{a_2 b_2}{1+a_2 t}+\cdots+\frac{a_r e_r}{1+a_r t}+\frac{k'(t)}{k(t)}$$
$$=\frac{a_1 d_1}{1+a_1 t}+\frac{a_2 d_2}{1+a_2 t}+\cdots+\frac{a_r d_r}{1+a_r t}+\frac{h'(t)}{h(t)}.$$

This equation, of course, is shorthand for the equation you get by multiplying it up; a priori this is an equation in $k[[t]]$, but it happens that every term lies in the polynomial ring $k[t]$. So we are now working with rational functions. Multiply (3.4) by $(1+a_i t)$ and substitute $t=-1/a_i$; we get

$$a_i b_i + a_i e_i = a_i d_i,$$

so that

$$b_i = d_i - e_i$$

and the exponents b_i are integers.

If we assume that the given series

$$(1+a_1 t)^{b_1}(1+a_2 t)^{b_2}\ldots(1+a_r t)^{b_r}$$

is a polynomial, then the proof above applies with $g(t)=1$, so that $e_i=0$ and we get $b_i=d_i$; thus the exponents b_i are non-negative integers. This proves the lemma.

I shall apply Lemma 3.3 to give parallel proofs of Proposition 3.2 and the following result, which is the special case $G=Z_{p^e}$ of Theorem 1.16.

Lemma 3.5. *Let $G=Z_{p^e}$, and let $x\in K(BG)$ be an element such that $\lambda_t(x)$ is a rational function of x; then $x\in \operatorname{Im}\alpha$.*

Proofs of (3.2), (3.5). Let $G=Z_{p^e}$, and let $\zeta=\exp(2\pi i p^{-e})$. Then the homomorphism

$$\chi_g\colon RG\to C$$

takes values in the ring of algebraic integers $Z[\zeta]\subset C$. Because $Z_p^\wedge$ is torsion-free, we have a monomorphism

$$Z_p^\wedge[\zeta]=Z_p^\wedge\otimes Z[\zeta]\to Z_p^\wedge\otimes C.$$

Let $Q_p^\wedge$ be the field of p-adic numbers; because $Z[\zeta]$ is torsion-free, we have a monomorphism

$$Z_p^\wedge[\zeta]=Z_p^\wedge\otimes Z[\zeta]\to Q_p^\wedge\otimes Z[\zeta]=Q_p^\wedge[\zeta].$$

It is easy to see that we can complete the triangle of monomorphisms

$$\begin{array}{ccc} Z_p^\wedge[\zeta] & \longrightarrow & Z_p^\wedge\otimes C, \\ & \searrow \quad \nearrow & \\ & Q_p^\wedge[\zeta] & \end{array}$$

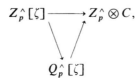

but I do not need this. The ring $Q_p^\wedge[\zeta]$ is a field; for if $e=f+1>0$, then the cyclotomic polynomial

$$\frac{z^{p^{f+1}}-1}{z^{p^f}-1}=1+z^{p^f}+z^{2p^f}+\cdots+z^{(p-1)p^f}$$

is irreducible over $Q_p^\wedge$, as may be seen by substituting $z=1+w$ and using Eisenstein's criterion. I shall take the field k in Lemma 3.3 to be $Q_p^\wedge[\zeta]$. In fact, it is clear from (2.1), (2.2)(i) that with $G=Z_{p^e}$ the maps $1\otimes\chi_g$ and $\chi_g^\wedge$ take values in $Z_p^\wedge\otimes Z[\zeta]\subset k$.

Let $q=p^e$, and let $\xi,\xi^2,\xi^3,\ldots,\xi^q=1$ be the irreducible representations of G, which are 1-dimensional; we may suppose ξ and a generator g for G chosen so that

$$\chi_g(\xi)=\zeta=\exp(2\pi i p^{-e}).$$

By (2.1), we can identify $K(BG;Z_p^\wedge)$ with $Z_p^\wedge\otimes RG$ under $\alpha^\wedge$, and consider the result as containing both $K(BG)$ and RG; let us write the given element $x\in K(BG)$ as

$$x=b_1\xi+b_2\xi^2+\cdots+b_q\xi^q$$

where $b_i\in Z_p^\wedge$ and $\sum_i b_i\in Z$. I claim that we have

$$\lambda_t(x)=(1+\xi t)^{b_1}(1+\xi^2 t)^{b_2}\ldots(1+\xi^q t)^{b_q}.$$

In fact, this is clearly true when the coefficients b_i lie in Z (so that x lies in RG). On the left the coefficient $\lambda^i x$ of t^i is a continuous function of x if we use the filtration topology, and on the right the coefficient of t^i is a continuous function of $b_1,b_2,\ldots,b_q$ if we use the p-adic topology; but these topologies agree by Lemma 2.1, and RG is dense in $K(BG)$. So by taking limits we obtain the general result, that is

$$\lambda_t(x)=(1+\xi t)^{b_1}(1+\xi^2 t)^{b_2}\ldots(1+\xi^q t)^{b_q}.$$

In Lemma 3.5 we assume that this is a rational function of t, and in Proposition 3.2 we assume that it is a polynomial in t. Apply the map $\chi_g^\wedge=1\otimes\chi_g$ into $k[[t]]$; we find that

$$(1+\zeta t)^{b_1}(1+\zeta^2 t)^{b_2}\ldots(1+\zeta^q t)^{b_q}$$

is a rational function or a polynomial as the case may be. Here $\zeta,\zeta^2,\ldots,\zeta^q$ are distinct non-zero elements of k; so we can apply Lemma 3.3 and conclude that the exponents b_i are integers in (3.5), or non-negative integers in (3.2). Thus

$$b_1\xi+b_2\xi^2+\cdots+b_q\xi^q$$

is a virtual representation in (3.5), or an honest representation in (3.2). This proves Proposition 3.2 and Lemma 3.5, which completes the proof of Proposition 3.1.

Proof of Lemma 1.5. Let G be a compact Lie group, and let $x \in K(BG)$ be an element such that $\lambda^i x = 0$ for i sufficiently large. Let $\varepsilon x = n$; then

$$\varepsilon(\lambda^i x) = \frac{n(n-1)(n-2)\dots(n-i+1)}{1.2.3\dots i};$$

if this is zero, then n must be one of $0, 1, 2, \dots, i-1$; that is, n must be a non-negative integer. We have $\lambda^n x \neq 0$ because $\varepsilon(\lambda^n x) = 1$.

Now let g be an element of p-power order in G, and let H be the cyclic subgroup generated by g. By Proposition 3.2, the image of x in $K(BH)$ has the form $\alpha \rho$ for some honest (positive) representation

$$\rho \colon H \to U(m);$$

and comparing augmentations, we see that

$$m = n.$$

Thus

$$\lambda^i(\alpha \rho) = 0 \quad \text{for } i > n,$$

and hence

$$\chi_g^\wedge(\lambda^i x) = 0 \quad \text{for } i > n.$$

Since this holds for all g of prime power order in G, we get

$$\lambda^i x = 0 \quad \text{for } i > n$$

by Theorem 2.3. This proves Lemma 1.5.

Lemma 3.6. *Let G be a finite p-group, and let $x \in K(BG)$ be an element such that $\lambda_t(x)$ is a rational function of t; then $x \in \operatorname{Im} \alpha$.*

Proof. By (2.1) (again) we can identify $K(BG; Z_p^\wedge)$ with $Z_p^\wedge \otimes RG$, under $\alpha^\wedge$, and consider the result as containing both $K(BG)$ and RG. Let us write the given element x in the form

$$x = \sum_i \mu_i \rho_i$$

where $\rho_1, \rho_2, \dots, \rho_r$ are the irreducible representations of G and $\mu_1, \mu_2, \dots, \mu_r$ are coefficients in $Z_p^\wedge$. To determine the coefficients μ_i we can write

$$\mu_i = \frac{1}{|G|} \sum_{g \in G} \overline{\chi_g(\rho_i)}\, \chi_g^\wedge(x)$$

in $Z_p^\wedge \otimes C$, because the corresponding formula holds for representations and we extend it by linearity over $Z_p^\wedge$.

We assume that $\lambda_t(x)$ is a rational function of t. Then the restriction of x to any cyclic subgroup lies in $\operatorname{Im} \alpha$ by Lemma 3.5. Since the character $\chi_g^\wedge(x)$ depends only on the restriction of x to the cyclic subgroup generated by g, we see that

$$\chi_g^{\wedge}(x) \in Z[\zeta],$$

where $Z[\zeta]$ is the subring of C generated by $\zeta = \exp(2\pi i p^{-e})$, and e is chosen so that the order of every element $g \in G$ divides p^e. Certainly we have

$$\overline{\chi_g(\rho_i)} \in Z[\zeta],$$

so it follows that

$$|G|\,\mu_i \in Z[\zeta].$$

Now $Z[\zeta]$ is a free Z-module, generated by the powers ζ^r with $0 \leqq r < (p-1)p^{e-1}$; so $Z_p^{\wedge} \otimes Z[\zeta]$ is a free $Z_p^{\wedge}$-module generated by the powers ζ^r with $0 \leqq r < (p-1)p^{e-1}$; and it is monomorphically embedded in $Z_p^{\wedge} \otimes C$. It follows that in $Z_p^{\wedge} \otimes C$ we have

$$(Z_p^{\wedge}) \cap (Z[\zeta]) = Z.$$

Since $|G|\,\mu_i$ lies both in $Z_p^{\wedge}$ and in $Z[\zeta]$, we have $|G|\,\mu_i \in Z$. Since $|G|$ is a power of p and μ_i is a p-adic integer, this proves $\mu_i \in Z$. Thus

$$\sum_i \mu_i \rho_i \in RG$$

and $x \in \operatorname{Im} \alpha$. This proves Lemma 3.6.

Proof of Theorem 1.16. Suppose that G is a finite group, that $x \in K(BG)$ and that $\lambda_t(x)$ is a rational function of t. Let us temporarily use the notation $\chi(\rho, g)$ for characters; by Proposition 3.1, the character $\chi^{\wedge}(x, g)$ lies in C for all g for which it is defined. I now construct a function $f\colon G \to C$ as follows. For any $g \in G$, let H be the cyclic subgroup of G generated by g; it is the direct sum of its Sylow subgroups H_p. Let the component of g in H_p be g_p; then g_p is of p-power order and $\chi^{\wedge}(x, g_p)$ is defined. Define

$$f(g) = d + \sum_p (\chi^{\wedge}(x, g_p) - d)$$

where d is the augmentation of x, or equivalently, $d = \chi^{\wedge}(x, 1)$.

It is clear that f is a complex-valued class function on G, and agrees with $\chi^{\wedge}(x, g)$ whenever g is an element of prime-power order. I claim that f is a (virtual) character. To prove this it is sufficient, by Brauer's "characterisation of characters" [14, 18], to show that the restriction of f to any "elementary" subgroup is a character. But an "elementary" subgroup E is, among other things, a direct product of p-groups: $E = \bigtimes_p E_p$. Let $\pi_p\colon E \to E_p$ be the projection map. By Lemma 3.6, the values of $\chi^{\wedge}(x, g)$ for $g \in E_p$ are the character of a virtual representation $\xi_p \in RE_p$. Then $f|E$ is the character of the virtual representation

$$d + \sum_p ((\pi_p^* \xi_p) - d).$$

It now follows that f is the character of a virtual representation $\xi \in RG$; then $\chi_g^{\wedge}(x) = \chi_g(\xi)$ for all g of prime-power order, and $x = \alpha\xi$ by Theorem 2.3. This proves Theorem 1.16, which completes the proof of Theorems 1.10 and 1.3.

§4. Some Examples

In this section I shall deal with the examples promised in §1.

First I consider Example 1.1, so I must prove Proposition 1.18, beginning with part (i). The element αi is of formal dimension 2, so since Ψ^k commutes with exterior powers the elements $\Psi^k(\alpha i)$ are of formal dimension 2; I have to prove there are no more.

In $G = SL(2,5)$, every element of prime power order is conjugate to an element in one of the following three subgroups:

$$Z_3 \text{ generated by } \begin{bmatrix} 0 & -1 \\ 1 & -1 \end{bmatrix},$$

$$Z_4 \text{ generated by } \begin{bmatrix} 0 & -1 \\ 1 & 0 \end{bmatrix},$$

$$Z_5 \text{ generated by } \begin{bmatrix} 1 & 1 \\ 0 & 1 \end{bmatrix}.$$

I shall write these subgroups Z_q, where $q = 3, 4, 5$. Let the irreducible representations of Z_q be $(\eta_q)^j$, where j runs over the residue classes $\bmod q$.

Let $x \in K(BSL(2,5))$ be an element of formal dimension 2. The image of x in $K(BZ_q)$ has the same property, and so by Proposition 3.2 has the form $\alpha\rho$ for some honest (positive) representation ρ, which necessarily has the form $(\eta_q)^r + (\eta_q)^s$ for some $r = r(q)$, $s = s(q)$. I claim that here we must have $r + s \equiv 0 \bmod q$; for example, since $H^2(BSL(2,5); Z) = 0$, the first Chern class $c_1 x$ is zero.

If we work in representation-theory rather than K-theory, the same considerations apply even more easily to the two standard embeddings $i, j\colon SL(2,5) \to SU(2)$. In this case we must find that r and s are prime to q, for otherwise the restriction of a standard embedding to Z_q would have a non-trivial kernel, which is ridiculous. So we can normalise the choice of η_q so that i restricts to $\eta_q + \eta_q^{-1}$. Then $\Psi^k i$ restricts to $(\eta_q)^k + (\eta_q)^{-k}$. By the Chinese remainder theorem we can choose $k \bmod 60$ so that

$$k \equiv r(3) \bmod 3,$$
$$k \equiv r(4) \bmod 4,$$
$$k \equiv r(5) \bmod 5.$$

Then $x = \alpha(\Psi^k i) = \Psi^k(\alpha i)$, for they have the same character $\chi_g^{\wedge}$ at all points g of prime-power order. This proves part (i).

As for part (ii), we know that $H^4(BSL(2,5)) \cong Z_{120}$, and that the second Chern class of $\Psi^k(\alpha i)$ is k^2 times a generator. So if $\Psi^k(\alpha i) = \Psi^l(\alpha i)$, then their second Chern classes are equal, and $k^2 \equiv l^2 \bmod 120$. I claim that the converse follows from the work above. In fact, the residue class of $k^2 \bmod 3$ determines the pair $(k, -k) \bmod 3$; the residue class of $k^2 \bmod 8$ determines the pair $(k, -k) \bmod 4$; and the residue class of $k^2 \bmod 5$ determines the pair $(k, -k) \bmod 5$. This proves part (ii).

Part (iv) also follows easily. There are three homomorphisms θ to be considered, namely i, j and the map constant at 1; and only j needs a calculation. For j we must have

$$\pm r(3) = \pm 1 \bmod 3,$$
$$\pm r(4) = \pm 1 \bmod 4,$$
$$\pm r(5) = \pm 2 \bmod 5.$$

Clearly $k=7$ is a solution, leading to $k^2 \equiv 49 \bmod 120$.

There remains part (iii). One way to construct maps f: $BSL(2,5) \to BSU(2)$ is to form the composite

$$BSL(2,5) \xrightarrow{Bi} BSU(2) \xrightarrow{\Psi^k} BSU(2),$$

where Ψ^k is a map of the sort constructed by Sullivan [22]. This certainly constructs a map f with

$$f^*\alpha'(1) = \Psi^k(\alpha i),$$

but the construction only works when k is odd (or zero). It is therefore more effective, and perhaps more elementary, to proceed as follows.

Since the group $SL(2,5)$ is perfect, we can form the plus construction $BSL(2,5)^+$ in the sense of Quillen [16] (there is an exposition in [6]). Since $BSU(2)$ is simply-connected, we have a $(1-1)$ correspondence

$$[BSL(2,5), BSU(2)] \longleftrightarrow [BSL(2,5)^+, BSU(2)].$$

Since $BSL(2,5)^+$ is a torsion space, it is equivalent to the wedge-sum of its localisations,

$$BSL(2,5)^+ \simeq BSL(2,5)^+_{(2)} \vee BSL(2,5)^+_{(3)} \vee BSL(2,5)^+_{(5)}.$$

On $BSL)^+_{(2)}$ we can take the constant map and also the map corresponding to Bi; thus we can realise Chern classes c_2 congruent to $0, 1 \bmod 8$. On $BSL(2,5)^+_{(3)}$ we can take the constant map and also the map corresponding to Bi; thus we can realise Chern classes congruent to $0, 1 \bmod 3$. On $BSL(2,5)^+_{(5)}$ we can take the constant map and also the maps corresponding to Bi, Bj; thus we can realise Chern classes c_2 congruent to 0, 1 and $4 \bmod 5$. We can make a map of the wedge-sum by combining these components in any way, and so realise Chern classes congruent to $k^2 \bmod 120$ whenever $k^2 \not\equiv 4 \bmod 8$.

It remains to show that we cannot have $f^*\alpha'(1) = \Psi^k(\alpha i)$ with $k^2 \equiv 4 \bmod 8$, that is, with $k \equiv 2 \bmod 4$. The obvious way is to use symplectic K-theory. As most of this paper concentrates on the systematic use of complex K-theory, I will not try to prove general results about symplectic K-theory (though I presume it can be done); I will simply make an ad hoc calculation. Recall that a

Sylow 2-subgroup of $SL(2,5)$ is isomorphic to the quaternion group $Q_8=\{\pm 1, \pm i, \pm j, \pm k\}$. Let

$$\gamma\colon KSp(BQ_8)\to K(BQ_8)$$

be the complexification map.

Lemma 4.1. *Let $x\in RQ_8$. Then $\alpha x\in \mathrm{Im}\,\gamma$ if and only if $x\in RSpQ_8$.*

Proof. It is clear that if $x\in RSpQ_8$, then $\alpha x\in \mathrm{Im}\gamma$; we have to prove the converse.

First I claim that by changing x suitably, we may assume without loss of generality that $\alpha x=\gamma y$ where $y\in KSp(BQ_8)$ and y is of filtration 8 in the Atiyah-Hirzebruch spectral sequence for $KSp(BQ_8)$. For this we must simply see that there are enough elements in $RSpQ_8$ to provide generators for the E_∞ term of the Atiyah-Hirzebruch spectral sequence in filtrations 0, 4, 5 and 6.

Let me name some generators. Let the 1-dimensional representations of Q_8 be $1, I, J$ and K, and let the irreducible 2-dimensional representation be H. The representations 1, I, J and K are real, while H is symplectic. In the Atiyah-Hirzebruch spectral sequence for $KO(BQ_8)$, we may use $I-1$ and $J-1$ as generators for the subquotient in filtration 1 (which is Z_2+Z_2); we may use $2(I-1)$ and $2(J-1)$ as generators for the subquotient in filtration 2 (which is also Z_2+Z_2). In the Atiyah-Hirzebruch spectral sequence for $KSp(BQ_8)$, we may use 2 as a generator for the subquotient in filtration 0 (which is Z); $H-2$ serves for the subquotient in filtration 4 (which is Z_8); $(I-1)(H-2)$ and $(J-1)(H-2)$ serve for the subquotient in filtration 5 (which is Z_2+Z_2); and $2(I-1)(H-2)$, $2(J-1)(H-2)$ serve for the subquotient in filtration 6 (which is Z_2+Z_2). All these elements lie in $RSpQ_8$, so the claim in the last paragraph is established.

Let us now write x in the form

$$x=a1+bI+cJ+dK+eH$$

and consider the position of $\alpha x=\gamma y$ in the Atiyah-Hirzebruch spectral sequence for $K(BQ_8)$; this is the only spectral sequence I consider for the rest of the proof. Clearly αx must map to zero successively in the subquotients of filtration 0, 2, 4 and 6.

The subquotient in filtration 0 is Z, and we find that

$$x=b(I-1)+c(J-1)+d(K-1)+e(H-2).$$

The subquotient in filtration 2 is Z_2+Z_2, with generators $I-1$ and $J-1$. Since $H-2$ has filtration 4, it maps to zero in this subquotient. Moreover, $(H-2)^2$ has filtration 8, and since $H^2=1+J+K$ we get

$$(H-2)^2=1+I+J+K-4H+4;$$

thus $I+J+K-3$ maps to zero in our subquotient. So we find that

$$b\equiv c\equiv d \bmod 2.$$

Using the relation above, we may now write x in the form

$$x=2b'(I-1)+2c'(J-1)+d(H-2)^2+e'(H-2).$$

The subquotient in filtration 4 is Z_8, generated by $(H-2)$, and the term $d(H-2)^2$ maps to zero in it since $(H-2)^2$ has filtration 8. We need to know about the first two terms. The product $(I-1)(H-2)$ by filtration 6, and since $IH=H$, we get

$$(I-1)(H-2)=-2(I-1);$$

thus $2(I-1)$ has filtration 6, and similarly for $2(J-1)$. We find that

$$e' \equiv 0 \bmod 8.$$

We may now write

$$x=2b'(I-1)+2c'(J-1)+d(H-2)^2+8e''(H-2).$$

The subquotient in filtration 6 is Z_2+Z_2, generated by

$$(I-1)(H-2)=-2(I-1),$$
$$(J-1)(H-2)=-2(J-1).$$

The term $d(H-2)^2$ maps to zero in it, since $(H-2)^2$ has filtration 8; we need to know about the last term. The element $(H-2)^3$ has filtration 12, and expanding it out, we find that

$$6(H-2)^2+8(H-2)$$

has filtration 12. Thus $8(H-2)$ has filtration 8. We find that

$$b' \equiv 0 \bmod 2, \qquad c' \equiv 0 \bmod 2.$$

We may now write

$$x=4b''(I-1)+4c''(J-1)+d(H-2)^2+8e''(H-2).$$

The subquotient in filtration 8 is Z_8, generated by $(H-2)^2$. Moreover, since αx comes from an element y of filtration 8 in the spectral sequence for $KSp(BQ_8)$, it follows that αx is an even multiple of the generator in this subquotient. Since $2(I-1)$ is of filtration 6, $2(I-1)(H-2)$ is of filtration 10, that is $4(I-1)$ is of filtration 10, and similarly for $4(J-1)$; so the terms $4b''(I-1)$, $4c''(J-1)$ yield zero in our subquotient. The term $8e''(H-2)$ yields $-6e''(H-2)^2$, which is an even multiple of the generator. We conclude that d is even. This shows that $x \in RSpQ_8$, and proves the lemma.

I can now return to the proof of Proposition 1.18, part (iii). Suppose then that $f\colon BSL(2,5) \to BSU(2)$ is a map and $f^*\alpha'(1)=\Psi^k(\alpha i)=\alpha(\Psi^k i)$. Since $\alpha'(1)$ comes from $KSp(BSU(2))$, $f^*\alpha'(1)$ is symplectic and its image in $K(BQ_8)$ is also symplectic. Now Lemma 4.1 shows that the element $\Psi^k i \in RQ_8$ lies in $RSpQ_8$. But for $k \equiv 2 \bmod 4$ we have

$$\Psi^k i = I+J+K-3$$

which does not lie in $RSpQ_8$; so this case is excluded. This proves Proposition 1.18, and completes the treatment of Example 1.1.

Secondly I turn to Example 1.4. The disctinctive feature of this example is that the map $f\colon BG \to BG'$ is constructed by a "Mayer-Vietoris" method from representations of two subgroups of G.

I begin with an auxiliary result about a suitable finite group. Let Q_8 be the quaternion group with 8 elements $\{+1, \pm i, \pm j, \pm k\}$. Then Q_8 contains the subgroup $Z_2 = \{\pm 1\}$; so $Q_8 \times Z_3$ contains the subgroup $Z_2 \times Z_3 \cong Z_6$. This has 6 irreducible representations of degree 1, say η^i where i runs over the residue classes mod 6. I claim that $\eta + \eta^5$ is not the restriction of any virtual representation of $Q_8 \times Z_3$.

In fact, we may identify $R(Z_3)$ with its image in $R(Z_6)$, so that the irreducible representations of Z_3 are η^{2j} where j runs over the residue classes mod 3; then the irreducible representations of $Q_8 \times Z_3$ are of the form $\xi\eta^{2j}$, where ξ runs over the 5 irreducible representations of Q_8. If ξ is one of the 4 irreducible representations of degree 1 then the image of $\xi\eta^{2j}$ in $R(Z_6)$ is η^{2j}, while if ξ is the irreducible representation of degree 2 (which is the usual representation of Q_8 on the quaternions) then the image of $\xi\eta^{2j}$ in $R(Z_6)$ is $2\eta^{2j+3}$. This proves the claim.

I now proceed to embed $Q_8 \times Z_3$ in a Lie group G so that the subgroup $Z_2 = \{\pm 1\}$ embeds in the identity component G_1. Consider $Q_8 \times Z_3$ as an extension with kernel $Z_2 = \{\pm 1\}$ and quotient $Z_2 \times Z_2 \times Z_3$. Embed $Z_2 = \{\pm 1\}$ in $U(1) = S^1$, and make $Z_2 \times Z_2 \times Z_3$ act trivially on S^1. Since extensions are covariant in this variable, I get the following diagram of extensions, which defines G.

$$\begin{array}{ccc} Z_2 & \longrightarrow & S^1 \\ \downarrow & & \downarrow \\ Q_8 \times Z_3 & \longrightarrow & G \\ \downarrow & & \downarrow \\ Z_2 \times Z_2 \times Z_3 & \xrightarrow{1} & Z_2 \times Z_2 \times Z_3 \end{array}$$

We can realise BG so that it takes part in the following fibering.

$$BS^1 \to BG \to B(Z_2 \times Z_2) \times (BZ_3).$$

I now prepare to construct a map of BG. Let $G(2)$, $G(3)$ be the subgroups of G over the Sylow subgroups $Z_2 \times Z_2$, Z_3 of $Z_2 \times Z_2 \times Z_3$; then $G(2) \cap G(3) = S^1$. Also $G(3)$ is a product extensions $S^1 \times Z_3$ (for example, because it is induced from a product extension $Z_2 \to Z_2 \times Z_3 \to Z_3$); so there are 3 homomorphisms

$$\beta_1, \beta_3, \beta_5\colon G(3) \to U(1)$$

which extend the identity map $\iota\colon S^1 \to S^1$; I number them so that the restriction of β_i to $Z_2 \times Z_3$ is η^i for $i = 1, 3, 5$. I now take

$$\beta = \beta_1 + \beta_5\colon G(3) \to U(2).$$

This is a representation of $G(3)$ whose restrictions to $Z_2 \times Z_3, S^1$ are $\eta + \eta^5$, 2ι.

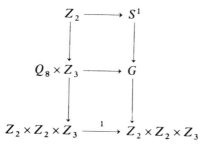

I claim that the representation 2ι of S^1 extends to a representation of $G(2)$. Let H be the subgroup of G over a subgroup Z_2 of $Z_2 \times Z_2$. Since the quotient Z_2 acts trivially on S^1, the extension $S^1 \to H \to Z_2$ must be a product extension, and there is a homomorphism γ: $H \to U(1)$ extending ι. Since H is of index 2 in $G(2)$ we can obtain from γ an induced representation δ: $G(2) \to U(2)$, and the restriction of δ to S^1 is 2ι.

It is now clear that the maps

$$B\delta\colon BG(2) \to BU(2),$$

$$B\beta\colon BG(3) \to BU(2)$$

agree on BS^1 and define a map which may be written

$$B\delta \cup B\beta\colon BG(2) \cup_{BS^1} BG(3) \to BU(2).$$

But $BG(2) \cup_{BS^1} BG(3)$ may be identified as the part of the fibering

$$BS^1 \to BG \to B(Z_2 \times Z_2) \times B(Z_3)$$

over $B(Z_2 \times Z_2) \vee B(Z_3)$. The relative homology groups

$$H_*(B(Z_2 \times Z_2) \times B(Z_3), B(Z_2 \times Z_2) \vee B(Z_3))$$

are zero, as is well known; the same goes for relative cohomology with any untwisted coefficients. Using the Serre spectral sequence and noting that the fundamental group of the base acts trivially on the fibre by construction, we see that the relative cohomology groups

$$H^*(BG, BG(2) \cup_{BS^1} (BG(3); \pi)$$

are zero for any untwisted coefficient group π. Since the space $BU(2)$ is simply-connected, obstruction-theory now shows that the map

$$B\delta \cup B\beta\colon BG(2) \cup_{BS^1} BG(3) \to BU(2)$$

extends to a map

$$f\colon BG \to BU(2).$$

This completes the construction; it remains to show that f has the property stated. First I claim that

$$\alpha\colon RG(3) \to K(BG(3))$$

is mono. In fact, elements of 3-power order are dense in $G(3) = S^1 \times Z_3$; that is, points g where $\chi_g^\wedge$ is defined are dense in $G(3)$. So if $\xi \in RG(3)$ and $\alpha\xi = 0$, then the character of ξ is 0 at points which are dense – so it is 0 by continuity. This proves the claim.

Now suppose that the element $x \in K(BG)$ determined by f lies in $\operatorname{Im}\alpha$; say $x = \alpha\xi$ for some $\xi \in RG$. Then the restriction of ξ to $G(3)$ is β, by the paragraph above. Therefore the restriction of ξ to $Z_2 \times Z_3$ is $\eta + \eta^5$. So the image of ξ in

$R(Q_8 \times Z_3)$ restricts to $\eta + \eta^5$, contradicting the result with which I began this discussion. This contradiction shows that $x \notin \operatorname{Im} \alpha$.

Thirdly I turn to Example 1.17. The distinctive feature of this example is that $\chi_g^\wedge(x)$ is complex-valued but not continuous on the set of points g where it is defined.

Take $G = S^1 \times Z_3$; then we may realise BG as $CP^\infty \times BZ_3$. Since this space has CP^∞ as a retract, $K(BG)$ splits as the direct sum of $K(CP^\infty)$ and a complentary summand $K(BG, CP^\infty)$.

Define a homomorphism $\theta: G \to G$ by $\theta(g) = g^2$. This homomorphism gives an automorphism of the 3-torsion subgroup, but the inverse of this automorphism is not continuous. From θ we get the following diagram of fiberings.

$$\begin{array}{ccc} CP^\infty & \xrightarrow{B\theta} & CP^\infty \\ \downarrow & & \downarrow \\ BG & \xrightarrow{B\theta} & BG \\ \downarrow & & \downarrow \\ BZ_3 & \xrightarrow{B\theta} & BZ_3 \end{array}$$

The induced map of $\tilde{H}^*(BZ_3; H^*(CP^\infty))$ is iso. By the Serre spectral sequence, the induced map $(B\theta)^*$ of $H^*(BG, CP^\infty)$ is iso. By the Atiyah-Hirzebruch spectral sequence, the induced map $(B\theta)^*$ of $K(BG, CP^\infty)$ is iso.

Now let $\xi \in K(CP^\infty)$ correspond to the identity representation of $U(1)$, and let $\eta \in K(BZ_3)$ correspond to a representation $Z_3 \to U(1)$ which identifies Z_3 with the subgroup $\{1, \omega, \omega^2\}$ where $\omega = \exp(2\pi i/3)$. The element $\xi(\eta - 1)$ restricts to 0 in $K(CP^\infty)$ and gives an element of $K(BG, CP^\infty)$. Since $(B\theta)^*$ is iso on this summand, there is an element $x \in K(BG, CP^\infty)$ such that

$$(B\theta)^* x = \xi(\eta - 1).$$

Clearly $\lambda_t(\xi(\eta - 1))$ is a rational function; in fact,

$$\lambda_t(\xi(\eta - 1)) = \frac{1 + t\xi\eta}{1 + t\xi}.$$

Multiply top and bottom by $1 - t\xi$; we get

$$\lambda_t(\xi(\eta - 1)) = \frac{1 + t\xi(\eta - 1) - t^2\xi^2\eta}{1 - t^2\xi^2},$$

that is,

$$(B\theta)^* \lambda_t(x) = (B\theta)^* \left(\frac{1 + tx - t^2\xi\eta^2}{1 - t^2\xi} \right).$$

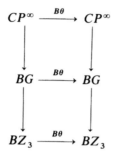

Here the restriction of

$$\frac{1+tx-t^2\xi\eta^2}{1-t^2\xi}$$

to CP^∞ is

$$\frac{1-t^2\xi}{1-t^2\xi}=1,$$

so

$$\frac{1+tx-t^2\xi\eta^2}{1-t^2\xi}$$

is a formal power-series with first term 1 and remaining coefficients in $K(BG, CP^\infty)$. The same goes for $\lambda_t(x)$; since $(B\theta)^*$ is iso on $K(BG, CP^\infty)$, we conclude that

$$\lambda_t(x)=\frac{1+tx-t^2\xi\eta^2}{1-t^2\xi},$$

and $\lambda_t(x)$ is a rational function of t.

Consider now the point $g=(1,\omega)$. This point has 3-power order, and since it has the form $g=\theta h$ for $h=(1,\omega^2)$, we may calculate $\chi_g^\wedge(x)$ as follows.

$$\begin{aligned}\chi_g^\wedge(x)&=\chi_{\theta h}^\wedge(x)\\&=\chi_h^\wedge((B\theta)^* x)\\&=\chi_h^\wedge(\xi(\eta-1))\\&=\omega^2-1.\end{aligned}$$

On the other hand, for any neighbourhood U of $g=(1,\omega)$, the set $\theta^{-1}U$ contains a small neighbourhood of $(-1,\omega^2)$, and in this we can choose elements k of 3-power order. For these we have

$$\begin{aligned}\chi_{\theta k}^\wedge(x)&=\chi_k^\wedge((B\theta)^* x)\\&=\chi_k^\wedge(\xi(\eta-1)),\end{aligned}$$

and as θk tends to g this tends to $-(\omega^2-1)$. Therefore $\chi_g^\wedge(x)$ is not a continuous function of g.

On the other hand, if we had $nx\in\operatorname{Im}\alpha$ for some non-zero integer n – say $nx=\alpha y$ – then $\chi_g^\wedge(x)=\frac{1}{n}\chi_g(y)$ would be a continuous function of g. This shows that we do not have $nx\in\operatorname{Im}\alpha$.

Finally I turn to Example 1.15. This example serves to show the difference between monogenic groups and groups which are not monogenic; its only distinctive feature is its utter triviality.

Let G be the group $Z_2\times Z_2$. The map

$$\alpha: RG\to K(BG)$$

is mono (e.g. because $\chi_g^\wedge$ is defined for all $g \in G$) and we will identify RG with its image in $K(BG)$. The group G has 4 irreducible representations $\xi_1, \xi_2, \xi_3, \xi_4$ (of degree 1); let x be the sum of any three minus the fourth, say

$$x = \xi_1 + \xi_2 + \xi_3 - \xi_4.$$

On any cyclic subgroup of $Z_2 \times Z_2$, the restriction of ξ_4 is equal to the restriction of one of ξ_1, ξ_2, ξ_3; so x restricts to an honest (positive) representation (of degree 2). It follows that for any $i > 2$, $\lambda^i x$ restricts to zero on each cyclic subgroup; so its character must be zero, and $\lambda^i x = 0$ for $i > 2$.

On the other hand, x is not an honest (positive) representation. This completes my discussion of the four examples.

§5. Discrete Approximation to a Monogenic Group

In §6 we shall study the special case in which our group G is a monogenic group $T \times Z_n$. As in §2, we shall approximate this group by finite subgroups. More precisely, let G_∞ be the p-torsion subgroup of $T \times Z_n$, taken with the discrete topology; then we have first to obtain corresponding results for the group G_∞. That is the object of this section. Of course it takes us outside the class of compact Lie groups, but there is no harm in that.

In $T^1 = S^1 = R/Z$, the p-torsion subgroup has the form

$$T^1_\infty = S^1_\infty = Z[1/p]/Z;$$

it is the union of an increasing sequence of cyclic subgroups

$$Z_p \subset Z_{p^2} \subset Z_{p^3} \subset \cdots.$$

This union is often written "Z_{p^∞}". In $T^d = (S^1)^d$, the p-torsion subgroup is

$$T^d_\infty = \mathop{\times}_{i=1}^{d} Z_{p^\infty}.$$

In $T^d \times Z_n$, the p-torsion subgroup is of the form

$$T^d_\infty \times Z_{p^f} = \left(\mathop{\times}_{i=1}^{d} Z_{p^\infty} \right) \times Z_{p^f}.$$

Proposition 5.1. *Let $A = T^d_\infty \times Z_{p^f}$, and let $x \in K(BA)$ be an element such that $\lambda^k x = 0$ for k sufficiently large. Then there are (finitely many) 1-dimensional representations*

$$\rho_i : A \to U(1)$$

such that

$$x = \alpha(\sum_i \rho_i);$$

moreover the ρ_i are unique (up to the order in which they are numbered).

This is the only result of this section which will be quoted later.

Proof. I begin with the uniqueness statement. Let A be a discrete abelian group which can be expressed as the union of an increasing sequence of finite p-groups A_i (as in our application). Then we can construct the classifying space BA as the union of an increasing sequence of subspace BA_i. Each map

$$RA_i \leftarrow RA_{i+1}$$

is epi; using (2.1), each map

$$K(BA_i) \leftarrow K(BA_{i+1})$$

is epi; so the map

$$K(BA) \to \varprojlim_i K(BA_i)$$

is iso. We may interpret the set of 1-dimensional representations of A as

$$\mathrm{Hom}(A, U(1)) = \varprojlim_i \mathrm{Hom}(A_i, U(1)).$$

We may interpret RA as the free abelian group on these 1-dimensional representations as generators; we still have a map

$$\alpha\colon RA \to K(BA).$$

Since each map

$$\alpha\colon RA_i \to K(BA_i)$$

is mono, we deduce that

$$\alpha\colon RA \to K(BA)$$

is mono. This proves the uniqueness statement.

For the rest of this section I shall omit α from the notation, and consider RA as a subgroup of $K(BA)$.

To prove the existence of the ρ_i, I proceed by induction over d. For $d=0$ the result is true by Proposition 3.2, so as an inductive hypothesis I assume that the result is true for $T_\infty^{d-1} \times Z_{p^f}$.

First I infer the result for T_∞^d, that is, I consider the special case $f=0$. We can consider the group $A = T_\infty^d$ as the union of the increasing sequence of subgroups

$$A_f = T_\infty^{d-1} \times Z_{p^f}.$$

Let $x \in K(BA)$ be as in the statement; then by the inductive hypothesis, the restriction of x to each BA_f has the form

$$\sigma_{f1} + \sigma_{f2} + \cdots + \sigma_{fn}$$

where each σ_{fi} is a 1-dimensional representation of A_f. Since $\varepsilon x = n$, the integer n is independent of f. The restriction of

$$\sigma_{f+1,1} + \sigma_{f+1,2} + \cdots + \sigma_{f+1,n}$$

to BA_f must be

$$\sigma_{f1}+\sigma_{f2}+\cdots+\sigma_{fn};$$

so after reordering the $\sigma_{f+1,i}$ if necessary, we can assume that $\sigma_{f+1,i}$ restricts to σ_{fi}. Then for each i the sequence σ_{fi} defines a 1-dimensional representation σ_i of A, and we have

$$x=\sigma_1+\sigma_2+\cdots+\sigma_n.$$

This proves the special case $f=0$.

Now I turn to the group $A=T^d_\infty\times Z_{p^f}$. Let $x\in K(BA)$ be as in the statement. Then by the special case just proved, the restriction of x to BT^d_∞ has the form

$$\sigma_1+\sigma_2+\cdots+\sigma_n$$

where $n=\varepsilon x$ and each σ_i is a 1-dimensional representation of T^d_∞. We keep this as standing notation for the rest of the proof.

We now choose an integer e and consider the following exact sequence.

$$\begin{array}{ccc}
T_e=\displaystyle\mathop{\times}_{i=1}^{d} Z_{p^e} & \xrightarrow{\subset} \displaystyle\mathop{\times}_{i=1}^{d} Z_{p^\infty} & =T^d_\infty \\
\downarrow & & \\
A_e=\left(\displaystyle\mathop{\times}_{i=1}^{d} Z_{p^{e+f}}\right)\times Z_{p^f} & \xrightarrow{\subset} \left(\displaystyle\mathop{\times}_{i=1}^{d} Z_{p^\infty}\right)\times Z_{p^f}=A & \\
\downarrow & & \\
\bar{A}_e=\left(\displaystyle\mathop{\times}_{i=1}^{d} Z_{p^f}\right)\times Z_{p^f}. & &
\end{array}$$

There are p^{de} 1-dimensional representations of T_e; for each such representation $\rho\colon T_e\to U(1)$, let us choose a fixed extension $\tilde{\rho}\colon A_e\to U(1)$ of ρ. Then these extensions $\tilde{\rho}$ form a system of representatives for the cosets of $\mathrm{Hom}(\bar{A}_e, U(1))$ in $\mathrm{Hom}(A_e, U(1))$. Therefore the elements $\tilde{\rho}$ form a base for $R(A_e)$ considered as a module over $R(\bar{A}_e)$. Taking the tensor product with $Z_p^\wedge$, we see that the elements $\tilde{\rho}$ form a base for $K(BA_e;Z_p^\wedge)$ considered as a module over $K(B\bar{A}_e;Z_p^\wedge)$.

Let x_e be the restriction of x to BA_e. By the work just done, we can write x_e in the form

$$x_e=\sum_\rho \tilde{\rho}\, y_{\rho e},$$

where ρ runs over the 1-dimensional representations of T_e and $y_{\rho e}$ lies in $K(B\bar{A}_e; Z_p^\wedge)$.

Consider now the restriction of x_e to BT_e. From the formula above, this restriction is

$$\sum_\rho \rho\, \varepsilon(y_{\rho e}),$$

where ε is the augmentation. On the other hand, the restriction is also that of $\sum_1^n \sigma_i$. We conclude that $\varepsilon(y_{\rho e})=0$ unless ρ is the restriction of at least one of the representations $\sigma_i\colon T_\infty^d \to U(1)$.

Let us now assume for the moment that ρ is not the restriction of any of the representations σ_i. Under this assumption I will prove that the character $\chi_g^\wedge(y_{\rho e})$ vanishes at all points $g \in \bar{A}_e$, except perhaps on the subgroup

$$\left(\bigtimes_{i=1}^{d} Z_{p^{f-1}}\right) \times Z_{p^f}.$$

To this end, choose an element $\bar{a} \in \bar{A}_e$ which is not in the subgroup

$$\bigtimes_{i=1}^{d} Z_{p^{f-1}} \times Z_{p^f}.$$

Writing $\bar{a}$ in the form

$$\bar{a}=(\bar{a}_1, \bar{a}_2, \ldots, \bar{a}_d, \bar{a}_{d+1}) \in \left(\bigtimes_{i=1}^{d} Z_{p^f}\right) \times Z_{p^f},$$

we see that at least one of the coordinates $\bar{a}_1, \bar{a}_2, \ldots, \bar{a}_d$ is a generator for Z_{p^f}; without loss of generality we may assume for definiteness that $\bar{a}_d$ is a generator for Z_{p^f}. Then there is a homomorphism

$$h\colon Z_{p^f} \to Z_{p^f}$$

such that $h(\bar{a}_d)=\bar{a}_{d+1}$. We now construct the subgroups displayed in the following diagram.

$$\begin{array}{ccl}
T_e & \xrightarrow{\;1\;} & T_e = \bigtimes_{i=1}^{d} Z_{p^e} \\
\downarrow & & \downarrow \\
A'_e & \longrightarrow & A_e = \left(\bigtimes_{i=1}^{d} Z_{p^{e+f}}\right) \times Z_{p^f} \\
\downarrow & & \downarrow \\
\bar{A}'_e & \longrightarrow & \bar{A}_e = \left(\bigtimes_{i=1}^{d} Z_{p^f}\right) \times Z_{p^f}.
\end{array}$$

Here $\bar{A}'_e$ is to be $\left(\bigtimes_{i=1}^{d-1} Z_{p^f}\right) \times \bar{\Gamma}$, where $\bar{\Gamma}$ is the graph of the homomorphism $h\colon Z_{p^f} \to Z_{p^f}$; that is, $\bar{\Gamma}$ is to be the set of pairs (z, hz). Thus $\bar{A}'_e$ contains the element $\bar{a}$. The group A'_e is to be the counterimage of $\bar{A}'_e$ in A_e, that is, $\left(\bigtimes_{i=1}^{d-1} Z_{p^{e+f}}\right) \times \Gamma$, where Γ is the counterimage of $\bar{\Gamma}$; Γ may also be described as the graph of the homomorphism

$$Z_{p^{e+f}} \to Z_{p^f} \xrightarrow{\;h\;} Z_{p^f},$$

and therefore it is cyclic. For the same reasons as before, the elements $\tilde{\rho}$ form a base for $K(BA'_e; Z_p^\wedge)$ considered as a module over $K(B\bar{A}'_e; Z_p^\wedge)$.

Since the restriction of x to BA_e has been written

$$x_e = \sum_\rho \tilde{\rho}\, y_{\rho e},$$

the restriction of x to BA'_e must be

$$x'_e = \sum_\rho \tilde{\rho}\, y'_{\rho e},$$

where $y'_{\rho e}$ is the restriction of $y_{\rho e}$ from $B\bar{A}_e$ to $B\bar{A}'_e$.

Since Γ is cyclic, our inductive hypothesis applies to the group

$$\left(\mathop{\times}_{i=1}^{d-1} Z_{p^\infty}\right) \times \Gamma,$$

and the restriction of x to this group is a sum of 1-dimensional representations. A fortiori the restriction of x to the group

$$A'_e = \left(\mathop{\times}_{i=1}^{d-1} Z_{p^{e+f}}\right) \times \Gamma$$

is a sum of 1-dimensional representations. That is, it has the form

$$x'_e = \sum_\rho \tilde{\rho}\, z'_{\rho e}$$

where each $z'_{\rho e}$ is a sum of 1-dimensional representations. Comparing this with the former expression, we find

$$y'_{\rho e} = z'_{\rho e};$$

that is, $y'_{\rho e}$ is a sum of 1-dimensional representations. But now from $\varepsilon(y'_{\rho e}) = 0$ we can conclude $y'_{\rho e} = 0$; that is, $y_{\rho e}$ restricts to zero on a subgroup $\bar{A}'_e$ containing the given element $\bar{a}$. This proves that the character $\chi_g^\wedge(y_{\rho e})$ vanishes at all the points g claimed.

Now let ϕ be any 1-dimensional representation of $\bar{A}_e$ which is 1 on the subgroup $\left(\mathop{\times}_{i=1}^{d} Z_{p^{f-1}}\right) \times Z_{p^f}$. If ρ is as above, then

$$\phi\, y_{\rho e} = y_{\rho e};$$

for at each point of $\bar{A}_e$, either $\chi_g^\wedge(y_{\rho e}) = 0$ or $\chi_g^\wedge(\phi) = 1$. We may thus write

$$y_{\rho e} = \sum_\tau a_\tau \tau$$

where τ runs over the 1-dimensional representations of $\bar{A}_e$, each coefficient a_τ is in $Z_p^\wedge$, and $a_{\phi\tau} = a_\tau$ when ϕ is as above.

It now follows that we may write the restriction of x to BA_e in the form

$$x_e = \sum_\theta b_\theta \theta$$

where θ runs over the 1-dimensional representations of A_e, and each coefficient b_θ is in $Z_p^\wedge$, and moreover

$$b_{\phi\theta} = b_\theta$$

whenever ϕ is a 1-dimensional representation of A_e which is 1 on $\left(\bigtimes_{i=1}^{d} Z_{p^{e+f-1}}\right) \times Z_{p^f}$ and θ is a 1-dimensional representation whose restriction to $T_e = \bigtimes_{i=1}^{d} Z_{p^e}$ is not the restriction of any of our original representations σ_i: $T_\infty^d \to U(1)$.

The same work is equally valid with e replaced by $e+1$, so we find

$$x_{e+1} = \sum_{\theta'} b'_{\theta'} \theta'$$

where the coefficients $b'_{\theta'}$ satisfy the appropriate condition. If we restrict this formula from BA_{e+1} to BA_e we must recover the former formula for x_e. But each 1-dimensional representation θ of A_e is the restriction of just p^d 1-dimensional representations of A_{e+1}, and if θ' is one of them then the others are exactly $\phi'\theta'$ where ϕ' runs over the p^d 1-dimensional representations of A_{e+1} which are 1 on $\left(\bigtimes_{i=1}^{d} Z_{p^{e+f}}\right) \times Z_{p^f}$; under the appropriate assumptions their coefficients $b_{\phi'\theta'}$ are all equal, and we find

$$b_\theta = p^d b'_{\theta'}$$

provided the restriction of θ' to T_{e+1} is not the restriction of any of the original representations σ_i.

But under the same assumption on θ' we can restrict from BA_{e+2} to BA_{e+1} and prove

$$b'_{\theta'} = p^d b''_{\theta''}$$

and so on; thus b_θ is divisible by arbitrarily high powers of p. Since b_θ is a p-adic integer, we conclude that $b_\theta = 0$.

It follows that we may write the restriction of x to BA_e in the form

$$x_e = \sum_\theta b_\theta \theta$$

where each coefficient b_θ lies in $Z_p^\wedge$, and the sum runs over those 1-dimensional representations θ of A_e whose restriction to $T_{e+1} = \bigtimes_{i=1}^{d} Z_{p^{e+1}}$ is the restriction of at least one of the original σ_i.

This result applies to BA_{e+f-1}, and we get an expansion of the form

$$x_{e+f-1} = \sum_\phi c_\phi \phi$$

where the sum runs over those 1-dimensional representations ϕ of A_{e+f-1} whose restriction to $T_{e+f}=\bigtimes_{i=1}^{d} Z_{p^{e+f}}$ is the restriction of at least one of the original σ_i.

Restricting this expression back on BA_e, we get an expression of the form

$$x_e=\sum_{\theta} d_\theta \theta$$

where each coefficient d_θ lies in $Z_p^\wedge$, and the sum runs over those 1-dimensional representations θ of A_e whose restriction to $T_{e+f}=\bigtimes_{i=1}^{d} Z_{p^{e+f}}$ is the restriction of at least one of the original σ_i.

Let η be the 1-dimensional representation

$$T_\infty^d \times Z_{p^f} \to Z_{p^f} \subset U(1),$$

where the first map is projection onto the second factor. Then the last formula can be written

$$x_e=\sum_{\sigma,r} a_{e,\sigma,r}\,\sigma\eta^r$$

where σ runs over the distinct members of the set $\{\sigma_1, \sigma_2, \ldots, \sigma_n\}$, r runs over the range $0 \leqq r < p^f$, and each coefficient $a_{e,\sigma,r}$ lies in $Z_p^\wedge$.

We may choose $e+f$ so large that whenever σ_i and σ_j are distinct, their restrictions to $T_{e+f}=\bigtimes_{i=1}^{d} Z_{p^{e+f}}$ are distinct. In this case the 1-dimensional representations $\sigma\eta^r$ of A_e are all distinct. Taking the formula for x_{e+1} and restricting it to BA_e, we see that

$$a_{e,\sigma,r}=a_{e+1,\sigma,r}.$$

Thus we may write

$$x_e=\sum_{\sigma,r} a_{\sigma,r}\,\sigma\eta^r$$

(independent of e) for all sufficiently large e; this formula is then true for all smaller e.

It remains to show that the coefficients $a_{\sigma,r}$ are ordinary integers, and non-negative. For this purpose we choose e so large that whenever σ_i and σ_j are distinct, their restrictions to $T_e=\bigtimes_{i=1}^{d} Z_{p^e}$ are distinct. We next form a subgroup A'_e as described above, but taking care that the homomorphism h is iso. Then the representations $\sigma\eta^r$ restrict to distinct 1-dimensional representations of A'_e. But using the group

$$\left(\bigtimes_{i=1}^{d-1} Z_{p^\infty}\right) \times \Gamma$$

as in the argument above, we see that the restriction of x to A'_e is the sum of n 1-dimensional representations. Thus the coefficients $a_{\sigma,r}$ are ordinary integers, and non-negative, and their sum is n. Since we have

$$x=\sum_{\sigma,r} a_{\sigma,r}\,\sigma\eta^r$$

this completes the induction and proves Proposition 5.1.

§ 6. Monogenic Groups

The object of this section is to prove Theorem 1.14. I shall begin with the case $G=S^1$, which can be handled by the essentially elementary argument due to Berstein [13]. First we need a lemma.

Lemma 6.1. *Suppose that $x\in K(BG)$ and $\lambda^i x=0$ for i sufficiently large. Then the Chern classes $c_i x$ vanish for i sufficiently large, more precisely, for $i>\varepsilon x$.*

Proof. I use the operation

$$\delta_t(x)=\gamma_t(x-\varepsilon x),$$

where γ_t is as in [9]. This may be characterised as the K-theory operation such that

$$\delta_t(y+z)=\delta_t(y)\cdot\delta_t(z),$$

and

$$\delta_t(L)\quad =1+(L-1)\,t$$

when L is a line bundle. Since we have

$$\gamma_t(x-\varepsilon x)=\gamma_t(x)/\gamma_t(\varepsilon x)$$

we can write $\gamma^m(x-\varepsilon x)$ as a linear combination of $\gamma^i(x)$ for $i\leqq m$, with integer coefficients depending on εx; and $\gamma^i(x)$ we can write as a linear combination of $\lambda^j(x)$ for $j\leqq i$. Thus $\delta^m(x)=\gamma^m(x-\varepsilon x)$ can be written as a linear combination of $\lambda^j(x)$ with $j\leqq m$, the coefficients depending on εx. If we assume $\varepsilon x=n\geqq 0$ and $m>n$, then the terms with $j\leqq n$ in this expansion are identically zero; for we can test the formula on the universal bundle over $BU(n)$, and here we have $\delta^m=0$ for $m>n$ and $\lambda^j=0$ for $j>n$, while $\lambda^0,\lambda^1,\lambda^2,\dots,\lambda^n$ are linearly independent.

In our application we have $\varepsilon x=n\geqq 0$ and $\lambda^j x=0$ for $j>n$, by Lemma 1.5. So we conclude $\delta^m x=0$ for $m>n$.

But in the Atiyah-Hirzebruch spectral sequence for BU, the universal δ^m is of filtration $2m$, and its class in the E_2-term is the universal Chern class c_m. So from $\delta^m(x)=0$ we infer $c_m(x)=0$ for $m>n$. This proves the lemma.

The method of Berstein involves adjoining complex coefficients in order to provide roots for a suitable polynomial, and I explain this construction next.

The Chern character may be considered as a natural transformation of cohomology theories

$$ch\colon K(X)\to\prod_n H^{2n}(X;Q).$$

This natural transformation may be induced by a map of spectra; and we can afflict spectra with coefficients in any abelian group, by a construction functorial for maps of spectra. In particular we can introduce coefficients in the rational numbers Q, and we see that ch extends to a Q-linear natural transformation

$$\begin{aligned} ch_Q\colon K(X,Q) &\to \prod_n H^{2n}(X;Q\otimes_Z Q) \\ &= \prod_n H^{2n}(X;Q). \end{aligned}$$

This transformation is iso, as is well known. We may now introduce coefficients in the complex numbers C; since $Q\otimes_Z C=C$, we get a C-linear natural isomorphism

$$ch_C\colon K(X;\ C)\to \prod_n H^{2n}(X;C).$$

In particular we get

$$ch_C\colon \tilde{K}(X;C)\xrightarrow{\cong} \prod_{n>0} H^{2n}(X;C)$$

when X is connected.

The standard formulae for converting Chern classes to Chern characters and back again make sense whenever the coefficient ring contains the rational numbers Q; they give an isomorphism from the additive group $\prod_{n>0} H^{2n}(X;C)$ to the multiplicative group $G(X;C)$ of formal series

$$1+x_1+x_2+\cdots, \qquad \text{where } x_i\in H^{2i}(X;C).$$

In this way we obtain the following commutative diagram.

$$\begin{array}{ccc} & & \prod_{n>0} H^{2n}(X;C) \\ & \nearrow^{ch_C}_{\cong} & \updownarrow\cong \\ \tilde{K}(X;\ C) & \xrightarrow{cc} & G(X;\ C) \\ \uparrow & & \uparrow \\ \tilde{K}(X) & \xrightarrow{c} & G(X;Z). \end{array}$$

Here I should point out that the isomorphism from $\prod_{n>0} H^{2n}(X;C)$ to $G(X;C)$ is not the standard exponential map; first one has to apply the additive isomorphism

$$\alpha\colon \prod_{n>0} H^{2n}(X;C)\to \prod_{n>0} H^{2n}(X;C)$$

which in degree $2n$ is multiplication by $(-1)^{n-1}(n-1)!$, and then one applies the standard exponential map

$$\exp(x)=\sum_{n\geqq 0} x^n/n!.$$

Similarly, to get back from $G(X;C)$ to $\prod_{n>0} H^{2n}(X;C)$, one first applies the standard logarithmic map

$$\log(1+x)=\sum_{n\geqq 1}(-1)^{n-1}x^n/n$$

and then applies the additive isomorphism

$$\alpha^{-1}\colon \prod_{n>0} H^{2n}(X;C)\to \prod_{n>0} H^{2n}(X;C)$$

which in degree $2n$ is multiplication by $(-1)^{n-1}/(n-1)!$. Nevertheless, the diagram is good if correctly understood.

Lemma 6.2. *Suppose that $G=S^1$, so that $BG=CP^\infty$; let $y\in K(CP^\infty)$ and suppose $\lambda^i x=0$ for i sufficiently large. Then $x=\alpha(\sum_j \rho_j)$ for a suitable finite set of 1-dimensional representations $\rho_j\colon S^1\to U(1)$.*

Proof. By Lemma 1.5 we have $\varepsilon x=n\geqq 0$. We have $H^*(CP^\infty)=Z[h]$, where $h\in H^2(CP^\infty)$ is the generator; by Lemma 6.1 the total Chern class $c(x)$ is a polynomial in h of degree $\leqq n$; in $H^*(CP^\infty;C)$ we can factorise it in the form

$$c(x)=(1+\mu_1 h)(1+\mu_2 h)\dots(1+\mu_n h)$$

where the scalars μ_i lie in C. The unique corresponding element of $\prod_{n>0} H^{2n}(X;C)$ is

$$\exp(\mu_1 h)+\exp(\mu_2 h)+\cdots+\exp(\mu_n h)-n;$$

since we know that $\varepsilon x=n$, we deduce that

$$ch\, x=\exp(\mu_1 h)+\exp(\mu_2 h)+\cdots+\exp(\mu_n h).$$

Let η be the canonical line bundle over CP^∞, and let $y=\eta-1$, so that

$$K(CP^\infty) \quad =Z[[y]],$$
$$K(CP^\infty;C)=C[[y]].$$

Then in $K(CP^\infty;C)$ we can use the binomial series to construct the element

$$\eta^\mu=(1+y)^\mu$$
$$=1+\mu y+\frac{\mu(\mu-1)}{1.2}y^2+\cdots$$

for any $\mu\in C$; and this element has

$$ch_C(\eta^\mu)=1+\mu\, ch\, y+\frac{\mu(\mu-1)}{1.2}(ch\, y)^2+\cdots$$

$$=1+\mu((\exp h)-1)+\frac{\mu(\mu-1)}{1.2}((\exp h)-1)^2+\cdots$$

$$=\exp(\mu h),$$

as expected. Since ch_C is iso, we have

$$x=\eta^{\mu_1}+\eta^{\mu_2}+\cdots+\eta^{\mu_n}.$$

From the formal properties of the binomial series, we have

$$\eta^{\lambda+\mu}=\eta^\lambda\cdot\eta^\mu.$$

In particular, for any integer m we have

$$\eta^m x=\eta^{m+\mu_1}+\eta^{m+\mu_2}+\cdots+\eta^{m+\mu_n}$$

$$=\eta^{\nu_1}+\eta^{\nu_2}+\cdots+\eta^{\nu_n},\quad \text{say;}$$

and here we can certainly choose m so that the real parts of $\nu_1, \nu_2, \ldots, \nu_n$ are positive. In this case, we have

$$\left|\frac{\nu-(r-1)}{r}\right|\leqq\frac{r-1}{r}$$

for $\nu=\nu_1, \nu_2, \ldots, \nu_n$ and r sufficiently large (because if $\nu=c+id$, then $c^2-2c(r-1)+d^2$ is ultimately negative). Thus we have

$$\left|\frac{\nu(\nu-1)\ldots(\nu-(r-1))}{1.2\ldots r}\right|=O(1/r),$$

and this binomial coefficient tends to O as $r\to\infty$. Thus the coefficient of y^r in

$$\eta^m x=(1+y)^{\nu_1}+(1+y)^{\nu_2}+\cdots+(1+y)^{\nu_n}$$

tends to zero as $r\to\infty$. But since $\eta^m x\in K(CP^\infty)$, these coefficients are integers; so they are ultimately zero. Thus $\eta^m x$ can be written as a polynomial in y (with integer coefficients). Since $y^r=(\eta-1)^r$, $\eta^m x$ can be written as a finite linear combination (with integer coefficients) of the power η^r (where $r=0, 1, 2, \ldots$). Thus x can be written as a finite Laurent series $x=\sum_r a_r\eta^r$ (where $r\in Z$, $a_r\in Z$). So the Chern class satisfies

$$c(x)=\prod_r(1+rh)^{a_r}.$$

Comparing this with the original expression for $c(x)$, we see that the original roots μ_i were integers. So the expression

$$x=\eta^{\mu_1}+\eta^{\mu_2}+\cdots+\eta^{\mu_n}$$

writes x in the form

$$x=\alpha(\sum_j \rho_j).$$

This proves Lemma 6.2.

Now we can turn to a more general case.

Lemma 6.3. *Suppose that G is the product $T^d \times Z_{p^e}$ of a torus and a finite cyclic p-group. If $x \in K(BG)$ and $\lambda^i x=0$ for i sufficiently large, then $x=\alpha(\sum_j \rho_j)$ for a suitable finite set of 1-dimensional representations ρ_j: $G \to U(1)$; moreover the ρ_j are unique up to the order in which they are numbered.*

Proof. The map

$$\alpha\colon RG \to K(BG)$$

is mono, because points g where $\chi_g^\wedge$ is defined are dense in G; this makes the uniqueness statement obvious. (Incidentally, this gives a uniqueness statement for the special case considered in Lemma 6.2.)

Now we prepare to exploit the work in §5. Let $G_\infty = T_\infty^d \times Z_{p^e}$ be the p-torsion subgroup of G, considered as a discrete group. The injection $G_\infty \to G$ induces a map i: $BG_\infty \to BG$, and so we obtain an element $i^* x \in K(BG_\infty)$ such that $\lambda^k(i^* x)=0$ for k sufficiently large. By Proposition 5.1, there are 1-dimensional representations

$$\rho_j\colon G_\infty \to U(1)$$

such that

$$i^* x=\alpha(\sum_j \rho_j).$$

(The number of these representations is of course εx.) Let us regard the torus T^{d+1} as the additive group R^{d+1}/Z^{d+1}, and let us embed $T^d \times Z_{p^e}$ in T^{d+1} as the subgroup of points $(t_1, t_2, \ldots, t_d, t_{d+1})$ such that $p^e t_{d+1} \in Z$; similarly, let us regard $U(1)$ as R/Z. Then a homomorphism ρ_j of G_∞ may be given by a linear function

$$\rho_j(t_1, t_2, \ldots, t_d, t_{d+1})=\sum_k \gamma_{jk} t_k$$

where the coefficient γ_{jk} are p-adic integers, $\gamma_{jk} \in Z_p^\wedge$, and $\gamma_{j,d+1}$ may be altered $\bmod p^e$ without affecting the answer (so we may take $\gamma_{j,d+1}$ to be an ordinary integer $\bmod p^e$ if we wish).

Now let $S^1(k) \subset T^d$ be the subtorus of points

$$(0, \ldots, 0, t_k, 0, \ldots, 0)$$

(where $1 \leqq k \leqq d$). Then the image of $x \in K(BG)$ in $K(B(S^1(k) \cap G_\infty))$ has a form

$$i^* x=\alpha(\sum_j \sigma_j)$$

like that give above, except that the homomorphism

$$\sigma_j\colon S^1(k)\cap G_\infty \to U(1)$$

is given by

$$\sigma_j(0,\dots,0,t_k,0,\dots,0)=\gamma_{jk}t_k.$$

On the other hand, by Lemma 6.2, the image of x in $K(B(S^1(k)))$ has the form $\alpha(\sum_j \rho'_j)$ for suitable 1-dimensional representations

$$\rho'_j\colon S^1(k)\to U(1).$$

Thus the image of x in $K(B(S^1(k)\cap G_\infty))$ has a form

$$i^* x=\alpha(\sum_j \sigma'_j)$$

like that above, except that the coefficients γ'_{jk} are ordinary integers. By the uniqueness clause in Proposition 5.1, we see that the original coefficients γ_{jk} are ordinary integers; this holds for each k in the range $1\leqq k\leqq d$. Then the formula

$$\bar\rho_j(t_1,t_2,\dots,t_d,t_{d+1})=\sum_k \gamma_{jk}t_k$$

defines a homomorphism

$$\bar\rho_j\colon G\to U(1)$$

such that x and $\alpha(\sum_j \bar\rho_j)$ have the same image in $K(BG_\infty)$. Using the final clause of Theorem 2.3, we conclude that

$$x=\alpha(\sum_j \bar\rho_j) \quad \text{in } K(BG).$$

This proves Lemma 6.3.

Proof of Theorem 1.14. Let G be a monogenic group $T\times Z_n$ and let $x\in K(BG)$ be an element such that $\lambda^i x=0$ for i sufficiently large, as in the statement of the theorem. Let p run over the primes which divide n; we may assume there is at least one such prime, for if there are none then the result is true by Lemma 6.3. For each such p, let Z_{p^e} be the p-component of the group Z_n; then Lemma 6.3 applies to the subgroup $T\times Z_{p^e}$. Each 1-dimensional representation of this subgroup may be written $\rho\sigma$, where ρ is a 1-dimensional representation of T and σ is a 1-dimensional representation of Z_{p^e}. Let $i_p^* x$ be the image of x in $K(B(T\times Z_{p^e}))$; by Lemma 6.3, we have

$$i_p^* x=\alpha(\sum_j \rho_j\sigma_j).$$

Thus the image of x in $K(BT)$ has the form

$$\alpha(\sum_j \rho_j),$$

since each σ_j restricts to 1. By the uniqueness clause in Lemma 6.3, applied to the group T, the representations ρ_j do not depend (up to order) on the prime p chosen; let us arrange the ordering so that ρ_j is independent of p. Then for each p we can write

$$i_p^* x = \alpha(\sum_j \rho_j \rho_{j,p}).$$

By the Chinese remainder theorem, we can choose a representation

$$\sigma_j \colon Z_n \to U(1)$$

which restricts to $\sigma_{j,p}$ on each subgroup Z_{p^e}. Then x and $\alpha(\sum_j \rho_j \sigma_j)$ have the same image in $K(B(T \times Z_{p^e}))$ for each p; so

$$x = \alpha \sum_j (\rho_j \sigma_j),$$

by Theorem 2.3. This proves Theorem 1.14.

§ 7. Structure and Representations of Disconnected Lie Groups

This section is entirely devoted to the structure and representations of Lie groups which need not be connected. Its object is to prove certain results which will be needed in § 8; these are numbered as (7.1) to (7.6); topologists who find they can believe these results may well want to skip their proofs.

The results we need later can be separated into three groups, and first we have a lemma about fixed-point sets which can be separated from the rest of the discussion. Let G be a compact Lie group and A a finite group of automorphisms of G; let H be the subgroup of elements $g \in G$ fixed under A.

Lemma 7.1. *If $\pi_0 G$ and A are p-groups, then $\pi_0 H$ is a p-group.*

Secondly we have four results which help us to obtain fuctions $f \colon G \to C$ and prove that they are virtual characters; here a "virtual character" on G means the character of a virtual representation, i.e. an element of RG. The first result is both easy and well-known.

Lemma 7.2. *Let $\pi \colon G \to H$ be an epimorphism of compact Lie groups. If $f \colon H \to C$ is a continuous class function and $f\pi \colon G \to C$ is a virtual character on G, then f is a virtual character on H.*

The next result is more substantial.

Proposition 7.3. *Let $f \colon G \to C$ be a continuous class function, and suppose that $f|F$ is a virtual character for every finite subgroup $F \subset G$; then f is a virtual character. More precisely, let T be a maximal torus in G_1, and let NT be its normaliser in G; then it is sufficient to assume that $f|F$ is a virtual character for every finite subgroup $F \subset NT$. If $\pi_0 NT$ is a p-group, then it is sufficient to assume that $f|F$ is a virtual character for every finite p-group $F \subset NT$.*

Proposition 7.3 will provide the last step in the proof of Theorem 1.8. We also need a result corresponding to Proposition 7.3 to use as the last step in the proof of Theorem 1.13, and this result will be stated as Proposition 7.5. For this purpose, however, we cannot suppose that our function f is given on the whole of G; for our characters $\chi_g^\wedge(x)$ are defined only for g of prime-power order, and if g is of p-power order in G, then its image $\pi_0 g$ is of p-power order in $\pi_0 G$. I therefore define $\Gamma \subset G$ to be the union of those components of G which lie over elements of prime-power order in $\pi_0 G$. If $\chi_g^\wedge(x)$ is defined, then $g \in \Gamma$; conversely, we have the following result.

Lemma 7.4. *Elements of p-power order in G are dense in those components of G over elements of p-power order in $\pi_0 G$; hence, points g where $\chi_g^\wedge(x)$ is defined are dense in Γ.*

This gives us hope that for suitable elements x we may perhaps extend $\chi_g^\wedge(x)$ by continuity so as to get a continuous function $f\colon \Gamma \to C$; and this will actually be done by Lemma 8.1. This will put us in a position to use the following result.

Proposition 7.5. *For each compact Lie group G there is a positive integer $n = n(G)$ with the following property. Suppose that $\phi\colon \Gamma \to C$ is a continuous function, that*

$$\phi(g\gamma g^{-1}) = \phi(\gamma)$$

for all $g \in G$ and $\gamma \in \Gamma$, and that $\phi | H$ is a virtual character of H for every monogenic subgroup $H \subset \Gamma$; then $n\phi$ is the restriction to Γ of some virtual character of G.

Here, when one write $\phi(g\gamma g^{-1})$, one should note that inner automorphisms of G preserve Γ.

The third sort of result needed later concerns certain monogenic subgroups $H \subset G$ which are particularly important because they are analogous to maximal tori in a connected group. I shall call them "*SS* subgroups" in honour of Segal [17] and de Siebenthal [19] who first studied them. (But I must admit that it is easier to read "*SS* subgroup" as "simply splendid subgroup".) Before explaining, it is best to fix some notation, for it will be clearer if our notation shows when a component of G is considered as a submanifold of G, and when it is considered as an element of $\pi_0 G$. I shall write β for a typical element of $\pi_0 G$, and G_β for the submanifold $\pi_0^{-1}\beta$ of G over β; this is consistent with the use of G_1 for the identity-component. Conversely, if G_β is declared to be a component of G (considered as a submanifold), then β will mean the same component considered as an element of $\pi_0 G$. Similarly if G is replaced by a subgroup H.

I shall say that H_δ is an "*SS* component" if it satisfies the following conditions.

(i) H_δ is one component of a monogenic subgroup $H \subset G$.

(ii) H_δ is maximal subject to (i).

Each *SS* component H_δ lies in some component G_β.

In condition (i), we can suppose without loss of generality that the subgroup H is generated by H_δ; for otherwise we replace H by the part of H over the cyclic subgroup of $\pi_0 H$ generated by δ.

I shall call H an "*SS* subgroup" if it is the monogenic subgroup generated by an *SS* component.

For example, a torus T in G_1 satisfies (i); and if we have $T \subset H_\delta$, then H_δ must be an identity-component, and hence a torus; so a maximal torus in G_1 satisfies (i) and (ii). Conversely, any monogenic subgroup $H \subset G_1$ lies in a torus, and hence the SS components in G_1 are exactly the maximal tori.

I do not aim to say much more than we need about SS subgroups, but the following results will be needed in § 8.

Proposition 7.6. (i) *Each point of G lies in an SS component; in particular, each component G_β of G contains an SS component.*

(ii) *If H_δ is an SS component in G_β, then the conjugates $g_1 H_\delta g_1^{-1}$ of H_δ by elements $g_1 \in G_1$ cover G_β.*

(iii) *If H_δ is an SS component, and two elements h, h' in H_δ are conjugate by an element $g \in G$, then they are conjugate by an element in NH, the normaliser of H in G.*

(iv) *If H_δ is an SS component in G_β, and β is of p-power order in $\pi_0 G$, then δ is of p-power order in $\pi_0 H$.*

This completes the statement of results needed in § 8. The following results about SS subgroups are not needed in § 8, but will be needed in the course of this section.

Proposition 7.7. (i) *Any SS subgroup H is of finite index in its normaliser.*

(ii) *Two SS components H_δ, K_ε which lie in the same component G_β are conjugate by an element $g_1 \in G_1$.*

(iii) *Let H_δ be an SS component in G_β, generating an SS subgroup H, and let H_ε be another component of H which also lies in G_β; then H_ε is also an SS component, and generates the same SS subgroup H as H_δ.*

(iv) *Let H be an SS subgroup, and let H_ε be a component of H which need not generate H; then H_ε is contained in a unique SS component. In particular, H_1 is contained in a unique maximal torus of G_1.*

After (7.6)(i), the basic result is (7.6)(ii), which is due to de Siebenthal; see [19] pp. 57–58. The results (7.6)(iii)(iv) and (7.7)(i)(ii)(iii) are due to Segal, who also proves (7.6)(ii); see [17] § 1.

I can now turn to the proofs. The following lemma will be used several times.

Lemma 7.8. *Let G be a compact connected Lie group, and θ: $G \to G$ an automorphism; then G contains a maximal torus T mapped to itself by θ.*

Proof. Let M be the space of all maximal tori in G; this is homeomorphic to G/NT_0, where T_0 is any one maximal torus and NT_0 is its normaliser; thus M is a manifold with the same rational cohomology as a point. The automorphism θ induces a continuous map θ_*: $M \to M$, and θ_* has a fixed point by the Lefschetz theorem. This proves the lemma.

Proof of Lemma 7.1. First I consider the case in which $\pi_0 G = 1$, so that G is connected, and $A = Z_p$, generated say by α: $G \to G$, where $\alpha^p = 1$. In this case, let $h \in H$; then h lies in some maximal torus T' in G. Let $Z(h)$ be the centraliser of h in G; then $T' \subset Z(h)$, and therefore h lies in the identity-component $Z(h)_1$. Since

α maps h to h it maps $Z(h)$ to $Z(h)$ and $Z(h)_1$ to $Z(h)_1$; so by Lemma 7.8, there is a maximal torus T of $Z(h)_1$ which is preserved by α. Then α acts on T, on its universal cover $\tilde{T}$ and on the integer lattice $I \subset \tilde{T}$ (that is, the kernel of the projection $\tilde{T} \to T$). Since h lies in the centre of $Z(h)$, it lies in every maximal torus of $Z(h)_1$ and in particular in T; so there is a vector $\tilde{t} \in \tilde{T}$ which maps to h. Since α fixes h we have

$$(\alpha - 1)\tilde{t} \in I.$$

By setting

$$\tilde{v} = (\alpha^{p-1} + \alpha^{p-2} + \cdots + \alpha + 1)\tilde{t},$$

$$\tilde{w} = (\alpha^{p-2} + 2\alpha^{p-3} + \cdots + (p-2)\alpha + (p-1))(\alpha - 1)\tilde{t}$$

we secure

$$(\alpha - 1)\tilde{v} = 0,$$

$$\tilde{w} \in I \quad \text{and}$$

$$\tilde{v} = \tilde{w} + p\tilde{t}.$$

So the 1-parameter subgroup determined by $\tilde{v}$ lies in H_1 and contains h^p. Thus every element of $\pi_0 H$ has order dividing p; so $\pi_0 H$ is a p-group. This completes the first case.

Secondly I consider the case in which $A = Z_p$, but $\pi_0 G$ is general. In this case the image of the induced map $\pi_0 \mathrm{H} \to \pi_0 G$ is a subgroup of $\pi_0 G$ and hence a p-group, while the kernel of this map is $(H \cap G_1)/H_1$, which is a p-group by the first case. So $\pi_0 H$ is a p-group in this case too.

Thirdly I consider the general case, and here the proof is by induction over the order of A. Since A is a p-group we can find a normal subgroup which is a cyclic group Z_p; let K be the subgroup of elements $g \in G$ fixed under Z_p. Then $\pi_0 K$ is a p-group by the second case; A/Z_p acts on K, and H is the subgroup of elements $k \in K$ fixed under A/Z_p; since we assume the result for A/Z_p as our inductive hypothesis, $\pi_0 H$ is a p-group. This completes the induction, and proves Lemma 7.1.

For completeness I shall prove Propositions 7.6 and 7.7, even though most of the parts of these propositions are known.

Proof of (7.6)(i). This can be done by essentially the same argument that one uses for maximal tori. Any element $g \in G$ generates a monogenic group H; if we take H_δ to be the component of H containing g, then H_δ satisfies condition (i) in the definition of an "*SS* component". Such components H_δ have bounded dimension; if $H_\delta \subset H'_{\delta}$; and $\dim H_\delta = \dim H'_{\delta'}$ then $H_\delta = H'_{\delta'}$; so we see that any such H_δ is contained in one which is maximal.

Proof of (7.7)(i). I repeat Segal's proof. Let H be an *SS* subgroup, and NH its normaliser in G. Since $\mathrm{Aut}\, H$ is discrete, $(NH)_1$ acts trivially on H. If we had $\dim(NH)_1 > \dim H$, then we could increase the dimension of H by adjoining a further 1-parameter subgroup, contrary to the maximality condition in the definition of an *SS* component. Thus $(NH)_1 = H_1$.

I shall return to (7.6)(ii) later, when I can pick it up for free.

I now explain that the results on representation-theory in this section will be proved by using integration formulae (just as Hermann Weyl used his classical integration formula to study the representation-theory of compact connected Lie groups). To begin with, let G_β be a component of G and H_δ an SS component in G_β; I shall reduce an integral over G_β to an integral over H_δ.

Lemma 7.9. *There is a positive integer d and a virtual representation $\eta \in RH$ (given more precisely below) such that*

$$d \int_{G_\beta} f = \int_{H_\delta} f \chi_\eta$$

for all continuous functions $f\colon G_\beta \to C$ which satisfy

$$f(g_1 g g_1^{-1}) = f(g)$$

for all $g_1 \in G_1, g \in G_\beta$.

Here the integrals are Haar integrals; for convenience we take them normalised so that

$$\int_{G_\beta} I = 1, \qquad \int_{H_\delta} 1 = 1$$

but the normalisation only affects the value of d. With this normalisation we have

$$d = |NH_\delta / H_1|$$

where NH_δ is the subgroup of elements $g_1 \in G_1$ such that $g_1 H_\delta g_1^{-1} = H_\delta$; this index is finite by (7.7)(i). It remains to give the weight-function χ_η. Let $L(G)$, $L(H)$ be the Lie algebras of G, H – that is, their tangent spaces at the identity. Then H acts by inner automorphisms on G, fixing H; so we get a representation Ad of H on $L(G)/L(H)$. (This is a representation over the real field R, but it gives one over C.) Then

$$\eta = \lambda_{-1}(Ad) = \sum_{i \geqq 0} (-1)^i \lambda^i(\mathrm{Ad}) \in RH,$$

where λ_{-1} is the well-known "Euler class in complex K-theory".

Proof. Let us define a function

$$\phi\colon (G_1/H_1) \times H_\delta \to G_\beta$$

by

$$\phi(g_1, h) = g_1 h g_1^{-1}.$$

This definition is legitimate because H is abelian. The space G_1/H_1 is an orientable manifold because H_1 is connected; the space $(G_1/H_1) \times H_\delta$ is an orientable manifold of the same dimension as G_β.

In order to apply to ϕ the usual naturality formula for integration, we must work out $\phi^* \omega_G$, where ω_G is the Haar volume form on G. But this sentence glosses over a small snag. Since our group G may be neither connected nor abelian, it is perfectly possible for an inner automorphism θ of G to reverse the orientation at the identity (think of $O(2)$). Since θ preserves the Haar measure, it must change the sign of the Haar volume form, because forms know about signs although measures don't. Therefore there is no chance of having a Haar volume form invariant both under right translation and under left translation; we shall have to choose. The choice only affects two signs which cancel in the final equation, and can be taken at our convencience. I choose to have ω_G invariant under right translation. Of course it is then also invariant under left translation *by elements* $g_1 \in G_1$.

I return to the task of working out $\theta^* \omega_G$. I claim that it is sufficient to do so at the point $(1, h) \in (G_1/H_1) \times H_\delta$. In fact, we have a commutative diagram of the following form.

$$\begin{array}{ccc} (G_1/H_1) \times H_\delta & \xrightarrow{\phi} & G_\beta \\ \Big\downarrow{\scriptstyle l \times 1} & & \Big\downarrow{\scriptstyle c} \\ (G_1/H_1) \times H_\delta & \xrightarrow{\phi} & G_\beta \end{array}$$

Here l is left translation by an element $g_1 \in G_1$ and c is conjugation by g_1. We have $c^* \omega_G = \omega_G$ because $g_1 \in G_1$; so if we can solve our problem at $(1, h)$ we can obtain the solution at any other point by left translation.

Let $T_{(1,h)}((G_1/H_1) \times H_\delta)$ be the tangent space to $(G_1/H_1) \times H_\delta$ at the point $(1, h)$; it has as a subspace the tangent space $T_h(H_\delta)$ at h; the tangent space $T_h(G_\beta)$ also has $T_h(H_\delta)$ as a subspace. I claim that on this subspace the induced map ϕ_* is the identity. In fact, if we vary the variable h in the formula $\phi = g_1 h g_1^{-1}$ when g_1 is held fixed at 1, we produce the same variation in the outcome. So to calculate $\phi^* \omega_G$ we need to see the induced map of quotient spaces corresponding to these subspaces. For this purpose I will identify the tangent space $T_h(G_\beta)$ at h with the tangent space $T_1(G)$ at 1, using right translation because this preserved ω_G. Then the induced map of quotient spaces comes as the sum of two parts; the first part arises because $\phi = g_1 h g_1^{-1}$ depends on g_1 via the first factor g_1, and the second part arises because $\phi = g_1 h g_1^{-1}$ depends on g_1 via the last factor g_1^{-1}. The first part is the identity map, for when we vary g_1 in the formula $\phi' = g_1 h 1$, the change in ϕ' is the right translate of the change in g_1. The second part may be written $-\mathrm{Ad}\, h$, where "$\mathrm{Ad}\, h$" is the action of h by conjugation on $L(G)/L(H) = T_1(G)/T_1(H)$; for when we vary g_1 by a vector v we vary g_1^{-1} by $-v$, and if we vary g_1 in the formula $\phi'' = 1 h g_1$ by a vector v, then we vary ϕ'' by the right translate of hvh^{-1}.

Therefore, we conclude that at $(1, h)$ we have

$$\phi^* \omega_G = \det(1 - \mathrm{Ad}\, h)\, \omega_P$$

where ω_P is the obvious product volume form on $(G_1/H_1) \times H_\delta$.

The usual naturality formula for integration now shows that

$$(\deg \phi) \int_{[G_\beta]} f \omega_G = \int_{[(G_1/H_1) \times H_\delta]} (f\phi)(\phi^* \omega_G),$$

where $\deg\phi$ is the degree of the map

$$\phi\colon (G_1/H_1)\times H_\delta \to G_\beta .$$

The integral on the right may be done in two steps; first integrate over each fibre $g_1 \times H_\delta$, and then integrate the result over G_1/H_1. In our diagram

$$\begin{array}{ccc} (G_1/H_1)\times H_\delta & \xrightarrow{\ \phi\ } & G_\beta \\ {\scriptstyle l\times 1}\downarrow & & \downarrow{\scriptstyle c} \\ (G_1/H_1)\times H_\delta & \xrightarrow{\ \phi\ } & G_\beta \end{array}$$

the integrand $f\omega_G$ is invariant under c, by our assumption on f; so the integrand $(f\phi)(\phi^*\omega_G)$ is invariant under the left translation map $l\times 1$. Thus we get the same integral over each fibre, namely

$$\int_{[H_\delta]} f(h)\det(1-\operatorname{Ad} h)\,\omega_H .$$

Since we have arranged for the measure of G_1/H_1 to be 1, we get

$$(\deg\phi)\int_{[G_\beta]} f\omega_G = \int_{[H_\delta]} f(h)\det(1-\operatorname{Ad} h)\,\omega_H .$$

Next we must identify $\det(1-\operatorname{Ad} h)$ with the character $\chi_\eta(h)$ given. Of course $\det(1-\operatorname{Ad} h)$ does not change if we complexify. But then since H is a compact group we can assume that h acts on $L(G)/L(H)$ with complex eigenvalues $\mu_1,\mu_2,\ldots,\mu_r$, where r is the dimension of G/H; then

$$\det(1-\operatorname{Ad} h) = \prod_{1\le i\le r}(1-\mu_i),$$

and this agrees with $\chi_\eta(h)$.

Finally I will show that $\deg\phi$ is as stated. We can find in H_δ a generator h for the monogenic group H. First I will show that $\phi^{-1}h$ is a set of d points, where $d=|NH_\delta/H_1|$; and then I will show that at each point the Jacobian of ϕ is positive; this will prove that $\deg\phi=d$.

It is clear that for any $n\in NH_\delta$ we get a point $(n,n^{-1}hn)$ in $\phi^{-1}h$, and that two such n give the same point of $(G_1/H_1)\times H_\delta$ if and only if they lie in the same coset of H_1. Conversely, suppose that $\phi(g_1,h')=h$, so that $g_1^{-1}hg_1=h'$; since h is a generator for H, we deduce that $g_1^{-1}Hg_1\subset H$; then since $h'\in H_\delta$ we must have $g_1^{-1}H_\delta g_1=H_\delta$, and $g_1\in NH_\delta$.

For any point (g_1,h') in $\phi^{-1}h$ the element h' is a generator for H (being conjugate to h by an element of NH). The Jacobian of ϕ at this point is

$$\chi_\eta(h') = \prod_{1\le i\le r}(1-\mu_i')$$

by the work above; and here the eigenvalues μ_i' which are not ± 1 have unit modulus and occur in pairs of complex conjugates. So the Jacobian will be positive if we can exclude the possibility that some eigenvalue μ_i' is 1. If it is, then when h' acts on $L(G)$ it fixes a subspace of larger dimension than $L(H)$, and so it fixes a vector v not in $L(H)$. Since h' is a generator the whole of H must fix v, and then by adjoining to H the 1-parameter subgroup corresponding to v we can increase the dimension of H, contrary to the maximality condition in the definition of an SS component. This proves that $\deg \phi = d$, and completes the proof of Lemma 7.9.

Proof of (7.6)(ii). By the work above, the map

$$\phi: (G_1/H_1) \times H_\delta \to G_\beta$$

has positive degree; so it must be onto.

Proof of (7.7)(ii). This can be done by essentially the same argument that one uses for maximal tori. Let H_δ, K_ε be two SS components which lie in the same component G_β, and let $k \in K_\varepsilon$ be a generator for K. By (7.6)(ii) there exists $g_1 \in G_1$ such that

$$k \in g_1 H_\delta g_1^{-1}.$$

Since k is a generator we have

$$K \subset g_1 H g_1^{-1},$$

and taking components

$$K_\varepsilon \subset g_1 H_\delta g_1^{-1}.$$

Since K_ε is maximal we must have

$$K_\varepsilon = g_1 H_\delta g_1^{-1}.$$

Proof of (7.6)(iii). I repeat Segal's proof. Suppose that $h, h' \in H_\delta$ and $h' = ghg^{-1}$. Consider the SS components H_δ and $gH_\delta g^{-1}$; they have the point h' in common. Let $Z(h')$ be the centraliser of h' in G; then H_1 is a maximal torus in $Z(h')$, for if we could find a larger torus there, we could enlarge H_δ. Similarly gH_1g^{-1} is a maximal torus in $Z(h')$, for if we could find a larger torus there we could enlarge $gH_\delta g^{-1}$. By (7.7)(ii), there exists $z_1 \in Z(h')_1$ such that $z_1 g H_1 g^{-1} z_1^{-1} = H_1$. Then $z_1 g$ conjugates h into h' and also conjugates H_1 into H_1, so it lies in NH.

Proof of (7.7)(iii). Suppose that H_ε is another component of H which also lies in G_β. Let K be the subgroup of H over the cyclic subgroup of $\pi_0 H$ generated by ε. Then K is monogenic, and has a generator k in the component K_ε. By (7.6)(ii) we can write

$$k = g_1 h g_1^{-1}$$

for some $g_1 \in G_1$ and $h \in H_\delta$. Conjugation with g_1^{-1} maps k into H and so maps the whole of K into H since k is a generator; say it defines a map $\psi: K \to H$. Since ψ carries k to h, the induced map $\psi_*: \pi_0 K \to \pi_0 H$ carries ε to δ; so it must be epi. Therefore $|\pi_0 K| \geqq |\pi_0 H|$; but by construction $\pi_0 K$ is a subgroup of $\pi_0 H$, so that $\pi_0 K = \pi_0 H$ and $K = H$. That is, H_ε generates the same subgroup as H_δ.

It is now clear that if we could enlarge H_ε that would enlarge H_δ, which is impossible; so H_ε is an SS component.

Proof of (7.6)(iv). Suppose that β is of p-power order; then we can certainly find ε in the p-torsion subgroup of $\pi_0 H$ which maps to $\beta \in \pi_0 G$; then ε generates $\pi_0 H$ by (7.7)(iii), and so $\pi_0 H$ must be a p-group.

This completes the proof of Proposition 7.6.

Proof of Lemma 7.4. Let G_β be the component of G over an element β of p-power order in $\pi_0 G$. Then every point g of G_β lies in an SS component H_δ, by (7.6)(i); the corresponding SS subgroup has the form $T \times Z_{p^e}$ by (7.6)(iv), and so elements of p-power order are dense in H.

The next lemma is needed for the proof of (7.7)(iv).

Lemma 7.10. *Let g be an element of a compact Lie group G. Then either* (i) *G_1 is a torus, or* (ii) *there is an element $g_1 \in G_1$ and a 1-parameter subgroup of G_1 which commutes with $g_1 g$.*

Proof. By Lemma 7.8, conjugation with g preserves a suitable maximal torus $T \subset G_1$. Let C be a fundamental Weyl chamber in the Stiefel diagram $\tilde{T}$; then gCg^{-1} is another Weyl chamber, and we can choose $g_1 \in G_1$ so that conjugation with g_1 preserves T and throws gCg^{-1} onto C.

Choose a vector v in the interior of C. We can average hvh^{-1} as h runs over the monogenic subgroup H generated by $g_1 g$ (this amounts to averaging over $\pi_0 H$, since the identity-component H_1 must act trivially on T). We obtain a vector u fixed under $g_1 g$.

If G_1 is not a torus then it has at least one root, and each root takes the same sign on all the vectors $(g_1 g)^n v (g_1 g)^{-n}$; from this we see that $u \neq 0$, and u determines a 1-parameter subgroup which commutes with $g_1 g$. This proves the lemma.

Proof of (7.7)(iv). Let H be the given SS subgroup of G, generated say by an SS component H_δ; and let $Z(H_1)$ be the centraliser of H_1 in G. Since H_1 is normal in $Z(H_1)$, we can form the quotient $K = Z(H_1)/H_1$; this is again a compact Lie group. Let $h \in H_\delta$; then $h \in Z(H_1)$, and h yields an element of K. Thus Lemma 7.10 applies to K, and yields two alternatives; let us first examine alternative (ii). In this case we obtain an element of K_1 and lift it to an element $z_1 \in Z(H_1)_1$; we also obtain a 1-parameter subgroup of K. Then $z_1 h$ lies in the same component G_β of G as h does, fixes H_1, and also fixes one more one-parameter subgroup; thus we obtain a monogenic group K of greater dimension than H and with a component K_ε in G_β. Then K_ε must be contained in an SS component of greater dimension than H_δ, contradicting (7.7)(ii). This contradiction shows that alternative (i) must apply; K_1 is a torus and therefore $Z(H_1)_1$ is a torus.

Now let h' lie in the given component H_ε. The subgroup of $Z(H_1)_1$ fixed under h' has as its identity-component a torus T. Let L_ϕ be any SS component containing H_ε; then L contains H_1, and hence lies in $Z(H_1)$; also L is fixed under h'; thus $L_1 \subset T$. But T and h' generate a monogenic subgroup M, so by the maximality of L_ϕ we have $T \subset M_1 \subset L_1$. Thus $L_1 = T$, and L_ϕ is unique. This proves (7.7) (iv).

We are now ready to go back to integration formulae, and give a formula which relates an integral over G to one over NT. Here, of course, T is a maximal torus in G_1, and NT is the normaliser of T in G. I need to give the weight-function in the formula. Let $L(G)$ and $L(T)$ be the Lie algebras of G and T. The group NT acts by inner automorphism on G and T, and this induces an action Ad of NT on $L(G)/L(T)$.

Let

$$\xi=\lambda_{-1}(\mathrm{Ad})=\sum_{i=0}^{\infty}(-1)^i\lambda^i(\mathrm{Ad})\in R(NT).$$

Proposition 7.11. *For any continuous class function f: $G\to C$ we have*

$$\int_{g\in G} f(g)=\int_{n\in NT} f(n)\,\chi_\xi(n).$$

Here the integrals are Haar integrals, normalised so that

$$\int_{g\in G} 1=1, \qquad \int_{n\in NT} 1=1.$$

This formula reduces to the Weyl integration formula when G is connected, because the integrand on the right is then zero on all components of NT except T, as we shall see later; while T has $1/|W|$ of the measure of NT, where $|W|$ is the order of the Weyl group.

Proof of Proposition 7.11. Let G_β be a component of G; first I claim that G_β contains at least one component $(NT)_\varepsilon$ of NT which contains an SS component H_δ for G_β.

In fact, let H_δ be an SS component in G_β, and let $h\in H_\delta$ be a generator for H. By Lemma 7.8, conjugation with h preserves some maximal torus T' in G_1, and we may write $T=g_1T'g_1^{-1}$; replacing H_δ by $g_1H_\delta g_1^{-1}$ and h by $g_1hg_1^{-1}$, we may suppose without loss of generality that conjugation with h preserves T; thus $h\in NT$ and, since h generates H, $H\subset NT$. We have only to take $(NT)_\varepsilon$ to be the component of NT containing H_δ.

Suppose then that we have

$$H_\delta\subset(NT)_\varepsilon\subset G_\beta.$$

Then H_δ is also an SS component for $(NT)_\varepsilon$; for if we cannot enlarge it in G_β, we cannot enlarge it in $(NT)_\varepsilon$. Thus Lemma 7.9 gives the following two equations.

$$|NH_\delta/H_1|\int_{g\in G_\beta} f(g)=\int_{h\in H_\delta} f(h)\,\chi_\eta(h)$$

$$|MH_\delta/H_1|\int_{n\in(NT)_\varepsilon} f(n)\,\chi_\xi(n)=\int_{h\in H_\delta} f(h)\,\chi_\xi(h)\,\chi_\zeta(h).$$

Here MH_δ is the analogue of NH_δ with G replaced by NT; that is, it is the subgroup of elements $t\in(NT)_1=T$ such that $tH_\delta t^{-1}=H_\delta$. Similarly, ζ is the analogue of η with G replaced by NT. Moreover, we must remember that the integrals are still normalised so that

$$\int_{g\in G_\beta} 1=1,$$

and similarly for the other groups.

Since $L(NT)=L(T)$, we have the following exact sequence of representations of H.

$$L(NT)/L(H)\to L(G)/L(H)\to L(G)/L(T)$$

Applying λ_{-1}, we find

$$\chi_\eta(h)=\chi_\xi(h)\,\chi_\zeta(h);$$

thus the two integrals over H_δ are equal, and we obtain

$$|NH_\delta/MH_\delta|\int_{g\in G_\beta} f(g)=\int_{n\in(NT)_\varepsilon} f(n)\,\chi_\xi(n).$$

In this equation, the coefficient $|NH_\delta/MH_\delta|$ is written in terms of H_δ, but the ratio of the two integrals cannot depend on the choice of H_δ inside $(NT)_\varepsilon$. I will show that

$$|NH_\delta/MH_\delta|=|\pi_0 K(\varepsilon)|,$$

where $K(\varepsilon)$ is the subgroup of elements $k\in G_1\cap NT$ such that $k(NT)_\varepsilon k^{-1}=(NT)_\varepsilon$.

First I will show that $NH_\delta\subset K(\varepsilon)$. By definition, NH_δ is the subgroup of elements $g_1\in G_1$ such that $g_1 H_\delta g_1^{-1}=H_\delta$. Conjugation with such an element g_1 preserves H, and so preserves T because T is the unique maximal torus of G containing H_1 by (7.7)(iv). Thus $g_1\in G_1\cap NT$; since conjugation with g_1 preserves H_δ it preserves $(NT)_\varepsilon$, and we have $NH_\delta\subset K(\varepsilon)$.

Secondly I note that the identity-component of $K(\varepsilon)$ is T, and we have $MH_\delta=T\cap NH_\delta$ by definition. Thus we get a monomorphism

$$NH_\delta/MH_\delta\to\pi_0 K(\varepsilon).$$

Thirdly I have to show that this monomorphism is epi. Take an element $k\in K(\varepsilon)$; then $kH_\delta k^{-1}$ is another SS component of G inside $(NT)_\varepsilon$, and so by (7.7)(i) applied to NT there is an element $t\in T$ such that

$$tkH_\delta k^{-1}t^{-1}=H_\delta.$$

Then $tk\in NH_\delta$ and tk represents the same element of $\pi_0 K(\varepsilon)$ as k. This completes the proof that

$$|NH_\delta/MH_\delta|=|\pi_0 K(\varepsilon)|.$$

At this stage we have the formula

$$|\pi_0 K(\varepsilon)|\int_{g\in G_\beta} f(g)=\int_{n\in(NT)_\varepsilon} f(n)\,\chi_\xi(n).$$

This equation holds for each component $(NT)_\varepsilon$ which lies in G_β and happens to contain an SS component H_δ for G_β. I will show that if G_β happens to contain any other components of NT, then the factor $\chi_\xi(n)$ is identically zero on such components. For let $(NT)_\varepsilon$ be a component on which $\chi_\xi(n)$ is not identically zero, and let h be a point at which it is not zero. Then h lies in an SS component

H_δ with respect to NT. Since $\chi_\xi(h) \neq 0$, the action of h on LG/LT has no eigenvalues $+1$; it follows that H_δ is also an SS component with respect to G. By contradiction, then, if $(NT)_\varepsilon$ does not contain an SS component for G, χ_ξ is identically zero on $(NT)_\varepsilon$.

The next step is to count how many components $(NT)_{\varepsilon'}$ in G_β happen to contain SS components H_δ for G_β, and I claim that the number is

$$|\pi_0(G_1 \cap NT)|/|\pi_0 K(\varepsilon)|.$$

Certainly, if NT_ε happens to contain an SS component H_δ, then its conjugate by an element in $G_1 \cap NT$ also does and lies in the same component G_β of G; and by definition of $K(\varepsilon)$, it is the subgroup of elements in $G_1 \cap NT$ which give us the same component $(NT)_\varepsilon$. So it remains to prove that all components $(NT)_{\varepsilon'}$ arise in this way. Suppose then that components $(NT)_\varepsilon, (NT)_{\varepsilon'}$ contain SS components H_δ, $H'_{\delta'}$ for G_β. By (7.7) (ii) there is an element $g_1 \in G_1$ such that $g_1 H_\delta g_1^{-1} = H'_{\delta'}$. Then

$$g_1 T g_1^{-1} \supset g_1 H_1 g_1^{-1} = H'_1 \subset T.$$

But by (7.7)(iv) H'_1 is contained in a unique maximal torus of G; so $g_1 T g_1^{-1} = T$ and $g_1 \in G_1 \cap NT$. This proves the claim.

At this stage, then, we have the formula

$$|\pi_0(G_1 \cap NT)| \int_{g \in G_\beta} f(g) = \int_{n \in G_\beta \cap NT} f(n)\,\chi_\xi(n).$$

Here we clearly have

$$|\pi_0(G_1 \cap NT)| = |\pi_0(NT)|/|\pi_0 G|;$$

and adding over all the components of G, we obtain

$$\frac{1}{|\pi_0 G|} \int_{g \in G} f(g) = \frac{1}{|\pi_0 NT|} \int_{n \in NT} f(n)\,\chi_\xi(n).$$

But this still refers to Haar integrals normalised so that

$$\int_{g \in G_1} 1 = 1, \qquad \int_{t \in T} 1 = 1;$$

if we change to Haar integrals normalised so that

$$\int_{g \in G} 1 = 1, \qquad \int_{n \in NT} 1 = 1$$

we get

$$\int_{g \in G} f(g) = \int_{n \in NT} f(n)\,\chi_\xi(n).$$

This completes the proof of Proposition 7.11.

We also need the following integration formula for use in proving Proposition 7.5.

Proposition 7.12. *For each compact Lie group G there is a positive integer $n=n(G)$, a set of SS subgroups $H_i \subset \Gamma$ and a set of virtual representations $\eta_i \in RH_i$ such that*

$$n \int_{\Gamma} f = \sum_i \int_{H_i} f \chi_{\eta_i}$$

for all continuous functions $f\colon \Gamma \to C$ such that $f(g_1 \gamma g_1^{-1}) = f(\gamma)$ for all $g_1 \in G_1$, $\gamma \in \Gamma$.

Here the integrals are Haar integrals, normalised so that

$$\int_G 1 = 1, \qquad \int_{H_i} 1 = 1.$$

Proof. This does not demand quite so much care as (7.11); and there is no great point in economising on n, so I won't.

Divide the elements β of prime-power order in $\pi_0 G$ into equivalence classes, setting $\beta \simeq \beta'$ if they generate the same cyclic subgroup of $\pi_0 G$. The indices i will correspond to the equivalence classes. Suppose that β is of p-power order in $\pi_0 G$ and lies in the ith equivalence class; let H_δ be an SS component in G_β, and let $H = H_i$ be the SS subgroup generated by H_δ; then δ is of p-power order in $\pi_0 H$ by (7.6) (iv). It follows that if $\beta' \simeq \beta$ and $H_{\delta'}$ is a component of H in $G_{\beta'}$, then δ' also generates $\pi_0 H$; consequently $H_{\delta'}$ generates H, and $H_{\delta'}$ is an SS component in $G_{\beta'}$. The factor

$$d' = |NH_{\delta'}/H_1|$$

in Lemma 7.9 is the same for all such δ'; therefore we may add the formula of Lemma 7.9 over all such pairs $H_{\delta'} \subset G_{\beta'}$, and we get a formula of the following form.

$$de \int_{U_i} f = \int_{U_i \cap H} f \chi_\eta .$$

Here U_i is the union of the components $G_{\beta'}$ as β' runs over the ith equivalence class, and e is the number of components $H_{\delta'}$ in any component $G_{\beta'}$; this is independent of β', being the order of the kernel of $\pi_0 H \to \pi_0 G$.

The next step is to modify χ_η so that it becomes zero over the components of H which are not in $U_i \cap H$. If $\beta = 1$, there is nothing to be done. Otherwise we have

$$H = T \times Z_{p^e}$$

with $e \geqq 1$. Now, it is easy to construct a virtual character on Z_p which is 0 at 1 and p elsewhere; just take $p - \rho$, where ρ is the regular representation of Z_p. By using the quotient map

$$H = T \times Z_{p^e} \to Z_{p^e} \to Z_p$$

we get from $p - \rho$ a virtual representation ζ of H whose character is zero over non-generators of $\pi_0 H$, p over generators of $\pi_0 H$. If $\beta = 1$ we interpret ζ as p.

Then we obtain

$$p\,de \int_{U_i} f = \int_H f\chi_{\eta\zeta}.$$

This still refers to Haar integrals normalised so that

$$\int_{G_1} 1 = 1, \quad \int_{H_1} 1 = 1;$$

if we change to Haar integrals normalised so that

$$\int_G 1 = 1, \quad \int_H 1 = 1$$

we get

$$p\,de|\pi_0 G| \int_{U_i} f = |\pi_0 H| \int_H f\chi_{\eta\zeta};$$

but here $|\pi_0 H|$ divides $e|\pi_0 G|$, so we get

$$n_i \int_{U_i} f = \int_{H_i} f\chi_{\eta\zeta}.$$

Let n be a common multiple for the integers n_i; then for each i we have

$$n \int_{U_i} f = \int_{H_i} f\chi_{\eta_i}$$

where $\eta_i = (n/n_i)\eta\zeta$; adding over the equivalence classes, we get

$$n \int_\Gamma f = \sum_i \int_{H_i} f\chi_{\eta_i}.$$

This proves the proposition.

We can now complete the proofs of the results in this section; this involves four proofs which are closely parallel.

Proof of Lemma 7.2. This is known, but I give it to introduce the argument.

Let $\pi: G \to H$ be an epimorphism. If $f: H \to C$ is a continuous class function, then we have an expansion

$$f = \sum_\rho \lambda_\rho \chi_\rho;$$

here ρ runs over the irreducible representations of H, $\lambda_\rho \in C$, and the expansion is convergent in mean square. We can determine the coefficients λ_ρ from the formula

$$\lambda_\rho = \int_H f(h)\bar{\chi}_\rho(h).$$

But this gives

$$\lambda_\rho = \int_G f(\pi g)\bar{\chi}_{\pi^*\rho}(g);$$

and since we assume that $f\pi$ is a virtual character on G, this integral gives an integer. Since the sum

$$\sum_{\rho} |\lambda_\rho|^2$$

is convergent, the integers λ_ρ must be zero for all but a finite number of ρ. Thus our initial expansion displays f as the character of the virtual representation

$$\sum_{\rho} \lambda_\rho \rho \in RH.$$

This proves Lemma 7.2.

Lemma 7.13. *Let $f: G \to C$ be a continuous class function, and suppose that $f|NT$ is a virtual character; then f is a virtual character.*

Proof. This is done by the same argument, but using Proposition 7.11 to do the integral. As in the previous proof, we can approximate f (in mean square) by an expansion

$$\sum_{\rho} \lambda_\rho \chi_\rho$$

and compute the coefficients λ_ρ from

$$\lambda_\rho = \int_G f \bar{\chi}_\rho.$$

By Proposition 7.11 this gives

$$\lambda_\rho = \int_{n \in NT} f(n) \bar{\chi}_\rho(n) \chi_\xi(n);$$

and since we assume that $f|NT$ is a virtual character, this gives an integer. The rest of the argument goes as before.

Proof of Proposition 7.5. This is done the same way, but using Proposition 7.12. Let n be as in Proposition 7.12, and let $\phi: \Gamma \to C$ be as given in (7.5). Define the function $f: G \to C$ to be $n\phi$ on Γ, 0 off Γ; then f is a continuous class function. As in the previous two proofs we can approximate f (in mean square) by an expansion

$$\sum_{\rho} \lambda_\rho \chi_\rho$$

and compute the coefficients λ_ρ from

$$\lambda_\rho = \int_G f \bar{\chi}_\rho.$$

This gives

$$\lambda_\rho = n \int_\Gamma \phi \bar{\chi}_\rho,$$

and using Proposition 7.12 we get

$$\lambda_\rho = \sum_i \int_{H_i} \phi \bar{\chi}_\rho \chi_{\eta_i}.$$

Here $\phi | H_i$ is a virtual character according to our data, and so each integral gives an integer. The rest of the argument goes as before.

Proof of Proposition 7.3. Let NT be as above. We can approximate NT by finite groups by using the construction in the proof of Lemma 2.6; we obtain the following diagram of extensions.

$$\begin{array}{ccccc}
\operatorname{Ker} m_1 & \longrightarrow & \operatorname{Ker} m_2 & \longrightarrow \cdots \longrightarrow & T \\
\downarrow & & \downarrow & & \downarrow \\
F(1) & \longrightarrow & F(2) & \longrightarrow \cdots \longrightarrow & NT \\
\downarrow & & \downarrow & & \downarrow \\
\pi_0 NT & \xrightarrow{1} & \pi_0 NT & \longrightarrow \cdots \longrightarrow & \pi_0 NT
\end{array}$$

Here the groups $F(r)$ are finite subgroups of NT; and if $\pi_0 NT$ is a p-group, we can arrange that each group $F(r)$ is a p-group.

Now let $f: G \to C$ be a continuous class function. As in the previous three proofs, we can approximate $f | NT$ (in mean square) by an expansion

$$\sum_\rho \lambda_\rho \chi_\rho$$

where ρ runs over the irreducible representations of NT. As before, we can compute the coefficients λ_ρ from

$$\lambda_\rho = \int_{NT} f \bar{\chi}_\rho.$$

But here the integrand is continuous, and it is clear that the points of $F(1)$, $F(2), \ldots$ are equidistributed in NT; so we can approximate this integral by taking

$$\int_{F(r)} f \bar{\chi}_\rho.$$

Here we assume that $f | F_r$ is a virtual character, and so this integral gives an integer. If we can approximate λ_ρ by integers, it must be an integer. The rest of the argument goes as before, and shows that $f | NT$ is a virtual character. But then f is a virtual character on G, by Lemma 7.13. This completes the proof of all the results in this section.

§8. Proofs for the General Case

The object of this section is to complete proving the main theorems. This will now take us only three lemmas. The fundamental tool is the first one. Let Γ be as in §7.

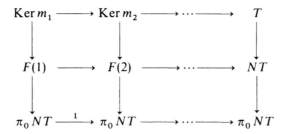

Lemma 8.1. *Let $x \in FF(BG)$; then $\chi_g^{\wedge}(x)$ extends to one and only one continuous function $f\colon \Gamma \to C$. Moreover this extension satisfies*

$$f(g\gamma g^{-1}) = f(\gamma)$$

for all $g \in G$, $\gamma \in \Gamma$; and its restriction to any monogenic subgroup $H \subset \Gamma$ is a virtual character.

Proof. Let $\beta \in \pi_0 G$ be an element of p-power order. By (7.6) (i) there is an SS component $H_\delta \subset G_\beta$; let H be the SS subgroup it generates. Since H is monogenic and $x \in FF(BG)$, the image of x in $K(BH)$ has the form $\alpha\xi$ for some $\xi \in RH$, by Theorem 1.14; thus on H the function $\chi_h^{\wedge}(x)$ extends to a continuous function $\chi_h(\xi)$.

Define functions

$$\phi\colon (G_1/H_1) \times H_\delta \to G_\beta$$
$$e\colon (G_1/H_1) \times H_\delta \to C$$

by

$$\phi(g_1, h) = g_1 h g_1^{-1}$$
$$e(g_1, h) = \chi_h(\xi).$$

I claim that e factors through ϕ. In fact, suppose that

$$\phi(g_1, h) = \phi(g'_1, h');$$

then h and h' are conjugate by an element of G_1; by (7.6) (iii), h and h' are conjugate by an element n in NH. But conjugation with such an element preserves the value of $\chi_h^{\wedge}(x)$; that is, it preserves the values of $\chi_h(\xi)$ at a set of point h which by using (7.6) (iv) we see to be dense in H; so it preserves $\chi_h(\xi)$ by continuity. Thus

$$e(g_1, h) = e(g'_1, h'),$$

and e factors through ϕ as claimed.

The map

$$\phi\colon (G_1/H_1) \times H_\delta \to G_\beta$$

is onto by (7.6) (ii), and is a quotient map since its source and target are compact Hausdorff spaces. So we get the following commutative diagram.

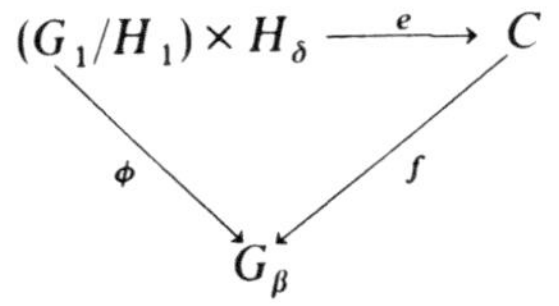

Here f is continuous and agrees with $\chi_g^{\wedge}(x)$ at all points in G_β where $\chi_g^{\wedge}(x)$ is defined.

This argument applies to each component G_β which is contained in Γ, and so we obtain the required extension. It is clearly unique because $\chi_g^{\wedge}(x)$ is defined for a set of points g dense in Γ, by (7.4).

An inner automorphism of G preserves $\chi_g^{\wedge}(x)$, and therefore preserves its unique continuous extension. This proves that

$$f(g\gamma g^{-1}) = f(\gamma).$$

Finally, let $H \subset \Gamma$ be a monogenic subgroup. Then a generator $h \in H$ lies in some SS component by (7.6)(i); and so we may assume without loss of generality that H is an SS subgroup. Using (7.6) (iv), we see that an SS subgroup in Γ has the form $T \times Z_{p^e}$ for some prime p; therefore points g where $\chi_g^{\wedge}(x)$ is defined are dense in H. According to Theorem 1.14, the image of x in $K(BH)$ has the form $\alpha\eta$ for some $\eta \in RH$; then the restriction of f to H agrees with χ_η, by continuity. This proves Lemma 8.1.

Proof of Theorem 1.13. Let $x \in FF(BG)$. By Lemma 8.1, $\chi_g^{\wedge}(x)$ extends to a continuous function $f: \Gamma \to C$ with the properties stated in (8.1). Let $n = n(G)$ be as in Proposition 7.5; then by Proposition 7.5, nf is the restriction to Γ of the character χ_η of some virtual representation $\eta \in RG$. Then $nx = \alpha\eta$ by Theorem 2.3. This proves Theorem 1.13.

The main aim of the next stretch of work is to prove Theorem 1.8. However, this demands some care because we have to go via various subgroups $G' \subset G$ such that $\pi_0 G'$ is *not* the union of its Sylow subgroups; so we shall have to obtain the values of our characters at some points $g' \in G'$ from their values at nearby points of G which are not in G'. The next two lemmas are special cases of Theorem 1.7.

Lemma 8.2. *Suppose that the identity-component G_1 is a torus T and that $\pi_0 G$ is a p-group; then $FF(BG) = \operatorname{Im}\alpha$.*

Proof. Let G be as given, and let $x \in FF(BG)$. By Lemma 8.1, we can extend $\chi_g^{\wedge}(x)$ to a continuous class function $f: G \to C$. Let F be a finite p-group contained in G; then the image of x in $K(BF)$ has the form $\alpha\eta$ for some $\eta \in RF$, by Theorem 1.10. Since $\chi_g^{\wedge}(x)$ is defined for all points $g \in F$, the function $f|F$ is a virtual character on F. By Proposition 7.3, f is the character of some virtual representation $\xi \in RG$. Then $x = \alpha\xi$ by Theorem 2.3.

Lemma 8.3. *Suppose that G is a quotient of the direct product of a connected group and a p-group; then $FF(BG) = \operatorname{Im}\alpha$.*

Proof. First I suppose that G is the direct product of a connected group G_1 and a p-group F. Let T be a maximal torus in G_1; then $G_1 \times F$ contains the subgroup $T \times F$. Let $x \in FF(BG)$; then by Lemma 8.2, the image of x in $K(B(T \times F))$ has the form $\alpha\eta$ for some $\eta \in R(T \times F)$. We may write η uniquely in the form

$$\eta = \sum_i \eta_i \zeta_i,$$

where $\zeta_1, \zeta_2, \ldots, \zeta_m$ are the irreducible representations of F, and the coefficients η_i lie in RT. Now the Weyl group W of G_1 acts on $G_1 \times F$ and $T \times F$ and fixes x; therefore it fixes each η_i. By the main theorem of Hermann Weyl – see for example [2] Theorem 6.20 p 153 – there are elements $\xi_i \in RG$ such that ξ_i restricts to $\eta_i \in RT$. Then the formula

$$\xi = \sum_i \xi_i \zeta_i$$

defines an element $\xi \in RG$, and we have

$$\chi_g^{\wedge}(x) = \chi_g^{\wedge}(\alpha\,\xi)$$

since every point $g \in G = G_1 \times F$ is conjugate to a point in $T \times F$. Thus $x = \alpha\,\xi$ by Theorem 2.3. This proves the special case in which G is a direct product.

Now suppose that we are given a quotient map $\pi\colon H \to G$, where H is the direct product of a connected group and a p-group. Let $x \in FF(BG)$; then $\chi_g^{\wedge}(x)$ extends to a continuous class function $f\colon G \to C$, by Lemma 8.1. Then $f\pi(h)$ is a continuous class function on H extending $\chi_h^{\wedge}(\pi^* x)$. Here $\pi^* x \in FF(BH)$, so by the result just proved for H, $\pi^* x \in \operatorname{Im}\alpha$ and we have

$$\chi_h^{\wedge}(\pi^* x) = \chi_h(\eta)$$

for some $\eta \in RH$. Points h where $\chi_h^{\wedge}$ is defined are dense in H by (7.4), so we get

$$f\pi(h) = \chi_h(\eta),$$

and $f\pi$ is a virtual character. So f is a virtual character by Lemma 7.2; that is,

$$f(g) = \chi_g(\xi)$$

for some $\xi \in RG$. Then $x = \alpha\,\xi$ by Theorem 2.3. This proves Lemma 8.3.

Proof of Theorem 1.8. Suppose that $\pi_0 G$ is the union of its Sylow subgroups, and that $x \in FF(BG)$. Let $f\colon G \to C$ be as in Lemma 8.1. First I aim to prove that for any finite subgroup $F \subset G$, the restriction $f|F$ of f to F is a virtual character.

By the Brauer induction theorem [14, 18] it is sufficient to prove this when F is an elementary subgroup E. Here I recall that an elementary subgroup E is, at least, a direct product of p-groups S_p for different primes p. I claim that all the factors S_p of E must be contained in G_1, except for one at most. In fact, let $s \in S_p$, $t \in S_q$ where p and q are distinct primes; then s and t commute in E. Therefore the images of s and t in $\pi_0 G$ commute, and generate a cyclic subgroup of order $p^\alpha q^\beta$ for some α, β. But we assume that $\pi_0 G$ is the union of its Sylow subgroups; so either $\alpha = 0$ or $\beta = 0$, that is, either $s \in G_1$ or $t \in G_1$. This proves the claim above.

It follows that we may write $E = D \times S_p$, where $D \subset G_1$ and the order $|D|$ of D is prime to p. Then S_p acts by inner automorphisms on G_1, fixing D; let H be the subgroup of G_1 consisting of elements $g_1 \in G_1$ fixed under S_p. By Lemma 7.1, $\pi_0 H$ is a p-group. Since the order $|D|$ of D is prime to p, the image of D in $\pi_0 H$ is 1, and we have $D \subset H_1$.

Since H_1 and S_p commute we get a map

$$\theta\colon H_1 \times S_p \to G;$$

let the image of θ be L. Then Lemma 8.3 applies to L; since $x \in FF(BG)$, we infer that the image of x in $K(BL)$ lies in $\operatorname{Im}\{\alpha_L\colon RL \to K(BL)\}$. Since points l where $\chi_l^{\wedge}$ is defined are dense in L, this shows that the restriction of f to L is a virtual character. Since we have $E \subset L$, we obtain the corresponding conclusion for E. This proves that the restriction of f to any finite subgroup F is a virtual character.

Proposition 7.3 now shows that f is the character of some virtual representation $\xi \in RG$. Then

$$\chi_g^\wedge(x) = \chi_g^\wedge(\alpha\xi)$$

for all g for which they are defined; so $x = \alpha\xi$ by Theorem 2.3. This completes the proof of Theorem 1.8, which in turn completes the proof of Corollary 1.9 and reproves Theorem 1.2.

Proof of Theorem 1.11. Suppose $x \in FF(BG)$. Then the image of x in $K(BG_p)$ lies in $FF(BG_p)$, and hence in $\mathrm{Im}\,\alpha_p$ by Theorem 1.8. This holds for all p, so x satisfies the condition given in the statement. The converse has already been proved as Corollary 2.4.

This completes the proof of all the results stated.

References

1. Adams, J.F.: Vector fields on spheres. Annals of Mathematics **75**, 603–622 (1962)
2. Adams, J.F.: Lectures on Lie groups. W.A. Benjamin 1969
3. Adams, J.F.: Stable homotopy and generalised homology. Chicago U.P. 1974
4. Adams, J.F.: Maps between classifying spaces, in the proceedings of a conference held in Evanston, March 1977; Springer Lecture Notes in Mathematics, vol. 657
5. Adams, J.F.: Maps between classifying spaces, in the proceedings of a conference held in Zürich, April 1977. l'Enseignement Mathématique, **24**, 79–85 (1978)
6. Adams, J.F.: Infinite loop spaces. "Annals of Mathematics Studies" vol. 90, Princeton U.P. (1978)
7. Adams, J.F., Mahmud, Z.: Maps between classifying spaces. Inventiones Math. **35**, 1–41 (1976)
8. Atiyah, M.F.: Characters and cohomology of finite groups. Publ. Math. de l'I.H.E.S. no. 9 (1961)
9. Atiyah, M.F.: *K*-theory, W.A. Benjamin Inc. 1967
10. Atiyah, M.F., Hirzebruch, F.: Vector bundles and homogeneous spaces. Proc. Symposia in Pure Math. no. 3. Amer. Math. Soc. 7–38, 1961
11. Atiyah, M.F., Segal, G.B.: Equivariant *K*-theory and completion. Jour. of Differential Geometry **3**, 1–18 (1969)
12. Becker, J.C., Gottlieb, D.H.: The transfer map and fibre bundles, Topology **14**, 1–12 (1975)
13. Berstein, I.: Bundles over products of infinite-dimensional complex projective spaces. Quart. J. Math. **19**, 275–279 (1968)
14. Brauer, R., Tate, J.: On the characters of finite groups. Annals of Math. **62**, 1–7 (1955)
15. Husemoller, D.: Fibre bundles. McGraw-Hill 1966
16. Quillen, D.: Cohomology of Groups. In: Proceedings of the International Congress of Mathematicians 1970, Gauthier-Villars, vol. **2**, 47–51, 1971
17. Segal, G.B.: The representation ring of a compact Lie group. Publ. Math. de l'I.H.E.S. **34** 113–128 (1968)
18. Serre, J.-P.: Représentations linéaires des groupes finis, 2° ed., §11.1, Hermann 1971
19. de Siebenthal, J.: Sur les groupes de Lie compacts non connexes, Comm. Math. Helv. **31**, 41–89 (1956)
20. Stasheff, J.D.: *H*-space problems. In: Lecture Notes in Mathematics no. 196, Berlin, Heidelberg, New York: Springer 1971; see especially problem no. 20, p. 127
21. Steenrod, N.E.: The topology of fibre bundles. Princeton U.P. 1951
22. Sullivan, D.: Geometric topology, part I. Localisation, periodicity, and Galois symmetry. Mimeographed notes, MIT, 1970; see especially Corollaries 5.10, 5.11
23. Thomas, C.B.: Free actions by finite groups on S^3, in the proceedings of a conference held in Stanford, July 1976. To appear in the "Proceedings of Symposia in Pure Mathematics"
24. Thomas, C.B.: Homotopy classification of free actions by finite groups on S^3, preprint

Received April 27, 1978

MAPS BETWEEN CLASSIFYING SPACES, III

J.F. Adams and Z. Mahmud

§1. Introduction. Let G and G' be compact connected Lie groups, and let $f\colon BG \longrightarrow BG'$ be a map. It is shown in [2] that the induced homomorphism $f^*\colon K(BG) \longleftarrow K(BG')$ carries the representation ring $R(G') \subset K(BG')$ into the representation ring $R(G) \subset K(BG)$. Moreover the induced map $R(G) \longleftarrow R(G')$ can also be induced by a homomorphism $\theta\colon T \longrightarrow T'$, where T, T' are the maximal tori in G, G'. Here of course one has to state that the behaviour of θ with respect to the Weyl groups W, W' is such that $\theta^*\colon R(T) \longleftarrow R(T')$ does indeed carry $R(G') \subset R(T')$ into $R(G) \subset R(T)$; in [2] maps θ with this behaviour are called "admissible maps".

Our present purpose is to see what further information can be obtained by using real and symplectic K-theory. Let $RO(G) \subset R(G)$ be the subgroup generated by real representations; similarly for $RSp(G) \subset R(G)$, using symplectic representations. Assume that the group G is semi-simple.

Proposition 1.1. For any map $f\colon BG \longrightarrow BG'$, the induced homomorphism $R(G) \longleftarrow R(G')$ preserves real elements, in the sense that it carries $RO(G')$ into $RO(G)$; similarly it preserves symplectic elements, in the sense that it carries $RSp(G')$ into $RSp(G)$.

The main problem, however, is to take this result and deduce useful, explicit conclusions about $\theta\colon T \longrightarrow T'$. For this purpose we must recall some representation-theory. In particular, if ρ is any irreducible (complex) representation of G, then under ρ any element z in the centre Z of G acts as a scalar. If $z^2 = 1$ in G, then of course the scalar $\rho(z)$ is ± 1.

<u>Lemma 1.2</u> (after Dynkin [5]). (a) Let G be a compact connected Lie group. Then there is a canonical element $\delta \in Z \subset G$ with the following properties. (i) $\delta^2 = 1$. (ii) For any irreducible self-conjugate representation ρ of G, $\rho(\delta)$ is +1 or -1 according as ρ is real or symplectic. (b) More explicitly, when G is simple δ is as follows. If G is SU(n) with n odd, Spin(m) with $m \equiv 0,1,2$ or 7 mod 8, G_2, F_4, E_6 or E_8 then $\delta = 1$. If G is SU(n) with n even or Sp(n) then δ is the matrix -1. If G is Spin(m) with $m \equiv 3,4,5$ or 6 mod 8 then δ is the non-trivial element in the kernel of $Spin(m) \longrightarrow SO(m)$. If G is E_7 then δ is the unique non-trivial element in the centre Z. (c) If $\zeta \in Z$ is any element of the centre such that $\zeta^4 = 1$, then $\delta\zeta^2$ also has properties (i) and (ii) of part (a).

We shall write $I \subset Z$ for the subgroup of elements ζ^2, where $\zeta \in Z$ and $\zeta^4 = 1$; and we shall regard I as an "indeterminacy" which affects elements satisfying (i) and (ii) of part (a).

Our object is now to conclude that if we take $f: BG \longrightarrow BG'$ and pass to an associated admissible map $\theta: T \longrightarrow T'$, then $\theta\delta \equiv \delta' \bmod I'$; in other words, θ preserves the Dynkin element (modulo indeterminacy). It is apparent from Lemma 1.2 that this condition is sufficient for $\theta^*: R(G) \longleftarrow R(G')$ to preserve real and symplectic elements; we aim to prove that this condition is also necessary, at least in certain cases.

This calls for two comments. First, we wish to emphasise that the condition "$\theta\delta \equiv \delta' \bmod I'$" is indeed useful and explicit. For example, consider the case $G = G' = Sp(1)$; the centre $Z = Z'$ is $\{\pm 1\}$, and the Dynkin element $\delta = \delta'$ is -1. The appropriate maps θ are those of the form $\theta_k(t) = t^k$, so the condition "$\theta\delta \equiv \delta'$ mod I'" becomes "k is odd".

Secondly we wish to emphasise that further assumptions will be needed. For example, in the case $G = G' = Sp(1)$ we know [6,7] that the correct conclusion is "k is odd <u>or zero</u>", and we know that the case $k = 0$ can actually occur (take the map $f: BG \longrightarrow BG'$ to be constant at the base-point). So in this case we shall need some assumption to exclude the case $k = 0$.

In fact, in general $\theta\delta$ need not even lie in the centre Z' of G'. (For an example, consider the usual injection $Sp(n) \longrightarrow Sp(n+1)$.) However, we have the following result.

<u>Proposition 1.3</u>. Let $f: BG \longrightarrow BG'$ correspond to an admissible map $\theta: T \longrightarrow T'$ which is irreducible in the sense of [2] p20; then θ carries Z into Z'.

Somewhat weaker assumptions should suffice to prove that $\theta\delta$ lies in Z'; for it is sufficient to verify that $\theta\delta$ lies in the kernel of each root of G', and apparently this could be done by methods similar to those given below. However, as the case of an irreducible admissible map is the most interesting one, (1.3) is probably enough.

For our main result, we shall make the assumptions which follow.

(i) G' is one of the simply-connected classical groups $Spin(m)$, $SU(n)$, $Sp(n)$.

(ii) $\theta: T \longrightarrow T'$ is an admissible map.

(iii) $\theta^*: R(G) \longleftarrow R(G')$ preserves real and symplectic elements.

(iv) $\theta\delta \in Z'$.

(v) If $G' = Spin(2n)$ with $2n \equiv 0 \bmod 4$, or if $G' = Sp(n)$, we assume that the map θ is irreducible in the sense of [2] p20.

To state our final assumption we need some notation. Let $x_1, x_2, \ldots, x_n$ be the basic weights in $G' = Spin(2n)$, $Spin(2n+1)$, $SU(n)$ or $Sp(n)$; and set $\theta_i = \theta^* x_i$, so that the map θ has components θ_i (except in the case $G' = SU(n)$, in which we have a relation $\sum_i \theta_i = 0$). Fix a Weyl chamber C in G, and let $\tau \in W$ be the element which carries C to $-C$.

(vi) If $G' = Spin(m)$ with $m \not\equiv 2 \bmod 4$ or $G' = SU(n)$ with $n \equiv 2 \bmod 4$ we assume that no θ_i is fixed by τ.

<u>Theorem 1.4</u>. Under these conditions we have $\theta\delta \equiv \delta' \bmod I'$.

We pause to comment on these assumptions. In (i), the only exceptional group we need to exclude is E_7, which would involve ad hoc calculations. We presume that for E_7 it would be appropriate to assume that θ is irreducible. Assumptions (ii) to (iv) seem acceptable. However assumption (iii) can sometimes be weakened, as we proceed to show. Lemma (1.2)(b) yields many cases in which $\delta' \in I'$, so that every self-conjugate representation of G' is real; in such cases θ automatically preserves symplectic elements, and so it is enough to assume that θ preserves real elements. The opposite happens when G' has enough basic symplectic representations. For example, if $G' = Sp(n)$ then it is sufficient to assume that $\theta^*\lambda^1$ is symplectic. For if so, then since θ^* commutes with λ^k, we see that $\theta^*\lambda^k$ is real or symplectic according to the parity of k, and similarly when λ^k is replaced by a monomial $(\lambda^1)^{e_1}(\lambda^2)^{e_2}\ldots(\lambda^n)^{e_n}$. Inspection of the proof of (1.4) will show what is actually used in each case.

Assumption (v) serves to rule out cases like the inclusion $SU(n) \subset Spin(2n)$ for $n \equiv 0 \bmod 2$ and the inclusion $SU(n) \subset Sp(n)$ for $n \equiv 1 \bmod 2$; in these cases the conclusion fails, although all the other assumptions hold. Assumption (vi) serves to rule out cases like the exterior powers $\lambda^{2i}: SU(n) \longrightarrow SU(m)$ where $m = n!/(2i)!(n-2i)!$ and

$m \equiv 2 \bmod 4$ (for example $\lambda^2 : SU(4) \longrightarrow SU(6)$). In these cases also the conclusion fails, although all the other assumptions hold. Assumption (vi) is not too restrictive; for many groups G we have $\tau = -1$, and then θ_i can be fixed by τ only if $\theta_i = 0$ - a possibility which is normally ruled out when θ is irreducible. In §5 we shall see how one can make use of assumption (vi).

The remainder of this paper is organised as follows. In §2 we prove Proposition 1.1. In §3 we deduce Lemma 1.2 from the statement originally given by Dynkin; we also prove Corollary 3.1, a useful result of representation-theory which however did not need to be stated in the introduction. In §4 we prove Proposition 1.3, and in §5 we prove the main result, Theorem 1.4.

The first author thanks the University of Kuwait for hospitality during the preparation of this paper.

§2. Preservation of real and symplectic elements. In this section we will prove Proposition 1.1.

We begin by studying the case in which the group G is simply-connected. In this case one of the main theorems of Hermann Weyl [1 p164] asserts that the representation ring R(G) is a polynomial algebra; and the proof of the theorem gives a preferred set of generators. More precisely, the weights in the closure of the fundamental dual Weyl chamber form a free (commutative) semigroup, with a unique set of generators, say $\omega_1, \omega_2, \ldots, \omega_\ell$ [1 p163]; to these weights there correspond irreducible representations $\rho_1, \rho_2, \ldots, \rho_\ell$; and the theorem says that R(G) is a polynomial algebra $Z[\rho_1, \rho_2, \ldots, \rho_\ell]$ on these generators.

The map $-1: L(T) \longrightarrow L(T)$ is conjugate to a map $-\tau$ (see §1) which preserves the fundamental Weyl chamber C. It follows that $-\tau$

permutes $\omega_1,\omega_2,\ldots,\omega_\ell$. Thus complex conjugation permutes the irreducible representations $\rho_1,\rho_2,\ldots,\rho_\ell$. Those of $\rho_1,\rho_2,\ldots,\rho_\ell$ which are self-conjugate are either real or symplectic, but not both [1 p64].

Complex conjugation also permutes the monomials $\rho_1^{i_1}\rho_2^{i_2}\ldots\rho_\ell^{i_\ell}$. Let us restrict attention to the self-conjugate monomials. We will call such a monomial "real" or "symplectic" according as the sum of the exponents of symplectic generators ρ_i is even or odd. Since a representation of the form $\rho\bar{\rho}$ is real [1 p166], it follows that each "real" monomial is a real representation, and similarly each "symplectic" monomial is a symplectic representation.

<u>Lemma 2.1</u>. The subgroup $RO(G)$ of $R(G)$ has a base consisting of the following elements.

(i) m, where m runs over the real monomials.

(ii) $m+\bar{m}$, where $(m,\bar{m})$ runs over pairs of distinct conjugate monomials.

(iii) $2m$, where m runs over the symplectic monomials.

The subgroup $RSp(G)$ of $R(G)$ has a base consisting of the following elements.

(i) $2m$, where m runs over the real monomials.

(ii) $m+\bar{m}$, where $(m,\bar{m})$ runs over pairs of distinct conjugate monomials.

(iii) m, where m runs over the symplectic monomials.

<u>Proof</u>. In view of the discussion preceding the lemma, it is clear that the subgroup generated by the elements listed is contained in $RO(G)$ or $RSp(G)$ as the case may be.

For the converse, we recall that it is easy to prove the corresponding descriptions of $RO(G)$ and $RSp(G)$ in terms of real irreducible representations, pairs of distinct complex conjugate irreducible representations, and symplectic irreducible representations; see for example [1 p66]. Any irreducible representation ρ can be written as a Z-linear combination of monomials, $\rho = \sum_I a_I m_I$, and therefore $\rho+\bar{\rho}$ can be written as $\rho+\bar{\rho} = \sum_I a_I(m_I+\bar{m}_I)$. Consider now a monomial $m = \rho_1^{i_1}\rho_2^{i_2}\ldots\rho^{i_\ell}$. Let σ be the irreducible representation corresponding to the "highest weight" $i_1\omega_1+i_2\omega_2+\ldots+i_\ell\omega_\ell$; then σ occurs in m with multiplicity 1 [1 pp161-164]. Thus m is self-conjugate or not with σ, and in the self-conjugate case, m is real or symplectic with σ. It now follows that when we write a self-conjugate irreducible representation σ' by induction as a Z-linear combination of monomials m', we use, apart from sums $m+\bar{m}$, only monomials m'' which are real or symplectic with σ'. This proves the lemma.

We now introduce the reduced generators $\sigma_i = \rho_i-(\varepsilon\rho_i)1$, where ε is the augmentation. These generators lie in the augmentation ideal I(RG). We still have $R(G) = Z[\sigma_1,\sigma_2,\ldots,\sigma_\ell]$. Complex conjugation permutes $\sigma_1,\sigma_2,\ldots,\sigma_\ell$ just as it permuted $\rho_1,\rho_2,\ldots,\rho_\ell$; σ_i is self-conjugate or not with ρ_i; and in the self-conjugate case, σ_i is real or symplectic with ρ_i. We may form monomials $\sigma_1^{i_1}\sigma_2^{i_2}\ldots\sigma_\ell^{i_\ell}$; such a monomial is self-conjugate or not with $\rho_1^{i_1}\rho_2^{i_2}\ldots\rho_\ell^{i_\ell}$; and in the self-conjugate case, we call $\sigma_1^{i_1}\sigma_2^{i_2}\ldots\sigma_\ell^{i_\ell}$ "real" or "symplectic" with $\rho_1^{i_1}\rho_2^{i_2}\ldots\rho_\ell^{i_\ell}$.

Lemma 2.2. The description of RO(G) and RSp(G) in Lemma 2.1 remains valid if we use monomials $m = \sigma_1^{i_1}\sigma_2^{i_2}\ldots\sigma_\ell^{i_\ell}$ in the reduced generators.

This follows immediately from Lemma 2.1.

Next we recall that by theorems of Atiyah and Segal [3 pp10, 14,17], we have canonical isomorphisms

$$R(G)^{\wedge} \xrightarrow{\cong} K(BG)$$
$$RO(G)^{\wedge} \xrightarrow{\cong} KO(BG)$$
$$RSp(G)^{\wedge} \longrightarrow KSp(BG).$$

Here $R(G)^{\wedge}$ means the completion of $R(G)$ with respect to the topology defined by powers of the augmentation ideal $I(RG)$. The topologies on $RO(G)$, $RSp(G)$ may be taken to be the restrictions of the topology on $R(G)$; see [3 p17], but note that the reference there to (5.1) should be to (6.1). The following diagrams commute.

$$\begin{array}{ccc} RO(G)^{\wedge} & \xrightarrow{\cong} & KO(BG) \\ \downarrow & & \downarrow \\ R(G)^{\wedge} & \xrightarrow{\cong} & K(BG) \end{array}$$

$$\begin{array}{ccc} RSp(G)^{\wedge} & \xrightarrow{\cong} & KSp(BG) \\ \downarrow & & \downarrow \\ R(G)^{\wedge} & \xrightarrow{\cong} & K(BG) \end{array}$$

We proceed to describe these completions explicitly in our case.

Corollary 2.3. $R(G)^{\wedge}$ is the ring of formal power-series $Z[[\sigma_1, \sigma_2, \ldots, \sigma_\ell]]$. An element of $R(G)^{\wedge}$ may be written uniquely as a formal Z-linear combination $\sum_I a_I m_I$ $(a_I \in Z)$ of monomials $m_I = \sigma_1^{i_1} \sigma_2^{i_2} \ldots \sigma_\ell^{i_\ell}$. Such an element is self-conjugate if and only if conjugate monomials appear with equal coefficients. If self-conjugate, it lies in the completion $RO(G)^{\wedge}$ of $RO(G)$ if and only if the coefficient of each symplectic monomial is even; it lies in the completion $RSp(G)^{\wedge}$ of $RSp(G)$ if and only if the coefficient of each real monomial is even.

This follows immediately from the work above.

Corollary 2.4. In $K(BG)$ we have

$$R(G) \cap KO(BG) = RO(G)$$

and

$$R(G) \cap KSp(BG) = RSp(G).$$

Proof. Take a typical element of $K(BG) = R(G)^{\wedge}$ as a formal Z-linear combination $\sum_I a_I m_I$. If it lies in $R(G)$ only finitely many of the coefficients a_I are non-zero. If it lies in $KO(BG) = RO(G)^{\wedge}$ then conjugate monomials occur with equal coefficients, and the coefficient of each symplectic monomial is even. If the element lies both in $R(G)$ and in $RO(G)^{\wedge}$, then it lies in $RO(G)$. The proof of the second statement is similar.

Proof of Proposition 1.1. By Corollary 1.13 of [2], an induced map

$$f^*: K(BG) \longleftarrow K(BG')$$

carries $R(G')$ into $R(G)$; it clearly carries $KO(BG')$ into $KO(BG)$, and $KSp(BG')$ into $KSp(BG)$. So when G is simply-connected, Proposition 1.1 follows from Corollary 2.4.

If G is merely semi-simple, let $\tilde{G}$ be its universal cover, so that we have a finite covering map $\pi: \tilde{G} \longrightarrow G$. Then the preceding result applies to the composite

$$B\tilde{G} \xrightarrow{B\pi} BG \xrightarrow{f} BG'.$$

So from $x \in RO(G')$ we infer $(B\pi)^* f^* x \in RO(\tilde{G})$. From this it follows that $f^*x \in RO(G)$. If we assume $x \in RSp(G')$, a similar argument applies. This proves Proposition 1.1.

§3. Dynkin elements. We begin by recalling the work of Dynkin [5]; the reader may also consult Bourbaki [4]. These authors assume that G is semi-simple, so we make this assumption for the moment and remove it later. Then the construction given for δ is as follows. Choose a Weyl chamber C; this determines a base of simple roots $\phi_1, \phi_2, \ldots, \phi_\ell$. Let us consider these roots as linear maps $L(T) \longrightarrow R$. Then there is a unique vector $\tilde{t} \in L(T)$ such that $\phi_i(\tilde{t}) = 1$ for $i = 1, 2, \ldots, \ell$. It follows that every root takes an integer value on $\tilde{t}$; so $\tilde{t}$ maps to the identity under the adjoint representation, and $\tilde{t}$ yields an element δ of the centre.

The authors cited guarantee the behaviour of $\rho(\delta)$ in the case they consider; we will discuss the other assertions of Lemma 1.2.

The element δ constructed above appears to depend on the choice of C; however, any other Weyl chamber C' may be obtained from C by the action of an element $w \in W$, which will carry $\delta(C)$ onto $\delta(C')$. Since W fixes central elements, $\delta(C) = \delta(C')$, and so the construction is canonical.

In particular, $-C$ is another Weyl chamber; it leads to simple roots $-\phi_1, -\phi_2, \ldots, -\phi_\ell$ and to the vector $-\tilde{t}$. Hence $\delta^{-1} = \delta$ and $\delta^2 = 1$.

This establishes assertions (i) and (ii) of the lemma when G is semi-simple. Moreover, the construction of δ yields the following further properties. (iii) If G is a product group $G_1 \times G_2$, then the Dynkin element δ in G is the product of the Dynkin elements δ_1, δ_2 in G_1, G_2. (iv) Let $\tilde{G} \longrightarrow G$ be a finite covering; then the Dynkin element $\tilde{\delta}$ in $\tilde{G}$ maps to the Dynkin element δ in G. If we insist on preserving these two properties, and define the Dynkin element of any torus to be 1, we get a canonical extension of δ from the class of compact connected semi-simple groups to the class of compact connected groups; and this extension has the properties (i),(ii) stated in the lemma.

Part (b) lists the value of δ in each simple Lie group, and this is an easy calculation.

It remains to prove part (c), about $\delta\zeta^2$. Since $\delta^2 = 1$ and we assume $\zeta^4 = 1$, and since δ and ζ are both central, it is clear that $(\delta\zeta^2)^2 = 1$. Suppose then that ρ is an irreducible self-conjugate representation of G, as in the lemma. Since ρ is self-conjugate we get $\overline{\rho(\zeta)} = \rho(\zeta)$, that is $\rho(\zeta)^{-1} = \rho(\zeta)$, so $\rho(\zeta^2) = 1$ and $\rho(\delta\zeta^2) = \rho(\delta)$. This completes the proof of Lemma 1.2.

Next we recall that weights ω may be considered as 1-dimensional representations of T, and symmetric sums $S(\omega)$ as representations of T. We note that a central element $z \in Z$ acts as a scalar under $S(\omega)$; in fact, since W fixes central elements we get $(\omega w)z = \omega(wz) = \omega(z)$ for each w.

We shall call a self-conjugate symmetric sum $S(\omega)$ "real" or "symplectic" according as $S(\omega)\delta$ is +1 or -1.

<u>Corollary 3.1</u>. The subgroup RO(G) of R(G) has a base consisting of the following elements.

(i) $S(\omega)$ where $S(\omega)$ runs over the real symmetric sums.

(ii) $S(\omega)+\overline{S(\omega)}$, where $(S(\omega),\overline{S(\omega)})$ runs over pairs of distinct conjugates.

(iii) $2S(\omega)$, where $S(\omega)$ runs over the symplectic symmetric sums.

The subgroup RSp(G) of R(G) has a base consisting of the following elements.

(i) $2S(\omega)$, where $S(\omega)$ runs over the real symmetric sums.

(ii) $S(\omega)+\overline{S(\omega)}$, where $(S(\omega),\overline{S(\omega)})$ runs over pairs of distinct conjugates.

(iii) $S(\omega)$ where $S(\omega)$ runs over the symplectic symmetric sums.

If we were allowed to replace the symmetric sum $S(\omega)$ by the irreducible representation $\rho(\omega)$ with ω as an extreme weight, this would become a standard result of representation-theory; see [1] p66. We can write each $\rho(\omega)$ in terms of the $S(\omega')$, and (by induction over ω) each $S(\omega)$ in terms of the $\rho(\omega')$; the work above assures us that in this process we only use weights ω' with a fixed value of $\omega'(\delta)$.

§4. Preservation of the centre. We interpret the Stiefel diagram $L(T)$ as the universal cover of T. As in [2 p22], let Γ be the extended Weyl group of G (generated by the reflections in all the planes of the Stiefel diagram, not just those which pass through the origin); and let Γ_o be the subgroup of translations in Γ.

Lemma 4.1. Let $f\colon BG \longrightarrow BG'$ correspond to $\theta\colon T \longrightarrow T'$; then $\tilde{\theta} = L(\theta)$ carries Γ_o into Γ_o'.

Proof. f induces a map from $\pi_1(G) = \pi_2(BG)$ to $\pi_1(G') = \pi_2(BG')$. Here we can suppose without loss of generality that $\pi_1(G')$ is free abelian, for we can arrange this by passing to finite covers of G and G' - a step which does not change $\tilde{\theta} = L(\theta)$. The diagram

$$\begin{array}{ccccccc} \pi_1(T) & = & \pi_2(BT) & \xrightarrow{(B\theta)_*} & \pi_2(BT') & = & \pi_1(T') \\ \downarrow & & \downarrow & & \downarrow & & \downarrow \\ \pi_1(G) & = & \pi_2(BG) & \xrightarrow{f_*} & \pi_2(BG') & = & \pi_1(G') \end{array}$$

is now commutative, for when $\pi_1(G')$ is free abelian this follows from the results on rational cohomology in [2]. We can interpret Γ_o as the kernel of $\pi_1(T) \longrightarrow \pi_1(G)$, and Γ_o' as the kernel of $\pi_1(T') \longrightarrow \pi_1(G')$; hence $\tilde{\theta} = L(\theta)$ carries Γ_o into Γ_o'.

Proof of Proposition 1.3. Take an element $z \in Z$, and lift it to a vector $\tilde{z} \in \tilde{T} = L(T)$. By Proposition 2.28 of [2] (applied with G' replaced by G) we have

$$w\tilde{z} \equiv \tilde{z} \bmod \Gamma_o$$

for all $w \in W$. Let $\alpha: W \longrightarrow W'$ be as in Theorem 2.29 of [2]; applying $\tilde{\theta} = L(\theta)$ and using Lemma 4.1, we get

$$(\alpha w)(\tilde{\theta}\tilde{z}) = (\tilde{\theta}\tilde{z}) \bmod \Gamma_o'.$$

Since in Proposition 1.3 we assume θ irreducible, Theorem 2.29(ii) of [2] yields

$$\tilde{\theta}\tilde{z} \quad \epsilon \ \tilde{Z}'.$$

So θ carries Z into Z'.

§5. Location of self-conjugate symmetric sums. The pattern of proof of Theorem 1.4 is as follows. We first select some irreducible self-conjugate representations ρ_i' of G'. We take care to choose enough representations ρ_i' so that the conditions

$$z' \in Z', \quad (z')^2 = 1 \text{ and } \rho_i'(z') = 1 \text{ for all } i$$

imply $z' \in I'$; we shall see that this can be done in each case. It is then sufficient to prove that

$$\rho_i'(\theta\delta) = \rho_i'(\delta') \text{ for each } i.$$

Since we assume that θ^* preserves real and symplectic elements, $\theta^*\rho_i'$ is real or symplectic with ρ_i'. The crucial step is now to prove that $\theta^*\rho_i'$ contains at least one self-conjugate symmetric sum $S(\omega_i)$ with odd multiplicity. If so, then we can apply Corollary 3.1 to the expression for $\theta^*\rho_i'$ in terms of symmetric sums, and conclude that $S(\omega_i)$ is real or symplectic with $\theta^*\rho_i'$ and ρ_i'; that is,

$$S(\omega_i)(\delta) = \rho_i'(\delta').$$

But since $S(\omega_i)$ occurs in $\theta^*\rho_i'$, we also have

$$\rho_i'(\theta\delta) = S(\omega_i)\delta.$$

Thus

$$\rho_i'(\theta\delta) = \rho_i'(\delta')$$

and this completes the proof.

To fill in this outline, we must first show that we can find enough representations ρ_i'. Let us dismiss as trivial the cases

$G' = Spin(n)$ with $n \equiv 2 \bmod 4$ and

$G' = SU(n)$ with $n \not\equiv 2 \bmod 4$.

In these cases the conditions

$$z' \in Z', \ (z')^2 = 1$$

already imply $z' \in I'$, so we need choose no representations ρ_i'.

Next we take the cases

$Spin(n)$ with $n \equiv 1 \bmod 2$

$SU(n)$ with $n \equiv 2 \bmod 4$ and

$Sp(n)$.

In each case we have just 2 elements $z' \in Z'$ such that $(z')^2 = 1$, and one representation ρ_i' will do. It is sufficient to take the spin-representation Δ on $Spin(n)$ with $n \equiv 1 \bmod 2$, the exterior power λ^m on $SU(2m)$ where $2m \equiv 2 \bmod 4$, and the fundamental representation λ^1 on $Sp(n)$.

Finally we take the case

$Spin(n)$ with $n \equiv 0 \bmod 4$.

In this case the centre is $Z_2 \times Z_2$. It is sufficient to take two representations ρ_i', namely the fundamental representation λ^1, and either one of the two half-spin representations Δ^+, Δ^-. This completes the choice of the ρ_i'.

The essential step is now to show in each case that $\theta^*\rho_i'$ contains at least one self-conjugate symmetric sum $S(\omega_i)$ with odd multiplicity. We consider first the cases

$$\begin{array}{l} Spin(2n) \text{ with } 2n \equiv 0 \bmod 4 \\ Sp(n) \end{array}$$

in which we have to consider $\rho_i' = \lambda^1$. In both cases W' permutes the 2n weights $\pm x_1, \pm x_2, \ldots, \pm x_n$; therefore W permutes the 2n elements $\pm\theta_1, \pm\theta_2, \ldots, \pm\theta_n$, and they fall into orbits under W. If $(\phi_1, \phi_2, \ldots, \phi_r)$ is an orbit, then $(-\phi_1, -\phi_2, \ldots, -\phi_r)$ is also an orbit. Suppose to begin with that one of the orbits other than $(\phi_1, \phi_2, \ldots, \phi_r)$ has the same elements as $(-\phi_1, -\phi_2, \ldots, -\phi_r)$. Then we can factor the admissible map θ; if $G' = Sp(n)$ we use the subgroup $U(r)\times Sp(n-r)$, while if $G' = Spin(2n)$ we use the pull back in the following diagram.

$$\begin{array}{ccccc} G'' & & \longrightarrow & & Spin(2n) \\ \downarrow & & & & \downarrow \\ U(r)\times SO(2n-2r) & \longrightarrow & SO(2r)\times SO(2n-2r) & \longrightarrow & SO(2n) \end{array}$$

This contradicts the assumption that θ is irreducible; therefore, it does not happen. We conclude (firstly) that $(-\phi_1, -\phi_2, \ldots, -\phi_r)$ is the same orbit as $(\phi_1, \phi_2, \ldots, \phi_r)$; therefore, each orbit has the form $(\pm\psi_1, \pm\psi_2, \ldots, \pm\psi_s)$. We also conclude that no orbit is repeated, for if the orbit $(\phi_1, \phi_2, \ldots, \phi_r)$ were repeated, the second copy of $(\phi_1, \phi_2, \ldots, \phi_r)$ would have the same elements as $(-\phi_1, -\phi_2, \ldots, -\phi_r)$. This says that when we decompose $\theta^*\lambda^1$ into symmetric sums, corresponding to the orbits $(\pm\psi_1, \pm\psi_2, \ldots, \pm\psi_s)$, each symmetric sum is self-conjugate and occurs with multiplicity 1.

In the remaining cases we have to consider $\rho_i' = \Delta$, λ^m and either Δ^+ or Δ^-. We use the same argument in all cases. Let

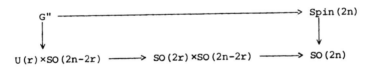

$\phi_1, \phi_2, \ldots, \phi_{2m}$ be

(i) $\pm\frac{1}{2}\theta_1, \pm\frac{1}{2}\theta_2, \ldots, \pm\frac{1}{2}\theta_m$ in the cases $G' = \mathrm{Spin}(2m)$ and $G' = \mathrm{Spin}(2m+1)$

(ii) $\theta_1, \theta_2, \ldots, \theta_{2m}$ in the case $G' = SU(2m)$.

In either case we have

$$\sum_1^{2m} \phi_j = 0.$$

By assumption (vi), ϕ_j is not fixed by τ, so the condition $\phi_j(\tau-1)v \neq 0$ is satisfied by an open dense set of vectors $v \in L(T)$. It follows that we can find a vector v such that $\phi_j(\tau-1)v \neq 0$ for all j. Now, τ permutes $\phi_1, \phi_2, \ldots, \phi_{2m}$, and for the element $\phi_j\tau$ we have

$$\begin{aligned}\phi_j\tau(\tau-1)v &= \phi_j(1-\tau)v \\ &= -\phi_j(\tau-1)v.\end{aligned}$$

So the elements $\phi_1, \phi_2, \ldots, \phi_{2n}$ fall into pairs $(\phi, \phi\tau)$, with one member of each pair being positive on $(\tau-1)v$ and one member of each pair being negative. Without loss of generality, we can suppose the ϕ's renumbered so that $\phi_1, \phi_2, \ldots, \phi_m$ are positive on $(\tau-1)v$. We claim that $\omega_i = \phi_1+\phi_2+\ldots+\phi_m$ is one of the weights of $\theta^*\rho_i'$. This is clear in the case $G' = SU(2m)$, $\rho_i' = \lambda^m$ since ω_i is merely the sum of some m of the θ_i. In the cases $G' = \mathrm{Spin}(2m)$ and $\mathrm{Spin}(2m+1)$ the ϕ's already fall into pairs

$$\pm\frac{1}{2}\theta_1,\ \pm\frac{1}{2}\theta_2, \ldots,\ \pm\frac{1}{2}\theta_m$$

taking opposite values on $(\tau-1)v$; we must have selected one from each pair and obtained

$$\omega_i = \frac{1}{2}(\varepsilon_1\theta_1+\varepsilon_2\theta_2+\ldots+\varepsilon_m\theta_m)$$

where each ε_i is ± 1. If $G = \mathrm{Spin}(2m+1)$ this is one of the weights of

$\theta^*\Delta$, because $\frac{1}{2}(\varepsilon_1 x_1+\varepsilon_2 x_2+\ldots+\varepsilon_m x_m)$ is a weight of Δ. If $G = \mathrm{Spin}(2m)$ then ω_i is either one of the weights of $\theta^*\Delta^+$ or one of the weights of $\theta^*\Delta^-$.

Next we claim that this weight occurs with multiplicity 1 in $\theta^*\rho_i'$. In fact, among the weights of $\theta^*\rho_i'$, ω_i is by construction the one with the maximum value at $(\tau-1)v$.

Finally we claim that the symmetric sum $S(\omega_i)$ is self-conjugate. In fact, by construction we have

$$\begin{aligned}\omega_i\tau &= \phi_1\tau+\phi_2\tau+\ldots+\phi_m\tau \\ &= \phi_{m+1}+\phi_{m+2}+\ldots+\phi_{2m} \\ &= -(\phi_1+\phi_2+\ldots+\phi_m) \quad (\text{since } \sum_1^{2m} \phi_j = 0) \\ &= -\omega_i .\end{aligned}$$

This proves the required result for $\theta^*\Delta$ and for $\theta^*\lambda^m$; and for $G' = \mathrm{Spin}(n)$ with $n \equiv 0 \bmod 4$ it shows that the required result holds either for $\theta^*\Delta^+$ or for $\theta^*\Delta^-$. This is enough, and it completes the proof.

References

[1] J.F. Adams, "Lectures on Lie Groups", W.A. Benjamin 1969; to be reprinted by the University of Chicago Press.

[2] J.F. Adams and Z. Mahmud, "Maps between classifying spaces", Inventiones Math. 35 (1976) 1-41.

[3] M.F. Atiyah and G.B. Segal, "Equivariant K-theory and completion", Jour. Differential Geometry 3 (1969) 1-18.

[4] N. Bourbaki, "Groups et algebras de Lie" Chap.VIII (especially pages 131-133), Hermann, Paris 1975.

[5] E.B. Dynkin, "Maximal subgroups of the classical groups", in American Math. Soc. translations, series 2, volume 6, American Math. Soc. 1957, 245-378.

[6] J. Hubbuck, "Homotopy homomorphisms of Lie groups", in "New developments in topology", L.M.S. Lecture Note Series no.11, C.U.P. 1974, 33-41, especially pages 33-34.

References (cont.)

[7] Z. Mahmud, "The maps BSp(1) ⟶ BSp(n)", Proc. Amer. Math. Soc. 52 (1975) 473-478.

Proceedings of the Royal Society of Edinburgh, **112A**, 231–235, *1989*

Maps between p-completed classifying spaces

J. Frank Adams*

Department of Pure Mathematics and Mathematical Statistics, 16 Mill Lane, Cambridge CB2 1SB, U.K.

and

Zdzisław Wojtkowiak

Universitat Autònoma de Barcelona, Departament de Matemàtiques, 08193 Bellaterra, Barcelona, Spain

(MS received 11 November 1988. Revised MS received 16 February 1989)

Synopsis

Let G and G' be two connected compact Lie groups with maximal tori T and T'. For a space X, let X_p be the p-completion of X. We will associate to each topological map $f: (BG)_p \to (BG')_p$ an "admissible map" $\varphi: \pi_1(T) \otimes_Z Z_p \to \pi_1(T') \otimes_Z Z_p$. We then show that the study of "admissible maps" in the p-complete case may be reduced to their study in the p-local case.

1. Introduction

In this paper, we will show that the results of [**1**] carry over to the p-complete case. More precisely (with notations and assumptions explained below), we will associate to each topological map

$$f: (BG)_p \to (BG')_p,$$

an "admissible map"

$$\phi: \pi_1(T) \otimes_Z Z_p \to \pi_1(T') \otimes_Z Z_p$$

(compare [**1**]). We then show that the study of "admissible maps" in the p-complete case may be reduced to their study in the p-local case, for which see [**1**, §2].

We assume throughout that G, G' are compact connected Lie groups with maximal tori T, T' and Weyl groups W, W'. We suppose given a map $f: (BG)_p \to (BG')_p$, where X_p is the p-completion of X.

THEOREM 1.1. *Then there is a map $\bar{f}: (BT)_p \to (BT')_p$ such that the diagram*

$$\begin{array}{ccc} (BG)_p & \xrightarrow{f} & (BG')_p \\ \uparrow & & \uparrow \\ (BT)_p & \xrightarrow{\bar{f}} & (BT')_p \end{array}$$

commutes up to homotopy. Moreover, $\bar{f}$ is unique up to the action of W'.

* Professor J. Frank Adams died in a car accident on 7 January 1989. The second author would like to express his profound regret and sorrow at this tragic death. He is very grateful to Professor Adams for his wonderful letters which during the last few years were for the second author the only light of hope.

This is the analogue of (1.1), (1.5) and (1.7) of [**1**]. Since $(BT)_p$ and $(BT')_p$ are Eilenberg–MacLane spaces $K(\pi, 2)$, the classification of maps $\bar{f}$ is given by their induced homomorphisms

$$\bar{f}_*: \pi_1(T) \otimes_Z Z_p \to \pi_1(T') \otimes_Z Z_p$$

where Z_p is the ring of p-adic integers.

More generally, we need to study R-linear maps

$$\phi: \pi_1(T) \otimes_Z R \to \pi_1(T') \otimes_Z R$$

for various commutative rings R; in the first place we have $R = Z_p$ in the p-complete case and $R = Z_{(p)}$ in the p-local case, but in [**1**], Q and $\mathbb{R}$ also occur. Such a map ϕ will be called "admissible" if for each w in W there exists w' in W' such that $\phi \circ w = w' \circ \phi$. The induced homomorphisms $\bar{f}_*$ above are admissible (apply the uniqueness clause of (1.1) to the map $\bar{f} \circ w$).

For each function $\alpha: W \to W'$, and R-linear map

$$\phi: \pi_1(T) \otimes_Z R \to \pi_1(T') \otimes_Z R$$

will be called "α-admissible" if $\phi \circ w = \alpha(w) \circ \phi$ for each w in W. Thus each admissible map is α-admissible for some function α. We write $V_\alpha(R)$ for the set of α-admissible maps; clearly it is an R-module.

PROPOSITION 1.2. *Assume $R \subset R'$ and R' is torsion-free over Z. Then the obvious map $V_\alpha(R) \to V_\alpha(R')$ induces an isomorphism*

$$V_\alpha(R) \otimes_R R' \cong (R').$$

This is a very simple result. With $R = Z_{(p)}$ and $R' = Z_p$, it suggests that by passing from the p-local case to the p-complete case we obtain "nothing essentially new". We obtain something new, of course; for example, there are more maps Ψ^k with k in Z_p than with k in $Z_{(p)}$; but the process is only "extension of scalars". Proposition 1.2 allows us to take results for the p-local case and deduce results for the p-complete case.

PROPOSITION 1.3. *Suppose given an admissible map*

$$\phi: \pi_1(T) \otimes_Z R \to \pi_1(T') \otimes_Z R$$

(where R is torsion-free over Z); then ϕ is α-admissible for some α which is a homomorphism from W to W'.

This generalises [**1**, Theorem 2.21].

At this point it follows that all the results about admissible maps in [**1**, §2] carry over to the p-complete case; we have only to express them in terms of spaces V_α and use (1.2). Here is a sample.

EXAMPLE 1.4. Assume $G = G'$ simple. Then the account of [**1**, (2.13)] carries over as follows. Let A be the automorphism group of the Dynkin diagram of $G = G'$, so that A is Σ_3 for $G = D_4$, Σ_2 for A_n $(n \geqq 2)$, D_n $(n \geqq 5)$, B_2, G_2, F_4 and E_6, otherwise 1. Let B be the subgroup corresponding to $-1: \pi_1(T) \to \pi_1(T)$, so that B is Σ_2 for A_n $(n \geqq 2)$, D_{2m+1} and E_6, otherwise 1. Form A/B, obtaining Σ_3

for $G=D_4$, Σ_2 for D_{2m} $(2m \geqq 6)$, B_2, G_2 and F_4, otherwise 1. Then the relevant V_α are in (1–1) correspondence with the elements of A/B, and each is 1-dimensional.

These results leave two problems open.

(i) How far is the original map $f:(BG)_p \to (BG')_p$ determined by $\bar{f}$? One might hope that it is determined up to homotopy. But a proof must await the development of technique; we have only preliminary results.

(ii) How many admissible maps ϕ are realised by topological maps f? One might expect that the set of realisable maps in $V_\alpha(Z_p)$ is the closure of the set of realisable maps in $V_\alpha(Z_{(p)})$. This seems more attainable.

Theorem 1.1 will be proved in Section 2, and Propositions 1.2 and 1.3 will be proved in Section 3.

2. Proof of Theorem 1.1

The solutions in T of $t^{p^n}=1$ make up a subgroup $T(n)$; let $T(\infty)=\bigcup_n T(n)$. Suppose given a map f as in (1.1), and consider the composite

$$BT(n) \to (BT)_p \to (BG)_p \xrightarrow{f} (BG')_p.$$

Since $T(n)$ is a finite p-group, it makes no difference whether we map $BT(n)$ into BG' or $(BG')_p$; the theorem of Dwyer and Zabrodsky [**2**, Theorem 1.1] assures us that the composite is homotopic to

$$BT(n) \xrightarrow{B\rho(n)} BG' \to (BG')_p$$

for some homomorphism $\rho(n): T(n) \to G'$, unique up to conjugacy in G'. We can easily choose the homomorphisms $\rho(n)$ by induction over n, so that $\rho(n+1) \mid T(n) = \rho(n)$; then the $\rho(n)$ yield a homomorphism $\rho(\infty): T(\infty) \to G'$. Let H' be the closure $Cl \operatorname{Im} \rho(\infty)$; we will show H' is a torus in G'. Clearly H' is a closed abelian subgroup. Since H' is compact, its group of components $\pi_0(H')$ is finite. Every element of $\pi_0(H')$ has p-power order, so $\pi_0(H')$ is a finite abelian p-group. But every element of $\pi_0(H')$ is p-divisible, so $\pi_0(H')=0$. Thus H' is a connected abelian Lie group, i.e. a torus.

We may thus assume, without loss of generality, that $\rho(\infty)$ maps $T(\infty)$ into T'. Thus we obtain the following diagram:

$$\begin{array}{ccc}
(BG)_p & \xrightarrow{f} & (BG')_p \\
\uparrow & & \uparrow \\
BT(\infty) & \xrightarrow{B\rho(\infty)} & BT'
\end{array}$$

This diagram commutes on any finite subcomplex of $BT(\infty)$, for any such finite subcomplex is contained in $BT(n)$ for some n. But since $(BG')_p$ is p-complete, we have a bijection

$$[X, (BG')_p] \to \varprojlim_\alpha [X_\alpha, (BG')_p]$$

where X_α runs over the finite subcomplexes of X. Thus the diagram is commutative. It remains to remark that $(BT(\infty))_p \cong (BT)_p$; this shows that there is a diagram, as stated in Theorem 1.1.

Now suppose given two maps $\bar{f}_1, \bar{f}_2$, such that the composites

$$(BT)_p \underset{\bar{f}_2}{\overset{\bar{f}_1}{\rightrightarrows}} (BT')_p \to (BG')_p$$

are homotopic. Maps $\bar{f}_i$ are classified as described in Section 1, and correspond to homomorphisms

$$T(\infty) \underset{\rho_2(\infty)}{\overset{\rho_1(\infty)}{\rightrightarrows}} T'(\infty)$$

such that the composites

$$BT(\infty) \underset{B\rho_2(\infty)}{\overset{B\rho_1(\infty)}{\rightrightarrows}} BT'(\infty) \to (BG')_p$$

are homotopic. Let $\rho_i(n)$ be the restriction of $\rho_i(\infty)$ to $T(n)$. By the uniqueness clause of the theorem of Dwyer and Zabrodsky [**2**, Theorem 1.1], there is for each n an element g'_n in G' which conjugates $\rho_1(n)$ into $\rho_2(n)$. Let E'_n be the set of g'_n in G' which conjugate $\rho_1(n)$ into $\rho_2(n)$; then the E'_n form a decreasing sequence of non-empty closed subsets in G', which is compact; so there is a point g'_∞ common to all the E'_n, and this conjugates $\rho_1(\infty)$ into $\rho_2(\infty)$. Hence it conjugates $H'_1 = Cl(\mathrm{Im}\,\rho_1(\infty))$ into $H'_2 = Cl(\mathrm{Im}\,\rho_2(\infty))$. Since H'_1 is a torus, it has a generator h'_1. Then g'_∞ conjugates h'_1 into a point h'_2 in $H'_2 \subset T'(\infty) \subset T'$. By a well-known theorem of Lie theory, there is an element g' of G' which conjugates h'_1 into h'_2 and T' into T'. That is, there is an element $w' \in W'$ which has the same effect as g'_∞ on h'_1 and hence on the whole of H'_1; that is, w' carries $\rho_1(\infty)$ into $\rho_2(\infty)$. Thus w' carries $\bar{f}_1$ into $\bar{f}_2$. This completes the proof of Theorem 1.1.

3. Proofs of Propositions 1.2 and 1.3

Proof of Proposition 1.2. Let $L(R)$ be the R-module of R-linear maps

$$\phi: \pi_1(T) \otimes_Z R \to \pi_1(T') \otimes_Z R.$$

Then $V_\alpha(R)$ can be defined by an exact sequence

$$0 \to V_\alpha(R) \to L(R) \xrightarrow{\delta} \prod_w L(R),$$

where the wth component of δ is the map

$$\phi \to \phi \circ w - \alpha(w) \circ \phi.$$

Standard techniques of commutative algebra now show that the map

$$V_\alpha(Z) \otimes_Z R \to V_\alpha(R)$$

is iso whenever R is flat over Z, i.e. torsion free over Z. With the assumptions of (1.2), this throws the map

$$V_\alpha(R) \otimes_R R' \to V_\alpha(R')$$

onto the map

$$(V_\alpha(Z) \otimes_Z R) \otimes_R R' \to V_\alpha(Z) \otimes_Z R',$$

which is of course iso.

Proof of Proposition 1.3. If ϕ is admissible then $\phi \in V_\beta(R)$ for some function β. By [**1**, (2.21)] in its original form, every element of $V_\beta(Q)$ is also in $V_\alpha(Q)$ for some homomorphism α; that is, $V_\beta(Q)$ is the union of finitely many vector subspaces $V_\alpha(Q) \cap V_\beta(Q)$. By [**1**, (3.1)] there is at least one α for which $V_\alpha(Q) \supset V_\beta(Q)$. Then $V_\alpha(Z) \supset V_\beta(Z)$. By (1.2), $V_\alpha(R) \supset V_\beta(R)$. Thus $\phi \in V_\alpha(R)$ for some homomorphism α.

Acknowledgment

Whilst writing this paper, the second author was supported by a grant from the Ministry of Education and Science of Spain.

References

1 J. F. Adams and Z. Mahmud. Maps between classifying spaces. *Invent. Math.* **35** (1976), 1–41.

2 W. Dwyer and A. Zabrodsky. Maps between classifying spaces. In *Algebraic Topology, Barcelona* 1986, Lecture Notes in Mathematics 1298 pp. 106–119 (Berlin: Springer-Verlag 1987).

(*Issued* 12 *September* 1989)

Reprinted from the *Proceedings of the Cambridge Philosophical Society*,
Volume 53, Part 4, pp. 922–923, 1957.

AN EXAMPLE IN HOMOTOPY THEORY

BY J. F. ADAMS

Communicated by S. WYLIE

Received 12 April 1957

The following query is due to Prof. J. H. C. Whitehead. *If two CW-complexes are of the same n-type for all n, are they necessarily of the same homotopy type?*

The answer to this query is 'No', and we will prove it by exhibiting a suitable counter-example. In order to account for the construction adopted, we will preface it with a weak result in the affirmative direction.

THEOREM. *If X, Y are two connected CW-complexes, of the same n-type for all n, and if $\Pi_m(X)$ is finite for all m, then X, Y are of the same homotopy type.*

Proof. Under the given conditions we know that $\Pi_m(Y)$ is finite for all m. Moreover, we may suppose that X has finitely many cells in each dimension; for if not, we may replace it by its minimal complex.

We now construct a homotopy equivalence $f: X \to Y$ by induction over the skeletons of X. Suppose $f^{n-1}: X^{n-1} \to Y^{n-1}$ so constructed that for any $m \geqslant n$ it can be extended to some m-equivalence $g^m: X^m \to Y^m$. (This is clearly possible for $n = 1$, taking any 0-map $f^0: X^0 \to Y^0$.) Then f^{n-1} can be extended to maps of X^n, and (by our first two remarks) these fall into finitely many homotopy classes. At least one of this finite number of classes must consist of maps with extensions to m-equivalences for arbitrarily large m (and thus for all m). Let $f^n: X^n \to Y^n$ lie in such a class.

The map $f: X \to Y$ thus inductively constructed induces homotopy isomorphisms in all dimensions, and is thus a homotopy equivalence.

It is clear that this proof depends on finding a chain of n-equivalences which induce one another; this is necessary and sufficient for the existence of a homotopy equivalence. We proceed to construct two complexes whose n-equivalences do not contain such a chain.

Counter-example. Let S be a sphere of dimension $d \geqslant 2$. Let R_m $(m > d)$ be a countable CW-complex of the same m-type as S, but with $\Pi_r(R_m) = 0$ for $r \geqslant m$. Define

$$X = \prod_{m>d} R_m, \quad Y = S \times X,$$

X being a restricted Cartesian product with the weak topology. Then X, Y clearly have the same n-type, which (for $n > d$) is that of

$$\Big(\prod_{d<m<n} R_m\Big) \times (R_n^{\aleph_0}).$$

This is true for all n. On the other hand, suppose (for a contradiction) that X, Y are related by homotopy equivalences $f: X \to Y$, $g: Y \to X$. Then $g(S)$ is contained in a finite subcomplex of X, hence in $\prod_{d<m<k} R_m$ for some k; call this Z. Since $fg \sim 1$, and $\Pi_r(Z) = 0$ for $r \geqslant k$, we see that $\Pi_r(S) = 0$ for $r \geqslant k$, in contradiction with a well-known result of J.-P. Serre.

Therefore X, Y are of the same n-type for all n, but not of the same homotopy type.

I wish to acknowledge my indebtedness to Prof. J. H. C. Whitehead, both on general grounds and for encouraging me to publish this answer to his query.

REFERENCES

(1) WHITEHEAD, J. H. C. Combinatorial homotopy. I. *Bull. Amer. Math. Soc.* 55 (1949), 213–45.
(2) WHITEHEAD, J. H. C. Combinatorial homotopy. II. *Bull. Amer. Math. Soc.* 55 (1949), 453–96.

TRINITY COLLEGE
CAMBRIDGE

Proc. Camb. Phil. Soc. (1964), **60**, 699
Printed in Great Britain

An example in homotopy theory

BY J. F. ADAMS AND G. WALKER
University of Manchester

Communicated by C. T. C. WALL

(*Received* 19 *July* 1963)

The following proposition answers a question of P. Olum (conveyed to us in a private communication).

PROPOSITION. *There exist two CW-complexes X, Y and a map $f: X \to Y$ such that* (i) *the map f is essential,* (ii) *the restriction of f to any finite-dimensional skeleton X^n is inessential, and* (iii) *Y is finite-dimensional.*

In our example, X is the suspension of infinite complex projective space CP^∞. Y has the homotopy type of a countable wedge of 4-spheres, but we shall present Y in a different way, so as to facilitate description of the map f. We first make a 'telescope' T as follows. Take a countable union of copies of the product of the 3-sphere S^3 with the unit interval $I = [0, 1]$. Glue the end $S^3_i \times 1$ of the ith copy to the end $S^3_{i+1} \times 0$ of the $(i+1)$th copy by identifying each point $(x, 1)$ with $(g_i(x), 0)$, where $g_i: S^3_i \to S^3_{i+1}$ is a map of degree $(i+1)$. The resulting space T is thus a Moore space $Y(Q, 3)$, where Q is the additive group of rationals; if s_i is the homology class of S^3_i, s_1 generates the integer subgroup $Z \subset Q$ and $s_i = (i+1)\, s_{i+1}$.

LEMMA. *T is also an Eilenberg–MacLane space $K(Q, 3)$.*

Assuming this lemma, we can complete the construction. Our complex Y is obtained from T by attaching a cone C_i across each 3-sphere S^3_i. Since T is a $K(Q, 3)$, the homotopy classes of maps from X to T are in 1-1 correspondence with the elements of $H^3(X; Q) \cong Q$. We choose a map $f: X \to T$ corresponding to a non-zero cohomology class; this, regarded as a map from X to Y, is the map we undertook to construct.

Proof of the lemma. Let T_m be the finite subcomplex of T obtained by cutting off the telescope at the mth sphere. The image of a map of S^r into T must be contained in the finite subcomplex T_m for some m ((3), page 225, (D)). Hence

$$\pi_r(T) \cong \operatorname*{dir\,lim}_{m \to \infty} \pi_r(T_m),$$

where the homomorphisms of the direct system are induced by inclusion maps. Now T_m has the sphere S^3_m as a deformation-retract. Hence $\pi_r(T) \cong \operatorname*{dir\,lim}_{m\to\infty} \pi_r(S^3_m)$, where the homomorphisms are now induced by the maps $g_i: S^3_i \to S^3_{i+1}$ of degree $(i+1)$. The fact that S^3 is an H-space implies that the homomorphism

$$(g_i)_* : \pi_r(S^3_i) \to \pi_r(S^3_{i+1})$$

is given by $(g_i)_* \theta = (i+1)\,\theta$. Now for $r \neq 3$, the group $\pi_r(S^3)$ is finite ((2), page 494), and so the direct limit in question must be zero; for $r = 3$ the direct limit is (of course) Q. This completes the proof. (For an alternative proof, see (1), page 54.)

Proof of the proposition. Our example obviously satisfies condition (iii). It is easy to see that (ii) is also satisfied. For X^n is a finite CW-complex and so, by the argument of (3) already referred to, its image under f is contained in some finite subcomplex T_m. Since T_m has S^3_m as a deformation-retract, and S^3_m bounds the cone C_m in Y, we can make a null-homotopy of $f|X^n$.

Turning to property (i), let us suppose that $f: X \to Y$ is inessential. Then f can be extended to a map $F: CX, X \to Y, T$, where CX is a cone on X. Let g be a generator of $H_4(CX, X)$, and let c_i be the homology class of the cone C_i, so that the c_i form a Z-base for $H_4(Y, T)$. Then $F_*(g) = \sum_{i=1}^{\infty} n_i c_i$ for some integers n_i, of which only a finite number are non-zero. Since $dc_i = s_i$ and $s_i = (i+1)\, s_{i+1}$, we have on taking boundaries

$$F_*(dg) = \left(\sum_{i=1}^{\infty} \frac{n_i}{i\,!}\right) s_1.$$

By our choice of f, $F_*(dg) \neq 0$. Hence there is an integer k such that $n_k \neq 0$, and we can choose a prime p so that $n_k \not\equiv 0 \bmod p$. Then if c_k^* is the class in $H^4(Y, T;\, Z_p)$ dual to c_k, we have $F^*(c_k^*) \neq 0$. Now the Steenrod power

$$P^1: H^2(CP^\infty;\, Z_p) \to H^{2p}(CP^\infty;\, Z_p)$$

is an isomorphism, and so therefore is

$$P^1: H^4(CX, X;\, Z_p) \to H^{2p+2}(CX, X;\, Z_p).$$

Hence $P^1F^*(c_k^*) \neq 0$. But $P^1(c_k^*) = 0$, since it lies in a trivial group. So we reach a contradiction, and it follows that the map f is essential.

REFERENCES

(1) Adams, J. F. The sphere, considered as an H-space mod p. *Quart. J. Math. Oxford Ser.* (2), 12 (1961), 52–60.

(2) Serre, J.-P. Homologie singulière des espaces fibrés: applications. *Ann. of Math.* **54** (1951), 425–505.

(3) Whitehead, J. H. C. Combinatorial Homotopy I. *Bull. American Math. Soc.* **55** (1949), 213-245.

Topology Vol. 10, pp. 185–198. Pergamon Press, 1971.

A VARIANT OF E. H. BROWN'S REPRESENTABILITY THEOREM

J. F. ADAMS

(Received 4 August 1970)

§1. INTRODUCTION AND STATEMENT OF RESULTS

I WILL begin by summarizing the relevant work of E. H. Brown [2, 3]. In what follows, all our complexes will be connected CW-complexes with base-point; maps and homotopies will preserve the base-point. If X and Y are such complexes, $[X, Y]$ will mean the set of such homotopy classes of such maps from X to Y (as usual). If we keep Y fixed, then $[X, Y]$ gives a contravariant functor of X; it is defined on the category in which the objects are CW-complexes and the morphisms are homotopy classes of maps; it takes values in the category of sets and functions. A contravariant functor H from CW-complexes to sets is said to be "representable" if we have a natural isomorphism

$$T : H(X) \cong [X, Y]$$

(for some fixed Y).

Brown remarks that if H is representable, it satisfies two conditions which he formulates. In order to state the first, let X be the wedge-sum $\bigvee_\alpha X_\alpha$, and let $i_\alpha : X_\alpha \to X$ be the injection of X_α. We have an induced function

$$i_\alpha{}^* = H(i_\alpha) : H(X) \to H(X_\alpha);$$

let

$$\theta : H(X) \to \prod_\alpha H(X_\alpha)$$

be the function whose αth component is $i_\alpha{}^*$. Brown's first condition is:

1.1. THE WEDGE AXIOM. *The function θ is an isomorphism of sets.*

In order to state the second condition, we consider the following diagram, in which the functions are induced by the obvious inclusions.

$$\begin{array}{ccc} H(X \cap Y) & \xleftarrow{a} & H(X) \\ {\scriptstyle b}\uparrow & & \uparrow{\scriptstyle c} \\ H(Y) & \xleftarrow{d} & H(X \cup Y) \end{array}$$

Brown's second condition is:

1.2. THE MAYER–VIETORIS AXIOM. *Suppose given $x \in H(X)$ and $y \in H(Y)$ such that $ax = by$; then there exists $z \in H(X \cup Y)$ such that $cz = x$ and $dz = y$.*

When we say that H satisfies the Wedge Axiom, we mean of course that (1.1) holds for all wedge-sums $\bigvee_\alpha X_\alpha$; and similarly for (1.2).

Brown shows that if a functor H satisfies these two axioms, then it is representable.

However, in some of the applications we have to consider a functor H which is defined not on all CW-complexes, but on some subcategory. I give two examples.

First, the reduced K-cohomology group $\tilde{K}(X)$ can be defined when X is a finite-dimensional complex by the usual construction in terms of vector-bundles [1]. If we were to apply the same construction when X is an infinite-dimensional complex it would lead to a functor which does not satisfy the wedge axiom, and does not agree with the "correct" functor $\tilde{K}(X)$.

We therefore have to consider the case of a functor H which is defined only on finite-dimensional complexes. In this case the theorems of Brown are perfectly satisfactory.

Secondly, consider the representability theorem of G.W.Whitehead [6]. In this case we suppose given a generalized homology theory H_*, and define a generalized cohomology theory on finite CW-complexes by S-duality:

$$\tilde{H}^n(X) = \tilde{H}_{-n}(DX).$$

We therefore have to consider the case of a functor H which is defined only on finite CW-complexes. In this case the theorems of Brown contain an inconvenient technical restriction: one has to assume that the sets $H(X)$ are countable [2, 3, 6].

However, in the example given (and in all other applications I know) the restriction to finite complexes arises from the fact that one uses S-duality. Therefore one really has to deal with stable homotopy theory rather than with unstable homotopy theory.

This makes it reasonable to look for a variant of Brown's Theorem which has the following features.

(i) The functor H is supposed given only on finite CW-complexes. (In what follows, the letters K, L, M will be reserved for finite CW-complexes.)

(ii) No assumption is made on the cardinality of the sets $H(K)$.

(iii) Instead, one makes some mild assumption of the sort that is satisfied when one is dealing with stable problems. For example, one might assume that the functor H has the form $H(K) = H'(SK)$ for some other functor H', where SK is the suspension of K; or one might assume that the functor H takes its values in the category of groups and homomorphisms (compare Brown [2, p. 471]). Neither assumption is too restrictive; both are satisfied in all the applications I know.

The object of this paper is to prove such a variant of Brown's Theorem, and to give some related results.

THEOREM 1.3 *Let H be a contravariant functor, defined on the category of finite CW-complexes K and homotopy classes of maps, and taking values in the category of groups and* *homomorphisms. Suppose that H satisfies the Wedge Axiom and the Mayer–Vietoris Axiom. Then there is a CW-complex Y and an isomorphism of sets*

$$T : [K, Y] \cong H(K)$$

defined and natural for all finite complexes K.

Such a complex Y, equipped with such a natural isomorphism T, may be called a "representing complex" for H.

The statement of Theorem 1.3 is incomplete in two respects. First, T is asserted to be an isomorphism of sets; it is desirable to know that Y is an H-space of a suitable kind, and that T is an isomorphism of groups. This point will be dealt with by Addendum 1.4 below.

Secondly, suppose that we apply the theorem twice, so that we have natural isomorphisms of sets

$$T : [K, Y] \cong H(K)$$
$$T' : [K, Y'] \cong H'(K).$$

It is desirable to know that any natural transformation from H to H' is induced by a map from Y to Y'. This point will be dealt with by Addendum 1.5 below.

In order to complete the theorem, we introduce the considerations which follow. We will say that two maps $f, g : X \to Y$ are "weakly homotopic", and write $f \sim_w g$, if $fh \sim gh$ for every map h from any finite complex K to X (compare [4]). Weak homotopy is an equivalence relation; composition of maps passes to weak homotopy classes. We write $[X, Y]_w$ for the set of weak homotopy classes of maps from X to Y; if X happens to be finite, then $[X, Y]_w = [X, Y]$. We say that Y is a "weak H-space" if it is provided with a product map $\mu : Y \times Y \to Y$ and an inverse map $\iota : Y \to Y$ which satisfy the usual identities, such as the associative law, up to weak homotopy. (Here $Y \times Y$ means the product in the category of CW-complexes.) If Y is a weak H-space, then $[X, Y]_w$ is a functor of X taking values in the category of groups and homomorphisms; in particular, if K runs through finite complexes, then $[K, Y]$ is a functor of K taking values in the category of groups and homomorphisms.

ADDENDUM 1.4. *If Y is a representing complex for H, then Y may be made into a weak H-space so that the natural transformation*

$$T : [K, Y] \cong H(K)$$

is an isomorphism of groups.

ADDENDUM 1.5. *Let X be any CW-complex, let Y be a representing complex for H, and let*

$$U : [K, X] \to [K, Y]$$

be a transformation of sets defined and natural for finite CW-complexes K. Then there is a map $f : X \to Y$ inducing U, and it is unique up to weak homotopy.

From these results we draw the obvious conclusions.

THEOREM 1.6. *Let H^* be a generalized cohomology theory defined on finite CW-complexes (without any assumption on the cardinality of the coefficient groups). Then H^* is the generalized cohomology theory corresponding to an Ω-spectrum E.*

This extends a result of Brown [2, p. 480].

In order to state the next result, we recall the expected behaviour of a generalized homology theory with respect to limits. Let H_* be a generalized homology theory defined on CW-complexes X; and let X_α run over the finite subcomplexes of X; then we have a canonical map

$$\theta : \underset{\overrightarrow{\alpha}}{\text{Lim}}\, H_*(X_\alpha) \rightarrow H(X).$$

We assume, as our axiom on limits, that the map θ is an isomorphism for all X.

THEOREM 1.7. *If H_* satisfies these assumptions, it is the generalized homology theory corresponding to a spectrum E.*

This extends a result of Whitehead [6].

The next result is about the behaviour of generalized cohomology theories with respect to limits. Let H be a contravariant functor from the category of all CW-complexes to the category of groups, satisfying the Wedge and Mayer–Vietoris Axioms. Let X run over the finite subcomplexes of X; then we have a canonical map

$$\theta : H(X) \rightarrow \underset{\overleftarrow{\alpha}}{\text{Lim}}\, H(X_\alpha).$$

THEOREM 1.8. *The canonical map*

$$\theta : H(X) \rightarrow \underset{\overleftarrow{\alpha}}{\text{Lim}}\, H(X_\alpha)$$

is an epimorphism.

This result is closely related to (1.5). If X is countable then the result is well known; but in the general case it seems to be new. We also know, of course, that θ need not be an isomorphism, even if X is countable [5].

The functor $\underset{\overleftarrow{\alpha}}{\text{Lim}}\, H(X_\alpha)$ can be introduced when H is given only on finite CW-complexes. It was introduced by Brown [2, 3]; it is useful as a technical device (see §4), and it has some intrinsic interest.

Formally, let H be a functor of the sort considered in Theorem 1.3; that is, H is a functor from finite complexes to groups, satisfying the Wedge and Mayer–Vietoris Axioms. Then we define the functor $\hat{H}$ on all complexes X by setting

$$\hat{H}(X) = \underset{\overleftarrow{\alpha}}{\text{Lim}}\, H(X_\alpha),$$

where X_α runs over the finite subcomplexes of X. If X happens to be finite, then $\hat{H}(X) = H(X)$. The properties of $\hat{H}$ are considered more closely in §3. For the moment we remark that if $f, g : X \rightarrow Y$ are weakly homotopic, then

$$f^* = g^* : \hat{H}(Y) \rightarrow \hat{H}(X).$$

Our next result shows that $\hat{H}$ is representable, in a suitable sense. We assume that Y is a representing complex for H.

THEOREM 1.9. (i) *There is one and only one natural transformation*

$$\hat{T} : [X, Y]_w \to \hat{H}(X)$$

which reduces to T when X is finite.

(ii) *The transformation $\hat{T}$ is an isomorphism of sets for all X.*

(iii) *The complex Y may be made into a weak H-space so that $\hat{T}$ is an isomorphism of groups.*

This result also is closely related to (1.5). It is clear that it includes (1.4).

This completes the statement of results. It is a pleasure to acknowledge the stimulus of correspondence with B. Bollobas and A. Heller. My original result in this direction was a stable analogue of Theorem 1.3; Heller obtained a different proof of this result, and this helped me to analyse what was essential in the situation.

§2. LIMITS

Our main proof involves a certain inverse system of sets, and difficulties would arise if the inverse limit of this inverse system happened to be empty. To avoid this difficulty we introduce the considerations in this section.

We will actually work with something slightly more general than an inverse system in the usual sense. Given any category C of sets and functions, we can form the set $\varprojlim C$; an element of this set is a function e which assigns to each object X of C an element e_X in X, in such a way that for every morphism $f: X \to Y$ in C we have

$$fe_X = e_Y.$$

We will restrict attention to categories C which have the following two properties.

(2.1) For any X and Y in C there is at most one morphism $f: X \to Y$ in C.

(2.2) For any two objects X and Y in C there is an object Z in C which admits morphisms $Z \to X$, $Z \to Y$ in C.

The difference between these axioms and those for an inverse system is slight, but perhaps deserves comment. In an ordered set $\alpha \leq \beta$, $\beta \leq \alpha$ usually implies $\alpha = \beta$. In our category C two objects X and Y can admit morphisms $X \to Y$, $Y \to X$ in C without X and Y being equal. This extra generality is crucial for our purposes.

If our category C happens to be an inverse system, our definition of $\varprojlim C$ gives the usual inverse limit.

We say that a subcategory C is cofinal in D if for every X in D there is a Y in C which admits a morphism $Y \to X$ in D.

LEMMA 2.3. *If C is cofinal in D, the restriction function*

$$\varprojlim D \to \varprojlim C$$

is an isomorphism of sets.

The proof is trivial.

An (infinite) sequence S is a category with objects X_n for $n = 1, 2, 3, \ldots$ and with a morphism $X_n \to X_m$ wherever $n \geq m$. In order not to exclude certain trivial cases it is desirable to consider also finite sequences; such a sequence is indexed over the integers $1, 2, 3, \ldots, N$.

LEMMA 2.4. *Let S be a sequence in which the objects are non-empty sets and the maps are epimorphisms of sets. Then* $\varprojlim S$ *is non-empty.*

This lemma is both easy and well-known.

LEMMA 2.5. *Let C be a category whose objects fall into countably many equivalence classes. Then C contains a cofinal sequence.*

The proof is trivial.

COROLLARY 2.6. *Let C be a category in which the objects are non-empty sets, the morphisms are epimorphisms, and the objects fall into countably many equivalence classes. Then* $\varprojlim C$ *is non-empty.*

This follows immediately from Lemmas 2.3–2.5.

In the applications we will face a category C whose objects may fall into uncountably many equivalence classes. Our response will be to adjoin new morphisms to C, so that objects which were not equivalent in C may become equivalent in the new category.

More precisely, suppose given a category C (satisfying (2.1) and (2.2), as always); and suppose that the object of C are non-empty sets, and the morphisms are epimorphisms. We define a new category $\bar{C}$ as follows. The objects of $\bar{C}$ are the objects of C. A function $f: X \to Y$ lies in $\bar{C}$ if we can find a commutative diagram

$$\begin{array}{ccc} Z & & \\ {\scriptstyle a}\downarrow & \searrow{\scriptstyle b} & \\ X & \xrightarrow{f} & Y \end{array}$$

with a and b in C.

LEMMA 2.7. *This construction defines a category $\bar{C}$ satisfying* (2.1) *and* (2.2); *$\bar{C}$ contains C as a cofinal subcategory; the objects of $\bar{C}$ are non-empty sets and the morphisms of $\bar{C}$ are epimorphisms.*

The proof is easy.

COROLLARY 2.8. *Suppose that the object of $\bar{C}$ fall into countably many equivalence classes. Then* $\varprojlim C$ *is non-empty.*

This follows immediately from Corollary 2.6 applied to $\bar{C}$, by using Lemma 2.3.

§3. PROPERTIES OF H AND $\hat{H}$

In this section we assume throughout that H is as in (1.3), so that H is a functor from finite CW-complexes to groups, satisfying the Wedge Axiom and the Mayer–Vietoris Axiom. We assume throughout that $\hat{H}$ is as defined in §1. We derive sufficient properties of H and $\hat{H}$ to permit the use of Brown's methods.

LEMMA 3.1. *For any cofibering*

$$K \xrightarrow{f} L \xrightarrow{i} L \cup_f CK$$

of finite complexes, the sequence

$$H(K) \xleftarrow{f^*} H(L) \xleftarrow{i^*} H(L \cup_f CK)$$

is exact.

This follows immediately from the axioms, as in Brown [2, 3].

For the next lemma, we assume that K is a finite complex, containing subcomplexes L and M.

LEMMA 3.2. *There is an exact sequence*

$$H(L) \times H(M) \xleftarrow{(i_1^*, i_2^*)} H(L \cup M) \leftarrow H(S(L \cap M)) \xleftarrow{g^*} H(S(L \vee M)),$$

natural for maps $K, L, M \to K', L', M'$, and in which the homomorphism g^* is induced by a map $g: S(L \cap M) \to S(L \vee M)$.

Proof. Consider the obvious map

$$L \vee M \to L \cup M,$$

and use it to start a cofibre sequence. We get

$$L \vee M \to L \cup M \to (L \cup M) \cup C(L \vee M) \to S(L \vee M).$$

The third term is homotopy-equivalent to $S(L \cap M)$, by a homotopy-equivalence which is natural. Applying Lemma 3.1, we get an exact sequence

$$H(L \vee M) \leftarrow H(L \cup M) \leftarrow H(S(L \cap M)) \xleftarrow{g^*} H(S(L \vee M))$$

which is still natural. We apply the Wedge Axiom to rewrite $H(L \vee M)$, and we obtain the required result.

LEMMA 3.3. *$\hat{H}$ satisfies the Wedge Axiom.*

The proof is trivial.

For the next lemma, let X be a CW-complex, and let $\{X_\alpha\}$ be any directed set of subcomplexes of X whose union is X.

LEMMA 3.4. *The canonical map*

$$\hat{H}(X) \to \varprojlim_{\alpha} \hat{H}(X_\alpha)$$

is an isomorphism.

The proof is trivial.

The next result is crucial. We suppose given a CW-complex X containing subcomplexes U and V.

PROPOSITION 3.5. *If $U \cap V$ is a finite complex, then the square*

$$\begin{array}{ccc} H(U \cap V) = \hat{H}(U \cap V) & \longleftarrow & \hat{H}(U) \\ \uparrow & & \uparrow \\ \hat{H}(V) & \longleftarrow & \hat{H}(U \cup V) \end{array}$$

satisfies the Mayer–Vietoris Axiom.

In order to make this result seem reasonable, one should perhaps observe that it follows from Theorem 1.9. In fact, suppose given elements $u \in \hat{H}(U)$, $v \in \hat{H}(V)$ whose restrictions to $U \cap V$ coincide. If we assume Theorem 1.9, then u and v can be represented by maps $f: U \to Y$, $g: V \to Y$ whose restrictions to $U \cap V$ are weakly homotopic, and hence homotopic since $U \cap V$ is a finite complex. Using the homotopy extension theorem in the usual way, we see that there is a map $h: U \cup V \to Y$ whose restrictions to U and V are homotopic to f and g. This gives the required element in $\hat{H}(U \cup V)$.

Of course, we propose to prove Proposition 3.5 directly in order to use it in the proof of Theorem 1.9.

Proof of Proposition 3.5. Let U_α, V_β run over those finite subcomplexes of U, V which contain $U \cap V$. Suppose given elements $u \in \hat{H}(U)$, $v \in \hat{H}(V)$ whose restrictions to $U \cap V$ coincide; that is, we are given elements $u_\alpha \in H(U_\alpha)$ for all α, $v_\beta \in H(V_\beta)$ for all β which behave in the obvious way under inclusion maps, and all have the same image in $H(U \cap V)$. We have to construct an element in $\hat{H}(U \cup V)$ which restricts to u in $\hat{H}(U)$ and to v in $\hat{H}(V)$. For each α and β, let $H_{\alpha,\beta}$ be the set of elements $w \in H(U_\alpha \cup V_\beta)$ such that the restriction of w to U_α is u_α and the restriction of w to V_β is v_β. Such elements exist by the Mayer–Vietoris Axiom for H, so $H_{\alpha,\beta}$ is non-empty. Whenever $U_\alpha \subset U_\gamma$ and $V_\beta \subset V_\delta$ we have a function

$$i^* : H_{\gamma,\delta} \to H_{\alpha,\beta}$$

induced by the inclusion $i: U_\alpha \cup V_\beta \to U_\gamma \cup V_\delta$.

We propose to apply the considerations of §2, taking C to be the category in which the objects are the sets $H_{\alpha,\beta}$ and the morphisms are the functions i^*. It will be sufficient to show that $\varprojlim C$ is non-empty, because an element of $\varprojlim C$ is an element of $\hat{H}(U \cup V)$ with the required properties. We proceed to check that the conditions of §2 apply; we begin with the fact that $i^* : H_{\gamma,\delta} \to H_{\alpha,\beta}$ is an epimorphism.

Take the exact sequence of Lemma 3.2 and apply it to the complex $U_\alpha \cup V_\beta$. We see that $H(S(U \cap V))$ acts transitively on $H_{\alpha,\beta}$ (on the left, say). Each map $i^* : H_{\gamma,\delta} \to H_{\alpha,\beta}$ commutes with this action (as we see from the fact that the sequence of Lemma 3.2 is natural). Since $H_{\gamma,\delta}$ is non-empty, it follows that $i^* : H_{\gamma,\delta} \to H_{\alpha,\beta}$ is an epimorphism of sets.

We note that in saying that $H(S(U \cap V))$ acts transitively on $H_{\alpha,\beta}$, we have begun to use the assumption that H takes values in the category of groups. At this point we could still avoid this assumption, if we were willing to restrict attention to a special case and replace $U \cup V$ by a mapping-cone $U \cup CK$, where K is a finite complex. However, the assumption that H takes values in the category of groups becomes essential in proving Lemma 3.6 below.

We construct $\bar{C}$ from C as in §2, and we pause to state its properties.

LEMMA 3.6. (i) *There is a morphism* $H_{\theta,\phi} \to H_{\alpha,\beta}$ *in* $\bar{C}$ *if and only if the image of*

$$g^*_{\theta,\phi} : H(S(U_\theta \vee V_\phi)) \to H(S(U \cap V))$$

is contained in the image of

$$g^*_{\alpha,\beta} : H(S(U_\alpha \vee V_\beta)) \to H(S(U \cap V))$$

where the maps g^* *are as in Lemma* 3.2.

(ii) $H_{\theta,\phi}$ *and* $H_{\alpha,\beta}$ *are equivalent in* $\bar{C}$ *if and only if the images of* $g^*_{\theta,\phi}$ *and* $g^*_{\alpha,\beta}$ *are equal.*

The proof is easy.

We resume the proof of Proposition 3.5. We can find a countable set of finite CW-complexes K containing at least one representative from each homotopy type. (For example consider the finite simplicial complexes.) For each K the maps $g : S(U \cap V) \to K$ fall into countably many homotopy classes. (This may be proved, for example, by replacing $S(U \cap V)$ and K with equivalent finite simplicial complexes and using the simplicial approximation theorem.) So there are in all only a countable number of images of homomorphisms

$$g^* : H(K) \to H(S(U \cap V)).$$

By Lemma 3.6, the objects in $\bar{C}$ must fall into countably many equivalence classes. Corollary 2.8 applies, and shows that $\varprojlim C$ is non-empty. This completes the proof of Proposition 3.5.

§4. APPLICATION OF BROWN'S METHOD

Now that we have proved Proposition 3.5, the results in §1 can be proved by following the methods of Brown [2, 3]; and in this section we will prove them. We will however use a variant of Brown's proof. Brown's original method allows one to construct the representing complex Y in a fairly economical way, without using an inordinate number of cells. The variant to be given shows how crude and wasteful one can be and still prove the required results.

We begin by explaining the use of $\hat{H}$ in Brown's method of proof.

Given a complex Y and an element $y \in \hat{H}(Y)$, we can construct a transformation

$$\hat{T} : [X_1\ Y]_w \to \hat{H}(X)$$

defined and natural for all complexes X; in fact, given $f : X \to Y$, we define

$$\hat{T}(\alpha) = f^* y.$$

By restricting X to run over finite complexes K, we can obtain a natural transformation

$$T : [K, Y] \to H(K).$$

Let us write Nat Trans (A, B) for the set of natural transformations from A to B. Then it is easy to check the following lemma, of which the most important part is due to Brown [2, p. 478].

LEMMA 4.1. *The construction above gives* (1–1) *correspondences*

$$\hat{H}(Y) \cong \text{Nat Trans}([X, Y]_w, \hat{H}(X))$$
$$\cong \text{Nat Trans}([K, Y], H(K)).$$

These correspondences are natural for maps of Y.

This lemma explains the importance of the functor $\hat{H}$ in Brown's method. In proving Theorem 1.3, you construct Y by induction; and if at some stage you have constructed an infinite complex X which you hope will be a subcomplex of the representing complex Y, you may plausibly assume as part of your inductive hypothesis that you have a natural transformation

$$T : [K, X] \to H(K).$$

Such a natural transformation is more easily handled by considering the corresponding element $x \in \hat{H}(X)$.

We proceed with the work.

LEMMA 4.2. *Let Y_n be a CW-complex provided with an element $y_n \in \hat{H}(Y_n)$. Then there is an embedding $i : Y_n \to Y_{n+1}$ and an element $y_{n+1} \in \hat{H}(Y_{n+1})$ which restricts to $y_n \in H(Y_n)$, so that the following condition is satisfied. If $f, g : K \to Y_n$ are any two maps from a finite complex K to Y, such that $f^*y_n = g^*y_n$,* then *if $\cong ig$ in Y_{n+1}.*

Proof. As in the proof of (3.5), it is sufficient to let K run over a countable set of representatives. For each such K, and each pair of homotopy classes of maps $f, g : K \to Y_n$ such that $f^*y_n = g^*y_n$, choose a pair of representative maps. We write these representative maps

$$f_\alpha, g_\alpha : K_\alpha \to Y_n,$$

where α runs over a suitably large set A of indices.

We now form

$$Y_{n+1} = Y_n \cup \bigcup_{\alpha \in A} (I \times K_\alpha / I \times pt),$$

where the reduced cylinder $I \times K_\alpha / I \times pt$ is attached to Y_n by the map f_α at one end and g_α at the other. We have certainly arranged that $if_\alpha \sim ig_\alpha$ for each α; it remains to construct y_{n+1}.

The construction of y_{n+1} is by Zorn's Lemma. We consider pairs (B, h) in which B is is a subset of A and h is an element in

$$\hat{H}(Y_n \cup \bigcup_{\beta \in B} (I \times K_\beta / I \times pt))$$

which restricts to y_n in $\hat{H}(Y_n)$. We order these pairs by writing $(B, h) \leq (B', h')$ if $B \subset B'$ and h' restricts to h.

The set of pairs is non-empty, since it contains the pair with $B = \emptyset$, $h = y_n$. It is inductive, for if we have a chain of pairs (B, h) we can define $B' = \bigcup B$ and construct h' by Lemma 3.4; the pair (B', h') provides an upper bound for the chain. Therefore our set of pairs contains a maximal element. Now, a pair (B, h) with $B \neq A$ is not maximal; for if $\alpha \notin B$, we can use Proposition 3.5 to extend h over

$$(Y_n \cup \bigcup_{\beta \in B} (I \times K_\beta / I \times pt)) \cup (I \times K_\alpha / I \times pt).$$

So a maximal element has the form (A, h), and we can take $y_{n+1} = h$. This completes the proof.

PROPOSITION 4.4. *Let Y_0 be a CW-complex, provided with an element $y_0 \in \hat{H}(Y_0)$. Then there is an embedding $i: Y_0 \to Y$ and an element $y \in \hat{H}(Y)$, which restricts to y_0 in $\hat{H}(Y_0)$, such that the corresponding natural transformation*

$$T : [K, Y] \to H(K)$$

is an isomorphism of sets for all finite complexes K.

It is clear that this result includes Theorem 1.3 (take Y_0 to be a point).

Proof. Let K run over a countable set of representatives, as in the proof of (3.5), and for each K let h run over $H(K)$. Form

$$Y_1 = Y_0 \vee \bigvee_{K, h} K;$$

and using Lemma 3.3, let $y_1 \in \hat{H}(Y_1)$ be the element which restricts to y_0 on Y_0 and to h on the (K, h)th summand of $\bigvee_{K, h} K$. The construction ensures that the natural transformation

$$T_1 : [K, Y_1] \to H(K)$$

corresponding to y_1 is an epimorphism of sets for all K.

Now construct complexes

$$Y_1 \subset Y_2 \subset Y_3 \subset \cdots \subset Y_n \subset \cdots$$

and elements $y_n \in \hat{H}(Y_n)$ by induction over n, using Lemma 4.2. Take $Y = \bigcup_n Y_n$ and (using Lemma 3.4) let $y \in \hat{H}(Y)$ be the element which restricts to y_n in $H(Y_n)$ for each n. The corresponding natural transformation

$$T : [K, Y] \to H(K)$$

is still epimorphic, for maps into Y_1 suffice to give all the elements of $H(K)$. But it is also a monomorphism of sets; for let $f, g : K \to Y$ be any two maps such that $f^*y = g^*y$; since K is a finite complex, f and g must map into Y_n for some n, so that $f^*y_n = g^*y_n$ and $f \sim g$ in Y_{n+1} by Lemma 4.2. This proves Proposition 4.3.

We are now ready to prove the results in §1.

Proof of Theorem 1.9. Our standing hypothesis is that Y is a representing complex for H, so that there is an isomorphism of sets

$$T : [K, Y] \to H(K)$$

defined and natural for all finite complexes K. Part (i), on the existence and uniqueness of

$$\hat{T} : [X, Y]_w \to \hat{H}(X)$$

follows immediately from Lemma 4.1. We turn to part (ii).

It is easy to show that

$$\hat{T} : [X, Y]_w \to \hat{H}(X)$$

is monomorphic, as follows. Let $f, g : X \to Y$ be two maps such that $\hat{T}f = \hat{T}g$, let K be any finite complex, and let $h : K \to X$ be any map. Then $h^*\hat{T}f = h^*\hat{T}g$, that is, $T(fh) = T(gh)$. Since T is an isomorphism, we have $fh \sim gh$. This holds for all K and h, so $f \sim_w g$.

It remains to show that $\hat{T}$ is epimorphic; we use an argument due to Brown [3]. By Lemma 4.1, T corresponds to an element $y \in \hat{H}(Y)$, and the natural transformation

$$\hat{T} : [X, Y]_w \to \hat{H}(X)$$

is given by

$$T(f) = f^*y.$$

Suppose given a complex X and a class $x \in \hat{H}(X)$. Apply Proposition 4.3, taking Y_0 to be the complex $X \vee Y$ and $y_0 \in \hat{H}(Y_0)$ to be the element which restricts to x in $\hat{H}(X)$, y in $\hat{H}(Y)$. By Proposition 4.3 there is an embedding (say) $Y_0 \to Y'$ and an element y' in $\hat{H}(Y')$, which restricts to y_0 in $\hat{H}(Y_0)$, such that the corresponding natural transformation

$$T' : [K, Y'] \to H(K)$$

is an isomorphism of sets. We have the following commutative diagram.

$$\begin{array}{ccc} [S^n, Y] & \xrightarrow{T} & \\ \downarrow i_* & & H(S^n) \\ [S^n, Y'] & \xrightarrow{T'} & \end{array}$$

We see that $i : Y \to Y'$ induces an isomorphism of homotopy groups; by the theorem of J. H. C. Whitehead, it is a homotopy equivalence. If we compose the injection $X \to Y'$ with a homotopy-inverse for i, we obtain a map $f : X \to Y$ such that $f^*y = x$. This shows that $\hat{T}$ is epimorphic, and proves part (ii).

We turn to part (iii). The product in $H(X)$ gives a natural transformation

$$\hat{H}(X) \times \hat{H}(X) \to \hat{H}(X);$$

applying the isomorphism $\hat{T}$, we get a natural transformation

$$[X, Y \times Y]_w \to [X, Y]_w.$$

Since this is defined for all X, it must be induced by a map

$$\mu : Y \times Y \to Y,$$

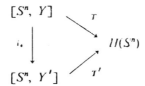

unique up to weak homotopy. Similarly for the existence of the inverse map $\iota : Y \to Y$, and for the identities which they satisfy. The construction ensures that

$$\hat{T} : [X, Y]_w \to \hat{H}(X)$$

is an isomorphism of groups.

This completes the proof of Theorem 1.9.

Proof of Addendum 1.5. The fact that f is unique up to weak homotopy is trivial; if

$$f_* = g_* : [K, X] \to [K, Y],$$

then we have $fh \sim gh$ for each map h from a finite complex K to X, that is, $f \sim_w g$. It remains to show that f exists. Let

$$U : [K, X] \to [K, Y]$$

be a transformation of sets defined and natural for any finite complex K. Consider the composite

$$[K, X] \xrightarrow{U} [K, Y] \xrightarrow{T} H(K).$$

By Lemma 4.1 it corresponds to an element $x \in \hat{H}(X)$, while T corresponds to an element $y \in \hat{H}(Y)$. By Theorem 1.9 (ii) there is a map $f : X \to Y$ such that $f^*y = x$. Using the naturality clause of Lemma 4.1, we see that $Tf_* = TU$. Since T is an isomorphism we have $f_* = U$. This completes the proof of Addendum 1.5.

Proof of Theorem 1.8. Let H be as in Theorem 1.8, so that H is representable by Brown's Theorem; we have a complex Y and a natural isomorphism

$$T : [X, Y] \to H(X)$$

valid for all X. So Y is a representing complex in the sense of (1.3), and Theorem 1.9 applies. Consider the following diagram.

$$\begin{array}{ccc} [X, Y] & \xrightarrow{T} & H(X) \\ \downarrow & & \downarrow \theta \\ [X, Y]_w & \dashrightarrow & \hat{H}(X) \end{array}$$

It is easy to check that θT factors through $[X, Y]_w$ so as to yield the lower arrow; this arrow is a natural transformation which reduces to T when X is finite, so by Theorem 1.9 it must be $\hat{T}$ and it must be an isomorphism of sets. Since $[X, Y] \to [X, Y]_w$ is trivially epi, it follows that $\theta : H(X) \to \hat{H}(X)$ is an epimorphism. This proves Theorem 1.8.

The deduction of Theorem 1.6 and 1.7 from the results already proved is done as in Brown [2, 3] and G. W. Whitehead [6].

REFERENCES

1. M. F. ATIYAH and F. HIRZEBRUCH: Vector bundles and homogeneous spaces, *Proc. Symp. Pure Math.* Vol. 3, pp. 7–38. Am. Math. Soc. (1961).
2. E. H. BROWN: Cohomology theories, *Ann. Math.* **75** (1962), 467–484; with a correction, *Annal. Math.* **78** (1963), 201.
3. E. H. BROWN: Abstract homotopy theory, *Trans. Am. Math. Soc.* **119** (1965), 79–85.
4. Séminaire H. Cartan, 1959/60, exposé 11, p. 11.01.

5. J. MILNOR: On axiomatic homology theory, *Pacific J. Math.* **12** (1962), 337–341.
6. G. W. WHITEHEAD: Generalised homology theories, *Trans. Am. Math. Soc.* **102** (1962), 227–283.

University of Manchester
Manchester

Idempotent Functors in Homotopy Theory

J. F. Adams

§ 1.

Work of Sullivan [6] has emphasised the importance in homotopy-theory of localisation and completion functors. The object of the present work is to begin an axiomatic investigation of the number and properties of such functors. Since beginning it, I have learnt of the independent but overlapping work of Deleanu, Frei and Hilton [4].

I suggest that the examples we wish to construct and investigate will admit the following formal description.

(i) There is a category C in which we work.

(ii) There is a functor $E: C \to C$ which assigns to each object its "localisation" or "completion".

(iii) There is a natural transformation $\eta : 1 \to E$.

Example 1.1. (Localisation in commutative algebra.) We suppose given a commutative ring R and a multiplicative subset T of R. Then C is the category of R-modules M. The functor $E: C \to C$ assigns to each module M its localisation $T^{-1}M$; this may be constructed so that the elements of $T^{-1}M$ are fractions m/t with m in M, t in T. The map $\eta_M : M \to T^{-1}M$ carries m to $m/1$.

Example 1.2. (Completion in metric spaces.) Here C is the category in which the objects are metric spaces and the morphisms are uniformly continuous maps. The functor $E: C \to C$ assigns to each metric space X its completion $\hat{X}$; this may be constructed so that the points of $\hat{X}$ are equivalence classes of sequences $\{x_1, x_2, \ldots, x_n, \ldots\}$ with $x_n \in X$. The map $\eta_X : X \to \hat{X}$ carries x to the constant sequence $\{x, x, \ldots, x, \ldots\}$.

As a homotopy-theorist, however, I wish to work in a different category C.

Definition 1.3. We define CW to be the category in which the objects are connected CW-complexes with base-point, and the morphisms are homotopy classes of maps. (Of course, maps and homotopies are required to preserve the base-point.)

§ 2.

Let us return to the general case. I suggest that the appropriate axioms to impose on E and η are as follows.

Axiom 2.1. $\eta_{EX} = E\eta_X : EX \to E^2X$ (for each object X in C).

Axiom 2.2. The map $\eta_{EX} = E\eta_X$ is an equivalence from EX to E^2X (for each X in C).

These axioms are satisfied by Examples 1.1 and 1.2. Therefore, we may expect these simple formal properties to hold in any case presenting a valid analogy with 1.1 or 1.2. Equivalently, if for some C, E and η either 2.1 or 2.2 is found to fail, then that by itself will tend to discredit any analogy with 1.1 or 1.2, to the point where one might hesitate to use the word "localisation" or "completion" for such a functor E.

Axioms 2.1 and 2.2 say that the functor E is "idempotent" (in a particular way). A pair (E, η) satisfying the axioms is called by the categorists an *idempotent monad* or *idempotent triple*. For our immediate purposes, it is sufficient to know that for each X, the map $\eta_X: X \to EX$ enjoys two distinct universal properties, either of which is sufficient to characterise it. (See 3.2 below.) In order to state these universal properties, we associate to any pair (E, η) a subset S of the morphisms in C, and also a subset D of the objects in C.

Definition 2.3. A morphism $f: X \to Y$ lies in S if and only if $Ef: EX \to EY$ is an equivalence.

Definition 2.4. (a) An object X lies in D if and only if X is equivalent to EY for some Y.

(b) An object X lies in D if and only if $\eta_X: X \to EX$ is an equivalence.

The two forms of Definition 2.4 are equivalent.

Example 2.5. We suppose given an integer $n \geq 1$. Let X be an object in CW. When we construct the "Postnikov system" of X, we construct an object EX and a map $\eta_X: X \to EX$ with the following properties.

(i) $(\eta_X)_*: \pi_r(X) \to \pi_r(EX)$ is an isomorphism for $r \leq n$.

(ii) $\pi_r(EX) = 0$ for $r > n$.

Then E can be made into a functor in such a way that η is a natural transformation; after that, Axioms 2.1 and 2.2 hold. In this case, a map $f: X \to Y$ lies in S if and only if

$$f_*: \pi_r(X) \to \pi_r(Y)$$

is an isomorphism for $r \leq n$; an object X lies in D if and only if $\pi_r(X) = 0$ for $r > n$.

The universal properties of the map $\eta_X: X \to EX$ will be given in §3. It follows from them that the subset S determines the pair (E, η) up to equivalence, and similarly for D. Here the appropriate equivalence relation is as follows: a pair (E, η) is equivalent to a pair (E', η') if there is a natural equivalence $\varepsilon: E \to E'$ such that $\eta' = \varepsilon\eta: 1 \to E'$.

The basic result in this area now gives an axiomatic characterisation of those subsets S which arise from pairs (E, η). This result will be stated as Theorem 4.7.

The remainder of this account is organised as follows. In §3 we give the characterisation of $\eta_X: X \to EX$ by universal properties. In §4 we give the axiomatic characterisation of sets S, and state the main theorem. In §5 we give an auxiliary axiom. In §6 and §7 I suggest that the properties of idempotent functors E deserve further investigation on axiomatic lines.

§ 3.

In this section we continue to work in a general category C. We assume given (E, η) satisfying 2.1 and 2.2. The first lemma shows how to characterise S in terms of D, and vice versa.

Lemma 3.1. (a) *Suppose given $f: X \to Y$ in C. Then f lies in S if and only if*

$$f^*: [X, Z] \leftarrow [Y, Z]$$

is bijective for all Z in D.

(b) *Suppose given Z in C. Then Z lies in D if and only if*

$$f^*: [X, Z] \leftarrow [Y, Z]$$

is bijective for all $f: X \to Y$ in S.

The second lemma characterises (E, η) in terms of S or D.

Lemma 3.2. *The following conditions on a map $f: X \to Y$ are equivalent.*

(a) *There is a commutative diagram*

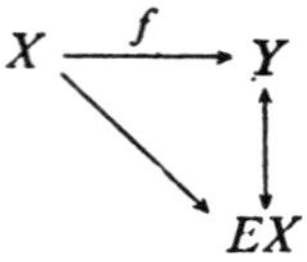

in which the vertical arrow is an equivalence.

(b) *f lies in S and Y lies in D.*

(c) *f lies in S and is couniversal with this property; more precisely, if $f': X \to Y'$ lies in S, then there is a unique map $g: Y' \to Y$ such that $gf' = f$.*

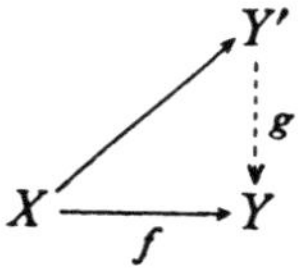

(d) *Y lies in D and f is universal with this property; more precisely, if $f': X \to Y'$ is a map with Y' in D, then there is a map $g: Y \to Y'$ such that $gf = f'$.*

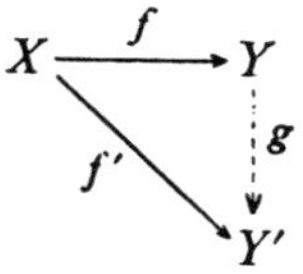

Example 3.3. In Example 2.5, we originally introduced the "n-type functor" EX and the map η_X by a statement of type (b): η_X lies in the relevant class S and EX lies in the relevant class D.

Example 3.4. The six characterisations of Sullivan's "localisation functor" may be obtained by substituting into 3.2(b), (c) and (d) either a homological account of the relevant classes S and D, or else a homotopical account. To make this precise, we work in a convenient category; let C be the full subcategory of CW, consisting of complexes X whose fundamental group $\pi_1(X)$ is nilpotent and acts trivially on the higher homotopy groups $\pi_i(X)$, $i > 1$. It is convenient to introduce a convention; for $\alpha = 2, 3, \cdots$ we let $\pi_\alpha(X)$ be the appropriate homotopy group, but we also let α run over a sequence of further indices, for which $\pi_\alpha(X)$ is defined to run over the subquotients of the lower central series of the fundamental group $\pi_1(X)$. Then $\pi_*(X) = \sum_\alpha \pi_\alpha(X)$ becomes a graded abelian group. We suppose given a subring R of the rational numbers Q. The situation then is that

$$\begin{aligned} f \text{ lies in } S \quad & \text{iff } f_*: H_*(X;R) \to H_*(Y;R) \text{ is iso} \\ & \text{iff } f_* \otimes 1 : \pi_*(X) \otimes R \to \pi_*(Y) \otimes R \text{ is iso,} \\ X \text{ lies in } D \quad & \text{iff } H_*(X) \text{ is an } R\text{-module} \\ & \text{iff } \pi_*(X) \text{ is an } R\text{-module.} \end{aligned}$$

§ 4.

We now suppose that C is a category with coproducts; since we have in mind the category CW, we write the coproducts as $\bigvee_\alpha X_\alpha$. If we are given (E, η) satisfying (2.1) and (2.2), and if S is defined by (2.3), then it can be deduced that S satisfies the following six axioms.

Axiom 4.1. S is closed under finite compositions.

Axiom 4.2. Given any diagram

$$\begin{array}{ccc} X & & \\ \uparrow{\scriptstyle s} & & \\ W & \xrightarrow{f} & Y \end{array}$$

with $s \in S$, there is a diagram

$$\begin{array}{ccc} X & \xrightarrow{g} & Z \\ \uparrow{\scriptstyle s} & & \uparrow{\scriptstyle t} \\ W & \xrightarrow{f} & Y \end{array}$$

with $t \in S$ and $gs = tf$.

Axiom 4.3. Given any diagram

$$W \xrightarrow{s} X \overset{f}{\underset{g}{\rightrightarrows}} Y$$

with $s \in S$ and $fs = gs$, there exists a diagram

$$W \xrightarrow{s} X \overset{f}{\underset{g}{\rightrightarrows}} Y \xrightarrow{t} Z$$

with $t \in S$ and $tf = tg$.

Axiom 4.4. If

$$W \xrightarrow{f} X \xrightarrow{g} Y \xrightarrow{h} Z$$

is a diagram in C such that $gf \in S$ and $hg \in S$, then $f \in S$.

Axiom 4.5. If $s_\alpha : X_\alpha \to Y_\alpha$ lies in S for each α, then

$$\bigvee_\alpha s_\alpha : \bigvee_\alpha X_\alpha \to \bigvee_\alpha Y_\alpha$$

lies in S.

Axiom 4.6. For any object Y in C, there is a *set* of morphisms $s_\alpha : Y \to Y_\alpha$ in S which is cofinal, in the sense that given a map $s: Y \to Y'$ in S there is a commutative diagram

$$\begin{array}{ccc} & Y' & \\ {\scriptstyle s}\nearrow & & \searrow \\ Y & \xrightarrow{s_\alpha} & Y_\alpha \end{array}$$

for some α.

For example, Axiom 4.5 follows from Lemma 3.1 (a).

We now comment on the meaning of these axioms. First we recall that modulo set-theoretic difficulties, we can construct from C and S the category of fractions $S^{-1}C$ [5], and there is a canonical functor $Q: C \to S^{-1}C$. If $f \in S$, then Qf is invertible in $S^{-1}C$. Axioms (4.1), (4.2) and (4.3) are standard axioms which ensure that $S^{-1}C$ and Q can be described conveniently; we have a "calculus of left fractions". Axiom (4.4) is equivalent to the statement that if Qf is invertible in $S^{-1}C$, then $f \in S$. Axiom (4.6) ensures that there are in fact no set-theoretic difficulties in constructing $S^{-1}C$ and Q; we can define

$$[QX, QY]_{S^{-1}C} = \varinjlim_{\alpha} [X, Y_\alpha]_C .$$

This axiom, (4.6), is comparable with the "solution-set condition" in category-theory. In particular, it is clearly necessary; for Lemma 3.2 (c) shows that it is sufficient to take a single map s_α, namely $\eta_Y: Y \to EY$.

Theorem 4.7. *Take C to be the category CW—see* (1.3). *Then the construction* (2.3) *gives a* (1-1) *correspondence between* (*equivalence classes of*) *pairs* (E, η) *satisfying* (2.1), (2.2) *and subsets* S *satisfying* (4.1) *to* (4.6).

Sketch proof. The essential step is to assume given a subset S satisfying (4.1) to (4.6), and reconstruct (E, η). This is done as follows. Given objects X in C and Y in $S^{-1}C$, we can form

$$[QX, Y]_{S^{-1}C} .$$

If we fix Y and vary X in C, we get a contravariant functor from C to sets. We check that this satisfies the hypotheses of Brown's Representability Theorem [2, 3]. (Needless to say, (4.5) is used in an essential way in verifying Brown's Wedge Axiom.) Therefore, there is an object RY in C and a (1-1) correspondence

$$[QX, Y]_{S^{-1}C} \longleftrightarrow [X, RY]_C$$

natural for maps of X in C. It is now standard to make R into a functor from $S^{-1}C$ to C, adjoint to Q. We take $E = RQ: C \to C$ and let $\eta_X: X \to RQX$ be the map corresponding under the adjunction to $1 \in [QX, QX]_{S^{-1}C}$. Using (4.4), we show that this pair (E, η) satisfies (2.1), (2.2) and gives rise to the subset S.

In this proof, the fact that we are working in the category CW is needed only to apply Brown's Representability Theorem. Therefore, Theorem 4.7 remains true in other categories where we have a similar theorem—for example, spectra of CW-complexes. In categories such as those usually considered by category-theorists, where we can apply one of the standard adjoint-functor theorems, the result is presumably easier —modulo appropriate restatement of the axioms.

§ 5.

In the applications, it is generally very easy to check axioms 4.1, 4.4 and 4.5. Rather than check 4.2 and 4.3 directly, it is usually easier to check a further axiom. To state it, we assume that $U \cup V$ is a complex which is the union of two subcomplexes U and V.

Axiom 5.1. If $i: U \cap V \rightarrow U$ is in S, then $j: V \rightarrow U \cup V$ is in S.

This is a sort of excision axiom.

Proposition 5.2. *Axioms* 4.1, 4.4, 4.5 *and* 5.1 *imply* 4.2 *and* 4.3.

Remark. Even for $C = CW$, there exist pairs (E, η) satisfying (2.1), (2.2) but such that S does not satisfy (5.1).

In the applications, it now becomes very easy to verify (4.1) to (4.5). For example, suppose given a subring R of the rational numbers Q, and define S as follows: a map $f: X \rightarrow Y$ lies in S if and only if $f_*: H_*(X; R) \rightarrow H_*(Y; R)$ is iso. Then (4.1), (4.4), (4.5) and (5.1) are easily verified.

By contrast, it is usually quite unclear whether (4.6) holds. However, for the subset S defined in the last paragraph, Bousfield [1] has obtained a proof that the pair (E, η) exists.

Axiom 5.1 carries the following further benefit.

Proposition 5.3. *If S satisfies* (4.1) *to* (4.6) *and* (5.1), *then D is closed under homotopy pull-backs and homotopy inverse-limits.*

Axiom 5.1 is also inherited by intersections $S_1 \cap S_2$, whereas axioms 4.2, 4.3 are not (see §7).

§ 6.

The example of Sullivan's localisation functor might lead us to expect an idempotent functor E to commute with sums, suspensions, cofiberings, homotopy pushouts, products, loop-spaces, fiberings, and homotopy pull-backs. However, a single pathological example is sufficient to show that an idempotent functor E may do none of these things. Therefore, we need further axioms to ensure that it does. For example, the axiom which ensures that E commutes with products is very simple.

Axiom 6.1. If f and g lie in S, then $f \times g$ lies in S.

§ 7.

Let R_1 and R_2 be two subrings of the rational numbers Q. Then the corresponding localisation functors E_1 and E_2 commute; their product $E_1E_2 = E_2E_1$ is the localisation functor corresponding to the subring of Q generated by R_1 and R_2. We have a diagram with good properties

$$\begin{array}{ccc} E_0X & \longrightarrow & E_1X \\ \downarrow & & \downarrow \\ E_2X & \longrightarrow & E_3X \end{array}$$

where E_0X is the localisation functor corresponding to $R_1 \cap R_2$.

Two general idempotent functors E_1 and E_2 do not commute. It is possible to introduce a good ordering relation between idempotent functors: one orders the corresponding subsets S or D by inclusion. One might seek to make the idempotent functors into a lattice, so as to obtain idempotent functors $E_1 \cap E_2$, $E_1 \cup E_2$ generalising the functors E_0, E_3 above. However, such lattice operations will be of little use unless the intersection is given by $S_1 \cap S_2$ and the union by $D_1 \cap D_2$. We should expect to introduce further axioms to make such a programme work.

References

[1] Bousfield, A. K., Private communication.
[2] Brown, E. H., Cohomology theories, Ann. of Math., **75** (1962), 467–484.
[3] ———, Abstract homotopy theory, Trans. Amer. Math. Soc., **119** (1965), 79–85.
[4] Deleanu, A., Frei, A. and Hilton, P. J., Generalised Adams completion, to appear.
[5] Gabriel, P. and Zisman, M., Calculus of Fractions and Homotopy Theory, Springer, Berlin, 1967.
[6] Sullivan, D., Geometric topology, Part I, Mimeographed notes, M.I.T., Cambridge, Mass., 1970.

Department of Pure Mathematics
and Mathematical Statistics
University of Cambridge
16, Mill Lane
Cambridge CB2 1SB
England

Proc. Camb. Phil. Soc. (1973), **73**, 45 45
PCPS 73–7

The Kahn–Priddy theorem

BY J. F. ADAMS
D.P.M.M.S., Cambridge

(*Received* 12 *May* 1972)

1. *Introduction.* Let $\pi_r^S(X)$ be the stable homotopy group

$$\operatorname*{Lim}_{n\to\infty} \pi_{n+r}(S^n X),$$

where $S^n X$ means the n-fold suspension of X. For example, the groups $\pi_r^S(S^0)$ are the stable homotopy groups of spheres. Let

$$O = \operatorname*{Lim}_{n\to\infty} O(n)$$

be the 'infinite-dimensional' orthogonal group. Then topologists are familiar with the 'stable J-homomorphism'

$$J: \pi_r(O) \to \pi_r^S(S^0).$$

G. W. Whitehead observed that J factors through an 'even more stable' J-homomorphism

$$J': \pi_r^S(O) \to \pi_r^S(S^0);$$

he conjectured that J' is epi (for $r > 0$).

It is well known that there is a canonical embedding of real projective space RP^∞ in the orthogonal group O. On the basis of calculations, M. E. Mahowald conjectured that $\pi_r^S(RP^\infty)$ maps onto the 2-primary component of $\pi_r^S(S^0)$ (for $r > 0$). Of course this implies the 2-primary part of G. W. Whitehead's conjecture. Mahowald's conjecture has been proved by Kahn and Priddy (4), and their work has several features of interest which I will not mention here. I am grateful to D. S. Kahn for a number of letters on this subject.

Two further points should be mentioned. The statement above refers to the prime 2, but Kahn and Priddy realized that there is an analogous statement for any odd prime p. The statement also refers to a particular map $RP^\infty \to O$, and this map induces a 'stable map' from RP^∞ to S^0. (The notion of a 'stable map' will be made precise below.) However, it is rather hard to construct a proof which uses the fact that the stable map from RP^∞ to S^0 is this particular one and no other. Kahn and Priddy quite rightly constructed a general proof, which applies to all stable maps from RP^∞ to S^0 with a given property. More precisely, they consider stable maps ϕ from RP^∞ to S^0 which induce an isomorphism of the first stable homotopy group π_1^S; this group is of course Z_2 both for RP^∞ and for S^0.

I have recently given a treatment of the Kahn–Priddy theorem in lectures. One of the key features of this treatment was as follows. By working in a suitable category of spectra, one can make the Kahn–Priddy theorem assert not only the existence and properties of a stable map ϕ from RP^∞ to S^0, but also the essential uniqueness of ϕ. Because of the uniqueness clause, it then becomes sufficient to do all the hard work not for a general map ϕ, but for one particular map of one's own choice. Now it was already known to Kahn and Priddy that there is available a 'convenient' choice for ϕ which simplifies the proof and reduces the hard work; one therefore makes this choice and uses the simpler proof.

As a digression, it is not obvious that the 'convenient' choice for ϕ is the same as the one obtained via O. Indeed, when one replaces the prime 2 by an odd prime, one still has a 'convenient' choice of ϕ; but no choice of ϕ will factor through O, even stably. It may be that the role of O in all this is that of a red herring. Perhaps one should replace it by the space '$\operatorname{Im} J$'.

To continue, my treatment included four points.

(i) The formulation of a 'strong' form of the Kahn–Priddy theorem, including a uniqueness clause. This formulation will be given as (2·3) below.

(ii) The formulation of a 'weak' form of the Kahn–Priddy theorem, in which it is sufficient to prove the result for one map of one's own choice. Equivalent formulations will be given as (2·4) and (2·5) below.

(iii) The proof of the 'weak' form of the Kahn–Priddy theorem. This I did only for $p = 2$, and therefore $p = 2$ is the only case for which I personally can swear that the result is true. There is no doubt, however, that the result is true for all primes. My proof followed Kahn and Priddy with simplifications, and so I will not give it here. (After taking the favourable choice of ϕ, I used different subgroups in the transfer argument. Instead of the Sylow subgroup of the symmetric group Σ_{2^n} on 2^n letters, it is sufficient to consider a 'wreath product' subgroup of Σ_{2m}, namely an extension with kernel $(Z_2)^m$ and quotient Σ_m.)

(iv) The proof of the following result.

THEOREM 1·1. *The 'weak' form of the Kahn–Priddy theorem,* (2·4) *or* (2·5), *is equivalent to the 'strong' form* (2·3).

This can be proved by standard techniques, and it is the object of this paper to record the proof.

The remainder of this paper is arranged as follows. Section 2 explains the various forms of the Kahn–Priddy theorem. Section 3 contains a few easy proofs postponed from Section 2. Section 4 contains the proof of Theorem 1·1.

2. *Formulation of the statements.* The object of this section is to explain the 'weak' and 'strong' forms of the Kahn–Priddy theorem.

When we see a result about homotopy groups, it is natural to ask what geometrical result about spaces underlies the result about homotopy groups. For example, a well-known theorem of James(3) asserts that if n is odd, we have an exact sequence

$$\ldots \to \pi_r(S^n) \xrightarrow{E} \pi_{r+1}(S^{n+1}) \xrightarrow{H} \pi_{r+1}(S^{2n+1}) \xrightarrow{\Delta} \pi_{r-1}(S^n) \to \ldots.$$

The underlying geometrical fact is that there is a fibering $F \to E \to B$ in which the spaces F, E and B are (up to weak equivalence) S^n, ΩS^{n+1} and ΩS^{2n+1} respectively. The exact sequence due to James is the exact homotopy sequence of this fibering. It is arguable that in such cases the geometrical fact should be treated as the main theorem; the consequent results about homotopy groups (or other invariants) should be treated as corollaries. We will apply this philosophy to the theorem of Kahn and Priddy.

As a matter of algebra, the cheapest way to prove that a homomorphism α is epi is to construct a one-sided inverse for it, say β, so that $\alpha\beta = 1$ and α is a split epimorphism. The proof of Kahn and Priddy may be put in this form. In this case, the groups which occur are stable homotopy groups; they may be considered either as homotopy groups of spectra, or as homotopy groups of spaces. The homomorphism α may be induced by a map of spectra or by a map of spaces. The homomorphism β can be induced by a map of spaces, but not by a map of spectra. The underlying geometrical fact is that one space is a retract in another. Thus, although the problem solved by Kahn and Priddy appears to lie in stable homotopy-theory, their solution uses methods which in a sense are unstable; this is one reason for interest in it.

To make all this precise, we need some details on spaces and spectra. Let $\mathscr{C}$ be the category in which the objects are CW-complexes with base-point, and the morphisms are homotopy classes of maps (where maps and homotopies preserve the base-point). Let $\mathscr{S}$ be Boardman's category of spectra (**1**, **6**), so that the objects are spectra $\mathbf{E}$ and the morphisms are homotopy classes (in a suitable sense.) (The reader is assured that we need only the simplest properties of this category.) $[X, Y]$ will mean the set of morphisms from X to Y in the appropriate category. The functor $S^\infty: \mathscr{C} \to \mathscr{S}$ assigns to each complex the corresponding suspension spectrum; for example, $S^\infty S^0$ is the sphere-spectrum. If X and Y are complexes and X is finite-dimensional, then

$$[S^\infty X, S^\infty Y]_{\mathscr{S}} = \operatorname*{Lim}_{n\to\infty} [S^n X, S^n Y]_{\mathscr{C}}.$$

(Here the limit is attained for n sufficiently large.) The words 'stable map from X to Y' in section 1 may be interpreted as meaning a morphism $S^\infty X \to S^\infty Y$ in $\mathscr{S}$.

The functor S^∞ has an adjoint $\Omega^\infty: \mathscr{S} \to \mathscr{C}$, so that

$$[S^\infty X, \mathbf{E}]_{\mathscr{S}} = [X, \Omega^\infty \mathbf{E}]_{\mathscr{C}},$$

where $X \in \mathscr{C}$, $\mathbf{E} \in \mathscr{S}$. Crudely speaking, a spectrum $\mathbf{E}$ is a sequence of spaces E_n, and the complex $\Omega^\infty \mathbf{E}$ is $\operatorname*{Lim}_{n\to\infty} \Omega^n E_n$. Alternatively, it is easy to show (using Brown's Representability Theorem (**2**)) that S^∞ has an adjoint; one can then take this adjoint as the definition of Ω^∞, and show that under suitable hypotheses, $\Omega^n E_n$ approximates to $\Omega^\infty \mathbf{E}$ in a suitable sense.

If $r \geqslant 0$, the rth homotopy group of a spectrum is given by

$$\begin{aligned} \pi_r(\mathbf{E}) &= [S^\infty S^r, \mathbf{E}]_{\mathscr{S}} \\ &= [S^r, \Omega^\infty \mathbf{E}]_{\mathscr{C}}. \end{aligned}$$

The stable homotopy groups of a complex X are the homotopy groups of the corresponding suspension spectrum. That is, if $r \geqslant 0$ we have

$$\begin{aligned} \pi_r^S(X) &= \operatorname*{Lim}_{n\to\infty} [S^n S^r, S^n X] \\ &= [S^\infty S^r, S^\infty X]_{\mathscr{S}} \\ &= [S^r, \Omega^\infty S^\infty X]_{\mathscr{C}}. \end{aligned}$$

In particular, the stable homotopy groups of spheres are the homotopy groups of the space $\Omega^\infty S^\infty S^0$. This space has pathwise-components in (1–1) correspondence with the integers; we write $(\Omega^\infty S^\infty S^0)_0$ for the component of the base-point. If W is a connected CW-complex, we have

$$[W, \Omega^\infty S^\infty S^0] = [W, (\Omega^\infty S^\infty S^0)_0].$$

Suppose given a prime p. We say that a complex W is p-primary if the homology groups $H_r(W)$ are p-primary for $r > 0$. Up to equivalence, we can decompose $(\Omega^\infty S^\infty S^0)_0$ as a product $X \times Y$, where the homotopy groups of X are p-primary and those of Y are of order prime to p. We see by obstruction-theory that if W is p-primary, then

$$[W, Y] = 0.$$

(Y is not simply connected if $p \neq 2$, but its fundamental group acts trivially on its higher homotopy groups.) So (if W is p-primary) we have

$$[W, X \times Y] = [W, X].$$

Let us write ${}_p(\Omega^\infty S^\infty S^0)_0$ for the p-primary factor X; then (if W is p-primary) we have

$$[W, (\Omega^\infty S^\infty S^0)_0] = [W, {}_p(\Omega^\infty S^\infty S^0)_0].$$

For the prime 2, then, the basic geometrical fact is that the complex ${}_2(\Omega^\infty S^\infty S^0)_0$ is a retract of $\Omega^\infty S^\infty RP^\infty$. We must now explain what space replaces RP^∞ when we replace the prime 2 by an odd prime p.

Let Z_p be a cyclic group of order p, embedded as a Sylow subgroup in the symmetric group Σ_p on p letters; and let BG be the classifying space of G. Then the inclusion $i: Z_p \to \Sigma_p$ induces $Bi: BZ_p \to B\Sigma_p$. When we specialize to the case $p = 2$, both BZ_p and $B\Sigma_p$ specialize to RP^∞. We might consider generalizing RP^∞ either to BZ_p or to $B\Sigma_p$. It turns out that although Z_p is a smaller group than Σ_p, $B\Sigma_p$ is better than BZ_p because $B\Sigma_p$ is 'smaller' than BZ_p in a homological sense. More precisely, let x be the generator in $H^1(BZ_p; Z_p)$, and let β be the Bockstein boundary; then we know that $y = \beta x$ is the generator in $H^2(BZ_p; Z_p)$, and that $H^*(BZ_p; Z_p)$ has a base consisting of the monomials $x^\epsilon y^n$, where $\epsilon = 0, 1$ and $n = 0, 1, 2, 3, \ldots$.

LEMMA 2·1. *The induced homomorphism*

$$(Bi)^*: H^*(B\Sigma_p; Z_p) \to H^*(BZ_p; Z_p)$$

is mono, and a base for its image is given by the monomials $x^\epsilon y^{m-\epsilon}$ *with* $\epsilon = 0, 1$ *and* $m \equiv 0 \bmod (p-1)$.

In order not to interrupt the exposition in this section, we defer the proofs of even the most trivial results, such as this one.

If $p > 2$, the space $B\Sigma_p$ is better than BZ_p, but it is not p-primary; for example, $H_1(B\Sigma_p) = Z_2$. We must repair this defect; first we need a definition. Let Q_p be the ring of fractions a/b with b prime to p; a map $f: X \to Y$ is said to be a p-equivalence if the map

$$f_*: H_*(X; Q_p) \to H_*(Y; Q_p)$$

is iso. If X and Y are connected and the homology groups $H_r(X)$, $H_r(Y)$ are finite for $r > 0$, it is equivalent to ask that the p-components of $H_r(X)$ and $H_r(Y)$ be isomorphic under f_* (for each $r > 0$).

LEMMA 2·2. (i) *If $p > 2$ there is a CW-complex $L = L(p)$ which is 1-connected, p-primary and admits a p-equivalence $f: B\Sigma_p \to L$.*

(ii) *Such a complex L is unique up to canonical equivalence.*

(iii) *The first non-zero homotopy group of $S^\infty L$ is Z_p in dimension $2p-3$.*

It follows immediately that we can realize L as a CW-complex, having one cell in each dimension n such that $n \geqslant 0$ and $n \equiv 0$ or $-1 \bmod 2(p-1)$, and having no cells in any other dimension. (Since L is simply-connected, we can use a 'homology decomposition' of L.) (The attaching map for the cell of dimension $2(p-1)n$ has degree p if $n > 0$.) We then write L^n for the skeleton of L consisting of the cells of dimension $\leqslant 2(p-1)n$. We interpret L^∞ as L.

If $p = 2$, we interpret L as RP^∞ and L^n as RP^{2n}.

Let ${}_pG$ be the p-primary component of the finite Abelian group G. When we go from the case $p = 2$ to the case $p > 2$, we have to replace the group $\pi_1^S(S^0) = Z_2$ by the p-component ${}_p\pi_{2p-3}^S(S^0) = Z_p$.

We can now formulate a strong form of the Kahn–Priddy theorem.

Formulation 2·3. (i) *For $n = 1, 2, 3, \ldots$ and for $n = \infty$ there is a morphism*

$$\phi_n: S^\infty L^n \to S^\infty S^0$$

(in $\mathscr{S}$) which induces an isomorphism of ${}_p\pi_{2p-3}$.

(ii) *Such a morphism ϕ_n is unique up to composition with an equivalence*

$$\epsilon: S^\infty L^n \to S^\infty L^n$$

in $\mathscr{S}$.

(iii) *For any such morphism $\phi = \phi_\infty$, the induced map*

$$\Omega^\infty\phi: \Omega^\infty S^\infty L \to {}_p(\Omega^\infty S^\infty S^0)_0$$

is, up to equivalence of spaces, the projection $X \times Y \to X$ of a product on one factor.

Various comments are in order. In part (ii), it is clear that we cannot hope to do better; if ϕ_n induces an isomorphism of ${}_p\pi_{2p-3}$ and if ϵ is an equivalence, then $\phi_n\epsilon$ induces an isomorphism of ${}_p\pi_{2p-3}$. In part (iii), the map $\Omega^\infty\phi$ runs in the first instance from $\Omega^\infty S^\infty L$ to $\Omega^\infty S^\infty S^0$; but since $\Omega^\infty S^\infty L$ is connected and p-primary, we may regard $\Omega^\infty\phi$ as mapping into ${}_p(\Omega^\infty S^\infty S^0)_0$.

In the original work of Kahn and Priddy, the functors Ω^∞ and S^∞ do not occur separately, but only as the composite $Q = \Omega^\infty S^\infty$. One justification for introducing the category $\mathscr{S}$ and the separate functors Ω^∞, S^∞ is that this allows one to formulate the uniqueness statement (2·3ii) in a simple and conceptual form. I concede that many of the benefits which one obtains from the pair of adjoint functors Ω^∞, S^∞ can also be obtained by exploiting the natural transformations $X \to QX$ and $QQX \to QX$; but by the time one gets to $QQQX$ such manipulations tend to obscure the real issues.

We can also formulate a weak form of the Kahn–Priddy theorem.

Formulation 2·4. *For $n = 1, 2, 3, \ldots$ there is a morphism $\phi_n: S^\infty L^n \to S^\infty S^0$ in $\mathscr{S}$ such that the induced homomorphism*

$$\phi_{n*}: [S^\infty W, S^\infty L^n]_{\mathscr{S}} \to [S^\infty W, S^\infty S^0]_{\mathscr{S}}$$

is epi whenever W is a connected p-primary CW-complex of dimension $< 2(p-1)n$.

Formulation 2·5. *For* $n = 1, 2, 3, \dots$ *and for* $m > 2(p-1)n$ *there is a map*

$$\phi_n : S^m L^n \to S^m$$

in $\mathscr{C}$ *such that the induced homomorphism*

$$\phi_{n*} : [S^m W, S^m L^n]_{\mathscr{C}} \to [S^m W, S^m]_{\mathscr{C}}$$

is epi whenever W *is a connected* p*-primary* CW*-complex of dimension* $< 2(p-1)n$.

Various comments are in order. The two formulations are strictly equivalent; if we take (2·4) and translate the statements about spectra into elementary terms, we obtain (2·5). The track groups $[X, Y]$ which occur in (2·5) are stable. As remarked above, the main advantage of (2·4) or (2·5) is that in order to prove it, it is sufficient to consider one map ϕ_n, which can be chosen to simplify the proof. It is not even necessary to have any relation between the choices of ϕ_n for different values of n.

Some readers may be willing to believe that

$$\phi_{n*} : [S^m S^r, S^m L^n] \to {}_p[S^m S^r, S^m]$$

is epi for $r < 2(p-1)n < m$, without being willing to replace the sphere S^r by a complex W. The answer is to look at the final steps of the Kahn–Priddy proof. Here we find the following diagram.

$$\begin{array}{ccccc} S^m X & \xrightarrow{f} & S^m \Omega^m S^m L^n & \xrightarrow{S^m \Omega^m \phi_n} & S^m(\Omega^m S^m)_0 \\ & & \downarrow{\scriptstyle \epsilon} & & \downarrow{\scriptstyle \epsilon} \\ & & S^m L^n & \xrightarrow{\phi_n} & S^m \end{array}$$

The natural transformation ϵ is defined by 'evaluation'. It is not asserted that the map f is an m-fold suspension; but the complex X and the map f are so arranged that the composite

$$S^m X \xrightarrow{(S^m \Omega^m \phi_n) f} S^m(\Omega^m S^m)_0$$

is a p-equivalence up to dimension $m + 2(p-1)n$ approximately. Suppose given a connected p-primary complex W and a map

$$g : S^m W \to S^m.$$

By considering the adjoint

$$g' : W \to (\Omega^m S^m)_0$$

we can factor g in the form

$$S^m W \xrightarrow{S^m g'} S^m(\Omega^m S^m)_0 \xrightarrow{\epsilon} S^m.$$

By the choice of X we can lift $S^m g'$ to $S^m X$, and so factor g through ϕ_n. This proves that ϕ_{n*} is epi; and in this proof, a complex W gives us no more trouble than a sphere.

3. *Some easy proofs.* This section contains a few proofs postponed from section 2.

Proof of Lemma 2·1. This induced homomorphism

$$(Bi)^* : H^*(B\Sigma_p; Z_p) \to H^*(BZ_p; Z_p)$$

is mono because the transfer provides a one-sided inverse for it. It can easily be shown that the image is contained in the subspace spanned by the monomials $x^\epsilon y^{m-\epsilon}$ with $\epsilon = 0, 1$ and $m \equiv 0 \bmod (p-1)$; see (5).

Let ρ be the permutation representation of Σ_p; then its Chern class $c_{p-1}(\rho)$ is an element of $H^{2(p-1)}(B\Sigma_p)$ which restricts to a non-zero multiple of y^{p-1} in

$$H^{2(p-1)}(BZ_p; Z_p).$$

So all the powers y^m with $m \equiv 0 \bmod (p-1)$ lie in the image. Finally, since $H_r(B\Sigma_p)$ has no elements of order p^2 and is zero for $r = 2s > 0$, the Bockstein boundary

$$\beta: H^{2s-1}(B\Sigma_p; Z_p) \to H^{2s}(B\Sigma_p; Z_p)$$

is iso for $s > 0$. This proves (2·1).

Proof of Lemma 2·2. (i) Assume $p > 2$. Let Z_2 be the subgroup of Σ_p generated by a single transposition, say (12). The inclusion $i: Z_2 \to \Sigma_p$ induces $Bi: BZ_2 \to B\Sigma_p$; form the mapping-cone

$$X = B\Sigma_p \cup_{Bi} CBZ_2.$$

Then X is simply-connected and its homology groups are finite. It is a standard deduction that X is equivalent to a product $L \times M$, where the homotopy groups of X are p-primary and those of M are of order prime to p. The composite map

$$B\Sigma_p \to X \to L$$

is a p-equivalence.

(ii) Suppose L, L' are 1-connected and p-primary, while $B\Sigma_p \to L$, $B\Sigma_p \to L'$ are p-equivalences. Without loss of generality we may assume $B\Sigma_p \subset L$. Then obstruction-theory shows that we can extend the map $B\Sigma_p \to L'$ over L.

(iii) Finally, the first non-zero homology group of L is Z_p in dimension $2p-3$. This proves (2·2).

Next we wish to prove that the 'strong' form of the Kahn–Priddy theorem does indeed imply the 'weak' one. For later use we record the first step as a lemma.

LEMMA 3·1. *Assume the statement* (2·3iii). *Then for any such* ϕ *the induced homomorphism*

$$\phi_*: [S^\infty W, S^\infty L] \to [S^\infty W, S^\infty S^0]$$

is epi whenever W *is a connected* p*-primary* CW*-complex.*

Proof. Assuming (2·3iii), the function

$$(\Omega^\infty \phi)_*: [W, \Omega^\infty S^\infty L] \to [W, \Omega^\infty S^\infty S^0]$$

is a surjection of sets. By adjointness,

$$\phi_*: [S^\infty W, S^\infty L] \to [S^\infty W, S^\infty S^0]$$

is epi. This proves (3·1).

If we assume further that W is of dimension $< 2(p-1)n$, then $[S^\infty W, S^\infty L]$ can be replaced by $[S^\infty W, S^\infty L^n]$, and ϕ can be replaced by its restriction to $S^\infty L^n$. This yields (2·4).

4. *Proof of the main theorem.* In this section we will see that the 'weak' form of the Kahn–Priddy theorem implies the 'strong' form. We build up the proof in steps.

LEMMA 4·1. *There is a morphism* $\phi_1: S^\infty L^1 \to S^\infty S^0$ (*in* $\mathscr{S}$) *which induces an isomorphism from* $\pi_{2p-3}(S^\infty L^1) = Z_p$ *to* ${}_p\pi_{2p-3}(S^\infty S^0) = Z_p$.

This is immediate; a map $f: S^{2p} \to S^3$ representing a generator of ${}_p\pi_{2p}(S^3) = Z_p$ can be extended to $S^{2p} \cup_p e^{2p+1}$.

From this point on, we assume (2·4).

LEMMA 4·2. *The existence statement* (2·3i) *is true for* $n < \infty$.

Proof. L^1 is a connected p-primary complex. So assuming 2·4, the morphism

$$\phi_1: S^\infty L^1 \to S^\infty S^0$$

of (4·1) can be factored through the map

$$\phi_n: S^\infty L^n \to S^\infty S^0$$

of (2·4). So

$$\phi_{n*}: \pi_{2p-3}(S^\infty L^n) \to {}_p\pi_{2p-3}(S^\infty S^0)$$

is epi. Since both groups are Z_p, this homomorphism ϕ_{n*} is iso.

LEMMA 4·3. *Let*

$$\ldots \leftarrow E_n \leftarrow E_{n+1} \leftarrow \ldots$$

be an inverse system of non-empty finite sets. Then $\underleftarrow{\operatorname{Lim}}\, E_n$ *is non-empty.*

This is a standard triviality.

LEMMA 4·4. *The existence statement* (2·3(i)) *is true for* $n = \infty$.

Proof. Let E_n be the subset of elements in $[S^\infty L^n, S^\infty S^0]$ which induce an isomorphism of ${}_p\pi_{2p-3}$. The sets E_n form an inverse system; they are non-empty by (4·2), and finite because $[S^\infty L^n, S^\infty S^0]$ is finite. So $\underleftarrow{\operatorname{Lim}}\, E_n$ is non-empty by (4·3). But

$$[S^\infty L^\infty, S^\infty S^0] \to \underset{\substack{\leftarrow \\ n}}{\operatorname{Lim}}\, [S^\infty L^n, S^\infty S^0]$$

is epi, so we can lift any element of $\underleftarrow{\operatorname{Lim}}\, E_n$ to a morphism $\phi^\infty: S^\infty L^\infty \to S^\infty S^0$. This morphism induces an isomorphism of ${}_p\pi_{2p-3}$.

LEMMA 4·5. *Let* $n = 1, 2, 3, \ldots$ *or* $n = \infty$. *Then a morphism* $f: S^\infty L^n \to S^\infty L^n$ *which induces an isomorphism of* π_{2p-3} *is an equivalence in* $\mathscr{S}$.

Proof. Let us use the expression $x^\epsilon y^{m-\epsilon}$ from (2·1) to mean also the corresponding element in $H^*(S^\infty L^n; Z_p)$. If $f: S^\infty L^n \to S^\infty L^n$ induces an isomorphism of π_{2p-3}, then $f^*x = cx$ for some $c \neq 0$ in Z_p. Now f^* commutes with the Steenrod operations $Q_0 = \beta$ and $Q_1 = P_1\beta - \beta P_1$ (where P_1 is interpreted as Sq^2 if $p = 2$). We have the formulae

$$Q_0 xy^{m-1} = y^m,$$

$$Q_1 xy^{m-1} = y^{m+(p-1)}.$$

We see by induction that

$$f^*: H^*(S^\infty L^n; Z_p) \to H^*(S^\infty L^n; Z_p)$$

is multiplication by c (at each step we use Q_1 to go up $2p-1$ dimensions and Q_0 to go down 1). So f is an equivalence, by the theorem of J. H. C. Whitehead in $\mathscr{S}$ (1, 6).

LEMMA 4·6. *The uniqueness statement* (2·3(ii)) *is true for* $n < \infty$.

Proof. Let $\psi: S^\infty L^n \to S^\infty S^0$ be any map inducing an isomorphism of ${}_p\pi_{2p-3}$. Let $\phi_{n+1}: S^\infty L^{n+1} \to S^\infty S^0$ be as in (2·4). Then by (2·4), we can factor ψ in the form

$$S^\infty L^n \xrightarrow{f} S^\infty L^{n+1} \xrightarrow{Q_{n+1}} S^\infty S^0.$$

Here f induces a monomorphism of ${}_p\pi_{2p-3}$, and therefore an isomorphism since both $\pi_{2p-3}(S^\infty L^n)$ and $\pi_{2p-3}(S^\infty L^{n+1})$ are Z_p. We can also assume, of course, that f maps $S^\infty L^n$ into $S^\infty L^n$. Then f is an equivalence from $S^\infty L^n$ to $S^\infty L^n$, by (4·5).

LEMMA 4·7. *Let $\phi: S^\infty L^\infty \to S^\infty S^0$ be any map as in* (2·3i). *Then the induced map*

$$\phi_*: [S^\infty W, S^\infty L^\infty] \to [S^\infty W, S^\infty S^0]$$

is epi where W is connected, p-primary and finite-dimensional.

Proof. Choose n finite but large compared with the dimension of W. Then by (2·4), there is a map

$$\phi_n: S^\infty L^n \to S^\infty S^0$$

such that

$$\phi_{n*}: [S^\infty W, S^\infty L^n] \to [S^\infty W, S^\infty S^0]$$

is epi. Let $\psi = \phi | S^\infty L^n$. Then by (4·6), $\psi = \phi_n f$, where f is an equivalence. Therefore

$$\psi_*: [S^\infty W, S^\infty L^n] \to [S^\infty W, S^\infty S^0]$$

is epi. This proves (4·7).

LEMMA 4·8. *We may assume without loss of generality that the complex $X = {}_p(\Omega^\infty S^\infty S^0)_0$ is the union of an increasing sequence of subcomplexes X_n which are finite, connected and p-primary.*

Proof. If $p > 2$ then ${}_p(\Omega^\infty S^\infty S^0)$ is 1-connected, and we can simply use a homology decomposition. It remains to consider the case $p = 2$. In this case we have a map

$$RP^\infty \to \Omega^\infty S^\infty S^0$$

(adjoint to ϕ) inducing an isomorphism of π_1; we may suppose it maps into

$${}_2(\Omega^\infty S^\infty S^0)_0 = \Gamma,$$

say. Let $\tilde\Gamma$ be the universal cover of Γ. Since Γ is an H-space, we can construct the map

$$RP^\infty \times \tilde\Gamma \to \Gamma \times \Gamma \to \Gamma.$$

It is an equivalence, by J. H. C. Whitehead's Theorem. Now filter RP^∞ by using the subcomplexes RP^{2n} and use a homology decomposition of the 1-connected space $\tilde\Gamma$.

LEMMA 4·9. *Let Z be a connected CW-complex with finite homotopy groups, X a CW-complex which is the union of an increasing sequence of finite subcomplexes X_n, and $f: Z \to X$ a map such that each diagram*

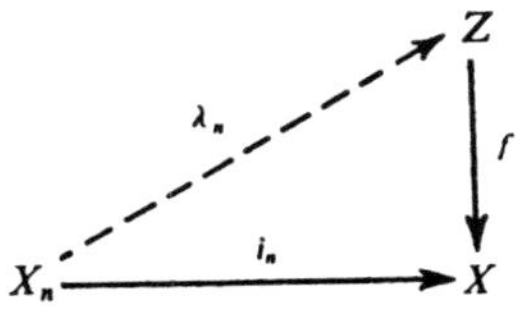

can be completed up to homotopy. Then the diagram

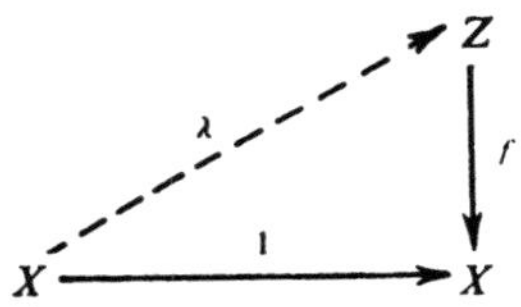

can be completed up to homotopy.

Proof. Let E_n be the subset of $[X_n, Z]$ consisting of maps λ_n such that $f\lambda_n \sim i_n$.

By hypothesis, it is non-empty, and it is finite since $[X_n, Z]$ is finite. By (4·3), $\operatorname{Lim}_{\leftarrow} E_n$ is non-empty. But

$$[X, Z] \to \operatorname*{Lim}_{\substack{\leftarrow \\ n}} [X_n, Z]$$

is onto; so we can choose a map $\gamma: X \to Z$ which projects to an element of $\operatorname{Lim}_{\leftarrow} E_n$. Then the composite $f\gamma$ induces the identity map of homotopy groups, because each map $S^r \to X$ factors through some X_n. By the theorem of J. H. C. Whitehead, $f\gamma$ is an equivalence. Then $\lambda = \gamma(f\gamma)^{-1}$ has $f\lambda \sim 1$. This proves (4·9).

LEMMA 4·10. (2·3(iii)) *is true.*

Proof. In (4·9) we take Z to be $\Omega^\infty S^\infty L$, X to be as in (4·8), and f to be $\Omega^\infty \phi$. By (4·7), the homomorphism

$$(\Omega^\infty \phi)_*: [X_n, \Omega^\infty S^\infty L] \to [X_n, \Omega^\infty S^\infty S^0]$$

is epi, so we can complete the following diagram.

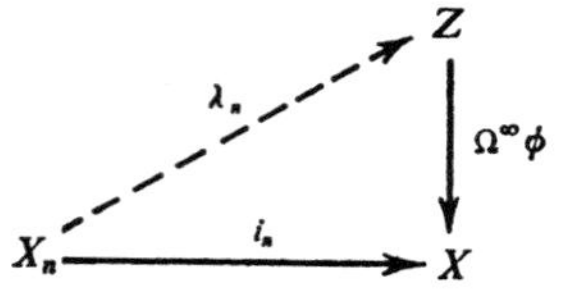

By (4·9), we can complete the following diagram.

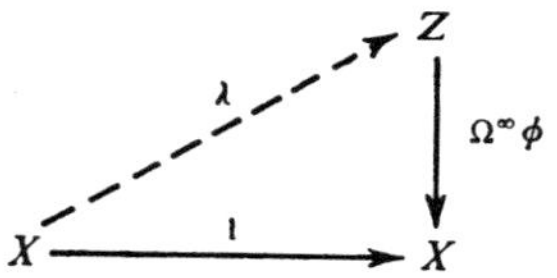

Let Y be a CW-complex weakly equivalent to the fibre of $\Omega^\infty\phi$. Since Z and X are H-spaces and $\Omega^\infty\phi$ is an H-map, we can construct the following diagram, in which x_0 is the base-point.

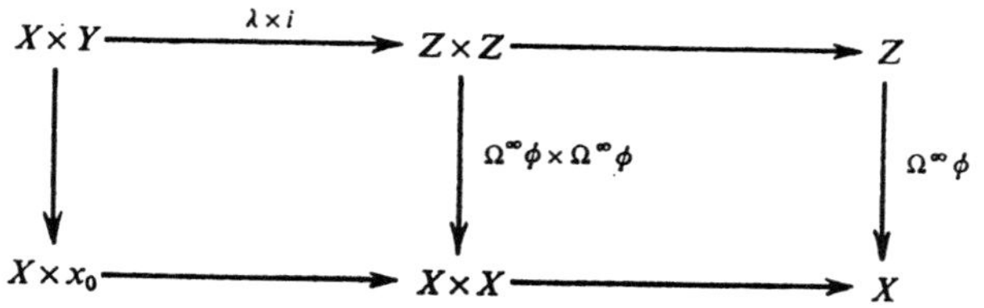

The top line is an equivalence by the theorem of J. H. C. Whitehead, and the bottom line is homotopic to the identity map. This proves (4·10).

At this stage we have the data for Lemma 3·1, and therefore we have its conclusion.

LEMMA 4·11. *The uniqueness statement* (2·3(ii)) *is true for* $n = \infty$.

Proof. Let $\phi, \psi: S^\infty L^\infty \to S^\infty S^0$ be two maps both inducing an isomorphism of ${}_p\pi_{2p-3}$. By (3·1) applied (say) to ϕ, we can express ψ as a composite

$$S^\infty L \xrightarrow{\epsilon} S^\infty L \xrightarrow{\phi} S^\infty S^0.$$

Here ϵ must induce an isomorphism of π_{2p-3}, and so must be an equivalence by (4·5).

This completes the deduction of all the parts of (2·3), and so proves Theorem 1·1.

REFERENCES

(1) Adams, J. F. *Stable Homotopy*, lecture notes, University of Chicago 1972.
(2) Brown, E. H. Cohomology theories. *Ann. of Math.* (2) **75** (1962), 467–484.
(3) James, I. M. On the suspension sequence. *Ann. of Math.* (2) **65** (1957), 74–107.
(4) Kahn, D. S. and Priddy, S. B. Applications of the transfer to stable homotopy theory (preprint).
(5) Steenrod, N. E. Cyclic reduced powers of cohomology classes. *Proc. Nat. Acad. Sci. USA* **39** (1953), pp. 217–223, especially Theorem 4.8, pp. 218–219.
(6) Vogt, R. *Boardman's Stable Homotopy Category*, Aarhus University (Lecture Notes series No. 21), 1970.

Math. Proc. Camb. Phil. Soc. (1976), **80**, 475
MPCPS 80–47

Uniqueness of *BSO*

BY J. F. ADAMS AND S. B. PRIDDY
University of Cambridge and Northwestern University

(*Received* 7 *November* 1975)

1. *Introduction.* This paper will show that after localization at any given prime p, the infinite loop space structure on the space BSO is essentially unique. If the word 'localization' is replaced by 'completion', the result continues to hold; and both results continue to hold if the space BSO is replaced by the space BSU.

In order to state this result formally, it is natural to suppose given a connected Ω-spectrum $\mathbf{X}$ whose 0th term X_0 is equivalent to the localization or completion at p of BG, where $G = SO$ or SU according to the case. One should then state and prove that this spectrum $\mathbf{X}$ is equivalent to some fixed spectrum $\mathbf{Y}$. Provided we arrange for $\mathbf{Y}$ to be an Ω-spectrum, this conclusion shows that there is an equivalence of infinite loop spaces from X_0 to the fixed infinite loop space Y_0.

Our proof that $\mathbf{X} \simeq \mathbf{Y}$ falls into two parts. The first step determines the mod p cohomology of $\mathbf{X}$ as a module over the mod p Steenrod algebra. The second step starts from a knowledge of the mod p cohomology of $\mathbf{X}$, and constructs an equivalence of spectra $\mathbf{X} \to \mathbf{Y}$.

The second step is valid not only for the cases $G = SO$ and $G = SU$, but also for the cases $G = O$ and $G = U$; but in the latter cases the first step is not valid in the form we have discussed so far. That is, in these cases, we require information about X_0 not only as a space, but as an H-space.

We therefore begin formal work by considering the second step, and for this purpose we first construct the 'obvious' fixed spectrum $\mathbf{Y}$.

Let $\mathbf{K_R}$ be the spectrum which represents classical (periodic) real K-theory; similarly for $\mathbf{K_C}$ in the complex case. Let d be a fixed integer; let $\mathbf{bg}$ be the spectrum obtained from $\mathbf{K_R}$ or $\mathbf{K_C}$ by killing homotopy groups in degrees $< d$, while retaining the homotopy groups in degrees $\geqslant d$. The spectrum $\mathbf{bg}$ therefore represents $(d-1)$-connected K-theory (real or complex). The notation $\mathbf{bg}$ is chosen to reflect the usual notation for connective K-theory; one obtains $\mathbf{bo}$ and $\mathbf{bso}$ from $\mathbf{K_R}$ by taking $d = 1$ and $d = 2$, while one obtains $\mathbf{bu}$ and $\mathbf{bsu}$ from $\mathbf{K_C}$ by taking $d = 2$ and $d = 4$.

Let p be a fixed prime. Let Λ be either the ring $\mathbf{Z}_p$ of p-adic integers, or the ring $\mathbf{Z}_{(p)}$ of integers localized at p (that is, the ring of fractions a/b with a, b integers and b prime to p). We can introduce coefficients Λ into any spectrum $\mathbf{W}$ by setting $\mathbf{W}_\Lambda = \mathbf{M}\Lambda \wedge \mathbf{W}$, where $\mathbf{M}\Lambda$ is a Moore spectrum for the group Λ. We take our fixed spectrum $\mathbf{Y}$ to be

$$\mathbf{Y} = \mathbf{bg}_\Lambda = \mathbf{M}\Lambda \wedge \mathbf{bg}.$$

So the spectrum $\mathbf{Y}$ represents $(d-1)$-connected K-theory with coefficients in Λ. We write $\mathbf{F}_p$ for the field with p elements.

THEOREM 1·1. *Let* $\mathbf{X}$ *be a spectrum whose homotopy groups are finitely generated modules over* Λ, *and bounded below. Suppose given an isomorphism*

$$\theta\colon H^*(\mathbf{X}; \mathbf{F}_p) \leftarrow H^*(\mathbf{Y}; \mathbf{F}_p)$$

of modules over the mod p *Steenrod algebra. Then there exists an equivalence* $f\colon \mathbf{X} \to \mathbf{Y}$ *of spectra such that* $f^* = \theta$.

If for simplicity we disregard the subsidiary hypotheses, this theorem shows that the 'standard' spectrum $\mathbf{Y}$ is characterized by its cohomology.

Theorem 1·1 uses the assumption that certain groups are finitely generated modules over Λ. It may be reassuring to remark that an abelian group can be made into a finitely generated Λ-module in at most one way (see Lemma 5·3).

We now turn to the first step. Let us arrange for $\mathbf{bg}_\Lambda$ to be an Ω-spectrum, and let us define BO_Λ, BSO_Λ, BU_Λ and BSU_Λ to be the 0th terms of the spectra $\mathbf{bo}_\Lambda$, $\mathbf{bso}_\Lambda$, $\mathbf{bu}_\Lambda$ and $\mathbf{bsu}_\Lambda$. With this definition, Theorem 1·2 can be understood without familiarity with the theory of localization or completion. However, we state the facts. For any spectrum $\mathbf{W}$ we have a map

$$\mathbf{W} \simeq \mathbf{S}^0 \wedge \mathbf{W} = \mathbf{MZ} \wedge \mathbf{W} \to \mathbf{M}\Lambda \wedge \mathbf{W},$$

where $\mathbf{S}^0$ is the sphere-spectrum and the final map is induced by the injection $\mathbf{Z} \to \Lambda$. In particular, we have a map $\mathbf{bg} \to \mathbf{bg}_\Lambda$. If we arrange for both spectra to be Ω-spectra and pass to their 0th terms, we have a map of infinite loop-spaces $BG \to BG_\Lambda$ (for $G = O, SO, U, SU$). If $\Lambda = \mathbf{Z}_{(p)}$, then this map displays BG_Λ as the localization of BG in the sense of Sullivan (13); if $\Lambda = \mathbf{Z}_p$, it displays BG_Λ as the completion of BG. In any case, it is useful to note that the map $BG \to BG_\Lambda$ induces an isomorphism of mod p cohomology.

To state Theorem 1·2, we recall that the equivalence $X_0 \simeq \Omega X_1$ determines an H-space structure on X_0 and a Pontryagin product in $H_*(X_0; \mathbf{F}_p)$.

THEOREM 1·2. *Let* $\mathbf{X}$ *be a connected* Ω-*spectrum. Suppose given a homotopy-equivalence of spaces* $X_0 \simeq Y_0$, *where* $Y_0 = BO_\Lambda$, BSO_Λ, BU_Λ *or* BSU_Λ *according to the case. If* $Y_0 = BO_\Lambda$ *and* $p = 2$, *assume further that the square* x^2 *of the generator* x *in* $H_1(X_0; \mathbf{F}_2)$ *is non-zero; if* $Y_0 = BU_\Lambda$, *assume further that the pth power* x^p *of the generator* x *in* $H_2(X_0; \mathbf{F}_p)$ *is non-zero. Then there is an isomorphism of A-modules*

$$\theta\colon H^*(\mathbf{X}; \mathbf{F}_p) \leftarrow H^*(\mathbf{Y}; \mathbf{F}_p).$$

COROLLARY 1·3. *Under the hypotheses of Theorem* 1·2, *there is an equivalence of spectra* $f\colon \mathbf{X} \to \mathbf{Y}$.

Proof. This follows by combining Theorems 1·1 and 1·2; it is only necessary to check that under the hypotheses of Theorem 1·2, the homotopy groups $\pi_r(\mathbf{X})$ of $\mathbf{X}$ are finitely generated modules over Λ. In fact, they are zero for $r < 0$, while for $r \geqslant 0$ we have

$$\pi_r(\mathbf{X}) \cong \pi_r(X_0) \cong \pi_r(Y_0) \cong \pi_r(\mathbf{Y}) \cong \pi_r(\mathbf{bg}) \otimes \Lambda.$$

We note that certain cases of Theorem 1·1 remain unused in Corollary 1·3, notably the real cases with $d \equiv 0$ or 4 mod 8. For example, one might seek to characterize the infinite loop space structure on $\mathbf{Z} \times BO$ (subject to localization or completion at the prime p, as usual); and it will become clear from our proof that this could be done by specifying rather less than the n-fold loop structure for some definite value of n, which indeed could be fairly small. However we leave such extensions of Theorem 1·2 to those readers who may have a use for them.

We have a natural example of a spectrum $\mathbf{X}$ to which Corollary 1·3 applies. For $G = O$, SO, U or SU we can consider BG as the classifying space for G-bundles of virtual dimension 1; then the tensor product of bundles gives BG the structure of an H-space, which we write $BG_{\otimes}$. Recent work of Segal and May (8, 12) shows that this H-space is the 0th term of a connected Ω-spectrum, which we write $\mathbf{bo}_{\otimes}$, $\mathbf{bso}_{\otimes}$, $\mathbf{bu}_{\otimes}$ or $\mathbf{bsu}_{\otimes}$ according to the case. It is then natural to write $\mathbf{bg}_{\oplus}$ for the spectrum we formerly called $\mathbf{bg}$, to show that there the H-space structure given to BG corresponds to the Whitney sum of bundles.

COROLLARY 1·4. *After localization at any prime p, the spectra $\mathbf{bso}_{\oplus}$ and $\mathbf{bso}_{\otimes}$ become equivalent; similarly for $\mathbf{bsu}_{\oplus}$ and $\mathbf{bsu}_{\otimes}$.*

This follows immediately from Corollary 1·3.

The corresponding statement clearly fails for $\mathbf{bu}$, and for $\mathbf{bo}$ at the prime 2, because the Pontryagin products in $BU_{\otimes}$ and $BO_{\otimes}$ do not behave as described in the assumptions of Theorem 1·2, and these assumptions are (of course) necessary, as well as sufficient, for the conclusion $\mathbf{X} \simeq \mathbf{Y}$.

Without localization the statement also fails, although this is harder to see.

Another application of Corollary 1·3 arises as follows. According to Boardman and Vogt (5) the spaces F/PL and F/Top are infinite loop spaces; and according to Sullivan ((13), p. 24) the spaces F/PL and F/Top become equivalent to BO upon localization at any odd prime p.

We add some historical remarks. Peterson (11) has proved that at any odd prime p, $BSO_{\oplus}$ and $BSO_{\otimes}$ are equivalent as two-fold loop spaces. Atiyah and Segal (4) have established an algebraic isomorphism

$$\widetilde{KSO}(\ )^{\wedge} \xrightarrow{\cong} (1+\widetilde{KSO}(\ ))^{\wedge}$$

of p-adic completions. Sullivan (14) has observed that at a regular (odd) prime p, a suitable characteristic class ρ^k induces an equivalence of H-spaces from $BSO_{\oplus}$ to $BSO_{\otimes}$. The second author (10) computed the Dyer–Lashof operations in $H_*(BSO; \mathbf{F}_p)$ which arise from the infinite loop space structures $\oplus$ and $\otimes$; he found that the resulting homology algebras are isomorphic (by an isomorphism very different from the identity). These facts, together with insights of J. P. May, led to the formulation of Corollary 1·4 as a conjecture.

We next attempt to prove this conjecture, in collaboration with J. P. May, by applying May's machinery (7) to specific maps constructed by representation theory. It turns out, however, that in this special case it is better to exploit the special good properties of the spaces BSO and BSU, rather than to rely on machinery which by its nature is adapted to the difficulties of the general case.

After our result was proved, it provided an essential input for the interesting work of Madsen, Snaith and Tornehave (6). In this connexion we note that the methods of the present paper can be pushed further, so as to yield an explicit description of $[\mathbf{Y}; \mathbf{Y}]$, the ring of endomorphisms of the standard spectrum $\mathbf{Y} = \mathbf{bg}_\Lambda$. However, we omit this from the present paper.

The remainder of this paper is organized as follows. In section 2 we take the 'first step', and prove Theorem 1·2. It remains to prove Theorem 1·1. In order to construct a map $f: \mathbf{X} \to \mathbf{Y}$ of spectra such that $f^* = \theta$, we use the Adams spectral sequence for computing the group $[\mathbf{X}, \mathbf{Y}]$ of maps from $\mathbf{X}$ to $\mathbf{Y}$. For this purpose we have to compute the relevant Ext groups

$$\mathrm{Ext}_A^{s,t}(H^*(\mathbf{Y}; \mathbf{F}_p), H^*(\mathbf{X}; \mathbf{F}_p)).$$

This will be done in section 4; it depends on some structure theory for modules over small subalgebras of A, and this will be given in section 3. At this stage we face a difficulty, for the spectra $\mathbf{X}$ and $\mathbf{Y}$ certainly do not satisfy any standard set of conditions known to be sufficient for the convergence of the Adams spectral sequence. In section 5 we will overcome this difficulty and prove Theorem 1·1.

We would like to thank J. P. May for many conversations and numerous letters; although his name does not appear at the head of this paper, he should be considered as a prime mover in this area. In particular, we owe to him the suggestion that we should cover the p-complete case; and for the adaptation of our proof to this case, we have in the end (after considering a variant of our own) preferred to follow his suggestions.

2. *Cohomology of* $\mathbf{X}$. In this section we will determine $H^*(\mathbf{X}; \mathbf{F}_p)$, where $\mathbf{X}$ is as in Theorem 1·2; in particular, $X_0 \simeq BG_\Lambda$, where $G = O$, SO, U or SU according to the case.

PROPOSITION 2·1. *If* $G = SU$, *then*

$$H^*(\mathbf{X}; \mathbf{F}_p) \cong \bigoplus_{r=2}^{p} \Sigma^{2r}(A/(AQ_0 + AQ_1)).$$

We pause to explain the notation. We write A for the mod p Steenrod algebra. We have $Q_0 = \beta_p$ and $Q_1 = P^1\beta_p - \beta_p P^1$, as usual; if $p = 2$ we interpret Q_1 as

$$Sq^{01} = Sq^1Sq^2 + Sq^2Sq^1.$$

The graded module ΣM is defined by regrading M so that an element of degree δ in M appears as an element of degree $\delta + 1$ in ΣM; for example, $\Sigma^{2r}A$ is a free module on one generator of degree $2r$.

PROPOSITION 2·2. *If* $G = U$, *then*

$$H^*(\mathbf{X}; \mathbf{F}_p) \cong \bigoplus_{r=1}^{p-1} \Sigma^{2r}(A/(AQ_0 + AQ_1)).$$

PROPOSITION 2·3. *If* $G = O$ *or* SO *and* $p > 2$, *then*

$$H^*(\mathbf{X}; \mathbf{F}_p) \cong \bigoplus_{s=1}^{\frac{1}{2}(p-1)} \Sigma^{4s}(A/(AQ_0 + AQ_1)).$$

PROPOSITION 2·4. *If* $G = SO$ *and* $p = 2$, *then*

$$H^*(\mathbf{X}; \mathbf{F}_p) \cong \Sigma^2(A/(ASq^3)).$$

PROPOSITION 2·5. *If* $G = O$ *and* $p = 2$, *then*

$$H^*(\mathbf{X}; \mathbf{F}_p) \cong \Sigma(A/(ASq^2)).$$

Theorem 1·2 will follow from these propositions, since they all apply to $\mathbf{Y}$ as well as they do to $\mathbf{X}$. The proofs of these propositions all follow the same pattern. We know the homotopy groups of $\mathbf{X}$; in fact, we are given $\pi_r(\mathbf{X}) = 0$ for $r < 0$, and for $r \geqslant 0$ we have

$$\pi_r(\mathbf{X}) \cong \pi_r(X_0) \cong \pi_r(Y_0) \cong \pi_r(BG) \otimes \Lambda.$$

We may now filter $\mathbf{X}$ by considering its Postnikov system. By applying $H^*(\ ; \mathbf{F}_p)$ to this Postnikov system, we obtain a spectral sequence for computing $H^*(\mathbf{X}; \mathbf{F}_p)$, as in (1). The E_1 term of this spectral sequence consists of the cohomology of those Eilenberg–MacLane spectra which appear in the Postnikov system. Thus for every homotopy group $\pi_r(BG)$ isomorphic to $\mathbf{Z}$ we obtain in our E_1 term a module $\Sigma^r(A/A\beta_p)$; and in the case $p = 2$, for every homotopy group $\pi_r(BG)$ isomorphic to $\mathbf{Z}/(2)$ we obtain in our E_1 term a module $\Sigma^r A$. (Of course homotopy groups isomorphic to $\mathbf{Z}/(2)$ arise only for $G = O$ and $G = SO$.) We note explicitly that what we have said applies just as well to the p-complete case $\Lambda = \mathbf{Z}_p$ as to the p-local case $\Lambda = \mathbf{Z}_{(p)}$.

To proceed further we need to know something about the differentials in our spectral sequence; this means that we need to know something about the k-invariants of the spectrum $\mathbf{X}$. We begin with the simplest case, $G = SU$.

LEMMA 2·6. (*a*) *The k-invariant* k^{2p+3} *of the space* BSU_Λ *is non-zero.*

(*b*) *If* $G = SU$, *and* $\mathbf{X}$ *is as above, then the k-invariant* k^{2p+3} *of the spectrum* $\mathbf{X}$ *is non-zero.*

Proof. The k-invariant k^{2i+1} of the spectrum $\mathbf{X}$ gives by suspension the k-invariant k^{2i+1} of the space X_0, that is, of BSU_Λ. So it is clear that part (*a*) of the lemma implies part (*b*). It is also clear that the k-invariant k^{2i+1} of the spectrum $\mathbf{X}$ is zero for $i \leqslant p$, for it lies in a zero group. So the k-invariant k^{2i+1} of X_0 or BSU_Λ is zero for $i \leqslant p$.

We now introduce notation following (1); if W is a space, then $W(m, \ldots, n)$ will mean that term in the Postnikov system of W whose homotopy groups π_r are the same as those of W for $m \leqslant r \leqslant n$, and zero for other values of r. We write $EM(\pi, n)$ for an Eilenberg–MacLane space of type (π, n).

If the k-invariant k^{2p+3} of BSU_Λ were zero, we would have

$$BSU_\Lambda(4, \ldots, 2p+2) \simeq \mathop{\times}_{r=2}^{p+1} EM(\Lambda, 2r);$$

so the operation $Q_1 = P^1\beta_p - \beta_p P^1$ would be non-zero on $H^4(BSU_\Lambda(4, \ldots, 2p+2); \mathbf{F}_p)$, and hence on $H^4(BSU_\Lambda; \mathbf{F}_p)$. This is a contradiction; the operation Q_1 is zero on $H^*(BSU; \mathbf{F}_p)$ since BSU is torsion-free, and hence it is zero on $H^*(BSU_\Lambda; \mathbf{F}_p)$. This contradiction shows that the k-invariant k^{2p+3} of BSU_Λ is non-zero, and proves the lemma.

Proof of Proposition 2·1. Suppose $G = SU$, and consider the spectral sequence mentioned above for computing $H^*(\mathbf{X}; \mathbf{F}_p)$. The differentials are A-module maps, and they are necessarily zero until we come to d_{2p-2}, which must be given on each module

$\Sigma^{2r}(A/A\beta_p)$ by some multiple of $a \mapsto aQ_1$. We claim that this multiple is non-zero. For the lowest differential

$$d_{2p-2}\colon \Sigma^{2p+2}(A/A\beta_p) \to \Sigma^4(A/A\beta_p)$$

this is equivalent to Lemma 2·6 (*b*), which we have just proved. For the other differentials we argue as follows. Let us use notation $\mathbf{W}(m, \ldots, n)$ for spectra analogous to that which we have used for spaces. Consider the spectrum $\mathbf{X}(2t+4, \ldots, \infty)$. Its 0th term $\mathbf{X}(2t+4, \ldots, \infty)_0$ is equivalent to the 0th term of $\mathbf{Y}(2t+4, \ldots, \infty)$; for we have

$$\mathbf{X}(2t+4, \ldots, \infty)_0 \simeq X_0(2t+4, \ldots, \infty),$$
$$\mathbf{Y}(2t+4, \ldots, \infty)_0 \simeq Y_0(2t+4, \ldots, \infty),$$

where the right-hand sides are constructed from the equivalent spaces X_0 and Y_0, by the method of killing homotopy groups as applied to spaces. *A fortiori*, the $(-2t)$th term $\mathbf{X}(2t+4, \ldots, \infty)_{-2t}$ is equivalent to the corresponding term $\mathbf{Y}(2t+4, \ldots, \infty)_{-2t}$. But by the Bott periodicity theorem the spectrum $\mathbf{bsu}(2t+4, \ldots, \infty)$ is equivalent, after reindexing its terms, to $\mathbf{bsu}$; this conclusion persists after we introduce coefficients Λ; therefore $\mathbf{Y}(2t+4, \ldots, \infty)_{-2t}$ is equivalent to BSU_Λ. So the work we have already done determines the lowest differential for $\mathbf{X}(2t+4, \ldots, \infty)$, which gives the differential

$$\Sigma^{2t+2p+2}(A/A\beta_p) \to \Sigma^{2t+4}(A/A\beta_p)$$

for $\mathbf{X}$.

Now, the sequence

$$\ldots \to \Sigma^{2t+4p-4}A/A\beta_p \to \Sigma^{2t+2p-2}A/A\beta_p \to \Sigma^{2t}A/A\beta_p,$$

in which every map is given by $a \mapsto aQ_1$, is exact. The simplest way to see this is as follows. Let B be the exterior algebra generated by $Q_0 = \beta_p$ and Q_1; then the sequence

$$\ldots \to \Sigma^{2t+4p-4}B/B\beta_p \to \Sigma^{2t+2p-2}B/B\beta_p \to \Sigma^{2t}B/B\beta_p,$$

in which every map is given by $b \mapsto bQ_1$, is exact. The previous sequence comes from this by applying the functor $A \otimes_B$, and this functor preserves exactness since A is free as a right module over B.

It follows that the spectral sequence we are studying becomes trivial after the differential d_{2p-2}, and we find

$$H^*(\mathbf{X}; \mathbf{F}_p) \cong \bigoplus_{r=2}^{p} \Sigma^{2r}(A/(AQ_0 + AQ_1)).$$

This completes the proof of Proposition 2·1.

We turn to the case $G = U$.

LEMMA 2·7. *If* $G = U$, *and* $\mathbf{X}$ *is as in Theorem* 1·2, *then the k-invariant* k^{2p+1} *of the spectrum* $\mathbf{X}$ *is non-zero.*

Proof. If this k-invariant were zero, then we would have an equivalence of spectra

$$\mathbf{X}(2, \ldots, 2p) \simeq \mathop{\times}_{r=1}^{p} \mathbf{EM}(\Lambda, 2r)$$

(where $\mathbf{EM}(\pi, n)$ means an Eilenberg–MacLane spectrum of type (π, n)). This would yield an equivalence of H-spaces

$$X_0(2, \ldots, 2p) \simeq \mathop{\times}_{r=1}^{p} EM(\Lambda, 2r).$$

But if x is the generator in $H_2(EM(\Lambda, 2); \mathbf{F}_p)$ then we have $x^p = 0$, and this contradicts the assumption in Theorem 1·2. This contradiction shows that $k^{2p+1} \neq 0$ and proves the lemma.

Proof of Proposition 2·2. We use the same spectral sequence as before. The lowest differential

$$d_{2p-2}\colon \Sigma^{2p} A/A\beta_p \to \Sigma^2 A/A\beta_p$$

is a non-zero multiple of $a \mapsto aQ_1$, by Lemma 2·7. The remaining differentials are determined by what we have already done, for the spectrum $\mathbf{X}(4, \ldots, \infty)$ has 0th term equivalent to BSU_Λ. The rest of the proof goes as for Proposition 2·1.

Proof of Proposition 2·3. If p is odd then on introducing coefficients Λ we have $\mathbf{bo}_\Lambda \simeq \mathbf{bso}_\Lambda$, so that $BO_\Lambda \simeq BSO_\Lambda$; thus the two cases are equivalent. We use the same spectral sequence as before. It behaves like that used in proving Propositions 2·1 and 2·2, except that we now have homotopy groups Λ in degrees $4s$ instead of in degrees $2r$. As for the differential d_{2p-2}, we now dispose of the relevant information about the space BSU_Λ (in any case, the information about BSU was already on record in (1)). We obtain the corresponding information about BSO_Λ by naturality, using either the map $BSO_\Lambda \to BSU_\Lambda$ or the map $BSU_\Lambda \to BSO_\Lambda$. The information about the space X_0 implies the required information about the spectrum $\mathbf{X}$. The rest of the proof goes as for Propositions 2·1 and 2·2.

We turn to the case $G = SO$, $p = 2$.

LEMMA 2·8. *Let n be a positive integer divisible by* 4. *Then the k-invariant k^{n+1} of the space BSO_Λ is non-zero.*

We actually need only the cases $n = 4$ and $n = 8$; but the proof is the same in general.

Proof. For brevity, we write W for BSO_Λ. Suppose (for a contradiction) that $k^{n+1} = 0$. Then we have

$$W(0, \ldots, n) \simeq W(0, \ldots, n-1) \times EM(\Lambda, n).$$

Suppose also that the indecomposable quotient of $H^*(W(0, \ldots, n-1); \mathbf{F}_2)$ in degree n has dimension δ over $\mathbf{F}_2$. Then (by the Künneth formula) the indecomposable quotient of $H^*(W(0, \ldots, n); \mathbf{F}_2)$ in degree n has dimension $\delta + 1$ over $\mathbf{F}_2$, and the same conclusion holds for W. But the indecomposable quotient of $H^*(BSO; \mathbf{F}_2)$ in degree n has dimension 1 over $\mathbf{F}_2$; so we infer that $\delta = 0$, and all elements of $H^n(W(0, \ldots, n-1); \mathbf{F}_2)$ are decomposable.

Since we have

$$W(0, \ldots, n) \cong W(0, \ldots, n-1) \times EM(\Lambda, n)$$

we can calculate $H^n(W(0, \ldots, n); \mathbf{F}_2)$ by the Künneth formula; we can calculate Sq^1 on it by the Cartan formula; and using the fact that Sq^1 annihilates $H^n(EM(\Lambda, n); \mathbf{F}_2)$, it follows that the image of

$$Sq^1\colon H^n(W(0, \ldots, n); \mathbf{F}_2) \to H^{n+1}(W(0, \ldots, n); \mathbf{F}_2)$$

consists entirely of decomposable elements. Therefore the same conclusion holds in W. But this contradicts the known relation

$$Sq^1 w_n = w_{n+1}$$

between the Stiefel–Whitney classes in $H^*(BSO; \mathbf{F}_2)$. This contradiction proves the lemma.

Proof of Proposition 2·4. We use the same spectral sequence as before. We need to know the first differentials, and we claim that they are as follows.

$$\begin{array}{l} \Sigma^{8t+8}A/ASq^1 \\ \quad\downarrow a \mapsto aSq^5 \\ \Sigma^{8t+4}A/ASq^1 \\ \quad\downarrow a \mapsto aSq^3 \\ \Sigma^{8t+2}A \\ \quad\downarrow a \mapsto aSq^2 \\ \Sigma^{8t+1}A \\ \quad\downarrow a \mapsto aSq^2 \\ \Sigma^{8t}A/ASq^1 \\ \quad\vdots \\ \Sigma^{8}A/ASq^1 \\ \quad\downarrow a \mapsto aSq^5 \\ \Sigma^{4}A/ASq^1 \\ \quad\downarrow a \mapsto aSq^3 \\ \Sigma^{2}A. \end{array} \tag{2·9}$$

In fact, for dimensional reasons, each differential must be a multiple of the one shown; we have to check that the multiple is non-zero.

In the cases where the differential is given by Sq^2, the result is easy; for composition with the essential map

$$\eta\colon S^{m+1} \to S^m$$

gives a homomorphism

$$\pi_m(BSO) \to \pi_{m+1}(BSO),$$

which is non-zero if $m \equiv 0$ or $1 \bmod 8$ and $m \geqslant 8$; so the same conclusion holds in the space X_0 and in the spectrum $\mathbf{X}$.

Next we take the final differential

$$\Sigma^4 A/A\,Sq^1 \to \Sigma^2 A.$$

This is zero if and only if the first k-invariant k^5 of the spectrum $\mathbf{X}$ is zero. But if this k-invariant were zero, then the k-invariant k^5 of the space X_0 would be zero, contradicting Lemma 2·8.

Note that in this argument it is essential to use cohomology with coefficients in $\mathbf{Z}_{(2)}$ or $\mathbf{Z}_2$ (as we have implicitly done by using k^5) rather than coefficients $\mathbf{F}_2$; for in the

Postnikov system of X_0, Sq^3 annihilates $H^2(EM(\mathbf{Z}/(2), 2); \mathbf{F}_2)$. The same remark applies to the next argument.

Next we take the penultimate differential

$$\Sigma^8 A/ASq^1 \to \Sigma^4 A/ASq^1.$$

This is related to the second k-invariant k^9 of the spectrum $\mathbf{X}$; here k^9 lies in

$$H^9(\mathbf{X}(2, \ldots, 4); \Lambda).$$

More precisely, consider the following exact sequence.

$$H^9(\mathbf{X}(2); \Lambda) \xrightarrow{j^*} H^9(\mathbf{X}(2, \ldots, 4); \Lambda) \xrightarrow{i^*} H^9(\mathbf{X}(4); \Lambda).$$

The differential in question is zero if and only if $i^*k^9 = 0$. If so, then k^9 lies in $\operatorname{Im} j^*$; in other words, k^9 is a linear combination of

$$\delta_2 Sq^6 b \quad \text{and} \quad \delta_2 Sq^4 Sq^2 b.$$

(Here δ_2: $H^m(\ ; \mathbf{F}_2) \to H^{m+1}(\ ; \Lambda)$ is the Bockstein boundary, and $b \in H^2(\mathbf{X}(2, \ldots, 4); \mathbf{F}_2)$ is the fundamental class.) But we will prove that both these classes suspend to zero in $H^9(X_0(2, \ldots, 4); \Lambda)$. To begin with, the fundamental class in $H^2(X_0(2, \ldots, 4); \mathbf{F}_2)$ can be identified with the Stiefel–Whitney class $w_2 \in H^2(X_0; \mathbf{F}_2) \cong H^2(BSO; \mathbf{F}_2)$, and of course Sq^6 annihilates it for dimensional reasons. To continue, we argue that

$$H^4(X_0(2, \ldots, 4); \mathbf{F}_2) \cong H^4(X_0; \mathbf{F}_2),$$

and that the element $Sq^2 w_2 = (w_2)^2$ in it is the reduction of a class in

$$H^4(X_0(2, \ldots, 4); \Lambda) \cong H^4(X_0; \Lambda).$$

In fact, in BSO the first Pontryagin class P_1 reduces to $(w_2)^2$; so it is sufficient to take the class corresponding to P_1 under the isomorphism

$$H^4(BSO; \Lambda) \leftarrow H^4(BSO_\Lambda; \Lambda).$$

(In the p-local case the fact that this map is an isomorphism in general is well known; in the p-complete case it is perhaps shortest to establish this particular case ad hoc, as is easily done, rather than set up general theory.) In any case, the element $Sq^4 Sq^2 w_2 = (w_2)^4$ in $H^8(X_0(2, \ldots, 4); \mathbf{F}_2)$ is also the reduction of a class defined over Λ, and $\delta_2 Sq^4 Sq^2 w_2 = 0$.

So the hypothesis that the penultimate differential is zero implies that the k-invariant k^9 of the space $X_0 \simeq BSO_\Lambda$ is zero. This contradicts Lemma 2·8.

For the higher differentials we argue as before. Consider the spectrum $\mathbf{X}(8t+2, \ldots, \infty)$. Its 0th term $X_0(8t+2, \ldots, \infty)$ is already equivalent to the 0th term $Y_0(8t+2, \ldots, \infty)$ of $\mathbf{Y}(8t+2, \ldots, \infty)$. *A fortiori*, its $(-8t)$th term $\mathbf{X}(8t+2, \ldots, \infty)_{-8t}$ is equivalent to the corresponding term $\mathbf{Y}(8t+2, \ldots, \infty)_{-8t}$. By the Bott periodicity theorem the spectrum $\mathbf{bso}(8t+2, \ldots, \infty)$ is equivalent, after reindexing its terms, to $\mathbf{bso}$; this conclusion persists after we introduce coefficients Λ; therefore $\mathbf{Y}(8t+2, \ldots, \infty)_{-8t}$ is equivalent to

BSO_Λ. So the work we have already done gives the last two differentials for

$$\mathbf{X}(8t+2, \ldots, \infty);$$

this gives the differentials

$$\Sigma^{8t+8}A/ASq^1 \to \Sigma^{8t+4}A/ASq^1 \to \Sigma^{8t+2}A$$

for $\mathbf{X}$. This completes the proof that the differentials are as shown in (2·9).

Now the sequence (2·9) is exact. To prove it we argue as before. Let B be the subalgebra of A generated by Sq^1 and Sq^2; then it is elementary to check that the following sequence is exact.

$$\begin{array}{l}
\Sigma^{8t+8}B/BSq^1 \\
\quad\downarrow b \mapsto bSq^2Sq^2 \\
\Sigma^{8t+4}B/BSq^1 \\
\quad\downarrow b \mapsto bSq^2 \\
\Sigma^{8t+2}B \\
\quad\downarrow b \mapsto bSq^1 \\
\Sigma^{8t+1}B \\
\quad\downarrow b \mapsto bSq^1 \\
\Sigma^{8t}B/BSq^1 \\
\quad\vdots \\
\quad\downarrow \\
\Sigma^2 B
\end{array}$$

The sequence (2·9) is obtained from this by applying the functor $A \otimes_B$, and this functor preserves exactness since A is free as a right module over B.

We conclude that the spectral sequence becomes trivial after the differentials shown in (2·9), and

$$H^*(\mathbf{X}; \mathbf{F}_2) \simeq \Sigma^2 A/ASq^3.$$

This completes the proof of Proposition 2·4.

We turn to the final case $G = O$, $p = 2$.

LEMMA 2·10. *If $G = O$, $p = 2$ and $\mathbf{X}$ is as in Theorem 1·2, then the k-invariant k^3 of the spectrum $\mathbf{X}$ is non-zero.*

Proof. This is parallel to the proof of Lemma 2·7. If this k-invariant were zero, then we would have an equivalence of spectra

$$\mathbf{X}(1,2) \simeq \mathbf{EM}(\mathbf{Z}/(2), 1) \times \mathbf{EM}(\mathbf{Z}/(2), 2).$$

This would yield an equivalence of H-spaces

$$X_0(1,2) \simeq EM(\mathbf{Z}/(2), 1) \times EM(\mathbf{Z}/(2), 2).$$

But if x is the generator in $H_1(EM(\mathbf{Z}/(2), 1); \mathbf{F}_2)$ then we have $x^2 = 0$, and this contradicts the assumption in Theorem 1·2. This contradiction shows that $k^3 \neq 0$ and proves the lemma.

Proof of Proposition 2·5. We use the same spectral sequence as before. This time, however, we claim that the first differentials are as follows.

$$
\begin{array}{l}
\Sigma^{8t+8}A/ASq^1 \\
\quad\downarrow a \mapsto aSq^5 \\
\Sigma^{8t+4}A/ASq^1 \\
\quad\downarrow a \mapsto aSq^3 \\
\Sigma^{8t+2}A \\
\quad\downarrow a \mapsto aSq^2 \\
\Sigma^{8t+1}A \quad a \mapsto aSq^2 \\
\quad\downarrow \\
\Sigma^{8t}A/ASq^1 \\
\quad\vdots \\
\quad\downarrow \\
\Sigma^{8}A/ASq^1 \\
\quad\downarrow a \mapsto aSq^5 \\
\Sigma^{4}A/ASq^1 \\
\quad\downarrow a \mapsto aSq^3 \\
\Sigma^{2}A \\
\quad\downarrow a \mapsto aSq^2 \\
\Sigma A
\end{array}
$$

In fact, the lowest differential $d_1: \Sigma^2 A \to \Sigma A$ is determined by Lemma 2·10, and all the higher ones are determined by our previous work, since $\mathbf{X}(2, \ldots, \infty)$ is a spectrum whose 0th term is BSO_Λ. The rest of the proof goes as for Proposition 2·4.

3. *Structure theory for modules.* In this section we will record some of the structure theory of modules over small subalgebras of the Steenrod algebra. The results are originally due to the first author.

Let B be a connected graded Hopf algebra of finite dimension over the ground field k. We have in mind the following two examples.

(3·1) $B = E[x, y]$, the exterior algebra over k on two primitive generators x and y of distinct degrees. In our applications, k is $\mathbf{F}_p$, and B is the subalgebra of the mod p Steenrod algebra generated by $x = Q_0$ and $y = Q_1$.

(3·2) $B = A_1$, the subalgebra of the mod 2 Steenrod algebra generated by Sq^1 and Sq^2.

We must now explain that we actually want to discuss stable structure theory rather than structure theory. Let L, M be (say) left B-modules, and let $f_0, f_1: L \to M$ be B-linear maps; for definiteness we may consider only maps which preserve the grading (leaving maps which change the grading to be introduced later by considering $\mathrm{Hom}_B(\Sigma^t L, M)$ or $\mathrm{Hom}_B(L, \Sigma^t M)$). We say that f_0 and f_1 are *homotopic* if $f_0 - f_1$ factors through a free module F. (If L is finitely generated we may take F to be finitely generated, for $f_0 - f_1$ must map into a finitely generated free submodule of F.) This notion of homotopy corresponds both to projective and to injective homotopy, which in general are distinct. Homotopy is an equivalence relation, for if $f_0 - f_1$ factors through F and $f_1 - f_2$ factors through F', then $f_0 - f_2$ factors through $F \oplus F'$. Composition of maps passes to homotopy classes, so we get a category of homotopy classes. Two modules L, M are *stably equivalent* if they are equivalent in the category of homotopy classes; Lemma 3·4 will show that this term has its usual meaning. We are in fact interested in the classification of modules rather than maps, so we fix on the adjective 'stable'; we speak of 'stable classes of maps' rather than 'homotopy classes', and write '$S\operatorname{hom}_B(L, M)$' for the group of stable classes of maps from L to M. The following results may be taken as justifying these definitions for our purposes.

LEMMA 3·3. (*a*) *For* $s > 0$, $\operatorname{Ext}^s_B(L, M)$ *is a bifunctor on the category of stable maps.*

(*b*) *Let*
$$0 \to L' \xrightarrow{i} L \to L'' \to 0$$
be an exact sequence in which L *is free; then*
$$\operatorname{Ext}^1_B(L'', M) \cong S\operatorname{hom}_B(L', M).$$

(*c*) *Let*
$$O \to M' \to M \xrightarrow{j} M'' \to O$$
be an exact sequence in which M *is free; then*
$$\operatorname{Ext}^1_B(L, M') \cong S\operatorname{hom}_B(L, M'').$$

LEMMA 3·4. *L and M are stably equivalent if and only if we have* $L \oplus F \cong M \oplus G$ *for some free modules F and G, which may be taken finitely generated if L and M are so.*

Proof of Lemma 3·3. (*a*) If F is free, then $\operatorname{Ext}^s_B(F, M) = 0$ for $s > 0$ because F is projective, and $\operatorname{Ext}^s_B(L, F) = 0$ for $s > 0$ because (under our hypotheses) F is injective.

(*b*) We have the following exact sequence.
$$0 \leftarrow \operatorname{Ext}^1_B(L'', M) \leftarrow \operatorname{Hom}_B(L', M) \xleftarrow{i^*} \operatorname{Hom}_B(L, M).$$
Here the image of i^* certainly consists of maps $L' \to M$ which factor through L, which is free, so this image maps to zero in $S\operatorname{hom}_B(L', M)$. Conversely, if we have a composite
$$L' \xrightarrow{j} F \xrightarrow{k} M$$
with F free, then since F is injective the map j factors through $i: L' \to L$, and kj lies in the image of i^*. We thus obtain an isomorphism
$$\operatorname{Ext}^1_B(L'', M) \cong S\operatorname{hom}_B(L', M).$$

(*c*) The proof of (*c*) is precisely dual to the proof of (*b*).

Proof of Lemma 3·4. If $L \oplus F \cong M \oplus G$ then L and M are stably equivalent, trivially. We have to prove the converse. Take a map $f: L \to M$ whose stable class is a stable equivalence. By adding to L a suitable free module F we can suppose that f is epi (and here we can suppose that F is finitely generated if M is so). Let K be the kernel of f; then for any module N we have an exact sequence
$$\dots \leftarrow \operatorname{Ext}^{s+1}_B(M, N) \leftarrow \operatorname{Ext}^s_B(K, N) \leftarrow \operatorname{Ext}^s_B(L, N) \xleftarrow{f^*} \operatorname{Ext}^s_B(M, N) \leftarrow \dots$$
in which f^* is iso for $s > 0$ by Lemma 3·3(*a*). Thus $\operatorname{Ext}^s_B(K, N) = 0$ for $s > 0$, and this for every module N; so K is projective. Hence K is free (note that under our strong

assumptions on B this follows without assuming K bounded above or below). Since K is free it is injective, and so $L \cong M \oplus K$. Here K is finitely generated if L is so. This proves the lemma.

Various constructions on modules are functorial in the sense that they carry stable maps into stable maps. First, obviously, we have the direct sum $L \oplus M$. Secondly we have the tensor product $L \otimes M$; since B is a Hopf algebra, we can make B act on $L \otimes_k M$ in the usual way; that is, if

$$\psi b = \sum_i b'_i \oplus b''_i,$$

we define

$$b(l \otimes m) = \sum_i (-1)^{|b''_i||l|} b'_i l \otimes b''_i m$$

(where $|l|$ means the degree of l, as usual). In order to see that the tensor product of maps passes to stable classes, we have to remark that if F is free, then $L \otimes F$ and $F \otimes M$ are free. Thirdly, we have the vector-space dual $M^* = \mathrm{Hom}_k(M, k)$. This is graded so that

$$|\langle m^*, m\rangle| = |m^*| + |m|,$$

where k is graded so that all of k is in degree 0. In other words, an element m^* is of degree d if it annihilates all homogeneous elements m except perhaps those of degree $-d$. The obvious way to make M^* a B-module is to make it a right B-module, so that

$$\langle m^* b, m\rangle = \langle m^*, bm\rangle.$$

However, we are willing to assume that B has a conjugation map c, and then we can make M^* into a left B-module by setting

$$bm^* = (-1)^{|b||m^*|} m^*(cb).$$

In order to see that the duals of maps pass to stable classes, we have to remark that if F is free, then (under our hypotheses on B) F^* is free.

The module k (graded so that all of k is in degree 0) is a unit for the tensor product. We call a module 'invertible' if its stable equivalence class is invertible (under the tensor product). Such invertible stable equivalence classes form a group, and we will calculate this group when B is as in (3·1) and (3·2); or more precisely, we will calculate the group of invertible classes which can be represented by finitely generated B-modules. First we need to characterize such B-modules, and for this purpose we need invariants of modules.

We can associate to a module M over the exterior algebra $E[x, y]$ the homology groups

$$H(M; x) = \mathrm{Ker}\, x/\mathrm{Im}\, x, \quad H(M; y) = \mathrm{Ker}\, y/\mathrm{Im}\, y.$$

We can also use these homology groups when $B = A_1$, by taking $x = Sq^1$,

$$y = Sq^{01} = Sq^1 Sq^2 + Sq^2 Sq^1.$$

These homology groups are functorial on the category of stable maps, for we have

$$H(F; z) = 0 \quad \text{when } F \text{ is free and } z = x \text{ or } y.$$

(We keep z as a letter which stands for x or y.) We have

$$H(L \oplus M; z) \cong H(L; z) \oplus H(M; z),$$
$$H(L \otimes M; z) \cong H(L; z) \otimes H(M; z)$$

(by the Künneth formula) and

$$H(M^*; z) \cong (H(M; z))^*.$$

So these homology groups commute with the three constructions considered above.

LEMMA 3·5. *Assume $B = E[x, y]$ or $B = A_1$. (a) If M is invertible, then $H(M; x)$ and $H(M; y)$ are of dimension 1 over k.*

(b) Suppose $H(M; x)$ and $H(M; y)$ are of dimension 1 over k, and M is finitely generated; then M is invertible, and its inverse is M^*.

It can be shown by examples that in clause (*b*), the assumption that M is finitely generated cannot be omitted.

Proof. First suppose that M is invertible; say $M \otimes N \simeq k$. Then by the remarks above,

$$H(M; z) \otimes H(N; z) \cong k,$$

so that $H(M; z)$ has dimension 1 over k.

Conversely, assume that M is finitely generated, and $H(M; x)$ and $H(M; y)$ are of dimension 1 over k. Consider the evaluation map

$$M^* \otimes M \to k;$$

by construction, it is a map of B-modules. We see from the Künneth formula that it induces an isomorphism of $H(\ ; x)$ and an isomorphism of $H(\ ; y)$. Therefore

$$M^* \otimes M \simeq k,$$

by the theorem of Adams and Margolis (3) (see theorem 4·2, and note that the proof given remains valid when A is replaced by B). This proves the lemma.

From this point up to and including (3·11) all B-modules are assumed to be finitely generated.

We now give examples of invertible modules.

(i) We define Σ to be the module which is k in degree 1 and zero in other degrees. Its inverse Σ^{-1} is the module which is k in degree -1 and zero in other degrees. Thus multiplying a module M by Σ simply regrades it; this is consistent with our use of the notation ΣM in section 2.

(ii) We define I to be the augmentation ideal of B. Then, we claim, $H(I; z)$ has dimension 1 over k for $z = x$ and for $z = y$; this follows immediately from the exact homology sequence which one obtains from the short exact sequence

$$0 \to I \to B \to k \to 0.$$

Thus I is invertible by Lemma 3·5, provided $B = E[x, y]$ or $B = A_1$.

A proof that I is invertible for more general B is given in (2), pp. 343–44.

(iii) When $B = A_1$, we define J to be the module $\Sigma^{-2}(B/BSq^3)$. It has a base over $\mathbf{F}_2$ indicated by the nodes in the following diagram. This module J satisfies the criteria

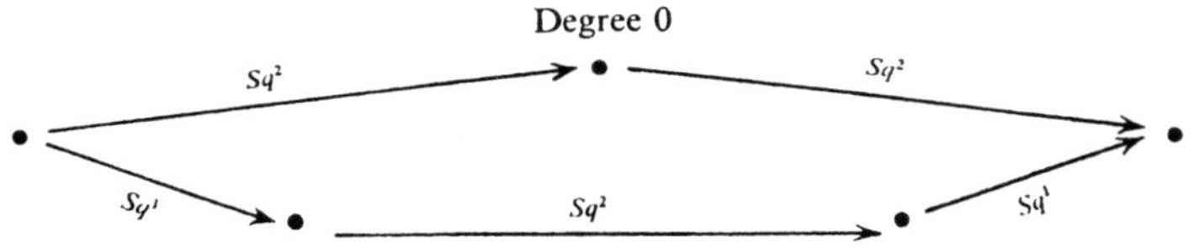

of Lemma 3·5; so it is invertible. Moreover, we have $J^* \cong J$; so J will serve as J^{-1} and we have $J^2 \simeq 1$.

For the rest of this section we reserve the letters I and J for the modules just defined.

THEOREM 3·6. *If* $B = E[x, y]$, *as in* (3·1), *then the group of invertible stable equivalence classes is* $\mathbf{Z} \oplus \mathbf{Z}$, *generated by* Σ *and* I.

THEOREM 3·7. *If* $B = A_1$, *as in* (3·2), *then the group of invertible stable equivalence classes is* $\mathbf{Z} \oplus \mathbf{Z} \oplus \mathbf{Z}/(2)$, *generated by* Σ, I *and* J.

Proof of Theorems 3·6 *and* 3·7. First we will prove that in Theorem 3·6, the stable equivalence classes $\Sigma^a I^b$ are all distinct. In fact, $H(\Sigma^a I^b; x)$ is k in degree $a + b|x| = c$, say, and zero in other degrees; while $H(\Sigma^a I^b; y)$ is k in degree $a + b|y| = d$, say, and zero in other degrees. Since $|x| \neq |y|$, c and d determine a and b.

Next we will prove that in Theorem 3·7, the stable equivalence classes $\Sigma^a I^b J^c$ (with $c = 0$ or $1 \bmod 2$) are all distinct. In fact, the homology groups of $\Sigma^a I^b J^c$ determine a and b, as above; we need one more invariant to determine c. The simplest is to consider $\dim_k M \bmod 8$, where $k = \mathbf{F}_2$. Using Lemma 3·4, we see that this is an invariant of the stable class of M; it sends the tensor-product of classes to the product of integers mod 8; and we have

$$\begin{aligned} \dim_k(\Sigma) &= 1 \bmod 8, \\ \dim_k(I) &= -1 \bmod 8, \\ \dim_k(J) &= 5 \bmod 8. \end{aligned}$$

So this invariant gives us c (and the residue class of $b \bmod 2$).

It remains to show that every invertible stable class has the form $\Sigma^a I^b$ or $\Sigma^a I^b J^c$, as the case may be.

First we assume $B = E[x, y]$, as in Theorem 3·6. Without loss of generality we may assume that $|x| < |y|$. Let M be an invertible module. Multiplying by some power of Σ, we may assume without loss of generality that $H(M; x)$ is k in degree 0. Let g_0 be a cycle representing the generator, so that $xg_0 = 0$ but $g_0 \notin \operatorname{Im} x$. Consider yg_0. We have $xyg_0 = -yxg_0 = 0$, so $yg_0 \in \operatorname{Ker} x$; since $H(M; x)$ is zero in this degree, we conclude that $yg_0 = xg_1$ for some g_1. Consider yg_1; we have

$$xyg_1 = -yxg_1 = -yyg_0 = 0.$$

Continuing in this way by induction, we find a sequence of elements

$$g_0, g_1, g_2, \ldots, g_n, \ldots$$

such that $\qquad xg_0 = 0 \quad \text{and} \quad yg_i = xg_{i+1} \quad$ for all i.

Now M is finitely generated over B, so we must have $g_{n+1} = 0$ for some n. Thus $yg_n = 0$.

As a first case, we consider the possibility that g_n represents a generator of $H(M; y)$. In this case, let $L(n)$ be the module presented by generators $l_0, l_1, l_2, \ldots, l_n$ and relations

$$xl_0 = 0, \quad yl_i = xl_{i+1}, \quad yl_n = 0.$$

$L(n)$:

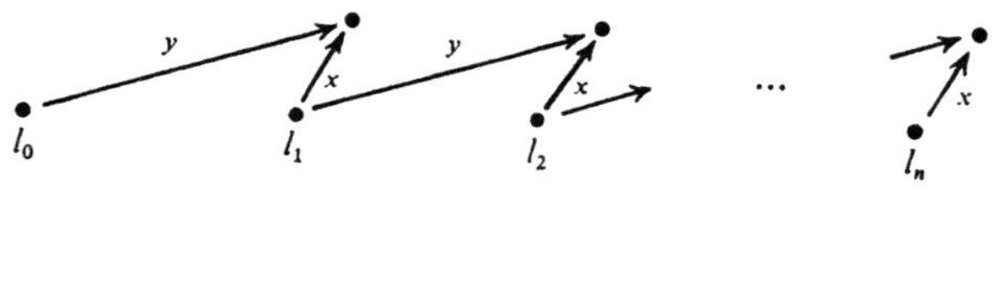

We have constructed a map $L(n) \to M$ (given by $l_i \mapsto g_i$) which induces an isomorphism of $H(\ ;x)$ and $H(\ ;y)$; so $M \simeq L(n)$ by the theorem of Adams and Margolis. We will show

$$L(n) \simeq (\Sigma^{-|x|}I)^n.$$

The obvious minimal resolution of $L(n)$ over B gives an exact sequence

$$0 \to \Sigma^{|x|}L(n+1) \to F \to L(n) \to 0$$

with F free. On the other hand, by taking the sequence

$$0 \to I \to B \to k \to 0$$

and tensoring with $L(n)$, we get

$$0 \to I \otimes L(n) \to B \otimes L(n) \to L(n) \to 0,$$

and here $B \otimes L(n)$ is free. By Schanuel's Lemma, we get

$$I \otimes L(n) \simeq \Sigma^{|x|}L(n+1).$$

By induction over n, starting with $L(0) \cong k$, we see

$$L(n) \simeq (\Sigma^{-|x|}I)^n.$$

As a second case, we consider the possibility that g_n represents the zero class in $H(M; y)$. In this case there is an element $h_{n-1} \in M$ such that $g_n = yh_{n-1}$. Replacing g_{n-1} by $g'_{n-1} = g_{n-1} + xh_{n-1}$, we recover our original situation with n replaced by $n-1$.

We may continue this process by induction downwards over n. Either at some stage we encounter the 'first case' and prove that $M \simeq \Sigma^a I^b$ for some $b \geq 0$, or else the induction continues right down to $n = 0$. There is no objection to replacing g_0 by $g'_0 = g_0 + xh_0$, for that does not alter its class in $H(M; x)$. We may thus suppose that $xg_0 = 0$, $yg_0 = 0$ and (unless $M \simeq k \simeq \Sigma^0 I^0$) that $g_0 = yh_{-1}$.

Suppose now that we have constructed $h_{-1}, h_{-2}, \ldots, h_{-n}$ with

$$yh_i = xh_{i+1}, \quad yh_{-1} = g_0.$$

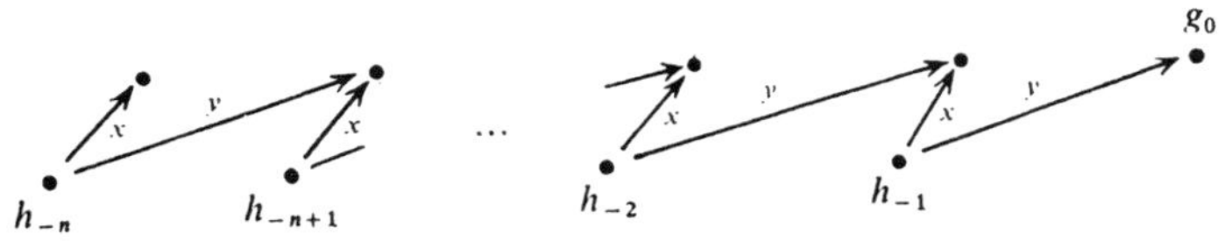

Consider xh_{-n}. If $n = 1$ we have

$$yxh_{-1} = -xyh_{-1} = -xg_0 = 0;$$

otherwise we have

$$yxh_{-n} = -xyh_{-n} = -xxh_{-n+1} = 0.$$

As a first case, we consider the possibility that xh_{-n} represents a generator of $H(M; y)$. In this case we can proceed precisely as before, but the module we map to M is now $L(n)^*$, and we obtain

$$M \simeq L(n)^* \simeq (\Sigma^{|x|}I^{-1})^n.$$

As a second case, we consider the possibility that xh_{-n} represents the zero element in $H(M; y)$. In this case there is an element h_{-n-1} in M such that $yh_{-n-1} = xh_{-n}$, and the induction continues. Either at some stage we encounter the 'first case' and prove that $M \simeq \Sigma^a I^b$ for some $b < 0$, or else the induction constructs h_n for all n. Now M is finitely generated over B, so we must have $h_{-n-1} = 0$ for some n. Then $xh_{-n} = 0$. Since $H(M; x)$ is zero in this degree, we conclude that $h_{-n} = xk_{-n}$ for some k_{-n}. Then if $n > 1$, we may replace h_{-n+1} by $h'_{-n+1} = h_{-n+1} + yk_{-n}$; we have $xh'_{-n+1} = 0$, and we continue the induction. This induction finally proves that

$$g_0 = yh_{-1} = yxk_{-1} = -x(yk_{-1}),$$

contradicting the choice of g_0. Therefore the 'first case' must have arisen at some stage, and we have $M \simeq \Sigma^a I^b$ for some a, b. This proves Theorem 3·6.

Next we assume $B = A_1$, as in Theorem 3·7. We will prove that every invertible class is of the form $\Sigma^a I^b J^c$. If M is an invertible module over $B = A_1$, then by neglecting some of its structure we obtain an invertible module over $B' = E[Sq^1, Sq^{01}]$. The module Σ over A_1 yields the module Σ' over B'; the module I over A_1 yields the module $I' \oplus \Sigma^2 B'$ over B'. Assume that over B' we have $M \simeq (\Sigma')^a (I')^b$. Take a B-module N in the stable class $\Sigma^{-a} I^{-b} M$; then over B' we have $N \simeq k$, and we have to prove that over A_1 we have either $N \simeq k$ or $N \simeq J$.

We work simultaneously on N and N^*. Since we know the stable class of N over B', it follows that we can find in N an element g of degree 0 such that $Sq^1 g = 0$, $Sq^{01} g = 0$ and g is simultaneously a generator for $H(N; Sq^1)$ and $H(N; Sq^{01})$. Similarly, in N^* we can find an element g^* of degree 0 such that $Sq^1 g^* = 0$, $Sq^{01} g^* = 0$ and g^* is simultaneously a generator for $H(N^*; Sq^1)$ and $H(N^*; Sq^{01})$. We must have $\langle g^*, g\rangle = 1$.

First suppose that g is indecomposable, so that $g \notin Sq^1 N + Sq^2 N$. Then we can find a map of graded $\mathbf{F}_2$-modules $\theta \colon N \to k$ which annihilates $Sq^1 N + Sq^2 N$ and maps g to 1. Then θ is an A_1-map from N to k which induces an isomorphism of $H(\ ; Sq^1)$ and $H(\ ; Sq^{01})$, so $N \simeq k$. Similarly, if g^* is indecomposable we find $N^* \simeq k$ and $N \simeq k$.

The only remaining possibility is to suppose

$$g = Sq^1 n + Sq^2 m,$$
$$g^* = Sq^1 n^* + Sq^2 m^*$$

for suitable elements n, m, n^*, m^*. Then

$$Sq^3 m = Sq^1 g + Sq^1 Sq^1 n = 0,$$
$$Sq^3 m^* = Sq^1 g^* + Sq^1 Sq^1 n^* = 0.$$

So m defines an A_1-map $J \to N$, and m^* defines an A_1-map $J \to N^*$, or equivalently $N \to J^* \simeq J$. Also we have

$$\begin{aligned}\langle Sq^2 m^*, Sq^2 m\rangle &= \langle g^* + Sq^1 n^*, g + Sq^1 n\rangle \\ &= \langle g^*, g\rangle + \langle n^*, Sq^1 g\rangle + \langle Sq^1 g^*, n\rangle \\ &= 1.\end{aligned}$$

This shows that the composite $J \to N \to J$ is the identity; so N contains J as a direct summand. Let the complementary direct summand be P; then $H(P; Sq^1) = 0$,

$H(P; Sq^{01}) = 0$ and so P is free. Thus $N \simeq J$. This proves that every invertible class over A_1 is of the form $\Sigma^a I^b J^c$, and completes the proof of Theorem 3·7.

We now turn to the application of these results in computing groups $\mathrm{Ext}^s_B(L, M)$.

LEMMA 3·8. *If $s > 0$ and N is invertible, then*

$$\mathrm{Ext}^s_B(L, M) \cong S\,\mathrm{hom}_B(I^s \otimes L \otimes N, M \otimes N).$$

Proof. If we take the exact sequence

$$0 \to I \to B \to k \to 0$$

and tensor with L, we get an exact sequence

$$0 \to I \otimes L \to B \otimes L \to L \to 0$$

in which $B \otimes L$ is free. So by the classical exact sequence we get

$$\mathrm{Ext}^s_B(L, M) \cong \mathrm{Ext}^{s-1}_B(I \otimes L, M)$$

for $s > 1$, and by induction over s we get

$$\mathrm{Ext}^s_B(L, M) \cong \mathrm{Ext}^1_B(I^{s-1} \otimes L, M).$$

But now using Lemma 3·3 (*b*) we get

$$\mathrm{Ext}^1_B(I^{s-1} \otimes L, M) \cong S\,\mathrm{hom}_B(I^s \otimes L, M).$$

And if N is invertible we clearly have an isomorphism

$$S\,\mathrm{hom}_B(I^s \otimes L, M) \cong S\,\mathrm{hom}_B(I^s \otimes L \otimes N, M \otimes N).$$

This proves the lemma.

The effect of this lemma is that we can read off the groups $\mathrm{Ext}^s_B(L, M)$ for all invertible L and M from comparatively few tables. In the case $B = E[x, y]$ it is sufficient to tabulate

$$S\,\mathrm{hom}_B(I^s, \Sigma^t);$$

in the case $B = A_1$ it is sufficient to tabulate

$$S\,\mathrm{hom}_B(I^s, \Sigma^t) \quad \text{and} \quad S\,\mathrm{hom}_B(I^s J, \Sigma^t).$$

In the case $B = E[x, y]$ we label the table as if the degrees of x and y are 1 and $2p-1$, as in the applications.

Table 3·9. $S\,\mathrm{hom}_{E[x,y]}(I^s, \Sigma^t)$

	2	—	—	—	—	—	—	—	—	k	—	k	—	k	—
	1	—	—	—	—	—	—	—	—	k	—	k	—	—	—
s	0	—	—	—	—	—	—	—	—	k	—	—	—	—	—
	-1	—	—	—	—	—	k	—	—	—	—	—	—	—	—
	-2	—	—	—	k	—	k	—	—	—	—	—	—	—	—
	-3	—	k	—	k	—	k	—	—	—	—	—	—	—	—
			$-6p+5$		$-4p+3$		$-2p+1$			0		$2p-2$		$4p-4$	

$t-s$

Table 3·10. $S\,\mathrm{hom}_{A_1}(I^s, \Sigma^t)$

s \ $t-s$	−15	−14	−13	−12	−11	−10	−9	−8	−7	−6	−5	−4	−3	−2	−1	0	1	2	3	4	5	6	7	8	9	10	11	12
6	—	—	—	—	—	—	—	—	—	—	—	—	—	—	—	k	—	—	—	k	—	—	—	k	—	k	—	—
5	—	—	—	—	—	—	—	—	—	—	—	—	—	—	—	k	—	—	—	k	—	—	—	k	k	—	—	—
4	—	—	—	—	—	—	—	—	—	—	—	—	—	—	—	k	—	—	—	k	—	—	—	k	—	—	—	—
3	—	—	—	—	—	—	—	—	—	—	—	—	—	—	—	k	—	—	—	k	—	—	—	—	—	—	—	—
2	—	—	—	—	—	—	—	—	—	—	—	—	—	—	—	k	—	k	—	—	—	—	—	—	—	—	—	—
1	—	—	—	—	—	—	—	—	—	—	—	—	—	—	—	k	k	—	—	—	—	—	—	—	—	—	—	—
0	—	—	—	—	—	—	—	—	—	—	—	—	—	—	—	k	—	—	—	—	—	—	—	—	—	—	—	—
−1	—	—	—	—	—	—	—	—	—	—	k	—	—	—	—	—	—	—	—	—	—	—	—	—	—	—	—	—
−2	—	—	—	—	—	—	—	—	—	k	k	—	—	—	—	—	—	—	—	—	—	—	—	—	—	—	—	—
−3	—	—	—	—	—	—	—	—	k	—	k	—	—	—	—	—	—	—	—	—	—	—	—	—	—	—	—	—
−4	—	—	—	—	—	—	k	—	—	—	k	—	—	—	—	—	—	—	—	—	—	—	—	—	—	—	—	—
−5	—	—	k	—	—	—	k	—	—	—	k	—	—	—	—	—	—	—	—	—	—	—	—	—	—	—	—	—
−6	—	k	k	—	—	—	k	—	—	—	k	—	—	—	—	—	—	—	—	—	—	—	—	—	—	—	—	—

Table 3·11. $S\,\mathrm{hom}_{A_1}(I^sJ, \Sigma^t)$

s \ $t-s$	−15	−14	−13	−12	−11	−10	−9	−8	−7	−6	−5	−4	−3	−2	−1	0	1	2	3	4	5	6	7	8	9	10	11	12
6	—	—	—	—	—	—	—	—	—	—	—	—	—	—	—	k	—	—	—	k	—	—	—	k	—	—	—	k
5	—	—	—	—	—	—	—	—	—	—	—	—	—	—	—	k	—	—	—	k	—	—	—	k	—	—	—	—
4	—	—	—	—	—	—	—	—	—	—	—	—	—	—	—	k	—	—	—	k	—	k	—	—	—	—	—	—
3	—	—	—	—	—	—	—	—	—	—	—	—	—	—	—	k	—	—	—	k	k	—	—	—	—	—	—	—
2	—	—	—	—	—	—	—	—	—	—	—	—	—	—	—	k	—	—	—	k	—	—	—	—	—	—	—	—
1	—	—	—	—	—	—	—	—	—	—	—	—	—	—	—	k	—	—	—	—	—	—	—	—	—	—	—	—
0	—	—	—	—	—	—	—	—	—	—	—	—	—	k	—	—	—	—	—	—	—	—	—	—	—	—	—	—
−1	—	—	—	—	—	—	—	—	—	—	—	—	k	—	—	—	—	—	—	—	—	—	—	—	—	—	—	—
−2	—	—	—	—	—	—	—	—	—	—	k	—	—	—	—	—	—	—	—	—	—	—	—	—	—	—	—	—
−3	—	—	—	—	—	—	k	—	—	—	k	—	—	—	—	—	—	—	—	—	—	—	—	—	—	—	—	—
−4	—	—	—	—	—	k	k	—	—	—	k	—	—	—	—	—	—	—	—	—	—	—	—	—	—	—	—	—
−5	—	—	—	—	k	—	k	—	—	—	k	—	—	—	—	—	—	—	—	—	—	—	—	—	—	—	—	—
−6	—	—	k	—	—	—	k	—	—	—	k	—	—	—	—	—	—	—	—	—	—	—	—	—	—	—	—	—

In each table, groups not indicated are zero; in Tables 3·10 and 3·11, k means $\mathbf{F}_2$. In each quadrant of each table, the obvious periodicity continues.

These tables are the result of simple and obvious calculations.

The observations above become applicable because certain modules which arise in the applications can be written as sums of invertible modules. At this point we have to consider modules which are not finitely generated over B, so we relax that assumption. We can now form infinite sums of modules; infinite sums pass to stable classes, because an infinite sum of free modules is free.

PROPOSITION 3·12. *Let* $B = E[Q_0, Q_1]$, *as in* (3·1). *Then the stable class of*

$$A/(AQ_0 + AQ_1) \cong A \otimes_B \mathbf{F}_p$$

is

$$\prod_{r=0}^{\infty} (1 + K_r + K_r^2 + \ldots + K_r^{p-1}),$$

where K_r *is the invertible class* $\Sigma^{a(r)} I^{b(r)}$ *with*

$$a(r) + b(r) = 2(p-1)p^r,$$

$$b(r) = \frac{p^r - 1}{p-1}.$$

PROPOSITION 3·13. *Take* $p = 2$, *and let* $B = A_1$, *as in* (3·2). *Then the stable class of*

$$A/(ASq^1 + ASq^2) \cong A \otimes_B \mathbf{F}_2$$

is

$$(1 + \Sigma^3 IJ)(1 + \Sigma^5 I^3)(1 + \Sigma^9 I^7) \ldots (1 + \Sigma^{2^r+1} I^{2^r-1}) \ldots.$$

In each proposition the infinite product is to be interpreted by expanding it as an infinite sum; for example, in Proposition 3·13 it means

$$1 + \Sigma^3 IJ + \Sigma^5 I^3 + \Sigma^8 I^4 J + \Sigma^9 I^7 + \ldots.$$

Proposition 3·13 is a reformulation by the first author of a lemma of Mahowald.

Proof of Proposition 3·12. We propose to proceed in the dual, and calculate the stable class of $(A/(AQ_0 + AQ_1))^*$. According to our principles, we should give this dual space the structure of a left B-module, letting b act by the dual of the map $a \mapsto (cb)a$ on $A/(AQ_0 + AQ_1)$. However, by using the conjugation c of A we can throw $A/(AQ_0 + AQ_1)$ onto $A/(Q_0 A + Q_1 A)$, and use the map $a' \mapsto a'b$ on $A/(Q_0 A + Q_1 A)$. This is convenient for purposes of calculation. In fact, assuming p odd, the dual A^* of A is the tensor product of an exterior algebra $E[\tau_0, \tau_1, \tau_2, \ldots]$ and a polynomial algebra $\mathbf{F}_p[\xi_1, \xi_2, \xi_3, \ldots]$. The dual of the quotient $A/(Q_0 A + Q_1 A)$ is the subalgebra

$$E[\tau_2, \tau_3, \ldots] \oplus \mathbf{F}_p[\xi_1, \xi_2, \xi_3, \ldots].$$

We must calculate its homology for the boundaries obtained by dualizing the maps $a \mapsto aQ_0$ and $a \mapsto aQ_1$ of $A/(Q_0 A + Q_1 A)$.

Under the first boundary our complex may be expressed as a tensor product of chain complexes. Here the first factor is the polynomial algebra $\mathbf{F}_p[\xi_1]$, with the zero boundary; while the rth factor for $r \geq 2$ has a base of monomials

$$\tau_r \xi_r^i \quad \text{and} \quad \xi_r^i \quad (i = 0, 1, 2, \ldots)$$

with the boundary

$$\tau_r \xi_r^i \to \xi_r^{i+1}, \quad \xi_r^i \mapsto 0.$$

Thus for $r \geqslant 2$ the homology of the rth factor is $\mathbf{F}_p$, generated by 1 in degree 0. We conclude that the homology of the tensor product is a polynomial algebra on one generator ξ_1.

Under the second boundary our complex may also be expressed as a tensor product of chain complexes. This time the rth factor has a base of monomials $\tau_{r+1}\xi_r^i$ and ξ_r^i $(i = 0, 1, 2, \ldots)$ with the boundary

$$\tau_{r+1}\xi_r^i \mapsto \xi_r^{i+p}, \quad \xi_r^i \mapsto 0.$$

We conclude that the homology of the tensor product is a truncated algebra, given by generators $\xi_1, \xi_2, \ldots$ and relations $\xi_1^p = 0$, $\xi_2^p = 0, \ldots$.

We now seek appropriate B-modules to map into $A/(Q_0A + Q_1A)^*$, so that eventually we may obtain a map inducing isomorphisms of $H(\ ;Q_0)$ and $H(\ ;Q_1)$, and so apply the theorem of Adams and Margolis to deduce that we have a stable equivalence of B-modules.

We take L_0 to be the submodule generated by ξ_1. We take L_1 to be the following submodule.

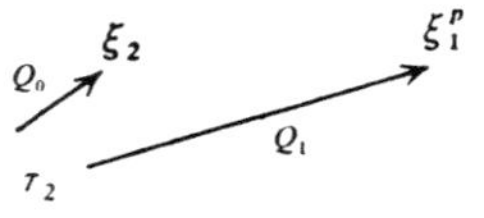

Suppose now, as an inductive hypothesis, that we have constructed L_r and mapped it to $(A/(Q_0A + Q_1A))^*$, so that L_r contains a submodule L_r' which maps isomorphically to the B-submodule which has the k-base ξ_{r+1}. Then using the algebra structure of $(A/(Q_0A + Q_1A))^*$, we can map the tensor power $(L_r)^p$ to $(A/(Q_0A + Q_1A))^*$ so that the submodule $(L_r')^p$ maps isomorphically to the B-submodule which has the k-base ξ_{r+1}^p. We now construct L_{r+1} by adjoining to $(L_r)^p$ the following submodule.

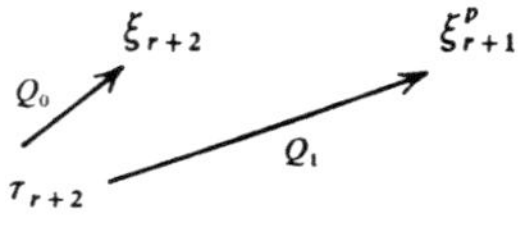

(Of course we identify $(L_r')^p$ with the part of this new submodule which has the k-base ξ_{r+1}^p.) We take L_{r+1}' to be the part of the new submodule which has the k-base ξ_{r+2}. This completes the induction.

The result of this induction is that $H(L_r; Q_0)$ has dimension 1 over $\mathbf{F}_p$, and its generator maps to the homology class of $\xi_1^{p^r}$, while $H(L_r; Q_1)$ has dimension 1 over $\mathbf{F}_p$, and its generator maps to the homology class of ξ_{r+1}.

By using the algebra structure of $(A/(Q_0A + Q_1A))^*$, we can now construct a map

$$\prod_{r=0}^{\infty} (1 + L_r + L_r^2 + \ldots + L_r^{p-1}) \to (A/(Q_0A + Q_1A))^*;$$

this map induces an isomorphism of $H(\ ;Q_0)$ and $H(\ ;Q_1)$, and is therefore a stable equivalence by the theorem of Adams and Margolis. Moreover, as we have said, $(A/(Q_0A + Q_1A))^*$ is isomorphic to $(A/(AQ_0 + AQ_1))^*$ with its correct structure as a B-module.

Now, the infinite product

$$\prod_{r=0}^{\infty} (1+L_r+L_r^2+\ldots+L_r^{p-1})$$

represents an infinite sum which is locally finite, and duality commutes with locally finite sums. Moreover, the modules L_r are invertible by Lemma 3·5; so all the summands in the infinite sum are invertible; but for an invertible module M we have $M^* \simeq M^{-1}$. Therefore we can write the dual formula in the form

$$A/AQ_0+AQ_1 \simeq \prod_{r=0}^{\infty} (1+L_r^{-1}+L_r^{-2}+\ldots+L_r^{-p+1}),$$

or as

$$A/AQ_0+AQ_1 \simeq \prod_{r=0}^{\infty} (1+K_r+K_r^2+\ldots+K_r^{p-1})$$

if we set $K_r = L_r^{-1}$. Finally, by calculating the invariant given in the proof of Theorem 3·6, we see that $K_r = L_r^{-1} \simeq \Sigma^{a(r)}I^{b(r)}$ where

$$a(r)+b(r) = 2(p-1)p^r,$$
$$b(r) = \frac{p^r-1}{p-1}.$$

The proof is valid for $p = 2$ if suitably interpreted (interpret τ_i as ζ_{i+1} and ξ_i as ζ_i^2). Alternatively, the result for $p = 2$ is given in (2), p. 335, line 1. This completes the proof of Proposition 3·12.

Proof of Proposition 3·13. This is wholly parallel to the proof of Proposition 3·12. Again, we proceed in the dual and calculate the stable class of $(A/(ASq^1+ASq^2))^*$. We use the conjugation map c to throw $A/(ASq^1+ASq^2)$ on $A/(Sq^1A+Sq^2A)$. The dual A^* of A is the polynomial algebra $\mathbf{F}_2[\zeta_1, \zeta_2, \ldots, \zeta_n, \ldots]$, and the dual of the quotient $A/(Sq^1A+Sq^2A)$ is the subalgebra $\mathbf{F}_2[\zeta_1^4, \zeta_2^2, \zeta_3, \zeta_4, \ldots, \zeta_n, \ldots]$. We must calculate its homology for the boundary obtained by dualizing the maps $a \mapsto aSq^1$ and $a \mapsto aSq^{01}$ of $A/(Sq^1A+Sq^2A)$.

Under the first boundary we get a tensor product of chain complexes, where the first factor is the polynomial algebra $\mathbf{F}_2[\zeta_1^4]$ with the zero boundary, and the rth factor for $r \geqslant 2$ has a base of monomials ζ_r^{2i} and $\zeta_r^{2i}\zeta_{r+1}$ $(i = 0, 1, 2, \ldots)$ with the boundary

$$\zeta_r^{2i}\zeta_{r+1} \mapsto \zeta_r^{2i+2}, \quad \zeta_r^{2i} \mapsto 0.$$

We conclude that the homology of the tensor product is a polynomial algebra on one generator ζ_1^4.

Under the second boundary our complex may also be expressed as a tensor product of chain complexes. This time the first factor has a base of monomials ζ_1^{4i} and $\zeta_1^{4i}\zeta_3$ $(i = 0, 1, 2, \ldots)$ with boundary

$$\zeta_1^{4i}\zeta_3 \mapsto \zeta_1^{4i+4}, \quad \zeta_1^{4i} \mapsto 0;$$

the rth factor for $r \geqslant 2$ has a base of monomials ζ_r^{2i} and $\zeta_r^{2i}\zeta_{r+2}$ with boundary

$$\zeta_r^{2i}\zeta_{r+2} \mapsto \zeta_r^{2i+4}, \quad \zeta_r^{2i} \mapsto 0.$$

We conclude that the homology of our tensor product is an exterior algebra on generators $\zeta_2^2, \zeta_3^2, \zeta_4^2, \ldots$.

We now (again) seek appropriate B-modules to map into $(A/(Sq^1A+Sq^2A))^*$. We take L_1 to be the following submodule.

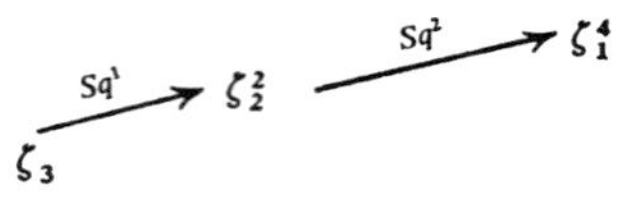

Suppose, as an inductive hypothesis, that we have constructed L_r and mapped it to $(A/(Sq^1A+Sq^2A))^*$ so that L_r contains a submodule L_r' which maps isomorphically to the following submodule.

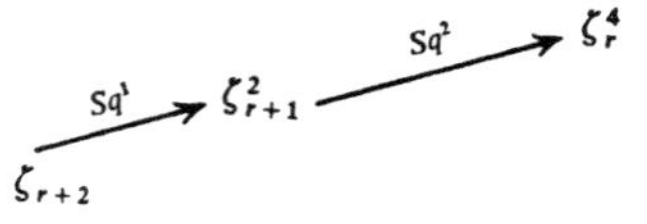

Then using the algebra structure of $(A/(Sq^1A+Sq^2(A))^*$, we can map the tensor square $(L_r)^2$ to $(A/(Sq^1A+Sq^2A))^*$. Now $(L_r')^2$ does not map isomorphically to its image; the kernel K is a submodule which (with an obvious notation) we may display as follows.

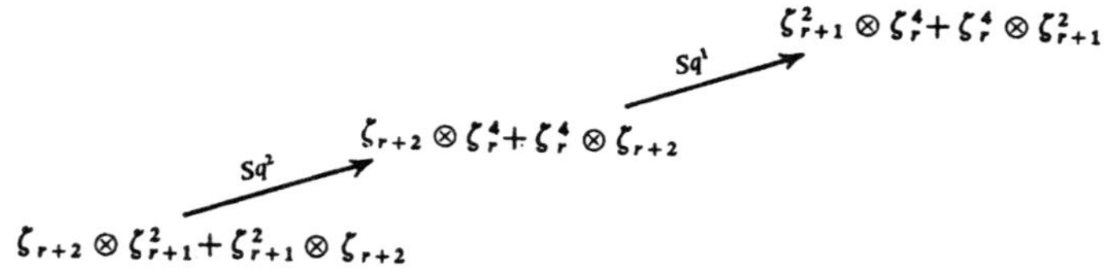

Let us form the quotient $(L_r)^2/K$; this quotient maps to $(A/(Sq^1A+Sq^2A))^*$, and has the following submodule M.

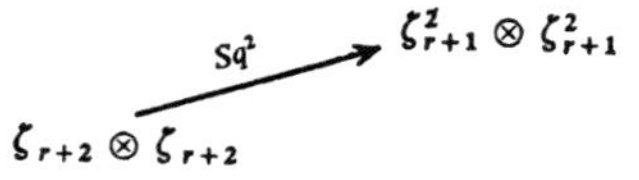

We now construct L_{r+1} by adjoining to $(L_r)^2/K$ the following submodule.

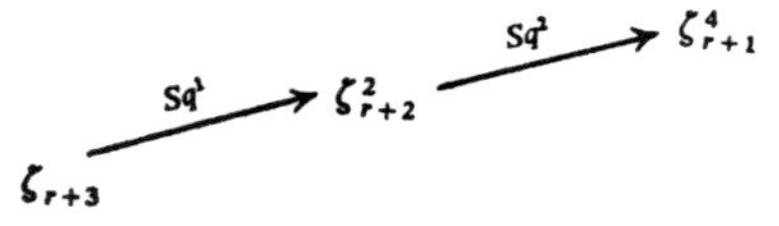

(Of course we identify M with the part of the new submodule which has the k-base $\zeta_{r+2}^2, \zeta_{r+1}^4$.) We take L_{r+1}' to be the new submodule we adjoined. This completes the induction.

The result of this induction is that $H(L_r; Sq^1)$ has dimension 1 over $\mathbf{F}_2$, and its generator maps to the homology class of $\zeta_1^{2^{r+1}}$, while $H(L_r; Sq^{01})$ has dimension 1 over $\mathbf{F}_2$, and its generator maps to the homology class of ζ_{r+1}^2.

By using the algebra structure of $(A/(Sq^1A+Sq^2A))^*$, as before, we can construct a map

$$(1+L_1)(1+L_2)(1+L_3)\ldots \to (A/(Sq^1A+Sq^2A))^*;$$

this map induces an isomorphism of $H(\ ;Sq^1)$ and $H(\ ;Sq^{01})$, and is therefore a stable equivalence by the theorem of Adams and Margolis. Dualizing exactly as in the proof of Theorem 3·12, we find

$$A/(ASq^1+ASq^2) \simeq (1+K_1)(1+K_2)(1+K_3)\ldots,$$

where $K_r = L_r^{-1}$. Finally, by calculating the invariants given in the proof of Theorem 3·7, we see that

$$K_1 = L_1^{-1} \simeq \Sigma^3 IJ$$

while

$$K_r = L_r^{-1} \simeq \Sigma^{2^r+1} I^{2^r-1} \quad \text{for} \quad r \geqslant 2.$$

This completes the proof of Proposition 3·13.

4. *Calculation of* Ext. In this section we will prove the following two results.

PROPOSITION 4·1. $\mathrm{Ext}_A^{s,t}(A/AQ_0+AQ_1,\ A/AQ_0+AQ_1) = 0$ *provided* $s>0$, $t-s$ *is odd and* $t-s > -2p+1$.

PROPOSITION 4·2. *Assume that* $p=2$, *and* M *is one of the following four* A-*modules:*

$$A/(ASq^1+ASq^2),\quad A/ASq^2,\quad A/ASq^3,\quad A/(ASq^1+ASq^2Sq^3).$$

Then
$$\mathrm{Ext}_A^{s,t}(M,M) = 0 \quad for \quad s>0,\quad t-s=-1.$$

Proof of Proposition 4·1. Let $B = E[Q_0,Q_1]$, as in (3·1). Then A is free as a right B-module, and we have

$$A/(AQ_0+AQ_1) \cong A\otimes_B \mathbf{F}_p.$$

By a standard change-of-rings theorem, we have

$$\mathrm{Ext}_A^{s,t}(A\otimes_B\mathbf{F}_p, M) \cong \mathrm{Ext}_B^{s,t}(\mathbf{F}_p, M).$$

If we take $M = A\otimes_B\mathbf{F}_p$, then by Proposition 3·12 it is stably equivalent to a sum of modules $\Sigma^a I^b$ with $a+b$ even and $a+b\geqslant 0$. By Lemma 3·8, we have

$$\mathrm{Ext}_B^{s,t}(\mathbf{F}_p, \Sigma^a I^b) \simeq S\,\mathrm{hom}_B(I^{s-b}, \Sigma^{t+a}).$$

By Table 3·9 (or by the calculations implicit in it), this is zero if $t+a-s+b$ is odd and greater than $-2p+1$. This proves Proposition 4·1.

The reader may notice how little use we have made of the precise details in Proposition 3·12, and may perhaps wonder if we could not rearrange the proof so as to omit much of section 3. So far as Proposition 4·1 goes this would be possible; in order to prove Proposition 4·1, it is sufficient to know merely the following about

$$M = A/AQ_0+AQ_1;$$

M is bounded below, $H(M;Q_0)=0$ in even negative degrees and $H(M;Q_1)=0$ in odd degrees. Unfortunately, similar remarks do not apply to Proposition 4·2. It can be shown by counter-examples that to prove Proposition 4·2 we need to know, at least, that M is stably equivalent to a sum of modules each finitely generated over B. For this we have to use the proof of Proposition 3·13; and if we have to use the proof, it seems unnecessarily obscure not to state what the proof proves.

We next prove Proposition 4·2, so far as it concerns the module

$$M = A/(ASq^1 + ASq^2).$$

This is parallel to the proof of Proposition 4·1. Let B be A_1, as in (3·2). Then A is free as a right module over B, and we have

$$A/(ASq^1 + ASq^2) \cong A \otimes_B \mathbf{F}_2.$$

By the same change-of-rings theorem, we have

$$\mathrm{Ext}_A^{s,t}(A \otimes_B \mathbf{F}_2, M) \cong \mathrm{Ext}_B^{s,t}(\mathbf{F}_2, M).$$

If we take $M = A \otimes_B \mathbf{F}_2$, then by Proposition 3·13 it is stably equivalent to a sum of modules $\Sigma^a I^b J^c$ with $a+b \equiv 0 \bmod 4$ and $a+b \geqslant 0$. By Lemma 3·8, we have

$$\mathrm{Ext}_B^{s,t}(\mathbf{F}_2, \Sigma^a I^b J^c) \cong S\,\mathrm{hom}_B(I^{s-b}J^{-c}, \Sigma^{t+a}).$$

By Tables 3·10, 3·11 (or by the calculations implicit in them) this is zero if

$$t+a-s+b \equiv -1 \bmod 4 \quad \text{and} \quad t+a-s+b \geqslant -1.$$

This proves that

$$\mathrm{Ext}_A^{s,t}(A/ASq^1 + ASq^2, A/ASq^1 + ASq^2) = 0$$

for $s > 0$, $t-s = -1$.

We next wish to deduce from this result the other three cases of Proposition 4·2. For this we use algebraic arguments analogous to the topological arguments one would use if one wished to prove that (with the notation of section 1) $[\mathbf{Y}, \mathbf{Y}]$ is essentially independent of d. First we need a subsidiary result.

LEMMA 4·3.

(i) $\mathrm{Ext}_A^{s,t}(A/ASq^1 + ASq^2, A/ASq^1) = 0$ *for* $s > 0, t-s > -5$.

(ii) $\mathrm{Ext}_A^{s,t}(A/ASq^1, \Sigma^2 A/ASq^2) = 0$ *for* $s > 0, t-s > -5$.

Proof. (i) As before, the usual change-of-rings theorem gives

$$\mathrm{Ext}_A^{s,t}(A/ASq^1 + ASq^2, A/ASq^1) \cong \mathrm{Ext}_B^{s,t}(\mathbf{F}_2, A/ASq^1).$$

Now of course we have an exact sequence

$$0 \to A/ASq^1 \to \Sigma^{-1}A \to \Sigma^{-1}A/ASq^1 \to 0$$

in which $\Sigma^{-1}A$ is free over B, and therefore injective. Proceeding dually to the proof of Lemma 3·8, and in particular applying Lemma 3·3(*c*), we get

$$\mathrm{Ext}_B^{s,t}(\mathbf{F}_2, A/ASq^1) \cong S\,\mathrm{hom}_B(\mathbf{F}_2, \Sigma^{t-s}A/ASq^1).$$

But by direct calculation,

$$S\,\mathrm{hom}_B(\Sigma^u, A/ASq^1) = 0$$

for $u < 5$. This proves part (i).

(ii) Let A_0 be the subalgebra of A generated by Sq^1; its augmentation ideal I_0 is Σ. The usual change-of-rings theorem gives

$$\mathrm{Ext}_A^{s,t}(A/ASq^1, \Sigma^2 A/ASq^2) \cong \mathrm{Ext}_{A_0}^{s,t}(\mathbf{F}_2, \Sigma^2 A/ASq^2).$$

Now a trivial analogue of Lemma 3·8 shows that

$$\mathrm{Ext}_A^{s,t}(\mathbf{F}_2, \Sigma^2 A/ASq^2) \cong S\,\mathrm{hom}_{A_0}(\mathbf{F}_2, \Sigma^{t-s+2} A/ASq^2).$$

But by direct calculation,

$$S\,\mathrm{hom}_{A_0}(\Sigma^u, A/ASq^2) = 0$$

for $u < 3$. This proves the lemma.

We now use the following exact sequence, which arises in the mod 2 cohomology of the Postnikov system for **bo**, as in section 2.

$$0 \to \Sigma^2 A/ASq^2 \to A/ASq^1 \to A/ASq^1 + ASq^2 \to 0.$$

From this, we get the following two exact sequences.

$$\mathrm{Ext}_A^{s,t}\left(\frac{A}{ASq^1+ASq^2}, \frac{A}{ASq^1+ASq^2}\right) \to \mathrm{Ext}_A^{s+1,t}\left(\frac{A}{ASq^1+ASq^2}, \Sigma^2\frac{A}{ASq^2}\right)$$
$$\to \mathrm{Ext}_A^{s+1,t}\left(\frac{A}{ASq^1+ASq^2}, \frac{A}{ASq^1}\right),$$

$$\mathrm{Ext}_A^{s,t}\left(\frac{A}{ASq^1}, \Sigma^2\frac{A}{ASq^2}\right) \to \mathrm{Ext}_A^{s,t}\left(\Sigma^2\frac{A}{ASq^2}, \Sigma^2\frac{A}{ASq^2}\right)$$
$$\to \mathrm{Ext}_A^{s+1,t}\left(\frac{A}{ASq^1+ASq^2}, \Sigma^2\frac{A}{ASq^2}\right).$$

In the first exact sequence, using our previous result on the left-hand group and Lemma 4·3 (i) on the right-hand group, we see that the middle group is zero for $s > 0, t-s = -1$. Now the second exact sequence, using Lemma 4·3 (ii) on the left-hand group, shows that

$$\mathrm{Ext}_A^{s,t}(\Sigma^2 A/ASq^2, \Sigma^2 A/ASq^2) = 0$$

for $s > 0, t-s = -1$.

Now we use the following exact sequence.

$$0 \to \Sigma^4 A/ASq^3 \to \Sigma^2 A \to \Sigma^2 A/ASq^2 \to 0.$$

Since $\Sigma^2 A$ is both projective and injective, this shows that for $s > 0$ we have

$$\mathrm{Ext}_A^{s,t}\left(\Sigma^2\frac{A}{ASq^2}, \Sigma^2\frac{A}{ASq^2}\right) \cong \mathrm{Ext}_A^{s+1,t}\left(\Sigma^2\frac{A}{ASq^2}, \Sigma^4\frac{A}{ASq^3}\right)$$
$$\cong \mathrm{Ext}_A^{s,t}\left(\Sigma^4\frac{A}{ASq^3}, \Sigma^4\frac{A}{ASq^3}\right).$$

This group is therefore zero for $s > 0, t-s = -1$.

Finally we use the exact sequence

$$0 \to \Sigma^7\frac{A}{ASq^1+ASq^2Sq^3} \to \Sigma^4 A \to \Sigma^4\frac{A}{ASq^3} \to 0,$$

and this shows similarly that

$$\mathrm{Ext}_A^{s,t}\left(\Sigma^4\frac{A}{ASq^3}, \Sigma^4\frac{A}{ASq^3}\right)$$
$$\cong \mathrm{Ext}_A^{s,t}\left(\Sigma^7\frac{A}{ASq^1+ASq^2Sq^3}, \Sigma^7\frac{A}{ASq^1+ASq^2Sq^3}\right)$$

for $s > 0$. This completes the proof of Proposition 4·2.

Similar arguments work for the exact sequence

$$0 \to \Sigma^{12}\frac{A}{ASq^1+ASq^2} \to \Sigma^7\frac{A}{ASq^1} \to \Sigma^7\frac{A}{ASq^1+ASq^2Sq^3} \to 0,$$

but for our purposes this is not necessary.

5. *Proof of Theorem* 1·1. In this section we will prove Theorem 1·1; so we assume that **X** and **Y** are as in Theorem 1·1. To deal with the difficulties over the convergence of the Adams spectral sequence, our plan is to approximate **X** by finite spectra $\mathbf{W}^n$, and construct maps $f^n\colon \mathbf{W}^n \to \mathbf{Y}$. In order to make the maps f^n compatible as we vary n, we plan to use the known endomorphisms of the fixed spectrum **Y**. We take these points in order.

PROPOSITION 5·1. *Let* **X** *be a spectrum whose homotopy groups are finitely generated modules over* Λ, *and bounded below. Then* X *is equivalent to a smash-product* $\mathbf{M}\Lambda \wedge \mathbf{W}$, *where* $\mathbf{M}\Lambda$ *is a Moore spectrum for the group* Λ, *and* **W** *is a spectrum whose skeletons* $\mathbf{W}^n$ *are all finite.*

Broadly speaking, this result says that any p-local spectrum **X** is the localization of a global spectrum **W**, and any p-complete spectrum **X** is the completion of a global spectrum **W**. We should perhaps only sketch the proof, since the result should appear in any complete treatment of localization and completion (compare (9, 13)). But for completeness we give at least a sketch.

It clearly follows from Proposition 5·1 that if **X** is as assumed there, then **X** is a module spectrum over the ring spectrum $\mathbf{M}\Lambda$. It is convenient to begin by proving at least part of this.

As in section 1, we define the map $\mathbf{X} \to \mathbf{M}\Lambda \wedge \mathbf{X}$ to be the composite

$$\mathbf{X} \simeq \mathbf{S}^0 \wedge \mathbf{X} \simeq \mathbf{MZ} \wedge \mathbf{X} \to \mathbf{M}\Lambda \wedge \mathbf{X},$$

where the last map is induced by the injection $\mathbf{Z} \to \Lambda$.

LEMMA 5·2. (i) *Let* **X** *be a spectrum whose homotopy groups are finitely generated modules over* Λ, *and bounded below. Then there is a map* $\nu\colon \mathbf{M}\Lambda \wedge \mathbf{X} \to \mathbf{X}$ *such that the composite*

$$\mathbf{X} \to \mathbf{M}\Lambda \wedge \mathbf{X} \xrightarrow{\nu} \mathbf{X}$$

is homotopic to the identity.

(ii) *Moreover, the induced map of homotopy groups*

$$\nu_*\colon \Lambda \otimes \pi_r(\mathbf{X}) \to \pi_r(\mathbf{X})$$

is the Λ*-module action map.*

Much more is true, but we state just what is needed for the proof of Proposition 5·1.

It is convenient to continue with subsidiary results. For example, part (ii) of Lemma 5·2 will follow immediately from part (i) by using the following result.

LEMMA 5·3. *Let A be a Λ-module, and B a finitely generated Λ-module; then any homomorphism $\theta: A \to B$ of abelian groups is a homomorphism of Λ-modules.*

Proof. If $\Lambda = \mathbf{Z}_{(p)}$ this follows by trivial algebra; if $\Lambda = \mathbf{Z}_p$ it follows because θ is continuous for the p-adic topology.

LEMMA 5·4. *Let V be a vector-space over the field $\mathbf{Q}$ of rational numbers; and let A be an abelian group which is complete and Hausdorff for the p-adic topology (e.g. a finitely generated module over $\mathbf{Z}_p$). Then*

$$\mathrm{Hom}_{\mathbf{Z}}(V, A) = 0, \quad \mathrm{Ext}_{\mathbf{Z}}(V, A) = 0.$$

Proof. The assertion about Hom is trivial; and since V is a direct sum of copies of $\mathbf{Q}$, it is sufficient to prove $\mathrm{Ext}_{\mathbf{Z}}(\mathbf{Q}, A) = 0$. We construct a $\mathbf{Z}$-free resolution of $\mathbf{Q}$

$$0 \to C_1 \xrightarrow{d} C_0 \xrightarrow{\epsilon} \mathbf{Q} \to 0$$

as follows. Let C_1, C_0 be $\mathbf{Z}$-free on bases $\{b_i\}, \{c_i\}$, $i \geqslant 1$; define d, ϵ by

$$d(b_i) = c_i - (i+1)\, c_{i+1},$$
$$\epsilon(c_i) = 1/i\,!.$$

Applying $\mathrm{Hom}_{\mathbf{Z}}(\ , A)$ we get an exact sequence

$$0 \leftarrow \mathrm{Ext}_{\mathbf{Z}}(\mathbf{Q}, A) \leftarrow \prod_{i=1}^{\infty} A \xleftarrow{\delta} \prod_{i=1}^{\infty} A,$$

where
$$\delta\{a_i\} = \{a_i - (i+1)\, a_{i+1}\}.$$

Our problem, then, is to take a vector $\{\alpha_i\} \in \prod_{i=1}^{\infty} A$ and solve the equations

$$\alpha_i = a_i - (i+1)\, a_{i+1} \quad (i = 1, 2, 3, \ldots).$$

The solution is
$$a_i = \sum_{j \geqslant i} \frac{j!}{i!} \alpha_j;$$

the series converges to a unique limit in A because we assume the p-adic topology on A is complete and Hausdorff. This proves the lemma.

Proof of Lemma 5·2. The homotopy groups of $\mathbf{X}$ are in any case modules over $\mathbf{Z}_{(p)}$; so the map $\mathbf{X} \to M\mathbf{Z}_{(p)} \wedge \mathbf{X}$ is an equivalence, because the induced map of homotopy maps

$$\pi_r(\mathbf{X}) \to \mathbf{Z}_{(p)} \otimes \pi_r(\mathbf{X})$$

is iso. Thus the map $\mathbf{X} \to M\mathbf{Z}_{(p)} \wedge \mathbf{X}$ has an inverse $M\mathbf{Z}_{(p)} \wedge \mathbf{X} \to \mathbf{X}$, unique up to homotopy.

In the p-complete case $\Lambda = \mathbf{Z}_p$, we note that the quotient $\mathbf{Z}_p/\mathbf{Z}_{(p)}$ is a vector-space $\mathbf{Q}$; therefore $H_*(M\mathbf{Z}_p \wedge \mathbf{X}, M\mathbf{Z}_{(p)} \wedge \mathbf{X})$ is a vector-space over $\mathbf{Q}$. In view of Lemma 5·4, the universal coefficient theorem yields

$$H^*(M\mathbf{Z}_p \wedge \mathbf{X}, M\mathbf{Z}_{(p)} \wedge \mathbf{X}; \pi_*(\mathbf{X})) = 0.$$

Now obstruction theory shows that there is a unique map $\mathbf{MZ}_p \wedge \mathbf{X} \to \mathbf{X}$ extending the map $\mathbf{MZ}_{(p)} \wedge \mathbf{X} \to \mathbf{X}$ obtained above.

This proves part (i) of Lemma 5·2, and part (ii) follows as remarked above.

Proof of Proposition 5·1. Alas, we have to divide cases. First we consider the p-local case $\Lambda = \mathbf{Z}_{(p)}$. We proceed by induction over n. Suppose constructed a finite n-skeleton $\mathbf{W}^n$ and a map $e^n\colon \mathbf{W}^n \to \mathbf{X}$ such that the homomorphism

$$\Lambda \otimes \pi_r(\mathbf{W}^n) \xrightarrow{1 \otimes e^n_*} \Lambda \otimes \pi_r(\mathbf{X}) \to \pi_r(\mathbf{X})$$

is iso for $r < n$ and epi for $r = n$. Consider the homomorphism

$$e^n_*\colon \pi_n(\mathbf{W}^n) \to \pi_n(\mathbf{X}).$$

Its kernel is a finitely generated module over $\mathbf{Z}$, and therefore we can find a finite number of maps $f_i\colon \mathbf{S}^n \to \mathbf{W}^n$ which generate it. Construct $\mathbf{V}$ by attaching to $\mathbf{W}^n$ stable $(n+1)$-cells $\mathbf{E}^{n+1}_i$, using the maps f_i as attaching maps; since the composites $e^n f_i\colon \mathbf{S}^n \to \mathbf{X}$ are null homotopic, we can extend the map $e^n\colon \mathbf{W}^n \to \mathbf{X}$ over $\mathbf{V}$. The induced homomorphism

$$\pi_n(\mathbf{V}) \to \pi_n(\mathbf{X})$$

is now monomorphic. Since localization preserves exactness, we infer that

$$\Lambda \otimes \pi_n(\mathbf{V}) \to \Lambda \otimes \pi_n(\mathbf{X}) \xrightarrow{\simeq} \pi_n(\mathbf{X})$$

is also mono. (At this point there seems to be no such simple argument for the p-complete case.)

We now take a finite number of maps $g_j\colon \mathbf{S}^{n+1} \to \mathbf{X}$ which generate $\pi_{n+1}(\mathbf{X})$ as a Λ-module. We construct

$$\mathbf{W}^{n+1} = \mathbf{V} \vee \bigvee_j \mathbf{S}^{n+1}_j,$$

and we extend the map $\mathbf{V} \to \mathbf{X}$ over $\mathbf{W}^{n+1}$ by using the map g_j on $\mathbf{S}^{n+1}_j$. This ensures that the homomorphism

$$\Lambda \otimes \pi_{n+1}(\mathbf{W}^{n+1}) \to \pi_{n+1}(\mathbf{X})$$

is epi. This completes the induction, and constructs a spectrum $\mathbf{W}$ with finite skeletons $\mathbf{W}^n$ and a map $e\colon \mathbf{W} \to \mathbf{X}$ such that

$$\Lambda \otimes \pi_r(\mathbf{W}) \xrightarrow{1 \otimes e_*} \Lambda \otimes \pi_r(\mathbf{X}) \to \pi_r(\mathbf{X})$$

is iso for all r. Let $\nu\colon \mathbf{M}\Lambda \wedge \mathbf{X} \to \mathbf{X}$ be as in Lemma 5·2; then the composite

$$\mathbf{M}\Lambda \wedge \mathbf{W} \xrightarrow{1 \wedge e} \mathbf{M}\Lambda \wedge \mathbf{X} \xrightarrow{\nu} \mathbf{X}$$

induces on homology groups the isomorphism just mentioned, and so is an equivalence. This completes the p-local case.

Secondly we consider the p-complete case $\Lambda = \mathbf{Z}_p$. The homotopy groups $\pi_r(\mathbf{X})$ are finitely generated modules over $\mathbf{Z}_p$. We can find subgroups $\sigma_r \subset \pi_r(\mathbf{X})$ which are finitely generated modules over $\mathbf{Z}_{(p)}$ and such that the composite

$$\mathbf{Z}_p \otimes \sigma_r \xrightarrow{1 \otimes i} \mathbf{Z}_p \otimes \pi_r(\mathbf{X}) \to \pi_r(\mathbf{X})$$

is iso for each r. (Take the whole of the p-torsion subgroup of $\pi_r(\mathbf{X})$, and one summand $\mathbf{Z}_{(p)}$ for each summand $\mathbf{Z}_p$ in $\pi_r(\mathbf{X})$.) The quotient $\pi_r(\mathbf{X})/\sigma_r$ is a direct sum of copies of $\mathbf{Z}_p/\mathbf{Z}_{(p)}$, and is a vector space over $\mathbf{Q}$. We can find a generalized Eilenberg–MacLane spectrum $\mathbf{E}$ such that $\pi_r(\mathbf{E}) \cong \pi_r(\mathbf{X})/\sigma_r$ for each r, and we can find a map $f\colon \mathbf{X} \to \mathbf{E}$ such that f induces on homotopy groups the projection $\pi_r(\mathbf{X}) \to \pi_r(\mathbf{X})/\sigma_r$. Let $\mathbf{F}$ be the fibre of f; then the injection $i\colon \mathbf{F} \to \mathbf{X}$ induces on homotopy groups the injection $\sigma_r \to \pi_r(\mathbf{X})$.

Alternatively, instead of using a fibering to realize the exact sequence

$$0 \to \sigma_r \to \pi_r(\mathbf{X}) \to \pi_r(\mathbf{X})/\sigma_r \to 0,$$

it is slightly more in line with general theory to use a pullback diagram to realize the following Cartesian square.

$$\begin{array}{ccc} \sigma_r = \mathbf{Z}_{(p)} \otimes \sigma_r & \to & \mathbf{Z}_p \otimes \sigma_r \cong \pi_r(\mathbf{X}) \\ \downarrow & & \downarrow \\ \mathbf{Q} \otimes \sigma_r & \to & \mathbf{Q}_p \otimes \sigma_r \end{array}$$

(Here $\mathbf{Q}_p$ is the field of p-adic numbers.) But really it makes no difference.

In any case, let $\nu\colon \mathbf{MZ}_p \wedge \mathbf{X} \to \mathbf{X}$ be as in Lemma 5·2; then the composite

$$\mathbf{MZ}_p \wedge \mathbf{F} \xrightarrow{1 \wedge i} \mathbf{MZ}_p \wedge \mathbf{X} \xrightarrow{\nu} \mathbf{X}$$

induces an isomorphism of homotopy groups, and so is an equivalence.

By the p-local case, which we have already established, we can find a spectrum $\mathbf{W}$ with finite skeletons such that $\mathbf{F} \simeq \mathbf{MZ}_{(p)} \wedge \mathbf{W}$. Then we have

$$\mathbf{X} = \mathbf{MZ}_p \wedge \mathbf{F} \simeq \mathbf{MZ}_p \wedge \mathbf{MZ}_{(p)} \wedge \mathbf{W}.$$

But since $\mathbf{Z}_p \otimes \mathbf{Z}_{(p)} \cong \mathbf{Z}_p$, $\mathbf{MZ}_p \wedge \mathbf{MZ}_{(p)}$ is again a Moore spectrum $\mathbf{MZ}_p$. This completes the proof.

One benefit which we obtain from having $\mathbf{W}^n$ finite is the following result, which is more or less standard.

LEMMA 5·5. *Let $\mathbf{W}^n$ be a finite spectrum and $\mathbf{Y}$ a spectrum whose homotopy groups $\pi_r(\mathbf{Y})$ are finitely generated modules over Λ. Consider maps $f\colon \mathbf{W}^n \to \mathbf{Y}$ such that*

$$f_*\colon \pi_r(\mathbf{W}^n) \otimes \mathbf{Q} \to \pi_r(\mathbf{Y}) \otimes \mathbf{Q}$$

is zero for all r. Then such maps fall into finitely many homotopy classes.

Proof. Let $\mathbf{Y}_{\mathbf{Q}}$ be the rationalization of $\mathbf{Y}$. Since $\mathbf{Y}_{\mathbf{Q}}$ is a generalized Eilenberg–MacLane spectrum, the hypothesis

$$f_* = 0\colon \pi_r(\mathbf{W}^n) \otimes \mathbf{Q} \to \pi_r(\mathbf{Y}) \otimes \mathbf{Q} \quad \text{for all } r$$

implies that the composite $\mathbf{W}^n \xrightarrow{f} \mathbf{Y} \to \mathbf{Y}_{\mathbf{Q}}$ is nullhomotopic. Now the obvious map

$$[\mathbf{W}^n, \mathbf{Y}] \otimes \mathbf{Q} \to [\mathbf{W}^n, \mathbf{Y}_{\mathbf{Q}}]$$

is iso when $\mathbf{W}^n$ is finite ((2), Proposition 6·7, p. 202). Therefore such maps f lie in the torsion subgroup of $[\mathbf{W}^n, \mathbf{Y}]$. But under the given hypotheses $[\mathbf{W}^n, \mathbf{Y}]$ is a finitely generated Λ-module, so its torsion subgroup is finite. This proves the lemma.

We will now discuss the endomorphisms of the fixed spectrum $\mathbf{Y}$.

LEMMA 5·6. *Let m be an integer for which $\pi_m(\mathbf{Y}) \simeq \Lambda$. Then there is a map $\phi(m): \mathbf{Y} \to \mathbf{Y}$ with the following properties*:

(i) *$\phi(m)^*: H^*(\mathbf{Y}; \mathbf{F}_p) \leftarrow H^*(\mathbf{Y}; \mathbf{F}_p)$ is zero.*

(ii) *$\phi(m)_*: \pi_t(\mathbf{Y}) \to \pi_t(\mathbf{Y})$ is a multiple of p for all t.*

(iii) *$\phi(m)_*: \pi_{2r}(\mathbf{Y}) \to \pi_{2r}(\mathbf{Y})$ is zero for $2r < m$, while for $2r = m$ it is multiplication by a non-zero scalar $\delta_m \in \Lambda$.*

The proof we give is not intended to lead to the best possible value for δ_m. Such a value would be relevant if we wished to consider $[\mathbf{Y}, \mathbf{Y}]$, and we do know the right numbers, but they are not relevant for our present purposes. This modest aim allows us to use a shorter proof. In particular, the reader will see that while the proof below may appear to be constructed with the complex case in mind, it is true word for word in the real case also, though in a rather wasteful way.

Proof of Lemma 5·6. The spectrum $\mathbf{Y}$ admits a map

$$\psi^k: \mathbf{Y} \to \mathbf{Y}$$

for each integer k prime to p (or even for each unit k in $\mathbf{Z}_{(p)}$). The induced map

$$(\psi^k)_*: \pi_{2r}(\mathbf{Y}) \to \pi_{2r}(\mathbf{Y})$$

is multiplication by k^r. If we take a Λ-linear combination $\sum_i \lambda_i \psi^{k_i}$, the induced map

$$(\sum_i \lambda_i \psi^{k_i})_*: \pi_{2r}(\mathbf{Y}) \to \pi_{2r}(\mathbf{Y})$$

is multiplication by $\sum_i \lambda_i (k_i)^r$. Let r run over the range $d \leqslant 2r \leqslant m$, where d is as in section 1. By taking as many i's as there are r's, and suitable coefficients $\lambda_i \in \mathbf{Z}_{(p)}$, we can ensure that

$$\sum_i \lambda_i (k_i)^r = 0 \quad \text{for} \quad d \leqslant 2r < m,$$

while for $2r = m$ we have

$$\sum_i \lambda_i (k_i)^r = \Delta,$$

where Δ is the determinant of the matrix whose (i, r)th entry is $(k_i)^r$. After removing from Δ a power of $\prod_i k_i$ which is a unit in $\mathbf{Z}_{(p)}$, we obtain a Vandermonde determinant, and this can be made non-zero by choosing the k_i distinct.

We can now satisfy all the conditions by taking

$$\phi(m) = p \sum_i \lambda_i \psi^{k_i}.$$

This proves the lemma.

Let m run over the integers for which $\pi_m(\mathbf{Y}) \simeq \Lambda$. By Lemma 5·3 we have

$$\mathrm{Hom}_{\mathbf{Z}}(\pi_m(\mathbf{X}), \pi_m(\mathbf{Y})) = \mathrm{Hom}_{\Lambda}(\pi_m(\mathbf{X}), \pi_m(\mathbf{Y})),$$

which, by our standing assumptions, is a finitely generated Λ-module. If we take it mod δ_m (where δ_m is as in Lemma 5·6) we get a finite group; let

$$\alpha_{m1}, \alpha_{m2}, \ldots: \pi_m(\mathbf{X}) \to \pi_m(\mathbf{Y})$$

be a finite set of homomorphisms containing one representative from each residue class mod δ_m.

We suppose given an isomorphism

$$\theta\colon H^*(\mathbf{X}; \mathbf{F}_p) \leftarrow H^*(\mathbf{Y}; \mathbf{F}_p)$$

as in Theorem 1·1.

LEMMA 5·7. *For each n there is a map $f = f^n\colon \mathbf{W}^n \to \mathbf{Y}$ with the following properties.*

(i) *The induced map f^* of mod p cohomology is the composite*

$$H^*(\mathbf{W}^n; \mathbf{F}_p) \leftarrow H^*(\mathbf{X}; \mathbf{F}_p) \xleftarrow{\theta} H^*(\mathbf{Y}; \mathbf{F}_p).$$

(ii) *The map* $\Lambda \otimes \pi_r(\mathbf{W}^n) \xrightarrow{1\otimes f_*} \Lambda \otimes \pi_r(\mathbf{Y}) \to \pi_r(\mathbf{Y})$

is iso for $r < n$, epi for $r = n$.

(iii) *For each m such that $\pi_m(\mathbf{Y}) \simeq \Lambda$ and $m < n$ the isomorphism*

$$\pi_m(\mathbf{X}) \xleftarrow{\cong} \Lambda \otimes \pi_m(\mathbf{W}^n) \xrightarrow{1\otimes f_*} \Lambda \otimes \pi_m(\mathbf{Y}) \to \pi_m(\mathbf{Y})$$

is one of the representatives α_{mi} chosen above.

Proof. First recall that $\mathbf{Y}$ is the spectrum $\mathbf{bg}_\Lambda$ obtained by introducing coefficients Λ into a spectrum $\mathbf{bg}$, and that the map $\mathbf{bg} \to \mathbf{bg}_\Lambda$ induces an isomorphism

$$H^*(\mathbf{bg}; \mathbf{F}_p) \leftarrow H^*(\mathbf{bg}_\Lambda; \mathbf{F}_p).$$

We define the A-module map ϕ so that the following diagram is commutative.

$$\begin{array}{ccc} H^*(\mathbf{X}; \mathbf{F}_p) & \xleftarrow{\theta} & H^*(\mathbf{Y}; \mathbf{F}_p) \\ \downarrow & & \downarrow \cong \\ H^*(\mathbf{W}^n; \mathbf{F}_p) & \xleftarrow{\phi} & H^*(\mathbf{bg}; \mathbf{F}_p) \end{array}$$

We begin by showing that there is a map $\mathbf{W}^n \to \mathbf{bg}$ whose induced map of mod p cohomology is ϕ.

In fact, the spectrum $\mathbf{bg}$ admits an Adams resolution of the conventional sort. By mapping any spectrum $\mathbf{X}$ into this resolution we obtain an 'Adams spectral sequence'; of course we assert nothing about the convergence of this spectral sequence. However, the spectral sequence is functorial for maps of $\mathbf{X}$; and the usual theorem for identifying its E_2 term is valid, for this depends only on hypotheses of finite generation in the Adams resolution, and the Adams resolution of $\mathbf{bg}$ is conventional. The isomorphism

$$H^*(\mathbf{X}; \mathbf{F}_p) \xleftarrow{\theta} H^*(\mathbf{Y}; \mathbf{F}_p) \xrightarrow{\cong} H^*(\mathbf{bg}; \mathbf{F}_p)$$

gives an element of the term E_2^{00} of the spectral sequence for $\mathbf{X}$. All differentials d_r are zero on this element, by Proposition 4·1 or 4·2 according to the case. By functoriality using the map $\mathbf{W}^n \to \mathbf{X}$, all differentials d_r are zero on the element ϕ in the term E_2^{00} of the spectral sequence for $\mathbf{W}^n$. But the spectral sequence for $\mathbf{W}^n$ is convergent in the usual way, because $\mathbf{W}^n$ is a finite complex. So there is a map $\mathbf{W}^n \to \mathbf{bg}$ which induces the map ϕ of cohomology. By taking the composite

$$\mathbf{W}^n \to \mathbf{bg} \to \mathbf{bg}_\Lambda = \mathbf{Y},$$

we get a map $f\colon \mathbf{W}^n \to \mathbf{Y}$ which satisfies clause (i) of the conclusion.

Next we show that any such map f satisfies clause (ii) of the conclusion. The map

$$\mathbf{W} \to \mathbf{M}\Lambda \wedge \mathbf{W} \xrightarrow{\cong} \mathbf{X}$$

induces isomorphisms of mod p cohomology; the map

$$H^r(\mathbf{W}^n; \mathbf{F}_p) \leftarrow H^r(\mathbf{W}; \mathbf{F}_p)$$

is iso for $r < n$, mono for $r = n$. It follows that the induced map

$$f^*: H^r(\mathbf{W}^n; \mathbf{F}_p) \leftarrow H^r(\mathbf{Y}; \mathbf{F}_p)$$

is iso for $r < n$, mono for $r = n$. Let $\bar{f}$ be the composite

$$\mathbf{M}\Lambda \wedge \mathbf{W}^n \xrightarrow{1 \wedge f} \mathbf{M}\Lambda \wedge \mathbf{Y} \xrightarrow{\nu} \mathbf{Y}$$

where ν is as in Lemma 5·2; then the restriction of $\bar{f}$ to $\mathbf{W}^n$ is homotopic to f. It follows that

$$\bar{f}^*: H^r(\mathbf{M}\Lambda \wedge \mathbf{W}^n; \mathbf{F}_p) \leftarrow H^r(\mathbf{Y}; \mathbf{F}_p)$$

is iso for $r < n$, mono for $r = n$. By the usual device of a mapping-cylinder, we can assume that $\bar{f}$ is an embedding; actually we do this only in order to write relative groups, and we could equally well use the corresponding groups of the map $\bar{f}$. In any case, we have

$$H^r(\mathbf{Y}, \mathbf{M}\Lambda \wedge \mathbf{W}^n; \mathbf{F}_p) = 0 \quad \text{for} \quad r \leqslant n.$$

By duality, we have

$$H_r(\mathbf{Y}, \mathbf{M}\Lambda \wedge \mathbf{W}^n; \mathbf{F}_p) = 0 \quad \text{for} \quad r \leqslant n.$$

We now argue by induction over r; suppose $\pi_s(\mathbf{Y}, \mathbf{M}\Lambda \wedge \mathbf{W}^n) = 0$ for $s < r$, where $r \leqslant n$. Then the Hurewicz theorem gives

$$\pi_r(\mathbf{Y}, \mathbf{M}\Lambda \wedge \mathbf{W}^n) \otimes \mathbf{F}_p = 0.$$

The homotopy sequence of the pair $(\mathbf{Y}, \mathbf{M}\Lambda \wedge \mathbf{W}^n)$ gives a short exact sequence of groups

$$0 \to A \to \pi_r(\mathbf{Y}, \mathbf{M}\Lambda \wedge \mathbf{W}^n) \to B \to 0$$

in which A and B are finitely generated modules over Λ. We have $B \otimes \mathbf{F}_p = 0$, so $B = 0$ (either by Nakayama's Lemma or by the structure theory for finitely generated modules). Hence the map $A \to \pi_r(\mathbf{Y}, \mathbf{M}\Lambda \wedge \mathbf{W}^n)$ is iso, and the argument of the last sentence shows that $\pi_r(\mathbf{Y}, \mathbf{M}\Lambda \wedge \mathbf{W}^n) = 0$. This completes the induction, which shows that $\pi_r(\mathbf{Y}, \mathbf{M}\Lambda \wedge \mathbf{W}^n) = 0$ for $r \leqslant n$, and establishes clause (ii) of the conclusion.

It remains to take our map f: $\mathbf{W}^n \to \mathbf{Y}$ and modify it so as to satisfy clause (iii). We do this by induction over m. Suppose that f_0: $\mathbf{W}^n \to \mathbf{Y}$ satisfies clauses (i) and (ii), and also satisfies clause (iii) in degrees $m' < m$. Let us replace f_0 by

$$f_0 + \lambda\phi(m) f_0,$$

where $\lambda \in \Lambda$ and $\phi(m)$ is as in Lemma 5·6. This process does not affect

$$f_0^*: H^*(\mathbf{W}^n; \mathbf{F}_p) \leftarrow H^*(\mathbf{Y}; \mathbf{F}_p).$$

It does not affect clause (ii), either because clause (ii) follows from clause (i), or because we alter

$$f_{0*}: \pi_r(\mathbf{W}^n) \to \pi_r(\mathbf{Y})$$

by a multiple of p. It does not affect the induced isomorphisms $\pi_{m'}(\mathbf{X}) \to \pi_{m'}(\mathbf{Y})$ for $m' < m$. However by varying λ we can make the induced isomorphism

$$\pi_m(\mathbf{X}) \to \pi_m(\mathbf{Y})$$

run over a residue class mod δ_m, so by a suitable choice of λ we can make it one of the representatives α_{mi}. This completes the induction and proves the lemma.

For the next lemma, we suppose that $\pi_n(\mathbf{Y}) \otimes \mathbf{Q} = 0$ (as happens, for example, when n is odd).

LEMMA 5·8. *The maps $f\colon \mathbf{W}^n \to \mathbf{Y}$ with the properties stated in Lemma 5·7 fall into finitely many homotopy classes.*

Proof. If f is as in Lemma 5·7, then the induced homomorphism

$$f_*\colon \pi_*(\mathbf{W}^n) \otimes \mathbf{Q} \to \pi_*(\mathbf{Y}) \otimes \mathbf{Q}$$

is wholly determined by the α_{mi} in Lemma 5·7. (In fact, $\pi_r(\mathbf{W}^n) \otimes \mathbf{Q} = 0$ for $r > n$, while $\pi_n(\mathbf{Y}) \otimes \mathbf{Q} = 0$ by assumption.) There are only a finite number of m's to be considered, and for each m only a finite number of α_{mi}, so there are only a finite number of possibilities for the induced homomorphism

$$f_*\colon \pi_*(\mathbf{W}^n) \otimes \mathbf{Q} \to \pi_*(\mathbf{Y}) \otimes \mathbf{Q}.$$

But for each induced homomorphism there are only finitely many homotopy classes of maps f, by Lemma 5·5. This proves the lemma.

LEMMA 5·9. *We can choose a sequence of maps $f^{2n+1}\colon \mathbf{W}^{2n+1} \to \mathbf{Y}$ with the properties stated in Lemma 5·7 so that $f^{2n+1}|\mathbf{W}^{2n-1} \simeq f^{2n-1}$ for each n.*

Proof. If we take a map $f\colon \mathbf{W}^N \to \mathbf{Y}$ with the properties stated in Lemma 5·7, then for any $n < N$ the restriction $f\,|\,\mathbf{W}^n$ also has these properties.

Suppose, as an inductive hypothesis, that we have chosen $f^{2n-1}\colon \mathbf{W}^{2n-1} \to \mathbf{Y}$ so that for an infinity of N, f^{2n-1} extends to some map $g^N\colon \mathbf{W}^N \to \mathbf{Y}$ with the properties stated in Lemma 5·7. (The induction starts when $2n-1 < d$; then there are maps g^N for all N by Lemma 5·7, and their restrictions to $\mathbf{W}^{2n-1}$ are all necessarily null homotopic.) Consider the restrictions $g^N|\mathbf{W}^{2n+1}$ of these maps g^N. They lie in a finite set of homotopy classes by Lemma 5·8. So at least one homotopy class arises for an infinity of g^N. Choose f^{2n+1} in such a homotopy class. This completes the induction and proves the lemma.

Proof of Theorem 1·1. The map

$$[\mathbf{W}, \mathbf{Y}] \to \underset{\overleftarrow{n}}{\mathrm{Lim}}\, [\mathbf{W}^{2n+1}, \mathbf{Y}]$$

is epi; therefore the sequence of maps f^{2n+1} of Lemma 5·9 arises from a map $f^\infty\colon \mathbf{W} \to \mathbf{Y}$. From clauses (i) and (ii) of Lemma 5·7 we see that the map of cohomology induced by f^∞ is the composite

$$H^*(\mathbf{W}; \mathbf{F}_p) \leftarrow H^*(\mathbf{X}; \mathbf{F}_p) \xleftarrow{\theta} H^*(\mathbf{Y}; \mathbf{F}_p),$$

and that (if ν is as in Lemma 5·2) the composite

$$\mathbf{M}\Lambda \wedge \mathbf{W} \xrightarrow{1\wedge f^\infty} \mathbf{M}\Lambda \wedge \mathbf{Y} \xrightarrow{\nu} \mathbf{Y}$$

is an equivalence. It only remains to take our map f: $\mathbf{X} \rightarrow \mathbf{Y}$ to be the composite

$$\mathbf{X} \xleftarrow{\simeq} \mathbf{M}\Lambda \wedge \mathbf{W} \xrightarrow{1\wedge f^\infty} \mathbf{M}\Lambda \wedge \mathbf{Y} \xrightarrow{\nu} \mathbf{Y}.$$

This completes the proof.

REFERENCES

(1) Adams, J. F. On Chern characters and the structure of the unitary group. *Proc. Camb. Philos. Soc.* **57** (1961), 189–199.

(2) Adams, J. F. *Stable homotopy and generalized homology* (Chicago University Press, 1974).

(3) Adams, J. F. and Margolis, H. R. Modules over the Steenrod algebra. *Topology* **10** (1971), 271–282.

(4) Atiyah, M. F. and Segal, G. B. Exponential isomorphisms for λ-rings. *Quart. J. Math.* **22** (1971), 371–378.

(5) Boardman, J. M. and Vogt, R. M. Homotopy-everything H-spaces. *Bull. Amer. Math. Soc.* **74** (1968), 1117–1122.

(6) Madsen, I., Snaith, V. and Tornehave, J. H^∞ endomorphisms of K-theory are infinite loop maps. (Manuscript.)

(7) May, J. P. *E_∞ spaces, group completions, and permutative categories*. London Math. Soc. Lecture Note, series no. 11 (Cambridge University Press, 1974).

(8) May, J. P. E_∞ ring spaces and E_∞ ring spectra. (To appear.)

(9) May, J. P. *Localisations, completions, and the stabilisation of homotopy theory* (London Math. Soc. Monograph). (To appear.)

(10) Priddy, S. B. Homology operations for the classifying spaces of certain matrix groups. *Quart. J. Math.* **26** (1975) 179–193.

(11) Peterson, F. P. The mod p homotopy type of BSO and F/PL. *Bol. Soc. Mat. Mexicana* **14** (1969), 22–27.

(12) Segal, G. B. Categories and cohomology theories. *Topology* **13** (1974), 293–312.

(13) Sullivan, D. Genetics of homotopy theory and Adams conjecture. *Ann. of Math.* **100** (1974), 1–79.

(14) Sullivan, D. Geometric topology. Part I. Localization, periodicity and Galois symmetry (mimeographed notes, M.I.T. 1970).

Proceedings of Symposia in Pure Mathematics
Volume 39 (1983), Part 1

Graeme Segal's Burnside Ring Conjecture

J. FRANK ADAMS

1. Introduction. My theme will be that algebraic topology still offers problems which reveal the present state of our art as inadequate. Such problems make us feel that there could be something good going on; but if there is, we have yet to understand it.

I shall devote §2 to explaining Segal's original conjecture. In §3–§6 I shall review what is proved about it so far. Broadly, for those few groups we have been able to handle the conjecture is found to be true, and there is no group for which the conjecture is known to be false. In §7–§9 I shall explain further conjectures related to the original one; the fact that these also are not yet disproved contributes to the impression that there could be something good going on. In §10 I shall comment briefly on our chances of going further.

2. Statement of the conjecture. In this section I shall explain Segal's original conjecture. I must begin by explaining cohomotopy.

Let X, Y be a good pair of spaces, for example, a finite-dimensional CW-pair. Then we have

$$\pi^n(X, Y) = \operatorname*{Lim}_{m \to \infty} [S^m X/Y, S^{m+n}].$$

Here maps and homotopies preserve the base-point; and if $Y = \phi$, I use Atiyah's convention that X/ϕ means X with a disjoint base-point. We can rewrite the right-hand side as

$$\operatorname*{Lim}_{m \to \infty} [X/Y, \Omega^m S^{m+n}]$$

or as

$$\left[X/Y, \operatorname*{Lim}_{m \to \infty} \Omega^m S^{m+n}\right].$$

The last expression gives us a definition of $\pi^n(X, Y)$ valid whether the pair X, Y is finite dimensional or not.

Cohomotopy is a generalised cohomology theory, namely the one corresponding to the sphere spectrum; and its coefficient groups are the stable homotopy groups of spheres. So it is like stable homotopy; if you could compute it, it would give you a lot of information, but unfortunately it is hard to compute.

However, there is one case in which we have a conjecture, and it is due to Graeme Segal. He asks for the analogue of a well-known theorem of Atiyah [3].

Reprinted from Bulletin Amer. Math. Soc. (N.S.) 6 (1982), 207–270.
1980 *Mathematics Subject Classification.* Primary 55Q55, 55R35.

Atiyah wished to compute the K-theory of BG, the classifying space of G, where G was originally a finite group. So Atiyah took the representation ring $R(G)$ and constructed a map

$$\alpha: R(G) \to K(BG).$$

He then completed $R(G)$ for the topology defined by the powers of the augmentation ideal. He extended α by continuity to get a map

$$\hat{\alpha}: R(G)\hat{} \to K(BG).$$

And finally he proved that this map $\hat{\alpha}$ is an isomorphism.

Segal proposed that we should replace K-theory by cohomotopy; so on the right we put $\pi^0(BG)$. On the left he proposed to replace the representation ring $R(G)$ by the Burnside ring $A(G)$ of G. To define this, instead of taking actions of G on vector spaces, we take actions of G on finite sets. (That is, assuming G is finite, of course.) We divide these actions into isomorphism classes, and make these classes into a semigroup under disjoint union. We take the corresponding Grothendieck group, and that is $A(G)$. We make it a ring so that the product of two finite G-sets is their Cartesian product. I emphasise that $A(G)$ should be regarded as something computable and known.

We have again a map

$$\alpha: A(G) \to \pi^0(BG).$$

I will not stop to define it here; the reader may consult §9 for a more general construction. We complete $A(G)$ for the topology defined by the powers of the augmentation ideal; and again α extends by continuity to give a map

$$\hat{\alpha}: A(G)\hat{} \to \pi^0(BG).$$

Segal's conjecture is that this map $\hat{\alpha}$ is an isomorphism.

For the trivial group $G = 1$ the conjecture is trivially true; both sides are Z and the map is an isomorphism. This seems to be the only group for which matters are trivial.

3. Results of Lin; the case $G = Z_2$. The group $G = Z_2$ already presents a problem. On the left, $A(Z_2)\hat{}$ comes to $Z \oplus \hat{Z}_2$, the direct sum of the integers and the 2-adic integers. On the right we can take real projective space RP^∞ as our model for BZ_2, and so the problem is to compute $\pi^0(RP^\infty)$; but for ten years nobody could get hold of it. This case was finally solved by W. H. Lin [**11, 12**], using a method which I once heard Graeme Segal call "that damn spectral sequence of yours". Lin found that the conjecture is true for $G = Z_2$.

Lin also found that $\pi^n(RP^\infty) = 0$ for $n > 0$. It seems fair to add to our list of conjectures the conjecture

$$\pi^n(BG) = 0 \quad \text{for } G \text{ finite, } n > 0.$$

Let me emphasise that this is not a statement which is true for trivial reasons, in the way that the homotopy groups of a space or spectrum vanish below the Hurewicz dimension. If you try to get hold of $\pi^n(BG)$ by obstruction-theory, or by an Atiyah-Hirzebruch spectral sequence, you find an infinity of nonzero groups

$$H^i(BG; \pi^j(pt)) \qquad (i+j=n).$$

What happens is that these groups all conspire to cancel each other out for $n>0$; for $G=Z_2$ this phenomenon was noticed long ago by Mahowald, and later by me [**1**], although neither of us had a proof.

Various sorts of further argument seem likely to need information about the groups $\pi^n(RP^\infty)$ also in the range $n<0$, where they are nontrivial. Lin obtained such information; but to describe the result will require some preliminary explanation.

4. Functional duals. In this section I will explain about functional duals, and so reach the results on $\pi^n(RP^\infty)$ for $n<0$.

Let us work in a suitable category of spectra where we can do stable homotopy theory. Consider $[W \wedge X, S^0]$, where X is fixed and W varies. This is a contravariant functor of W, and satisfies the hypotheses of Brown's Representability Theorem, so we get a natural (1-1) correspondence

$$[W \wedge X, S^0] \leftrightarrow [W, DX],$$

where the representing object DX is by definition the "functional dual" of X. In particular,

$$\pi^{-n}(X) = [S^n \wedge X, S^0] \leftrightarrow [S^n, DX] = \pi_n(DX).$$

So if we know DX, we get information about the cohomotopy of X.

Let's use the same symbol for a space with base-point and for its suspension spectrum. Then we want to know $D(RP^\infty/\phi)$, and we have

$$RP^\infty/\phi \simeq S^0 \vee (RP^\infty/pt),$$

so

$$D(RP^\infty/\phi) \simeq (DS^0) \vee D(RP^\infty/pt) \simeq S^0 \vee D(RP^\infty/pt).$$

Thus it is sufficient to know $D(RP^\infty/pt)$.

Let $M\hat{Z_2}$ be a Moore spectrum for the 2-adic integers $\hat{Z_2}$ (in dimension 0). Let

$$M\hat{Z_2} \wedge (RP^\infty/pt) \to S^0$$

be any (stable) map for which the functional Sq^2 is nonzero; since the left-hand side is equivalent to RP^∞/pt, such maps exist. Let

$$(RP^\infty/pt) \wedge (RP^\infty/pt) \to S^0$$

be any map for which the functional Sq^4 is nonzero; again, such maps exist. Then by adjointness we get a map

$$(M\hat{Z_2}) \vee (RP^\infty/pt) \to D(RP^\infty/pt).$$

PROPOSITION. *Any map*

$$(M\hat{Z_2}) \vee (RP^\infty/pt) \to D(RP^\infty/pt)$$

constructed in this way is an equivalence.

This is a result of Lin and myself which is not yet written up for publication; I sketch the proof. We have an induced map of homotopy

$$\pi_n\big((M\hat{Z_2}) \vee (RP^\infty/pt)\big) \to \pi_n\big(D(RP^\infty/pt)\big) \cong \pi^{-n}(RP^\infty/pt).$$

In fact we can obtain an induced map between the Adams spectral sequences which should converge to these groups; we have one spectral sequence converging to

$$\big[S^0, (M\hat{Z_2}) \vee (RP^\infty/pt)\big]_*$$

and one converging to

$$\big[(RP^\infty/pt), S^0\big]_*.$$

It is necessary to insist, however, that this map of spectral sequences carries $E_r^{s,t}$ to $E_r^{s+1,t+1}$; that is, it raises the filtration degree by 1. By quoting the algebraic result of [**12**], it is shown that this comparison map of spectral sequences is an isomorphism.

The reader will notice that the result about the functional dual implicitly contains the result that $\pi^n(RP^\infty) = 0$ for $n > 0$ and the result that $\pi^0(RP^\infty)$ is $Z \oplus \hat{Z_2}$, as well as information about the groups $\pi^n(RP^\infty)$ for $n < 0$. Of course, it is still necessary to take the isomorphism $\pi^0(RP^\infty) \cong Z \oplus \hat{Z_2}$ given by the proposition and reconcile it with the map $\hat{\alpha}$. Similar remarks will apply whenever we describe a functional dual $D(BG)$.

This completes my account of the results for $G = Z_2$.

5. Results of Gunawardena and Ravenel; the case $G = Z_n$. The next case should be the case $G = Z_p$, where p is an odd prime. Here Gunawardena has shown that with suitable care Lin's results go over [**9**].

The next case should be the case $G = Z_{p^e}$, p prime, $e \geqslant 1$. In this case D. C. Ravenel finds

$$D(BZ_{p^e}/pt) \simeq \Big(\bigvee_1^e M\hat{Z_p}\Big) \vee \Big(\bigvee_{i=1}^e BZ_{p^i}/pt\Big).$$

This is in agreement with the conjecture. An announcement of Ravenel's work is in course of publication [**14**]. The proof sketched lacks only a treatment of the convergence of a spectral sequence constructed for the purposes of the proof; it is plausible that this should not present an essential obstacle. The proof is by induction over e, and it relies on the previous work of Lin and Gunawardena to deal with the case $e = 1$ and start the induction.

The case of a general cyclic group Z_n follows immediately. After all, for a cyclic group Z_n we have a stable equivalence

$$BZ_n/pt \simeq \bigvee_p BS_p/pt,$$

where S_p is the Sylow p-subgroup of Z_n. This gives

$$D(BZ_n/pt) \simeq \bigvee_p D(B(S_p/pt)),$$

where the right-hand side is known by the work of Ravenel.

6. Other results. One should expect that suitable statements for a finite group G will follow from the corresponding statements about the Sylow subgroups S_p of G. For the conjecture "$\hat{\alpha}$ mono" this was proved by Laitinen [**10**]. For the conjectures "$\hat{\alpha}$ iso" and "$\pi^n(BG) = 0$ for $n > 0$" it has been proved by May and McClure; but I do not have details about publication. In particular, the conjectures should be true for any finite group G all of whose Sylow subgroups are cyclic.

Results which show that $\hat{\alpha}$ is mono are perhaps less impressive than ones which show that $\hat{\alpha}$ is iso. However, Laitinen [**10**] has proved that $\hat{\alpha}$ is mono when G is an elementary abelian p-group. This result was later extended to all finite abelian groups by Segal and Stretch. They use BP-theory; again I do not know details about publication.

At this point I more or less run out of positive results to report. Noncyclic groups present a substantial problem. To do calculations (corresponding to those which Lin did for $G = Z_2$) for so small a group as $G = Z_2 \times Z_2$ seems to be something contemplated only by a select few.

7. Could the functional duals be doing something good? At this point it is natural to step back and see if we can guess some larger pattern into which the pieces might fit. In this section I will do so for the functional duals.

Let G be a finite group, and let H be a subgroup of G. Then the coset space G/H is a finite G-set. We can also consider the set of G-maps $G/H \to G/H$, and this turns out to be $W_H = N_G(H)/H$, where $N_G(H)$ is the normaliser of H in G. So we have $W_H \times G$ acting on the finite set G/H. Thus we get a finite covering associated to the universal bundle over $B(W_H \times G) \simeq BW_H \times BG$, say

$$G/H \to E \xrightarrow{p} BW_H \times BG.$$

Take the unit element $1 \in \pi^0(E)$ and apply the transfer [**2, 5, 7, 8**]; we get an element

$$\begin{aligned} p_!1 \in \pi^0(BW_H \times BG) &= [(BW_H \times BG)/\phi, S^0] \\ &= [(BW_H/\phi) \wedge (BG/\phi), S^0]. \end{aligned}$$

This corresponds to a map

$$BW_H/\phi \to D(BG/\phi).$$

Conjugate subgroups H lead to the same outcome. So I get a map

$$\bigvee_{\{H\}} (BW_H)/\phi \to D(BG/\phi)$$

where in the wedge-sum I take one subgroup H from each conjugacy class. The left-hand side contains a summand S^0 which arises for $H = G$ ($W_G = N_G(G)/G = 1$). This maps to the obvious summand S^0 of $D(BG/\phi)$ (the constant map $G \to 1$ gives $BG/\phi \to B1/\phi$, which dualises to $D(B1/\phi) \to D(BG/\phi)$). As for the complementary summands, one may conjecture that this map gives an equivalence provided that G is a p-group and on the left you complete at the prime p. This appears consistent with what is known for cyclic p-groups and can be guessed for other groups. If G were not a p-group life would become more complicated.

8. Could equivariant cohomotopy be doing something good? In this section I will discuss some generalisations of the original conjecture which involve equivariant cohomotopy.

In a seminar in Oxford in February 1980 I proposed the conjecture in §7; and they were very polite and told me that this was indeed the obvious conjecture, and that it could be incorporated into further conjectures which were better and brighter and even more conjectural. These conjectures are again due to Segal, and again he follows an analogy.

Atiyah and Segal long ago generalised Atiyah's theorem

$$\hat{\alpha}: R(G)\hat{} \xrightarrow{\cong} K(BG)$$

in the following direction [**4**]. Let X be a suitable G-space, for example, a finite G-CW-complex in the sense of Matumoto [**13**]. Then there are two ways to define the equivariant K-theory of X. First, we can form $E_G \times_G X$ and take ordinary K of it, so as to get $K(EG \times_G X)$. This way works with K replaced by any generalised cohomology theory k; you can form $k(EG \times_G X)$. The second way only works if you have a good geometrical construction for k; then you go through the construction again with G acting on everything. In particular, Atiyah and Segal define $K_G(X)$ in terms of G-vector-bundles over the G-space X. Then they get

$$K_G(X)\hat{} \xrightarrow{\cong} K(EG \times_G X).$$

Taking X to be a point, they recover as a special case the theorem

$$R(G)\hat{} \xrightarrow{\cong} K(BG).$$

When one comes to replace K-theory by cohomotopy, one replaces $K(EG \times_G X)$ on the right by $\pi^n(EG \times_G X)$. On the left we should replace $K_G(X)$ by the equivariant cohomotopy of X according to some direct definition. For simplicity I will begin with the case $G = Z_2$. In this case we have two ways to suspend a given G-space: one, say S, in which Z_2 preserves the suspension coordinate, and another, say T, in which Z_2 reverses the suspension coordinate. So let X, Y be a finite-dimensional G-CW-pair, where $G = Z_2$. We can form

$$\pi_G^{i,j}(X, Y) = \operatorname*{Lim}_{m,n \to \infty} \left[S^m T^n X/Y, S^{m+i} T^{n+j}\right]_G.$$

Here maps and homotopies are G-maps and G-homotopies preserving the base-point. Now just as I did in §2, I can rewrite this in the following form.

$$\operatorname*{Lim}_{m,n\to\infty}\left[X/Y, \Omega_S^m\Omega_T^n S^{m+i}T^{n+j}\right]_G.$$

Here the function-space on the right has the obvious action of G; an element $g \in G$ acts on a function f to give $g \circ f \circ g^{-1}$. I can rewrite things again in the following form.

$$\left[X/Y, \operatorname*{Lim}_{m,n\to\infty} \Omega_S^m\Omega_T^n S^{m+i}T^{n+j}\right]_G$$

In this form the definition is valid whether the pair X, Y is finite-dimensional or not.

For a general finite group G of course I have to allow for suspension using any finite-dimensional real representation of G, and I get groups $\pi_G^\rho(X, Y)$ indexed by elements ρ in the real representation ring $RO(G)$. It is clear enough how the generalisation goes.

On the left of our conjecture we can now put the groups $\pi_G^\rho(X, Y)$. But on the right we still have ordinary cohomotopy groups

$$\pi^n(EG \times_G X, EG \times_G Y)$$

which are indexed over the integers. What is to be done?

It seems the best move is the following. EG is usually considered as a space with G acting on its right, but we can make G act on its left by using $e.g.^{-1}$. Then we can form

$$\pi_G^\rho(EG \times X, EG \times Y).$$

The constant map $c\colon EG \to pt$ of course induces

$$(c \times 1)^*\colon \pi_G^\rho(X, Y) \to \pi_G^\rho(EG \times X, EG \times Y).$$

If ρ happens to be an integer n, we can prove

$$\pi_G^n(EG \times X, EG \times Y) \cong \pi^n(EG \times_G X, EG \times_G Y).$$

The proof is not quite the same as it was for K-theory, but still it can be done. Whether ρ is an integer or not, we can conjecture that $(c \times 1)^*$ becomes an isomorphism if we complete the left-hand side. (Notice that $c\colon EG \to pt$ may be a homotopy equivalence, but it is not a G-homotopy-equivalence.)

On the face of it, it is not clear how the left-hand side $\pi_G^\rho(X, Y)$ just considered is related to the sort of left-hand side, involving groups W_H, which we saw in the last section. Experts in this area see a relation, but I am not ready to report on it with confidence.

I believe that when G is cyclic the conjecture in this section can probably be deduced from the results I have described above.

9. Could something make sense for compact Lie groups in general? Up to this point G has been a finite group; I now seek to generalise so that G can be any compact Lie group. Let $(BG)^m$ be the m-skeleton of BG; then we have the following exact sequence due to Milnor.

$$0 \to \operatorname*{Lim^1}_{\overleftarrow{m}} \pi^*((BG)^m) \to \pi^*(BG) \to \operatorname*{Lim^0}_{\overleftarrow{m}} \pi^*((BG)^m) \to 0.$$

When G is finite the Lim^1 term is zero, but this need no longer happen when G has positive dimension. It seems advisable to replace $\pi^n(BG)$ in our conjecture by $\underset{\overleftarrow{m}}{\mathrm{Lim}}^0\, \pi^n((BG)^m)$. Once this is done, hope does not seem to be ruled out.

At this point I should give briefly some perspective on the Burnside ring in this context. Take some good class of compact G-spaces X, for example, finite G-CW-complexes. By a G-Euler characteristic, I shall mean a pair (A, χ) of the following sort. A is a given abelian group; χ is a function which assigns to each finite G-CW-complex X an element $\chi(X)$ in A; and χ satisfies the following axioms.

(0) (Zero) $\chi(\phi) = 0$.

(i) (Invariance). If X and Y are G-homotopy-equivalent, then $\chi(X) = \chi(Y)$.

(iii) (Mayer-Vietoris) Under the obvious assumptions,

$$\chi(X_1 \cup X_2) + \chi(X_1 \cap X_2) = \chi(X_1) + \chi(X_2).$$

(Axiom (0) seems to be needed; one can replace $\chi(X)$ by $\chi(X) + c$, where c is a constant, without affecting axioms (i) and (ii).)

Among such G-Euler-characteristics, there is one which is universal. In this, the group A is a free abelian group with one generator $a_{\{H\}}$ for each conjugacy class $\{H\}$ of closed subgroups $H \subset G$. The function χ is defined as follows. Given X, let $r(\{H\}, n)$ be the number of G-cells in X of type $(G/K) \times E^n$ with $K \in \{H\}$; set

$$\chi(X) = \sum_{\{H\},n} (-1)^n r(\{H\}, n) a_{\{H\}}.$$

This is the most naive generalisation of the Euler characteristic; we count the G-cells with a sign $(-1)^n$ depending on their dimension, but we keep accounts separately for G-cells of the different symmetry types. For all this, see [**6**].

We now proceed as follows. For any such G-space X we can form the bundle with fibres X associated to the universal G-bundle, that is

$$EG \times_G X \xrightarrow{p} BG.$$

We can take the unit element

$$1 \in \pi^0(EG \times_G X)$$

and apply transfer [**5, 7, 8**] getting an "index"

$$p_! 1 \in \pi^0(BG).$$

We check that this satisfies the axioms for a G-Euler-characteristic of X; by the universal property, we get a homomorphism

$$\alpha \colon A \to \pi^0(BG).$$

(If we use a version of the transfer which is only defined when the base is finite dimensional, or even when the base is a finite complex, we just replace $\pi^0(BG)$ by a suitable inverse limit.)

At this point the general case has one new twist which doesn't show up when G is finite. Let H be a closed subgroup of G; then by definition, $\alpha a_{\{H\}}$ is $p_!(1)$ in the following fibering.

$$G/H \to EG \times_H pt \to BG.$$

Let N be the normaliser of H, and assume that N/H is of positive dimension. Then we can find a 1-parameter subgroup θ_t in N which leads from the identity element to a point not in H. Acting on $EG \times_H pt$ by right translation with θ_t, we can deform the identity map

$$1: EG \times_H pt \to EG \times_H pt$$

to a map without fixed-points. So the fixed-point index $p_! 1$ is zero in this case. Thus we can pass to a quotient of our previous universal A in which we map to zero each generator $a_{\{H\}}$ such that N/H is of positive dimension; we leave only the generators $a_{\{H\}}$ such that N/H is finite. We thus reach tom Dieck's version of the Burnside ring $A(G)$ [**6**], and we still have a homomorphism

$$\alpha: A(G) \to \pi^0(BG).$$

The one case which seems accessible to checking at the present time is the case $G = S^1$, for one can approximate to S^1 by its cyclic subgroups Z_n. As our model for BS^1 we can take the complex projective space CP^∞. When we audit the conjectures about

$$\operatorname*{Lim}_{\substack{\leftarrow \\ m}} \pi^n(CP^m)$$

for $n \geqslant 0$, they seem to stand up.

10. What are our chances? Finally, I owe you some comments about our chances of going further. To begin with, it will be prudent to allow for a chance of say 5% or 10% that the conjectures fail when you go beyond cyclic groups. If so, then "calculation is the way to the truth"; one can always hope that a sufficiently heroic effort will settle one more group. On the other hand, if the conjectures are true, then we must ask what sort of a proof to seek. From the attractive nature of some of the formulations, you might be encouraged to hope for a fairly conceptual proof. I don't think it's all that likely, myself. After all, the original formulation was copied from K-theory, and in that case the best proof we have is the proof by Atiyah and Segal [**4**], which works by slowly building up the results for a succession of well-chosen particular groups. It would seem more reasonable to look for a proof like that, and the question is, how to build up.

The first obvious suggestion is to follow the approach used by Atiyah in [**3**]; try to prove something for p-groups G by induction, applying the inductive hypothesis either to a suitable normal subgroup $H \subset G$, or to the corresponding quotient G/H, or to both. This is hard and nobody can do it yet; the reader will find that the difficulty is already present in the case $G = Z_2 \times Z_2$.

The second obvious suggestion is to follow the approach used by Atiyah and Segal in [**4**]; go via compact Lie groups, even if you only want a conclusion about finite groups. This approach also does not seem realistic yet.

My conclusion is that this area deserves further study. But in order to work in it, those who love conceptual theories had better not scorn calculation, while those who love horrendous calculations should not neglect such help and guidance as may be had from conceptual theory.

References

1. J. F. Adams, *Operations of the nth kind in K-theory, and what we don't know about* RP^∞, London Math. Soc. Lecture Notes No. 11, Cambridge Univ. Press, 1974, pp. 1–9.

2. ______, *Infinite loop spaces*, Princeton Univ. Press, Princeton, N.J., 1978, especially Chapter 4.

3. M. F. Atiyah, *Characters and cohomology of finite groups*, Inst. Hautes Études Sci. Publ. Math., No. 9 (1961), 23–64.

4. M. F. Atiyah and G. B. Segal, *Equivariant K-theory and completion*, J. Differential Geometry **3** (1969), 1–18.

5. J. C. Becker and D. Gottlieb, *Transfer maps for fibrations and duality*, Compositio Math. **33** (1976), 107–133.

6. T. tom Dieck, *Transformation groups and representation theory*, Lecture Notes in Math., vol. 766, Springer-Verlag, Berlin and New York, 1979.

7. A. Dold, *The fixed-point index of fibre-preserving maps*, Invent. Math. **25** (1974), 281–297.

8. ______, *The fixed point transfer of fibre-preserving maps*, Math. Z. **148** (1976), 215–244.

9. J. H. C. Gunawardena, *Segal's conjecture for cyclic groups of (odd) prime order*, J. T. Knight Prize Essay, Cambridge, 1980.

10. E. Laitinen, *On the Burnside ring and stable cohomotopy of a finite group*, Math. Scand. **44** (1979), 37–72.

11. W. H. Lin, *On conjectures of Mahowald, Segal and Sullivan*, Math. Proc. Cambridge Philos. Soc. **87** (1980), 449–458.

12. W. H. Lin, D. M. Davis, M. E. Mahowald and J. F. Adams, *Calculation of Lin's Ext groups*, Math. Proc. Cambridge Philos. Soc. **87** (1980), 459–469.

13. T. Matumoto, *On G-CW-complexes and a theorem of J. H. C. Whitehead*, J. Fac. Sci. Univ. Tokyo **18** (1971), 363–374.

14. D. C. Ravenel, *The Segal conjecture for cyclic groups*, Bull. London Math. Soc. **13** (1981), 42–44.

Pure Mathematics Department, University of Cambridge, Cambridge, England

Topology Vol. 27, No. 1, pp. 7 21, 1988

A GENERALIZATION OF THE SEGAL CONJECTURE

J. F. Adams, J.-P. Haeberly, S. Jackowski and J. P. May

(Received 13 May 1986)

§1. INTRODUCTION

Theorem 1.4 below is a generalization of the Segal conjecture about equivariant cohomotopy. It asserts an invariance property of the G-cohomology-theory $S^{-1}\pi_G^*(—)_I^\wedge$ obtained from equivariant cohomotopy π_G^* by first localizing with respect to a general multiplicatively-closed subset S in the Burnside ring $A(G)$, and then completing with respect to a general ideal $I \subset A(G)$. We first explain how we place previous "localization theorems" and "completion theorems" in one setting by formulating suitable invariance statements.

Let G be a finite group; all our G-spaces will be G-CW complexes [23]. Let $\mathscr{H}$ be some class of subgroups and let $f: X \to Y$ be a G-map. We will say that f is an "$\mathscr{H}$-equivalence" if the induced map of fixed-point-sets $f^H: X^H \to Y^H$ is an ordinary homotopy equivalence for each $H \in \mathscr{H}$. (Thus we may assume without loss of generality that $\mathscr{H}$ is closed under passing to conjugate subgroups.) Let h be a functor defined on G-spaces and G-maps; we will say that h is "$\mathscr{H}$-invariant" if it carries each $\mathscr{H}$-equivalence to an isomorphism in the target category of h. The same property was previously introduced in [34] and studied further in [35].

In particular, let $\mathscr{H}$ be the class of all subgroups $H \subset G$; then an $\mathscr{H}$-equivalence is just a G-homotopy-equivalence, and every G-cohomology-theory is $\mathscr{H}$-invariant.

To place "localization theorems" in this setting, we assume that $\mathscr{H}$ is closed under passing to conjugate subgroups and larger subgroups. Then for any X we have an $\mathscr{H}$-fixed-point subcomplex

$$X^{\mathscr{H}} = \cup\{X^H : H \in \mathscr{H}\},$$

and the inclusion $i: X^{\mathscr{H}} \to X$ is an $\mathscr{H}$-equivalence.

Remark 1.1. In this case, h is $\mathscr{H}$-invariant iff $h(i): h(X) \to h(X^{\mathscr{H}})$ is iso for each X. ("Only if" is clear; and we will explain the converse in §7.)

"Localization theorems" usually state that $h(i)$ is iso when h is a functor obtained by localization, $h = S^{-1}k$, and $S, \mathscr{H}$ are suitably related. Such theorems go back to Segal [29, Prop. 4.1].

We place "completion theorems" in this setting. A class $\mathscr{H}$ which is closed under passing to conjugate subgroups and smaller subgroups is called a "family". A G-space Y qualifies as a universal space $E\mathscr{F}$ for the family $\mathscr{F}$ if Y^H is contractible for $H \in \mathscr{F}$ and empty for $H \notin \mathscr{F}$. For background on spaces $E\mathscr{F}$, see [27, 10, 11 p. 175, 13]. For any X the projection

$$p: E\mathscr{F} \times X \to X$$

is an $\mathscr{F}$-equivalence.

Remark 1.2. In this case, h is $\mathscr{F}$-invariant iff $h(p):h(X)\to h(E\mathscr{F}\times X)$ is iso for each X. ("Only if" is clear; and if $f:X\to Y$ is an $\mathscr{F}$-equivalence, then $1\times f: E\mathscr{F}\times X\to E\mathscr{F}\times Y$ is a G-homotopy-equivalence.)

"Completion theorems" usually state that $h(p)$ is iso when h is a functor obtained by completion, $h=k(-)_I^{\wedge}$, and I, $\mathscr{F}$ are suitably related. Such theorems go back to Atiyah and Segal [6].

We will show that it makes sense to look for a "best possible" invariance result.

THEOREM 1.3. *For each G-cohomology-theory h^* satisfying the axioms given in* §7, *there is a unique minimal class $\mathscr{H}$ such that h^* is $\mathscr{H}$-invariant.*

[Note that as $\mathscr{H}$ decreases, the $\mathscr{H}$-invariance property gets stronger, because less data on f suffice to prove $h(f)$ iso.]

We seek specific invariance results (preferably best possible) for particular functors. The functors we consider are progroup-valued. The role of progroups in this subject has been recognized ever since the work of Atiyah and Segal [6]. Let h be a functor from finite G-CW complexes to R-modules. Then h yields a progroup-valued functor $\mathbf{h}$ defined on all G-CW complexes X; we define $\mathbf{h}(X)$ to be the inverse system $\{h(X_\alpha)\}$, where X_α runs over the finite G-CW subcomplexes of X. Localization of promodules over R (with respect to a multiplicative set $S\subset R$) is done termwise: $S^{-1}\{M_\alpha\}=\{S^{-1}M_\alpha\}$. To complete promodules (with respect to an ideal $I\subset R$) we define $\{M_\alpha\}_I^{\wedge}$ be the inverse system $\{M_\alpha/I^rM_\alpha\}$, where α runs as before and r runs over the non-negative integers. In particular, even if X is a finite complex, the completion $\mathbf{h}(X)_I^{\wedge}$ is a progroup.

We take h to be equivariant cohomotopy—see [1] or [30].

THEOREM 1.4. *The theory $S^{-1}\pi_G^*(-)_I^{\wedge}$ (progroup-valued equivariant cohomotopy localized at $S\subset A(G)$ and completed at $I\subset A(G)$) is $\mathscr{H}$-invariant, where*

$$\mathscr{H}=\cup\{Supp(P): P\cap S=\varnothing\ \&\ P\supset I\}.$$

Here P runs over prime ideals of $A(G)$, and Supp(P) is the support of P, which we define following Dress [12]. [$H\in$Supp(P) if P comes from H via the restriction map $A(G)\to A(H)$ and P does not come from any $K<H$. Dress shows that Supp(P) is a single conjugacy class of subgroups H.]

Our companion paper on K-theory [2] shows that a theorem precisely analogous to (1.4) holds for equivariant K-theory; one just replaces the Burnside ring $A(G)$ by the representation ring $R(G)$, and "supports" in the sense of Dress [12] by "supports" in the sense of Segal [28].

Originally we sought the special case $S=\{1\}$ of (1.4); this goes as follows.

THEOREM 1.5. *For any family $\mathscr{F}$ the theory $\pi_G^*(-)_{I(\mathscr{F})}^{\wedge}$, equivariant cohomotopy completed at*

$$I(\mathscr{F})=\bigcap_{H\in\mathscr{F}}\operatorname{Ker}(A(G)\to A(H)),$$

is $\mathscr{F}$- invariant.

COROLLARY 1.6. *There is a pro-isomorphism*

$$\pi_G^*(X)\hat{}_{I(\mathscr{F})} \leftrightarrow \pi_G^*(E\mathscr{F} \times X)$$

natural in the G-space X.

On the right of (1.6) we can omit the completion at $I(\mathscr{F})$, because $\pi_G^*(E\mathscr{F} \times X)$ is already complete (see §6). Given this, the result follows from (1.5) and (1.2).

We refer to our companion paper [2] for the application of (1.6) to calculate the equivariant cohomotopy of equivariant classifying spaces.

We may pass from the inverse systems in (1.6) to their inverse limits. We assume that X is a finite G-CW complex; then the inverse system $\pi_G^*(X)\hat{}_{I(\mathscr{F})}$ is Mittag–Leffler; therefore the pro-isomorphic inverse system $\pi_G^*(E\mathscr{F} \times X)$ is Mittag–Leffler; therefore its inverse limit is the representable G-cohomotopy of $E\mathscr{F} \times X$. All this goes back to [6].

The classical case is that in which $\mathscr{F} = \{1\}$, $E\mathscr{F}$ becomes EG and the completion is done using the augmentation ideal $\mathrm{Ker}(\varepsilon: A(G) \to Z)$. In this case (1.6) becomes the Segal conjecture, which has been proved by the combined efforts of a number of mathematicians, by far the greatest contribution being due to Carlsson [8].

Compared with the special case $\mathscr{F} = \{1\}$, the general case (1.5), (1.6) has more flexibility, and (1.4) has more flexibility still. By adjusting S and I, we can obtain results about functors closer to cohomotopy, at the price of using stronger hypotheses on our spaces and maps. Conversely, (1.4) shows what price (in terms of S and I) will pay for a given level of invariance (every class $\mathscr{H}$ arises for suitable S and I, usually for many).

One of us [25] has obtained a further generalization of (1.4). In this he replaces the representing spectrum for cohomotopy, that is the sphere spectrum, by the suspension spectrum of a suitable classifying space. (See appendix.)

As for history: completion theorems of the general form of (1.6) were proposed by one of us [17, 18]. For equivariant K-theory (over a compact Lie group G), such a theorem was proved independently, using different approaches, by two of us [16, 19]. The analogy between K-theory and cohomotopy led to the starting-point of the present work, an attempt to prove (1.6). The statement (1.4) grew out of our attempts to explain our proof of (1.6); in order to prove completion theorems in cohomotopy, we were driven to use intermediate results which involved localization as well as completion, and involved classes $\mathscr{H}$ which were not families.

The rest of this paper is organised as follows. Necessary preliminaries about progroups come in §2, and necessary preliminaries about the Burnside ring come in §3. §4 and §5 go to proving (1.4); §6 deduces (1.5) and (1.6); and finally, §7 covers (1.1) and (1.3).

The proof of (1.4) may be summarized as follows. We assemble the result from information "over the rationals", which is easy to come by, and p-adic information, which we derive ultimately from Carlsson [8]. The assembly job is done by (2.3), which is our main algebraic weapon. Carlsson proceeds from his p-adic result to the I-adic statement of the Segal conjecture by quoting the work of May and McClure [26]; our main proof, in §5, subsumes and generalises that part of the proof of the Segal conjecture. (Note that even for p-groups (1.4) gives some new information, because its proof builds in "rational" information.) The steps of our main argument prove special cases of (1.4) which grow successively more general.

In the course of upgrading our information in §5, we need a relation between equivariant cohomotopy over a group G and equivariant cohomotopy over a quotient group G/H. We prepare this result in §4. The difference between the proof of (1.4) and that in [2] is explained by the fact that this relation works much better in cohomotopy than in K-theory, while the

Euler class is much more accessible in K-theory than in cohomotopy. Otherwise the only topological ingredient worth mentioning in §5 is the use of "transfer" in (5.4).

§2. PROGROUPS

In this section we will summarize what we need about progroups. The language of progroups is due to Grothendieck [15] and may be found in [4, 6] and later references.

Inverse systems of Abelian groups, indexed on directed sets, qualify as progroups. The progroups which arise in the examples given in §1 are of this form. However, at the end of §7 we assume that h^* carries any direct limit of G-spaces to an inverse limit in the category of progroups. To construct an inverse limit in the category of progroups, you take all the data contained in your inverse system of progroups, and interpret it as a single progroup [4]. To make this idea work as stated, one generalizes the allowable indexing systems to "filtering categories".

If $\{M_\alpha\}$ and $\{N_\beta\}$ are progroups, one defines

$$\mathrm{Prohom}(\{M_\alpha\},\{N_\beta\}) = \lim_{\overleftarrow{\beta}} \lim_{\overrightarrow{\alpha}} \mathrm{Hom}(M_\alpha, N_\beta),$$

where both limits are taken in the category of groups. There is a unique sensible definition for the composite of prohomomorphisms. The progroups and prohomomorphisms make up a category. A prohomomorphism $\{M_\alpha\} \to \{N_\beta\}$ is a pro-isomorphism if it is an isomorphism in this category.

In §1 we introduced a progroup-valued functor $\mathbf{h}$, giving the definition on objects as $\mathbf{h}(X) = \{h(X_\alpha)\}$. It is easy to supply the definition of h on maps.

The main use of the language of progroups is to make statements about inverse systems which cannot be expressed as statements about their limits. These are mostly statements about exactness. In fact, the category of progroups is an Abelian category, in which one can conduct exactness arguments.

LEMMA 2.1. *The functor $S^{-1}\pi_G^*(-)_I^\wedge$ of* (1.4) *carries pairs and cofiberings to pro-exact sequences.*

Of course, the assertion about "cofiberings" assumes that one introduces the reduced theory $\tilde{\pi}_G^*$ and uses it in the usual way.

It may be reassuring, and help in checking lemmas and details, if we make the definition of "pro-exact" utterly explicit. Let

$$\mathbf{L} \xrightarrow{\mathbf{f}} \mathbf{M} \xrightarrow{\mathbf{g}} \mathbf{N}$$

be a sequence of two prohomomorphisms whose composite is the zero prohomomorphism. By definition, the element

$$\mathbf{f} \in \lim_{\overleftarrow{\beta}} \lim_{\overrightarrow{\alpha}} \mathrm{Hom}(L_\alpha, M_\beta)$$

is a system of compatible elements

$$f_\beta \in \lim_{\overrightarrow{\alpha}} \mathrm{Hom}(L_\alpha, M_\beta),$$

and each f_β is an equivalence class of representatives

$$f_{\alpha\beta} \in \mathrm{Hom}(L_\alpha, M_\beta).$$

The sequence is pro-exact at **M** if for each such representative

$$L_\alpha \xrightarrow{f_{\alpha\beta}} M_\beta$$

there is a diagram

$$\begin{array}{ccc} & & M_\gamma \xrightarrow{g_{\gamma\delta}} N_\delta \\ & & \downarrow m \\ L_\alpha & \xrightarrow{f_{\alpha\beta}} & M_\beta \end{array}$$

in which m is a map of **M**, $g_{\gamma\delta}$ is a representative for some component g_δ of g, and

$$m(\operatorname{Ker} g_{\gamma\delta}) \subset \operatorname{Im} f_{\alpha\beta}.$$

Cultural aside: inverse limits in the category of progroups preserve pro-exactness.

Proof of (2.1). If X_α is a finite G-CW complex, then $\pi_G^n(X_\alpha)$ is a finitely generated Z-module [1] and therefore finitely generated over $A(G)$. Thus $S^{-1}\pi_G^n(X_\alpha)$ is finitely generated over the Noetherian ring $S^{-1}A(G)$. The Artin–Rees lemma [5] may now be used to show that if $X_\alpha \subset Y_\beta$ is a finite pair, the sequence

$$\cdots \to \left\{\frac{S^{-1}\pi_G^n(Y_\beta, X_\alpha)}{(S^{-1}I)^r S^{-1}\pi_G^n(Y_\beta, X_\alpha)}\right\} \to \left\{\frac{S^{-1}\pi_G^n(Y_\beta)}{(S^{-1}I)^r S^{-1}\pi_G^n(Y_\beta)}\right\} \to \left\{\frac{S^{-1}\pi_G^n(X_\alpha)}{(S^{-1}I)^r S^{-1}\pi_G^n(X_\alpha)}\right\} \to \cdots$$

is proexact. Varying the finite pair, we get enough to prove the required proexactness statement for a general pair $X \subset Y$. Similarly for cofiberings.

LEMMA 2.2. *In order to prove that a G-cohomology theory h^* is $\mathscr{H}$-invariant, it is sufficient to verify the following special case: if Z is a pointed G-space such that Z^H is contractible for $H \in \mathscr{H}$, then $\tilde{h}^*(Z) = 0$.*

The proof of (2.2) would be clear if h^* were group-valued. We would assume given an $\mathscr{H}$-equivalence $f: X \to Y$, and apply the assumed property of h^* to the mapping-cone $Z = Y \cup_f CX$. We would then use the exact cohomology sequence of a cofibering (which is the only significant assumption on h^* we need) to show that $h^*(f)$ is an isomorphism.

Of course, this proof carries over to progroup-valued functors, and it is for this purpose that we have stated (2.1) explicitly. The equation "$\tilde{h}^*(Z) = 0$" should now be read "$\tilde{h}^*(Z)$ is prozero". Here a progroup $\{M_\alpha\}$ is prozero if it is a zero object in the category of progroups, and this is equivalent to the following explicit condition: for each of its objects M_α, the progroup has a zero map

$$M_\beta \xrightarrow{0} M_\alpha.$$

Now we need a result for proving that progroups are prozero, and what follows is our main algebraic weapon. Let $\mathbf{M} = \{M_\alpha\}$ be a pro-object of finitely generated modules over a Noetherian ring R; let S be a multiplicative subset of R, and let I be an ideal in R.

LEMMA 2.3. *$S^{-1}\mathbf{M}_I^\wedge$ is prozero iff $S_P^{-1}\mathbf{M}_P^\wedge$ is prozero for each prime ideal $P \subset R$ such that $P \cap S = \varnothing$ and $P \supset I$.*

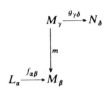

Here S_P^{-1} means "localization at P"; that is, the multiplicative set S_P is the complement of P.

Proof. It is immediate that if $S^{-1}\mathbf{M}_I^\wedge$ is prozero then so are all the other $S_P^{-1}\mathbf{M}_P^\wedge$; we have to argue in the other direction.

First we note that it is enough to prove the special case $S=\{1\}$, in which data are given for all $P\supset I$ and the conclusion is $\mathbf{M}_I^\wedge=0$. For then to prove (2.3) in the generality given, we apply the special case to the promodule $S^{-1}\mathbf{M}$ over $S^{-1}R$; the primes Q of $S^{-1}R$ for which we require data correspond to the primes P of R for which we have data.

Assuming $S=\{1\}$, we take a typical term in $\mathbf{M}_I^\wedge$, say $T=M_\alpha/(I^rM_\alpha)$. We will find a finite number of prime ideals $P_1, P_2, \ldots, P_n$ containing I and integers $s(i)$ such that the map

$$T=M_\alpha/(I^rM_\alpha)\to\bigoplus_i S_{P_i}^{-1}(M_\alpha/(I^r+P_i^{s(i)})M_\alpha)$$

is mono.

In fact, we take $P_1, P_2, \ldots, P_n$ to be the associated prime ideals of T, which are finite in number by a standard result [22]. These prime ideals contain I^r, and therefore contain I. Let $L_i\subset T$ be the submodule annihilated by P_i. By the Artin–Rees lemma [5] there exists $s(i)$ such that

$$L_i\cap P_i^{s(i)}T\subset P_iL_i=0.$$

We will show that the kernel K_i of the map

$$T\to S_P^{-1}(T/P_i^{s(i)}T)$$

does not have P_i as an associated prime.

For suppose it did, and for convenience write P, L, s, K instead of $P_i, L_i, s(i), K_i$. Then we would have a monomorphism $R/P\to K$, which must map into L. Since $L\to T/P^sT$ is mono by the choice of s, we would get a monomorphism $R/P\to T/P^sT$. Since localization preserves exactness, we would get the following commutative diagram.

$$\begin{array}{ccc} R/P & \xrightarrow{\text{mono}} & T/P^sT \\ {\scriptstyle\text{mono}}\downarrow & & \downarrow \\ S_P^{-1}(R/P) & \xrightarrow{\text{mono}} & S_P^{-1}(T/P^sT) \end{array}$$

But the diagonal is zero because we assumed R/P mapped into K. This contradiction shows that K_i does not have P_i as an associated prime.

But then the kernel of

$$T\to\bigoplus_i S_{P_i}^{-1}(T/P_i^{s(i)}T)$$

has no associated primes, and must be zero as claimed.

Given $T=M/(I^rM_\alpha)$, we now have the following commutative diagram for any map $m\colon M_\beta\to M_\alpha$ in $\mathbf{M}$.

$$\begin{array}{ccc} M_\beta/(I^rM_\beta) & \longrightarrow & M_\alpha/(I^rM_\alpha)=T \\ \downarrow & & \downarrow{\scriptstyle\text{mono}} \\ \bigoplus_i S_{P_i}^{-1}(M_\beta/(I^r+P_i^{s(i)})M_\beta) & \to & \bigoplus_i S_{P_i}^{-1}(M_\alpha/(I^r+P_i^{s(i)})M_\alpha) \end{array}$$

For each P_i our hypotheses allow us to choose m so that $S_{P_i}^{-1}M_\beta$ maps to zero in $S_{P_i}^{-1}(M_\alpha/P_i^{s(i)}M_\alpha)$. We can do this for a finite number of i, and so ensure that the lower horizontal arrow is zero; then m must be zero. This proves (2.3).

§3. THE BURNSIDE RING

In this section we will say what we need about the Burnside ring.

The Burnside ring $A(G)$ is the Grothendieck group constructed from (finite) G-sets [12]. For each subgroup $H \subset G$ there is a homomorphism of rings

$$\phi_H: A(G) \to Z$$

which carries a G-set W to $|W^H|$; ϕ_H depends only on the conjugacy class of H. With these maps as components, we obtain a map

$$\Phi: A(G) \to \prod_{(H)} Z,$$

where the product runs over all conjugacy classes (H); Φ is mono. By the going-up theorem [5], each prime ideal P of $A(G)$ is the restriction of a prime ideal of $\prod_{(H)} Z$; that is, it may be written in the form

$$q(H,p) = \phi_H^{-1}(p),$$

for some H and some prime ideal (p) in Z. Here (p) is clearly determined by P; however, we may still get the same ideal $q(H,p)$ for different choices of H. Fix a prime $p > 0$; for each subgroup $H \subset G$, let H_p be the smallest normal subgroup of H such that H/H_p is a p-group. Then $(H_p)_p$ is a characteristic subgroup of H_p, and hence normal in H, so $(H_p)_p = H_p$; thus H_p is "p-perfect", meaning that any quotient of it which is a p-group is trivial. Dress [12] says that H and K are p-equivalent, and writes $H \sim_p K$, if H_p is conjugate to K_p; he shows that $q(H,p) = q(K,p)$ iff $H \sim_p K$. The "support" of $q(H,p)$ is then the conjugacy class of H_p. For $p = 0$ we can interpret this discussion in the same way as for any other prime which does not divide $|G|$; H_0 becomes H, and 0-equivalence becomes conjugacy.

In the rest of this paper we shall make free use of localization with respect to prime ideals in $A(G)$. Integer denominators are sometimes more convenient than general elements of $A(G)$, and we can reduce to that case. Let P be a prime in $A(G)$, and let (p) be its counter-image under $Z \to A(G)$; we write $S_{(p)}^{-1}$ for localization over Z at (p).

LEMMA 3.1. *The map $S_{(p)}^{-1}A(G) \to S_P^{-1}A(G)$ is epi.*

To prove this conveniently, we discuss the idempotents in $S_{(p)}^{-1}A(G)$. Such idempotents have been used by several authors [11 p8, 14, 3, 31]. We continue to write ϕ_H after localizing at (p). If $H \sim_p K$ then $\phi_H(x) \equiv \phi_K(x) \bmod p$ for any x; in particular, if e is idempotent then $\phi_H(e)$ must be constant at 0 or 1 as H runs over a p-equivalence class. By a standard result of commutative algebra [7] the Boolean algebra of idempotents in $S_{(p)}^{-1}A(G)$ is canonically isomorphic to the Boolean algebra of open-and-closed sets in spec $S_{(p)}^{-1}A(G)$. This spectrum has been explicitly described by Dress [12]; it is the disjoint union of finitely many open sets, each containing just one of the ideals $q(H,p)$. There is therefore just one primitive idempotent e_H in $S_{(p)}^{-1}A(G)$ for each conjugacy class of p-perfect subgroups H, given by

$$\phi_K(e_H) = \begin{cases} 1 \text{ if } K \sim_p H \\ 0 \text{ if } K \not\sim_p H. \end{cases}$$

These idempotents can also be obtained by more elementary methods. For $(p)=0$ we can interpret this discussion in the obvious way.

Proof of (3.1). Let $H\in \mathrm{Supp}(P)$ and let $e=e_H$ be the corresponding idempotent. Consider the map

$$A(G)\to S_{(p)}^{-1}A(G)\to eS_{(p)}^{-1}A(G).$$

This is a map of rings which carries every element of S_P to an invertible element. (The target $eS_{(p)}^{-1}A(G)$ is a local ring because it has only one maximal ideal, and the counter-image of that maximal ideal is P.) It is also universal among such maps. [Any such map carries e to an invertible element and $(1-e)$ to zero.] This characterizes the target as $S_P^{-1}A(G)$. But clearly the map

$$S_{(p)}^{-1}A(G)\to eS_{(p)}^{-1}A(G)$$

is epi.

It may be helpful to know that the localized cohomology theory $S_P^{-1}h^*$ is the same as that obtained by first localizing over Z to get $S_{(p)}^{-1}h^*$, and then taking the summand $e_H S_{(p)}^{-1}h^*$; compare [20].

§4. TOPOLOGICAL PRELIMINARIES

In this section we prove a topological result needed for the main proof. Let G be a finite group and (p) a given prime; let $H=G_p$ and let P be the corresponding prime ideal $q(G,p)=q(H,p)$ in $A(G)$, as in §3.

PROPOSITION 4.1. *Then there is an isomorphism*

$$S_P^{-1}\tilde{\pi}_G^n(X)\cong S_{(p)}^{-1}\tilde{\pi}_{G/H}^n(X^H)$$

natural as X runs over finite pointed G-spaces.

Results of this sort were known to Araki [3] McClure [21] and probably others. We separate off the first part of the proof.

LEMMA 4.2. *Restriction gives a natural isomorphism*

$$S_P^{-1}\{X,Y\}^G\xrightarrow{\cong}S_P^{-1}\{X^H,Y\}^G.$$

Here X runs over finite pointed G-spaces; Y runs over pointed G-spaces which may be infinite; and $\{X,Y\}^G$ means stable G-homotopy-classes of stable G-maps.

Sketch proof of (4.2). $S_P^{-1}\{X,Y\}^G$ is one group of a G-cohomology-theory which is zero on all the G-cells

$$(G/K)\times E^m,(G/K)\times S^{m-1}$$

of X which are not in X^H. See [20], Theorem 4.8.

Proof of (4.1). By suspending X if necessary we can assume $n\geqslant 0$.

First we construct the natural transformation. Restriction on H-fixed-point-sets gives a natural map

$$\tilde{\pi}_G^n(X)\to\tilde{\pi}_{G/H}^n(X^H).$$

This is a map of $A(G)$-modules, provided we make $A(G)$ act on $\tilde{\pi}^n_{G/H}(X^H)$ via the homomorphism $\theta: A(G) \to A(G/H)$ which carries a G-set W to W^H. Notice now that G/H is a p-group, $S_{(p)}^{-1} A(G/H)$ is a local ring, and the counter-image of its unique maximal ideal in $A(G)$ is P. Thus θ carries an element of $A(G)$ not in P to an element invertible in $S_{(p)}^{-1} A(G/H)$. So we get an induced map

$$\phi: S_P^{-1} \tilde{\pi}^n_G(X) \to S_{(p)}^{-1} \tilde{\pi}^n_{G/H}(X^H).$$

We show that ϕ is epi. The map

$$S_{(p)}^{-1} \tilde{\pi}^n_G(X^H) \to S_{(p)}^{-1} \tilde{\pi}^n_{G/H}(X^H)$$

is split epi because any representative (G/H)-map is also a G-map. *A fortiori*,

$$S_P^{-1} \tilde{\pi}^n_G(X^H) \to S_{(p)}^{-1} \tilde{\pi}^n_{G/H}(X^H)$$

is epi. The map

$$S_P^{-1} \tilde{\pi}^n_G(X) \to S_P^{-1} \tilde{\pi}^n_G(X^H)$$

is epi by (4.2) applied to $Y = S^n$.

We show that ϕ is mono. Take an element of $S_P^{-1} \tilde{\pi}^n_G(X)$; using (3.1), we may write it $[f]/d$, where d is an integer prime to p and f is a representative G-map

$$S^V \wedge X \xrightarrow{f} S^V \wedge S^n$$

for a suitable representation V of G. Now assume $[f]/d \in \operatorname{Ker} \phi$. Then after increasing both d and V, we may assume that the restriction of f to $S^{V^H} \wedge X^H$ is G-nullhomotopic. Thus $[f]$ maps to zero in

$$S_P^{-1} \{S^{V^H} \wedge X^H, S^V \wedge S^n\}^G.$$

But then $[f]$ maps to zero in

$$S_P^{-1} \{S^V \wedge X, S^V \wedge S^n\}^G$$

by (4.2) applied to $Y = S^V \wedge S^n$. Thus $[f]/d = 0$.

§5. THE MAIN PROOF

In this section we will prove Theorem 1.4. By (2.3) it is sufficient to consider $S_P^{-1} \pi^*_G(X)^\wedge_P$; in this case the only relevant assumption is the contractibility of X^H for one conjugacy class of H.

LEMMA 5.1. *Let G be a p-group. If X^1, the underlying space of X, is contractible, then $\tilde{\pi}^*_G(X)^\wedge_{(p)}$ is prozero.*

The result remains true in a trivial way if we take (p) to be the prime ideal (0), for we have to interpret it so that $G = 1$ is the only group which qualifies.

Proof of (5.1). Carlsson [8] proves that the inverse limit of the inverse system

$$\left\{ \frac{\tilde{\pi}^n_G(X_\alpha)}{p^r \tilde{\pi}^n_G(X_\alpha)} \right\}$$

is zero. Since the groups of this inverse system are finite groups, it follows that the inverse system is prozero.

Now let G be a finite group and (p) a given prime; let $H=G_p$ and $P=q(G,p)=q(H,p)$ be as in §4.

LEMMA 5.2. *If X^H is contractible, then $S_P^{-1}\tilde{\pi}_G^*(X)_P^\wedge$ is prozero.*

Proof. By (4.1) we have an isomorphism of progroups

$$\left\{\frac{S_P^{-1}\tilde{\pi}_G^n(X_\alpha)}{p^r S_P^{-1}\tilde{\pi}_G^n(X_\alpha)}\right\} \overset{\cong}{\longleftrightarrow} \left\{\frac{S_{(p)}^{-1}\tilde{\pi}_{G/H}^n(X_\alpha{}^H)}{p^r S_{(p)}^{-1}\tilde{\pi}_{G/H}^n(X_\alpha{}^H)}\right\}.$$

Since G/H is a p-group and X^H is contractible, the right-hand side is prozero by (5.1). That is, $S_P^{-1}\tilde{\pi}_G^*(X)_{(p)}^\wedge$ is prozero. Since completion at P is more drastic than completion at (p), the result follows.

Now let P be a general prime ideal $q(H,p)$ in $A(G)$, where H is p-perfect.

PROPOSITION 5.3. *If X^H is contractible, then $S_P^{-1}\tilde{\pi}_G(X)_P^\wedge$ is prozero.*

The proof involves a construction. Let N be the normalizer of H in G, and let F/H be a Sylow p-subgroup of N/H, where F/H is interpreted as H/H if $(p)=0$.

LEMMA 5.4. *$S_P^{-1}\tilde{\pi}_G^n(X)_P^\wedge$ is a direct summand in*

$$\{S_P^{-1}\pi_G^n((G/F)\times X_\alpha, G/F)\}_P^\wedge.$$

Here $\{S_P^{-1}\pi_G^n((G/F)\times X_\alpha, G/F)\}$ is a progroup in which X_α runs over the finite subcomplexes of X.

LEMMA 5.5. *If X^H is contractible, then $\{S_P^{-1}\pi_G^n((G/F)\times X_\alpha, G/F)\}_P^\wedge$ is prozero.*

Proof of 5.4. For each finite G-space X_α we have a G-covering-map

$$(G/F)\times X_\alpha, G/F \xrightarrow{\varpi} X_\alpha, pt$$

natural in X_α. This gives the following commutative diagram, in which "Tr" means transfer—see, for example, [1] or [21].

$$\begin{array}{ccc} & \pi_G^n((G/F)\times X_\alpha, G/F) & \\ {}^{\varpi^*}\nearrow & & \searrow^{\mathrm{Tr}} \\ \pi_G^n(X_\alpha, pt) & \xrightarrow{\quad [G/F] \quad} & \pi_G^n(X_\alpha, pt) \end{array}$$

The horizontal arrow is multiplication by the class of G/F in $A(G)$. Using the fact that H is p-perfect and F/H is a p-group, we find

$$\phi_H(G/F)=|N/F|$$

which is prime to p by the choice of F. Thus $[G/F]$ does not lie in P, and on localizing at P we get the following commutative diagram, which is natural for maps of X_α.

$$\begin{array}{ccc} & S_P^{-1}\pi_G^n((G/F)\times X_\alpha, G/F) & \\ {}^{\varpi^*}\nearrow & & \searrow^{\mathrm{Tr}} \\ S_P^{-1}\pi_G^n(X_\alpha, pt) & \xrightarrow{\quad \cong \quad} & S_P^{-1}\pi_G^n(X_\alpha, pt) \end{array}$$

The conclusion follows.

Proof of (5.5). Of course we use the natural isomorphism

$$\pi_G^n((G/F)\times X_\alpha, G/F)\cong \pi_F^n(X_\alpha, pt).$$

This isomorphism is a map of $A(G)$-modules, if we make $A(G)$ act on $\pi_F^n(X_\alpha, pt)$ via the restriction map i^*: $A(G)\to A(F)$. It follows that

$$\{S_P^{-1}\pi_G^n((G/F)\times X_\alpha, G/F)\}\hat{}_P=\{S^{-1}\pi_G^n(X_\alpha, pt)\}\hat{}_I,$$

where on the right-hand side localization and completion are done over $A(F)$, taking

$$S=i^*S_P, \qquad I=(i^*P)A(F).$$

We now wish to prove that $S^{-1}\tilde{\pi}_F^n(X)\hat{}_I$ is prozero. By (2.3), it is sufficient to prove that $S_Q^{-1}\tilde{\pi}_F^n(X)\hat{}_Q$ is prozero for each prime ideal Q of $A(F)$ such that $Q\cap S=\varnothing$ and $Q\supset I$. Equivalently, we have to consider prime ideals Q whose counter-image in $A(G)$ is P. We will show there is only one such ideal, namely the ideal $q(H,p)$ of $A(F)$.

Any such Q has to be an ideal $q(K,p)$ of $A(F)$ for the same p and some $K\subset F$ which is p-perfect and conjugate to H in G; it follows that $K=H$.

Thus (5.2) applies and shows $S_Q^{-1}\tilde{\pi}_F^n(X)\hat{}_Q=0$. This proves (5.5).

Proposition 5.3 follows from (5.4) and (5.5). Theorem 1.4 follows immediately by assembling (2.2), (2.3) and (5.3).

§6.

In this section we deduce (1.5) and (1.6).

Proof of (1.5). We deduce (1.5) from (1.4) by taking the ideal I in (1.4) to be the ideal $I(\mathscr{F})$ in (1.5). For this it is enough to show that if $\mathscr{F}$ is a family, then the class

$$\mathscr{H}=\{\text{Supp}(P): P\supset I(\mathscr{F})\}$$

in (1.4) is contained in $\mathscr{F}$.

Since $\mathscr{F}$ is a family, the ideal

$$I(\mathscr{F})=\bigcap_{H\in\mathscr{F}}\text{Ker}(A(G)\to A(H))$$

is the intersection of the prime ideals $q(H,0)$ over $H\in\mathscr{F}$. If $P\supset I(\mathscr{F})$, then P must contain one of these ideals $q(H,0)$. According to Dress [12], this means that $P=q(H,p)$ for some H and some p. Since $H_p\subset H\in\mathscr{F}$ and $\mathscr{F}$ is a family, we have $\text{Supp}(P)\subset\mathscr{F}$. This holds for each $P\supset I(\mathscr{F})$, so $\mathscr{H}\subset\mathscr{F}$.

[Of course, $\mathscr{H}=\mathscr{F}$ since every $H\in\mathscr{F}$ is the support of an ideal $q(H,0)$.]

We turn to (1.6). Let $\mathscr{F}$ be a family.

LEMMA 6.1. *Let Y be a finite G-CW complex such that Y^H is empty for $H\notin\mathscr{F}$. Then $\pi_G^*(Y)$ is annihilated by some power of $I(\mathscr{F})$.*

LEMMA 6.2. *Let Y be a G-space such that Y^H is empty for $H\notin\mathscr{F}$. Then the canonical pro-map*

$$\pi_G^*(Y)\to\pi_G^*(Y)\hat{}_{I(\mathscr{F})}$$

is a pro-isomorphism.

We omit the proof of (6.1) and the proof of (6.2) from (6.1); both are sufficiently well known, and the ideas go back to [6].

Lemma 6.2 applies to $Y = E\mathscr{F} \times X$ and shows that in (1.6), the right-hand side $\pi_G^*(E\mathscr{F} \times X)$ is already complete.

§7

In this section we explain (1.1) and prove (1.3).

We say that a G-CW complex X is an "$\mathscr{H}$–complex" if its G-cells are all of the form $G/H \times E^m$ with $H \in \mathscr{H}$. It is easy to prove the appropriate generalization of the "theorem of J. H. C. Whitehead"; this says that if X is an $\mathscr{H}$–complex and $f\colon Y \to Z$ is an $\mathscr{H}$-equivalence, then the induced map

$$f_*\colon [X, Y]^G \to [X, Z]^G$$

is a bijection. In particular, an $\mathscr{H}$-equivalence between $\mathscr{H}$–complexes is a G-homotopy-equivalence; this was certainly known to previous authors [34, §1].

Assume, as in (1.1), that $\mathscr{H}$ is closed under passing to larger subgroups. Then an $\mathscr{H}$-equivalence $f\colon X \to Y$ induces a map $f^{\mathscr{H}}\colon X^{\mathscr{H}} \to Y^{\mathscr{H}}$ which is an $\mathscr{H}$-equivalence between $\mathscr{H}$–complexes, and therefore a G-homotopy-equivalence by the remarks above. Now (1.1) follows.

One may also deduce the result from [21, II, 9.3].

We turn to (1.3). Our assumptions on h^* are as follows. It is Z-graded and satisfies Eilenberg–Steenrod Axioms 1–6, with the words "exact" and "isomorphism" interpreted as "pro-exact" and "pro-isomorphism" if h^* is progroup-valued. No axiom of "suspension with respect to arbitrary representations" is required.

Proof of (1.3). First we define the required class $\mathscr{H}$. For any subgroup K, let $\mathscr{C}(K)$ be the complement of the conjugacy class (K), i.e., the class of subgroups not conjugate to K. We lay down that K is not in $\mathscr{H}$ if and only if h^* is invariant with respect to $\mathscr{C}(K)$. It follows that if h^* if $\mathscr{L}$-invariant, then $\mathscr{L} \supset \mathscr{H}$; for if $K \notin \mathscr{L}$, then $\mathscr{C}(K) \supset \mathscr{L}$, so the $\mathscr{L}$-invariance of h^* implies the $\mathscr{C}(K)$-invariance, and $K \notin \mathscr{H}$. This justifies the words "unique minimal" in (1.3). It remains to prove that h^* is $\mathscr{H}$-invariant.

Let $\mathscr{F}$ be a family, and let $\mathscr{F}' = \mathscr{F} \cup (H)$ be the "adjacent" family obtained by adjoining the conjugacy class of a subgroup H all of whose proper subgroups lie in $\mathscr{F}$ Consider the map

$$i\colon X \wedge (E\mathscr{F} \sqcup P) \to X \wedge (E\mathscr{F}' \sqcup P).$$

Then i^K is an equivalence for $K \not\sim H$, so if $H \notin \mathscr{H}$ it follows that $h^*(i)$ is iso..If $H \in \mathscr{H}$, the same conclusion follows trivially if we assume X^H contractible. Suppose then that X^H is contractible for all $H \in \mathscr{H}$. We can get from the empty family to the family of all subgroups by a finite number of the steps considered above; so $h^*(i)$ is iso for the map

$$i\colon X \wedge P \to X \wedge S^0.$$

That is, $\tilde{h}^*(X) = 0$. Now (2.2) shows that h^* is $\mathscr{H}$-invariant.

We remark that this proof carries over when G becomes a compact Lie group. We need one more assumption on h^*: it carries any direct limit of G-spaces to an inverse limit in the category of progroups (see §2). The finite induction implicit in the proof above is replaced by an appeal to Zorn's Lemma, using the class $\mathscr{C}$ of families $\mathscr{F}$ such that

$$\tilde{h}^*(X \wedge (E\mathscr{F} \sqcup P)) = 0.$$

The induction starts because this class contains the empty family. We have to show that for any totally ordered subset $\{\mathscr{F}_\alpha\}$ of families in $\mathscr{C}$, the union $\mathscr{F} = \bigcup_\alpha \mathscr{F}_\alpha$ will serve as an upper bound in $\mathscr{C}$. For this we construct the homotopy-limit $\operatorname{Holim}_{\vec{\alpha}} E\mathscr{F}_\alpha$. The construction of this limit involves extending G-maps $(\partial\sigma^n) \times E\mathscr{F}_\alpha \to E\mathscr{F}_\beta$ over $\sigma^n \times E\mathscr{F}_\alpha$ by induction over n, but this is certainly possible in view of the properties of $E\mathscr{F}_\alpha$, $E\mathscr{F}_\beta$ and the fact that we always have $\mathscr{F}_\alpha \subset \mathscr{F}_\beta$. We observe that $\operatorname{Holim}_{\vec{\alpha}} E\mathscr{F}_\alpha$ qualifies as $E\mathscr{F}$, so

$$X \wedge (E\mathscr{F} \sqcup P) = \operatorname{Holim}_{\vec{\alpha}}(X \wedge (E\mathscr{F}_\alpha \sqcup P))$$

and the assumed property of h^* gives

$$\tilde{h}^*(X \wedge (E\mathscr{F} \sqcup P)) = 0.$$

Zorn's lemma now shows that $\mathscr{C}$ has a maximal element $\mathscr{F}$. If $\mathscr{F}$ were not the family of all subgroups, then there would be a subgroup H minimal among subgroups not in $\mathscr{F}$, and the argument above applied to $\mathscr{F}' = \mathscr{F} \cup (H)$ would yield a contradiction. Thus we conclude $\tilde{h}^*(X) = 0$, as before.

Acknowledgements—J.F.A. and S.J. thank the University of Chicago and Northwestern University, and S.J. similarly thanks also the Forschungsinstitut für Mathematik ETH Zurich, for hospitality during discussions on the successive versions of this paper.

REFERENCES

1. J. F. Adams: Prerequisites for Carlsson's lecture, *Lecture Notes in Mathematics*, Springer, Berlin **1051** (1984), 483–532.
2. J. F. Adams, J.-P. Haeberly, S. Jackowski and J. P. May: A generalization of the Atiyah–Segal completion theorem, *Topology* **27** (1988), 1–6.
3. S. Araki: Equivariant stable homotopy theory and idempotents of Burnside rings, preprint (1983).
4. M. Artin and B. Mazur: Etale homotopy, *Lecture Notes in Mathematics*, Springer, Berlin **100** (1969), 154–166.
5. M. F. Atiyah and I. G. Macdonald: *Introduction to commutative algebra*, Addison–Wesley U.K. (1969), 62–107.
6. M. F. Atiyah and G. B. Segal: Equivariant K-theory and completion, *J. Diff. Geom.* **3** (1969), 1–18.
7. H. Bass: Algebraic K-theory, W. A. Benjamin U.S.A. (1968), p. 103.
8. G. Carlsson: Equivariant stable homotopy and Segal's Burnside ring conjecture, *Ann. of Math.* **120** (1984), 189–224.
9. J. Caruso and J. P. May: Completions in equivariant cohomology theory. To appear.
10. T. tom Dieck: Orbitypen und aquivariante Homologie I, *Arch. Math.* **23** (1972), 307–317.
11. T. tom Dieck: Transformation groups and representation theory, *Lecture Notes in Mathematics*, Springer, Berlin **766** (1979).
12. A. Dress: A characterisation of solvable groups, *Math. Zeit.* **110** (1969), 213–217.
13. A. D. Elmendorff: Systems of fixed point sets, *Trans. Am. Math. Soc.* **277** (1983), 275–284.
14. D. Gluck: Idempotent formula for the Burnside algebra with applications to the p-subgroup simplicial complex, *Illinois J. Math.* **25** (1981), 63–67.
15. A. Grothendieck: Technique de descente et théorems d'existence en géometrie algébrique II, Seminaire Bourbaki 12ième année, (1959–60), exp. 195.
16. J.-P. Haeberly: Completions in equivariant K-theory, Thesis, University of Chicago (1983).
17. S. Jackowski: Localisation and completion theorems in equivariant cohomology theories, Thesis, University of Warsaw (1976).
18. S. Jackowski: Equivariant K-theory and cyclic subgroups, *Lecture Notes Lond. Math. Soc. C.U.P.* **26** (1977), 76–92.

19. S. Jackowski: Families of subgroups and completion, *J. Pure and Appl. Algebra* **37** (1985), 167–179.
20. C. Kosniowski: Equivariant cohomology and stable cohomotopy, *Math. Ann.* **210** (1974), 83–104.
21. L. G. Lewis, Jr, J. P. May and M. Steinberger (with contributions by J. E. McClure): Equivariant stable homotopy theory, *Lecture Notes in Mathematics*, Springer, Berlin, **1213** (1986).
22. H. Matsumara: Commutative Algebra, W. A. Benjamin, U.S.A. (1970), p. 52.
23. T. Matumoto: on *G-CW*-complexes and a theorem of J. H. C. Whitehead, *J. Fac. Sci. Univ. Tokyo, Sect. IA Math.* **18** (1971/72), 363–374.
24. J. P. May: The completion conjecture in equivariant cohomology, *Lecture Notes in Mathematics*, Springer, Berlin **1051** (1984), 620–637.
25. J. P. May: A further generalization of the Segal conjecture, in preparation.
26. J. P. May and J. E. McClure: A reduction of the Segal conjecture, *Can. Math. Soc. Conference Proceedings*, Vol. 2 (1982), 209–222.
27. R. S. Palais: The classification of *G*-spaces, *Memoirs Am. Math. Soc.* **36** (1960).
28. G. B. Segal: The representation ring of a compact Lie group, *Inst. Hautes Etudes Sci. Publ. Math.* **34** (1968), 113–128.
29. G. B. Segal: Equivariant *K*-theory, *Inst. Hautes Etudes Sci. Publ. Math.* **34** (1968), 129–151.
30. G. B. Segal: Equivariant stable homotopy theory, in *Proceedings of the International Congress of Mathematicians* 1970, Gauthier–Villars, Paris, **2** (1971), 59–63.
31. T. Yoshida: Idempotents in Burnside rings and Dress induction theorem, *J. Algebra* **80** (1983), 90–105.
32. L. G. Lewis, Jr., J. P. May and J. E. McClure: Classifying G-spaces and the Segal conjecture, *Can. Math. Soc. Conference Proceedings*, **2** (1982), 165–179.
33. J. P. May: Stable maps between classifying spaces, *Contemp. Math.* **37** (1985), 121–129.
34. R. M. Seymour: On *G*-cohomology theories and Kunneth formulae, *Can. Math. Soc. Conference Proceedings*, **2** (1982), 251–271.
35. R. M. Seymour: A Kunneth formula for *RO*(*G*)-graded equivariant cohomology theories, preprint.

APPENDIX: BY J. P. MAY

In [32] the Segal conjecture was generalized to the assertion that equivariant cohomotopy with coefficients in equivariant classifying spaces is $\{1\}$-invariant. This generalization specializes to give a calculation of the stable maps between classifying spaces in the non-equivariant world [33]. In a later paper [25], I will use the theorem below to prove that equivariant cohomotopy with coefficients in equivariant classifying spaces satisfies the $\mathscr{H}$-invariance property analogous to (1.4) above. This generalization of (1.4) specializes to give a calculation of the equivariant stable maps between equivariant classifying spaces.

The proof of (1.4) takes given information about p-groups and p-adic completion, namely that supplied by Carlsson and quoted as (5.1), and derives from it the strongest implications. This idea works in considerable generality.

Let h^*_G be a Z-graded cohomology theory defined on G-CW complexes. We want h^*_G to take values in modules over $A(G)$, and the natural way to ensure this is to require h^*_G to be $RO(G)$-gradable, or equivalently, representable. We also require h^*_G to be of finite type, in the sense that each $h^q_G(G/H)$ is finitely generated, and this ensures that each $h^q_G(X)$ is finitely generated when X is a finite G-CW complex. We obtain a Z-graded progroup-valued cohomology theory on general G-CW complexes by setting $\mathbf{h}^*_G(X)=\{h^*_G(X_\alpha)\}$ as in §1.

We need a relation like (4.1), and this requires us to construct representable theories $h^*_{H/K}$ on H/K-CW complexes for subquotient groups H/K of G. There is a sensible way to do this [24, p. 626; 9;3; 21, II, §9], the evident analog of (4.1) holds [3; 21, V, §6] and so does the analog of (5.4). When h^*_G is stable G-cohomotopy with coefficients in a G-space Y, the theory associated to H/K is just stable H/K-cohomotopy with coefficients in the H/K-space Y^K. In this case, the proofs above of (4.1) and (5.4) apply with only notational changes.

From here, one can argue exactly as in §5 to reach the following conditional conclusion.

Theorem. *Suppose that $h^*_{H/K}$ is of finite type for each subquotient H/K of G and that $\mathbf{h}^*_{H/K}(X)^{\wedge}_{(p)}=0$ whenever H/K is a p-group and X is a nonequivariantly contractible H/K-space. Then the theory $S^{-1}\mathbf{h}^*_G(-)^{\wedge}_I$ is $\mathscr{H}$-invariant, where*

$$\mathscr{H}=\cup\{\mathrm{Supp}(P)|P\cap S=\varnothing\,\&\,P\supset I\}.$$

A comparable reduction from general p-groups to elementary Abelian p-groups can be obtained by Carlsson's methods [24, 9].

Department of Pure Mathematics,
University of Cambridge,
16 *Mill Lane,*
*Cambridge CB*2 1*SB, U.K.*

Dept. of Mathematics
University of Washington
Seattle, WA 98195
U.S.A.

Mathematical Institute
University of Warsaw
Palac Kultury i Nauki IXp
00–901 *Warszawa*
Poland

Dept. of Mathematics
University of Chicago
Chicago, IL 60637
U.S.A.

Topology Vol. 27, No. 1, pp. 1–6, 1988

A GENERALIZATION OF THE ATIYAH–SEGAL COMPLETION THEOREM

J. F. Adams, J.-P. Haeberly, S. Jackowski and J. P. May

(*Received* 9 *June* 1986)

§1. INTRODUCTION

We shall give a geodesic path from the equivariant Bott periodicity theorem to the generalized completion theorem in equivariant K-theory.

Let G be a compact Lie group. Our G-spaces are understood to be G-CW complexes and we let K_G^* denote the progroup valued G-cohomology theory specified by

$$K_G^n(X) = \{K_G^n(X_\alpha)\},$$

where X_α runs over the finite subcomplexes of X. For a subgroup H of G, we have a restriction homomorphism $r_H^G\colon R(G) \to R(H)$ and we let I_H^G be its kernel. (Subgroups are understood to be closed.) A set $\mathscr{J}$ of subgroups of G closed under subconjugacy is called a family. We let $(K_G^*)\hat{}_{\mathscr{J}}$, the $\mathscr{J}$-adic completion of K_G^*, denote the progroup valued G-cohomology theory specified by

$$K_G^n(X)\hat{}_{\mathscr{J}} = \{K_G^n(X_\alpha)/JK_G^n(X_\alpha)\},$$

where J runs over the finite products of ideals I_H^G with $H \in \mathscr{J}$. (The relevant information about progroups is summarized in [2, §2].)

Theorem 1.1. *If a G-map $f\colon X \to Y$ restricts to a homotopy equivalence $f^H\colon X^H \to Y^H$ for each $H \in \mathscr{J}$, then $(f^*)\hat{}_{\mathscr{J}}\colon (K_G^* Y)\hat{}_{\mathscr{J}} \to (K_G^* X)\hat{}_{\mathscr{J}}$ is an isomorphism.*

The same assertion holds with K_G and $R(G)$ replaced by KO_G and $RO(G)$.

Theorem 1.1 was first conjectured in 1976 [8] and was first proven, independently, by two of us in 1983 [7, 9]. The case $\mathscr{J} = \{1\}$ is the Atiyah–Segal completion theorem of [4], and the proof in [7] follows [4] in outline. The proof in [9] contained the key idea of proceeding by direct induction rather than giving unitary groups and tori a privileged role. Our variant of this idea exploits an argument due to Carlsson [5] in cohomotopy to obtain an immediate reduction to quotation of Bott periodicity for the equivariant K-theory of G-spheres. It is to be emphasized that our argument, like that of [9], includes a new proof of the original Atiyah–Segal theorem.

We use (1.1) to compute equivariant K-theory characteristic classes in §2. We prove (1.1) in §3 and make a few remarks on it in §5. In §4, we use (1.1) to prove the following mixed localization and completion theorem. Its cohomotopy analog was the main result of our paper [2], and more discussion of such invariance theorems may be found there. Pro-$R(G)$-modules are localized termwise, $S^{-1}\{M_\alpha\} = \{S^{-1}M_\alpha\}$.

Theorem 1.2. *Let $S \subset R(G)$ be a multiplicative set, let $I \subset R(G)$ be an ideal, and define*

$$\mathscr{H} = \cup\{\mathrm{Supp}(P) \mid P \cap S = \phi \text{ and } P \supset I\}.$$

If a G-map f: $X \to Y$ restricts to a homotopy equivalence f^H: $X^H \to Y^H$ for all $H \in \mathscr{H}$, then $S^{-1}(f^*)_I^\wedge$: $S^{-1}K_G^*(Y)_I^\wedge \to S^{-1}K_G^*(X)_I^\wedge$ is an isomorphism. The same assertion holds with K_G and $R(G)$ replaced by KO_G and $RO(G)$.

Here P runs over prime ideals of $R(G)$ and Supp(P) is the support of P as defined by Segal [13]: $H \in$ Supp(P) if P comes from H via the restriction map $R(G) \to R(H)$ and P does not come from any $K \subset H$. Segal shows that Supp(P) is a single conjugacy class of (topologically) cyclic subgroups H. The theorem has content even when $S = \{1\}$ and $I = 0$.

Corollary 1.3. *If a G-map f: $X \to Y$ restricts to a homotopy equivalence f^H: $X^H \to Y^H$ for all cyclic subgroups H, then f^*: $K_G^*(Y) \to K_G^*(X)$ is an isomorphism.*

For finite groups G, this result goes back to [8].

§2. EQUIVARIANT K-THEORY OF CLASSIFYING SPACES

The main motivation for Theorem 1.1 comes from the following consequence (which is actually equivalent to the theorem).

Let $E\mathscr{J}$ be a universal $\mathscr{J}$-free G-space, so that $(E\mathscr{J})^H$ is contractible if $H \in \mathscr{J}$ and is empty if $H \notin \mathscr{J}$. For any G-space X, the projection $E\mathscr{J} \times X \to X$ restricts to a homotopy equivalence $(E\mathscr{J} \times X)^H \to X^H$ for each $H \in \mathscr{J}$, so (1.1) gives an isomorphism $K_G^*(X)_{\mathscr{J}}^\wedge \to K_G^*(E\mathscr{J} \times X)_{\mathscr{J}}^\wedge$. For a G-space Y, such as $E\mathscr{J} \times X$, all of whose isotropy groups are in $\mathscr{J}$, the groups of the inverse system $K_G^*(Y)$ are $\mathscr{J}$-adically complete. For a finite G-CW complex X, the inverse system $K_G^*(X)_{\mathscr{J}}^\wedge$ satisfies the Mittag–Leffler condition. These facts imply that the algebraic completion $K_G^*(X)_{\mathscr{J}}^\wedge$ is isomorphic to the topological completion $K_G^*(E\mathscr{J} \times X)$.

Corollary 2.1. *If X is a finite G-CW complex, then the projection $E\mathscr{J} \times X \to X$ induces an isomorphism $K_G^*(X)_{\mathscr{J}}^\wedge \to K_G^*(E\mathscr{J} \times X)$.*

McClure has obtained interesting applications [12]. For example, he has shown that $K_G^*(X)$ is detected by the family of finite subgroups of G, so that a G-vector bundle is stably trivial if it is stably trivial when regarded as an H-vector bundle for each finite subgroup H of G.

With X a point, the original Atiyah–Segal completion theorem specializes to a calculation of the K-theory of classifying spaces in terms of completions of representation rings. There is an analogous specialization of (2.1) to the calculation of the K_G-theory of classifying G-spaces. To see this, let Π be a normal subgroup of a compact Lie group Γ with quotient group G. The orbit projection $q: Y \to Y/\Pi$ of a Π-free Γ-space is a kind of equivariant bundle, and there is a universal bundle $E(\Pi; \Gamma) \to B(\Pi; \Gamma)$ of this sort. Classically, $\Gamma = G \times \Pi$, and q is then called a principal (G, Π)-bundle. For example, a smooth G-n-plane bundle has an associated principal $(G, O(n))$-bundle. The universal Π-free Γ-space $E(\Pi; \Gamma)$ is just $E\mathscr{J}$, where $\mathscr{J} = \mathscr{J}(\Pi; \Gamma)$ is the family of subgroups Λ of Γ such that $\Lambda \cap \Pi = e$, and we have the following calculation of $K_G(B(\Pi; \Gamma))$.

Corollary 2.2. *The projection $E(\Pi; \Gamma) \to \{pt\}$ induces an isomorphism*

$$R(\Gamma)_{\mathscr{J}(\Pi;\Gamma)}^\wedge \xrightarrow{\cong} K_\Gamma(E(\Pi; \Gamma)) \cong K_G(B(\Pi; \Gamma)).$$

Parenthetically, we insert the analogous specialization of [2, (1.6)]. The case $\Pi = 1$ is Segal's original version of the Segal conjecture.

COROLLARY 2.3. *Let Γ be finite. The projection $E(\Pi;\Gamma)\to\{pt\}$ induces an isomorphism*

$$A(\Gamma)\hat{}_{\mathscr{J}(\Pi;\Gamma)}\xrightarrow{\cong}\pi^0_\Gamma(E(\Pi;\Gamma))\cong\pi^0_G(B(\Pi;\Gamma)).$$

The right-hand change of groups isomorphisms in (2.2) and (2.3) are standard; see [14, 2.1] for K-theory, [1, 5.3] for cohomotopy, and [11, II§8] for general theories.

§3. PROOF OF THEOREM 1.1

Since $(K^*_G)\hat{}_{\mathscr{J}}$ is a progroup valued cohomology theory (as explained in [2]), exact sequences derived from cofibre sequences imply that (1.1) is equivalent to the following vanishing theorem.

THEOREM 3.1. *$\tilde{K}^*_G(X)\hat{}_{\mathscr{J}}$ is pro-zero for every based G-space such that X^H is contractible for each $H\in\mathscr{J}$, and similarly for $(KO^*_G)\hat{}_{\mathscr{J}}$.*

We deduce this from a special case. Let U be the sum of countably many copies of each of a countable set of non-trivial representations V_i of G such that each $V_i^G=0$ and some $V_i^H\neq 0$ if H is a proper subgroup of G. For K^*_G, we restrict attention to complex representations. For KO^*_G, we restrict attention to Spin representations with dimension divisible by eight. Since the arguments are otherwise identical, we concentrate on the complex case from now on. Let Y be the colimit of the one-point compactifications S^V of the finite dimensional subrepresentations V of U. Since each $V_i^G=0$, $Y^G=S^0$. If $V\subset W$ and $(W-V)^H\neq 0$, where $W-V$ is the complement of V in W, then the inclusion $S^V\to S^W$ is null H-homotopic. It follows that Y^H is contractible and Y is H-contractible for $H\neq G$.

LEMMA 3.2. *If $\mathscr{J}$ is proper ($G\notin\mathscr{J}$), then $\tilde{K}^*_G(Y)\hat{}_{\mathscr{J}}$ is pro-zero.*

To deduce (3.1) from (3.2), we need to know the behavior of $(\tilde{K}^*_G)\hat{}_{\mathscr{J}}$ with respect to restriction to subgroups. For $H\subset G$, let $\mathscr{J}|H$ be the family of subgroups of H which are in $\mathscr{J}$.

LEMMA 3.3. *For any based G-space X,*

$$\tilde{K}^*_G((G/H)_+\wedge X)\hat{}_{\mathscr{J}}\cong\tilde{K}^*_H(X)\hat{}_{\mathscr{J}|H}.$$

Proof of (3.1). Since $\mathscr{J}$ is a family, the equivariant Whitehead theorem shows that X is H-contractible and thus $\tilde{K}^*_H(X)$ is pro-zero for $H\in\mathscr{J}$. We must show that $\tilde{K}^*_G(X)\hat{}_{\mathscr{J}}$ is pro-zero. To avoid triviality, we assume that $G\notin\mathscr{J}$. Since the descending chain condition on subgroups allows induction, we may assume that $\tilde{K}^*_H(X)\hat{}_{\mathscr{J}|H}$ is pro-zero for all proper subgroups $H\subset G$. Since $Y^G=S^0$, we have a cofibre sequence

$$S^0\to Y\to Y/S^0.$$

Taking smash products with X, we obtain a cofibre sequence

$$X\to X\wedge Y\to X\wedge(Y/S^0).$$

It suffices to prove that $\tilde{K}^*_G(X\wedge Y)\hat{}_{\mathscr{J}}$ and $\tilde{K}^*_G(X\wedge(Y/S^0))\hat{}_{\mathscr{J}}$ are both pro-zero. We claim first that $\tilde{K}^*_G(W\wedge Y)\hat{}_{\mathscr{J}}$ is pro-zero for any G-CW complex W. Since the zero skeleton W^0 and the skeletal quotients W^n/W^{n-1} for $n>0$ are wedges of G-spaces of the form $(G/H)_+\wedge S^n$ and since we may as well assume that W is finite, we need only verify this for $W=(G/H)_+\wedge S^n$ and thus, by suspension, for $W=(G/H)_+$. Here (3.2) gives the conclusion if $H=G$ and (3.3) and the H-contractibility of Y give the conclusion of $H\neq G$. We claim next that $K^*_G(X\wedge Z)^\wedge_{\mathscr{J}}$ is pro-zero for any G-CW complex Z, such as Y/S^0, such that Z^G is a point. Arguing as above, we need only verify this when $Z=(G/H)_+$ for a proper subgroup H, and here the conclusion holds by (3.3) and the induction hypothesis.

The argument just given is an adaptation of the preliminary steps in Carlsson's proof of the Segal conjecture [5].

Proof of (3.2). By Bott periodicity [3, 14], $\tilde{K}^*_G(S^V)$ is the free $\tilde{K}^*_G(S^0)$-module generated by the Bott class $\lambda_V\in\tilde{K}^0_G(S^V)$. Moreover, λ_V restricts to the Bott class in $\tilde{K}^0_H(S^V)$ for each $H\subset G$. The Euler class $\chi_V\in R(G)=\tilde{K}^0_G(S^0)$ is $e^*(\lambda_V)$, where $e\colon S^0\to S^V$ is the evident inclusion. If $H\neq G$ and $V^H\neq 0$, then e is null H-homotopic and $X_V\in I^G_H$. If $V\subset W$, then the inclusion $i\colon S^V\to S^W$ is $l\wedge e$, $e\colon S^0\to S^{W-V}$. Since $\lambda_W=\lambda_{W-V}\lambda_V$, the homomorphism $i^*\colon \tilde{K}^*_G(S^W)\to\tilde{K}^*_G(S^V)$ is given by the formula

$$i^*(x\lambda_W)=x\chi_{W-V}\lambda_V,$$

for $x\in\tilde{K}^*_G(S^0)$; that is, i^* is multiplication by χ_{W-V}.

We may view $\tilde{K}^*_G(Y)^\wedge_{\mathscr{J}}$ as the inverse limit in the category of progroups of the $\tilde{K}^*_G(Y)/J\tilde{K}^*_G(Y)$, where J runs over the finite products of ideals I^G_H with $H\in\mathscr{J}$ (see [2, §2]). So it suffices to prove that $\tilde{K}^*_G(Y)/J\tilde{K}^*_G(Y)$ is pro-zero for each such J. This means that, for each V, there exists $W\supset V$ such that

$$i^*\colon \tilde{K}^*_G(S^W)/J\tilde{K}^*_G(S^W)\to\tilde{K}^*_G(S^V)/J\tilde{K}^*_G(S^V)$$

is zero. If $J=I^G_{H_1}\dots I^G_{H_n}$ and we choose $W-V$ to be the sum of representations W_i such that $W_i^{H_i}\neq 0$, then i^* is zero since it is multiplication by $\chi_{W_1}\dots\chi_{W_n}\in J$.

Since $\tilde{K}^*_G((G/H)_+\wedge X)\cong\tilde{K}^*_H(X)$ as pro-$R(G)$-modules, where $R(G)$ acts on $\tilde{K}^*_H(X)$ through $r^G_H\colon R(G)\to R(H)$, the following algebraic fact implies (3.3).

LEMMA 3.4. *The $\mathscr{J}$-adic and $(\mathscr{J}|H)$-adic topologies coincide on $R(H)$.*

This follows from Segal's results on $R(G)$ [13, §3]. The key point is the following observation about supports of prime ideals, which can be derived from [13, 3.5 or 3.7].

LEMMA 3.5. *If $S\subset H$ is a support of a prime ideal $Q\subset R(H)$ and if $P=(r^G_H)^{-1}(Q)\subset R(G)$, then S is a support of P.*

Proof of (3.4). If $L\in\mathscr{J}|H$, then $r^G_H(I^G_L)\,R(H)\subset I^H_L$ since $r^H_L r^G_H=r^G_L$. Conversely, if $K\in\mathscr{J}$ and if $I=r^G_H(I^G_K)R(H)$, then I contains some product of ideals I^H_L with $L\in\mathscr{J}|H$. To see this, note that some product of prime ideals $Q\supset I$ is contained in I and that any prime ideal $Q\subset R(H)$ contains I^H_S, where $S\subset H$ is a support of Q. So it suffices to check that S is in $\mathscr{J}$ when Q contains I. If $P=(r^G_H)^{-1}(Q)\subset R(G)$, then S is a support of P and P contains I^G_K. Since $R(K)$ is finitely generated and thus integral over $R(G)/I^G_K$ [13, 3.2], $P=(r^G_K)^{-1}(P')$ for some prime ideal $P'\subset R(K)$. Therefore P has a support $S'\subset K$. Since any two supports of a given prime ideal are conjugate [13, 3.7] and S' is in $\mathscr{J}$, S is in $\mathscr{J}$.

Remark 3.6. The previous two lemmas remain valid for $RO(G)$. The essential points are that any prime ideal Q of $RO(G)$ is the restriction of a prime ideal P of $R(G)$ and that if P is also the restriction of $P'\neq P$, then P' is the complex conjugate of P.

§4. PROOF OF THEOREM 1.2

Again, (1.2) is equivalent to the following vanishing theorem.

THEOREM 4.1. *$S^{-1}\tilde{K}^*_G(X)^\wedge_I$ is pro-zero for every based G-space such that X^H is contractible for each $H \in \mathscr{H}$, and similarly for $S^{-1}(KO^*_G)^\wedge_I$.*

As a matter of algebra [2, 2.3], it suffices to prove that $S_P^{-1}\tilde{K}^*_G(X)^\wedge_P$ is pro-zero for each prime ideal $P \subset R(G)$ such that $P \cap S = \phi$ and $P \supset I$. Here S_P^{-1} means "localization at P"; that is, the multiplicative set S_P is the complement of P. Let $H \in \mathrm{Supp}(P)$ and let $\mathscr{J}$ be the family of subgroups of G subconjugate to H. By (3.1), $\tilde{K}^*_G(Y)^\wedge_{\mathscr{J}}$ is pro-zero if Y^K is contractible for all $K \in \mathscr{J}$. Since P contains I^G_H, it follows that $\tilde{K}^*_G(Y)^\wedge_P$ is pro-zero, and a fortiori $S_P^{-1}\tilde{K}^*_G(Y)^\wedge_P$ is pro-zero. For X as in (4.1), X^H is contractible but X^K need not be contractible for $K \subset H$. However, we can embed X as a subcomplex of a G-CW complex Y such that $Y^K = X^K$ for all K which contain a conjugate of H and Y^K is contractible for all other K. For example, we can take $Y = X \wedge \tilde{E}\mathscr{G}$, where $\mathscr{G}$ is the family of subgroups of G which do not contain a conjugate of H and $\tilde{E}\mathscr{G}$ is the unreduced suspension of $E\mathscr{G}$ with one of the cone points as basepoint; the inclusion of S^0 in $\tilde{E}\mathscr{G}$ induces the inclusion of X in Y. The classical localization theorem [14, 4.1] implies that $S_P^{-1}\tilde{K}^*_G(Y) \to S_P^{-1}\tilde{K}^*_G(X)$ is a pro-isomorphism; *a fortiori* $S_P^{-1}\tilde{K}^*_G(Y)^\wedge_P \to S_P^{-1}\tilde{K}^*_G(X)^\wedge_P$ is a pro-isomorphism and $S_P^{-1}\tilde{K}^*_G(X)^\wedge_P$ is pro-zero. In more detail, let $\{Y_\alpha\}$ run over the finite subcomplexes of Y and let $X_\alpha = X \cap Y_\alpha$. Then $S_P^{-1}\tilde{K}^*_G(Y_\alpha) \to S_P^{-1}\tilde{K}^*_G(X_\alpha)$ is an isomorphism for each α by induction up the finitely many cells of Y_α not in X_α since these cells are of orbit type G/K with $K \in \mathscr{G}$ and since $S_P^{-1}R(K) = 0$ for such K by [13.3.7].

Remark 4.2. Every collection $\mathscr{H}$ of cyclic subgroups of G can be realized, generally in several ways, as

$$\mathscr{H} = \cup\{\mathrm{Supp}(P) \mid P \cap S = \phi \text{ and } P \supset I\}$$

for some multiplicative set S and ideal I.

§5. REMARKS

We conclude with three unrelated comments.

Remark 5.1. Clearly (2.1) remains valid when X is finitely dominated but not necessarily finite. The extra generality is significant because locally linear compact topological G-manifolds are compact G-ENR's and are therefore finitely dominated, but they need not have the homotopy types of finite G-CW complexes (even stably). It is also possible to rework our proofs for general compact G-spaces such that each $K^n_G(X)$ is finitely $R(G)$-generated.

Remark 5.2. Let $\mathscr{J}$ be a family in G. As one of us observed [9], (3.4) and Segal's results in [13, p. 121] imply that $\mathscr{J}$ contains all topologically cyclic subgroups of G iff the $\mathscr{J}$-adic topology on $R(G)$ is complete. McClure [12] has proven that $\mathscr{J}$ contains all finite cyclic subgroups of G iff the $\mathscr{J}$-adic topology is Hausdorff on all finitely generated $R(G)$-modules. (This fact is the key to the proof of his result cited in §2.)

Remark 5.3. One might expect the Real K-theory case to work equally well. However, in the presence of a non-trivial involution on G, use of families other than $\{1\}$ leads to difficulty. A Real G-space is the same thing as a $\tilde{G}$-space, where $\tilde{G}$ is the semi-direct product of G and Z_2

determined by the involution, and $KR_{\tilde{G}}^*$ is a cohomology theory on $\tilde{G}$-spaces. For a general subgroup L of G, we do not have a good description of $KR_{\tilde{G}}^*(G/L)_+ \wedge X)$; if $L=\tilde{H}$ for a Real subgroup H of G, then this is $KR_{\tilde{H}}^*(X)$. This suggests that we should restrict attention to Real families in G, but some of our arguments require use of actual families in $\tilde{G}$.

REFERENCES

1. J. F. Adams: Prerequisites for Carlsson's lecture. *Lecture Notes in Mathematics*, Springer **1051** (1984), 483–532.
2. J. F. Adams, J.-P. Haeberly, S. Jackowski and J. P. May: A generalization of the Segal conjecture. *Topology* **27** (1988), 7–21.
3. M. F. Atiyah: Bott periodicity and the index of elliptic operators. *Quart. J. Math. Oxford Ser.* **19** (1983), 113–140.
4. M. F. Atiyah and G. B. Segal: Equivariant K-theory and completion. *J. Diff. Geom.* **3** (1969), 1–18.
5. G. Carlsson: Equivariant stable homotopy and Segal's Burnside ring conjecture. *Ann. Math.* **120** (1984), 189–224.
6. T. tom Dieck: Faserbundel mit Gruppenoperation. *Archiv der Math.* **20** (1969), 136–143.
7. J.-P. Haeberly: Completions in equivariant K-theory. Thesis. University of Chicago (1983).
8. S. Jackowski: Equivariant K-theory and cyclic subgroups. *London Math. Soc. Lecture Notes Cambridge Univ. Press* **26** (1977), 76–92.
9. S. Jackowski: Families of subgroups and completions. *J. Pure Appl. Algebra* **37** (1985), 167–179.
10. R. Lashof: Equivariant bundles. *Illinois J. Math.* **26** (1982), 257–271.
11. L. G. Lewis, Jr, J. P. May, and M. Steinberger: Equivariant stable homotopy theory. *Lecture Notes in Mathematics*. Springer, in press.
12. J. E. McClure, Jr: Restriction maps in equivariant K-theory. *Topology* **25** (1986), 399–409.
13. G. B. Segal: The representation ring of a compact Lie group. *Inst. Hautes Etudes Sci. Publ. Math.* **34** (1968), 113–128.
14. G. B. Segal: Equivariant K-theory. *Inst. Hautes Etudes Sci. Publ. Math.* **34** (1968), 129–151.

D.P.M.M.S., 16 *Mill Lane,*
*Cambridge CB*2 1*SB, England.*

Dept. of Mathematics, University of Washington,
Seattle, WA 98195, *U.S.A.*

Dept. of Mathematics, University of Warsaw,
Palac Kultury i Nauki IXp,
00–901 *Warsaw, Poland.*

Dept. of Mathematics, University of Chicago,
Chicago, IL 60637, *U.S.A.*

Proceedings of the Edinburgh Mathematical Society (1989) **32**, 473–481 ©

ATOMIC SPACES AND SPECTRA

by J. F. ADAMS* and N. J. KUHN

(Received 28th July 1988)

1. The subject-matter of this paper is in some sense known; but we will try to organise, explain and reprove it, and to give examples.

In essence, a space or spectrum X is "atomic" if a map $f\colon X \to X$ may be proved to be an equivalence by a simple, computable test applied in one dimension; this goes back to [**4**] (published as [**5**]) and first appeared in print in [**12**]. That it is useful to prove X atomic and then apply the fact has been amply shown, beginning with [**3**].

This notion is related to two others. Unique factorisation results for spaces and spectra have been considered in [**6, 9, 14**]. Here one needs the notion of an "irreducible" or "indecomposable" object X, and a slightly stronger notion of "prime".

We first show that the case of "spaces" and the case of "spectra" can be considered together, by concentrating on the fact that the hom-set $[X,X]$ is (under suitable assumptions) a profinite monoid. In this case we show that the "weaker" condition implies the "stronger", as follows.

(a) If X is indecomposable then its hom-set $[X,X]$ is "good", and

(b) if $[X,X]$ is "good" then X is both "atomic" and "prime".

We give some illustrative examples, including some which arise "in nature" as stable summands of classifying spaces BG. We conclude with the proofs.

Related results have been obtained by M. C. Crabb and J. R. Hubbuck; we are grateful to them for letters, and also to F. R. Cohen and F. P. Peterson.

2. First we unify the two cases to be considered.

Proposition 2.1. *The hom-set $[X,X]$ is a profinite monoid with zero in both the following cases.*

(2.2) *X is a p-complete CW-complex of finite type and $[X,X]$ means homotopy classes of pointed maps.*

(2.3) *X is a p-complete spectrum of finite type and $[X,X]$ means maps in the homotopy category of spectra.*

We will comment in Section 3.

*The Society is saddened by the sudden death on 7 January 1989 of Professor J. F. Adams, F.R.S.

Proposition 2.4. *Suppose M is a profinite monoid with zero. Then either*

(*a*) *M contains a non-trivial idempotent, or*

(*b*) *M is "good" in that the sense that each $f \in M$ is either invertible or topologically nilpotent.*

In (a), an idempotent is "non-trivial" if it is neither 0 nor 1.

In (b), "f is topologically nilpotent" means that as $n \to \infty$, so $f^n \to 0$ in the profinite topology on M.

The proof of (2.4), which is elementary, will be given in Section 3.

When $[X, X]$ contains a non-trivial idempotent, X is "reducible" or "decomposable". For spaces this means that X has a non-trivial retract; that is, there is a diagram

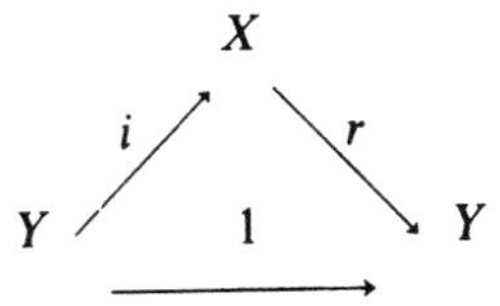

in which Y is not contractible and i, r are not equivalences. For spectra we can go on to infer a non-trivial decomposition as a wedge-sum, $X \simeq Y \vee Z$.

If X is "irreducible" or "indecomposable" then the possibility of a non-trivial idempotent is excluded, and we conclude that $[X, X]$ is "good".

Thus (2.1) plus (2.4) is an analogue, for homotopy-theorists, of a well-known algebraic result: under suitable finiteness conditions, if an R-module X is indecomposable, then every map $f\colon X \to X$ is either invertible or nilpotent. We were surprised to find that such a result survives in a context with no addition.

We will sketch the argument that if $[X, X]$ is good, then X is atomic. In the applications, a simple computable test will dismiss the possibility that f is topologically nilpotent. For example, suppose we choose any dimension n where $H_n(X; F_p) \neq 0$. If f is an equivalence, then $f_*\colon H_n(X; F_p) \to H_n(X; F_p)$ must be iso. But conversely, if $f_*\colon H_n(X; F_p) \to H_n(X; F_p)$ is iso, then it cannot be nilpotent, so f cannot be topologically nilpotent, and f must be an equivalence (assuming $[X, X]$ is good). We conclude that X is atomic.

Of course, many functors other than $H_n(-; F_p)$ would serve as well.

We will sketch the argument that if $[X, X]$ is good, then X is "prime".

If f and g are both topologically nilpotent, then the equation

$$f + g = 1 \neq 0$$

cannot hold even after passing to a finite quotient M_α of M and embedding M_α in a ring R (where we can add). In fact, in the finite quotient M_α we would have $f^m = 0$, $g^n = 0$; in R, f would commute with $1 - f = g$; so we would have $(f + g)^{m+n-1} = 0$.

We now assume that "X divides YZ". For spectra this means that we assume given a retraction

$$X \to Y \vee Z \to X.$$

We take f, g to be the composites

$$X \to Y \to X$$

$$X \to Z \to X.$$

We then have $f+g=1$ in $[X, X]$; assuming $[X, X]$ is good, we deduce that either f or g is an equivalence, and X is a retract either of Y or of Z. That is, "if X divides YZ, then X divides either Y or Z".

For spaces, $Y \vee Z$ should become $Y \times Z$. We cannot argue in quite the same way because we cannot add in $[X, X]$; but we can obtain the equation $f+g=1$ after passing to a suitable ring $R = \operatorname{End} h(X)$, where h is a suitable functor $\pi_r(-) \otimes F_p$ or $QH^r(-; F_p)$ with $h(X) \neq 0$.

We turn to the examples.

Example 2.5 There is a p-local spectrum X of finite type which is indecomposable (but becomes decomposable on completion) and for which $[X, X]$ is not good.

This justifies the assumption of p-completeness above. The construction will be given in Section 4.

In the case of spectra, $[X, X]$ is a profinite ring R. When R is good it is local: the topologically-nilpotent elements make up the unique maximal ideal $\operatorname{rad}(R)$. (Given the indications above, the proof may be left to the reader; the result is due to [**9**].)

Remark 2.6. In this case the quotient $R/\operatorname{rad}(R)$ is a finite field.

The proof is easy, but this too is postponed to Section 4.

This raises the question, which finite fields occur as $R/\operatorname{rad}(R)$. Here we present two examples; one involves infinite spectra which "arise in nature", and the other involves finite spectra constructed by hand.

Example 2.7. Each finite field arises as $R/\operatorname{rad}(R)$ for a suitable X which is an indecomposable stable summand of a classifying space BG. Indeed, if the field is of characteristic p, then G can be a p-group.

For simplicity we now take $p=2$.

Example 2.8. Each finite field of characteristic 2 arises as $R/\operatorname{rad}(R)$ for a suitable X which is a finite spectrum

The constructions will be given in Section 4.

3. We begin by commenting on (2.1).

The cases of spaces, (2.2), is presumably known; but we sketch a proof avoiding certain difficulties.

First we set up some finite quotients of the monoid $N = [X, X]$. Let X_α be a space whose homotopy groups $\pi_r(X_\alpha)$ are finite p-groups, and zero except for a finite number of r. Then $[X, X_\alpha]$ is a finite set and $M=[X,X]$ acts on it from the right; let M_α be the image of M in the monoid $\operatorname{End}([X, X_\alpha])$. Then M_α is a finite monoid; and the map $M \to M_\alpha$ is continuous because the map $m \mapsto i_\alpha m$: $[X,X] \to [X, X_\alpha]$ is continuous for each of the finitely many i_α in $[X, X_\alpha]$.

Secondly we note that the map $M \to \prod_\alpha M_\alpha$ is mono. In fact, according to Sullivan [**13**] we can arrange an isomorphism

$$[X,X] \xrightarrow{\{m \mapsto i_\alpha m\}} \lim_{\substack{\leftarrow \\ i_\alpha}} [X, X_\alpha].$$

Thus $i_\alpha m' = i_\alpha m''$ for all i_α implies $m' = m''$.

Thirdly we order the M_α by considering diagrams

$$\begin{array}{ccc} & & M_\alpha \\ & \nearrow & \downarrow \\ M & & \\ & \searrow & \\ & & M_\beta \end{array}$$

(without requiring any relation between the spaces X^α, X_β). We show that the map.

$$M \to \lim_{\leftarrow} M_\alpha$$

is an epi by a standard compactness argument.

This completes the sketch proof that $[X,X]$ is a profinite monoid.

The case of spectra, (2.3), is due to [**9**, Proposition 4, p. 155}.

We turn to the proof of (2.4).

Suppose given $f \in M$. Let M_α be a finite quotient of M. We show first that some power f^n of f (with $n \geqq 1$) becomes idempotent in M_α.

In fact, if infinitely many powers f^n lie in the same finite set M_α, then two of them must be equal, say $f^a = f^{a+b}$ for some $a \geqq 1, b \geqq 1$. Applying f and iterating we get $f^c = f^{c+bd}$ for $c \geqq a$. Taking $bd \geqq a$ we get $f^{bd} = f^{2bd}$ is idempotent in M_α.

Next let F be $\{f^n | n \geqq 1\}$, the set of powers of f, and let $\bar{F}$ be the closure of F in M. We show that $\bar{F}$ contains an idempotent (possibly 0 or 1). In fact, for each finite quotient M_α of M, let E_α be the set of elements in $\bar{F}$ which map to idempotents in M_α. Then E_α is closed and non-empty (for by the last paragraph it contains some power of f). Indeed, the sets E_α have the finite intersection property, for any finite intersection $E_\alpha \cap E_\beta \cap \ldots \cap E_\delta$ contains another E_ε (consider the pull-back of $M_\alpha, M_\beta, \ldots, M_\delta$). M is compact, so there is an element common to all the E_α, i.e. an idempotent in $\bar{F}$.

It is possible that the idempotent in $\bar{F}$ is 1; in this case we will show that f is invertible. In fact, assume $1 \in \bar{F}$; then in each finite quotient M_α of M we have $1 = f^n$ for some $n \geqq 1$, so f has an inverse $1 = f^{n-1}$ in M_α. This inverse is unique, and these inverse elements give an element of $\lim_{\leftarrow} M_\alpha$, providing an inverse for f in M.

It is possible that the idempotent in $\bar{F}$ is 0; in tnis case we will show that f is topologically nilpotent. In fact, assume $0 \in \bar{F}$, and let M_α be a finite quotient of M. Then some f^n maps to 0 in M_α; hence f^r maps to 0 in M_α for $r \geqq n = n(\alpha)$. Thus $f^r \to 0$ in the profinite topology.

This completes the proof of (2.4).

4. We begin with (2.5).

Our spectrum X will have $H_0(X) = Z_{(p)} \oplus Z_{(p)}$, so that $\mathrm{End}(H_0(X))$ is a ring of 2×2 matrices. We will construct X so that the image of

$$[X, X] \to \mathrm{End}(H_0(X))$$

is a ring

$$\frac{Z_{(p)}[A]}{(A^2 - A + p)}.$$

The proposed minimum polynomial $x^2 - x + p$ has no real roots (because $b^2 - 4ac < 0$), and *a fortiori* no roots in $Z_{(p)}$. It does have p-adic roots α_0, α_1 congruent to $0, 1 \bmod p$ (e.g. by Hensel's Lemma). A convenient matrix with this minimum polynomial is

$$A = \begin{bmatrix} 0 & 1 \\ -p & 1 \end{bmatrix};$$

this has $(1, \alpha_0)^T$, $(1, \alpha_1)^T$ as eigenvectors with eigenvalues α_0, α_1.

Next we need a sequence of elements $g_i \in \pi^S_{n_i - 1}(S^0)$, $i = 1, 2, \ldots$, such that g_i has order p^i in $\pi_*^S(S^0)$ and still has order p^i in $\pi_*^S(S^0)^*/I_i$, where I_i is the ideal generated by the g_j with $j \neq i$. These conditions can easily be satisfied by suitable elements in the image of the J-homomorphism.

We now take

$$X = (S^0 \vee S^0) \cup \left(\bigcup_i e^{n_i} \right),$$

where we work localised at p but omit the notation for it, and where the attaching map for e^{n_i} has components $(g_i, \alpha_i g_i)$. (Here α_i means α_0 or α_1 according as i is even or odd.) This has the following effect. A self-map f of $S^0 \vee S^0$, given by a matrix B, extends over e^{n_i}, with degree d_i on e^{n_i}, if and only if $(1, \alpha_i)^T$ is an eigenvector for $B \bmod p^i$, with eigenvalue $d_i \bmod p^i$; it extends over X if and only if this condition holds for all i, that is, if and only if $(1, \alpha_0)^T$ and $(1, \alpha_1)^T$ are p-adic eigenvectors for B. By construction A satisfies this condition, so A comes from a map $X \to X$. Conversely, suppose B satisfies it; then B is a p-adic linear combination of I and A; here the coefficients of I, A are the entries B_{11}, B_{12} in B, so they lie in $Z_{(p)}$. This proves that the image of

$$[X, X] \to \mathrm{End}(H_0(X))$$

is

$$\frac{Z_{(p)}[A]}{(A^2 - A + p)}.$$

We show that X is indecomposable (over $Z_{(p)}$). In fact,

$$\frac{Z_{(p)}[A]}{(A^2 - A + p)}$$

is in integral domain; so for any idempotent $e \in [X, X]$, either e or $1-e$ maps to 0 in $\mathrm{End}\, H_0(X)$, and the other maps to 1; the one which maps to 1 maps each cell e^{n_i} with degree congruent to $1 \bmod p^i$, and must be an equivalence.

We show that $[X, X]$ is not good. In fact, the map which realises A is neither an equivalence nor topologically nilpotent, for on $H_0(X; F_p)$ it induces an idempotent of rank 1.

Proof of (2.6). Consider the quotient map q from R to a finite quotient ring $R_\alpha \neq 0$. Under q invertible elements map to invertible elements, and topologically nilpotent elements map to nilpotent ones; thus $q^{-1}(\mathrm{rad}(R_\alpha)) = \mathrm{rad}(R)$ and

$$R/\mathrm{rad}(R) \cong R_\alpha/\mathrm{rad}(R_\alpha).$$

So $R/\mathrm{rad}\, R$ is finite; being a finite division algebra, it must be a finite field.
We turn to (2.7). Here we need some hold on the ring of stable maps $\{BG_+, BG_+\}$.

Lemma 4.1. *Let G be a finite p-group. Then the group ring $F_p[\mathrm{Out}(G)]$ is a quotient of the ring $\{BG_+, BG_+\}$.*

The obvious map is in the direction

$$Z[\mathrm{Out}(G)] \to \{BG_+, BG_+\};$$

but we definitely need a quotient of $\{BG_+, BG_+\}$.

Sketch proof. The ring $\{BG_+, BG_+\}$ is known [**11**, p. 397, Corollary 2.3; **10**, p. 128, Corollary 15] as a consequence of the Segal conjecture. In particular, we have

$$F_p \otimes \{BG_+, BG_+\} \cong F_p \otimes A(G,G)$$

where $A(G,G)$ is a ring that plays the same role here that the Burnside ring $A(G)$ does in studying $\{BG_+, S^0\}$ [**1, 10**]. (In [**11**] the present $A(G,G)$ is written $F(G,G)$.) As a Z-module, $A(G,G)$ is free, with a base of elements which may be written $\theta_* i^*$. Here i runs over the inclusions of subgroups $i: H \to G$, and i^* corresponds to the transfer map $Tr: BG_+ \to BH_+$; θ runs over homomorphisms $\theta: H \to G$, and θ_* corresponds to the induced map $B\theta_+: BH_+ \to BG_+$ [**1**, Section 9; **7**, p. 433]. If θ is epi, then we must have $H = G$ and $i = 1$, and θ must be iso. Let I be the Z-submodule of $A(G,G)$ generated by the remaining elements $\theta_* i^*$, in which θ is not epi. We claim I is an ideal.

Consider a product $\theta_* i^* \phi_* j^*$, and assume first that θ is not epi. The product $i^* \phi_* j^*$ can be reduced to a sum of terms $\sum_\alpha (\psi_\alpha)_* k^*$, so we obtain a sum of terms $(\theta\psi_\alpha)_* k^*$ in which $\theta\psi_\alpha$ is not epi.

Assume secondly that ϕ is not epi. By the last paragraph it is sufficient to consider the case in which θ is iso and $i = 1$; but then we get $(\theta\phi)_* j^*$, in which $\theta\phi$ is not epi. Thus I is an ideal.

We now see that the quotient ring $A(G,G)/I$ is $Z[\mathrm{Out}(G)]$. (Two automorphisms θ of G give the same basis element in $A(G,G)$ if and only if they differ by conjugation in G). Thus

$$A(G,G)/(I + (p)) \cong F_p[\mathrm{Out}(G)],$$

and this proves the lemma.

Lemma 4.2. *The finite field F_q, where $q = p^n$, may be obtained as a quotient of the ring $\{BG_+, BG_+\}$ for a suitable p-group G.*

Proof. By a theorem of Bryant and Kovacs [**2, 8**, p. 403, Theorem 13.5] there is a p-group G whose abelianisation $G/[G,G]$ is the additive group $(Z/p)^n$ of F_q and whose automorphism group $\mathrm{Aut}\, G$ acts on $G/[G,G]$ as the multiplicative group $F_q^\times$ of F_q. Clearly this action factors through $\mathrm{Out}(G)$, so we get epimorphisms

$$F_p[\mathrm{Out}(G)] \to F_p[F_q^\times] \to F_q.$$

Using (4.1), we get a map of rings from $\{BG_+, BG_+\}$ onto F_q.

(2.7) now follows. If we take a complete decomposition of 1 into orthogonal idempotents in $\{BG_+, BG_+\}$, then just one of the idempotents maps to 1 in F_q; if X is the corresponding summand of BG_+, then $\{X, X\}$ maps onto F_q and $R/\mathrm{rad}(R) \cong F_q$, as in the proof of (2.6).

We turn to (2.8). In order to realise the finite field F_{2^q}, we begin with $W = \bigvee_1^q S^0$. (We work completed, but for simplicity we omit the notation for it.) We next form

$$X = W \cup_f CS^8 W$$

where the attaching map f has to be described. For any $w \in \pi_0(W)$, f is to carry $S^8 w$ to

$$w \cdot \bar{v} + \phi(w) \cdot \varepsilon;$$

here $\bar{v}, \varepsilon$ are the two generators for $\pi_8^S(S^0) = Z_2 \oplus Z_2$, and ϕ has to be described. Since the result depends only on $\phi(w) \bmod 2$, we may interpret ϕ as an endomorphism of

$$V = \pi_0(W) \otimes F_2 = H_0(W; F_2) = H_0(X; F_2).$$

We take $\phi: V \to V$ to be some linear map whose minimum polynomial is an irreducible polynomial P of degree q over F_2.

An endomorphism of $H_*(X; F_2)$ is now given by a linear map $\lambda: V \to V$ in degree 0 and a linear map $\mu: V \to V$ in degree 9. Such a pair (λ, μ) is induced by a map $g: X \to X$ if and only if it commutes with the boundary map, that is

$$\lambda w \cdot \bar{v} + \lambda \phi w \cdot \varepsilon = \mu w \cdot \bar{v} + \phi \mu w \cdot \varepsilon.$$

Equivalently, $\lambda = \mu$ and $\lambda\phi = \phi\lambda$, that is, λ commutes with ϕ.

Multiplication by ϕ gives V the structure of a module over $F_2[\phi]/P \cong F_{2^q}$; this structure is of course a 1-dimensional vector space over F_{2^q}. The possible maps λ are the endomorphisms of this structure, i.e. multiplication by the elements of F_{2^q}. This shows that the image of

$$R = [X, X] \to \mathrm{End}(V)$$

is F_{2^q}.

It is now clear that X is indecomposable; and $R/\mathrm{rad}(R) \cong F_{2^q}$, as in the proof of (2.6).

Acknowledgement. The second author gratefully acknowledges the support of the Sloan Foundation, the S.E.R.C. and the N.S.F.

REFERENCES

1. J. F. Adams, J. H. Gunawardena and H. Miller, The Segal conjecture for elementary abelian p-groups, *Topology* **24** (1985), 435–460.

2. R. M. Bryant and L. G. Kovacs, Lie representations and groups of prime power order, *J. London Math. Soc.* **17** (1978), 415–421.

3. F. R. Cohen and M.E. Mahowald, A remark on the self-maps of $\Omega^2 S^{2n+1}$, *Indiana Univ. Math. J.* **30** (1981), 583–588.

4. F. R. Cohen, J. C. Moore and J. A. Neisendorfer, Moore spaces have exponents, preprint, circa 1981.

5. F. R. Cohen, J. C. Moore and J. A. Neisendorfer, Exponents in homotopy theory, in *Algebraic Topology and Algebraic K-Theory* (Annals of Mathematics Studies no. **113,** Princeton University Press 1987), 3–34.

6. P. J. Freyd, Stable homotopy, in *Proceedings of the Conference on Categorical Algebra, La Jolla 1965* (Springer 1966), 121–172.

7. J. C. Harris and N. J. Kuhn, Stable decompositions of classifying spaces of finite abelian p-groups, *Math. Proc. Cambridge Philos. Soc.* **103** (1988), 427–449.

8. B. Huppert and N. Blackburn, *Finite Groups II* (Springer 1982).

9. H. R. Margolis, *Spectra and Steenrod Algebra* (North-Holland 1983).

10. J. P. May, Stable maps between classifying spaces, *Contemp. Math.* **37** (1985), 121–129.

11. G. Nishida, Stable homotopy type of classifying spaces of finite groups, in *Algebraic and Topological Theories—to the Memory of Dr Takehiko Miyata* (Kinokuniya, Tokyo 1985), 391–404.

12. P. S. Selick, On the indecomposability of certain sphere-related spaces (Canadian Math. Soc. Conference Proceedings Vol. 2, Part 1, 1982), 359–372.

13. D. Sullivan, Genetics of homotopy theory and the Adams conjecture, *Ann. of Math.* **100** (1974), 1–79.

14. C. W. Wilkerson, Genus and cancellation, *Topology* **14** (1975), 29–36.

Department of Mathematics
University of Virginia
Charlottesville, VA 22903
USA

Two Theorems of J. Lannes

J. F. Adams

§1. This paper does not correspond to anything I said in Seattle; it represents material from seminars in Cambridge later. These seminars were devoted to work of J. Lannes, which seems to offer an elegant and efficient way to build up a body of algebra relevant to the solutions of the Segal conjecture and the Sullivan conjecture, and potentially useful for other purposes.

Lannes' formulation rests on the solution of the following problem:

$$\mathrm{Hom}_U(M, N \otimes P) \xleftarrow{\cong} \mathrm{Hom}_U(?, P). \tag{1.1}$$

Here U is the category of unstable modules over the mod p Steenrod algebra A^* (with details as for cohomology modules), and M, N, P are three objects in U. Lannes writes the solution "?" as $(M/N)_U$ or $T(M)$.

Comparing this idea with the "formal function-objects" of [**1**] (which may appear similar in spirit) one notes important differences. (i) The formalism of [**1**] was conceived as an exercise in stable algebra; Lannes' formalism is conceived as an exercise in unstable algebra. This fits it for calculating "unstable Ext groups"; and in stable applications, it allows one to exploit the fact that the relevant objects in the stable world come from the unstable world. (ii) To show that something is a "formal function-object" in the sense of [**1**], one has to verify statements which involve Ext groups; so that formalism is more useful for processing information about Ext groups than obtaining it in the first place. By contrast, to show that something solves the problem (1.1), one only has to verify a statement about Hom groups; it turns out later that the solution has useful properties with respect to Ext groups.

In fact, it is easy to display a theoretical solution to the problem (1.1). Lannes assumes that in each degree d the vector-space N^d is finite-dimensional; he then sets

$$(M/N)_u = UQ(M \otimes N^*),$$

where N^* is the dual of N (made into an A^*-module in the usual way), and $UQ(L)$ is the universal unstable quotient of the A^*-module L.

Of course, in so far as the inspired choice of the problem (1.1) makes it easier to show the existence of a solution, it throws more work onto showing that the solution has good properties. Lannes specialises to the case $N = H^*(BZ_p; F_p)$, and shows that in this case the functor T has remarkable properties; the following two are fundamental to his method.

Theorem 1.2. (J. Lannes) *The functor T is exact.*

Theorem 1.3. (J. Lannes) *The functor T preserves tensor products.*

In order to make (1.3) precise, one must explain the natural map

$$T(M' \otimes M'') \xrightarrow{\gamma} T(M') \otimes T(M'')$$

which is proved to be an isomorphism. Its existence depends on the fact that $N = H^*(BZ_p; F_p)$ is an algebra. The product map $N \otimes N \to N$ is dual to a coproduct map $N^* \to N^* \otimes N^*$, which induces the following map.

$$\begin{array}{ccc} M' \otimes M'' \otimes N^* & \longrightarrow & M' \otimes N^* \otimes M'' \otimes N^* \\ & & \downarrow \\ & & UQ(M' \otimes N^*) \otimes UQ(M'' \otimes N^*) \end{array}$$

Since the target is unstable, this map factors through $UQ(M' \otimes M'' \otimes N^*)$ to give the required canonical map

$$T(M' \otimes M'') \xrightarrow{\gamma} T(M') \otimes T(M'').$$

I will not discuss the theory which Lannes erects on these foundations; in my view it provides good reason to study his work, but I have nothing to add to it. I will discuss (at least for $p = 2$) an alternative approach to Theorems 1.2 and 1.3, which aims to make $T(M)$ more accessible to direct calculation.

The main difficulty in getting a practical grasp on the theoretical solution $UQ(M \otimes N^*)$ lies in giving usable necessary and sufficient conditions for an element of $M \otimes N^*$ to map to zero in $UQ(M \otimes N^*)$. Propositions 2.1 and 2.2 of §2 will offer such conditions. As soon as they are stated, I will seek to justify them by deducing Theorem 1.2.

The proof I offer for Theorem 1.3 is similar; but this proof works best if one can refer to elements which emerge naturally from the proof of (2.2); so §3 proves (2.2) and §4 proves (1.3). Finally, §5 proves (2.1), which reduces to an exercise on identities in the Steenrod algebra.

§**2.** The necessary and sufficient conditions mentioned in §2 involve particular Steenrod operations, which we must now define. We restrict attention to the case $p = 2$.

Each Milnor operation [**2**]

$$Sq^K = Sq^{k_1 k_2 \cdots k_s}$$

has a degree defined by

$$d(K) = \sum_i k_i(2^i - 1)$$

and an "excess" defined by

$$e(K) = \sum_i k_i.$$

One can then define a second grading by

$$f(K) = \sum_i k_i 2^i$$

so that

$$d(K) = f(K) - e(K).$$

We define the element $a(e, f) \in A^{f-e}$ by

$$a(e, f) = \sum Sq^K \mid e(K) = e, f(K) = f.$$

Let H be $H_*(BZ_2; F_2)$, which is the dual N^* of $N = H^*(BZ_2; F_2) = F_2[x]$. Let $\{h_i\}$ be the base in H dual to the base $\{x^i\}$ in $F_2[x]$. We define a homomorphism

$$\theta_f : (M \otimes H)^d \to M^{d+f}$$

by

$$\theta_f \left(\sum_i m(i) \otimes h_i \right) = \sum_i a(i, f) m(i).$$

PROPOSITION 2.1. *If an element $x \in (M \otimes H)^d$ maps to zero in $UQ(M \otimes H)$, then there exists t such that $\theta_f(x) = 0$ for all $f \equiv 0 \mod 2^t$.*

This necessary condition is also sufficient. Suppose given $x \in (M \otimes H)^d$ of the form

$$x = \sum_{i \leq L} m(i) \otimes h_i.$$

The condition $i \leq L$ serves to set an explicit upper bound on the i which occur.

PROPOSITION 2.2. *Then there exists $t_0 = t_0(d, L)$ such that x maps to zero in*
$UQ(M \otimes H)$ if $\theta_0(x) = 0$
and $\theta_f(x) = 0$ for one
value $f = 2^t$ with $t \geq t_0$.

The condition $\theta_0(x) = 0$ simply means $m(0) = 0$ (since $a(0,0) = 1$ and $a(i,0) = 0$ for $i > 0$). An explicit estimate for the condition $t \geq t_0(d, L)$ would be

$$2^t \geq \mathrm{Max}(2^L, (d+1)2^{L-1})$$

PROOF OF THEOREM 1.2, ASSUMING (2.1) AND (2.2)

The fact that $T(M)$ arises as the solution of the problem (1.1) leads to an easy proof that T is right exact for any N. It is thus sufficient to show that T carries any monomorphism $\varphi : M \to M'$ to a monomorphism $T(\varphi)$.

Suppose then that

$$x = \sum_{i \leq L} m(i) \otimes h_i$$

is an element in $(M \otimes H)^d$ such that $(T\varphi)(x) = 0$ in $T(M')$. By (2.1) there exists t such that

$$\sum_{i \leq L} a(i, f)\varphi m(i) = 0 \qquad \text{in } M'$$

for all $f \equiv 0 \mod 2^t$. Since φ is mono, this shows that

$$\sum_{i \leq L} a(i, f) m(i) = 0 \qquad \text{in } M$$

for all $f \equiv 0 \mod 2^t$. By (2.2), x maps to zero in $T(M)$. This proves (1.2).

§3. The proof of (2.2) depends on a sequence of simple ideas, given as Lemmas 3.1, 3.2 and 3.4 below. At the same time, we prove (3.3) and (3.5); these easy results will be used in §4 to prove Theorem 1.3, but can be ignored till then.

The dual A_* of A^* is a polynomial algebra $F_2[\zeta_1, \zeta_2, \cdots]$ [**2**]. Let $A(r)_*$ be the quotient

$$A_*/(\zeta_1, \zeta_2, \cdots, \zeta_{r-1});$$

this is still a Hopf algebra. Its dual $A(r)^*$ is a sub-Hopf-algebra of A^*; more precisely, it has as a base the Milnor operations

$$Sq^{k_1 k_2 \cdots k_s}$$

such that $k_i = 0$ for $i < r$.

LEMMA 3.1. *$am \otimes n^* = m \otimes (\chi a)n^*$ in $UQ(M \otimes N^*)^d$ provided $a \in A(r)^*$ and $2^r - 1 \geq d$.*

Here χ is the canonical anti-automorphism in A^*. In effect, the lemma says that in degrees $d \leq 2^n - 1$, the quotient map

$$(M \otimes N^*)^d \to UQ(M \otimes N^*)^d$$

factors through $(N^* \otimes_{A(r)^*} M)^d$.

PROOF: By induction over $|a|$. The result is true when $|a| = 0$, and the algebra $A(r)^*$ is zero in degrees between 0 and $2^r - 1$, so we may assume $|a| \geq 2^r - 1$. Thus $m \otimes n^*$ lies in degree ≤ 0; so in $UQ(M \otimes N^*)$ we have $a(m \otimes n^*) = 0$. With the usual notation $\psi a = \sum_i a_i' \otimes a_i''$, we get

$$\begin{aligned} 0 &= \sum_i a_i' m \otimes a_i'' n^* \\ &= am \otimes n^* + \sum_{|a_i'|<|a|} m \otimes (\chi a_i') a_i'' n^* \\ &\qquad \text{(by the inductive hypothesis)} \\ &= am \otimes n^* - m \otimes (\chi a) n^*. \end{aligned}$$

This completes the induction and proves (3.1).

In order to exploit (3.1), we need to show that the elements "a" we use lie in $A(r)^*$ and satisfy $(\chi a)h_i = h_j$ for suitable i, j. Lemmas 3.2 and 3.5 will address the first point, 3.4 and 3.5 the second.

LEMMA 3.2. *The operation $a(e, f)$ lies in $A(r)^*$ if $f \equiv 0 \mod 2^{e+r-1}$.*

PROOF: Assume $f \equiv 0 \mod 2^{e+r-1}$. Consider an index K with $e(K) = e$, $f(K) = f$, that is

$$\sum_i k_i = e, \qquad \sum_i k_i 2^i = f.$$

If $k_i > 1$ we get

$$(k_i - 1)2^i + \sum_{j \neq i} h_j 2^j = f - 2^i.$$

On the right we have a number whose binary expansion has at least $e + r - 1 - i$ nonzero digits and on the left it is written as a sum of $e - 1$ powers of 2; this is only possible if $e - 1 \geq e + r - 1 - i$, that is, $i \geq r$. So any such Sq^K lies in $A(r)^*$.

COROLLARY 3.3. *If $f \equiv 0 \mod 2^{e+r-1}$, then the coproduct $\psi(a(e, f))$ can be written*

$$\psi(a(e, f)) = \sum a(e', f') \otimes a(e'', f'')$$

where the sum runs over $e' + e'' = e$, $f' + f'' = f$, $f' \equiv 0 \mod 2^r$ $f'' \equiv 0 \mod 2^r$.

PROOF: A priori the formula would be

$$\psi(a(e, f)) = \sum a(e', f') \otimes a(e'', f'')$$

where the sum runs over all solutions of $e' + e'' = e$, $f' + f'' = f$. However, some of these terms may be zero. If $f \equiv 0 \mod 2^{e+r-1}$, then by (3.2) $a(e, f)$ lies in $A(r)^*$, and so $\psi(a(e, f))$ lies in $A(r)^* \otimes A(r)^*$; all Sq^K in $A(r)^*$ have second grading $f(K) \equiv 0 \mod 2^r$.

LEMMA 3.4. *If $e \geq 1$ and $f = 2^t$ with $t \geq e$ then*

$$(\chi a(e,f))h_f = h_e.$$

PROOF: The proof falls into two parts: first

$$(\chi Sq^{f-e})x^e = x^f,$$

and secondly, $a(e,f) = \chi Sq^{f-e}$ modulo terms of excess $> e$, which must annihilate x^e. If so, then

$$a(e,f)x^e = x^f,$$

and the result follows.

To prove that $(\chi Sq^{f-e})x^e = x^f$ when $e \geq 1$ and $f = 2^t > e$ is a simple exercise in $F_2[x, x^{-1}]$ which may be left to the reader. It is also easy to show by induction over r that

$$\chi\zeta_r = \zeta_1^{2^r-1} \quad \text{modulo other monomials,}$$

whence

$$\chi(Sq^n) = \sum Sq^K \Big| d(K) = n.$$

It is now sufficient to show that when $f \equiv 0 \mod 2^e$ there is no K with $d(K) = f - e$, $e(K) < e$. This is easily proved by the same method as (3.2), but replacing the k_i used there by a k_0 chosen so that

$$\sum_{i \geq 0} k_i = e.$$

PROOF OF (2.2): Since we are given $m(0) = 0$, we may take $x \in (M \otimes H)^d$ in the form

$$x = \sum_{1 \leq i \leq L} m(i) \otimes h_i$$

and assume $\theta_f(x) = 0$, that is

$$\sum_{1 \leq i \leq L} a(i,f)m(i) = 0 \quad \text{in } M^{d+f}.$$

Assuming $f = 2^t$ with $t \geq L$ and $2^{t-L+1} \geq d+1$, we have

$$\begin{aligned} x &= \sum_{1 \leq i \leq L} m(i) \otimes h_i \\ &= \sum_{1 \leq i \leq L} m(i) \otimes \chi a(i,f)h_f \qquad \text{by (3.4)} \\ &= \sum_{1 \leq i \leq L} a(i,f)m(i) \otimes h_f \qquad \text{in } UQ(M \otimes H) \end{aligned}$$

by (3.1), since by (3.2) $a(i,f) \in A(r)^*$ with $r = t - L + 1$ and $2^r - 1 \geq d$. Thus x maps to zero in $UQ(M \otimes H)$. This proves (2.2).

LEMMA 3.5. *There is in $A(r)^*$ a finite subalgebra $B(r)^*$ such that for any $i \not\equiv 0 \mod 2^r$ we have*

$$h_i = (\chi b)h_j$$

for a suitable b in the augmentation ideal I of $B(r)^*$ and a suitable $j \equiv 0 \mod 2^r$.

PROOF: Consider the quotient

$$A_*(\zeta_1, \cdots, \zeta_{r-1}, \zeta_r^{2^r}, \zeta_{r+1}, \cdots).$$

This quotient is a Hopf algebra, and its dual is a subalgebra $B(r)^*$ of $A(r)^*$ which has as a base the Milnor operations $Sq^{0\cdots 0\,k_r}$ with $0 \le k_r < 2^r$. We have

$$Sq^{0\cdots 0k_r}x^i = x^j$$

where k_r is the residue of $i \mod 2^r$, and $j = i + k_r(2^r - 1)$. This is equivalent to

$$h_i = (\chi Sq^{0\cdots 0k_r})h_j.$$

§4. Proof of Theorem 1.3, assuming (2.1) The first half of the proof will show that the map

$$\gamma : T(M' \otimes M'') \to T(M') \otimes T(M'')$$

is epi. It is sufficient to consider a fixed degree d. Choose r with $2^r - 1 \ge d$, and let I be the augmentation ideal in the algebra $B(r)^*$ of (3.5). We will show that elements $(m' \otimes h_{i'}) \otimes (m'' \otimes h_{i''})$ of degree d lie in $\operatorname{Im}\gamma$ for successively larger classes of $m' \in M'$ and $m'' \in M''$; more precisely, we will show that an element $(m' \otimes h_{i'}) \otimes (m'' \otimes h_{i''})$ of degree d lies in $\operatorname{Im}\gamma$ provided $m' \in I^{s'}M'$ and $m'' \in I^{s''}M''$. Since the ideal I is nilpotent, this is certainly true if either s' or s'' is sufficiently large (in fact, if $s' \ge r+1$ or $s'' \ge r+1$). We may thus argue by induction downwards over $s' + s''$.

We may suppose $m' \otimes h_{i'}$ is of degree ≥ 0 (since otherwise it is zero in $T(M')$ and the result is trivial); similarly, we may suppose $m'' \otimes h_{i''}$ is of degree ≥ 0. This ensures that $m' \otimes h_{i'}$ and $m'' \otimes h_{i''}$ are of degree $\le d$; in other words, we may assume

$$|m'| - d \le i' \le |m'|, \quad |m''| - d \le i'' \le |m''|.$$

Suppose $i' \not\equiv 0 \mod 2^r$. Then by (3.5) there is a $b \in I$ with

$$h_{i'} = (\chi b)h_j,$$

and

$$\begin{aligned} m' \otimes h_{i'} &= m' \otimes (\chi b)h_j \\ &= bm' \otimes h_j \qquad \text{by (3.1)}. \end{aligned}$$

If $m' \in I^{s'}M'$ and $m'' \in I^{s''}M''$ then $bm' \in I^{s'+1}M'$ and

$$(bm' \otimes h_j) \otimes (m'' \otimes h_{i''}) \in \operatorname{Im}\gamma$$

by the inductive hypothesis. Similarly if $i'' \not\equiv 0 \mod 2^r$.

It therefore remains to consider the case $i' \equiv 0 \mod 2^r$, $i'' \equiv 0 \mod 2^r$. In this case we have

$$\gamma(m' \otimes m'' \otimes h_{i'+i''}) = \sum_{j'+j''=i'+i''} (m' \otimes h_{j'}) \otimes (m'' \otimes h_{j''})$$

and for the reasons given in the second paragraph it is sufficient to run this sum over the range

$$|m'| - d \leq j' \leq |m'|, \quad |m''| - d \leq j'' \leq |m''|.$$

But since $d \leq 2^r - 1$, these inequalities admit only one solution with $j' \equiv 0 \mod 2^r$, $j'' \equiv 0 \mod 2^r$, namely the solution $j' = i'$, $j'' = i''$. Thus

$$\gamma(m' \otimes m'' \otimes h_{i'+i''}) \equiv (m' \otimes h_{i'}) \otimes (m'' \otimes h_{i''})$$

modulo terms which lie in $\operatorname{Im}\gamma$ by the third paragraph.

This completes the induction. The induction ends when $s' = 0$ and $s'' = 0$, and this proves that the map γ is epi.

The second half of the proof will show, assuming (2.1), that the map

$$\gamma : T(M' \otimes M'') \to T(M') \otimes T(M'')$$

is mono. Take in $(M' \otimes M'' \otimes H)^d$ an element

$$x = \sum_\alpha m'_\alpha \otimes m''_\alpha \otimes h_{i(\alpha)}$$

which maps to zero in $T(M') \otimes T(M'')$. Then its image in $(M' \otimes H) \otimes (M'' \otimes H)$, that is

$$\overline{x} = \sum_{\alpha, i'+i''=i(\alpha)} (m'_\alpha \otimes h_{i'}) \otimes (m''_\alpha \otimes h_{i''}),$$

can be written in the form

$$\overline{x} = \sum_\beta y'_\beta \otimes y''_\beta + \sum_\gamma z'_\gamma \otimes z''_\gamma$$

where the sums are finite and each y'_β maps to zero in $T(M')$, while each z''_γ maps to zero in $T(M'')$. By (2.1) it follows that there exists t such that

$$(\theta_{f'} \otimes \theta_{f''})\overline{x} = 0 \quad \text{in } M' \otimes M''$$

for all $f' \equiv 0 \mod 2^t$, $f'' \equiv 0 \mod 2^t$. Explicitly, this gives

$$\sum_{\alpha, i'+i''=i(\alpha)} (a(i', f')m'_\alpha) \otimes (a(i'', f'')m''_\alpha) = 0$$

in $(M' \otimes M'')^{d+f'+f''}$ whenever $f' \equiv 0 \mod 2^t$, $f'' \equiv 0 \mod 2^t$.

Provided we choose f divisible by a sufficiently high power of 2, (3.3) now shows that

$$\begin{aligned}\theta_f(x) &= \sum_\alpha a(i(\alpha), f)(m'_\alpha \otimes m''_\alpha) \\ &= \sum a(i', f')m'_\alpha \otimes a(i'', f'')m''_\alpha\end{aligned}$$

where the sum runs first over α and then over $i' + i'' = i(\alpha)$, $f' + f'' = f$, $f' \equiv 0 \mod 2^t$, $f'' \equiv 0 \mod 2^t$. So the last paragraph gives $\theta_f(x) = 0$. Now (2.2) shows that x maps to zero in $UQ(M \otimes H)$. This proves that γ is mono, and completes the proof of (1.3), assuming (2.1).

§**5.** It remains only to prove some identities in A^*. Let

$$\sigma_k : A(2)^* \to A^*$$

be the "shift" map given on the generators by

$$\sigma_k(Sq^{0k_2k_3\cdots}) = Sq^{kk_2k_3\cdots} ;$$

it increases both degree and excess by k.

LEMMA 5.1. *Assume $f \equiv 0 \mod 2^{e+1}$. Then*

$$a(e, f)Sq^k = \sum_{0 \le j \le \min(e,k)} \lambda_{e,j}\sigma_{k-j}a(e-j, f)$$

where $\lambda_{e,j}$ is the coefficient in

$$Sq^j x^{e-j} = \lambda_{e,j}x^e;$$

in particular, $\lambda_{e,0} = 1$, $\lambda_{e,e} = 0$.

All the arguments $a(e-j, f)$ do lie in $A(2)^*$ by (3.2) (using the hypothesis on f).

PROOF: Every element $b \in A^*$ has a unique expression as a finite sum $b = \sum_i \sigma_i b(i)$ with $b(i) \in A(2)^*$. For the element

$$b = a(e, f)Sq^k$$

it is sufficient to run the sum over $i \le k$, so we may write

$$a(e, f)Sq^k = \sum_{0 \le j \le k} \sigma_{k-j}b(j, k)$$

It remains to determine the coefficients $b(j, k)$.

First we note that they are independent of k. In fact, multiplication by ζ_1 gives a map $A_* \to A_*$ which preserves the structure of a left comodule over $A(2)_*$; transposing, we get a map $A^* \to A^*$ which preserves the structure of a left module over $A(2)^*$, carries Sq^k to Sq^{k-1}, and carries σ_i to σ_{i-1} (or to zero if $i = 0$). Applying this map to the formula

$$a(e, f)Sq^k = \sum_{0 \le j \le k} \sigma_{k-j}b(j, k)$$

we get

$$a(e, f)Sq^{k-1} = \sum_{0 \le j \le k-1} \sigma_{k-j-1}b(j, k).$$

Thus $b(j, k-1) = b(j, k)$.

We may now write

$$a(e, f)Sq^k = \sum_{0 \le j \le k} \sigma_{k-j}b(j)$$

and determine $b(j)$ from the final term in $a(e,f)Sq^j$. (In particular, for $j = 0$ we get $b(0) = a(e,f)$.) To this end, we introduce the following diagram.

$$\begin{array}{ccc} F_2[\zeta_2, \zeta_3, \cdots] & \xrightarrow{\psi} & A_* \otimes A_* \\ \downarrow{\scriptstyle\varphi} & & \downarrow{\scriptstyle\varphi\otimes q} \\ F_2[x] & \xrightarrow[\psi]{} & F_2[x] \otimes F_2[\zeta_1] \end{array}$$

Here φ is the homomorphism of rings such that

$$\varphi(\zeta_i) = x \qquad \text{for each } i.$$

The map q is just the quotient map

$$A_* \to A_*/(\zeta_2, \zeta_3, \cdots) = F_2[\zeta_1];$$

it is dual to the injection into A^* of the coalgebra on the Sq^k as base ($k = 0, 1, 2, \cdots$). The lower horizontal arrow ψ expresses the usual action of the Sq^k on $F_2[x]$:

$$\psi x = x \otimes 1 + x^2 \otimes \zeta_1.$$

It is easy to check that the diagram commutes.

Now, if we restrict the map φ to elements $a \in A_{f-e}$, and take the coefficient of x^e in $\varphi(a)$, we get the linear function $A_{f-e} \to F_2$ which defines the element $a(e,f) \in A^{f-e}$. Therefore this diagram calculates the value of $a(e,f)Sq^j$ on $F_2[\zeta_2, \zeta_3, \cdots]$, and it comes out as stated.

LEMMA 5.2. *Assume $f \equiv 0 \mod 2^{e+1}$. Then*

$$\sigma_k a(e,f) = \sum_{0 \le j \le \min(e,k)} \mu_{e,j} a(e-j, f) Sq^k$$

where $\mu_{e,j}$ is the coefficient in

$$Sq^j h_e = \mu_{e,j} h_{e-j};$$

in particular $\mu_{e,0} = 1$, $\mu_{e,e} = 0$.

PROOF: Invert the previous result. That is, take the right-hand side of (5.2) and substitute from (5.1); one comes out with $\sigma_k a(e,f)$.

LEMMA 5.3. *If $f \equiv 0 \mod 2^{L+1}$, then the map*

$$\theta_f : (M \otimes H)^d \to M^{d+f}$$

annihilates all elements $Sq^k(m \otimes h_i)$ with $2k > d$, $i \le L$.

PROOF: We have

$$\begin{aligned} \theta_f Sq^k(m \otimes h_i) &= \theta_f \sum_{0 \le j \le \min(i,k)} Sq^{k-j} m \otimes Sq^j h_i \\ &= \theta_f \sum_{0 \le j \le \min(i,k)} Sq^{k-j} m \otimes \mu_{i,j} h_{i-j} \\ &\qquad\qquad \text{(with the notation of (5.2))} \\ &= \sum_{0 \le j \le \min(i,k)} \mu_{i,j} a(i-j, f) Sq^{k-j} m \\ &= (\sigma_k a(i,f)) m \qquad\qquad \text{(by (5.2))}. \end{aligned}$$

But here $\sigma_k a(i,f)$ is an operation of excess $k+i$, and m is a class of degree $d+i-k$; since we assume $2k > d$, we have

$$k + i > d + i - k.$$

Thus

$$(\sigma_k a(i,f))m = 0.$$

PROOF OF PROPOSITION 2.1: Assume that an element $x \in (M \otimes H)^d$ maps to zero in $UQ(M \otimes H)$. Then by the construction of the universal unstable quotient, x can be written as a finite linear combination of elements $Sq^k(m \otimes h_i)$ with $2k > d$. Let L be an upper bound for the i which occur; if $f \equiv 0 \mod 2^{L+1}$ then the map

$$\theta_f : (M \otimes H)^d \to M^{d+f}$$

annihilates all these elements $Sq^k(m \otimes h_i)$ by (5.3), and so it annihilates x. This completes all the proofs.

REFERENCES

[1] J. F. Adams, J. H. C. Gunawardena and H. Miller. *The Segal conjecture for elementary Abelian p-groups–I*, Topology **24** (1985), 435-460.
[2] J. Milnor. *The Steenrod algebra and its dual*, Annals of Math. **67** (1958), 150–171.

The Work of M.J. Hopkins

J. F. Adams

The work I shall report has the following significance. At one time it seemed as if homotopy theory was utterly without system; now it is almost proved that systematic effects predominate.

I shall begin historically, with one piece of evidence for system. Let p be an odd prime; take a "Moore space" $V(0) = S^n \cup_p e^{n+1}$ with $n \geq 3$. it is shown in [**1**], [**5**] that there is a map

$$S^q V(0) \xrightarrow{A} V(0)$$

where $q = 2(p-1)$ such that the induced map of K-theory

$$K^*(S^q V(0)) \xleftarrow{A^*} K^*(V(0))$$

is an isomorphism. Here $K^*(V(0))$ is non-zero. It follows that the composite of any number i of factors

$$S^{iq}V(0) \to \cdots \to S^{2q}V(0) \xrightarrow{S^q A} S^q V(0) \xrightarrow{A} V(0)$$

induces a non-zero map of K-theory, and is therefore essential. In the diagram

$$\begin{array}{ccccccccc} S^{iq}V(0) & \longrightarrow & \cdots & \longrightarrow & S^{2q}V(0) & \xrightarrow{S^q A} & S^q V(0) & \xrightarrow{A} & V(0) \\ \uparrow & & & & & & & & \downarrow \\ S^{n+iq} & & & & & & & & S^{n+1} \end{array}$$

The initial and final maps are the obvious ones; the composite is (essentially) the map which Toda calls $\alpha_i : S^{n+iq} \to S^{n+1}$. This is the prime example of a "systematic family" in the stable homotopy of spheres.

Miller, Ravenel and Wilson [**7**] found evidence that this sort of systematic behaviour went much further. They were using a spectral sequence invented by Novikov [**9**], of the form

$$\mathrm{Ext}^{s,t}_{MU_*MU}(MU_*X, MU_*Y) \Longrightarrow \{X, Y\}_*.$$

They were trying to compute the Ext group, at least for the case $X = Y = S^0$, knowing that this algebraic work had a guaranteed application to topology; and in their algebra they found a hierarchy of periodic phenomena.

The question thus arose, what if any was has the correspondence between these periodic phenomena in algebra, and anything that might be happening in homotopy theory. Indeed, that question is not yet fully settled. But Ravenel wrote a paper on the subject[**10**], and ended it with about ten conjectures which if true would tend to establish a very good relationship between the periodicity phenomena in the algebra and periodicity phenomena in homotopy theory. And here one must give Ravenel credit for his insight and leadership; for what Hopkins and his collaborators have done is to prove most of Ravenel's conjectures.

Ravenel's first conjecture asked for a far-reaching generalisation of a theorem of Nishida. In [**8**], Nishida shows that any element $x \in \pi^S_q(S^0)$, $q > 0$, is nilpotent: some power x^m of x is zero. To generalise that, you must give sense to the iterated product x^m. In view of examples such as A, one probably begins by thinking of the composition product. So it is natural to suppose given a map

$$S^qX \xrightarrow{f} X$$

and iterate f under composition:

$$S^{mq}X \to \cdots \to S^{2q}X \xrightarrow{S^qf} S^qX \xrightarrow{f} X.$$

But this is not the only sort of product one can use. Indeed, Barratt and Hilton [**2**] proved that the composition product in $\pi_*^S(S^0)$ is commutative, in the graded sense, by remarking that it agrees with the smash product. So one's second thought is to take a map

$$X \xrightarrow{f} Y$$

and form the iterated smash product

$$\bigwedge_1^m X \xrightarrow{\bigwedge_1^m f} \bigwedge_1^m Y.$$

The third option is to suppose given a ring-spectrum R. Then π_*R will be a ring, so it makes sense to ask if $\alpha \in \pi_q(R)$ is nilpotent in that sense. If we take $R = S^0$, we recover the usual product.

Hopkins and his collaborators, Devinatz and Smith, have theorems to deal with all three versions of the problem.

THEOREM 1. (Devinatz, Hopkins and Smith) *Let X be a finite spectrum and $f : S^qX \to X$ a map. In order that f be nilpotent under composition, it is necessary and sufficient that $MU_*f : MU_*X \to MU_*X$ be nilpotent.*

This is a very striking result; we feel that we can compute with MU_*, but the condition on MU_*f is necessary and sufficient for a homotopy-theoretic problem which previously seemed out of reach. With $X = S^0$ we recover Nishida's result; if $q > 0$ then $MU_*f = 0$, so f is nilpotent.

THEOREM 2. (Devinatz, Hopkins and Smith). *Let R be a ring-spectrum (say homotopy-associative). Then $\alpha \in \pi_q(R)$ is nilpotent if and only if its image in $MU_q(R)$ is nilpotent.*

Comments similar to those on Theorem 1 would apply to Theorem 2. Notice also that Theorem 2 needs no finiteness assumption on R; it need not even be bounded below.

THEOREM 3. (Devinatz, Hopkins and Smith) *Let X be a finite spectrum and Y a spectrum bounded below. Then a map $f : X \to Y$ is nilpotent under smash-product if and only if the induced map of Morava K-theory*

$$K(n)_*f : K(n)_*X \to K(n)_*Y$$

is zero for all p and n.

For Morava K-theory, see [**6**], [**10**]. It depends on a prime p, which is not displayed, as well as on n, which is. Two extreme cases are allowed: for $n = \infty$, $K(n)_*$ means ordinary homology with mod p coefficients; for $n = 0$, $K(n)_*$ means rational homology (for each p).

Morava K-theory satisfies a Künneth formula of essentially the same form as holds for ordinary homology with field coefficients. When one deals with smash products, this is too convenient to pass up.

The applications demand a version of Theorem 3 "with parameters". Let W be a finite spectrum where we keep our parameters, and let $f : X \to Y$ be as in Theorem 3.

THEOREM 4. (Devinatz, Hopkins and Smith) *Suppose that*

$$K(n)_*(1 \wedge f) : K(n)_*(W \wedge X) \to K(n)_*(W \wedge Y)$$

is zero for all p *and* n. *Then there exists* m *such that*

$$1 \wedge \bigwedge_1^m f : W \wedge \bigwedge_1^m X \to W \wedge \bigwedge_1^m Y$$

is null homotopic.

The proof of these theorems is a work of considerable virtue; it involves a great many lemmas and intermediate steps. I have presented it in a course of 24 graduate lectures in Cambridge, and I have checked everything I wish to check about it.

The next theorem asserts the existence of periodicity maps like the map A with which I started. Let X be a finite spectrum; for simplicity, let us fix a prime p and assume X is a p-torsion spectrum. Suppose given a map $f : S^q X \to X$, $q > 0$. Hopkins says f is a "v_n-map" if the induced map of Morava K-theory

$$K(i)_* f : K(i)_* X \to K(i)_* X$$

is

$$\begin{cases} \text{an isomorphism} & \text{for } i = n \\ \text{nilpotent} & \text{for } i \neq n. \end{cases}$$

For the next result, we assume $K(i)_* X = 0$ for $i < n$.

THEOREM 5. (Devinatz, Hopkins and Smith)

(i) *X has a v_n-map f.*

(ii) *After replacing f by a sufficiently high power, you may assume $K(i)_* f$ is*

$$\begin{cases} \textit{multiplication by a unit in } K(i)_*(pt) & \textit{for } i = n \\ \textit{zero} & \textit{for } i \neq n. \end{cases}$$

(iii) *f is unique if you are willing to replace f by a sufficiently high power; that is, if f and g are two v_n-maps, then there exists $r > 0$, $s > 0$ such that $f^r \sim g^s$.*

(iv) *After replacing f by a sufficiently high power you may assume that f is central in $[X, X]_*$.*

(v) *Indeed, given a v_n-map f for X and a v_n-map g for Y, you may after replacing them by sufficiently high powers assume that the diagram*

$$\begin{array}{ccc} S^? X & \xrightarrow{h} & S^? Y \\ \downarrow f & & g \downarrow \\ S^? X & \xrightarrow{h} & Y \end{array}$$

commutes for all h.

The strategy of this proof is interesting. First you prove that if you are given v_n-maps, then they have all the good properties you want, leading up to (v). Given Theorems 1–4, this is fairly straightforward; but of course it does nothing to prove the existence of any v_n-maps.

Secondly, you consider the class $\mathcal{C}$ of finite p-torsion spectra X, with $K(i)_*X = 0$ for $i < n$, which do carry a v_n-map. This class $\mathcal{C}$ is closed under the following operations.

(i) If $X \in \mathcal{C}$ and $X' \simeq X$ then $X' \in \mathcal{C}$ (trivially).

(ii) If $X \in \mathcal{C}$ then $S^n X \in \mathcal{C}$ for $n \in Z$ (trivially).

(iii) If two terms of a cofibering $X \to Y \to Z$ lie in $\mathcal{C}$ then so does the third (use (5) (v)).

(iv) If $X \vee Y \in \mathcal{C}$ then $X \in \mathcal{C}$ (use (5)(ii)).

Thirdly, one can prove , using Theorem 1–4, that if such a class $\mathcal{C}$ contains even one spectrum X with $K(n)_*X \neq 0$, then it contains all spectra of the sort considered. So the problem reduces to constructing one spectrum X which carries a periodicity map; it must be a finite spectrum but it does not matter if it is horribly large. Well, the technology is available to construct a periodicity map by the classical Adams spectral sequence if $H^*(X; F_p)$ is extraordinarily favourable. Moreover, such X can be constructed. If $p = 2$ you take a finite approximation to RP^∞, take the smash-product of a large number of such factors, and then split it using an idempotent you get from the symmetric group. Similarly for p odd.

This suggests to me that one of our next tasks is to get more insight into the following question: how many finite complexes are there to which we can apply Theorem 5? In other words, can we set up an appropriate classification scheme for finite complexes?

Let $\mathcal{C}$ be some class of finite spectra closed under the operations (i), (ii), (iii) above. I have in mind the class $\mathcal{C}_n$ of finite p-torsion spectra such that $K(i)_*X = 0$ for $i < n$, but what follows will make sense for other classes.

Let (X_α) be a set of finite spectra: I say it "generates" $\mathcal{C}$ if the smallest class containing (X_α) and closed under (i), (ii), (iii) is exactly $\mathcal{C}$.

QUESTION 1: Find a set of generators for $\mathcal{C}$.

Here I don't want to use operation (iv), because if you do you lose control over what you generate.

The second question will concern the structure of $\mathcal{C}$.

An "Euler characteristic" defined on $\mathcal{C}$ will be a function χ from $\mathcal{C}$ to some abelian group A with the following properties.

(i) If $X \simeq X'$ then $\chi(X) = \chi(X')$.

(ii) if $X \to Y \to Z$ is a cofibering then $\chi(Y) = \chi(X) + \chi(Z)$.

EXAMPLE: If $X \in \mathcal{C}_n$, then $BP_*(X)$ has a finite filtration whose subquotients are cyclic modules

$$S^m(\pi_*(BP)/(p, v_1, v_2, \dots, v_{i-1}))$$

with $i \geq n$. Count the number of subquotients of the most significant sort, i.e., with $i = n$. Let $\chi_n(X)$ be the number with m even, minus the number with m odd, while m is, as above, the degree in which you find the generator for your cyclic module. This χ_n has the required properties.

In any case, I can consider the universal Euler characteristic $\chi : \mathcal{C} \to K_0(\mathcal{C})$.

QUESTION 2: What is the target $K_0(\mathcal{C})$ of the universal Euler characteristic?

EXAMPLE 0: Let $\mathcal{C}$ be the class of all finite spectra. Then S^0 is a generator. The usual Euler characteristic $\chi : \mathcal{C} \to Z$ is the universal one.

The interpretation is as follows. Every finite spectrum can be constructed from cells by iterated cofiberings. χ counts how many cells you use in the even degrees, minus the number you use in odd degrees.

EXAMPLE 1: $\mathcal{C}_1$ is the class of p-torsion spectra. $V(0) = S^0 \cup_p e^1$ is a generator. The Euler characteristic $\chi_1 : \mathcal{C}_1 \to Z$ defined above is the universal one.

The interpretation is similar. A p-torsion spectrum can be constructed from copies of $V(0)$ by iterated cofiberings. χ_1 counts how many $V(0)$'s you use in even degrees, minus the number you use in odd degrees.

The implication is that for $n = 1$, $V(0)$ is the canonical best choice for X in Theorem 5. Notice that any Euler characteristic defined on S^0 will give $\chi(V(0)) = 0$; but if we restrict to a smaller class of spectra, we can study more subtle questions about them.

EXAMPLE 2: $\mathcal{C}_2$ is the class of all p-torison spectra X whose K-theory K_*X is zero.

For p odd, it is a reasonable conjecture that $V(1)$ is a generator and the Euler characteristic $\chi_2 : \mathcal{C}_2 \to Z$ defined above is the universal one.

For $p = 2$, it is perhaps a semi-reasonable conjecture that the universal Euler characteristic is $\mathcal{C}_2 \xrightarrow{\chi_2} 2Z$.

Beyond that I think nobody has much confidence; I would be glad if anyone could tell us more.

REFERENCES

1. J.F. Adams, *On the groups $J(X)$ –IV*, Topology **5** (1966), 21–71.
2. M. G. Barratt and P.J. Hilton, *On join operations in homotopy groups*, Proc. London Math. Soc. **3** (1953), 430–445.
3. A. K. Bousfield, *The Boolean algebra of spectra*, Comment. Math. Helvetica **54** (1979), 368–377.
4. A. K. Bousfield, *The localisation of spectra with respect to homology*, Topology **18** (1979), 257–281.
5. F. R. Cohen and J. A. Neisendorfer, *Note on desuspending the Adams map*, Math. Proc. Camb. Phil. Soc. **99** (1986), 59–64.
6. D. C. Johnson and W. S. Wilson, *BP operations and Morava's extraordinary K-theories*, Math. Zeit. **144** (1975), 55–75.
7. H. R.Miller, D. C. Ravenel and W. S. Wilson, *Periodic phenomena in the Adams-Novikov spectral sequence*, Annals of Math **106** (1977), 469–516.
8. G. Nishida, *The nilpotency of elements of the stable homotopy groups of spheres*, Jour. Math. Soc. Japan **25** (1973), 707–732.
9. S. P. Novikov, *The methods of algebraic topology from the view point of cobordism theories*, Izvest. Akad. Nauk SSSR Ser. Mat. **31** (1967), 855–951.
10. D. C. Ravenel, *Localisation with respect to certain periodic homology theories*, American Jour. Math. **106** (1984), 351–414.

CPSIA information can be obtained at www.ICGtesting.com
262462BV00005B/4/P

9 780521 110686